Microstructure and Properties of High-Temperature Superconductors

I. A. Parinov

Microstructure and Properties of High-Temperature Superconductors

Second Edition

Springer

I. A. Parinov
Vorovich Research Institute of Mechanics
and Applied Mathematics
Southern Federal University
Rostov-on-Don
Russia

ISBN 978-3-642-43183-8 ISBN 978-3-642-34441-1 (eBook)
DOI 10.1007/978-3-642-34441-1
Springer Heidelberg New York Dordrecht London

Printed on acid-free paper

Springer is part of Springer Science+Business Media (www.springer.com)

To my wife, Nina

Preface

From the time of publication of the first edition of the monograph (2007), the rapidly changing world of superconductivity ensured a whole set of new discoveries and achievements in theoretical and experimental aspects, as technologies and applications. In my view, three main achievements generated constant interest of the scientific society to this area, namely (i) discovery of superconductive pnictides in 2008 with temperature of superconductive transition equal to 56 K that is near the liquid nitrogen temperature (77 K), allowing in many researches to relate these superconductors to HTSCs, (ii) active invasion of nanotechnology approaches and methods in processing and investigations of HTSCs that allowed one to significantly modernize the processing techniques and obtain HTSCs with improved structure-sensitive properties, and (iii) growing application of non-conventional superconductors with enough high superconductive properties (in particular, MgB_2) leading to considerable reducing price of the numerous superconductive goods.

All these achievements (but not only they) contributed to the modernization of the first edition. The second edition includes two additional chapters but a majority of the other chapters and appendices were either reworked or considerably extended. The main results of the book are devoted to traditional HTSCs of Y(*RE*)BCO and BSCCO families, and also superconductive pnictides demonstrating many common features with the pointed layered HTSCs, while the main trends in development of superconductivity also are discussed in detail. I am hoping the book can help many students to better understand and study this very interesting region of modern Physics and Material Science.

The author would be grateful for reports of typographical and other errors to be sent via the following e-mail: ppr@math.rsu.ru.

Rostov-on-Don, August 2012 I. A. Parinov

Preface to the First English Edition

In 2006, the scientific society was celebrating the twentieth anniversary of the discovery of the high-temperature superconductivity by George Bednorz and Alex Müller. Dynamically developing researches in this field give new and new scientific results. This caused a significant modernization of the English edition compared to the Russian one [806], which was written in 2003. Considerable changes have been introduced in Chaps. 1–3 and Appendix A, in particular, the new Sect. 3.1.2 is devoted to acoustic emission study of BSCCO/Ag tapes under bending. The new Chap. 4 is devoted to carbon problem in HTSC and includes as "old" text from Sects. 2.6 and 7.7 of the Russian edition, as "new" text (Sects. 4.3 and Appendix B), presenting mathematical modeling of the brittle carbonate formation and following fracture during interaction of YBCO with CO_2. The main aims of the monograph have been retained and connected with Material Science of HTSC and their mathematical modeling. Comparatively, lower attention has been devoted to Physics of HTSC. The main results as before have been related to the YBCO and BSCCO families, while the main trends in R&D of other superconductors have also been marked.

The author would be grateful for reports of typographical and other errors to be sent via the following web-page http://www.math.rsu.ru/niimpm/strl/welcome.en.html, where an up-to-date errata list will be maintained.

Rostov-on-Don, December 2006 I. A. Parinov

Preface to the Russian Edition

The discovery in 1986 of high-temperature superconductors (HTSC) on the base of copper oxides with the temperature of superconducting transition that is greater than the temperature of low-cost, nontoxic, and accessible liquid nitrogen (77 K), marked a qualitative jump in the development and application of new technical conductors, devices for energy transmission, transformation, and storage. Together with enough high-critical temperatures T_c, an intrinsic brittleness of oxide cuprates, the layered anisotropic structure, and the super-short ($\sim$1 nm) coherence length ξ, presenting itself as a spatial characteristic of superconducting electrons, are other main features defining HTSC microstructure and properties. Due to the above-mentioned peculiarities, even the existence of an intergranular boundary could be enough to suppress superconductivity, but the structure-sensitive properties of HTSC systems depend greatly on the weak links of intergranular boundaries, by manufacturing them in the polycrystalline form, demonstrating coexistence of inter- and transgranular currents. Also, superconductivity can be destroyed after the attainment of the critical value of the external magnetic field H_{cm}. The interfaces of the "superconductor-normal metal", "superconductor-insulator" and other types based on them are the localization places of different defects. The microstructure features, connected with phase composition, domain structure, crystallographic properties, and existence of structure defects, pores, microcracks, inclusions, etc., define directly useful properties of HTSC materials and composites. The main goal of this monograph is to study the microstructure, strength, electromagnetic, and superconducting properties. Another aim includes discussion of the optimization directions for the fabrication techniques, superconducting compositions, external loading, and thermal treatments to obtain HTSC, possessing improved and more controlled physical and mechanical properties. The link "composition-technique-experiment-theory-model" investigated in the book, assuming considerable HTSC defectiveness and structure heterogeneity, forms a whole picture of modern representations on the microstructure, strength and connected with them the structure-sensitive properties of the materials considered. Special attention in the book is devoted to Bi-Sr-Ca-Cu-O and Y-Ba-Cu-O families that today are most important for applications.

The monograph is addressed to students, post-graduate students, and specialists taking part in the development, preparation, and researching new materials. The author has the pleasure to thank the Russian Foundation for Basic Research, Russian Department of Education and Science, Soros Foundation and American International Program COBASE (Collaboration in Basic Science and Engineering), grants of which during the last decade have rendered considerable financial support and promoted to publish this book.

I am also grateful to colleagues and nearest scientific workers, who have directly or indirectly contributed to the book. In particular, I wish to thank V. P. Zatsarinny, D. N. Karpinsky, E. A. Dul'kin, E. M. Kaydashev, E. V. Rozhkov, A. A. Polyanskii, and D. C. Larbalestier.

Corrections and proposals by the book readers will be considered with thanks. They could be presented by E-mail: ppr@math.rsu.ru.

September 2003 I. A. Parinov

Contents

Periodic Table of the Chemical Elements

IA	IIA	IIB	IVB	VB	VIB	VIIB	VIII	VIII
1 - HEX *a* = 3.750 *c* = 6.120 $1s^{1}$ - **H**								
3 20.0* HEX *a* = 3.111 *c* = 5.039 $2s^{1}$ - **Li**	**4** 0.026 HEX *a* = 2.286 *c* = 3.584 $2s^{2}$ - **Be**							
11 - BCC *a* = 4.282 $3s^{1}$ - **Na**	**12** - HEX *a* = 3.202 *c* = 5.299 $3s^{2}$ - **Mg**							
19 - BCC *a* = 5.247 $4s^{1}$ - **K**	**20** 25.0* FCC *a* = 5.560 $4s^{2}$ - **Ca**	**21** 8.1* HEX *a* = 3.308 *c* = 5.265 $3d^{1}4s^{2}$ - **Sc**	**22** 0.39 HEX *a* = 2.951 *c* = 4.679 $3d^{2}4s^{2}$ 60 **Ti**	**23** 16.5* BCC *a* = 3.028 $3d^{3}4s^{2}$ 1400 **V**	**24** - BCC *a* = 2.884 $3d^{5}4s^{1}$ - **Cr**	**25** - CUB *a* = 8.912 $3d^{5}4s^{2}$ - **Mn**	**26** 2.1* BCC *a* = 2.866 $3d^{6}4s^{2}$ - **Fe**	**27** - HEX *a* = 2.510 *c* = 4.090 $3d^{7}4s^{2}$ - **Co**
37 - BCC *a* = 5.700 $5s^{1}$ - **Rb**	**38** 7.0* FCC *a* = 6.085 $5s^{2}$ - **Sr**	**39** 19.5* HEX *a* = 3.647 *c* = 5.731 $4d^{1}5s^{2}$ - **Y**	**40** 11.0* HEX *a* = 3.223 *c* = 5.147 $4d^{2}5s^{2}$ 54 **Zr**	**41** 9.9* BCC *a* = 3.300 $4d^{4}5s^{1}$ 1970 **Nb**	**42** 0.92 BCC *a* = 3.147 $4d^{5}5s^{1}$ 90 **Mo**	**43** 7.77 HEX *a* = 2.735 *c* = 4.391 $4d^{6}5s^{1}$ 1410 **Tc**	**44** 0.51 HEX *a* = 2.706 *c* = 4.281 $4d^{7}5s^{1}$ 69 **Ru**	**45** 0.0003 FCC *a* = 3.790 $4d^{8}5s^{1}$ - **Rh**
55 1.3* BCC *a* = 6.141 $6s^{1}$ - **Cs**	**56** 5.0* BCC *a* = 5.019 $6s^{2}$ - **Ba**	**57** 13.0* HEX *a* = 3.770 *c* = 12.159 $5d^{1}6s^{2}$ 808 **La**	**72** 8.6* HEX *a* = 3.195 *c* = 5.051 $5d^{2}6s^{2}$ 12.1 **Hf**	**73** 4.5* BCC *a* = 3.307 $5d^{3}6s^{2}$ 831 **Ta**	**74** 0.012 BCC *a* = 3.165 $5d^{4}6s^{2}$ 1.15 **W**	**75** 1.4 HEX *a* = 2.757 *c* = 4.463 $5d^{5}6s^{2}$ 200 **Re**	**76** 0.65 HEX *a* = 2.750 *c* = 4.319 $5d^{6}6s^{2}$ 70 **Os**	**77** 0.14 FCC *a* = 3.831 $5d^{7}6s^{2}$ 19 **Ir**
87 - BCC $7s^{1}$ - **Fr**	**88** - BCC $7s^{2}$ - **Ra**	**89** - FCC *a* = 5.311 $6d^{1}7s^{2}$ - **Ac**	**104** - - $6d^{2}7s^{2}$ - **Db**					

Lanthanides	**57** 13.0* HEX *a* = 3.770 *c* = 12.159 $5d^{1}6s^{2}$ 808 **La**	**58** - FCC *a* = 4.85 $4f^{1}5d^{1}6s^{2}$ - **Ce**	**59** - HEX *a* = 3.664 *c* = 11.867 $4f^{3}6s^{2}$ - **Pr**	**60** - HEX *a* = 3.658 *c* = 11.799 $4f^{4}6s^{2}$ - **Nd**	**61** - HEX *a* = 3.650 *c* = 11.650 $4f^{5}6s^{2}$ - **Pm**	**62** - ROM *a* = 3.626 α = 26.18° $4f^{6}6s^{2}$ - **Sm**
Actinides	**89** - FCC *a* = 5.311 $6d^{1}7s^{2}$ - **Ac**	**90** 1.37 FCC *a* = 5.086 $6d^{2}7s^{2}$ - **Th**	**91** 1.4 TET *a* = 3.631 *c* = 3.236 $5f^{2}6d^{1}7s^{2}$ - **Pa**	**92** 2.4* ORC *a* = 2.844 *b* = 5.869 *c* = 4.960 $5f^{3}6d^{1}7s^{2}$ - **U**	**93** 0.075 ORC *a* = 4.721 *b* = 4.888 *c* = 3.670 $5f^{4}6d^{1}7s^{2}$ - **Np**	**94** - MCL *a* = 6.183 *b* = 4.822 *c* = 10.963 β = 101.79° $5f^{6}7s^{2}$ - **Pu**

La are the elements used in superconducting compounds

P are the elements not used in superconducting compounds

Legend (example: La):

ATOMIC NUMBER	57
CRITICAL TEMPERATURE (K)*	4.88
STRUCTURE	HEX
LATTICE PARAMETERS (Å)	a = 3.770 c = 12.159
ATOMIC CONFIGURATION	$5d^1 6s^2$
CRITICAL MAGNETIC FIELD (Oe)	808
SYMBOL	La

* signs "under pressure"

VIII	IB	IIB	IIIA	IVA	VA	VIA	VIIA	VIIIA
								2 - BCC a = 4.110 $1s^2$ - **He**
			5 11.0* TET a = 8.800 c = 5.050 $2s^2p^1$ - **B**	6 - DIA a = 3.567 $2s^2p^2$ - **C**	7 - BCC a = 5.644 $2s^2p^3$ - **N**	8 0.6* MCL a = 5.403 b = 3.429 c = 5.086 β = 132.53° $2s^2p^4$ - **O**	9 - MCL a = 5.500 b = 3.280 c = 10.010 β = 134.66° $2s^2p^5$ - **F**	10 - FCC a = 4.462 $2s^2p^6$ - **Ne**
			13 1.14 FCC a = 4.050 $3s^2p^1$ 105 **Al**	14 8.2* DIA a = 5.430 $3s^2p^2$ - **Si**	15 13.0* CUB a = 7.180 $3s^2p^3$ - **P**	16 17.3* ORC a = 10.465 b = 12.866 c = 24.486 $3s^2p^4$ - **S**	17 - ORC a = 4.500 b = 6.290 c = 8.210 $3s^2p^5$ - **Cl**	18 - FCC a = 5.300 $3s^2p^6$ - **Ar**
28 - FCC a = 3.524 $3d^8 4s^2$ - **Ni**	29 - FCC a = 3.615 $3d^{10} 4s^1$ - **Cu**	30 0.9 HEX a = 2.664 c = 4.946 $3d^{10} 4s^2$ 55 **Zn**	31 7.0* ORC a = 4.526 b = 4.519 c = 7.657 $4s^2p^1$ 59.3 **Ga**	32 5.35* DIA a = 5.658 $4s^2p^2$ - **Ge**	33 2.4* ROM a = 4.129 α = 54.1° $4s^2p^3$ - **As**	34 8.0* HEX a = 4.363 c = 4.959 $4s^2p^4$ - **Se**	35 1.4* ORC a = 4.480 b = 6.670 c = 8.720 $4s^2p^5$ - **Br**	36 - FCC a = 5.706 $4s^2p^6$ - **Kr**
46 - FCC a = 3.882 $4d^{10}$ - **Pd**	47 - FCC a = 4.086 $4d^{10} 5s^1$ - **Ag**	48 0.56 HEX a = 2.960 c = 5.630 $4d^{10} 5s^2$ 29.6 **Cd**	49 3.4 TET a = 4.583 c = 4.936 $5s^2p^1$ 281.5 **In**	50 5.3* DIA a = 6.504 $5s^2p^2$ 305 **Sn**	51 3.9* ROM a = 6.085 α = 57.1° $5s^2p^3$ - **Sb**	52 7.5* HEX a = 4.457 c = 5.929 $5s^2p^4$ - **Te**	53 1.2* ORC a = 4.774 b = 7.250 c = 9.772 $5s^2p^5$ - **I**	54 - FCC a = 6.250 $5s^2p^6$ - **Xe**
78 0.05 FCC a = 3.200 $5d^9 6s^1$ 0.07 **Pt**	79 - FCC a = 4.070 $5d^{10} 6s^1$ - **Au**	80 4.16 TET a = 3.463 c = 6.740 $5d^{10} 6s^2$ 411 **Hg**	81 2.39 HEX a = 3.450 c = 5.514 $6s^2p^1$ 181 **Tl**	82 7.19 FCC a = 4.950 $6s^2p^2$ 803 **Pb**	83 8.5* ROM a = 3.475 α = 57.28° $6s^2p^3$ - **Bi**	84 - CUB a = 3.359 $6s^2p^4$ - **Po**	85 - $6s^2p^5$ - **At**	86 - FCC $6s^2p^6$ - **Rn**

63 - BCC a = 4.572 $4f^7 6s^2$ - **Eu**	64 - HEX a = 3.636 c = 5.783 $4f^7 5d^1 6s^2$ - **Gd**	65 - HEX a = 3.604 c = 5.698 $4f^9 6s^2$ - **Tb**	66 - HEX. a = 3.592 c = 5.655 $4f^{10} 6s^2$ - **Dy**	67 - HEX a = 3.577 c = 5.619 $4f^{11} 6s^2$ - **Ho**	68 - HEX a = 3.560 c = 5.595 $4f^{12} 6s^2$ - **Er**	69 - HEX a = 3.537 c = 5.558 $4f^{12} 5d^1 6s^2$ - **Tm**	70 - FCC a = 5.483 $4f^{14} 6s^2$ - **Yb**	71 2.5* HEX a = 3.505 c = 5.553 $4f^{14} 5d^1 6s^2$ 400 **Lu**
95 2.2* HEX a = 3.462 c = 11.223 $5f^7 7s^2$ - **Am**	96 - HEX a = 3.496 c = 11.331 $5f^7 6d^1 7s^2$ - **Cm**	97 - $5f^8 6d^1 7s^2$ - **Bk**	98 - $5f^{10} 7s^2$ - **Cf**	99 - $5f^{11} 7s^2$ - **Es**	100 - $5f^{12} 7s^2$ - **Fm**	101 - $5f^{13} 7s^2$ - **Md**	102 - $5f^{14} 7s^2$ - **No**	103 - $5f^{14} 6d^1 7s^2$ - **Lr**

Chapter 1
Superconductors and Superconductivity: General Issues

1.1 Superconductivity Discovery

The discovery of superconductivity, that is, the phenomenon by which current flow in material occurs without noticeable energy dissipation, has been recognized as one of the greatest scientific achievements of twentieth century. This phenomenon is accompanied by a sudden drop of the electrical resistance to zero[1] by cooling below the *critical temperature* (T_c), which is the temperature of superconducting transition that is defined for every specific material.

In 1911, Heike Kamerlingh-Onnes when researching the properties of some metals in the vicinity of liquid helium (4.2 K) had found that mercury cooled to ~4.25 K losing its electrical resistance that is transformed into superconductor [1268]. In the next years, the superconductivity of some other metals, several alloys and intermetallic compounds had also been discovered (Fig. 1.1).[2] However, they demonstrated very low critical temperatures (maximal value $T_c = 23.2$ K for Nb_3Ge), some increasing the liquid helium temperature. This circumstance impeded the practical applications of superconductors in tremendous degree due to the high cost of liquid helium (~ $25 per liter) and difficulties in its preparation. The long absence of noticeable successes in the increasing of critical temperature (last record was achieved in 1973 for compound Nb_3Ge) raised the highly restrained moods of scientists, who worked in this field in the middle of 1980. So, experimentalists discussed the issue of the prospective of the T_c increasing for Nb_3Sn up to a value of 30 K with the application of very exotic techniques [360], and theorists predicted a 40 K T_c ceiling [676].

The situation changed dramatically in 1986 with the discovery of the so-called high-temperature superconductivity (HTSC) in non-traditional compounds, namely cuprates. In spite of the long time and tremendous forces of world scientific society,

---

[1] On the basis of the sensitivity of modern equipment, it may be argued that the resistivity of superconductors is no more 8×10^{-25} Ω cm. For comparison, we note that the resistivity of high-purity copper is of the order 10^{-9} Ω cm at 4.2 K.

[2] The total classification of known superconductors is presented in Appendix A.

I. A. Parinov, *Microstructure and Properties of High-Temperature Superconductors*,
DOI: 10.1007/978-3-642-34441-1_1,

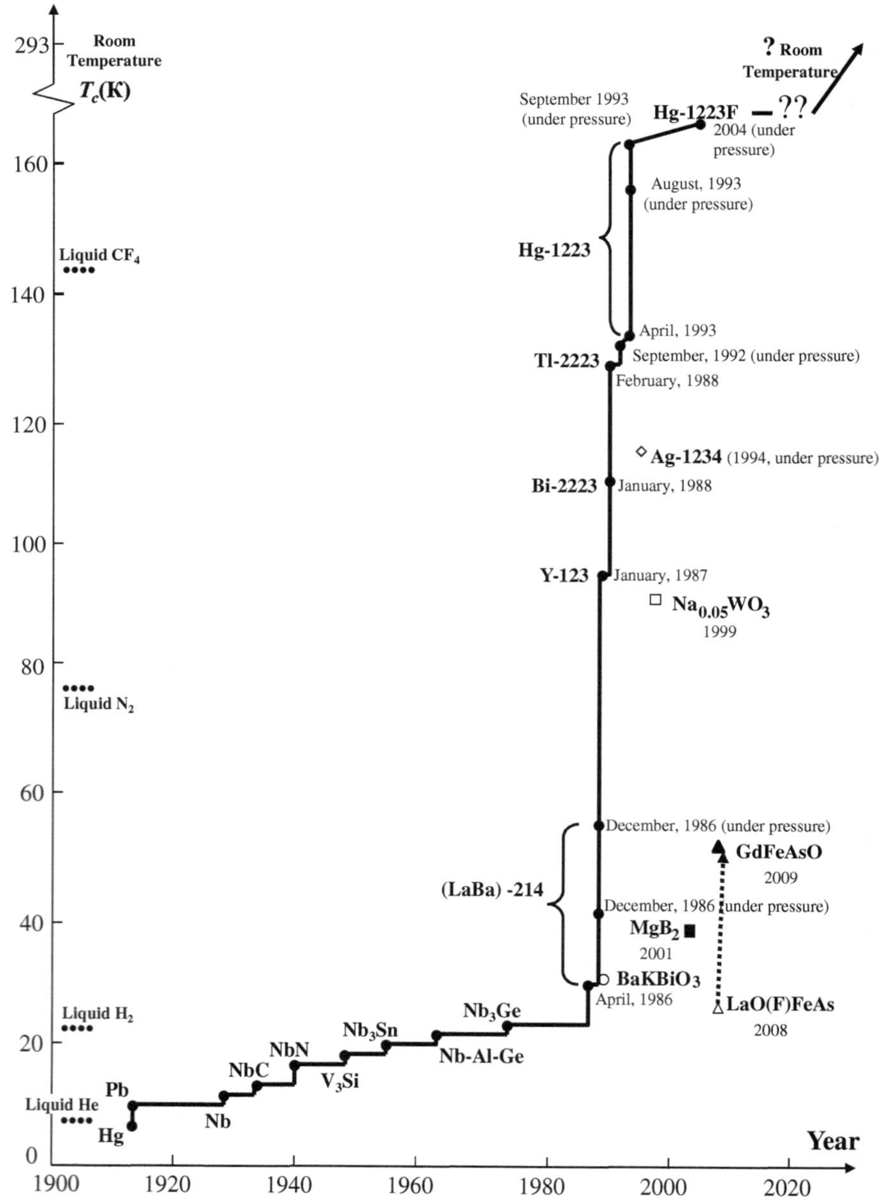

Fig. 1.1 History of superconductors' discovery and of T_c-increasing

the mechanism of superconductivity for these materials is not yet stated. The distinctive feature of HTSC is that all these compounds have atomic CuO_2 plane, playing key role as the origin of superconductivity. The first from HTSC was $La_{2-x}Ba_xCuO_4$ with $T_c = 30$ K (George Bednorz and Alex Müller, April 1986)

[117]. The main peak of discoveries took place between December 1986 and March 1987, when the existence of the critical temperature $T_c > 30$ K was confirmed by Chu's group and Kitazawa's group, independently [285], and then the superconductivity in $YBa_2Cu_3O_7$ at 93 K, attained jointly by Chu's group and Wu's group [1810], was published. These achievements opened a new epoch in the superconductivity investigations as they brought down the liquid nitrogen temperature barrier of 77 K. Low cost ($\sim$\$0.5 per liter), simple conditions for its preparation and utilization led to a considerable progress, in the next years, in the development, manufacture and initial application of high-temperature superconductors. The discovery of HTSC initiated real ecstasy in scientific society (the number of researchers in this field had increased more than order in 1987), and also the tremendous interest of press and public. In 1 year, the critical temperature increased on 70 K, whereas, for the previous 75 years of superconductivity researches the growth of T_c was 20 K only! Scientists, as simple people, called 1987 "the year of progress in physics". Public interest caused a sharp increase in financing of HTSC studies in the world. In the following years, other compounds were discovered (see Fig. 1.1 and Appendix A) with critical temperature also above the liquid nitrogen temperature, namely $Bi_2Sr_2Ca_2Cu_3O_{10+x}$ ($T_c = 110$ K) [1067] (Ag, Cu)$Ba_2Ca_3Cu_4O_{11-x}$ ($T_c = 117$ K) [737], $Tl_2Ba_2Ca_2Cu_3O_{10}$ ($T_c = 125$ K) [1554] and $HgBa_2Ca_2Cu_3O_8$ ($T_c = 134$ K) [1562]. The record critical temperature, 164 K, was achieved at high pressure (30 GPa) in $HgBa_2Ca_2Cu_3O_{8+x}$ family in September 1993 [287].

In 1988, Sheng, et al. presented results of measurements of the electrical resistance in $Tl_2Ba_2Ca_{x-1}Cu_xO_{2x+4}$ samples [1553, 1554]. The results and data obtained by Hazen, et al. [655] demonstrated that at increasing number x the critical temperature $T_c(x)$ grows by the next way: $T_c(1) = 90$ K, $T_c(2) = 110$ K, $T_c(3) = 125$ K. By the linear dependence on x, the temperature $T_c = 300$ K is attained at $x = 10$. However, the values of $T_c > 125$ K for thallium oxides were not defined. Similar predictions for the behavior of T_c on increasing x had been fulfilled for $Bi_2Sr_2Ca_{x-1}Cu_xO_{2x+4}$ family [706]. Regretfully, the dependence also destroys at $x > 3$.

Thus, numerous studies of HTSC have stated that maximal critical temperature, T_c, is reached in compounds with three CuO_2 layers per elementary cell. Moreover, it is necessary to carry out two conditions: (i) to reach the small distance, $d_{\text{Cu–O}} = a/2$, between atoms of copper and oxygen (where a is the period of two-dimensional lattice in the CuO_2 plane) and (ii) a hole concentration in CuO_2 layers should be near the optimal value of $p = 0.16$ (in account per copper atom). All these conditions have been realized in mercury HTSC Hg-1223 with fluorine additives (Hg-1223F) in 2004. The maximal $T_c = 138$ K (at $P = 0$) has been attained in samples with $a = 0.38496$ nm, and a record $T_c = (166 \pm 1.5)$ K at $P = 23$ GPa [1184]. In mercurial HTSC, a linear dependence of T_c on the lattice constant, a, is observed, namely: T_c increases with decreasing of a. The critical temperature $T_c \sim 100$ K in Hg-1201 at $a = 0.388$ nm, and $T_c = 138$ K in Hg-1223F at $a = 0.38496$ nm. Then, in order to reach $T_c = T_{\text{room}} = 293$ K, it is necessary that $a \approx 0.374$. However, a decreasing of T_c is observed at increasing number of the CuO_2 layers, $x > 3$, that obviously occurs due to a *buckling* of the

CuO_2 layers at diminishing a. All the above-mentioned cuprates are hole-doped. In 1989, single cuprate family was discovered to be electron-doped: (Nd, Pr, Sm)CeCuO with critical temperature $T_c = 24$ K.

Periodically appeared information on the sightings of superconductivity at room temperatures and even above 300 K today are doubtful because they do not satisfy the four criteria to determine the superconductivity existence, stated in 1987 by the researchers who discovered the HTSC [285]: (i) zero resistivity, (ii) Meissner's effect marked (when by decreasing temperature and magnetic field below the critical values, the total displacement of magnetic flux from conductor, transforming into superconductor, is observed), (iii) high reproducibility of results, and (iv) high stability of effect. Today, the mistakes of the superconductivity effect statement at temperatures above 130 K are linked usually with two general problems, namely:

(1) with experimental mistakes of non-practiced experimentalists having insufficient knowledge of the modern measuring methods or with *what* should be demonstrated to prove the superconductivity of material (the last cause is met very seldom today due to development of this field);
(2) with discovery of "new" superconductors really being known superconducting chemical compounds involving some additional components.

Note that the last problem is very difficult even for experienced researchers.

The main cause that decreased the interest of scientific society considerably to HTSC in the middle of 1990 and diminished considerably financial supporting (in particular, in USA), directed to these aims, concluded that the high-temperature superconductors could not replace low-temperature superconductors and attain sufficiently wide applications. The main obstacle is the material brittleness, which is intrinsic for HTSC compositions. It forms microstructure defects (voids, microcracks, damage, etc.) during sample preparation and loading that can be rapidly transformed to macrodefects and degrade the superconductivity properties. Moreover, the critical current density, J_c, was found to be the key parameter for engineering application than the value of T_c. Its value is the limit magnitude of direct undamped electric current in superconducting sample, above which the sample transforms to normal (i.e., non-superconducting) state. In particular, the (Bi, Pb)$_2Sr_2Ca_2Cu_3O_{10+x}$/Ag tapes possess critical current density up to 80 kA/cm^2 (at 77 K and 0 T) that changes in dependence on the conductor length and shape [487]. At a temperature of 4 K and in the absence of the magnetic field, their value of J_c increases three times. The volume critical current density, $j_c = (2–5) \times 10^6$ A/cm^2 in $Bi_2Sr_2CaCu_2O_{8+x}$ monocrystal, found on the basis of the measurement of the critical current in Cu_2O_4 surface layer (i.e., pair of parallel CuO_2 layers) at $T = 4.5$ K [1665], evidently, is also the limit for Bi-2212 thin films and tapes. Note also that the critical current density $J_c > 10^5$ A/cm^2 has been reached in Bi-2212 ceramics [329, 744–746].

Melt-processed Y(*RE*)BCO bulks (where *RE* is the rare-earth elements), the most prospective for applications, demonstrate values of $J_c > 100$ kA/cm^2 (at 77 K and 0 T) that decrease rapidly with increase of temperature and magnetic

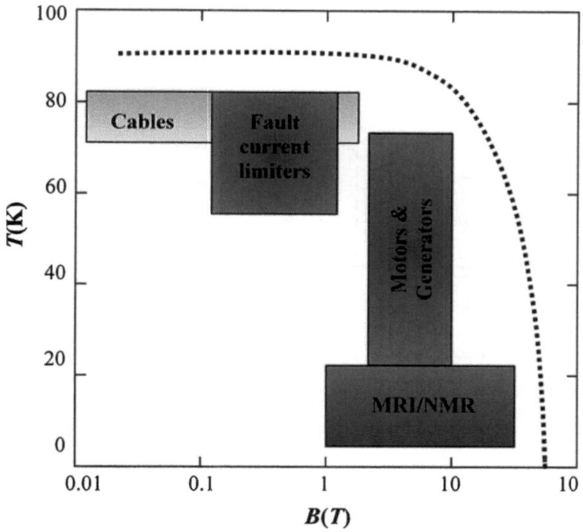

Fig. 1.2 Operating magnetic fields and temperatures where the HTSC tapes have performances adapted to the different applications. The irreversibility line (the upper field where dissipation-less currents can be achieved) in YBCO is shown by *dotted curve*

field [1461]. The best values achieved up to now in Y(*RE*)BCO coated conductors (CCs) are 200–300 A/(cm·width) at 77 K and self-field in kilometer length while in shorter lengths values as high as 1000 A/(cm·width) have been demonstrated by using vacuum deposition technology [654]. For comparison, the low-temperature superconductors (LTSCs) of NbTi family possess $J_c = 300$ kA/cm^2 (at 4 K and 5 T), but Nb_3Sn samples demonstrate $J_c = 100$–200 kA/cm^2 (at 4 K and 10 T) [978].

Remarkable discoveries were also made in another (no cuprate) superconducting systems. In particular, in 2001 in broadly accessible and very cheap MgB_2 system, $T_c = 39$ K [1224] was obtained. In 2008, the new era of iron based superconductors has started with the discovery of the layered pnictides and chalcogenides of iron. The superconducting state, invented in La(O, F)FeAs material at 26 K [806], very rapidly rose up to $T_c \sim 56$ K in *M*FFeAs (*M* = Ca, Sr, Ba, Eu) with superconductivity arising upon modification with Ln^{3+} ions [269, 1809]. These achievements again increased sharply an interest to superconductivity in the world.

1.2 Progress and Prognosis of Superconductivity Applications

At the beginning of 1990, the epoch of engineering applications of HTSC was started marked by considerable technical achievements that stated the use of HTSC in specific products and devices. In particular, achieved progress for different HTSC applications is presented in Fig. 1.2. A prospective for global market of superconductivity is depicted in Figs. 1.3, 1.4 and also in Tables 1.1, 1.2.

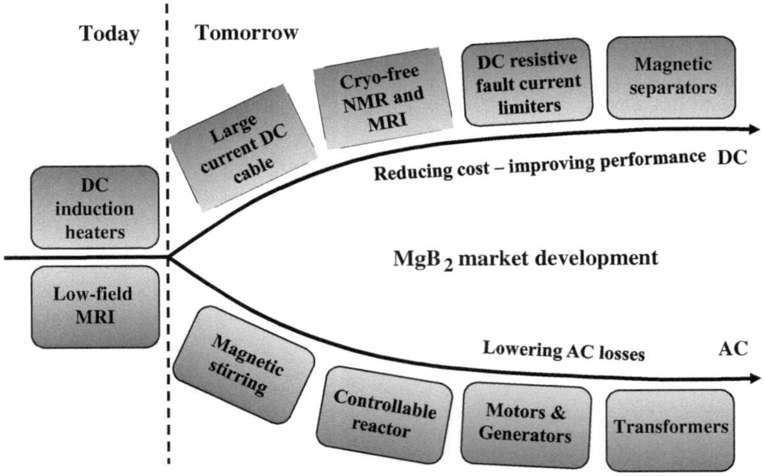

Fig. 1.3 Prospective of MgB_2 market development

Fig. 1.4 Prognosis for global market of superconductivity [303]

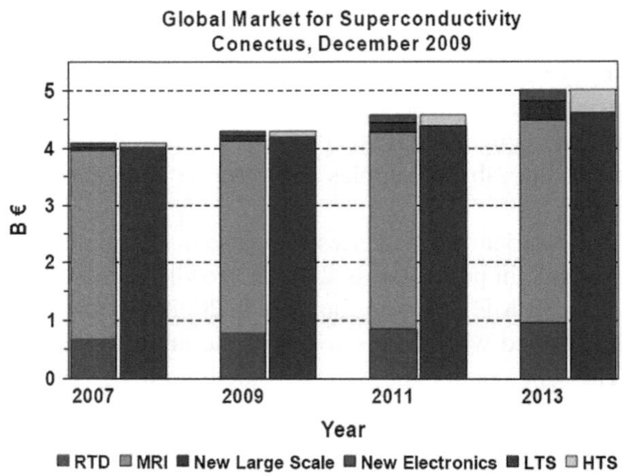

Historically, superconductors have been utilized in science, research and technological development (RTD), and in military fields. The next stage of the development included medical applications due to weakness of competition from non-superconductive devices. Today, niobium-based LTSC (NbTi and Nb_3Sn) are used in wire form in some applications, for example Magnetic Resonance Imaging (MRI), nuclear magnetic resonance (NMR), silicon crystal growers, different magnets (from small laboratory samples to enormous systems for large facilities), etc. They constituted a world market of superconductivity exceeding 4 billion euro in 2009 [303]. It should be noted, that the wider adoption of superconductors for power electric devices has been limited by either the cooling cost and complexity

Table 1.1 New markets for large scale and electronics applications of superconductivity [303]

Application	2009	2010	2011	2012	2013
Magnets for new medical applications	▲ ▲	▲ ▲	▲ ▼	▼ ▼	▼ ▼
Magnets for industrial processing	◊ ▲	▲ ▲	▲ ▲	▼ ▼	▼ ▼
Fault current limiters	◊ ◊	◊ ◊	▲ ▲	▲ ▲	▼ ▼
Power cables	◊ ◊	◊ ◊	▲ ▲	▲ ▲	▼ ▼
Rotating electric machines	◊ ◊	◊ ◊	▲ ▲	▲ ▲	▼ ▼
Magnetic bearings & Levitation		◊	◊ ◊	◊ ▲	▲ ▼
Magnetocardiography	Δ Δ	Δ ►	► ►	► ►	◄◄
Digital circuits	Δ Δ	Δ Δ	Δ Δ	Δ ►	► ►
Filter systems for wireless communication	► ►	► ►	►►	► ►	► ►
Other electronic devices & Sensors	Δ Δ	Δ Δ	Δ Δ	Δ ►	► ►

Large Scale & Electronics:
◊ Δ pre-commercial orders related to RTD activities, field tests and prototype operation
▲► emerging market
▼◄ established market

Table 1.2 Global market for superconductivity (in M€) [303]

Business field	2007	2009	2011	2013
Research and Technological Development (RTD)	660	765	845	955
Magnetic Resonance Imaging (MRI)	3300	3355	3435	3525
Total of RTD and MRI	3960	4120	4280	4480
New large scale applications	65	100	155	325
New electronics applications	60	80	125	180
Total of emerging new business	125	180	280	505
Total market	4085	4300	4560	4985
Market shares for LTSC	4025	4205	4385	4600
Market shares for HTSC	60	95	175	385

for LTSC operating at 4.2 K in liquid helium, or the fabrication cost and alternating current (AC) losses of HTSC at a much increased operating temperature of 77 K in liquid nitrogen.

New competing fields are based on new superconducting materials and system designs. The highly cost-competitive market of energy, information and communication, industrial processing and transportation is actively and steady developed. Based on estimate of [303] LTSC and HTSC contributions in total market will of 4.6 and 0.4 billion euro in 2013, respectively.

In whole, the areas of existing and potential applications of superconductivity (both HTSC and LTSC) broadly described in [1313] and could be depicted in the form of symbolic "tree" (Fig. 1.5).

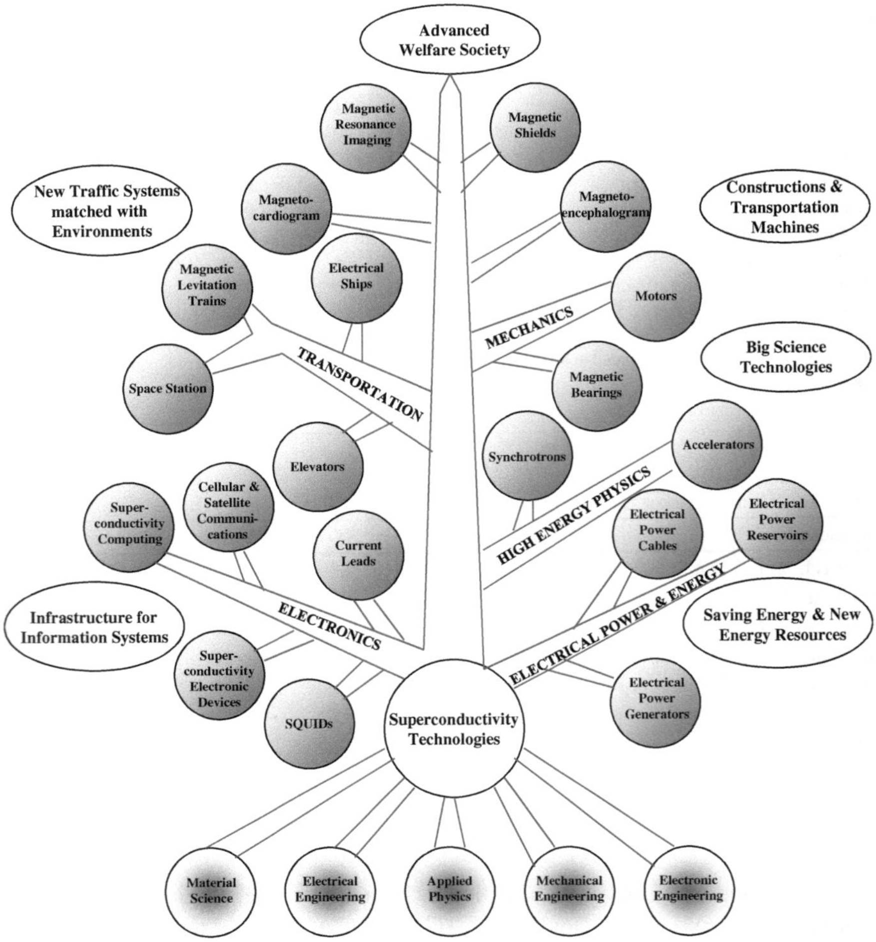

Fig. 1.5 Symbolic "tree" representing wide perspectives of already existing and potential applications of superconductivity [1656]

1.2.1 Technical Directions

These are present at the "roots" of the tree. However, the set of the disciplines pointed does not cover all possible applications. The experience shows that practically every scientific and technical direction—and also most social ones—can impact upon and be impacted positively by the future of superconductivity. Then we proceed clockwise, considering the branches of the tree.

1.2.2 Electronics

In this area, thin-film and SQUID-based systems,[3] as well as cellular and satellite communicating devices, are the most close to achieving commercial success. The next development of superconductivity applications in the area of electronic and information systems promises very fruitful competition between semiconductor and superconductor technologies, a competition that involves the state of development of cost-effective refrigeration and packaging circuits and devices. It may be assumed that existing advantages semiconductors will decrease more and more with solution of problems intrinsic for HTSC and taking into account unique properties of superconducting devices, namely, high rate of switching, small consumed power, weakened noise and high sensitivity to external electromagnetic radiation.

Superconductive electronics use three kinds of effects, namely: (i) low losses, almost non-dispersive and linear transport properties of LTSC and HTSC at frequencies which are lower than their frequencies of energy gap, (ii) quantum nature of magnetic flux, and (iii) two Josephson effects.

Today, the application of superconducting electronics, in particular, includes *antennas* (Josephson antenna lattices), passive super-high-frequency (SHF) *filters* for system of communications, hot electron *bolometers* (HEB), etc.

High-effective antennas for micro- and millimeter-waves are required due to growing needs in cellular, space and wireless communications. However, emission effectiveness of electric small antennas processed on the base of usual conductors is very small because of that the conductivity losses are greater than the emission losses. It could be supposed that superconductive antennas will be effective even for waves being shorter than the length of resonance wave. This circumstance is caused that the surface resistance of superconductive films in microwave range significantly lower than usual conductive films. Different types of HTSC antennas, in particular helical antennas [377, 761], meander patch [1630, 1631] and patch antennas [249, 1262] have been processed and studied. In [1632], it has been developed approach which did not demand of using lens and applied the tapered slot antenna (TSA) of type of the break line tapered slot antenna (BLTSA) proposed in [1857]. The method to determine two-dimensional distribution of critical current in superconductive thin-film devices in conditions of excitation THz-irradiation with focused femtosecond laser impulses have been developed in [631]. In this case, the bow-tie antenna was applied for mapping by using displacement current. The log-periodic toothed trapezoid antenna has been presented in [810] which introduced in Josephson junction (JJ) with aim to increase power itself Josephson emission. In [21], the micro-strip antennas with direct feeding energy and right-hand circular polarization (RHCP) have been developed. Table 1.3 [21] points the parameters used in design corresponding copper and superconductive (YBCO) antennas. Application of superconductors in

[3] SQUID is an acronym of superconducting quantum interference device.

Table 1.3 Designed parameters of micro-strip antennas

Parameters	Copper antenna	YBCO antenna
Substrate thickness (mm)	0.50	0.50
Dielectric constant (ε_r)	10.5	9.65 (at ~77 K)
Tangent of dielectric losses (tanδ)	0.011	6.1×10^{-6} (at ~77 K)
Film thickness (μm)	18	0.5
Surface resistance (Ω)	7.4×10^{-2}	1.15×10^{-5} (at 20 K)
Surface reactance (Ω)	7.4×10^{-2}	2.05×10^{-2} (at 20 K)
Penetration depth (nm)	–	219.27 (at 20 K)
$L = W$ (mm)	3.7	3.9
Wms11 (mm)	0.05	0.04
Wms12 (mm)	0.50	0.45

L and W are the length and width of square patch antenna, respectively; the feeding energy is directed by transducer with width Wms11 at width of micro-strip line of feeding energy equals to Wms12

patch antennas improves not only effectiveness of itself antenna but devices integrated together with it [422].

In order to manufacture small, light antenna lattices for radar and communication systems possessing low losses is required a superconductive device of phase shear. The receipt device of the phase shear processed by using S–N–S junctions (where S is the superconductor, and N is the normal metal) doped by Co has been proposed in [125]. The devices of the phase shear also were processed in conditions of low-temperature superconductivity by using magnetic connection (circuits) [412].

Several methods for combination of emission of the superconductive oscillators with open space have been demonstrated based on using arrays of quasi-optic junctions, two-dimensional arrays with a wave slot-line antenna [155], two-dimensional lattices with HTSC junctions [1089] and flux-flow oscillators with a tapered slot-line antenna [883]. One-dimensional arrays of superconductive oscillators could be used for off-chip applications due to they possess great required power and large impedance of source [884]. Josephson arrays of phase synchronization have been studied as oscillators into millimeter and sub-millimeter ranges.

In [951], it has been proposed and in [952] experimentally studied a new class of Josephson oscillators based on Cherenkov irradiation from Josephson flux lines (fluxons). As it has been shown in [253], the Cherenkov irradiation is perspective for improving existing Josephson oscillators of flux flow (flux-flow oscillators, FFO). They become to be competitive into frequency range of 200–600 GHz compared to usual non-superconducting electronic sources: for example, translation wave tubes (TWT) and back wave tubes (BWT).

Active antenna arrays applying semiconductive oscillators are used into millimeter and sub-millimeter wave ranges. Due to transition from single high-power micro-wave sources with electric and transmitting lines of the power distribution to the active arrays, in which each antenna has small oscillator, the complexities and losses, caused by transmission of radio-frequency power through the transmitting lines, may be eliminated. This common conception could be applied to the active antenna arrays in which oscillators to be arrays of small Josephson junctions (so-called, sub-arrays). The main case to use JJ-oscillators instead of semiconductive ones is determined that they can oscillate on extremely high frequencies (>1 THz) with high effectiveness [1023]. Triangle JJ-arrays working in external magnetic field have been proposed in [1858] in order to obtain high power of useful emission into millimeter and sub-millimeter ranges.

Based on HTSC approach, it is possible to manufacture filters possessing advantages compared to conventional ones. They demonstrate low losses of input signal, sharp bounds in the filter characteristics, super-low band-width, compactness and thermal stability. Ideal substrate for typical filter should demonstrate isotropic dielectric properties, very low tangent of dielectric losses ($<10^{-5}$), crystalline lattice and thermal extension factor which are similar with superconductive film, and also be not subjected to twinning. The superconductive films should have high critical temperature (T_c), high critical current density (J_c), low microwave surface resistance (R_s), and small nonlinearities at high level of microwave power.

The high-temperature superconducting filters for cellular stations present the first practical development of HTSCs attained now a wide application. HTSC filters increase in two times a capacity of phone channel. Potential advantages of HTSC applications in high-power filters consist in following: small size, light weight, very low losses in band-width, very high strip-off removal of hindrances and sharp bounds of characteristics. HTSC filters have more than one order of magnitude smaller sizes and weight compared to usual non-superconductive filters.

The superconducting filters of linear frequency modulation are capable to provide a width of band-width outside of limits of 1 GHz receipt and compression of the surface acoustic waves (SAWs) [181], or outside of limits of 2 GHz acoustic division multiplexing [34]. Only superconductivity is capable to support successively multi-Gigahertz band-width by using structure of extension filters. First, the superconducting filters of linear frequency modulation have been proposed in [1061]. They use two advantages of superconductors in systems of transmitting lines: their extremely low losses at microwave frequencies and without dispersion (that is independent on frequency) penetration depth. Significant progress has been achieved beginning since 1986 owing to application of HTSC thin films for passive analogous microwave filters. YBCO filter of linear frequency modulation with band-width of 3 GHz [1062] was one of the first demonstrated HTSC devices.

A bolometer is the heat non-selective receiver of radiation based on change of electric resistance thermal-sensitive element from metal, semiconductor or dielectric at its heating due absorption of the measured radiation flow. Bolometer has been invented by Langley in 1880 [95], and classic bolometer consisted of the radiation absorber and thermometer. First bolometers used metallic strip as absorber and in 1970 it has been replaced by superconductor. A bolometer is used for measuring summary power of radiation, and together with spectral device—for definition spectral composition of radiation. Superconductive bolometers working in depth cooling (3–5 K) are based on sharp change of electric resistance at transition of some metals and semiconductors from normal to superconducting state. In 1993, again took place a return to use of thin film of normal metal as the radiation absorber. These were HEBs in normal metal with Andreev mirrors for heat isolation of absorber from input electrodes. As thermometer, the tunnel junction "superconductor–insulator–superconductor" (S–I–S) has been used. The base of this idea has been constructed in [1225, 1226]. In [955] measured output signal from this bolometer not only on direct current, but on external noise signal in frequency of 300 GHz.

However, the Andreev mirrors demonstrate a set of shortcomings, therefore instead Andreev junctions, the tunnel junctions have been used [956], which for account of capacity form a good high-frequency contact between absorber and antenna. Moreover, the tunnel junctions carry out functions of thermometer and electronic cooler. So, it has been formed a new concept of cold electrons bolometer (CEB) in difference on HEB in usual bolometers. The advantages of CEB are obvious: instead heating the absorber film by displacement current and decreasing output signal and increasing phonon noise, in the bolometer with electronic cooling takes place decreasing electronic temperature, increasing output signal and decreasing contribution of noises of the following amplifier. In other bolometer construction [954], instead of normal metal film it has been applied superconductive absorber with trap from normal metal and superconductor. This led to effective pinning quasi-particles by normal metal due to proximity effect and also to decreasing effective volume of normal metal. As a result, significantly stronger electronic cooling of normal metal and additional rise of output signal were occurred.

The superconductive bolometers are one of most interesting applications of passive HTSC. In [1432], it has been shown that HTSC bolometric detectors can potentially on two orders of magnitude to be higher than sensitivity of existing analogues at or above temperature of liquid nitrogen for infrared waves with lengths more than 20 μm. By using superconductive bolometer as frequency transducer it has been observed transformation at the frequencies of >2.5 THz [815]. Traditionally, bolometers are used in radio-astronomy, where required high-sensitive and broad-strip detectors. They are also broadly applied in measured and laser techniques as receivers of infra-red radiation.

For superconductive mixers of frequencies, based on Josephson scheme of transformation, are required junctions demonstrating supercritical damping, so majority HTSC junctions naturally fit to this goal. Due to higher energetic gap, it could be waited they will act in the range of Terahertz frequencies. In contrast, possibilities of usual LTSC frequency transducers of the S–I–S type usually degrade above the frequency of 700 GHz. The advantages and shortcomings of HTSC Josephson frequency transducers are directly connected with the HTSC material properties. High energetic gap leads to high frequency of output signal upon to some Terahertz, and high superconductive transition temperature allows transducers to work in wide temperature range.

First, the concept of superconductive HEB-transducers has been proposed for receivers of sub-millimeter waves in [528]. As it has been pointed above, this technology for transducers is competitive with S–I–S transducers in frequencies near 1 THz and to be very prospective for applications on higher frequencies (at least, up to 100 THz). In [210], HTSC HEB has been designed and manufactured. The device used ultra-thin ($\leq$20 nm) YBCO-films passivated by YSZ-layer. In difference of LTSC, on the base of which have been processed high-qualitative S–I–S junctions with subcritical damping, the typical HTSC JJ demonstrates supercritical damping. Therefore, it is necessary to modify the concept of HTSC-receiver. In particular, in [130], the oscillator, based on discrete arrays of HTSC JJs, has been integrated with single HTSC JJ transducer.

Superconductive receivers of radiation are based on changing state(s) of superconductor (or system of superconductors) under action of radiation. Application of superconductors demonstrating a small noise level and strong non-linear properties allows one to attain a high sensitivity of the superconductive receiver close to theoretical (quantum) limit. Most spreading has the next kinds of superconductive receivers: the superconductive bolometers, receivers based on the Josephson effects (tunneling of coupled electrons) and receivers based on the single-particle tunneling. In [921], it has been developed completely superconductive integral receiver of sub-millimeter waves for space and Earth radio-astronomy, and also aeronomy. Important goal of investigations, directed to effective generation and sensitive detection of sub-millimeter radiation by using HTSC Josephson devices, is connected with manufacture integrated HTSC receivers. These devices are capable to detect radiation in Terahertz range and work above helium temperatures. In [919], the devices of integrated receivers, based on the LTSC technology and acting in the range of 500 GHz, have been developed. They used flux-flow oscillators and S–I–S transducers of frequencies. The progress achieved in development of superconductive integrated receivers (SIR) processed on quartz and demonstrating small noise [920], led to the next development of superconductive integrated array receivers (SIAR) for astronomy.

One of the main goals is RTD of superconducting computer. Superconducting Josephson qubits are two-level quantum systems, in which electric charge or magnetic flux is the degree of freedom. Josephson contacts, that is, two superconducting layers with dielectric buffer layer, present their basis. In 1999, the solid realization of the qubit in SQUID loop was proposed [757]. This idea is based on using "exotic" superconductors with symmetry of parameter of the superconducting order, D, which is lower than symmetry of crystalline lattice (e.g., HTSC $YBa_2Cu_3O_7$, possessing d-wave symmetry of D in superconducting CuO_2 layers). In 2000, new computer architecture, namely, *hybrid technology multi-threaded architecture* (HTMT), was started. The schemes of *rapid single flux quantum* (RSFQ) are the basis of the computer, and it is assumed that the superconducting supercomputer will carry out up to 10^{15} floating-point operations per second. The RSFQ conception is based on using shunted Josephson junctions. In 2003, a formation of the *entangled state* between two solid superconducting qubits [1339] has been demonstrated. In 2004, a coherent coupling of Josephson qubit with harmonic modes of transmitting line (for "charge" qubit [1761]) and with plasma oscillations (for "flux" qubit [278]) has been carried out. By disposing Josephson qubits on one chip, it is possible in principle to organize their coupling, by means of electromagnetic waves. In the case of actual HTSC applications, it is necessary to reach 5 % current spread per plate with 10^3 Josephson junctions.

Behavior of single Josephson fluxes has been observed in 2005 [1762] that opened a prospective their using as carriers of quantum information bits. In 2009 [363], by using the entangled quantum states, it has been demonstrated a two-qubit superconducting processor and the implementation of the Grover search and Deutsch—Jozsa quantum algorithms [357, 586].

While the proper time of de-coherence of superconductive qubits (~ 1 μs) is on some orders of magnitude lesser than for ions in catch, but at the same time the operations with them could be fulfilled significantly faster (in this case, in order to obtain final result were required only approximately ten operations). The qubit's number can be increased in many times by means of mechanic addition of superconductive circuits using a well-developed technique. However, a big problem is connected with super-low work temperatures (~ 1 μK).

Obviously, the moment of creation of the quantum computer will set when will be demonstrated a solution of any problem to be important with practical view and besides faster than on classical computer. Nevertheless, future superconductive quantum computer will allow one to form infrastructure for processing and following exploitation of cryogenic information systems, in particular will be solved the issues connected with requirement to use super-low work temperatures. This infrastructure will promote more wide application of superconductive computation devices. In this case, application areas of low-temperature electronics may be sufficiently broad and include analogous-digital and digital-analogous transducers, digital SQUIDs, digital autocorrelators, generators of pseudo-accidental signals and applications connected with high levels of radiation. One of the possible directions of the superconductive quantum computer application will be connected with existence of quantum algorithm which allow effective modeling different Hamiltonians describing coupling of electrons (for example, Bardeen–Cooper–Schrieffer (BCS) Hamiltonian) on quantum computer by using NMR. This algorithm will give a possibility to calculate spectrum of low-energetic excitations near energetic gap between ground and first excited states. In particular, it may be used for verification of applicability of the BCS model to mesoscopic superconductive systems (for example, ultra-small metallic grains).

In whole, the areas of application of HTSC products are divided into passive and magnetic ones based on the state (diamagnetic or ferromagnetic) of the sample in working regime (see Table 1.4). The capability of superconductor pushes off external magnetic field is used in the diamagnetic state (in this case, an existence of origin of the external direct field is assumed). On the other hand, superconductor itself becomes the origin of magnetic field in the ferromagnetic state (state with "freezed" magnetic flux).

Table 1.4 Passive and magnetic applications of HTSC

Passive applications	Magnetic applications
Magnetic bearings	Quasi-continuous magnets
Fly-wheels	Magnetic separation
Cryogenic pumps	MagLevs
Filling systems	Magnetic captures
Hysteresis motors (rotor)	Motors (stator)
Linear carrying capacitors	Magnetic dampers
Inertial transformers	Control of bundles of the charge particles
Electric drives	Magnetic apparatus
Magnetic screens	

1.2.3 Transportation

The next main directions of superconductivity applications in the aviation and astronautics have been formed, namely:

(1) *Systems of electromechanical start.*

In the construction of the newest single-pass motor of air- spacecraft, the necessary air inflow is attained at super-sound velocities of apparatus; it only demands initial acceleration of the craft. For this aim, it is possible to use the take-off superconducting platform does not leaving the airport limits, but racing aircraft up to sound velocity. The electromagnetic systems of start from Earth to its orbit, using HTSC magnet or linear synchronous motor, present themselves the high-speed electromagnetic guns for start to space of the heavy useful loads. Analogous starting systems are applied for replacement of the first step of the multi-step rockets.

(2) *Levitation systems.*

The passive superconducting magnetic suspenders for contactless suspension of shafts may be used to stabilize satellites, but electromechanical storages of energy to feed HTSC bearings.

(3) *Electric motors*, working on the principles of hysteresis and "freezed" magnetic flux.

The hysteresis motors developed consisting of cylindrical or disc HTSC rotors are based on the use of two types of bulk superconducting ceramics, namely (i) monocrystalline (YBCO) or bulk (BSCCO) elements and (ii) melt-textured YBCO samples. The constructive elements of motors of the synchronous–asynchronous type are the HTSC bulks, but *ac*-synchronous and *dc*-unipolar electric machines use HTSC wires (from BSCCO).

The main directions of superconductivity applications in various transport systems cover the following:

(1) *High-speed trains* and other kinds of ground transport on magnetic suspension.

The superconducting (SC) electromagnetic systems (using linear electric motors to speed up the truck) are applied in the development of high-speed ground transportation based on the SC windings placed in the route and SC magnets located in the vehicle. By this, in order to economically, accumulate and use energy effectively, the SC storages of electric energy that feed HTSC bearings are applied.

Magnetic levitation (MagLev) train, moved and controlled by magnetic forces, does not touch rail surface during motion in difference of traditional trains. Because a gap exists between train and surface movement, a friction is excluded and sole braking force is the force of aerodynamic resistance. MagLev, related to monorail transport, can attain speeds compared with aircraft velocity and allows one to compete with flies on small (for aviation) distances (up to 1000 km). Now MagLev could not use existing transport infrastructure, while there are projects with disposition of elements of the magnetic road between rails of usual rail-road or under motor way bed. There are three main technologies of magnetic suspension of trains, namely: (i) on superconductive magnets (electrodynamic suspension, EDS), (ii) on electric magnets (electromagnetic suspension, EMS), and (iii) on permanent magnets. The train levitates owing to repulsion of the same poles of magnets and on the contrary due to attraction of different poles. A big problem of designing is the great weight of sufficiently power magnets requiring strong magnetic field to keep in air the massive train.

Magnetic levitation systems based on the use of magnetic forces to "suspend" one object above another have been developed in Japan, the USA, Germany, France, Switzerland, China and Canada. HTSC systems have been demonstrated in China during realization of Germanic China project directed to create the levitated transportation mean. During the last 25 years, great efforts have been taken in Japan to develop high-speed trains on magnetic suspension by the application of superconducting systems. In particular, for the passenger train on magnetic suspension, caused by superconductivity, a speed of over 1000 km/h [1450] has been demonstrated! Today, an analogous train has been developed, called the Japanese Linear Chuo Shinkansen Project (MLX) with nominal speed of 500 km/h. At the same time, trains with superconducting magnetic suspension compete with high-speed trains (e.g., France train TGV and German Transrapid System with speed of 420 km/h). The Inductrack MagLev System is developed in the USA. The Swissmetro Project (speed of 320 km/h) should be noted. The German Transrapid System is based on "classical" electromechanical technologies, requiring small air gap (<12 mm) between the vehicle and the truck. The Japanese MagLev is based on active superconductivity magnetic systems, permitting large air gap (>80 mm). The USA MagLev is based on passive superconductive magnets, also permitting large air gap (>80 mm). Swissmetro presents a unique example of MagLev systems: it is designed to work under partial vacuum (<10 kPa). The vehicles run in tunnels of small inside diameter (5 m), which require partial vacuum in order to reduce the aerodynamic resistance. Independent magnetic systems permit to have medium air gap of 20 mm.

By using HTSC bulks in "freezed" magnetic flux regime in superconductive engines as a stator leads to sharp increase of rotation moment. These HTSC magnets can replace conventional NbTi coils which are used now in MagLevs.

The calculations show that to attain this aim the HTSC magnets should satisfy the following requirements at work temperature which are really-achieved in *RE*BCO systems: critical current density $J_c > 10^5$ A/cm^2, sample diameter is more than 10 cm and mechanical strength is more than 100 MPa [501].

Thus, based on the melted HTSC ceramic, the superconducting magnets are able to capture high magnetic fields (tens of Tesla) compared with the classical magnets, and allow to create ecologically pure high-speed ground transport systems with broadened air gap (the thickness of magnetic suspension). This defines the possibility to design rail-roads with lesser financial and operation expenses.

Finally note that in whole the advantages of MagLev are connected with (i) theoretically most high speed compared to the velocity of the usual ground (non-sport) transport kind, and (ii) low noise of the transport. The shortcomings of MagLev are determined by (i) high cost of construction and service of the route, (ii) big weight of magnets and electric energy consumption, (iii) the electromagnetic field created by magnetic suspension may be bad for the train service personal, near-disposed inhabitants and passengers who use cardio-stimulators, (iv) requirement of super-rapidly acting control systems which at high speed (hundreds kilometers per hour) could control effectively a gap between rail-road and train (some centimeters), (iv) complexity of route infrastructure (for example, MagLev points consist of two rail-road sites changing one of other in dependence on direction of the train turn; this determines very big difficulties in organization in the slightest degree branched rail-road nets with branching and crossing), (v) while the standard rail-roads reconstructed to high-speed transport remain to be accessible for usual passenger and suburban trains, at the same time MagLev route cannot be used in another way and this circumstance will require additional rail-roads for low-speed transport.

Moreover, the catastrophe of the improved train prototype on magnetic suspension by September 22, 2006 near Lathen (Germany) in which perished 23 passengers stated as one of the most problem, the problem of safety of the high-speed ground systems. In this case, one of the main problems of the safety is connected with comparatively lightness of the train weight on the magnetic suspension. As a result, any hindrance on track becomes a problem (the bigger weight of the hindrance is more danger).

(2) *Electromobiles and automobiles.*

HTSC electric motors for electromobiles and automobiles act on the principles of hysteresis and "freezed" magnetic flow. The superconducting wires, bulks and thick films are used in construction. The next electric devices and machines, using superconductivity, that are developed now:

- *ac*-synchronous and *dc*-unipolar electric motors, using HTSC wires (from BSCCO) and acting at the temperature of liquid hydrogen;
- hysteresis motors with cylindrical and disc YBCO rotors that are developed on the basis of monocrystalline YBCO- or BSCCO-ceramic, and also melt-textured YBCO bulks;

- motors of synchronous–asynchronous type, consisting of HTSC bulks;
- motors with rotor from layered composites, namely YBCO bulks (thick films) with intermediate ferromagnetic inserts (steel plates);
- HTSC levitated systems, realized in electromechanical storages of energy and levitated suspenders that use high-speed electromobiles.

(3) *High-speed ships and Navy.*

The linear synchronous electric motors are applied in starters of aircraft carriers. The superconducting magneto-hydrodynamic systems are used in Navy with the aim of decreasing the noise detection of motor in torpedoes and in the development of high-effective motors for fast-moving ships. There are researches directed to the use of superconductivity in the ship impulse source of energy, for example, for aircraft carrier. The Navy has intention of creating a warship with unique superconducting equipment and apparatus [588], including: (i) engines, (ii) magnetic mine-sweepings for fishing out of sea mines, (iii) board systems of feeding on the basis of inductive storage and (iv) radio-locator, using superconducting magnet in the protection system of the ship.

Superconducting electric engines have small sizes and weight, consume small capacity. Their important advantage consists in the absence of mechanical moving details that exclude acoustic noise at the ship movement. In 2003, it has been constructed and tested 5 MW HTSC engine for sea ships. Fast discharge is the advantage of superconducting storage of the electric energy, which could be important for applications in guns, catapults, torpedoes with electromagnetic launching. The clearance of the coastal waters from sea mines is another important problem of the Navy. The mechanism of sea mine launching is connected with their sensitivity to the magnetic image of ship (i.e., with perturbations of Earth's magnetic field under influence of magnetic mass of the ship). Small ship with superconducting magnet can imitate the magnetic image of moving ship, thus collecting the mines. Under large magnetic field (for this, the superconducting magnet is used) the mines can be blown up sufficiently far from the magnet. Today, HTSC magnet, which may be placed in helicopter and models fields, imitating moving ship, has been developed. With the aim to find and counteract small, low-flying rockets, the Navy develops powerful high-frequency radio-locator on the base of superconducting magnet. In this case, interaction of electronic ray with magnetic field creates electromagnetic radiation with frequency proportional to the magnetic field.

1.2.4 Medicine

At the top of our technological tree are seen benefits, connected with increase of human welfare and, especially, with development of medicine. The main problem is to use the uniquely useful, ecologically safe and energy-saving potential of superconducting materials and magnets. In particular, it relates to use of MRI. This method rapidly established itself as a new and virtually indispensable medical diagnostic tool. The perspectives of the method are caused by progress in the technique of electronic image and cryogenic refrigeration. The base application of MRI involves visualization of concentrations of "hydrogen molecules" or "liquid" content of the various organs in the body. The use of even higher field superconducting magnets (so-called *functional MRI*) allows one to state very accurately the distribution of other chemical elements in the human body that exist in much more limited concentrations than hydrogen. Another perspective direction is the use of SQUIDs that are the most sensitive detectors of flux or magnetic fields. SQUIDs are used to measure brain waves and brain functions. Similar excitement surrounds the use of SQUIDs in cardiology. Their other applications cover microbiology, biomedicine, high-energy physics, nanoparticle magnetism, non-destructive control, archeology, geology, and SQUID microscopy. Medical HTSC tomographs can be used to control the quality of goods, in particular of food products.

1.2.5 Mechanical Systems

The application of superconducting motors shows that such motors could be at least 20 % more efficient than typical present products. The total energy saving on use of this technology could be enormous. In particular, 5000 horsepower motors have been demonstrated [392]. Other examples of superconductivity applications are the devices of energy accumulation and storage based on the levitation phenomenon, high-field magnets for laboratory research, superconducting magnets for specialized fabrication processes of the chemical and pharmaceutical products, industrial growing silicon crystals, material separation and so on.

A restriction for wide application of superconductive magnets for growing monocrystals and magnetic separation of minerals is caused by complexity of the refrigeration systems. The situation is changed due to started in 2000 cryogenic cooler manufacture without application of liquid helium. Namely owing to these cryocoolers, it has been achieved a significant progress in development of HTSC current enters and LTSC solenoids (10 T in work gap of 100 mm, and 14 T at gap of 25 mm), and also demonstrated in Japan potential possibilities of HTSC-magnets of the Bi-superconductive family (7 T in gap of 80 mm). Application of magnetic field to melt is effective at growing large silicon monocrystal with high purity (development of HTSC magnet for this aim began in 1997). High magnetic fields prevent convection of melted silicon caused by large thermal gradients between walls and central part of crucible. Development of HTSC magnet for these goals began in 1997. For separation of minerals, in particular refining kaolin from metallic admixture is actively applied devices with LTSC magnets and USA projects of HTSC separators started in 2000. The magnetic separation into great gradients of magnetic fields is used for refining rocks from magnetic admixtures. Traditionally, with this aim the copper electromagnets with iron core and magnetic field up to 2 T are applied. However, these devices have high energy consumption.

The situation is simplified by using HTSC magnets, for example based on Bi-2223 tapes in Ag shell.

In 2002, in Japan, it has been designed based on superconductive magnets new system for industry purifying water-pipe waters and sewage from pollutions. The system removes upon to 85 % of phosphor containing in water. The system is compact and occupies only 10 % of the square required for modern systems, its manufacture cost on 30 %, and exploitation on 20 % cheaper than similar systems.

HTSC current enters is one of the high-temperature superconductor development which finds today a significant application. In 1998, in Japan, the HTSC current enters have been stated on superconductive magnet (Nb_3Sn) with 52 mm heat hole and field of 15.1 T. The length of the $Bi_2Sr_2Ca_2Cu_3O_{10+d}$ current enters was 180 mm, external and internal diameters were 23 mm and 20 mm, respectively. The current enter passed critical current up to 1 kA at 77 K in absence of magnetic field. In order to process HTSC current enters are used Bi-2212 rods, tubes and tapes and also Y-123 melt-textured rods. One of construction including HTSC current enters is current-restricting reactor intended for regulation of reactive power of energetic systems. In fact, the reactor is the coil which has constant inductivity, serving for restriction of short circuit currents and supporting a voltage on buses at accident regime.

1.2.6 Scientific Research

The application of superconducting magnets has probably been the most radical event for high-energy physics.

Before 1999, three coils (FNAL,[4] DESY[5] and Dubna nuclotron in Russia) for acceleration of hadrons existed in the world, in which the "magnetic path" has been fabricated based on superconductive magnets from NbTi and Nb_3Sn. In 1999, fourth superconductive coil of Relativistic Heavy Ion Collider (RHIC) began its work in Brookhaven National Laboratory (USA). In order to manufacture its 1800 superconductive magnets were required 21 millions meters of superconductors. The supplier of primary ion beam for RHIC became oldest Brookhaven synchrotron (Alternating Gradient Synchrotron, AGS) constructed in 1960.

Note, in particular, most powerful in the world an accelerator of high-energetic charged particles (Large Hadron Collider), constructed in Europe by CERN[6] consortium. In total, approximately 1200 tons (50,000 km) of NiTi superconducting wires and cables were required, which were supplied by several companies of Europe, Japan and the USA. 27-km tunnel was equipped by 1200 dipole (each of

[4] Fermi National Accelerator Laboratory (USA).

[5] Deutsches Elektronen Synchrotron (Germany).

[6] CERN is European Laboratory for Particle Physics (Switzerland).

which has length of 15 m) and 400 quadrupole (with length of 4 m) superconductive magnets (with fields up to 8.4 T) using NbTi cable with cooling down to 1.8 K.

The proton beams were successfully circulated in the main ring of the LHC by November 20, 2009. First recorded proton–proton collisions occurred 3 days later at the injection energy of 450 GeV per beam. After the 2009 winter shutdown, the LHC was restarted and the beam was ramped up to 3.5 TeV per beam (half its designed energy). First planned collisions took place by March 30, 2010 between two 3.5 TeV beams, a new world record for the highest-energy man-made particle collisions. The LHC will continue to operate at half energy until the end of 2012. It will not run at full energy (7 TeV per beam) until 2014.

High-temperature superconductive round wires from Bi-2212 (4.4 kA, 0 T, 4 K) is planned to use for magnets of "Very" Large Hadron Collider (VLHC) developed in Lawrence Berkeley Laboratory (USA). Note also, that there are other Collider projects, in particular in Russia.

Idea of TOroid CAmera with MAgnetic Coils (TOCAMAC) being a device for retention of high-temperature plasma with help of high magnetic field has been proposed by Tamm and Sakharov in 1950 and first experiments these systems began in 1956. On European TOCAMAC JET[7] and Japanese JT-60 (Japan Torus) output power approximates to input one. A record result has been attained on JET: maximal output power was equal to 16 MW. Re-equipment of fusion reactor JT-60 started in 2006 was caused by statement in it superconducting magnets. This will allow one carry out preliminary tests of possible work scenarios of controlled thermonuclear synthesis.

International Thermonuclear Experimental Reactor (ITER) is the fusion reactor of TOCAMAC type, which should prove that the process of regeneration of energy occurring on Sun may be launched on Earth. The objective of ITER is a demonstration of practicability of creation of the fusion reactor and solution of physical and technological problems stating in this way. The core of ITER is the giant vacuum vessel rounded by several types of giant superconductive coils (19 coils of torroidal field with total weight of 900 tons) which can retain hydrogen plasma heated to hundreds of millions degrees. Hydrogen isotopes in rapid reaction of synthesis in superhot plasma ($^2_1H + ^3_1H \rightarrow ^4_2He + ^1_0n + 17.6\,\text{MeV}$) will generate approximately in 10 times greater heat than is necessary to spend for the reactor work. The projected characteristics of ITER are present in Table 1.5. They guarantee that the future fusion electric station will work out excessive electricity. ITER uses carefully selected configuration of magnetic fields. In 2006, Evans from General Atomics (San Diego, CA) proposed to change this system. His research group demonstrated a method which was able to prevent potentially dangerous plasma discharge which is similar to flare and can damage key parts of the reactor. This discharge occurs when plasma "embeds" into weak regions of magnetic field retaining it. By modifying TOCAMAC reactor *DIII-D* of General Atomics, Evans group could solve the problem of gradual but no explosive transformation of

[7] JET is the Joint European Torus—Europe's largest Fusion Device (UK).

Table 1.5 The projected characteristics of ITER

General radius of construction	10.7 m
Height	30 m
Great radius of plasma	6.2 m
Small radius of plasma	2.0 m
Plasma volume	837 m^3
Magnetic field	5.3 T
Maximal current in plasma column	15 MA
Power of external heating plasma	40 MW
Fusion power	~500 MW
Factor of increasing power	In 10 times
Mean temperature	10^8 °C
Impulse duration	>400 s

plasma excess. This idea is very perspective for ITER, however for its realization may require statement of expansive superconducting coils near or even inside of bounded space of the torus in which generates plasma. The fusion reactor is safer than atomic reactor in relation to radiation due to comparatively small quantity of radioactive substances. An energy which can be emitted as a result of any accident also is small and cannot lead to the reactor fracture. Nevertheless, the safety problems state in the center of the ITER project. There are several sources of possible radioactive pollution, namely: (i) radioactive isotope of hydrogen—tritium, (ii) radioactivity introduced in materials of the device as a result of neutron irradiation, (iii) radioactive dust forming as a result of influence of plasma on first wall of ITER, (iv) radioactive products of corrosion which can initiate in cooling system. The estimates have shown that even in the accident case, radioactive outbursts will not present danger for inhabitants.

The participants financing ITER project (its cost is equal to $12 billions) are EU, India, China, Russia, USA, South Korea and Japan. The reactor will be constructed during 2007–2015 in Cadarache (France). After that, it is suggested fulfilment of experiments during 20 years with following finishing this project.

Scanning HTSC SQUID microscopes have highest sensitivity to weak magnetic fields in broad frequency diapason. They are able to register magnetic fields generated by vortex currents, magneto-tactile bacteria and currents of leakage in integral schemes. Electron-ray origins of multi-charged ions are used widely in laboratories. The parameters of HTSC origin [1229] are sufficient to create the multi-charged ions, namely, helium-like xenon and neon-like uranium. It is used to study a surface modification of different materials by the multi-charged ions, using the scanning probe microscope.

Nearest broad application of SQUIDs, as rule, is connected with medical diagnostics (magnetocardiography). However, in the last years, SQUIDs received interest as sensitive elements for non-destructive magnetic control of big metallic constructions (aircraft frames, bridge constructions, etc.). Modern methods of non-destructive control of these constructions, based on measurement of electric resistance change, fix a fact of existing fracture (for example, initiation of crack). The magnetic control, based on SQUID application, can registers formation of plastic strain which precedes the crack initiation. Typical SQUID system for non-destructive magnetic control consists of coils exciting low-frequency (usually of 80–130 Hz) currents in specimen and SQUID registering magnetic field of the excited currents. A simplified construction of this system can manage without application these coils of excitation. HTSC films, based on YBCO ($T_c \sim 90$ K), are used basically for film elements of the SQUID systems. In 2001, SQUIDs of direct current have been fabricated on the base of one-layer technology from epitaxial films of $Tl_2Ba_2CaCu_2O_x$ (TBCCO-2212, $T_c \sim 125$ K) on $SrTiO_3$ bi-crystal substrate [626]. These SQUID magnetometers are well base for electronic gradiometer intended for work without magnetic screen. In 2006, it has been constructed SQUID with Josephson junctions from one-wall carbon nanotube [299]. After precipitation of the nanotube on substrate, its exact disposition has been defined by using atomic force microscope, and then SQUID loops and junctions were formed by method of electronic-beam lithography. Due to small sizes of nanotube junctions their electronic energy spectra to be discrete as in quantum dots. This allows one to "switch in" and "switch off" for account of shear of size quantization levels of nanotube relatively Fermi level in superconductor supplying voltage on electric gates. This SQUID may be used, for example for studying the processes of re-magnetization of nanoparticle or even single molecular, if locate it on one of two junctions (cross-section sizes of molecular and nanotube are approximately the same, therefore the factor of couple of the molecular spin with circuit is far greater than in usual SQUIDs). Moreover, simple electric controlling direction of super-current in junctions becomes possible in these SQUIDs.

Many constructions of magnetic microscopes based on SQUIDs have been proposed and realized practically. First variants of these microscopes used LTSC SQUIDs working at liquid helium temperatures and intended for study of samples be under low temperature. Substitution of LTCS by HTSC SQUIDs significantly extended area of studies and stimulated their commerce applications. In this microscope, the specimen is scanned by closely disposed SQUID, and computer registers signal from SQUID in dependence on its location relatively of the sample. In first versions of SQUID microscopes, the specimen and scanning system were completely dipped into bath with liquid helium, liquid nitrogen or in evacuated volume rounded by liquid coolant. By using these systems, it has been attained spatial resolution of 10 μm at sensitivity to magnetic field of 20 $pT/Hz^{1/2}$ [1799]. The use of the SQUID microscope led to significant difficulties connected with matching and disposition of SQUID relatively of a sample and also at dipping and re-dipping of the specimen. In 1997, the microscope constructions in which a sample may be tested at the room temperature were developed. In these versions, evacuated volume with SQUID was separated from sample by thin window from non-magnetic material (for example, sapphire). Moreover, the microscopes with refrigerators of close cycle have been manufactured. For example, in SQUID microscope intended for non-destructive control of vortex currents, the Joule-Thompson refrigerator has been used in which absented moving parts on cold end.

SQUID microscope with refrigerator of close cycle has been also manufactured in which moving mechanical components carried out into compressor, that excluded noises caused by vibration near SQUID. In 2002 SQUID-microscopy has been replaced by laser SQUID-microscopy which allowed one to significantly extend a scanning area and the method sensitivity, and also its application in the micro-scheme processing [906]. Main advantage of the new approach consisted in possibility to control micro-schemes without connection to power supply that was general shortcoming of all previous methods. This allowed significantly improving technique of non-destructive control for searching defects in silicon substrates and integral schemes.

Superconductive detector of X-ray radiation has been based on creation of strongly non-equilibrium occupancy in comparatively small control current. X-ray photon, found oneself in superconductor, creates excessive quasi-particle excitations which then are fixed by using, for example superconductive tunnel junction. In order to compel this detector to work effectively, the idea of trap has been proposed in 1987 [156]. The detector device presented oneself S–I–S junction with small sizes jointed with absorber in the form of superconductive film having energy gap and physical sizes significantly greater than corresponding parameters of the S–I–S junction. Excessive quasiparticles, created in the absorber, diffused in one of the junction plates, rapidly sank down on the gap level and then sufficiently long time to be on it with following either recombination in Cooper pairs or crossing potential barrier revealing oneself in the form of tunneling current.

Superconductive detector of photons is based on study of hot electrons in ultra-thin superconductive films. Superconductive tunnel junctions are used for spectroscopy of single photon into range of the wave lengths from X-rays to almost infra-red irradiation. In 1999, it has been presented a single quasiparticle photon counter (SQPC) demonstrating combination of elements of superconductive and single electron transistor (SET) technologies for fabrication of detector with waited sensitivity of 10^{-20} W/Hz$^{1/2}$ and even higher [1508]. In 2001, the single-photon detector (photon counter) into diapason from ultraviolet to infrared waves was created [916]. This detector can register light flows from 1 photon per minute on the detection square up to 10^9–10^{10} photons per second. In 2002, superconductive single-photon detector begun to apply in micro-scheme tester for control of processed scheme in various stages their manufacture. The work of this tester is based on the picosecond imaging circuit analysis (PICA) [830]. The PICA method has a high resolution and allows one to collect data in parallel way from many individual transistors with sufficiently great area dividing them on time.

Record parameters of magnetic field transducers [1866]:

(1) Sensitivity of LTSC and HTSC *SQUIDs* is equal to 10^{-15} T/Hz$^{1/2}$ (1 Hz, 4 K, screened room) and 5×10^{-14} T/Hz$^{1/2}$ (1 Hz, 77 K, screened room), respectively.
(2) Sensitivity of giant magnetic resistance (GMR) *spin valves* is equal to 4×10^{-10} T/Hz$^{1/2}$ (1 Hz, 300 K, without screening) and 4×10^{-11} T/Hz$^{1/2}$ (1 Hz, 4.2 K, without screening).

(3) Sensitivity of *GMR spin valves with superconducting transformer of flux* is equal to 10^{-12} T/Hz$^{1/2}$ (1 Hz, 77 K, without screening) and 3×10^{-13} T/Hz$^{1/2}$ (1 Hz, 4.2 K, without screening).
(4) Sensitivity of *atomic magnetometer with sizes of microelectronic chip* (at minimal consumed capacity) is equal to 5×10^{-11} T/Hz$^{1/2}$ (10 Hz).

1.2.7 Electric Power

The superconductivity could render potentially enormous influence on the electric power industry, in particular, the power generation, power storage, power transformation and distribution and also the improvement and assurance of power quality. Even though the efficiency of power transmission lines, motor generators and especially electric transformers is already very impressible, energy losses due to use of ordinary *resistive* copper and aluminum conductors are enormous. Taking into account the basic influence that energy renders on all economic sectors the use of superconductivity in this field is connected with even more enormous perspectives. LTSC superconducting windings for electric power generator equipment to reduce energy losses have been carried out in a number of countries. While some very challenging engineering problems were successfully overcome, the economics, defined by using liquid helium and existence of high magnetic fields, pose extra challenges. In this case, the application of HTSC, which do not lose their properties at liquid nitrogen temperature, provides, in perspective, considerable advantages. For example, while the earlier approach involved the use of homopolar machine design—to have the windings exposed to *dc*-magnetic fields only—the present HTSC design involves *ac*-synchronous approaches, which also maintain *dc*-fields around the superconducting windings [393]. In addition to lower losses, through the utilization of superconducting windings instead of copper, the decrease of size and weight should also lead to lower overall cost of product. Finally, greater reliability and potentially longer equipment lifetimes can be expected due to a constant and much cooler operating temperature environment.

The problems of thermonuclear energetic are directly connected with the use of high-field superconducting magnets (the proposals of TOCAMAC, LHC, ITER, etc.). While low-temperature superconductors will continue to be employed initially, the potential applications of HTSC in this field have great perspectives.

A very attractive feature of superconducting magnets is the fact that they can be considered to be “electromagnetic batteries”. They can store enormous amounts of energy for long time sufficiently. Compared to electrochemical batteries, SMES[8] systems—although initially more costly—are considered to be more efficient, environmentally clean and less expensive for long term [198]. The SMES systems are integrated in autonomic complexes and intended for their continuous work

[8] SMES is superconducting magnetic energy storage.

during sufficiently long time. Usually, these systems include also refrigerator (cryocooler), HTSC current enters, powerful electronics with system of continuous feeding. SMESs are used, for example for bomb shelters and industry lines of electric transmission. Bad weather conditions, transport accidents and unpredicted breaking equipment work could lead to interruption in electric energy feeding or voltage jumps in electric net, equipment failure and breaking industrial processes. The situation in most degree could be changed at application of SMESs. Power electronics "sense" instantaneous re-distribution of power in electric net and during 2 s re-switch the net on feeding from the storage. After stabilization of electric power in the net, it is fulfilled back re-switching.

Successful feasibility demonstrations of superconducting transformers with HTSC windings have been carried out in Japan, Germany and the USA [1109]. The economic savings, operational reliability and environmental benefits are the main advantages of superconductors, used in this case. Because of HTSC capabilities, they can readily and rapidly replace existing transformers, while providing higher ratings, overload protection capability and smaller footprints for equivalent ratings of conventional transformers.

Superconducting cables are of considerable interest to the electric industry because they offer: (i) efficiency gains in transmission lines, reducing resistive losses, and (ii) opportunities to replace existing underground cables with much higher capacity cables, than in existing pipelines, that also lead to their quantitative decrease. HTSC cables one of the perspective developments high-temperature superconductivity for energetic applications. Now, two constructions, which to be principally different one of other, are developed, namely with heat and cold (at the cryogenic temperatures) dielectric. In the construction with cold dielectric the current-carrying element of cable is rounded by coaxial superconductive layer intended for screening magnetic field. The dielectric, impregnated with liquid nitrogen, disposes between the current-carrying element and external screening layer. The objective of this construction consists in excluding losses on alternating current, caused by influencing magnetic field initiated by current in adjacent phases and also by vortex currents induced in metallic parts of adjacent equipment. The cables with heat dielectric do not consist of this superconductive screening layer therefore their cost is significantly lower. A result of using this construction is a lower expense of superconductor and application of usual insulator materials in the contrast to new dielectrics which should be developed anew. Moreover, because the cable with heat dielectric is constructively similar to usual cable then in its processing, mounting and jointing could be used multi-times verified and reliable previous technologies. The narrower diameter of the superconductive cable with heat dielectric allows one to use it in existing lines of electric transmission. In many cases, the conductors of adjacent phases can be lightly removed in significant distance one of other. Due to the necessity of magnetic screening is eliminated. In spite of initial technological and economic advantages of the construction with heat dielectric, the construction with using superconductive screening element can provide most full utilization of the superconductor's properties. First HTSC cables have been installed in 2001, in Amager

Koblingsstation (Denmark), namely three pieces of superconductive cable, each with length of 30 m in electric net with voltage of 30 kV. In 2006, HTSC cable with length of 350 m carrying a current greater than 200 A has been stated in electric net of New-York.

The requirements to HTSC wires for different electric technical devices are presented in Table 1.6, and properties of main superconductors are shown in Table 1.7.

Another unique application of superconductivity in energetics is the development of approaches for single-phase and multi-phase systems having significant near-term commercial potential. In particular, fault current limiters are devices utilizing the ability of superconductors to act like a resistance "switch," capable of exhibiting zero resistance, when superconducting, but capable of returning to a higher resistance state, when critical temperature, current or magnetic field limits are exceeded. A need in the superconductive fault current limiters is very great because they allow considerable increasing net capacity and simplify requirements to circuit breakers staying for them. Many types of the superconductive fault current limiters have been proposed as with independent action as operating from the system of detecting short circuit. Existing circuit breakers to be slow and this is one of the cause for necessity of their replacement. In the both types of the HTSC fault current limiters, is used the transition of superconductor into normal state. This lowers quick-action of the current limiter and leads to slow restoration of superconductive state that is readiness to take new current impact. However, there is one type of the current limiters without these shortcomings, namely superconductive fault current limiters with biasing magnetic core. Superconducting winding holds the magnetic core in magnetized state. In waiting regime, an inductivity of the system is small till will be initiated short circuit and through power winding a current flows of high amplitude. In this case, the magnetic core goes out steady state, inductivity rises leading to restriction of alternating current in net. Advantageous of the construction with the magnetic core consists in that the winding remains always in superconductive state and actually work in the regime of direct current (it is not the least of the factors for superconductor no loving the alternating current). An operation of that current limiter occurs very rapidly during first quarter of period after coming short circuit. At the same time, the main shortcoming is connected with inconvenience caused by the magnetic core.

SuperFoam, synthesized from $YBa_2Cu_3O_z$ ceramic [1421], may be ideal material for the fault current limiters. Advantages of superconducting foam under "tape" and "bulk" devices from BSCCO include the following ones [1252]: (i) it endures critical currents at $T = 77$ K, which are significantly higher, compared to Bi-2212 phase, having problems with pinning centers at this "high" temperatures, (ii) it has sufficiently high electric resistance at room temperature in order to dissipate in heat an energy of supercurrent and (iii) it switches rapidly from and in superconducting state. The last two properties are forced in significant degree due to the foam includes open pores, in which there is cryogenic cooler. This provides continuously a directed contact with liquid nitrogen, compensating bad heat capacity of ceramics in the bulks.

Table 1.6 The requirements to HTSC wires for different electric technical devices [1235]

Device	J_c (A/cm^2)	Field (T)	Working temperature (K)	I_c (A)	Wire length (m)	Strain (%)	Twis-ting radius (m)	Cost ($/kA m)
Fault current limiters	10^4–10^5	0.1–3	20–77	10^3–10^4	1000	0.2	0.1	10–100
Big motors	10^5	4.5	20–77	500	1000	0.2–0.3	0.05	10
Generators	10^5	4–5	20–50	>1000	1000	0.2	0.1	10
SMES	10^5	5–10	20–77	10^4	1000	0.2	1	10
Cables	10^4–10^5	<0.2	65–77	100 per strand	100	0.4	2 (in cable)	10–100
Transformers	10^5	0.1–0.5	65–77	10^2–10^3	1000	0.2	1	10

Table 1.7 Main superconductors and their properties [1235]

Material	Crystalline structure	Anisotropy	T_c (K)	H_{c2} (T)	H^* (T)	ξ (nm)	λ (0) (nm)	De-pairing current density, at 4.2 K (A/cm^2)	Critical current density (A/cm^2)	ρT_c (μΩ cm)
NbTi (Nb, 47 wt.%)	b.c.c.	Neglected	9	12 (4 K)	10.5 (4 K)	4	240	3.6×10^7	4×10^5 (5 T)	60
Nb_3Sn	A-15, cubic	Neglected	18	27 (4 K)	24 (4 K)	3	65	7.7×10^8	$\sim 10^6$	5
MgB_2	P6/mmm, hexagonal	2–2.7	39	15 (4 K)	8 (4 K)	6.5	140	7.7×10^7	$\sim 10^6$	0.4
YBCO	Ortho-rhombic, layered perovskite	7	92	>100 (4 K)	5–7 (77 K)	1.5	150	3×10^8	$\sim 10^7$	~40–60
Bi-2223	Tetragonal layered perovskite	50–100	108	>100 (4 K)	0.2 (77 K)	1.5	150	3×10^8	$\sim 10^6$	~150–800

Of course, above review far not cover all superconductivity applications (in more details they have been considered, for example in Ref. [1313]) but it gives a representation on place of superconductivity in modern and future science and innovations.

1.3 Superconductivity Phenomena

1.3.1 Critical Field

Not only thermal fluctuations destroy superconductivity effect, but a sufficiently intense magnetic field can restore resistance and return the superconductor to its normal, that is, resistive state. This field is called *the critical field of bulk material*, H_{cm}, or the thermodynamic critical field, being a characteristic value of intensity of the constant magnetic field, exceeding of which lead to penetration of the magnetic field from surface in depth of superconductor on distance L which is greater the penetration depth of magnetic field λ (that defines the *Meissner effect* [1128]). The thermodynamic critical field depends on temperature and material properties of superconductor. In particular, the temperature dependence of H_{cm} is well described by the empirical formula:

$$H_{cm}(T) = H_{cm}(0)\left[1 - (T/T_c)^2\right]. \tag{1.1}$$

This dependence is shown in Fig. 1.6a, which essentially represents the H–T phase diagram of the superconducting state. Within area of S, any point in H–T plane corresponds to the superconducting state. In type-I superconductors the magnetic field penetration occurs in a result of first order phase transition, and thermodynamic critical field, H_{cm}, defines the difference of specific free energies in normal (non-superconductive, F_n) and superconductive (F_s) states.

$$H_{cm}^2/(8\pi) = F_n - F_s. \tag{1.2}$$

The critical field can be related to the critical temperature T_c. Since the critical field is zero at T_c, then in this case there is a second order phase transition. The entropy is continuous at T_c in zero magnetic field, but the specific heat is discontinuous. When $H \neq 0$, transition takes place at $T < T_c$ and becomes of first order.

The zero resistance state can also be destroyed by large enough electrical currents, stated by the Silsbee criterion [1572]. It defines an existence of *critical current*, the application of which through bulk superconductor causes initiation on its surface of the critical magnetic field, destroying superconductivity.

As we will see below, the oxide high-temperature superconductors relate to type-II superconductors and demonstrate high anisotropy of critical magnetic fields and critical current. In the specimens of small sizes (thin films, tapes, etc. with transverse sizes compared with value of λ), initiation and breaking

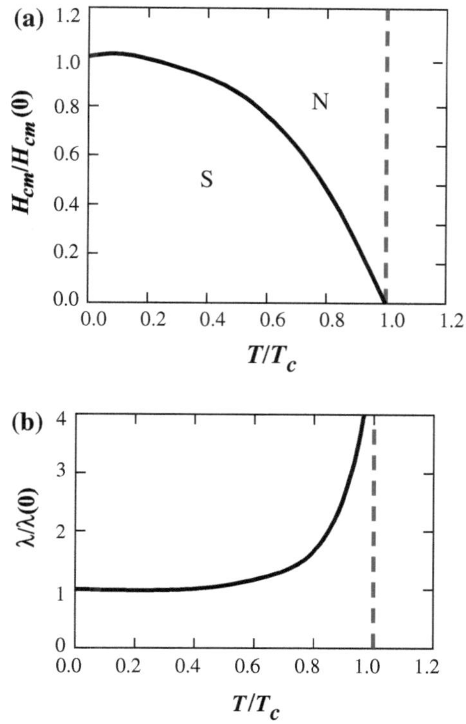

Fig. 1.6 Temperature dependence of the critical field, H_{cm} (**a**); temperature dependence of penetration depth, λ (**b**)

superconductivity in magnetic field do not directly connected with the value of the thermodynamic critical field, but in dependence on transverse sizes of specimen maybe realized through phase order as first as second order.

1.3.2 Critical Current

Critical current of superconductor, I_c, is a maximal value of direct electric current which can flow via superconductor without energy dissipation. If current exceeds the critical value then matter of superconductor transits in whole or partially in normal (non-superconducting) state and in the sample is initiated energy dissipation leading to its heating. The state of zero resistance may be fractured by sufficiently strong electric current, that it has been stated by *Silsbee rule* [1572]. In bulk type-I superconductor, the critical current, I_c, defines a current initiating the critical magnetic field, H_{cm}, at the sample surface. For instance, in the cylindrical superconductor $I_c = cRH_{cm}/2$, where c is constant, R is the radius of cylinder. At the current $I > I_c$, a type-I superconductor transits in *intermediate state* characterized by alternation of normal and superconductive areas (domain structure). In a type-II superconductor, the value of critical current is found by initiation into the sample of *vortices*[9] (non-superconductive vortex lines, at formation of which the superconductor transits in so-called *mixed state*). The vortices begin to form at the current initiating at the sample surface, so-called *first critical magnetic field* H_{c1}. The "core" of each vortex line presents oneself a quant of magnetic flux. Due to the Lorentz force acts on the vortices from the side of flowing current. However, in the actual samples, the vortices are pinned on defects of crystallite lattice and near the sample boundaries, therefore they be in equilibrium in sufficiently small current defining *magnetic flux pinning*. By increasing current the Lorentz force rises and at the current exceeding a critical value (the critical current of pinning) occurs frustration of vortices (at the current equal to the critical value is occurred so-called *critical state of pinning*). The vortices begin movement and as a result an electric field generates in the sample for account of induction mechanism and energy dissipation (*resistive state of superconductor*) occurs. In this stage, the sample matter in whole maintains superconductive properties (superconductivity is suppressed only into cores of the vortices).

There are soft and rigid type-II superconductors. In the soft superconductors the pinning force is small and the critical current in practice is equal to the current in which the critical magnetic field on surface H_{c1} creates on the sample surface. In the rigid superconductors, the pinning force is great the value of the critical current is determined by vortex frustration and may be very significant. Maximal critical currents demonstrate so-called *composite superconductors* in which owing to special technology is formed microstructure which maximally prevents to movement of the vortices. In these superconductors the critical current density, J_c, could attain 10^6 A/cm^2.

In thin superconductors with transverse sizes which are lesser than the penetration depth of the magnetic flux, the superconductivity failure and dissipation initiation occur for account of increasing velocity of superconducting electrons (*Cooper pairs or pairons*) at the current increasing. The critical current is the current of fracture of the Cooper pairs (see the Cooper effect). Magnetic field of the critical current in thin samples is small and matter of superconductor can transits in normal state either in whole or partially (resistive state). The concept of the critical current meets also in the *Josephson effects*.

1.3.3 The Meissner Effect

An external magnetic field which is lower than definite critical value, depended on the type of matter, does not penetrate into superconductor in the shape of infinite solid cylinder with axis parallel to the field direction and differs from zero only in

[9] These vortices are called by *Abrikosov vortices*. Moreover, there are *Josephson vortices* which form in Josephson junctions (for example, in sandwich from two superconductors divided by thin dielectric layer). At the tunneling Cooper pairs through dielectric, the circulating current occurs which creates the vortex.

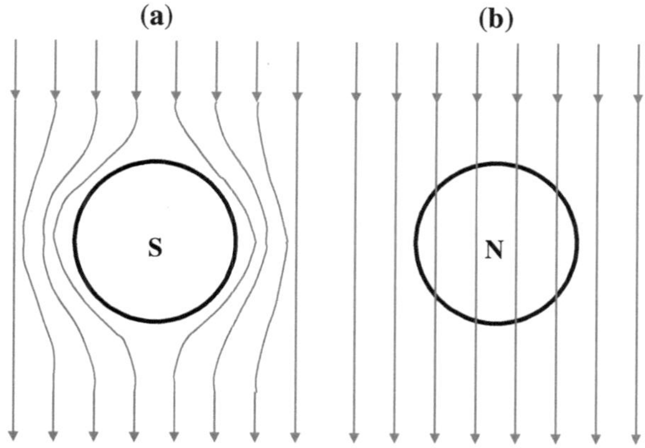

Fig. 1.7 Dependence of magnetic state of an ideal conductor on its history at $T < T_c$ and $H > 0$: (**a**) magnetic field applied to an ideal conductor at $T < T_c$; (**b**) field applied at $T > T_c$

thin surface layer. The displacement of the constant magnetic field from bulk conductor when the last becomes superconductor is one of the basic properties of superconductivity. So, not only is the superconductivity phenomenon a state of zero resistance, but a superconductor is not simply an ideal conductor to being a piece of metal with zero resistance.

Consider the behavior of an ideal conductor in an external magnetic field that is weak enough so as not to destroy the specimen's ideal conductivity. Suppose that initially the ideal conductor is cooled down below the critical temperature in zero external magnetic field. After that an external field is applied. From general considerations, it is easy to show that the field does not penetrate the interior of the sample (Fig. 1.7a). Let us prove this with the help of Maxwell's equations. As the induction $\boldsymbol{B}$ changes, an electric field $\boldsymbol{E}$ must be induced in the specimen due to the equation

$$\mathrm{curl}\, \boldsymbol{E} = -c^{-1} \partial \boldsymbol{B} / \partial t, \tag{1.3}$$

where c is the speed of light in vacuum.

In the ideal conductor $\boldsymbol{E} = 0$, since

$$\boldsymbol{E} = \boldsymbol{j} \rho, \tag{1.4}$$

where ρ is the resistivity (which in our case is zero) and $\boldsymbol{j}$ is the density of the induced current.

It follows that $\boldsymbol{B} = \mathrm{const}$ and taking into account that $\boldsymbol{B} = 0$ before applying the external field, we arrive $\boldsymbol{B} = 0$ in any point of the ideal conductor also after the field is applied. However, the same situation (an ideal conductor at $T < T_c$ in an external magnetic field) can be reached through another sequence of events, that is, by first applying the external field to a "warm" sample and then cooling it down to $T < T_c$. In that case, electrodynamics predicts an entirely different result. At $T > T_c$, the resistivity of the sample is finite and, therefore, the magnetic field penetrates into it. After cooling the specimen down through the superconducting transition, the field remains in it (see Fig. 1.7b).

The experiment by Meissner and Ochsenfeld [1128] showed that the superconductor is not only an ideal conductor. They revealed that at $T < T_c$ the field inside a superconducting specimen was always zero ($\boldsymbol{B} = 0$) in the presence of an external field, independent of which a procedure has been chosen to cool the superconductor below T_c. Because of expression $\boldsymbol{B} = \boldsymbol{H} + 4\pi\boldsymbol{M}$ between induction, $\boldsymbol{B}$, intensity of magnetic field, $\boldsymbol{H}$, and magnetization, $\boldsymbol{M}$, the last at the $\boldsymbol{B} = 0$ is equal to $M = -H/4\pi$, that is, superconductor in the Meissner state leads itself as ideal diamagnetic with negative magnetic susceptibility $\chi = -1/4\pi$ (i.e., ratio of magnetization to external magnetic field) and states zero field inside itself. So, if condition $\boldsymbol{B} = 0$ is independent of the specimen's history, the zero induction can be treated as an intrinsic property of the superconducting state at $H < H_{cm}$. Furthermore, it implies that we can treat a transition to the superconducting state as a phase transition, but the superconducting state obeys the equations

$$\rho = 0, \tag{1.5}$$

$$B = 0. \tag{1.6}$$

The Meissner effect disappears at complete transition superconductor in normal (non-superconducting) state. Due to a possibility of existing non-damped currents in superconductor, external magnetic field is screened by supercurrents flowing near the sample surface (the Meissner current) into layer with thickness of order of the magnetic field penetration depth. The state with non-decayed surface currents preventing penetration of magnetic field in interior of superconductor corresponds to minimum of free energy including the energy of magnetic field, kinetic energy of superconductive electrons and energy of superconductive condensation.
Total Meissner effect exists into range of intensities of magnetic field do not exceeding the critical magnetic field at which whole or partial failure of superconductivity takes place. Incomplete Meissner effect is observed as in type-I superconductors into intermediate state, when magnetic field penetrate the sample via areas of normal phase, as in type-II superconductors in the range from first critical field, H_{c1}, up to second critical field, H_{c2} (mixed state), when magnetic field penetrate the superconductor in the form of vortices carrying quant of magnetic flux. During cooling the specimen to be under magnetic field down to temperature, which is lower than critical temperature of superconductive transition, the displacement of the magnetic field from the sample is occured. In this process, the areas of normal phase, containing magnetic flux, or quantified vortices aspire go out the superconductor displacing from interior to the sample surface. In materials with defects of crystallite lattice, this movement of magnetic flux could be impeded, that will lead to "freezing" magnetic flux into sample.

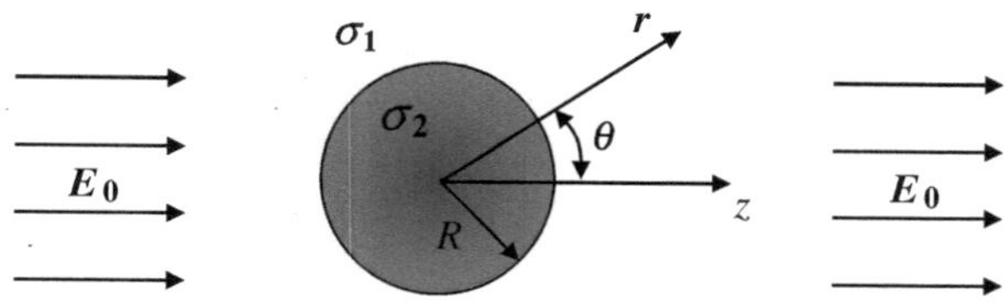

Fig. 1.8 Coordinate system for sphere of conductivity σ_2 and radius R in an infinite matrix of conductivity σ_1

1.3.4 Equipotential Lines for Superconductive Inclusion

In order to state mathematically strict picture of equipotential lines around superconductive inclusion, following [1697] consider the boundary value problem on the potential (temperature) field T for a spherical inclusion of radius R and conductivity σ_2 in an infinite matrix of conductivity σ_1 in which the applied intensity field is the constant vector $\boldsymbol{E}_0$. Let $\boldsymbol{r}$ be a position of vector emanating from the center of the sphere (see Fig. 1.8). Then, within and outside of the sphere, T must satisfy Laplace's equation:

$$\Delta T(r) = 0, \tag{1.7}$$

at the boundary conditions:

$$T_+ = T_-, \quad r = R; \tag{1.8}$$

$$\sigma_1 \boldsymbol{n} \cdot (\nabla T)_+ = \sigma_2 \boldsymbol{n} \cdot (\nabla T)_-, \quad r = R; \tag{1.9}$$

$$T = -\boldsymbol{E}_0 \cdot \boldsymbol{r}, \quad r \to \infty, \tag{1.10}$$

where T_+ and T_- are the limits of T on the sphere surface $r = R$ taken from the outside and inside, respectively. Similarly $(\nabla T)_+$ and $(\nabla T)_-$ are the limits of ∇T at $r = R$ from the outside and inside, respectively, where $\boldsymbol{n} = \boldsymbol{r}/r$ is the unit vector of normal. The solution of Eqs. (1.7–1.10) for T can be found by using the separation of variables technique [768].

The general form of the solution in the matrix is given as

$$T = -\boldsymbol{E}_0 \cdot \boldsymbol{r} + A\boldsymbol{E}_0 \cdot \nabla\left(\frac{1}{r}\right) = -\left(r + \frac{A}{r^2}\right) E_0\, cos\theta, \quad \text{at } r \geq R, \tag{1.11}$$

where A is an arbitrary constant, θ is the angle between vectors $\boldsymbol{E}_0$ and $\boldsymbol{r}$, $E_0 = |\boldsymbol{E}_0|$. Obviously, this axisymmetric field is the superposition of two elementary solutions of the Laplace's equation: the field due to a uniform intensity and the field due to a dipole located at the origin.

The general form of the solution inside the sphere is given as

$$T = -\boldsymbol{E}_0 \cdot \boldsymbol{r} + B\boldsymbol{E}_0 \cdot \boldsymbol{r} = (B - 1)E_0 r \cos\theta, \quad \text{at } r \leq R, \tag{1.12}$$

where B is an arbitrary constant.

By substituting the interface continuity conditions (1.8) and (1.9) in relations (1.11) and (1.12), we obtain two relations for definition of the unknown constants A and B, that leads to

$$A = -R^3 \frac{\sigma_2 - \sigma_1}{\sigma_2 + 2\sigma_1}; \quad B = \frac{\sigma_2 - \sigma_1}{\sigma_2 + 2\sigma_1}. \tag{1.13}$$

Substitution of these expressions into (1.11) and (1.12) gives

$$T = \begin{cases} -\boldsymbol{E}_0 \cdot \boldsymbol{r} + \beta_{21}\boldsymbol{E}_0 \cdot \boldsymbol{r}\left(\frac{R}{r}\right)^3, & r \geq R; \\ -\boldsymbol{E}_0 \cdot \boldsymbol{r} + \beta_{21}\boldsymbol{E}_0 \cdot \boldsymbol{r}, & r \leq R, \end{cases} \tag{1.14}$$

where

$$\beta_{21} = \frac{\sigma_2 - \sigma_1}{\sigma_2 + 2\sigma_1}. \tag{1.15}$$

The quantity β_{21} is termed the "polarizability" because the potential outside the sphere is equivalent to the applied field plus the field of a dipole at the origin with dipole moment $\beta_{21}R^3E_0$.

Then the corresponding expression for the intensity field $\boldsymbol{E} = -\nabla T$ is

$$\boldsymbol{E} = \begin{cases} \boldsymbol{E}_0 + \beta_{21}R^3\boldsymbol{t}(\boldsymbol{r}) \cdot \boldsymbol{E}_0, & r > R; \\ \boldsymbol{E}_0 - \beta_{21}\boldsymbol{E}_0, & r < R, \end{cases} \tag{1.16}$$

where

$$\boldsymbol{t}(\boldsymbol{r}) = \nabla\nabla\left(\frac{1}{r}\right) = \frac{3\boldsymbol{nn} - \boldsymbol{I}}{r^3} \tag{1.17}$$

is the dipole tensor. Figure 1.7a shows equipotential lines for superconducting inclusion ($\sigma_2/\sigma_1 = \infty$).

1.3.5 The Cooper Effect

Coupled pairs of particles can form in degenerate system of fermions at existence of any weakest attraction between them. Cooper by solution Schrödinger equation for two particles of degenerate fermions gas (gas of electrons) has shown that weak attraction between them led to so-called *particle pairing* which to be near Fermi surface that is to formation of coupled states of the two particles [310]. The Cooper effect presents oneself a basis of microscopic theory of superconductivity of BCS model very well explaining conventional low-temperature superconductivity. In ideal Fermi gas a superconductivity (that is a superfluidity of system of the charged particles) is no possible. In order to initiate superconductivity, it is required that a finite energy gap formed in energy spectrum of Fermi excitations above ground state. The Cooper pairing of particles with finite energy of binding namely leads to formation of that gap. So, for Fermi systems with pairing is satisfied Landau criterion of superfluidity. As a result of the Cooper effect any degenerate Fermi system with attraction between particles must possess superconductivity (superfluidity). In actual metals, an interaction between electrons is caused by screened (at great distances) Coulomb repulsion and attraction around electrons due to possible exchange by virtual phonons and polarization of crystal [496, 497]. An interrelation of these interaction types defines a possibility of superconductivity in metals. For initiation of Cooper pairing, it is enough that in the decomposition of the scattering amplitude of fermions one on other on Legendre polynomials at least one of terms of the decomposition to be negative (an attraction on corresponding harmonic). The Cooper pairs possess orbital moment to be equal number of this harmonic. As rule, a binding energy of couples and respectively the critical current of superconductive transition rapidly decrease with increasing the orbital moment. Therefore, the pairing is fulfilled with lowest permissible value of the moment. Summary spin of the couple is equal to zero at even orbital moment and unity at the odd one (in this case, the couples are bosons). In majority of known superconductors the Cooper pairs possess zero orbital moment.

The Cooper pair (pairon) is a quasiparticle describing coupled state of two electrons near Fermi surface caused by effective interelectronic attraction. At the decreasing temperature lower than the critical value T_c, an electric resistance of superconductor transforms in zero and magnetic flux expels from the specimen. In conventional LTSCs, the Cooper pairs, which are responsible for these phenomena, form at the $T = T_c$. For HTSCs (cuprate superconductors), it is safely stated the following properties:

(i) HTSC superconductivity is the sequence of formation of the Cooper pairs, each of which has charge $2e$ (or more strictly, the sequence of initiation at the $T = T_c$ non-diagonal far order); this is directly confirmed by experiments on quantization of magnetic flux and existence of Josephson effect,
(ii) copper-oxygen planes CuO_2 is the obligatory attribute of crystallite structure of all cuprate high-temperature superconductors,
(iii) it may be assumed in first approximation, that the Cooper pairs form independently in various layers of CuO_2 (or groups from several these adjacent layers) divided by "charge reservoirs"; on this witnesses a small energy of the electrons jumping between layers (in comparison with value of $k_B T_c$, where k_B is Boltzmann constant) defined by temperature dependence of electric resistance along c-axis,
(iv) mechanism of forming the Cooper pairs either to be no phonon, or includes the phonons much more complex manner than in BCS model; this is supported by very weak *isotope effect* or its total absence in HTSC,
(v) spin state of the Cooper pairs to be singlet,
(vi) orbital state of the Cooper pairs in tetragonal HTSC has d-wave symmetry,

(vii) proper sizes of the Cooper pairs are equal to 1–3 nm,
(viii) the Cooper pairs also as in the BCS model form from electrons being in converted to time states.

1.3.6 The Isotope Effect

The isotope effect was discovered by two groups in 1950, who discovered it independently working with different mercury isotopes [1105, 1426]. The isotope effect provided a crucial key to the development of the BCS superconductivity theory discussed in Sect. 1.6. It was found that for a given element, T_c was proportional to $M^{-1/2}$, where M is the isotope mass. The vibration frequency of a mass M on a spring is proportional to $M^{-1/2}$. The same relation holds for the characteristic vibration frequencies of the atoms in a crystal lattice. So, the existence of the isotope effect indicated that although superconductivity is an electronic phenomenon, it is nevertheless connected with the vibrations of the crystal lattice in which the electrons move. However, after the development of the BCS theory, for some conventional (i.e., obeying the BCS theory) superconductors, the exponent of M is not $-1/2$, but near zero (see Table 1.8). The studies also showed that together with T_c, the critical magnetic field H^0_{cm} (at the $T \to 0$) changes, however ratio H^0_{cm}/T_c remains constant for various isotopes of given superconductive metal.

In high-temperature superconductors, the above exponent is also near zero. This fact caused research of HTSC mechanisms that are distinct on the electron–phonon interactions. Afterwards, the magnetic isotope effect in HTSCs has been discovered [1665] connected with alteration of anomaly character of the electronic dispersion law (so-called, "*kink*") in high-temperature superconductors, namely in vicinity of Fermi level at partial substitution of Cu atoms by atoms of Zn and Ni, not much differing on mass but being in other spin states. In this case, the kink arises because of interaction of electrons with spin fluctuations but no with acoustic and optic phonons, that to be argument for no phonon mechanism of pairing in HTSC. Thus, existence of the magnetic isotope effect can point on main role of magnetic interactions for pairing carriers in HTSC.

1.3.7 Penetration Depth and Coherence Length

Detailed studies have shown that provided the magnetic field remains weak enough, it gradually diminishes at the superconductor surface over a depth λ of some hundreds of angstroms order (see the Meissner effect). This is called the

Table 1.8 Isotope effect ($T_c \propto M^{-\alpha}$) [940]

Element	Mg	Sn	Re	Mo	Os	Ru	Zr
α	0.5	0.46	0.4	0.33	0.21	0 (±0.05)	0 (±0.05)

penetration depth. When an external magnetic field is applied, a direct thermodynamic current appears at the superconductor surface in such a way as to screen the bulk from the applied field.

Mathematically, the penetration depth is determined as $\lambda = \frac{1}{H}\int_0^\infty B(x)dx$, where H is the intensity of the external magnetic field, directed as the vector of magnetic induction in interior of superconductor parallel to the superconductor surface occupying half-space $x > 0$. In the case of exponential decaying the magnetic field into depth of the superconductor $\boldsymbol{B} = \boldsymbol{H}\exp(-x/\lambda)$, where λ is determined by above integral representation. This exponential law is observed in so-called London's case [1045, 1046] when λ is much greater than *the coherence length* (see below).

The phenomenological theory of superconductivity, proposed by Gorter and Casimir and based on the assumption that there are two components of conducting electronic "fluid" in superconducting state, namely "normal" and "superconducting," gave the name *two-fluid model* [553]. The properties of the "normal" component are identical to system of electrons in the normal metal, but the "superconducting" component is responsible for the anomalous properties. In the framework of the two-fluid model of superconductor, the temperature dependence of the penetration depth λ is approximated well by [552]

$$\lambda(T) = \frac{\lambda(0)}{\left[1 - (T/T_c)^4\right]^{1/2}}. \tag{1.18}$$

The divergence of λ at transition ($T \to T_c$) shows that we pass continuously from the normal metal, in which $\lambda = \infty$, to the superconducting state (see Fig. 1.6b). In non-zero field, λ varies discontinuously from $\lambda = \infty$ to a finite value, and the transition is then first order.

Taking into account sharpness of the superconducting transition at the absence of magnetic field and also significant dependence of the observed depth λ on the admixture concentration, Pippard concluded that superconducting state should be characterized by a finite coherency length of electron impulse (Δp) [1365], but no infinite one as in the theory of the brothers London [1046]. Therefore, the order parameter changes smoothly at the distance ξ, called the coherence length. Based on the estimation made by Pippard for dependence λ on the field, the length of $\xi \sim 1$ μm. The coherence length is the distance between two electrons of Cooper pair, to be in essence spatial characteristic of superconducting electrons, on which its value could be judged by using the relation of uncertainties

$$\xi \approx \frac{\hbar}{\Delta p} \approx \frac{\hbar v_F}{\kappa T_c}, \tag{1.19}$$

where v_F is Fermi velocity and $\kappa = \lambda/\xi$ is Ginzburg–Landau parameter.

To explain the dependence of depth λ on the length of the free run of electrons l, Pippard proposed that an effective coherence length $\xi(l)$ is connected with corresponding value of ξ_0 for pure metal by the equation [1366]

$$\frac{1}{\xi(l)} = \frac{1}{\xi_0} + \frac{1}{Al}, \tag{1.20}$$

where A is the constant near unit.

In the two limit cases, the evident equations for λ may be obtained in the form:

$$\lambda = \lambda_L \left[\frac{\xi_0}{\xi(l)}\right]^{1/2} \quad \text{at } \xi << \lambda \text{ (London limit)}, \tag{1.21}$$

where

$$\lambda_L = \frac{c}{e}\left(\frac{m}{4\pi n_s}\right)^{1/2} \tag{1.22}$$

is the London penetration depth; m and e are the mass and charge of electron, respectively; n_s is the quantitative density of superconducting electrons. The condition $\xi \ll \lambda$ is carried out for pure metals near T_c (where $\lambda \to \infty$) and also for alloys and thin admixture films, where l and ξ are decreased or limited by the electronic scattering at defects, admixtures or at film boundaries, so that $\xi \to l$ at $l \to 0$.

The contrast limit case ($\xi \gg l$), corresponding to most part of pure bulk superconductors ($l \to \infty$), to be at the temperatures which are far from T_c,

$$\lambda_{l\to\infty} = \left(\frac{\sqrt{3}}{2\pi}\xi_0\lambda_L^2\right)^{1/3} \quad \text{at } \xi > > \lambda \text{ (Pippard limit)}. \tag{1.23}$$

The penetration depth depends also on the character of the electrons reflection from surface of superconductor and field frequency. Near the temperature of superconductive transition, T_c, into framework of BCS model (London case) $\lambda_L^2 = mc^2/[8\pi e^2 n(1 - T/T_c)]$, where n is the total density of electrons.

1.3.8 Weak Superconductivity and the Josephson Effects

The quantum nature of superconductivity is provided by the so-called *weak superconductivity*, or Josephson effects [796]. The weak superconductivity covers a set of phenomena, occurring in weakly-coupled superconductive systems (heterogeneous superconductive structures), which contain narrow (in direction of flowing current) areas with absent or strongly suppressed superconductivity. This term has been introduced by Anderson in 1964 [31] because the critical current and critical magnetic field in the weakly-coupled superconductors are significantly lower than in usual ones. The weak superconductivity is observed in the tunnel junctions (when two superconductors are divided by thin dielectric buffer layer of ~10 Å), junctions with buffer layer from normal (non-superconducting) metal or semiconductor, superconducting bridges with contraction, point contacts, granular superconductors consisting of numerous Josephson junctions, etc. In the tunnel junction, electrons can cross dielectric barrier that leads to arising one-particle tunnel current. Sharp changes of the one-particle current, linked with features in the state density of superconductors, display in volt-ampere characteristic (VAC) at the voltages, V, on junction $eV = \Delta_1 + \Delta_2$ and $eV = |\Delta_1 - \Delta_2|$, where e is the electron charge, Δ_1 and Δ_2 are the values of superconductive gaps forming the junction.

There are two Josephson effects predicted in 1962, namely stationary (*dc*) and non-stationary (*ac*).

dc-Josephson effect. Let us apply a current through a weak link (or so-called Josephson junction). If the current is small enough, it passes through the weak link without resistance, even if the material of the weak link itself is not superconducting (e.g., it is an insulator in a tunnel junction). Through the weak link, the electrons of the two superconductors merge into a single quantum body. In other words, all superconducting electrons on both sides of the weak link are described by the same wave function, the phase of which smoothly changes along superconductor at flowing current into it (the phase coherence of superconductive electrons). The presence of the weak link should not change significantly the wave functions on the two sides, compared to what they had been before the link was established.

ac-Josephson effect. Let us increase the *dc*-current through the weak link until a finite voltage appears across the junction. The corresponding value of current, I_c (critical Josephson current) depends on properties of the junction, temperature and magnetic field. The value of I_c is defined by the current of superconductive (coupled) electrons, which in this case becomes to be alternative (its frequency depends on the voltage on the junction), and the current caused by normal (non-superconductive) electrons crossing the buffer layer. At the flowing superconductive electrons through non-superconductive buffer layer, the phase coherence partially (in dependence on the ratio of the barrier layer thickness to coherence length) failure and the flowing Josephson current via the barrier layer is accompanied by phase jumping the wave function of the superconductive electrons on this buffer layer $\varphi = \varphi_2 - \varphi_1$, where φ_2 and φ_1 the phases of the wave function in the superconductors on the two sides of the buffer layer. In this case the current via junction is equal to $I = I_c \sin\varphi$. Hence, the Josephson current cannot exceed value of I_c.

In order to obtain the measured Josephson current, the thickness of insulating buffer layer must be of ~10–20 Å. At the non-stationary Josephson effect, the phase difference on junction depends on time:

$$\hbar(\partial\varphi/\partial t) = 2eV, \tag{1.24}$$

where V is the voltage on junction, $\hbar$ is Planck's constant. This equation is the sequence of Schrödinger equation for the wave function of couple of the superconductive electrons at existence of constant potential energy, $2eV$, and it is not connected with existence of buffer layer and has a general character. The angular frequency ω of superconductive current through junction is defined as

$$\hbar\omega = 2eV\,. \tag{1.25}$$

Equations (1.24) and (1.25) are called Josephson relationships.

At irradiation of Josephson junction by radiation of super-high frequency (SHF) Ω satisfying condition $n\hbar\Omega = 2eV$ (where n is integer number), the flowing superconductive electrons through junction occurs with absorption of photons of the external field. This leads to arising additional current via the junction that is formation of sites on VAC with zero differential resistance. The observation of these sites has to be first indirect discovery of the non-stationary Josephson effect [1547]. A direct observation of generating SHF-radiation by Josephson junction under direct voltage has been carried out in 1965 [1831].

If the junction width, L, is small compared to Josephson penetration depth $\lambda_J = c\Phi_0/(8\pi^2 j_c d)^{1/2}$, where j_c is the critical current density, d is the thickness of region in which penetrates magnetic field, Φ_0 is the quant of magnetic flux, c is the speed of light, then external field penetrates in the junction region uniformly, but dependence of critical current on the applied magnetic field is described by the function which is proper for Fraunhofer diffraction picture:

$$I_c(H) = I_c(0)\left|\frac{\sin(\pi\Phi/\Phi_0)}{\pi\Phi/\Phi_0}\right|, \tag{1.26}$$

where $\Phi = HLd$. If voltage is no zero, then in existence of magnetic field in the junction can propagate waves of current density with speed $v = cV/(Hd)$. These waves are observed on steps on VAC (so-called, Fiske steps [484]). If the junction width $L > \lambda_J$, then magnetic field penetrates into tunnel junction non-uniformly in the form of Josephson vortices (magnetic field in which decays exponentially on length λ_J). The Josephson vortices can displace along the junction due to the current influence. The weak superconductivity is strongly subjected to influence of fluctuations leading to two effects. From one hand, the accidental changes along the junction plane of Josephson phase or critical current density, connected with local heterogeneities of the tunnel junction (structural fluctuations), lead to distortions of the Fraunhofer dependence of critical current on magnetic field. From other hand, the voltage, connected with accidental change of the Josephson phase in time, can arise on the junction at the current which is lower than the critical one. The probability of these phase jumps increases with the temperature growth, however at the low temperatures, it is possible macroscopic quantum tunneling (there is no zero probability of changing the Josephson phase with time at the $T \to 0$).

The variety of effects allows one to use the weak superconductivity for physical studies (definition of superconductive gap by using VAC of one-particle current, investigation of heterogeneities, etc.). The Josephson effects are used in different cryogenic devices, in particular in SQUIDs. The Josephson junctions are applied as generators and detectors of SHF-range. The property of Josephson junctions to re-switch from zero to finite voltage at exceeding by current of a critical value together with small capacity allow their using as rapidly-acting logic elements of computers. The condition $n\hbar\Omega = 2eV$ could be used to more accurate definition of basic physical constants and designing standards of voltage. Based on the Josephson effects has been measured the ratio $2e/\hbar = 4.83594000 \times 10^{14}$ Hz/V with error of 2×10^{-8} that allowed one to process a volt standard with error of $\sim 10^{-9}$.

1.4 Type-I Superconductors

These superconductive materials constitute one of two classes (together with type-II superconductors) on which divide all superconductors in dependence on their behavior in magnetic field. A cylinder from type-I superconductor, placed in longitudinal magnetic field with intensity H which is less than critical value (thermodynamic critical field) H_{cm}, displays complete Meissner effect (if the sample has not heterogeneities) in difference of the type-II superconductors which demonstrate incomplete Meissner effect into definite range of magnetic fields. The type-I superconductor loses superconductivity with jump (first order phase transition) ether at attainment of critical temperature, T_c (H), corresponded to the given field, or at the increasing an external field upon to the critical value H_{cm} (T). Hidden specific heat of the phase transition is defined as

$$Q = T(S_n - S_s) = -T\frac{\partial(F_n - F_s)}{\partial T} = -\frac{T}{4\pi}H_{cm}\frac{\partial H_{cm}}{\partial T}, \tag{1.27}$$

where S_n and S_s are the specific entropies of the normal and superconductive phases, F_n and F_s are the specific free energies in normal and superconductive states. The jump of specific heat capacity is calculated at the $T = T_c$ as

$$\Delta C = C_s - C_n = \frac{T_c}{4\pi}\left(\frac{\partial H_{cm}}{\partial T}\right)^2. \tag{1.28}$$

In absence of external magnetic field $Q = 0$ at the $T = T_c$, that is occurs second order phase transition.

1.4.1 Magnetic Properties of Type-I Superconductors

Let us consider the magnetization curve of a superconductor representing a long cylinder in a longitudinal external magnetic field $\boldsymbol{H}$. When the field $\boldsymbol{H}$ increases, the induction inside the sample does not change at first; it remains at $\boldsymbol{B} = 0$.

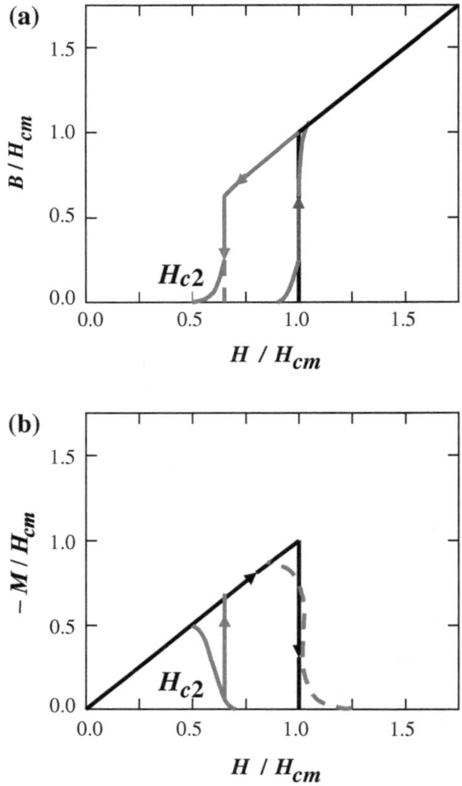

Fig. 1.9 Magnetic induction (**a**) and magnetization (**b**), as a function of applied field in type-I superconductors. When the magnetic field is reduced, superconductivity reappears at a field $H_{c2} < H_{cm}$. *Dotted lines* correspond to impure samples [1006]

As soon as H reaches the value of H_{cm}, the superconductivity is destroyed, the field penetrates into the superconductor and $\boldsymbol{B} = \boldsymbol{H}$. Therefore, the magnetization curve $B = B(H)$ has the form shown in Fig. 1.9a. The magnetic induction $\boldsymbol{B}$ and magnetic field $\boldsymbol{H}$ are related to each other by the expression:

$$\boldsymbol{B} = \boldsymbol{H} + 4\pi\boldsymbol{M}, \tag{1.29}$$

where $\boldsymbol{M}$ is the magnetic moment per unit volume (Fig. 1.9b). In type-I superconductors, the diamagnetism (i.e., negative value of magnetic moment and its contrast direction to magnetic induction vector) remains up to the value of field H_{cm}. This supercooling of normal phase is caused by the difficulty of local nucleating a superconducting region within a normal region in type-I superconductors. If there are defects (e.g., dislocations) in the crystal lattice, these can modify such theoretical behavior by smoothing out transitions. Main magnetic properties of type-I superconductors are led from (1.5) and (1.6) [970]:

(1) *Magnetic field lines outside a superconductor are always tangential to its surface*. Indeed, it is known from electrodynamics that magnetic field lines (i.e., lines of the magnetic induction) are continuous and closed. This can be

written as the equation div $\boldsymbol{B} = 0$. Then, the components of $\boldsymbol{B}$ that are normal to the surface must be equal on both sides of the surface that is inside and outside of the sample. However, the field in the interior of a superconductor is absent and, consequently, $B_n^{(i)} = 0$. It follows that the normal component $B_n^{(e)}$ at the outside of the superconductor's surface is zero, too: $B_n^{(e)} = 0$, that is, the magnetic field lines are tangential to the surface of the superconductor.

(2) *A superconductor in an external magnetic field always carries an electric current near its surface.* This is one of the consequences of the first property, because from Maxwell's equation curl $\boldsymbol{B} = (4\pi/c)\boldsymbol{j}$ and the requirement $\boldsymbol{B} = 0$, it follows that the volume current in the interior of the superconductor is zero ($\boldsymbol{j} = 0$) and only a surface current is possible. Then, from relation between the surface current $\boldsymbol{j}_s$ and the magnetic field at the surface of the superconductor $\boldsymbol{H}$

$$\boldsymbol{j}_s = (c/4\pi)[\boldsymbol{nH}], \tag{1.30}$$

where $\boldsymbol{n}$ is the unit vector along the normal to the surface, it is obvious that the surface current is completely defined by the magnetic field at the surface of a superconductor. In other words, the surface current assumes a value such that the magnetic field generated by it inside a superconductor is exactly equal in value and opposite in direction to the external field. This assumes zero total field in the interior: $\boldsymbol{B} = 0$.

(3) *In a simply connected superconductor (i.e., body inside of which an arbitrary closed path can be reduced to a point without crossing the boundaries of the body), surface currents can exist only when the superconductor is placed in an external magnetic field.* Indeed, if the surface current remains after switching off the external field, it would create its own field in the superconductor, which is impossible.

1.4.2 The Intermediate State

Consider the behavior of a superconducting sphere, placed in an external magnetic field (Fig. 1.10). Since the magnetic field lines are always tangential to the surface of a superconductor then in this case, it is obvious that the field lines have a higher density at the equator thereby producing a local increase of the magnetic field. At the same time, the field at the poles is absent. Far away from the sphere, where any perturbations are averaged, the uniform external field H is lower than that at the equator. Therefore, an attainment of the field at the equator of the critical value of H_{cm} the magnitude of $H < H_{cm}$ and it is not permitted for the whole sphere to revert to the normal state. On the other hand, it is not permitted for the whole sphere to be superconducting either, because the field at the equator has already reached the critical value. This contradiction involves a co-existence of alternating superconducting and normal regions (domains) within the sphere, which is called

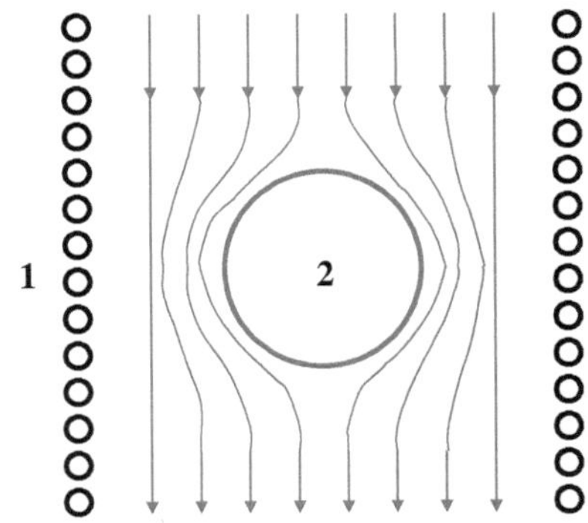

Fig. 1.10 Superconducting sphere in the homogeneous field of a solenoid. *Numbers* denote winding of the solenoid (*1*) and superconducting sphere (*2*)

the intermediate state. The interfaces between these regions are always parallel to the external field, while in the cross-section perpendicular to the field they may assume various intricate configurations.

Assume that before a superconducting body goes into the intermediate state, the maximum field at its surface (in the case of a sphere, at the equator) is H_m, and the external field far away from the body is H. Then, it is obvious that, on the one hand, $H_m > H$ and, on the other hand, H_m is proportional to H with the proportionality factor dependent on the exact shape of the body. It can be written in the form $H_m = H/(1-n)$. The values of the demagnetizing factor n for different forms are present in Table 1.9.

With the help of Table 1.9, it is possible to calculate the field H corresponding to the transition into the intermediate state for a body of a certain shape. The transition takes place when the field H_m reaches the value of H_{cm}. In particular, for a sphere it occurs at the external field $H = H_{cm}\ (1-1/3) = 2/3H_{cm}$.

Let us now consider the conditions for thermodynamic equilibrium in the intermediate state. Assume that the field in one of the normal regions exceeds H_{cm}. Then it must destroy the superconductivity of the adjacent superconducting regions. Conversely, if the field in a normal region is less than H_{cm}, this region must be superconducting. Therefore, a stable co-existence of the normal and superconducting regions is only possible, if the field in the normal regions equals to H_{cm}.

During cooling type-I superconductor placed in magnetic field, the magnetic flux is expelled due to movement of the domain boundaries. If this movement is impeded by structure heterogeneities, then occurs "freezing" the magnetic flux. In this non-equilibrium state, it is observed incomplete Meissner effect. The intermediate state and alternative phases in the type-I superconductor is only possible in proposition that the interface between these phases possesses positive surface

Table 1.9 The demagnetizing factor for different forms of superconductor

Sample geometry	Cylinder in parallel field	Cylinder in transverse field	Sphere	Thin plate in perpendicular field
n	0	½	1/3	1

energy, σ_{ns}, magnitude and sign of which depend on the Ginzburg–Landau parameter, $\kappa = \lambda/\xi$, where λ is the penetration depth, ξ is the coherence length. The sign of σ_{ns} (or value of κ) strictly define the type of superconductor: for type-I superconductor $\kappa < 1/\sqrt{2}$ and $\sigma_{ns} > 0$, for type-II superconductor $\kappa > 1/\sqrt{2}$и $\sigma_{ns} < 0$.

The term "intermediate state of superconductor" has been introduced by Peierls [1347], and structure of the intermediate state has been ascertained by Landau [969]. Under influence of current flowing in the sample, may be formed so-called dynamic intermediate state of superconductor in which interfaces continuously move through the sample (with speeds 10^{-5}–10^{-4} m/s) initiating on one its surface and disappearing on other. The normal and superconductive regions with size $d \gg \xi$ can co-exist in equilibrium only in the type-I superconductors, where the penetration depth $\lambda < \xi/\sqrt{2}$. In the type-II superconductors at the $H > H_{c1}$ (where H_{c1} is the lower critical field) arises a *mixed state*, in which is impossible to select normal and superconductive regions because the proper scale of microscopic structure of the mixed state $d_{ms} \approx \xi < \lambda\sqrt{2}$.

1.4.3 Intermediate State and Scaling Laws

As it has been shown, the intermediate state of type-I superconductor presents oneself fine-size mix of normal and superconductive regions (domains). Based on the principle of energy minimization, introduced by Landau [969], an actual disposition of the regions corresponds to local minima of non-convex variation problem. Investigation of scaling law of minimal energy allows one to state corresponding it qualitative properties of general picture for the normal and superconductive regions. The minimal energy may be presented as a sum $E = E_0 + E_1$, where E_0 is the value obtained by neglecting surface energy and E_1 is the addition caused by non-zero surface tension, σ. The scaling law states dependence of E_1 on the surface tension on interface between the normal and superconductive regions and also normalized value ($0 < b_a < 1$) of applied magnetic field. The principal term E_0, which describes the behavior in the limit of $\sigma \to 0$, corresponds to the thermodynamic theory (depicted by Landau) of the intermediate state defining, in particular volume fraction of the normal regions. However, in order to study actual domain structure, it is necessary to take into account correcting term E_1. Therefore, it is required to identify a scaling law for this addition and state microstructure features corresponding to this scaling law.

The different regimes related to concrete microstructure have been stated for the plate with thickness, L, being under action of applied transverse field [281, 282]:

(i) for intermediate values of b_a far disposed from 0 and 1, $E_1 \sim \sigma^{2/3}L^{1/3}$ (this scaling law is attained sufficiently simply by a picture of the magnetic flux at the main requirement of corresponding local scale of the length),

(ii) for relatively small values of b_a into range of $(\sigma/L)^{2/7} \leq b_a \ll 1$, $E_1 \sim b_a^{2/3}\sigma^{2/3}L^{1/3}$ (in order to attain this scaling, the magnetic flux must cross the sample by means of uniformly-distributed family of branching tubes of the magnetic field),

(iii) for smallest values of b_a, when $b_a \leq (\sigma/L)^{2/7} \ll 1$, $E_1 \leq b_a\sigma^{4/7}L^{3/7}$; this is less than scaling law (ii), because $b_a\sigma^{4/7}L^{3/7} \ll b_a^{2/3}\sigma^{2/3}L^{1/3}$, at the $b_a \ll (\sigma/L)^{2/7} \ll 1$ (in order to attain this scaling, the magnetic flux must cross the sample by means of non-uniformly-distributed family of branching tubes of the magnetic field),

(iv) for relatively great values of b_a, that is for b_a near to 1, the situation is similar to case (ii), however does not coincides with it: $E_1 \leq (1-b_a)|\log(1-b_a)|^{1/3}\sigma^{2/3}L^{1/3}$ (in order to attain this scaling with logarithmic factor, the magnetic flux must fill main part of the sample leaving only uniformly-distributed family of superconducting tunnels),

(v) for greatest values of b_a, the sample completely to be in normal state and $E_1 \sim (1-b_a)^2L$; this is fulfilled, when $(1-b_a)^2L \ll (1-b_a)|\log(1-b_a)|^{1/3}\sigma^{2/3}L^{1/3}$, that is, when $(\sigma/L)^{2/3} \gg (1-b_a)|\log(1-b_a)|^{-1/3}$.

Analysis of the scaling allows one to display some aspects of intermediate state of the superconductor:

(1) *Hysteresis*. The picture of magnetic flux distribution at the increasing from zero the applied magnetic field value strongly differs from the picture observed at decreasing field from critical value. This agrees with that the pictures, caused by regimes (ii–iv), quite differ one from others.

(2) *Complexity of plane picture*. The experimental data show very complex weave of normal and superconductive regions, in particular for intermediate values of applied field (when b_a far disposes from 0 and 1). Is the complexity to be advantageous energetically? The scaling law answers this question negatively. Actually, its lower bounds show that even most complex picture of interacting regions cannot attain better scaling law than simplest ordered structures.

(3) *Changing scale of length*. Landau was first who assumed that for $\sigma/L \ll 1$ the microstructure scale of length should be changed with depth. The lower scaling bound supports this assumption, showing that if the microstructure is independent on depth, then is impossible to attain the optimal scaling law $\sigma^{2/3}L^{1/3}$.

Type-I superconductors include all superconducting elements except niobium. Niobium, superconducting alloys and chemical compounds are type-II superconductors. The high-temperature superconductors also belong to this group. The main difference between the superconductors of both types lies in their different response to an external magnetic field. The Meissner effect and intermediate state of superconductor are observed only in type-I superconductors but no in type-II superconductors.

1.5 Type-II Superconductors

First, the idea on existence of two types of superconductors has been proposed by Abrikosov [3] and Zavaritski [1861] based on the experimental results of Schubnikow, et al. [1521] on the magnetization curves for superconducting alloys and Zavaritski's data on critical currents in thin superconductive films.

By introducing admixtures, the materials being in a "pure" state by type-I superconductors can transform in type-II superconductors. The coherence length in alloys $\xi \sim (\xi_0 l)^{1/2}$, where ξ is the coherence length of "pure" material, l is the mean free path of electrons in alloy. Even at the small (1 %) concentration of admixtures, it is possible that $\xi \ll \xi_0$. On the contrary, the penetration depth in alloys $\lambda \sim \lambda_0(\xi_0/l)^{1/2}$ (where λ_0 is the penetration depth for "pure" material) increases by introducing the admixtures, therefore for the alloys $\kappa = 0.75\lambda_0/l$. Thus, practically all alloys and non-ordered films are type-II superconductors.

As it has been noted, in contrast to type-I superconductors, type-II superconductors do not show Meissner effect and are characterized by the *mixed state*, but no intermediate state. In this case, electric resistance is absent, but a magnetic field penetrates into superconductor in a quite extraordinary way. Let us consider a type-II superconductor in the form of a long cylinder placed in a longitudinal magnetic field, which is increased from zero. First, the cylinder pushes out all field, causing the zero magnetic induction in the interior of the cylinder. This means that the Meissner effect is observed at this stage. However, beginning from a certain value of the field, there is a non-zero (i.e., finite) induction in the cylinder. This field is called *the lower critical field* and denoted by H_{c1}. With next increase of the external field H, the induction will build up until the average field in the cylinder becomes equal to the external field H, but the cylinder itself goes to the normal state. This will happen at the so-called *upper critical field* H_{c2}.

So, at the applied magnetic field H in type-II superconductors, the Meissner state is stable at $H < H_{c1}$, and mixed state is stable into range of $H_{c1} < H < H_{c2}$. At the same time, at the transition from the superconductive state to the mixed one there is superheating region. de Gennes [342] and Chapman [255] have shown that at the $H_{c1} < H_0 < H_{sh}$ the superconductive state remains only locally stable and its stability is violated at the *superheating field of type-II superconductor*, H_{sh}. The estimation of this critical field may be obtained by studying corresponding non-linear elliptic system.

Then, Saint-James and de Gennes have shown that the superconductivity will remain even at $H > H_{c2}$ in a thin surface layer, until $H \leq 2.392\kappa H_{cm}$, where κ is the Ginzburg–Landau parameter, H_c is the thermodynamic critical field, that is, $H \leq 1.695H_{c2}$ [1460]. This field $H = 1.695H_{c2}$, also destroying superconductivity in the surface layer has been named by *the third critical field* and denoted by H_{c3}. However, based on the measurements of magnetic susceptibility of the Nb cylinders under fields above H_{c2} [240, 926], it has been stated that the ratio of H_{c3}/H_{c2} depends on processing method of surface, but was above the theoretical value of 1.695 [1460] in all cases. The near-surface critical current was absent in the fields above $H_{c3}^{c}(T) = 0.81H_{c3}(T)$, where factor 0.81 was independent of the temperature, surface quality and admixtures. The last two parameters led to alteration of the ratio H_{c3}/H_{c2} from 1.86 to 2.57. It could be assumed [240, 926] that below H_{c3} down to H_{c3}^{c} there is only local near-surface superconductivity in the form of separate sites, which contribute to the magnetic susceptibility, but do not provide surface critical current. They couple together only in the field below H_{c3}^{c} and a single coherent superconducting state forms.

1.6 Theories of Superconductivity

The first theory which had success in describing the electrodynamics of superconductors was the phenomenological theory of brothers F. and H. London (1935). It had introduced two equations, in addition to Maxwell's ones, governing the electromagnetic field in a superconductor [1046]. These equations provided a correct description of the two basic properties of superconductors: absolute diamagnetism and zero resistance to a *dc*-current. The London theory did not resolve the microscopic mechanism of superconductivity on the level of electrons. Therefore, the question: "*Why* does a superconductor behave according to the London equations?" remained beyond its scope.

According to the London theory, electrons in a superconductor maybe considered as a mixture of superconducting and normal electrons. The quantitative density of the superconducting electrons, n_s, decreases with increasing of temperature and attains zero at $T = T_c$. On other hand, at $T = 0$ the value of n_s is equal to the total density of conductivity electrons. This postulates two-fluid model of superconductor proposed by Gorter and Casimir. A flux of superconducting electrons meets no resistance. Obviously, such a current cannot generate a direct electric field in a superconductor because, if it did, it would cause the superconducting electrons to accelerate infinitely. Therefore, under stationary conditions, corresponding to absence of an electric field, the normal electrons are at rest. In contrast, in the presence of an *ac*-electric field, both the normal and the superconducting components of the current are finite and the normal current obeys Ohm's law.

The London equations provided a description for the behavior of the superconducting component of the electronic fluid in both *dc*- and *ac*-electromagnetic field. They also helped to understand a number of aspects of the superconductors' behavior in total. However, by the end of the 1940, it was clear that one question at least was not answered in the framework of the London theory. For the interfaces between adjacent normal and superconducting regions, the theory predicted a negative surface energy: $\sigma_{ns} < 0$. This implied that a superconductor in an external magnetic field could decrease its total energy by turning into a mixture of alternating normal and superconducting regions. In order to make the total area of the interface within the superconductor as large as possible, the size of the regions must be as small as possible. This was supposed to be the case even for a long cylinder in a longitudinal magnetic field, in contradiction to experimental evidence existed at that time. Experiments showed that such a separation of the normal and superconducting regions took place only for samples with a non-zero demagnetizing factor (the intermediate state). Moreover, the layers were rather thick (~ 1 mm), which could only be the case if $\sigma_{ns} > 0$, which also contradicted the London theory.

This contradiction was reconciled by a theory proposed by Ginzburg and Landau, which was also phenomenological but took into account quantum effects [536]. In that moment, it became clear why it is so important to include quantum effects in considerations. Assume that there is a wave function (or order parameter) Ψ, describing the electrons quantum-mechanically. Then, the squared amplitude of this function (which is proportional to n_s) must be zero in a normal region, increase continuously through the normal-superconducting interface and finally reach a certain equilibrium value in a superconducting region. Therefore, a gradient of Ψ must appear at the interface. At the same time, as is known from quantum mechanics, $|\nabla\Psi|^2$ is proportional to the density of the kinetic energy. So, quantum effects taken into account lead to an additional positive energy stored at the interface, which creates the opportunity to obtain $\sigma_{ns} > 0$.

Tremendous importance of the Ginzburg–Landau theory consisted in that it introduced quantum mechanics into the description of superconductors. It assumes the description of the total number of superconducting electrons by a wave function, depending on spatial coordinates (or equivalently, a wave function of n electrons is a function of n coordinates, $\Psi(\boldsymbol{r}_1, \boldsymbol{r}_2,\ldots,\boldsymbol{r}_n)$). Based on this, the theory established *the coherent (coupled) behavior* of all superconducting electrons. Indeed, in quantum mechanics, a single electron in the superconducting state is described by a function $\Psi(\boldsymbol{r})$. If we now have n_s absolutely identical electrons (where n_s, the superconducting electron number density, is a macroscopically large number), and all these electrons behave coherently, it is obvious that the same wave function of a single parameter is sufficient to describe each of them. This idea permitted to involve in the description the superconductivity quantum effects, at the same time, retaining macroscopic features of the material.

It is interesting that analysis of experimental data on the basis of the Ginzburg–Landau theory permitted to estimate effective charge, e^*, which rendered approximately two times greater than electron charge. However, then nobody had any idea about coupling of electrons, while the Ogg's paper [1259][10] that was published some years ago (in 1946) contained the idea about coupling of the electrons with their further Bose–Einstein condensation.[11]

---

[10] In this paper, Ogg claimed that a solid sodium-ammonia solution may be a superconductor at temperatures as high as 190 K (the melting point of the solution) which created great interest in this metastable solid but not obtained (such a high superconductivity) in later experimental support.

[11] Bose–Einstein condensation is quantum phenomenon, consisting in that what in a system from great number of particles, obeying Bose–Einstein statistics at the temperature which is lower than the degeneracy temperature, the finite fraction of all particles of the system is found in the state

The Ginzburg–Landau theory was built on the basis of the theory of second-order phase transitions (the Landau theory) [971] and, therefore, it is valid only in the vicinity of the critical temperature. By applying the Ginzburg–Landau theory to superconducting alloys, Abrikosov in 1957 developed a theory of the so-called type-II superconductors [4]. It turned out that superconductors need not necessarily have $\sigma_{ns} > 0$. Materials that provide this condition are type-I superconductors and transition in external magnetic field from superconducting state to normal one for them is the phase transition of I type. However, the majority of superconducting alloys and chemical compounds demonstrate $\sigma_{ns} < 0$, and they are type-II superconductors. For type-II superconductors, there is no Meissner effect; magnetic field penetrates inside the material but in a very unusual way, that is, in the form of *quantized vortex lines* (quantum effect on the macroscopic scale!). Superconductivity in these materials can survive up to very high magnetic fields, and transition in external magnetic field from superconducting state to normal one for them is the phase transition of II type. Abrikosov found theoretically vortex structures in superconductors, and thus explained experiments of L.W. Schubnikow, who together with co-authors discovered them as long ago as 1937 by observing the unusual behavior of some superconductors in external magnetic field [1522]. Abrikosov assumed that mixed state of superconductor (or Schubnikow phase) is the vortex state, in which superconducting vortices form periodic lattice [4].

Neither the London nor the Ginzburg–Landau theory could answer the question: "What are those 'superconducting electrons,' whose behavior they were intended to describe?" It was 46 years since the discovery of superconductivity, but at the microscopic level a superconductor remained a mystery. This issue was finally resolved in 1957 by the work of Bardeen, Cooper and Schrieffer [91] in which the so-called BCS theory was presented. In 1958, an important contribution was also made by Bogoliubov [151], who developed mathematical methods which were widely used in studies of superconductivity. The BCS theory described the process of coupling of the conductivity electrons with formation of pairs and regrouping into one quantum state. The formed pairs of electrons are called *Cooper pairs (pairons)*. In superconductor, at the $T < T_c$, a part of electrons united in the Cooper pairs form Bose–Einstein condensate (BEC). All electrons being in BEC are described by the same coherent wave function. The rest of the electrons to be in excited over-condensate states (Fermi quasiparticles) and their energetic spectrum reform in comparison with the spectrum of electrons in normal metal. The BCS theory is based on idea on superfluidity of the Cooper pairs of electrons (*Cooper effect*). In contrast to ideal electrons, all pairs could be in the same quantum state, forming macroscopic quantum wave. In these conditions, electric current exists for account of a motion the all aggregate of *paired electrons* without scattering of

(Footnote 11 continued)
with zero impulse. In this case, below this temperature, the same particles with integer spin "condensate" in the same quantum state. Superconductivity could be considered as a consequence of the Bose–Einstein condensation of correlated Cooper pairs of the electrons with opposite impulses and spins.

energy of the single electrons, that is, electric resistance. In the BCS theory, the formation of pairs is explained by the existence of a certain type of indirect interactions between electrons. This interaction has a character of attraction and, therefore, is a contrast to Coulomb force that pushes away one electron on other. Displacing the conductivity electron in metal during its motion causes a local deformation of crystalline lattice on account of attraction forces, with which it acts on positive ions (cations of crystal). This deformation (one from the types of the lattice excitations) can interact with the second electron. Due to this process, there is an attraction of two electrons that render to be "coupled" for account of crystalline lattice. This interaction exists on account of link of the electrons with quanta of excitations, that is, with lattice phonons. An existence of interaction "electron–lattice–electron" leads to change of energetic spectrum of electrons. It is known that in the solid the energetic levels accessible for electrons form consequent zones. Conductor is characterized by the conductivity zone occupied partially up to the energy of E_F, called *the Fermi level*. In superconductor the BCS theory predicts at this energy E_F the beginning of a restricted zone (*energy gap*) that separates ground state of system of the coupled electrons from excited states occupied by usual (uncoupled) electrons. Width, Δ, of this zone is equal to an energy that it is necessary to apply in order to break the pair and consequently for the failure of superconductivity. One depends on a temperature, namely has maximum value at the zero temperature and is equal to zero at the $T = T_c$. In several cases, superconductivity can exist even in the absence of a restricted zone.

The BCS theory explained directly the isotope effect, discovered in 1950, and, it was very important to find the critical temperature, T_c, through phonon and electron characteristics. This opened a possibility to search sensibly new superconductors with higher T_c from materials with great values of Debye temperature and constant electron–phonon coupling.

In 1959, the microscopic theory of superconductivity was elaborated further by Gor'kov, who developed a method to solve the BCS problem, using Green's functions [554–556]. He applied this method, in particular, to find microscopic interpretations for all phenomenological parameters of the Ginzburg–Landau theory, as well as to define the theory's range of validity. The works of Gor'kov completed the development of the Ginzburg–Landau–Abrikosov–Gor'kov theory.

In 1964, Little [1029] and Ginzburg [535] expressed independently an idea about probable non-phonon mechanism of superconductivity in low-dimensional (quasi-1D or quasi-2D) systems. It was shown that substitution of phonons by excitons (excitations of sub-system of the coupled electrons) should permit, in principle, to increase T_c up to 50–500 K. However, search of these superconductors was unsuccessful.

The soliton (or bisoliton) model of superconductivity was considered for the first time by Brizhik and Davydov in 1984 [185] in order to explain the superconductivity in organic quasi-1D conductors discovered by Jérome, et al. in 1979 [778].

In 1986, trying to explain the superconductivity in heavy fermions discovered in 1979, Miyake and co-authors considered the mechanism of superconductivity based on the exchange of antiferromagnetic spin fluctuations [1168]. The calculations showed that the anisotropic even-parity couplings are assisted, and the odd-parity as well as the isotropic even-parity are impeded by antiferromagnetic spin fluctuations.

Interest in the research of superconductivity obtained powerful impulse in 1986 due to the discovery of oxide high-temperature superconductors,[12] made by Bednorz and Müller. The classical BCS theory was unable to explain many of their properties. The electron–phonon mechanism became questionable.

By analyzing the layered structure of cuprates, Krezin and Wolf proposed, in 1987, a model of high-temperature superconductivity based on the existence of two energy gaps, namely superconducting and induced [938, 939]. Indeed, different experiments performed after 1987 have demonstrated the existence of two gaps; however, they both have the same superconducting origin.

In 1987, Gor'kov and Sokol proposed existence of a new type of microscopic and dynamical phase separation [557] that was later discovered in other theoretical models

In this year, Anderson, et al. proposed a model of superconductivity in cuprates, separating the coupling mechanism and the mechanism for the establishment of phase coherence [38, 99]. The HTSC model, called a "resonating valence bond (RVB)" state, proposes that the magnetic interactions between electrons localized on different atoms leads to that what the electrons divide into pairs with opposite spins (spin liquid) and these pairs "resonate" between copper atoms similarly to Cooper pairs. At the doping CuO_2 plane (for instance, due to introducing admixture atoms of other valence), the number of electrons becomes lesser than number of copper atoms, that is the vacant sites (holes) arise, by means of which the electrons can displace and RVB state evolve in superconductive state. In other words the doping leads to spontaneous breaking symmetry, as a result of which the magnetic singlet pairs become charged superconductive couples. Key prediction of RVB model, namely a linear on concentration of admixture atoms increasing the density of charge carriers taking part in superconductivity well agree with experiments.

In 1988, Davydov suggested that high-temperature superconductivity occurs due to the formation of bisolitons, as it takes place in organic superconductors [338]. In 1990, he presented a HTSC theory based on the concept of a moderately strong electron–phonon coupling [339, 340]. The theory utilizes the concept of

[12] In fact, in the early 1970s, a compound of lanthanum and copper oxides was synthesized in Moscow (USSR). This research was not connected with superconductivity, but the researchers were looking for good and cheap conductors. At low temperatures, the conductivity of this new material showed an abnormal behavior. The scientists understood the significance of this abnormality, but, nevertheless, they could not continue the next experiments because in these years there were great difficulties with liquid helium. Moreover, they did not reveal persistence in defense their questionable results. The mysterious compound was put away in a cupboard and forgotten. Thus, in 1986, Bednorz and Müller discovered high-temperature superconductivity practically in the same compound that had been synthesized before in the USSR [1360]. So, the possibility of the superconductivity discovery in cuprates a few years earlier, than in heavy fermions and organic superconductors, was missed.

bisolitons or electron (hole) pairs coupled in a single state due to local deformation of the –O–Cu–O–Cu– chain in CuO_2 planes.

In the early 1990s, based on the Anderson's assumption, namely, that in cuprates the coupling mechanism and the mechanism for the establishment of phase coherence are different, some theorists autonomously proposed that independently of the origin of coupling mechanism, spin fluctuations cause the long-range phase coherence in cuprates. Several groups concluded that the lowest energy RVB is (in one representation) the "$s + id$" state with two orthogonal gaps. The identical state in a totally different representation could be created by Gutzwiller projecting[13] the single-particle state.

In 1994, Alexandrov and Mott showed that, in cuprates, it was necessary to distinguish the "internal" wave function of a Cooper pair and the order parameter of the Bose–Einstein condensate, which may have various symmetries [17].

In 1995, Emery and Kivelson emphasized that superconductivity requires coupling and long-range phase coherence [433]. They demonstrated that in cuprates, the coupling can occur above T_c without the phase coherence.

In the same year, Tranquada, et al. found the presence of coupled, dynamical modulations of charges (holes) and spins in Nd-doped $La_{2-x}Sr_xCuO_4$ (LSCO) by using neutron diffraction [1700].

In 1997, Emery, Kivelson and Zachar presented a theoretical model of HTSC based on the presence of charge stripes in CuO_2 planes [434].

In 1998, Chakraverty, et al. attempted to prove that the theory of bi-polaron superconductivity of HTSC contradicted with experiments and was theoretically discrepant [250]. In answer of Alexandrov [15], has been stated contrary opinion, namely: the negation of the bi-polaron superconductivity of HTSC is the result erroneous approximation for energetic spectrum of bi-polarons and erroneous application of the bi-polaron theory carried out by Chakraverty et al. Based on *two-zone* model, he obtained a formula for T_c, which was free from adjusted parameters and included besides basic constants the concentration of carriers, n, and penetration depths of magnetic flux λ_{ab} and λ_c along two mutually perpendicular crystallographic directions. The substitution of test values of n, λ_{ab} and λ_c (for Y-123) estimated $T_c \sim 100$ K. It proves self-consistency of the bi-polaron approach and testifies HTSC to be in the regime of Bose–Einstein condensation.

In 1999, analysis of tunneling and neutron scattering measurements, carried out by Mourachkine, showed that in $Bi_2Sr_2CaCu_2O_{8+x}$ (Bi-2212) and $YBa_2Cu_3O_{6+x}$ (YBCO), the phase coherence is established due to spin excitations [1196, 1197],

---

[13] The procedure of the Gutzwiller projection consists in the following: The sites are, for free electrons, randomly occupied ½ of the time by up electrons and ½ of the time by down ones, so the probability that any given site is empty is ¼, and the probability that it is doubly full is also ¼. Let us throw away every part of the wave function for which any site is either empty or doubly full, so that on every site there is exactly one electron with spin up or down. You realize that we have thrown away almost all of the wave function, leaving only $(1/2)^N$ of it; but what we have left has a spin on every site and thus will do as an approximate wave function for the Heisenberg antiferromagnetic chain [43].

which cause the appearance of the so-called magnetic resonance peak in inelastic neutron scattering spectra [1451].

In this year, Cronstrom and Noga determined a new solution of BCS equations in approximation of mean field, which pointed to the existence in thin superconducting films (or in superconducting bulks with layered structure) of type-III phase transitions [319]. The critical temperature of this transition increases at decreasing of the layer thickness and is independent of isotope mass. The electronic heat capacity is a continuous function of temperature, but has discontinuity of derivative.

In 1999, Leggett defined a very simple dependence, $T_c(n)$ for "calcium" HTSC (where n is the number of CuO_2 layers per elementary cell): $T_c(n) = T_c(1) + T_0(1-1/n)$, where T_0 is the own constant for each Bi, Hg and Tl family, in which the CuO_2 layers are separated by calcium layers [1003]. In particular, it is followed from this formula that $[T_c(3)-T_c(2)]/[T_c(2)-T_c(1)] = 1/3$, which agrees with test data 0.25–0.28 and 0.25–0.34 for HTSC on the basis of Hg and Tl, respectively.

In 2000, Tang measured the critical temperature, T_c, of ultra-thin HTSC films $YBa_2Cu_3O_7$ depending on their thickness d. It has been observed approximately linearly on d a diminishing of T_c with decreasing of the film thickness at $d < 10$ nm [1657]. The dependence $T_c(d)$ is well described by the empirical formula: $T_c = T_{c0}(1-d_m/d)$, where $T_{c0} = 90$ K and $d_m = 1.56$ nm. The critical thickness, d_m, is near to the thickness of one elementary cell along c-axis that supports quasi-2D nature of HTSC superconductivity.

In 2001, Kivelson proposed the following way for increasing of T_c: it is necessary to create multi-layer systems with different concentrations of carriers in various layers, so the layers with low concentration of carriers provide their coupling, but the layers with high concentration of carriers guarantee phase rigidity [881]. Maximal critical temperature of HTSC is found by competition of two effects: (i) the coupling interaction weakens at increasing of the charge carrier concentration, x, that is connected with properties of doped Mott's dielectric, but (ii) the density of super-fluid component, which controls the system rigidity in relation to phase fluctuations, increases with growth of x. Thus, optimal T_c is reached at the boundary of the region with prevalence of the phase alignment and region with prevalence of the coupling interaction.

In the same year, based on tunneling measurements, Mourachkine provided evidence that the quasiparticle peaks in tunneling spectra of Bi-2212 crystals are caused by condensed soliton-like excitations, which form the Cooper pairs [1198–1200].

In 2001, the superconducting system MgB_2 has been discovered by Nagamatsu, et. al. which demonstrated highest volume superconductivity (critical temperature, $T_c = 39$ K) on that moment among of non-copper oxide conductors combining simplicity and accessibility of this material with complexity of the superconductivity phenomenon [1224].

In 2002, Cui proposed a possible responsibility for superconductivity of relativistic attraction of electrons. At least, there are two types of collective movement in superconductors, which can suppress usual Coulomb repulsion of electrons as the attraction component becomes predominant. This movement is caused by combination of electron gas and phonons in conventional superconductors and by itself electron gas (or electron liquid)—in HTSC. The repulsion and attraction between electrons balance approximately each other in majority of matters; therefore, the theory of electron gas (i.e., non-interacting particles) works well. However, the repulsion predominates over attraction in some matters, and then electron sub-system demonstrates properties directly contrary to superconductivity [320].

In this year, Laughlin showed that cuprate HTSC in undoped state should be considered as no dielectrics, but superconductors with very great gap and extremely small super-fluid density [987]. Laughlin named these superconductors "gossamer superconductors". In practice, a brittleness of coupled state of this superconductor creates obstacles for stating superconductivity in all volumes. However, he assumed that the wave function of this coupled state can serve as a good starting point for understanding of correlations between Mott's dielectric and HTSC.

In 2003, Hussey, et al. proved an existence of 3D Fermi surface, based on investigations of angle oscillations of magnetic resistance in HTSC $Tl_2Ba_2CuO_{6+\delta}$ [734]. Thus, almost a 20-year-old argument about coherency or incoherency of electronic states along the *c*-axis was solved in favor of coherency.

In 2004, Homes, et al. obtained the next universal scaling relationship between physical values, characterizing normal and superconducting states of HTSC [701]: $\rho_s = A\sigma_{dc}T_c$ (where ρ_s is the density of super-fluid component, σ_{dc} is the static specific conductivity and T_c is the critical temperature), which should be carried out for *all* HTSC, without dependence on the value of T_c, type of carriers (holes or electrons), doping level, crystalline structure and current direction (parallel or perpendicular to CuO_2 planes). In this case, the proportionality factor, $A = 120 \pm 25$, if ρ_s is measured in s^{-2}, σ_{dc} in $(\Omega\ cm)^{-1}$ and T_c in K. The straight dependence, $\rho_s(\sigma_{dc}T_c)$, includes even the points for low-temperature superconductors Pb and Nb. It covers diapason above of five degree of magnitude on each of the coordinate axes. This empirical dependence is surprising, if taking into account the principally different character of current transfer in various crystallographic directions (i.e., coherent in *ab*-plane and incoherent along *c*-axis).

In this year, Alexandrov again explained HTSC physics on the basis of bi-polarons mechanism. HTSC properties weakly differ from usual metals, namely: there is standard BCS phenomenon, only Bogoliubov's quasiparticles (Cooper's pairs with *d*-symmetry) transfer current. The magnetic mechanism of superconductivity exists, but as subsequent of electron–phonon coupling. Because constants of electron–phonon coupling are greater than 1, polarons and super-light bi-polarons must arise. In this case, the critical temperature is found as [16]

$$T_c = 1.64\left[eR_{\mathrm{H}}/(\lambda_{ab})^4(\lambda_c)^2\right]^{1/3}, \qquad (1.31)$$

where e is electron charge, R_{H} is Hall constant, λ_{ab} and λ_c are the penetration depth of magnetic field in *ab*-plane and along *c*-axis, respectively.

In 2004, Anderson, et al. finished formation of the "plain vanilla" theory being the simple Hubbard model transformed into the *t–J* model but unembellished by esoteric complications [44]. The elements of which are: (i) the super-exchange vertex, resulting from (ii) the projective canonical transformation, which (iii) in turn renormalizes the kinetic energy and hence T_c and the superfluid density; (iv) the gauge ambiguity is resolved by the spin-charge locking mechanism [40]. As a result, the "plain vanilla" gap equations have been obtained which could explain some features of experimental data.

In 2005, Li, et al., in the course of transport studies of cuprate superconductors, came upon an extensive region of the (x, T)-phase diagram, in which these studies gave unequivocal evidence of quantized vorticity above the critical temperature T_c [1014]. So, they discovered a new phase of matter which it is appropriate to describe as a vortex fluid. The theoretical ground of existence of this state has been given by Anderson [43].

In 2006, Honma and Hor proposed to distinguish 2D and 3D hole concentration [704]. It is well known that the critical temperature of HTSC is a universal function of this parameter (the hole concentration is usually found by number of holes per Cu atom in CuO_2 layer). Their analysis shows that T_c is defined by 3D *concentration*, n_h, and universal bell-like dependence takes place just for $T_c(n_h)$.

In this year, Anderson derived for the actual shape of the Nernst or the equivalent diamagnetic response, a formula which is remarkably successful in describing the overall shape, though neither the experiments nor the theory are adequate to describe fine details, especially the precise behavior at low magnetic field [41].

In 2008, the superconductive oxypnictides were discovered by Kamihara, et al. in which the critical temperature very rapidly rose from 26 K [806] up to $T_c \sim 56$ K [269, 1809]. These superconductors possessing layered structure in many properties are similar to cuprate high-temperature superconductors. The high critical temperature of the oxypnictides which is near the temperature of liquid nitrogen (77 K) allows one to concern them to HTSC.

Obviously, the above list of mechanisms and superconductivity theories is not complete. Finally note that beginning from 1987, theorists proposed more than 100 models of high-temperature superconductivity, in particular, based on the representations about polarons, plasmons, excitons, solitons, super-exchange, direct interaction between electrons, etc. and also published more than 100,000 papers. Today, the intensive discussion of HTSC mechanisms and high critical temperatures, T_c, to being proper for these materials is continued. Some review results of different HTSC theories can be found in monographs [39, 197, 506, 507, 725, 849, 1027, 1201, 1569].

1.7 Connection of Superconductivity with Other Physical Researches

History of the superconductivity investigation and development its ideas and concepts are inseparably connected with many other fields of modern physics. Below in this section we will discuss these connections in some details.

1.7.1 Development of Common Research Methods of Superconductivity, Condensed Matter Physics, Electromagnetism and Gravitation

A prospective of developing these joint ideas and studies is based on the following moments [894].

Firstly, Dirac in 1931 made a proposition on that what owing to quantum mechanics, the Maxwell equations could be adapted to magnetic monopoles in spite of disappearance in this case of the magnetic field divergence [378]. Always, it is possible to form magnetic field arising from point supplying far field to this point by using thin magnetic vortex tube. However, owing to quantum mechanics this vortex tube cannot be found by known physical methods. It is a truth when carries out the known Dirac condition of charge quantization which states that all electric charges are integer multipliers of $2\pi\ \hbar c/\Phi$, where $\hbar$ is Planck constant, c is the speed of light in vacuum, Φ is the total magnetic flux via the tube. This physically indeterminable tube is called Dirac string. The magnetic field lines arise from finite point of this string in outside direction similarly to the lines of electric field arising from point electric charge that is this finite point acts as magnetic monopole. In this case, the string shape is arbitrary and has not any interest. The Dirac condition of quantization has been specified by Schwinger [1529]. In experiment, the magnetic monopole has not been found in spite of intensive studies and the Dirac theory was theoretical abstraction a long time. However, in last 40 years, it has been again added to arsenal in attempt to explain a phenomenon of quark confinement.

Secondly, Berezinski [123], Kosterlitz and Thouless [922–924] showed that the phase transition in film of superfluid helium could be understood on the base of statistical mechanics of vortices of the superfluid liquid. Their description relates to each point a phase angle of the condensate wave function belonging to range (0, 2π). After vortex passing of a whole circle in any point of space, this angle jumps on value of 2π. A line of this jump connects the vortex with antivortex and forms an analog of the Dirac string, whose precise shape to be unessential. If this idea to transfer on spatial superfluid helium in three-dimensional statement, then it leads to the statistical mechanics of vortex contours which interact with the same long-range forces as the electric circuits.

Thirdly, another basis of these investigations is the theory of plastic deformations applied to understand work hardening of metals and material fatigue. This theory has been developed by Orowan [1277], Polany [1384] and Taylor [1662] after discovery of dislocations in 1934. By using field methods this theory has been extended by Kleinert [890] on statistical mechanics of linear defects. At the same time, it has been shown importance of melting transitions at the condensation of linear defects.

Fourthly, Bilby [139], Kondo [910] and Kröner [942] have demonstrated that the linear defects could be described by geometrically. The crystal distortion types do not change geometry of defects, so playing the same role as Einstein coordinate transformations. The crystal with defects forms especial version of Riemann–Cartan space. The theory of these spaces has been developed by Cartan [235–237], which supplemented a distorted Riemann space with other geometrical property, namely torsion. This Cartan work inspired A. Einstein on creation of the gravitation theory in Riemann–Cartan space possessing the property of teleparallelism [238]. After 20 years Schrödinger [1518–1520] has attempted to link the torsion properties with electromagnetism. He noted that presence of torsion in Universe should make photons to be massive and restrict long-range action of magnetic fields caused by planets and stars. By using the observed in experiment ranges of changing magnitudes attained in this time, he obtained upper bound of photon mass. Then, Utiyama [1730], Sciama [1530] and Kibble [850] stated the dependence between torsion and spin density of gravitation field.

Application of the order–disorder theory to the linear defects in crystals has been fulfilled by Kleinert in 1980s [894]. Original description of these defects has been based on functions which to be discontinuous on surfaces with boundaries in the form of defect lines. The shape of these surfaces has been arbitrary, at the same time the boundaries fixed. As a result, it has been stated that deformations of these surfaces could be formulated as gauge transformations of the gauge fields of new type (so-called, defect gauge fields). Owing to so-called duality transformation, it may be reformulated the theory of defects and their interactions in more proper type of the gauge theory. The duality formulation could be considered as a linearized form of other Einstein–Cartan geometrical space, in which the gauge transformations to be a combination of the Einstein local translations and local generalization of the Lorentz invariance. The relationship between dual and actual description of defects in the terms of jump-like surfaces is completely to be similar to well-known relationship between Maxwell theory of magnetism, formulated into terms of the gauge field, vector potential and alternative formulation in which magnetic field is a gradient of multivalued scalar field.

In order to simply explain the phenomenon of the quark confinement by using electric color field, the physics of superconductivity becomes an important ground. With the time of appearance of the superconductivity theory proposed by London [1046], it is well-known that superconductors must trap (or restrict of movement) magnetic charges if they exist. The cause is explained by Meissner effect which leads to expelling magnetic flux lines from superconductor on its surface. Due to the lines arising from magnetic monopole compress in tubes of magnetic flux with fixed thickness. The energy of these tubes is proportional to their length suggesting that opposite magnetic charges retain together. From BCS theory (1957), it is well-known that this effect is caused by Bose–Einstein condensate of electric charges known as Cooper pairs of electrons. This phenomenon grounded a modern view on the quark confinement. As it is assumed, the vacuum state of Universe consists of a condensate of magnetic color monopoles. This condensate acts under influence of electric color fields the same way as the Cooper pairs act in superconductor under influence of magnetic field causing the Meissner effect and the confinement of electric color charges. The models which used this mechanism of the confinement have been first developed in 1974 by Polyakov [1386], Nambu [1232], t'Hooft [1673] and Wilson [1803]. During study of this phenomenon are observed closed mathematical analogies between Dirac magnetic monopoles and above-mentioned defect structures. Dirac used vector potential together with jump-like surfaces for considering infinitely thin tube of magnetic flux with magnetic point source on its end. Thus, a world line of monopole in the space–time could be considered as a type of "vortex line" into Maxwell field. After constructing the theory of disorder of the vortex lines, Kleinert [891–893] developed the theory of field disorder of the world line of monopoles which today serves as a simplest model of the quark confinement. At extension of statistical mechanics of the vortices on defect lines is used dual description of the defect lines, which defined by their linear approximation of geometrical description in the Riemann–Cartan space. This circumstance has motivated the development of the defect theory based on multi-valued fields to return to their geometrical description and reformulate the gravitation theory which is traditionally to be treated as a geometrical theory by alternative way by using jump-like surfaces of fields of translation and torsion. In the plasticity theory these singular transformations are used for transformation of ideal crystal in the crystals with translational and torsion defects. Their geometrical analogues filled the plane space–time interior of the space–time with curvature and torsion. Thus, the multi-valued fields are the mathematical ground defining a new geometry. Future physical and mathematical description of these fields is the prospective direction which could join together numerous features which are proper to, on first view, different physical phenomena as superconductivity, gravitation and crystal plasticity.

1.7.2 Spontaneous Symmetry Breaking or on Connection Between Superconductivity and Elementary Particle Physics

In 1933 Meissner wondered: Is the electrical current carried in superconductor by known charged particles electrons and ions? The tremendous achievement of BCS theory consisted in that what understanding the superconductivity phenomenon Bardeen, Cooper and Schrieffer in 1957 made without introducing any-body new forces or elementary particles. An idea of spontaneous breaking of symmetry first was borrowed by physics studying elementary particle from the BCS theory. One of these symmetries is an invariance of the nature laws under action of the symmetry group of three-dimensional rotations or, in order words an invariance of laws at the change of orientation of the measuring equipment.

When a physical system does not demonstrate all symmetries of laws on which it is controlled, then occurs a spontaneous breaking symmetries. As an example, we note a spontaneous magnetization. The laws controlling atoms in magnet are completely invariant at the three-dimensional rotations, but at the temperatures which are lower than the critical one, spins of these atoms spontaneously line up in one direction initiating magnetic field. Any superconductor (as high-temperature, as supporting to the BCS theory) presents oneself a material, in which especial symmetry of the nature laws (electromagnetic gauge invariance) to be spontaneously broken. Here, a symmetry group is a group of two-dimensional rotations. These rotations act on two-dimensional vector, whose two components to be real and imaginary parts of electron field. The quantum–mechanical operator annihilates electrons in the quantum theories of field. The rotation angle of group of the broken symmetry can change with the location in superconductor and then the symmetry transformations influence also on electromagnetic potentials. The symmetry, breaking in superconductor, remains unbroken in a result of rotation on 180°, which only changes a sign of electric field. Due to this spontaneous breaking symmetry, products of any even number of electron fields have non-vanishing values of expectation in a superconductor, at the same time the field of single electron does not [1787].

All proper (exact) properties of superconductors, namely: zero electric resistance, expelling magnetic field from a superconductor (Meissner effect), a quantization of magnetic flux in thick superconductive ring and Josephson formula for frequency of *ac*-current in Josephson junction between two superconductors with different voltage, follow from the assumption on this breaking electromagnetic gauge invariance. By this, there is need to state a breaking mechanism of this symmetry. It should be noted, that in the seminal paper [91], which presented the BCS theory has been described (while not mentioned) the breaking mechanism of the electromagnetic gauge invariance. The authors obtained properties of superconductors on the base of their dynamical model, but no fact of breaking symmetry. This dynamic model has been based on approximation that a pair of electrons interacts only when value of their moments very near some magnitude which is known as Fermi surface. In this case, there is a question how can be stated exact properties of superconductors, for instance zero electric resistance and quantization of magnetic flux on the base of approximate dynamic theory. Only one argument follows from exact principles of symmetry, which can in a whole explain remarkable exact properties of superconductors. While the spontaneous breaking symmetry has not been emphasized in the BCS theory, acknowledgement of this phenomenon led to revolution in elementary particle physics. The cause consists in that what at any spontaneous breaking symmetry, excitations of the systems arise with frequency which aspires to zero at the large wavelength. In elementary particle physics, this defines a particle with zero mass.

The first step in this direction has been made by Nambu [1231] which received Nobel Prize in Physics in 2008. He showed that these collective excitations in superconductors play a decisive role in agreement of corresponding breaking gauge invariance in a superconductor with exact gauge invariance of ground theory, controlling matter and electromagnetism. Y. Nambu speculated that these collective excitations to be necessary sequence of this exact gauge invariance. Latter, he put this idea in elementary particle physics. Due to nuclear β-decay, the electron and neutrino (or their anti-particles) are created by currents of two different kinds flowing in the nucleus which are known as vector and axial-vector currents. It has been known that the vector current conserved in the same sense as usual electric current. Could be this spoken on the axial current? The current conservation is usual symptom of some symmetry of basic theory and conserves independently of spontaneous symmetry breaking. This symmetry is electromagnetic gauge invariance for usual electric current. Moreover, the vector current during β-decay conserves due to isotropic spin symmetry in nuclear physics. It may be simply presented some different symmetries of the known chiral symmetries type, which could lead to conservation of axial vector current. However, in this case, any chiral symmetry should imply either zero mass of nucleon (that is not true), or that exists triplet of mass-less strongly interacting particles with zero spin and odd state (this also no truth). These both possibilities correspond to two possibilities related to symmetry, whatever it is, either is not, or if it exists then spontaneously broken not only in material similar to superconductor but even in empty space. Nambu proposed a requirement of existence of this symmetry which breaks spontaneously in empty space. However, a symmetry being spontaneously broken is not to be exact. So, a particle with zero spin and odd state (it is required by broken symmetry) is not massless and only much lighter than other strongly interacting particles.[14] In a subsequent paper [1233] Nambu and Jona-Lasinio presented a theory, in which by using drastic approximations they defined that a result of chiral symmetry breaking is arising light pion as coupled state of a nucleon and an anti-nucleon.

Many years after that there were not proofs that the broken exact symmetries always directly lead to mass-less particles. Only examples of approximate calculations existed into framework of special theories. Then, Goldstone presented some new instances and a proof that the phenomenon to be common result [541]. These mass-less particles have been called Goldstone bosons (or Nambu–Goldstone bosons). Soon after, Goldstone, Salam and Weinberg formulated this result in the form of rigorous general theorem [542]. However, first in elementary particle physics, this theorem has been seen as a disappointing result, because it was assumed that the phenomenon of spontaneous symmetry breaking would explain why symmetries discovered in physics of strong interactions were not exact. The predictions of new mass-less particles which have been discovered in experiments in the early 1960s seemed to close this hope. However, it was not this case.

[14] This light elementary particle has been called the pion, which is lightest and first particle discovered among mesons.

Excepting especial cases, the spontaneously broken symmetry does not look as an approximate unbroken symmetry. It manifests itself in masslessness of bosons with zero spin and in details of their interactions. Today, it is clear, that the approximate symmetries as isospin and chiral invariance are consequences that what masses of single quarks, for some unknown reason, happen to be relatively small. At the same time, these false hopes led to important consequence. Higgs [674, 675], Englert and Brout [442], and Guralnik, Hagen and Kibble [594] found an exception from the general theorem which is applied to theories, in which ground physics to be invariant at local symmetries. The transformations of these symmetries which to be similar electromagnetic gauge transformations can change from site to site in space and time.[15] The vector field must exist for each local symmetry, which is similar to electromagnetic field, whose quanta would be massless if symmetry was not spontaneously broken. The quanta of each that field are the particles with component of angle moment in direction of movement which is equal in natural units to +1 or −1. However, if symmetry is spontaneously broken, then these two states of helicity (chirality) join with state of zero helicity of Goldstone boson in order to form three-component states of massive particle of spin 1. So, when a local symmetry breaks spontaneously neither vector particles with which associates symmetry, nor Nambu–Goldstone particles created by the symmetry breaking have zero mass (it has been shown in [442, 674, 675]).

In actuality, this has been previously shown by Anderson on the example of the BCS theory [37]. However, the BCS theory is not relativistic, but Lorentz invariance which to be proper for special theory of relativity played a crucial role in the general Goldstone–Salam–Weinberg theorem. Therefore, the argumentation of Anderson was ignored by physicists studying elementary particles. In fact, Anderson was true because the cause for exception stated by Higgs, et al. consisted in that what it is impossible to introduce a quantization into theory with local symmetry by method which conserves both as manifested Lorentz invariance as usual rules of quantum mechanics including the requirement of positive probabilities. In fact, there are two ways to quantize theories with local symmetry: (i) the way conserving positive probabilities, but loses the manifested Lorentz invariance, and (ii) the way conserving the Lorentz invariance, but loses the positive probabilities. So, these theories, in fact agree with as the Lorentz invariance as the positive probabilities, however they contradicts to the general Goldstone–Salam–Weinberg theorem.

The appearance of mass for the quanta of vector bosons in the theory of local symmetry again discovered an old suggestion of Yang and Mills on that what strong interactions could be created by vector bosons associated with some kind of local symmetry to be more complex than the known electromagnetic gauge invariance [1828]. This possibility has been specially pointed by Englert and Brout [442].

---

[15] This sharply contrasts with chiral symmetry associated with axial vector current at β-decay, which is applied only when the symmetry transformations are the same in a whole space–time.

It took a few years for transformation of this theory into special theory, which was no theory of strong interactions. It is possible, that the delay has been caused by earlier idea of Nambu on that what the pion was almost massless boson associated with approximated chiral symmetry which was no local symmetry [1233]. The mathematical approaches developed in this period in physics were used by Polchinski [1385] and other physicists for explanation of approximations in their works on superconductivity which have been based on the BCS theory.

In 1967, Weinberg returned to idea of the theory of strong interactions based on the group theory of spontaneously broken local symmetry [1783]. Since, the subgroup consisting of usual isospin transformations is no spontaneously broken then in it the mass-less vector particle would exist, associated with these transformations. However soon he understood that it was related to the theory of weak nuclear interactions which similar to β-decay. In this case, there was only one natural selection for proper local symmetry corresponding to the symmetry group which has been proposed by Glashow [538], while no in context of spontaneously broken local symmetry. The same symmetry group has been independently considered by A. Salam and Ward [1467], and also Weinberg [1785]. Even in this exact case the spontaneous symmetry breaking would introduce massive vector particles (charged W-particles which have been subjected to theoretical speculation for decades) and neutral particle, called Z-particle which mediates weak interaction of "neutral current" which has not been observed in experiments. The same symmetry breaking creates mass of electron and other leptons and at the following extension of the theory, to the quarks. This symmetry group contained electromagnetic gauge invariance, and due to this subgroup is obviously no spontaneously broken (excepting the case of superconductors), then the theory requires existence of mass-less vector particle, namely photon (light quantum). Today, this theory is known as the theory of electroweak interaction it has been independently developed by Salam [1466] and Weinberg [1784]. The mathematical ground of this theory has been developed by 't Hooft [1672].[16] Their detail studies have demonstrated properties of the particles expected by the theory of weak interaction. Still is an issue: how is symmetry of the weak interaction broken?

In the BCS theory, the spontaneous breaking the electromagnetic gauge invariance arises because of attraction forces between electrons near the Fermi surface. These forces are no strong and symmetry could be broken by these weak forces. However, this feature exists only owing to presence of the Fermi surface, therefore in this sense the BCS theory does not agree with elementary particle physics. In absence of the Fermi surface, dynamical spontaneous symmetry breaking requires influence strong forces. Today, are unknown the forces influencing on quarks and leptons which be enough strong that to lead dynamically to the observed breaking local electric weak symmetry. Therefore, Salam and

---

[16] The weak interactions of neutral current have been discovered in 1973 by Lagarrigue, et al., W- and Z-particles also found at CERN a decade later (Rubbia and van der Meer awarded the Nobel Prize in Physics in 1984 for this discovery).

Weinberg did not assume a possibility of dynamic symmetry breaking. Instead, they introduced elementary scalar fields in the theory, whose values of vacuum expectation into framework of classical approximation would break symmetry. In this case, the only elementary scalar quanta which eliminate at spontaneous symmetry breaking in theory are those which become helicity-zero states of the vector *W*- and *Z*-particles. As the physical particles, the other elementary scalars appear which are known as Higgs bosons. These Higgs bosons, predicted by the Salam–Weinberg theory of electroweak interaction, now are the research objective of the CERN large hadron collider (LHC).

At the same time, there is another possibility proposed by independently by Weinberg [1786] and Susskind [373] in the late of 1970s. The electroweak symmetry could be dynamically broken similarly to as in the BCS theory. For this possibility, it is necessary to introduce in consideration forces of super-strong interaction known as technicolour forces which influence on novel particles differing of known quarks and leptons. Into framework of these assumptions, it may be simply obtain true masses for *W*- and *Z*-particles and large masses for all novel particles. However, there are significant difficulties in giving masses to the ordinary quarks and leptons. It is possible, that the experiments on LHS will not find the Higgs bosons, but in any case they discover numerous novel heavy particles, associated with the technicolour forces, and may be stated the mechanism of symmetry breaking electroweak interaction.

1.7.3 Field Theory and Study of Superconducting Phase Transition

As it has been noted in Sect. 1.6 the Ginzburg–Landau (GL) phenomenological theory of superconductivity was proposed in 1950 [536] for some years before the BCS microscopic theory [91]. In 1959, Gor'kov obtained the GL-theory from the BCS theory [554–556]. It has been made by using development of effective theory for the Cooper pairs which fulfils near the critical temperature T_c. The modern approaches use functional integrals that introduce a collective quantum field $\Delta(\boldsymbol{x}, t)$ for the Cooper pairs by using Hubbard–Stratonovich transformation [724, 1617] and integrating fermions in partition function. At the same time, the GL-theory conserved its actuality in our time. It is well suitable in description of HTSC, at the same time the BCS theory is inadequate in many details at description of these materials. The success of the GL-theory in study of modern problems of superconductivity is based on its universal effective character in which the details of microscopic behavior are no important. In vicinity of critical point, the GL-theory covers a broad set of applications, many of which to be outside of superconductivity.

It is possible its application to the phase transition from smectic to nematic phase in liquid crystals is the most known example. In elementary particle physics the Gor'kov result in the GL-theory initiated four-dimensional relativistic version of the BCS theory, the Nambu–Jona-Lasinio model [1233]. The GL-field describes in this GL-theory the quark-antiquark border states, mesons π, σ, ρ and A_1. This model widely uses in nuclear physics.

Another relativistic GL-model is required for explanation of effect of arising masses for vector bosons W and Z in common theory of electromagnetic and weak interaction in renormalized form. It is carried out by using a non-Abelian analog of the Meissner effect that in this context is called Higgs mechanism. In 1973, Coleman and Weinberg [302] showed that four-dimensional Abelian GL-model (or scalar electrodynamics, from viewpoint of elementary particle physics) demonstrates spontaneous arising mass in originally massless theory. From viewpoint of statistical mechanics, this implies type-I phase transition, when mass of scalar field crosses zero.

In 1974, Halperin, Lubensky and Ma [618] stated similar behavior in three-dimensional GL-theory of superconductivity (HLM-model). These works open a new epoch in study of the GL-model: firstly the renormalization group (RG) methods have been used for studying superconductive phase transition. The RG-results have been obtained on the version of Wilson of the RG-theory [1803], in which fast modes are integrated for statement of effective theory in terms of slow modes. The approach of field theory using Callan–Symanzik equation [226, 1635, 1636] leads to equivalent result in perturbation statement of the problem.

At the level of the mean field theory, the GL-model demonstrates type-II phase transition. However, the HLM-model makes a conclusion that fluctuations change the phase transition type to first one. Type-II phase transition would be characterized by absurdly high number of complex components of the order parameter. If number of the complex constants is equal to $N/2$, then the one-loop RG-analysis in the HLM-model leads to lower bound of $N > N_c = 365$ for the type-II phase transition. By this, the dictated infrared stable fixed point did not find for $N < N_c$.

Most popular non-perturbation approach of the field theory and statistical physics is connected with $1/N$-expansion. The critical exponents of the GL-model in the case of $O(1/N)$ have been calculated in the HLM-model for three-dimensional case. However, in 1981, Dasgupta and Halperin [334] expressed doubt in the pointed results of the HLM-model by using arguments of duality and Monte-Carlo simulation on lattice model in London's approximation. They demonstrated that the phase transition for $N = 2$ is type-II phase transition. Therefore, the result of the RG-theory which suggested that in this case the phase transition be always type-I phase transition turn out to be artifact of ε-approximation.

In 1982, Kleinert [888, 889] carried out quantitative duality on lattice model and grounded the field theory of disorder (FTD) which explained the disagreement between the RG-result of the HLM-model and the Dasgupta–Halperin computer simulation. It has been shown that there is tricritical point on phase diagram of superconductor at line dividing the regimes of type-I and type-II. Existence of this tricritical point allowed one to construct successive picture, in which the limit of strong couple relative to type-II phase transition and the limit of weak couple relative to weak type-I phase transition co-exist with normal phase meeting at the tricritical point. Based on FTD the GL-parameter has been calculated in the tricritical point: $\kappa = \lambda/\xi \approx 0.8/\sqrt{2}$, where λ is the penetration depth, and ξ is the coherence length. This result agreed with accuracy into bounds of 5 % with large-scale numerical experiments based on Monte-Carlo simulation fulfilled by Mo, et al. [1174]. Previous computer simulation based on the Monte-Carlo method, carried out by Bartholomew [98], led to the rougher estimate $\kappa \approx 0.42/\sqrt{2}$. FTD allowed one to also state that the value of critical exponent ν corresponds to the value of *XY-model* ($\nu \approx 0.67$). At the superconductive phase transition only critical exponents ν, ν' and α are measured. Here α is the exponent at the specific heat connected with ν through hyper-scale relationship $d\nu = 2-\alpha$.

Other critical exponent η for the present does not measured and even can doubt in existence its physical sense. It is well known arguments that superconductive order parameter could not be treated as physically measured value since its conjugate field has not physical sense. In ferromagnetic, the field, conjugated with magnetization, to be external magnetic field which can be experimentally controlled. Another problem connected with that what the order parameter $\langle\psi\rangle$ is not to be gauge-invariant. Therefore, a calculation of $\langle\psi(\boldsymbol{r})\psi^{+}(\boldsymbol{r}')\rangle$ will depend on gauge selection, then η will be also defined by gauge. Nogueira and Kleinert [1246] demonstrated a possibility to define the critical exponent η out of dependence on gauge. In the case of GL-model, the calculations gave a negative value of η $(-1 < \eta < 0)$ in three-dimensional case. The physical sense of the critical exponent η is stated on the base of consideration of, so-called Lifshits point in phase diagram for magnetic systems, at which co-exist paramagnetic, ferromagnetic and helical phases. In the case of superconductor, it corresponds to the point, in which co-exist superconductive phases of type-I and type-II, and also normal phase. In complete analogy with magnetic systems, this phase diagram is represented in plane $\kappa^2 - T$. It is necessary to note, that in the case of GL-model, behavior at the Lifshits point would be defined by thermal fluctuations. During last decade, the approach, based on renormalization groups to the GL-model, has been studied by several scientists with aim to improve the RG-analysis, so to obtain the prescribed fixed point for $N = 2$, predicted by dual approach. However, in spite of significant progress, understanding this problem remains far insufficient.

1.7.4 Optical Conductivity and Spatial Heterogeneity of HTSC

The dynamical conductivity $\sigma(q, \omega)$ is the linear function which couples current density with electric field (where q is the electric charge and ω is the angular frequency of oscillations). In the limit of $q \to 0$ the function $\sigma(q, \omega) = \sigma(\omega)$ is called optical conductivity because describes response of the medium on electromagnetic waves with the wave length, which is far longer than scales of proper sizes of condensed electronic systems. The real part of the optical conductivity,

$\sigma_1(\omega)$, describes dissipation of electromagnetic energy in the medium, and imaginary part, $\sigma_2(\omega)$, presents screening applied field. The measurement of $\sigma(\omega)$ in superconductor is used to estimate dynamics of quasiparticle excitations and size of energetic gap. In accordance to the BCS-theory, there are three dissipative processes defining $\sigma_1(\omega)$ in superconductor: (i) superplastic acceleration, (ii) creation of quasiparticle couple, and (iii) quasiparticle scattering. The first process presents a work required for acceleration of electrons for attaining by them of Meissner screening current. This contribution appears in the form of Dirac delta-function at zero frequency in $\sigma_1(\omega)$, whose spectral weight to be superfluid density ρ_s. Other two contributions appear above zero frequency. At the formation of quasiparticle couple, the electromagnetic energy dissipates, when photon excites the quasiparticle couple out of the BCS-vacuum. At the scattering quasiparticles, photon moves the quasiparticle preliminary excited out of the vacuum due to thermal excitation on higher energetic level.

Some proper properties of HTSC make the study of $\sigma(\omega, T)$ (where T is the temperature) especially important for these materials. Due to the wide energetic gap even relatively poor samples are to be in the limit in which the scattering speed is lesser than frequency of the gap oscillation or $1/\tau \ll \Delta$, where Δ is the gap wide. Owing to the spectral weight, associated with creation of Cooper pair, is very small and contribution from quasiparticle scattering can be exactly stated. Then, $\sigma(\omega)$ corresponding to the quasiparticle scattering can be absolutely clearly modeled by using Drude response for dilute gas of weakly interacting particles. It is not carried out for the BCS s-wave superconductors, where density of states defines a singularity of the gap energy, and the coherence parameters strictly cause the conductivity spectra. In d-wave superconductors, these effects are significantly weak due to existence of wide range of values of the gap wide stretching from 0 up to maximal magnitude of Δ_0. The combination of dynamics in the above-mentioned limit and d-wave density of states proposes that simple two-liquid model (TLM) is applied for $\sigma(\omega, T)$ in superconducting cuprates.

Due to that the creation of the Cooper pair has infinitely small spectral weight then the total conductivity will consist only of two components: condensate δ-function and the Drude-like peak associated with thermal quasiparticles. The normal (ρ_n) and superfluid (ρ_s) densities are spectral weights of these two components, respectively. The rule of summary conductivity requires that the value of $\rho_n(T) + \rho_s(T)$ did not depend on T. Actually, TLM provides a fine complete description of microwave properties of optimally-doped single $YBa_2Cu_3O_{7-\delta}$ (YBCO) crystals, as it has been shown by Hosseini, et al. [711], in particular for temperature above some kelvins. At the proper selection of $\sigma(\omega, T)$ by using TLM, $\rho_n(T)$, $\rho_s(T)$ and $1/\tau(T)$ could be determined.

However, the microwave properties of all other cuprate systems are very different clearly demonstrating far greater disorder than in YBCO. And what is more, $\sigma(\omega)$ in these systems does not agree with "dirty d-wave" picture which ensures mean-field description of disorder influence on conductivity. According to the "dirty d-wave" picture, the quasiparticle spectrum is sensitive to disorder lower of proper energetic level E^* (or temperature level $T^* = E^*/k_B$, where k_B is

Boltzmann constant). For the $E < E^*$ the density of states attains non-zero value of $N(0)$ at the chemical potential remaining linear at the $E > E^*$. Existence of non-zero density of states at the chemical potential causes low-temperature properties of ρ_n and quasiparticle conductivity.

Linear diminishing together with T is expected for "pure" *d*-wave superconductors: $\rho_n(T) \rightarrow \alpha T^* + O(T^2)$, when $T \rightarrow 0$, where α is the temperature factor of $\rho_n(T)$ in the "pure" limit. Residual quasiparticle spectral weight increases conductivity which attains the found by Lee [1001] "universal value" $\sigma_d = (v_F/v_\Delta)\sigma_Q/\pi^2$ at low temperatures T, where v_F is Fermi velocity, v_Δ is the "gap velocity", $\sigma_Q = e^2/(\hbar d)$, e is the electron charge, $\hbar$ is Planck constant, d is the distance between planes. The "dirty *d*-wave" picture implies a correlation between low-temperature behavior of $\rho_n(T)$ and $\rho_s(T)$, which is not observed in the measurements fulfilled on cuprates differed on YBCO. Orenstein studied disagreement of mean-field aspect of the "dirty *d*-wave" [1269], that is assumption on that what the disorder generates non-zero value of $N(0)$, which to be spatially homogeneous. In spite of, it has been shown that the density of states at the chemical potential $N(0)$, and hence ρ_n and ρ_s change with location in the medium. This shape of heterogeneity changes qualitatively a nature of optical properties in essence not satisfying the selection rule, which forbids joining longitudinal perturbations of the order parameter to $\sigma(\omega)$.

Explanation of anomalies in $\sigma_1(\omega)$, which are based on spatial heterogeneity, is confirmed by measurements of scanning tunneling microscopy (STM) [975, 1291]. These experiments demonstrated that the local density of states (LDS) of Bi_2Sr_2-$CaCu_2O_{8+\delta}$ (BSCCO) changed in space with the spatial fluctuations having minimal wave length of $\sim$50 Å. The LDS changes confirm a suppressed heterogeneity of local concentration of carriers and therefore local ρ_s. As it has been shown in [1047], the spatial changes defined directly by STM-spectroscopy, are concentrated near the sample surfaces. The measurements of specific heat, nuclear magnetic resonance, Knight shift and fluctuations near the temperature of superconductive transition, T_c, disagree with great amplitude fluctuations observed in STM. The majority measurements of optical conductivity affirm that the specific heat changes as T^2 at low temperatures not demonstrating linear component on T, which could be identified with non-zero value of $N(0)$ in superconductive state. However, this conclusion contradicts to other measurements made by Emerson, et al. in 1999 and showing that linear factor of the specific heat $\gamma \neq 0$ even in high-qualitative crystals. In actuality, the quantitative modeling of low-temperature specific heat in cuprates is difficult due to inclusion possible contributions from normal electrons, *d*-wave quasiparticles, paramagnetic centers and phonons. All these contributions disappear at the $T \rightarrow 0$, as it has been shown in [483]. The optical conductivity, compared with the specific heat, is far more directed probe of $N(0)$. By this, there is not only several possible contributions, but $N(0)$ arises directly as a component of spectral weight, aspiring to non-zero value at the $T \rightarrow 0$.

If interpret the combination of $\sigma(\omega)$ and STM-results as hard witness of existence of the volume heterogeneity in HTSC, then its source becomes central issue. This issue could be formulated in terms of internal (intrinsic) and external mechanisms, while difference between them is not always clear. Externally caused heterogeneity arises due to spatial change of defects or chemical additives which shear the carrier concentration in the side from the half-filled Mott isolated state. Internal heteroheneity arises, when the charge-density wave (CDW) and/or the spin-density wave (SDW) of the state compete with the *d*-wave superconductive order. As it is expected, the intrinsic heterogeneity changes in time in absence of disorder, while potentially at low-frequency level. However, presence of chemical or structural disorder could cause suppress low-frequency changes generating static spatial heterogeneity. "Universality" is the critical test for intrinsic heterogeneity. If the spatial changes arise by internal way owing to physical features of competing interactions, then they would arise in all superconductive cuprate families. As a test for universality, the measurements of the low-frequency optical conductivity in dependence on the carrier concentration would be very useful. Special interest is connected with $La_{2-x}Sr_xCuO_4$ system, in which suppressed stripe-like heterogeneities have been observed by using the neutron scattering. The great values of no condensed spectral weight for optimally doped specimens have been obtained in [1409, 1700].

In composition with universality the characteristic of super-pure and strongly-ordered YBCO system should be considered separately. In mathematics, one opposite examples is enough to refute a proposition. However, in HTSC, it is difficult to apply these strict logics since the experimental "truth" changes often during time. At the same time, it is obvious that majority tests defining the internal heterogeneity must state it in most pure HTSC, in spite of existing fluctuations. Interpretation of the optical conductivity in the YBCO crystal system does not require to state strong spatial heterogeneity. The microwave conductivity agrees with weak elastic scattering, at the same time, arising infrared conductivity is connected with inclusion of the Holstein joining to boson spectrum with energy of order 40 meV. It is associated either with phonons, or spin fluctuations that has been demonstrated in [100]. However, it has been then shown by experimentally that the optical conductivity characteristic in YBCO was not completed [1269]. The spectrum range, lost in the experiments, includes the region to be upper than the scattering velocity of carriers and lower than frequency of gap oscillations, where could exist fluctuations. Future measurements must fill this information vacuum for the YBCO system and extend acknowledge on other superconductive cuprates. This will has tremendous value for estimation of source and role of heterogeneity in HTSC.

1.7.5 Analogy Between Superconducting Current and Laser

Superconductive current arises as movement of state including great number of bosons. This multi-particle state cannot be failured by admixtures, phonons, etc. This situation is similar to river (macro-object), which cannot be stopped by small obstacles (micro-objects). In stricter formulation, the change of state of the Cooper pairs (pairons) could occur only in the case, when transition including great number of pairons takes place from one state of the system to other state. In this sense the supercurrent is very similar to laser. In particular, two supercurrents in SQUID dividing one of other by distance up to 1 mm can demonstrate interference.

Similar to laser, the supercurrent is described by wave function $A\ \exp[i(\boldsymbol{k}\cdot\boldsymbol{r}-\omega t)]$, where A is the amplitude, $i=\sqrt{-1}$, $\boldsymbol{k}$ is the wave vector, $\boldsymbol{r}$ is the coordinate vector, ω is the angular frequency and, t is the time. It corresponds to condensed bosons moving in accordance with linear dispersion relationship. This flux of bosons possesses self-focusing energy. In turn, the laser ray becomes to be self-focused after crossing glass plate (scattering element). Similarly, the flux of condensed pairons becomes to be monochromatic after crossing Josephson junction. So, both as laser ray as supercurrent can interfere on macroscale. However, only superconductive current can carry electric currents because the pairons have charges $\pm 2e$ (where e is the electron charge) and therefore interact with electromagnetic fields. Thus, the pairons can stop at zero angular momentum, at the same time, the photons moves with the light speed and cannot stop. By this the self-focusing energy, for the present is unknown for fermion fluxes.

Chapter 2
High-Temperature Superconductors. Overview

2.1 General Remarks on Type-II Superconductors

High-temperature superconductors placed in the center of our research are the type-II superconductors. Therefore, their properties and superconductivity mechanisms are considered in more detail. The term type-II superconductors was first introduced by Abrikosov in his classical paper [4], where he assumed a detailed phenomenological theory of these materials' behavior, based on the Ginzburg–Landau theory, and explained their magnetic properties. Initially, Abrikosov's theory was greeted with certain skepticism: so much out of the ordinary was in its predictions. However, at the next development of physics of superconductors this theory obtained numerous experimental supports. Finally, several years later it was accepted in total, when it consequently explained the complex behavior of superconducting alloys and compounds, in particular the very high critical fields of some materials. As it has been noted for type-II superconductors, the energy of an interface between a normal and a superconducting region $\sigma_{ns} < 0$. Total displacement of external field from superconductor does not lead to a state with the least energy, if the contribution of surface energy of the interface between two phases is significant. Therefore, in this case, the energetically favorable state is that at which superconductor of corresponding shape (any one, besides an infinitely long cylinder placed in a parallel magnetic field) is divided into great number of alternating superconducting and normal regions.

The magnetization curve of a type-II superconductor in the form of a long cylinder, placed in a parallel magnetic field, is shown schematically in Fig. 2.1. As long as the external field is $H < H_{c1}$, the field in the interior of the sample is absent ($B = 0$). At the attainment of the lower critical field $H_{c1} = H_{cm}\kappa^{-1}(\ln\kappa + 0.08)$

I. A. Parinov, *Microstructure and Properties of High-Temperature Superconductors*,
DOI: 10.1007/978-3-642-34441-1_2,

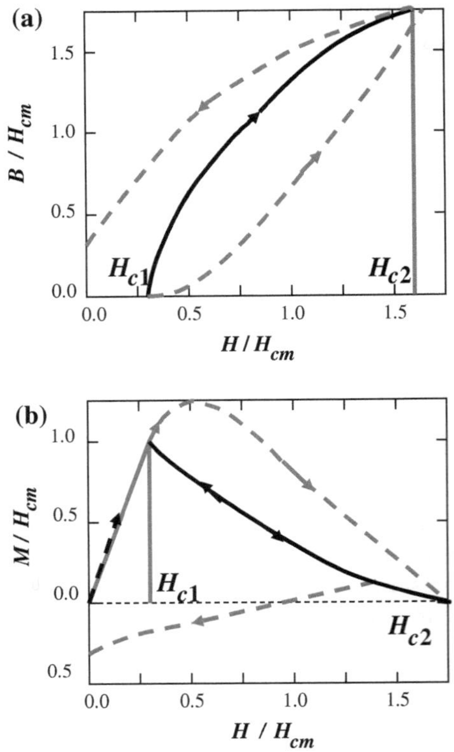

Fig. 2.1 Magnetic induction (**a**) and magnetization (**b**) as a function of applied field in type-II superconductors. *Dotted lines* correspond to impure samples [1006]

[1502], when the Ginzburg–Landau parameter $\kappa \gg 1$, then it becomes to be energetically advantageous the penetration of magnetic field into superconductor.

At the $H_{c1} < H < H_{c2}$, a steadily increasing field penetrates the superconductor in the form of flux lines (vortices). It remains below the external field H, and superconductivity of the sample is not destroyed. At a certain field $H = H_{c2} = \sqrt{2}\kappa H_{cm}$ [4], the field in the interior becomes equal to external field H, and the bulk superconductivity disappears. In contrast to the type-I superconductors, the superconductivity may easily originate in a heterogeneous way in these materials. In this case, the field of the superconductivity initiation H_{c2} may be well above the critical thermodynamic field H_{cm}. Between H_{c1} and H_{c2}, the material has no electric resistance and contains a lattice of flux lines, which can be simplistically treated as normal regions. This is the mixed state, also known as the Schubnikow phase [1570].

In existence of electric field, Lorentz force influences on the vortices. If the vortices are not pinned onto defects or material heterogeneities, then they move and induce electric field, in this case takes place energy dissipation and critical current is equal to zero. If the sample to be in no external magnetic field, then the critical current is defined by equality of the Lorentz force and the pinning force holding the vortices. The anchoring the vortices by impurities (or defects) can lead to significant hysteresis and even paramagnetization curves, as shown in Fig. 2.1. These material heterogeneities can be formed artificially so increasing the critical current of the flux pinning (the materials with the critical current of the flux pinning $\sim 10^5$ A/cm^2 relate to *rigid superconductors*).

Thus, the above H_{c1} type-II superconductors do not show the Meissner effect. Magnetic field penetrates into these materials in the form of quantized flux lines (vortices), each of which has a normal core, which can be approximated by a long thin cylinder with its axis parallel to the external magnetic field. Inside the cylinder, the order parameter, $\Psi = 0$. The radius of the cylinder is of the order ξ, the coherence length. The direction of the super-current, circulating around the normal core, is such that the direction of the magnetic field, generated by it, coincides with that of the external field and is parallel to the normal core. By this, the vortex current circulates into region with radius of the order λ. The size of this region is by far above a value ξ, because $\lambda \gg \xi$ for type-II superconductors [1502].

Each vortex carries one magnetic flux quantum. Penetration of vortices in the interior of a superconductor becomes thermodynamically favorable at $H > H_{c1}$. Inside the superconductor, the vortices arrange themselves at distances $\sim \lambda$, from each other, so that in the cross-section, they form a regular triangular or square lattice (see Fig. 2.2). This state of superconductor (at $H_{c1} < H < H_{c2}$) is the mixed state, because it is characterized by a partial penetration of the magnetic field in the interior of the sample. Once formed at H_{c1}, the vortex lattice persists at much higher fields. As the external field increases, the lattice period steadily decreases and the density of the vortices rises.

The picture of the mixed state remains before upper critical field H_{c2}, when superconductivity into the specimen volume (excepting only thin surface layer) eliminates completely. At this field occurs failure of Cooper pairs caused by their twisting, because the pair can exist as a single whole until the radius of Larmor precession exceeds the proper size of Cooper pair, ξ. At the field $H = H_{c2}$, the vortex lattice becomes so dense that the distance between the neighboring vortices, that is, the lattice period attains the order ξ. This means that the normal cores of the vortices come into contact with each other and the order parameter Ψ becomes zero over the volume of the superconductor, that is, a second-order phase transition occurs.

2.2 High-Temperature Superconductive Cuprates

2.2.1 Doping of Cuprates

The simplest copper oxide perovskites are insulators. In order to become superconducting, they should be doped by charge carriers. There are two ways to increase the number of charge carriers in cuprates chemically: (i) to substitute metallic atoms in the intermediate planes by higher-valence atoms and/or (ii) to

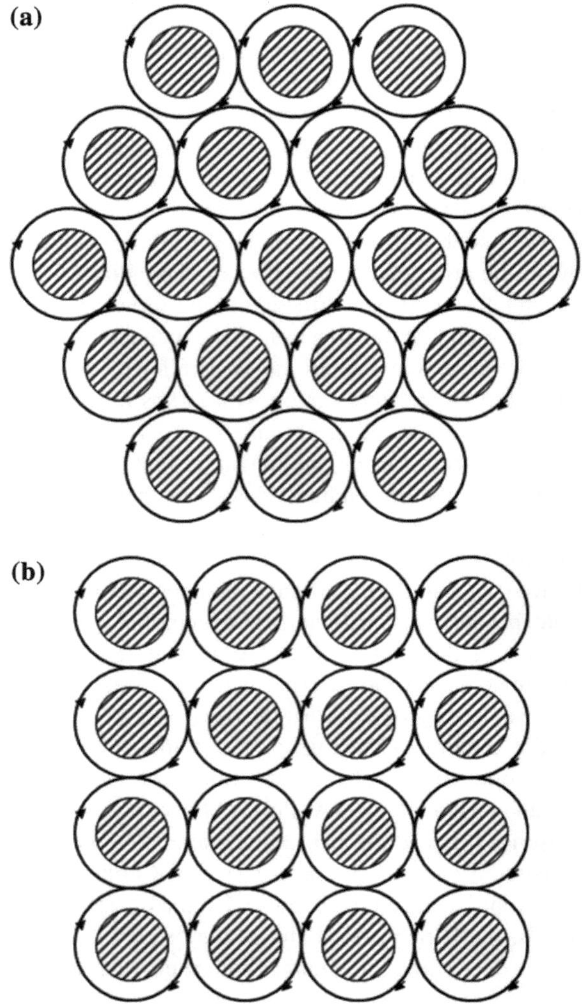

Fig. 2.2 Mixed state of a type-II superconductor. Superconducting vortices form a regular *triangular* (*hexagonal*) (**a**) or *square* (**b**) lattice. Vortex cores (*dashed regions*) are normal

change the number of oxygen atoms. Doping increases the number electrons or holes at the Fermi level. The concentration of charge carriers in HTSC is low ($\sim 5 \times 10^{21}$), in comparison with conventional superconductors ($\sim 5 \times 10^{22}$ – 10^{23}). However, due to the large coherence length in conventional superconductors, only a 10^{-4} part of the electrons, located near the Fermi surface, participate in coupling. At the same time, in cuprates, ~ 10 % of all conduction electrons (holes) form the Cooper pairs.

In conventional superconductors, the critical temperature increases monotonically with growth of charge carriers: $T_c(p) \propto p$. In cuprates, this dependence is non-monotonic. In most of hole-doped cuprates (but not in all) the $T_c(p)$ dependence has the bell-like shape and can be approximated as [1399]

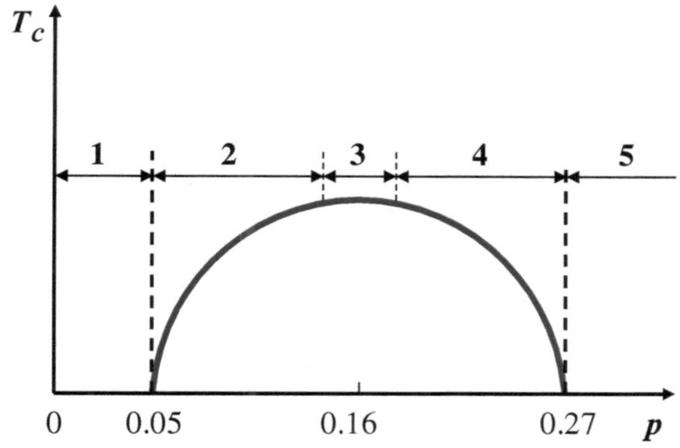

Fig. 2.3 Critical temperature as a function of doping [1201]. Numbers mark the next regions: *1* undoped ($p < 0.05$); *2* underdoped ($0.05 \le p \le 0.14$); *3* Optimally doped ($0.14 < p < 0.18$); *4* overdoped ($0.18 \le p \le 0.27$), and *5* metalic phase ($p > 0.27$)

$$T_c(p) \approx T_{c,\,max}[1 - 82.6(p - 0.16)^2], \tag{2.1}$$

where $T_{c,max}$ is the maximum critical temperature for a given compound.

Superconductivity occurs within the limits, $0.05 \le p \le 0.27$, which vary slightly in various cuprates. Thus, the different doping regions of the superconducting phase may be chosen such as *the underdoped, optimally doped and overdoped* regions (Fig. 2.3).[1] The insulating phase at $p < 0.05$ is called *the undoped* region, but above 0.27, cuprates become *metallic*.

The ratio between the maximum critical temperatures in hole-doped and electron-doped cuprates is 135 K/24 K = 5.6. Obviously, it may be assumed that the electron–hole asymmetry has fundamental character: superconducting hole-doped compounds will always have the critical temperature a few times higher than the same electron-doped superconductors.

In cuprates and in many other compounds with low dimensionality, the distribution of charge carriers is heterogeneous. Moreover, in cuprates, this distribution is heterogeneous in a micro-, as in a macroscopic scale (Fig. 2.4). In the undoped region ($p < 0.05$), doping holes are preferably distributed heterogeneously into CuO_2 planes, and they form dynamical one-dimensional charge stripes (so-called *charge-stripe phase*). In the undoped region, these stripes have diagonal shapes and locate not along –O–Cu–O–Cu– bonds, but along the diagonal –Cu–Cu–Cu– direction, as shown in Fig. 2.4. In undoped cuprates, the concentration of holes is low, but the distance between charge stripes, separated by

---

[1] For these regions various model representations have been developed. In particular, as one of the models of underdoped cuprates, it is considered so-called *granular superconductivity*. This state realized in the case, when microscopic superconductive grains are divided by non-superconductive regions, through which the grains interact one with others by means of *Josephson tunneling* (as a result, it is stated macroscopic superconducting state). While a structure of HTSC monocrystals is no granular, the theory speaks on that, what at low hole concentration of carriers (underdoped compounds), the holes concentrate into some regions leading to arising isolated between themselves superconductive domains. The granular superconductivity arises owing to the Josephson tunneling between these domains.

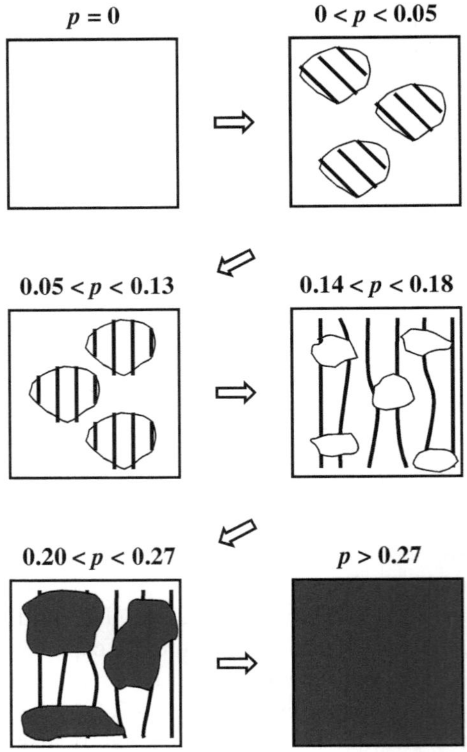

Fig. 2.4 Distribution of charges in CuO_2 planes as function of doping [1201]. The antiferromagnetic and metallic phases are shown in *white* and *gray*, respectively. The *lines* depict charge stripes

two-dimensional insulating antiferromagnetic domains, is large. The charge-stripe phase is distributed heterogeneously: there are two types of small islands, containing either the antiferromagnetic or the charge-stripe phase.

In the underdoped region ($0.05 < p < 0.13$), charge stripes are vertical (or horizontal) and located closer to each other. In this region, the average distance between charge stripes d_s is approximately proportional to $1/p$ and saturates at $p = 1/8$ (Fig. 2.5). Above $p = 1/8$, the distance between stripes is practically constant. As p increases, the concentration of antiferromagnetic regions decreases, but the two types of islands, containing the antiferromagnetic phase and the vertical

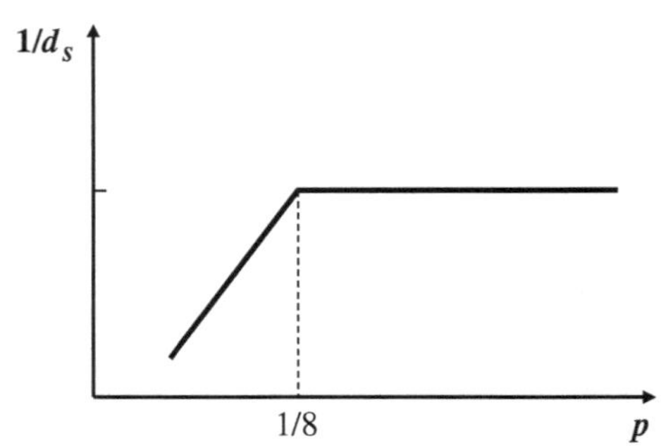

Fig. 2.5 Dependence of incommensurability δ ($\propto 1/d_s$) of spin fluctuations on doping level [1818]

charge-stripe phase, still co-exist. The dynamical charge stripes can move in the transverse direction, and they are *quasi-one-dimensional*. In the near optimally doped region ($p \sim 0.16$) and in the overdoped region ($0.2 < p < 0.27$), the average distance between charge stripes remains almost constant (see Figs. 2.4 and 2.5). Therefore, as the doping level increases, new doped holes cover antiferromagnetic islands, which completely vanish at $p = 0.19$. Above this value, small metallic islands start appearing. Above $p = 0.27$, the charge-stripe distribution becomes homogeneous in two-dimensional CuO_2 planes, and cuprates transform in non-superconducting (normal) metals.

2.2.2 Anderson Phase Diagram of Cuprate Superconductivity

Obviously, that the p–T_c phase diagram presented in Fig. 2.3 is very approximate and not take into account some effects which are characteristic for cuprate superconductivity. Therefore, in this section we consider more complex and probable p–T phase diagram developed by Anderson for superconductive cuprates [43] and presented in Fig. 2.6.

This phase diagram has been developed based on viewpoint on HTSC as a kind of perovskite, key concepts of Mott insulator and super-exchange, Jahn–Teller distortions, antiferromagnetism and *d*-shells. The "plain vanilla" theory—the simple Hubbard model transformed into *t–J* model, which was created by Anderson with co-authors (see in some details Sect. 1.6), allows one to understand a behavior of the cuprates in dependence on the doping degree. This theory which

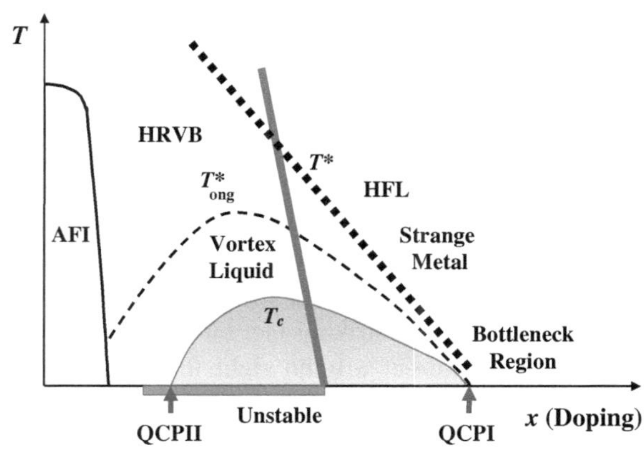

Fig. 2.6 Schematic phase diagram of cuprate superconductivity which is enough general for different materials [43]. Abbreviators design the following: *AFI* Anti-Ferromagnet Insulator, *HRVB* Hidden Resonating Valence Bond, *HFL* Hidden Fermi Liquid, *QCP* Quantum Critical Point; T^* is the pseudogap temperature at which a gap begins to open in the single-particle spectrum, T^*_{ong} is the temperature bound the role of which will be discussed bellow into text

is capable to explain crucial experimental anomalies, in particular tunnel spectra proper for different superconductive cuprates has dealing with the constrained Hilbert space which follows from the requirement of Gutzwiller projection.

In the considered problem, the projective *t*–*J* Hamiltonian has two terms, and there are three possible situations defined by doping and temperature changes, namely:

(i) *t* may dominate (the overdoped case),
(ii) both terms are relevant, but *J* may be considered as a perturbing interaction (the optimally doped case),
(iii) *J* is dominant, and determines the shape of the spectrum, but *t* is also relevant (the underdoped case).

Case (i) applies everywhere above and to the right of the pairing line T^* on the generalized phase diagram (Fig. 2.6) and imagines the energy gap vanishing. In that event, the mean field result of Anderson–Ong [45] implies a jump discontinuity at the Fermi energy. In order to treat this case Anderson used a self-consistent *ansatz* by the same arguments based on momentum and energy conservation which underpin the conventional Fermi liquid [43], while there is not rigorous proof that it is true, the best evidence that it coincides with experiment: fitting laser-excited Angle-Resolved Photoemission Spectroscopy (ARPES) energy distribution curves to shapes derived from this assumption [242]. Then the single-fermion operators acting within the unprojected Hillbert space are the "Hidden Fermi Liquid".[2] They represent true eigen excitations of the system but are not true quasiparticles (they are to be so-called, "pseudoparticles" being true excitations, but not continuously associated to the real electrons and holes of the system). The quasiparticles are three-pseudoparticle operators. They are the physical particles which are exchanged by tunneling or acted on by physical fields. It is the Green's functions of the true quasiparticles which underline all of the phenomena of the "Strange Metal" (sometimes referred to as the "Marginal Fermi Liquid") region of the superconductive cuprate phase diagram. The "linear *T*" relaxation rates are striking feature of the strange metal. The other idea, namely: the "bottleneck" completes our understanding of this case. A great deal of test data has supported the idea that there are two relaxation rates and two relaxation processes active in this region, characterized by linear *T* and quadratic *T* relaxation rates. By using these two concepts Casey has been able to understand the full range of data on the bulk properties available in the strange metal region [241]. The bottleneck process can explain one striking fact on the resistivity: that the "linear *T*" resistivity often passes without a hitch right through the Mott–Ioffe–Regel "maximum metallic resistivity" [39] without saturating as conventional resistivity in metals does.

Case (ii) develops a true energy gap and due to the optimally doped superconductor can be treated to a fair level of approximation as a simple modified BCS

---

[2] The general "Hidden" scheme has been presented in [43, 1830].

theory ignoring the incoherent terms caused by projection [45]. In discussing the Fermi liquid, it is introduced the above-pointed pseudoparticles and if they are directly used in deriving the tunneling spectrum, we would find it symmetric. However, the spectra observed for optimally doped HTSC [43] have perceptibly more prominent coherence peaks than the prediction, but otherwise the general course of the test results is in fair agreement with the mean field theory for doping not too far from optimal, and in excellent agreement as far as the unique features of the spectra are concerned: (i) the "hump and dip" occurring primarily on the hole side, at energies related to the gap and the doping percentage; (ii) their absence on the electron side; and (iii) the symmetry of the singularity at the gap energy. The asymmetry is incompatible with any perturbation method starting from weak coupling (as, for example spin fluctuation theory, or any phonon-based theory) and shows that the system must be described as a doped Mott insulator.

Case (iii) presents the underdoped case, in which it is expected a region of fluctuating "preformed pairs" between T_c and T^* that is the pseudogap is qualitatively predicted (see Fig. 2.6). The RVB state is initially an insulator on two causes: (i) at weak kinetic energy (that is, small effective t and domination of J), the spin-charge locking mechanism [40] does not work and the pairs are effectively neutral (ii) there is a commensurability energy which holds the structure in register with the crystal lattice (that is, the nodes are pinned at the $\pi/2$, $\pi/2$ points [1830]). As we add holes, the RVB structure does not change and provides the major proportion of the pseudoparticle self-energy. The hole excitations on the nodal lines will feel no or a small gap and show the putative Fermi surface for the ordinary kinetics energy, and such excitations form hole pockets. True superconductivity, when it develops, is confined at first to the pocket-arcs alone. This correlates with all the known experimental facts, including the "Fermi arc phenomenon" [811] and the mysterious simultaneous existence of strong Dirac nodes and hole pockets [1435]. Then, the correct approach to this region of the phase diagram is based in terms of a "Hidden RVB", as for the overdoped case it is the "Hidden Fermi Liquid".

The transport studies of the cuprate superconductors, carried out by Ong et al. [1774], discovered an extensive region of the p–T phase diagram in which these studies gave unequivocal evidence of quantized vorticity above the superconductivity temperature T_c. This discovery has been based on measurements of the Nernst effect and non-linear diamagnetism [1013, 1014]. The discovered equilibrium effect can be shown to be equivalent to the Nernst measurement if and only if the former is due to vortex motion. The region in which this occurs is shown for two sets of superconductive compounds in Fig. 2.7. The facts that these phenomena occur over a finite, in fact, extensive, region of phase space, and that one manifestation is a non-analyticity of an equilibrium property, argue that this is not critical fluctuations but a distinct new phase of matter. It is supported by models for T_c such as the two-dimensional Kosterlitz–Thouless one and XY model.

In the phase space region, discovered by Ong with co-authors, the electrons are apparently to be fairly strongly paired, superconductivity having been destroyed only by phase fluctuations of the order parameter, as demonstrated in [1470].

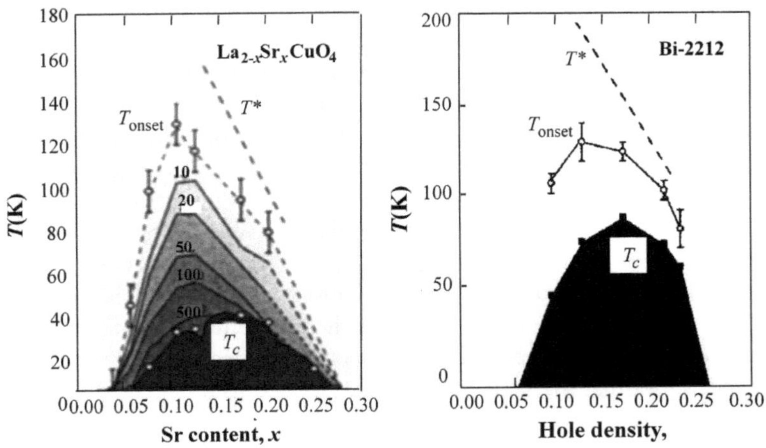

Fig. 2.7 The region of fluctuating phase ("vortex fluid") in which a vortex Nernst effect is measurable in two compounds with a range of doping [43]

Therefore, it is reasonable to think on the currents as being predominantly carried by paired electrons, i. e. bosons. If there is finite local pair amplitude above T_c, the pair wave function will have a time- and space-varying phase Φ and the pair current will be proportional to $\nabla\Phi$ and conserved. If so, Φ will be completely determined by a network of vortex lines—in three-dimensional case mostly vortex loops. Thus, it is appropriate to describe this phase as a "vortex liquid" [41, 42].

2.2.3 Influence of High Pressure on Superconductivity

In difference of majority superconductors, for which the critical temperature of superconductive transition T_c decreases with rise of pressure, in HTSC T_c increases usually. Studies of one-axis pressure witness that this increase in main degree is caused by decreasing a square, A, of CuO_2 planes ($T_c \propto A^{-2}$), but no distance between them. These results support theoretical models, which attribute superconductivity, in first order to intra-plane coupling interactions. Hence, the future experiments must state dependences on hydrostatic and one-axis pressure of the main parameters, namely: T_c, superconductive gap, pseudogap, carrier concentration and exchange interaction in all range of doping.

First studies of influence high pressures on superconductors have been carried out in 1925 by Sizoo and Onnes [1579], which discovered that for Sn and In, also as majority superconductors T_c decreased with growth of pressure. As it became to be clear afterwards, an explanation that diminishing was connected with the isotope effect. It is not be accidental, that many scientists apply the high-pressure techniques as important diagnostic instrument. It is defined that what the experiments with using a high pressure could ensure valuable information for study superconductors with higher values of T_c.

In difference of magnetic materials, tremendous technological importance of which is caused by that, what their magnetism is stable to temperatures, which are far higher of usual ones, the material cannot become to be superconductive until not cooled at least on 160 K lower than above-mentioned temperatures. It is a very complex and elaborated process for long-length specimens, therefore most aim of superconductivity studies oriented on technological applications is creation materials with room temperature of superconductive transition.

One of the ways to understand ability of new superconductor to attain higher value of T_c is definition of dependence of T_c on chemical composition and/or pressure p. The great value of $|dT_c/dp|$ allows one to hope on possibility of attainment of the higher values of T_c. For example, successive application this strategy in experiments with high pressure on cuprate $La_{2-x}Ba_xCuO_4$ (La-214) has allowed Wu et al. [1810] to substitute La^{3+} cation by smaller Y^{3+} cation at high values of dT_c/dp (+8 K/GPa) and discover known superconductive compound $YBa_2Cu_3O_{7-\delta}$ (Y-123) with T_c, exceeding the liquid nitrogen temperature ($\sim$77 K). Another example is connected with oxide $La_{2-x}Sr_xCuO_4$, for which as it has been stated by Gugenberger et al. [589], T_c increases at the compression of material along one direction, at the same time decreasing at the compression along other direction. This fact has been used by Locquet et al. [1042], which created proper deformations in thin films of this oxide during their epitaxial growth on substrate and due to increased in two times the value of T_c from 25 up to 49 K. Finally, Klehe et al. [885] observed increase of T_c in conditions of high pressure applied to samples of $HgBa_2Ca_2Cu_3O_8$ (Hg-1223) that allowed to Gao et al. [515] to increase T_c for this superconductor from 134 K upon to $\sim$160 K. Moreover, Ashcroft [63] stated that elementary hydrogen could become room-temperature superconductor under very high pressure.

The experiments with high pressure contribute in superconductivity in some directions:

(i) as it has been above noted dT_c/dp is the well indicator that what higher values of T_c could be possible for given superconductor at ordinary pressure owing to the chemical substitution or using the technique of epitaxial growth,
(ii) some superconductors could be synthesized with proper properties only in conditions of high pressure and high temperature,
(iii) many non-superconducting materials could become superconductive at enough high pressure: to existing 29 elementary superconductors at ordinary pressure (Be, Al, Ti, V, Zn, Ga, Zr, Nb, Mo, Tc, Ru, Rh, Cd, In, Sn, Hf, Ta, W, Re, Os, Ir, Hg, Tl, Pb, La, Th, Pa, U, Am) from Mendeleev Table of Chemical Elements are added 23 ones, which become to be superconductive at application of pressure (Li, B, O, Si, P, S, Ca, Sc, Fe, Ge, As, Se, Br, Sr, Y, Sb, Te, I, Cs, Ba, Bi, Ce и Lu),

(iv) main electronic and crystalline lattice properties of superconductor change at the decreasing temperature due to thermal compression of the crystalline lattice, then the high pressure directly can change the lattice parameters at any temperature and so it allows one to correct the effects of the thermal compression at ordinary pressure,

(v) determination of dependences of T_c and other superconductive properties on individual parameters of the crystalline lattice of single sample creates a quantitative test for theoretical models and give information on the coupling mechanism.

For instance, if HTSC is caused, in first order by interlayer couple, then it may be proposed arising strong change of T_c, if apply one-axial pressure along perpendicular direction to layers. However, all experiments with high pressure differ one of others. In the case of superconductivity, the dependence of T_c on pressure p could be defined by medium used for creation of pressure and other factors. For example, the curve $T_c(p)$ obtained by Thomasson et al. [1667] for Pb by using pressure in rigid medium of "methanol–ethanol" disposed significantly higher than corresponding curve obtained for the same pressure but in the medium of helium.

In ideal case, the tests would be either totally hydrostatical, or only one-axis. However, the totally hydrostatical tests are possible only in restricted range of change of the pressure and temperature, because all liquids harden under pressure, for instance liquid helium freezes at the pressure of 12 GPa in the conditions of room temperature. The solid helium is very soft therefore it supports only very weak shear stresses. The dense He is used in experiments with high pressure as a pressure medium. One of the practical way to define, are the test results to be sensitive to effects of shear stresses, consists in use of two different pressure media. Achievement of highest values of T_c demands careful optimization of proper electronic and structural (crystalline lattice) properties. This optimization is especially difficultly to be realized in elementary bodies. Here, maximal value T_c is restricted by temperature range of 9–20 K (for Nb under ordinary pressure and for Li, B, P, S, Ca, V, Y, Zr and La at the very high pressures).

It is not surprising, that the multi-atomic systems demonstrate higher values of T_c since their structural flexibility allows greater degree of optimization. Quasi-two-dimensional solids (for example, HTSC, oxypnictides and MgB_2) demonstrate most values of T_c. However, the superconductive curates show great structural and electronic complexity as in ordinary conditions as under high pressures. This fact initiates numerous attempts to attain main understanding physical mechanisms which are responsible for superconductive state. The experiments, presented by Schilling [1498], confirm a conclusion on that what sizes of CuO_2 planes in greater degree determine maximal value of T_c for given HTSC, than a distance between these planes. The CuO_2 planes are to be plane and quadratic but decreasing their surface leads to rising magnitude of T_c. Thus, following Wijngaarden et al. [1798], it may be noted that c-axis causes in main case a doping, at the same time a-axis defines intrinsic T_c. Hence, interior interactions in the CuO_2 planes are responsible in first order for superconductive pairing. Therefore, everywhere intrinsic dependence of T_c on volume ($T_c \propto V^{-1.2}$) for almost optimally doped HTSC is caused by rather compression of the CuO_2 planes than decreasing distance between them. Intrinsic dependence of T_c on crystalline lattice parameter, a, inside the plane for one-, two- and three-layer superconductive Hg-compounds is determined in the form, respectively: $d\ln T_c/d\ln a = -4.1, -4.8$ and -5.0. Hence, it may be obtained approximate dependence: $T_c \propto a^{-\delta}$, where $\delta = 4.5 \pm 0.5$. This dependence implies that an optimal doping is approximately inversely proportional to the square of region A of the CuO_2 planes, $T_c \propto A^{-2}$. Similar result has been obtained for other optimally doped HTSC. This one of most valuable results obtained in experiments on HTSC subjected to high pressure. In actuality, this information could not be obtained another means. Besides information on superconductivity mechanism, this dependence states additional strategy for following increasing T_c.

Thus, additionally to above-mentioned five rules (i–v) for optimization of T_c, it should be added sixth one, namely: (vi) it is required a search of structures applying maximal compression of the CuO_2 planes without their buckling.

In according to above relationships, if apply a pressure to optimally doped sample of Hg-1223 and compress its size in plane approximately on 20 % without addition of defects or increasing number of charge carriers, then T_c should rise from 134 K up to 304 K that is attain room-temperature superconductivity. In general case, HTSC systems with the same number of the CuO_2 planes at optimal doping have different values of $T_c^{\max}$.

It is interesting to elucidate, is this difference due to change of the lattice parameter, a, that is fulfills the dependence $T_c^{\max} \propto a^{-4.5}$. However, as it has been shown by tests there is not this simple correlation. The one-layer superconductor with highest T_c (Hg-1201), which has $T_c^{\max} \approx 98$ K, possesses most parameter a, but the compound with least value of T_c ($La_{1.85}Sr_{0.15}CuO_4$), which has $T_c^{\max} \approx 36$ K, demonstrates the least magnitude of a. It is possible, that $La_{1.85}Sr_{0.15}CuO_4$ shows anomalous low value of T_c due to excessive compression of the CuO_2 plane leading to strong structural distortions and buckling inside the plane that is to the effects which as it is well known decrease T_c. The values of a for other cuprate HTSC systems differs only into limits of 1.4 %, that corresponds to $\sim$5 GPa or change of T_c only on 7–8 K. The pressure level increasing from 1.4 upon 20 % is deserving, but difficultly attainable aim for material science scientists. Important remaining moment is proof that the above-mentioned dependence $T_c^{\max} \propto a^{-4.5}$ will lead to superconductive state of HTSC.

If suppose that two electrons couple into Cooper pair with help of electron–phonon, electron–electron, electron–magnon or anybody another effective interaction, V_{eff}, then the BCS theory carries out for weak interactions and $T_c \approx \langle\omega\rangle \exp\{-1/[V_{eff} N(E_f)]\}$, where $\langle\omega\rangle$ is the characteristic energy of intermediate bosons, $N(E_f)$ is the electron density of states, E_f is Fermi energy. Since, the both parameters V_{eff} and $N(E_f)$ are included in the expression for exponent then it should be supposed that their dependence on pressure is responsible for the critical temperature, T_c. The measurements of spin sensitivity at high pressures carried out for $La_{1.85}Sr_{0.15}CuO_4$ [23] and for Y-123 [22], and also calculations of structure of the energetic levels in Hg-1223 [1573] have stated that changes of $N(E_f)$ under pressure were lesser on 0.2, 0.1 and 0.5 %/GPa, respectively. In the case of $La_{1.85}Sr_{0.15}CuO_4$, this change of $N(E_f)$ was small due to fast increasing T_c under pressure. In order to make a similar estimate for Y-123 and Hg-1223, where dT_c/dp is far lesser, it is necessary to significantly increase an accuracy of definition of $dN(E_f)/dp$.

The remaining issue is: why $V_{eff}N(E_f)$ increases with pressure growth with the speed that $T_c \propto a^{-4.5}$. To regret, one only this dependence $T_c(a)$ does not give enough information, allowing simple identification of the pairing interaction. Intrinsic dependence of critical temperature on pressure dT_c/dp, which changes into approximate bounds from +1 up to 2 K/GPa for ideal HTSC decreases easily into broad limits, observed for superconductive transit metals, for which it has been stated the electron–phonon coupling. Based on analysis of results obtained for HTSC by using tests with high pressure, it has been concluded [1239] that the electron–phonon coupling is a possible superconductive mechanism also for HTSC.

To regret, the analysis leading to the intrinsic dependence of $T_c \propto a^{-4.5}$ is satisfactory only for HTSC near of optimal doping level. This restriction is introduced to exclude dependence of parameter n (the number of holes into CuO_2 plane) on pressure. In order to state intrinsic dependence $T_c^{\text{int}}(p)$ for wide range of doping, it is necessary first to carry out independent definition of dn/dp into the whole of this range. In order to understand the mechanism of the pairing interaction by using investigations fulfilled with application of high pressures, it is required to combine estimates obtained for $T_c(p)$ due to action of hydrostatic and one-axis pressures with simultaneous measurements (preferably on the same crystal) other important properties of superconductive and normal states, namely: superconductive gap, pseudo-gap into zone of underdoping, energy of superconductive condensation, magnetic sensitivity, electric resistance, Hall effect, etc. Then, it has been studied Raman scattering at high pressures in antiferromagnetic La_2CuO_4 and stated, that super-exchange interaction J increased approximately as $J \propto a^6$ [59].

These studies after spreading on HTSC could be used for testing a viability of spin-fluctuation theories. The measurements of magnetic sensitivity in dependence on pressure at the elevated temperatures should ensure similar information on $J(p)$. However, following studies on the La-214 system can lead to error conclusions because strong structural distortions and transitions, which are characteristic for this system, make quantitative analysis very complex. More real attempt is connected with these studies fulfilled on monocrystals of Hg- and Tl-superconductive families into the whole of range of the doping change. Especial attention should be given to the experiments with one-axis pressure, since they can possess necessary detail information.

2.2.4 *Specific-Heat Anomaly at Temperature of Superconducting Transition in HTSC*

Due to short coherence length in HTSC, the fluctuation effects in specific heat near transition temperature T_c at zero field which have observed in conventional low-temperature superconductors, to be also proper for HTSC. However, there is vagueness: are these fluctuations to be two- or three-dimensional and Gaussian or critical? For magneto-field measurements is also important an issue of applying set of the scaling relationships. Other fundamental moment connected with anomaly of heat capacity near T_c is change of anomaly shape in dependence on doping level. The experimental witnesses that what this change of the anomaly shape reflects alteration from the BCS-transition to one of the transition being a variant of the Bose–Einstein condensation, have been presented by Fisher et al. [483].The specific-heat anomaly at the transition temperature T_c is better found in YBCO and *RE*BCO compounds compared to other HTSC.

Ginzburg–Landau (GL) theory can well explain into framework of the mean field approximation the character of phase transition in superconductors satisfying the BCS-theory and predicts the transition behavior until the fluctuation effects are sufficiently small and can be described by Gaussian distribution. However, at the enough approximation of T to T_c, when the sample to be already in critical zone, in order to describe the fluctuations, the renormalization group theory is more available, which has been developed for description of phase transitions by Goldenfeld [540]. The temperature regions, in which two kinds of fluctuations expect, are compared in size with mean-field discontinuity $\Delta C_e(T_c)$, expressed in terms of the Ginzburg criterion [534] or lesser-known criterion of Brout [188]. Here, C_e is the contribution into specific heat of conductivity electrons. The both criteria could be expressed in terms of the normalized temperature $t = (T/T_c - 1)$. Due to these criteria the critical region of three-dimensional superconductor is situated into range $|t| < t_G = \{k_B/[\pi\Delta C_e(T_c)\xi^3]\}^2/32$, where $\xi = (\xi_a\xi_b\xi_c)^{1/3}$ is the GL coherence length of superconductive order parameter at $T = 0$. The parameter $\Delta C_e(T_c)$ is expressed in units of energy density, K^{-1}, k_B is Boltzmann constant.

The Gaussian fluctuations stated by Kapitulnik et al. [813] could be expressed into wider range $1 \gg t_B > |t| > t_G$, where $t_B \sim (t_G)^{1/2}$. The contribution of fluctuations into anomaly of specific heat near T_c, is determined by positive curvature of C_e/T lower of T_c and "tail" upper T_c. There is important witness that superconductivity in HTSC is characterized by a strong couple which increases $\Delta C_e(T_c)$, and also creates a positive curvature lower T_c. Therefore, the possibility of strong couple introduces further indefiniteness in addition to that which is associated with selection of the crystalline lattice contribution in analysis of specific heat near T_c. Strong dependence of both as t_B, as t_G on ξ make to be obvious that what the coherence length is defining factor without dependence on existence of the fluctuation effects. In the case of typical type-I superconductor, for example Sn with $\Delta C_e(T_c) \sim 1$ mJ/(cm^3 K) and $\xi \sim 2000$ Å, $t_G \sim 10^{-14}$ and $t_B \sim 10^{-7}$ [1203], the Brout criterion affirms that the contribution of the Gaussian fluctuations cannot be observed without making measurements of specific heat into range of 1 μK or lesser near T_c. Therefore, it is not surprising that the specific-heat anomalies near T_c of conventional superconductor are well described by standard mean-field theory. On contrary, in the case of HTSC, for instance YBCO with $\Delta C_e(T_c) \sim 30$ mJ/(cm^3 K), $\xi \sim 10$ Å and $T_c \sim 90$ K, $t_G \sim 0.0015$ and $t_B \sim 0.04$ [1203]. In this case, the Ginzburg criterion and Brout criterion propose that the Gaussian fluctuations would take place into range of ~5 K near T_c, at the same time the critical fluctuations would be into much more narrower range of ~0.1 K near T_c.

The calculations [50, 753] have shown that for *s*-wave pairing, the contribution of the Gaussian fluctuations C_f into C_e at zero magnetic field could be written as $C_f^{\pm} = A^{\pm}[k_B/(16\pi\xi^d)]|t|^{-\alpha}$, where $\alpha = (4 - D)/2$, D is the system dimensionality, $A^+ = 2$ and $A^- = 2^{D/2}$ for $t > 0$ and $t < 0$, respectively. The more earlier measurements of C_f [752] corresponded to $\alpha = ½$, as it was expected for three-dimensional systems, but in this moment there have not been obtained values of A^+ and A^- or their ratio A^+/A^-. Then, it has been shown that in critical region, C_f may be presented as $C_f^{\pm} = (A^{\pm}/\alpha)|t|^{-\alpha}$, where the "critical exponent" α and factors $A^{\pm}$ are constants, depending on "universality class" (three-dimensional *XY*, Ising, etc.) [1506]. By this, the value of α is to be near but no equal to zero. For very small α, the term $|t|^{-\alpha}$ could be replaced by $(1 - \alpha\ln|t|)$.

While the Ginzburg and Brout criteria are useful rules, but do not allow one to differ the regions of the Gaussian and critical fluctuations. For example, it has been supposed for type-II superconductors, that the bounds of critical region exceed the value of t_G in difference on that how this was defined above [480]. Moreover, there is divergence with nature of crossover from the Gaussian to critical fluctuations. Therefore, due to the pointed discrepancies as with definition of wide of the critical region, as form of the fluctuation term in each region of the crossover, it is useful to carry out a comparison with test data. Different research groups (see [801]) have shown that for optimally doped YBCO and DyBCO monocrystals the specific heat is well predicted by using three-dimensional *XY*-model [540, 1506], which well describes λ-transition of liquid ^{4}He [1026].

In the existence of field, the measurement results of optimally doped *RE*BCO in the case of critical representation coincide better with three-dimensional *XY*-model, compared to the Gaussian case. As it has been shown [993], the field and temperature dependence of C_f could be presented as $C_f = C_0 - B^{\alpha/2\nu}f(x)$, where $C_0 = 0$ for $t > 0$ and is to be constant for $t < 0$, $x = t/B^{1/2\nu}$, α is the heat capacity exponent, and ν is the coherence length exponent, for example $\xi(t) = \xi/t^{\nu}$. In the fluctuation equation arises scaled temperature x, but no normalized temperature t as before. The simplified form of the equation for fluctuation, as it has been shown in [1470] could be written as $[C(0, T) - C(B, T)]B^{\alpha/2\nu} = f(x)$, where $\nu = 0.669$ in the case of three-dimensional *XY*-model, and $\alpha = -0.013$ is the test value obtained for ^{4}He. Advantage of the simplified equation consists in that what it allows one to predict the scaling without assumptions stated for crystalline lattice specific heat C_{lat}, excepting that what it independents on magnetic field B. In practice, because of indefiniteness of estimate of the critical temperature T_c, the authors prefer to define it on the base of known values rather than based on measurements of $C(0, T)$.

The satisfactory of the three-dimensional *XY*-model for description of critical behavior of the optimally doped *RE*BCO has been stated [801]. Some authors tested predictions of the three-dimensional *XY*-scaling studying as derivatives of C change in dependence on field and temperature. Additionally, there have been studied predictions of scaling with aim to apply them to field dependence of magnetization. Moreover, it has been verified applicability another approach to scaling, namely the lowest Landau level (LLL-scaling) [1361]. While, there is discrepancy in opinions relatively most acceptable range of magnetic field B change for given scaling procedure, however, if it is suggested that the upper critical field $B_{c2}(T)$ changes as $|t|^{4/3}$, but not as $|t|$, then the both models (three-dimensional *XY* and LLL) ensure satisfactory description the specific heat and magnetization under magnetic fields from 1 up to 16 T.

The anomaly height of specific heat in optimally and weakly overdoped *RE*BCO could be of order 3–4 % from C. In the case of underdoped superconductive *RE*BCO, this height in general case is lower, at the same time it seldom exceeds 1.5 % from C in La-, Tl- Bi-, and Hg-HTSC. The measurements fulfilled on Pb-stabilized Bi-2212 [1266] and on Bi-2212 [1507] agreed with two-dimensional (but no three-dimensional) Gaussian fluctuations, since it has been stated, that C_f changed as $|t|^{-1}$, but not as $|t|^{-1/2}$. Existing experimental data for Tl-2201, Hg-1201 and Hg-1223 demonstrate the specific-heat anomaly which is similar to Bi-2212.

2.2.5 Coherence Length and HTSC Anisotropy

Despite the fact that there is no definite theory to explain high critical temperatures of HTSC, their magnetic and superconducting properties can be well described (as it has been noted above) into framework of the classical BCS/Ginzburg–Landau theory. They demonstrate a set of properties that are similar to conventional low-temperature superconductors. In particular, superconductivity in cuprates occurs due to coupling of electrons. Moreover, there is energy gap in a spectrum of electron excitations that is caused by electron coupling. Non-monotonous dependence $T_c(p)$ (see Fig. 2.3) is similar to non-monotonous behavior $T_c(p)$ of superconducting semiconductors. Finally, isotope effect also exists in cuprates, while it is directly found by the concentration of holes [1201].

The main difference from conventional superconductors is caused by intrinsic material properties, for example, the extremely short coherence length ξ (in conventional superconductors $\xi = 400 - 10^4$ Å). Short coherence length is a consequence of the big energy gap and the small Fermi velocity. Due to the extremely short coherence length, even a grain boundary can be sufficient to suppress superconductivity in cuprates. In particular, the grain boundaries can be

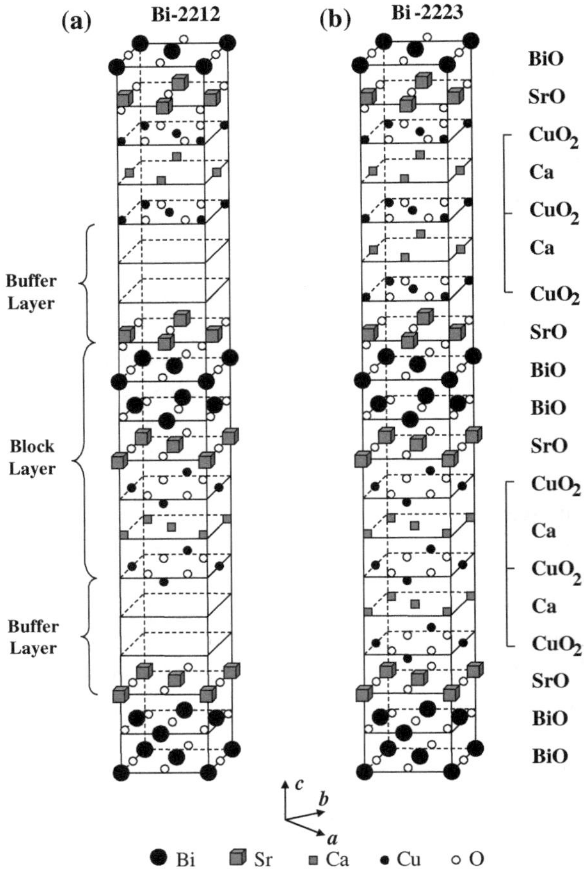

Fig. 2.8 Crystalline structures of Bi-2212 (**a**) and Bi-2223 (**b**). The layered structure of Bi-2212 can be divided into block layers and buffer layers for intercalation of additional Ca/CuO_2 plane, forming Bi-2223. The height shown of buffer layers in the structure of Bi-2212 is expanded to compare with real one for clearness of comparison of both structures [224]. Below, the principal axes *a*, *b* and *c* are shown

used to fabricate devices of Josephson type (in the form of epitaxial films on bi-crystalline substrates), based on the existence of weak links.

The second important property of high-temperature superconductors is their huge anisotropy caused by the layered crystalline perovskite structure.[3] For example, $Bi_2Sr_2CaCu_2O_8$ (Bi-2212) crystal, presented in Fig. 2.8a, consists of the sequence of CuO_2 planes, alternating with other oxide layers. The basic block is the CuO_2 double layer (intercalated by Ca). These blocks are separated by four oxide layers namely two SrO and two BiO ones. In Fig. 2.8b, a process of intercalation of the additional Ca/CuO_2 plane is present to form $Bi_2Sr_2Ca_2Cu_3O_{10}$ (Bi-2223) crystal.

Due to two-dimensional structure of cuprates, the coherence length depends on the crystallographic direction, namely along *c*-axis the value ξ_c is far lesser than in *ab*-plane (ξ_{ab}). In the different hole-doped cuprates, ξ_{ab} = 10–35 Å, at the same

---

[3] Crystalline structures of some HTSCs are present in Appendix A.

Table 2.1 Characteristics of optimally doped cuprates [1201]

Material	T_c (K)	ξ_{ab} (Å)	ξ_c (Å)	λ_{ab} (Å)	λ_c (Å)	$B_{c2\parallel}$ (T)	$B_{c2\perp}$ (T)
NCCO	24	70–80	~15	1200	260000	7	–
LSCO	38	33	2.5	2000	20000	80	15
YBCO	93	13	2	1450	6000	150	40
Bi-2212	95	15	1	1800	7000	120	30
Bi-2223	110	13	1	2000	10000	250	30
Tl-1224	128	14	1	1500	–	160	–
Hg-1223	135	13	2	1770	30000	190	–

time, $\xi_c = 1$–5 Å. As a rule, the coherence length of cuprates with low critical temperature is longer than in cuprates with high T_c (see Table 2.1). In electron-doped NCCO, the coherence length is in some times longer than in other hole-doped cuprates. Small values of ξ_c mean that transport along the c-axis is not coherent, even in the superconducting state. For example, $\xi_c \sim 1$ Å in Bi-2212 that is in some times shorter compared to the distance between layers.

Two critical fields $H_{c2\parallel}$ and $H_{c2\perp}$, directed parallel and perpendicular to the basic ab-plane, respectively, correspond to two principle axes (in ab-plane and along c-axis). The notation is as follows. The upper critical field perpendicular to the ab-plane, $H_{c2\perp}$, is determined by vortices (with magnetic flux Φ_0) whose screening currents flow parallel to this plane. Then, for the dependence between critical field and coherence length, we have the following equation [1502]:

$$H_{c2\perp} = \frac{\Phi_0}{2\pi\xi_{ab}^2}. \tag{2.2}$$

The indices ab or c of parameters λ and ξ show the directions of the screening currents.

Due to the high-temperature superconductors possess the layered crystal structure superconductivity in HTSC is confined to the CuO_2 planes. They are separated from neighboring planes by weakly superconducting, normal or even insulating regions of the crystal. Three-dimensional phase coherence is provided by the Josephson currents, which flow between above planes. If we assume a homogeneous order parameter and use description of anisotropy in the framework of the Ginzburg–Landau theory, then for parameter $H_{c2\parallel}$ we have

$$H_{c2\parallel} = \frac{\Phi_0}{2\pi\xi_{ab}\xi_c}; \tag{2.3}$$

then the anisotropy ratio is

$$\frac{H_{c2\parallel}}{H_{c2\perp}} = \frac{\xi_{ab}}{\xi_c}. \tag{2.4}$$

In the case of weakly anisotropic material, such as $YBa_2Cu_3O_{7-x}$, this representation is sufficient. However, for anisotropic material, such as $Bi_2Sr_2CaCu_2O_8$, the value of ξ_c then would be of the order of 0.1 nm, that is approaching atomic scales. In any case, this contradicts the assumption of a homogeneous order parameter. In cuprates, low critical fields $B_{c1\parallel}$ and $B_{c1\perp}$ are very small. For example, in YBCO $B_{c1\parallel} \sim 2 \times 10^{-2}$ T and $B_{c1\perp} \sim 5 \times 10^{-2}$ T. It is interesting that anisotropy of values B_{c1} has different sign than of B_{c2}, namely $B_{c2\perp} < B_{c2\parallel}$ and $B_{c1\perp} > B_{c1\parallel}$. In conventional superconductors, $B_{c2} \propto T_c^2$; at the same time, $B_{c2} \propto T_c^{\sqrt{2}}$ [1201] in cuprates with low T_c. The extreme anisotropy is also responsible for many particular effects associated with the flux line lattice in high-temperature superconductors.

2.2.6 Vortex Structure of HTSC and Magnetic Flux Pinning

The layered structure of the cuprate superconductors, with the superconductivity arising within the CuO_2 planes, causes the properties of a single vortex. The orientation of the CuO_2 planes is defined by the crystallographic *a*- and *b*-axes. The CuO_2 planes are coupled to each other by Josephson junctions. Lawrence and Doniach proposed the phenomenological model for this layered structure [992]. The Lawrence–Doniach theory contains the anisotropic Ginzburg–Landau and London theories as limiting cases, when the coherence length ξ_c in *c*-direction exceeds the distance between layers *s*. In this limit, the anisotropy may be considered in terms of the reciprocal mass tensor with the principal values $1/m_{ab}$, $1/m_{ab}$ and $1/m_c$. Here m_{ab} and m_c are the effective masses of Cooper pairs, moving into *ab* plane and along the *c*-axis, respectively. If the interlayer coupling is weak, then we have $m_{ab} \ll m_c$. Into framework of the anisotropic Ginzburg–Landau limit, the extended relations (2.4) may be obtained:

$$\left(\frac{m_c}{m_{ab}}\right)^{1/2} = \frac{\lambda_c}{\lambda_{ab}} = \frac{\xi_{ab}}{\xi_c} = \frac{H_{c2\parallel}}{H_{c2\perp}} = \frac{H_{c1\perp}}{H_{c1\parallel}}. \tag{2.5}$$

If magnetic field is oriented along the *c*-axis, the flux lines reduce to stacks of two-dimensional point vortices or pancake vortices. A detailed modeling of the layered cuprate superconductor in terms of a stack of thin superconducting films into framework of the Lawrence–Doniach theory has been carried out [295, 481]. Energetically, the perfect stacking of the pancake vortices along *c*-axis is favorable, than a more disordered structure. At the same time, compared to a continuous flux line, as it exists in the conventional superconductors, a stack of the pancake vortices has additional degrees of freedom for thermal excitations. As an example, we consider the displacement of a single pancake vortex, presented in Fig. 2.9. This displacement is equivalent to the excitation of a vortex–antivortex pair (Kosterlitz–Thouless transition), possessing the interaction energy [725]: $U(r) = \varphi_0^2/(\mu_0 r)$, where φ_0 is the quantum of magnetic flux, μ_0 is the vacuum permeability and r the distance between the vortex and antivortex. For two-dimensional screening length, Λ, we have the binding energy, $U(r) \approx \varphi_0^2/(\mu_0 \Lambda)$.

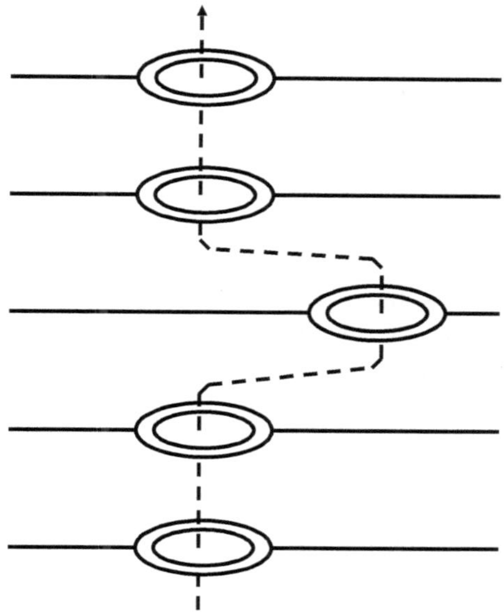

Fig. 2.9 Displacement of a single pancake vortex

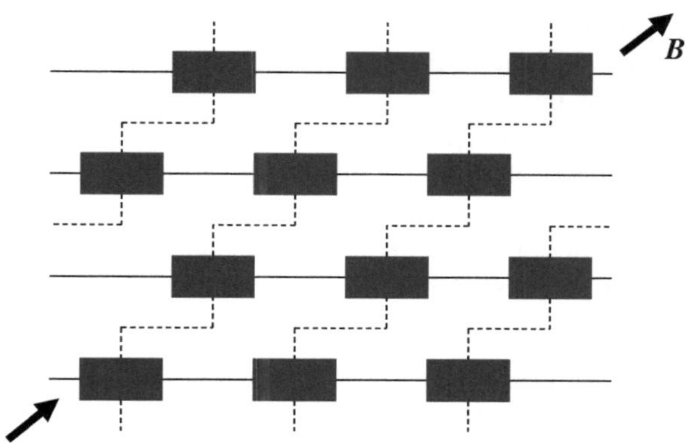

Fig. 2.10 Pancake vortices (*gray rectangles*) coupled by Josephson strings (*horizontal dashed lines*)

By interpreting the displacement of a single pancake vortex as an evaporation process, the evaporation temperature, T_D, has form [295]:

$$T_D \approx \frac{1}{k_B} \frac{\varphi_0^2}{\mu_0 \Lambda}. \tag{2.6}$$

When an external field is nearly parallel to the *ab*-plane, the vortex core preferably runs between the CuO_2 layers. When the coupling between layers is weak, vortex lines along the *ab*-plane are referred to as Josephson vortices or strings. For any magnetic field direction not parallel to the *ab*-plane, the pancake vortices existing in the CuO_2 planes are coupled by such Josephson strings, as shown in Fig. 2.10.

Investigation of the temperature and magnetic field dependence of the magnetization in the powder samples of Ba–La–Cu–O [1206] discovered *an irreversibility line* in the *H–T* phase space (*H* is the magnetic field, *T* is the temperature). Above this line, the magnetization is perfectly reversible with no detectable magnetic flux pinning. However, below this line the hysteresis of *magnetization* arises, and the equilibrium vortex distribution can no longer be established due to *the magnetic flux pinning*. Soon after this discovery, a similar line was found in YBCO single crystal [1836]. Due to these and similar observations, the concept of "vortex matter" may be stated, taking *a liquid, glassy or crystalline* state in the phase diagram. These features have important influence on the transport processes associated with vortex motion. The vortex lattice, originally proposed by Abrikosov [4], consisted of a regular configuration of the magnetic flux lines in the form of a triangle (hexagonal) or square lattice that minimized their interaction energy (see Fig. 2.2). In HTSC, thermal energies are large enough to melt the Abrikosov vortex lattice (at $H = H_m < H_{c2}$), forming a vortex liquid over a large part of the phase diagram. In addition to the high temperatures, there is the structure of magnetic flux lines, consisting of individual, more or less strongly coupled pancake vortices, which promote this *melting transition*. In order to avoid energy dissipation at existence of transport current, each vortex should be fixed at pinning center. In this case, linear and plane defects are most effective. Increasing the number of defects is capable of moving the line $H_m(T)$ at phase diagram into a region of greater values of *H* and *T*. There is a universal field, H_1, such that $H_1(T) < H_m(T) < H_{c2}(T)$, at which thermodynamic fluctuations of order parameter lead to tearing of vortices from lengthy pinning centers [478]. This field presents upper boundary of the irreversibility field, $H_{irr}(T)$, at which the dissipation begins.

The simplest theoretical description of a melting transition is based on the Lindemann criterion, according to which a crystal melts if the thermal fluctuations $\langle u^2 \rangle^{1/2} = c_L a$ of the atomic positions are of the order of the lattice constant *a*. The Lindemann parameter $c_L \approx 0.1$–0.2 depends only slightly on the specific material. This approach has been used to determine the melting transition of vortex lattices. In particular, expressions for the melting temperature in the two- and three-dimensional cases have been derived [146, 170, 172, 482, 1681]. A schematic phase diagram for melting of the solid vortex lattice for three-dimensional material such as YBCO with field applied parallel to the *c*-axis is shown in Fig. 2.11.

The various parts of the vortex phase diagram are caused by competition of four energies, namely thermal, vortex interaction, vortex coupling between layers and pinning. The thermal energy pushes the vortex structure towards the liquid state, the interaction energy favors lattice state, the coupling energy tends to align the pancake vortices in the form of linear stacks and the pinning energy generates disorder. An interaction of these energies, whose relative contributions vary strongly with magnetic field and temperature, results in the complex phase behavior defined by the vortex matter. Then, the irreversibility line may be interpreted as the melting line, above which the state of the vortex liquid is

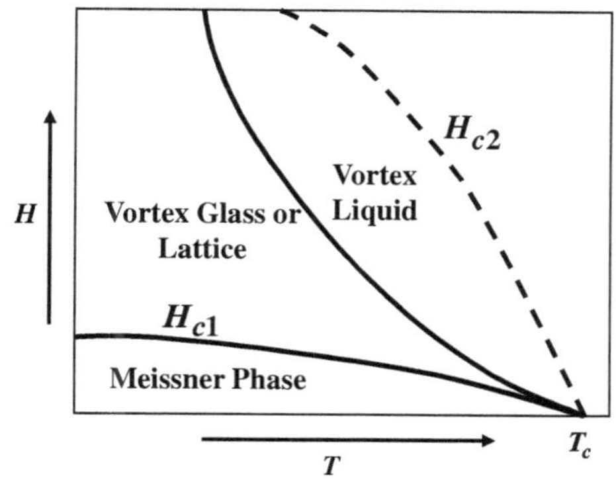

Fig. 2.11 Schematic phase diagram for a three-dimensional material such as YBCO [725]

attained, and below which there is the state of the vortex glass or vortex lattice. The vortex glass state is connected with magnetic flux pinning in the sample, disrupting any vortex motion. The first clear evidence for vortex lattice melting has been obtained from transport measurements (electrical resistance) for de-twinned single crystals of YBCO with $\boldsymbol{H}$ parallel to the c-axis. At a well-defined freezing temperature of the magnetic flux T_m, which depends on the magnetic field, a sudden drop to zero of the resistivity was observed, defining the onset of strong pinning in the vortex solid. The sharp drop of the resistivity at T_m demonstrates *a first-order freezing transition* [312]. *The first order vortex-lattice melting transition* has been observed in thermodynamic measurements, using a high-quality single crystal of BSCCO with $\boldsymbol{H}$, again parallel to the c-axis [1865]. Review, devoted to vortex matter and its melting transition, has been presented in [313].

Early investigations of transport processes in HTSC demonstrated the power dependence of volt–ampere I–V characteristic (I is the current, V is the voltage) [904, 1863] that in the following has been selected as a criterion for the freezing transition into limit of the superconducting vortex glass structure. In another interpretation, a distribution of the activation energy is used for this [575].

Weakening of flux pinning by melting of the vortex lattice is expected only when there are many more flux lines to compare with existing pinning centers. At the same time, in the opposite case, softening of the vortex lattice often leads to stronger pinning than in a rigid vortex lattice. This is explained by the concept that the atomic-scale defects (also as oxygen vacancies) can act as pinning centers for HTSC (the case is often realized in practice). Therefore, melting of vortex matter does not necessarily result in a reduction of pinning. Because of the complexity of this question, there is no simple answer (I see reviews [171, 172]). Flux pinning is caused by spatial heterogeneity of the superconducting material, leading to local depression in the Gibbs free energy density of the magnetic flux structure. Due to the short coherence length in HTSC, heterogeneities, even on an atomic scale, can act as pinning centers. As these important examples, we note deviations from stoichiometry, oxygen vacancies in the CuO_2 planes, and twin boundaries. The separation of a flux line into individual pancake vortices also promotes pinning caused by atomic size defects.

An original discussion of magnetic flux pinning caused by atomic defects in the superconducting CuO_2 planes (in this case by oxygen vacancies) has been carried out in [847, 1734]. By this, the elementary pinning interaction of vortices with the oxygen vacancies was calculated, and the vacancy concentration was related to the critical current density. The various structure defects in HTSC, acting as *pinning centers*, were considered in review [1807], but a detailed research of pinning effect on magnetic relaxation has been carried out in paper [1837]. An advance in solution of the problem of the statistical summation of pinning forces has been attained into framework of the Larkin–Ovchinnikov theory of collective pinning [984]. In this theory, the elastic deformation of the vortex lattice in the presence of a random spatial distribution of pinning centers plays a central role, but the increase of the elastic energy is balanced by the energy gained by passing the flux lines through favorable pinning sites. A discussion of corresponding physical basis is represented in monograph [1681]. We shall return to the discussion of the flux pinning problem in other chapters of our book.

2.2.7 Interactions of Vortices with Pinning Centers

As has been noted, for the attainment of high density of the critical current, it is necessary that microstructure of superconductor retained vortex lines of magnetic flux on the moving, caused by the Lorentz forces. It is reached only by pinning of vortices on the microstructure heterogeneities (or defects). However, no any defect can effectively interact with vortex lines. For example, in conventional superconductors, vacancies, individual atoms of secondary phases or other similar tiny defects are not effective pinning centers due to obvious causes: as a rule, a specific size of vortex (coherence length) is far greater than atomic size, that is, proper size of this defect. Therefore, vortex line simply "does not notice" them. On the contrary, the structure defects with size of $\sim\xi$ and greater become effective ones in this sense, and they can cause high density of critical current.

However, in the case of HTSC there is another situation. Here, the coherence length is extremely short, and point defects have sizes commensurable with ξ. Therefore, consider in more detail a situation arisen on example of vortex interaction with a cavity in superconductor.

Consider an infinite superconductor, containing a defect in the form of a cylindrical cavity. How will a single vortex parallel to the cavity interact with it? Assume that the diameter of the cavity, d, satisfies the inequality $d > \xi(T)$. If the vortex is far away from the cavity, its normal core (of diameter $\sim 2\xi$) stores a positive energy (relative to the energy of the superconductor without the vortex), because the free energy of the normal state exceeds that of the superconducting state by $H_{cm}^2/8\pi$ (per unit volume). Then, the energy of the normal core (per unit length) is

$$\frac{H_{cm}^2}{8\pi}\pi\xi^2. \tag{2.7}$$

On the other hand, if the vortex is trapped by the cavity, that is, passes through its interior, then it does not have a normal core and, accordingly, the energy of the system is reduced by the amount of (2.7). This means that the vortex is attracted to the cavity. The interaction force per unit length, f_p, can be found easily, if we recall that the energy changes by the value of (2.7), when the vortex changes its position near the edge of the cavity by $\sim\xi$

$$f_p \approx H_{cm}^2\xi^2/8. \tag{2.8}$$

For a spherical cavity of diameter d, the interaction force f_{pd} caused by the vortex can be obtained from (2.8) in the form

$$f_{pd} \approx H_{cm}^2\xi d/8 \tag{2.9}$$

To get an idea of how large this force is, let us find the current (j) that must be applied in the direction perpendicular to the vortex in order to produce the Lorentz force $f_L > f_{pd}$. It is known that the Lorentz force per unit length of a vortex is $j\Phi_0/c$ (Φ_0 is the magnetic flux of the vortex, c is the speed of light in vacuum) [1502]. Then, the force applied to the part of the vortex which actually interacts with the defect is $j\Phi_0 d/c$. By equating this to f_{pd}, from (2.9), we obtain

$$j = \frac{cH_{cm}^2}{8\Phi_0}\xi. \tag{2.10}$$

Furthermore, since $H_{cm} = \Phi_0/(2\sqrt{2}\pi\lambda\xi)$ [1502], then from (2.10) we obtain

$$j = \frac{cH_{cm}}{16\sqrt{2}\pi\lambda}. \tag{2.11}$$

It may be shown that j in the last expression is of the same order of magnitude as the Cooper pair-breaking current [1502]. Thus, in order to tear the vortex off a spherical void, it is necessary to apply the maximum possible current for a given superconductor.

The above discussion is also applied to a superconductor, containing tiny dielectric inclusions. It also remains valid (at least by the order of magnitude) for normal metal inclusions, provided the size of the inclusions is larger than ξ. The restriction is caused by the *proximity effect,*[4] which is essential only at distances of the order of the coherence length ξ from the interface. Therefore, the various types

[4] If thin layer of a superconductor is brought in contact with layer of a normal metal, then pairs of electrons with summary zero impulse, formed in the superconductor will stretch on both layers. Proximity effect depends on the interface nature and the relative thickness of both layers.

of inclusions present effective pinning centers in superconductors. This property is widely used in technical applications, when large critical currents and magnetic fields act.

As effective pining centers in superconductors we also note dislocations, dislocation walls, grain boundaries and interfaces between different superconductors. The above analysis of vortex-defect interactions forms the basis for interpretation of effects caused by the so-called *columnar defects* [1173], produced in high-temperature superconductors by irradiation. In this case, the normal cylindrical amorphous regions are formed in the superconductor. The nuclear tracks represent themselves as cylindrical regions with diameter being only 5–10 nm, that is, of the order of the coherence length ξ. As we have just seen, cavities of such sizes are effective pinning centers. This is one of the very few possibilities for producing artificial defects of sizes comparable with the coherence length in HTSC. In given materials, the coherence length is so short that any metallurgical methods used to produce pinning centers, such as precipitates or grain boundaries, have small effect, due to the disproportional size of defects, or not applicable at all, due to the brittleness of the material.

Each defect is capable of trapping approximately one vortex. Hence the optimum pinning efficiency can be expected at magnetic fields, for which the vortex lattice period is less than the average distance between the amorphous tracks. When the Lorentz force[5] caused by the applied field becomes greater than an interaction force with defect or heterogeneity, then a displacement of vortex flux leads to energy dissipation and initiation of finite electrical resistance. This state of vortex structure is called the *resistive state*.

As may be expected, vortex moving caused by the Lorentz force and corresponding resistance to magnetic flux strongly depend on three unique properties of HTSC, namely high critical temperature, small coherence length and layered anisotropic structure. Combining the above features strongly facilitates vortex moving and destroys superconductivity. Due to this, the *resistive transition* essentially expands in magnetic field. This is one from causes, stimulating tremendous efforts of material scientists directed to change of the HTSC manufacture technology with aim to decrease the vortex moving, making more active the magnetic flux pinning. Obviously, this problem will be complicated with the discovery of superconductors, possessing even greater temperatures of transition into superconducting state.[6] Composition features and modern manufacture techniques for basic HTSC systems will be considered in Chap. 3.

[5] In general case, the Lorentz forces can be initiated as transport current, as screening currents caused by sample magnetization.

[6] Nevertheless, it is not obligatory that superconductors with transition temperatures near to room temperature will possess oxide structure.

2.2.8 *Future Studies of Vortex State*

Study of vortices in superfluid directly connected with the problems its turbulence which due to Feynman is the state with disorder sequence of quantified vortices. Since the vortex structure is topologically stable object in superfluid and only idealization into framework of classic liquid, then the turbulence of the superfluid could help to attain a new understanding turbulence in general case. Moreover, this could help to investigate mutual links between superconductivity and ferro- (antiferro-) magnetism. The interests of study of the vortex state are not restricted only by study of vortex structures for several single vortices, but also expanded on investigation collective effects of great number of the vortices and vortex lattices, and also on their interaction with material structure and defects with account of the superconductor topology and geometry.

The studies of the Ginzburg–Landau (GL) equations for fullerene-like structures could lead to novel understanding on vortex state in Josephson junction arrays. Moreover, interaction of vortices with other objects, for instance vortex structure of the Bose–Einstein condensate in optical traps and their interaction with lasers are also interesting objects for future studies.

Note the following perspective mathematical directions of future investigations:

(1) *Vortex creep*. Since, the temperature fluctuations cause motion, for example vortex vibration, then this, in one's turn, leads to arising electric resistance, which can tear vortices from their flux pinning centers. While definite results of computer simulation of these processes have been obtained, the strict mathematical theory remains to be restricted. These problems are related to study of vortex glass and liquid states, and also critical fields and transition temperatures between glassy state (when vortices are to be stationary or pinned) and liquid one (when moving vortices possess definite electric resistance).
(2) *Critical currents*. Great applied currents or magnetic fields also fracture holding pinning forces and cause vortex motion introducing electric resistance. Preliminary studies of vortex states in presence of external fields have been fulfilled as well as computer simulation of movement of spatially-periodic vortices. However, complete mathematical description of these processes is to be very approximate.
(3) *Three-dimensional vortex structure*. Movement, entanglement, orientation, pinning, etc. of vortex lines in three-dimensional statement, as well as collective properties of great number of the vortices, their properties of elastic and plastic states require fulfillment of corresponding mathematical studies.

As problems connected with technical applications of superconductivity, note optimal design of composite superconductors, in particular definition most better distribution of pinning centers which are capable to support high currents, and also control the movement ways and speeds of the vortex lines that could be important at designing superconductive devices. From viewpoint of computational processes numerous questions remain which are related to strong analysis of computer algorithms, in particular to schemes of partition on time of non-linear GL equations depending on time. In the case of the vortex density models, the numerical analysis is still restricted in a whole by two-dimensional examples. Therefore, generalization to three-dimensional models is to be required and perspective stage of the following studies. With practical viewpoint, the computer simulation of enough great number of vortices based on the GL model is very interesting. However, it is still accompanied by great computation difficulties, in particular owing to insufficiently studied effect of the vortex quantization.

Speaking more concretely, for square sample with n internal vortices, the phase angle of complex order parameter will demonstrate change on value of $2n\pi$ near the specimen boundary. If m points are necessary to obtain numerical solution for this single period of phase change, then $mn/4$ points required for each side of the square specimen. Thus, the computational network will demand $m^2n^2/16$ nodes. In the case of great values of n and even moderate values of m this can lead to very significant computational difficulties even in two-dimensional statement of the problem no speaking on three-dimensional case.

It is natural that the adaptive schemes and parallel calculations could significantly decrease the computational difficulties. However, in the case of densely-located vortices, decreasing the network scale could be required practically everywhere, and effectiveness of these schemes will far lesser. In particular note that it is not enough to decrease the network only near the vortex cores, because correct study of phase change is also to be very important. Enough exact definition of high phase oscillations for correct statement of effect of the topological quantization will accompanied by computational difficulties which similar, for instance to definition of high-frequency solution of Helmholtz equation. Moreover, increasing in two times of diameter of considered specimen on networks with distributed vortices will lead to fourfold increasing the vortex number forcing also the computational problems.

On other hand, in the case of HTSC, computational schemes developed for mesoscopic GL models cannot be directly applied at designing corresponding devices because of existence of great number of vortices. The vortex network in superconductive specimen, for example of millimeter size will consist of great number of vortices due to typical distance of ~ 100 Å between single vortices. At the same time, computational and analytical results used for these media, as rule are based on simple homogenized or macroscopic models, for instance the mean-field models describing the vortex state in terms of the vortex density. Since, the superconductivity is phenomenon, in which quantum–mechanical effects demonstrate macroscopic behavior, then it is obvious that more effective computational schemes for the vortex states could be developed based on multi-scale approach combining, for example the GL model or laws of vortex dynamics with the models of mean field.

In relation to experimental studies, it is known that in the layered cuprate HTSC, the vortex lattice melts in wide range of fields and temperatures. As it has been above noted, at phase diagram of the magnetic states of HTSC are differed regions of Meissner phase, vortex crystal and vortex liquid. The properties of the magnetic vortex liquid in HTSC remain be studied insufficiently. In particular, the theoretical estimates of melting temperature of the vortex lattice, and also value of energetic barrier preventing to destruction of vortices at their crossing are changed into very wide bounds. The magnetic force microscope could be used for manipulation with single magnetic vortices. In particular, this allows one to "bind" two vortices in some similarity of double DNA spiral. Moreover, by using this microscope, it may be attempted to tear a vortex from pinning center. In a whole, it could be obtained in future by using this method new information on the vortex glass phase, which realized at low temperatures.

2.2.9 Energy Gap Properties of Superconductors

In conventional superconductors the temperature of superconductive transition T_c is directly defined by value of the superconductive gap, Δ, which presents itself an binding energy of two coupled electrons (Cooper pair or pairon) in superconductive state. However, this situation is quite no situation of HTSC, where is absent direct dependence between T_c and Δ. For example, variation of charge carrier concentration can lead to decreasing T_c, but increasing Δ. The value of Δ can be no zero at the $T > T_c$, at the same time in normal state of conventional superconductors always $\Delta = 0$.

The metals become superconductive, when their free electrons divide into the Cooper pairs, at the same time wave functions of all pairs have the same phase. The phase coherence is responsible, in particular for transformation of electric resistance in zero at the cooling lower than the superconductive transition temperature, T_c, and existence in superconductor of the coupled pairs leads to arising gap in spectrum of quasiparticle excitations, which is registered by different spectroscopic methods.

In conventional LTSC the phase coherence of pairs is stated practically simultaneously with their formation, therefore the electric resistance disappears at the $T = T_c$ simultaneously with arising the gap in spectrum. At the same time in HTSC, the gap is conserved also in absence of phase coherence that is in non-superconductive samples. Origin of this gap, called by pseudogap, up to date is subject of discussions. According to one of viewpoints, the pseudogap is caused by presence of coupled, but non-coherent pairs (macroscopic phase coherence is fractured by thermal fluctuations), that is in fact to be portent of superconductivity. Advocates of alternative interpretation assume that the pseudogap forms due to proximity of HTSC to some non-superconductive (possibly, magnetic) state which competes with superconductive one and is to be a sign of existence another type of ordering (for example, spin or charge one). According to two scenarios, in HTSC simultaneously could exist either only one gap (superconducting or pseudogap), or the both together.

At the temperature which is upper some temperature (the temperature of arising of the pseudogap) $T^* > T_c$, the "square-like" Fermi surface (the Fermi contour in *ab*-plane) is observed. Decreasing temperature lower than T^* leads to arising the pseudogap in the electron state density in the Fermi level: in some parts of the Fermi surface, the energy gap presences, at the same time it absences in other sites having shape of segments (so-called, "Fermi arcs") divided one of other. In this case, the electron number in the Fermi surface decreases sharply. At the following cooling the sample down to T_c, the transition in superconducting state occurs, that is in the Fermi surface already arises the superconducting gap. The attempts to experimentally solve this dilemma are complicated by strong anisotropy of the superconductive gap which is zero at "nodes" and maximal at "antinodes". Since, majority spectrographic methods give information either on maximal value of the gap, or its mean value, then distinguishing the great non-superconductive gap in antinode region and small superconductive gap in the rest of part of the Fermi surface is very difficult problem.

Nevertheless, the experiments witness on existence of two different types of the energetic gap in HTSC, namely: charge (it is displayed in spectrum of charge excitations, for instance when is studied single-particle tunneling and relaxation of quasiparticles with time resolution) and spin (it is displayed in spectrum of nuclear magnetic resonance, scattering spin-polarized neutrons and other methods which are to be sensitive to electron spin flip). The pseudogap state is characterized by existence of electron ordering in actual (coordinate) space. This ordering can arise because of pinning the fluctuating incommensurable wave of spin density. Thus, complex phase diagram of the cuprate HTSC (see Sect. 2.2.2) is a sequence of competition between various types of ordering. The energetic gaps are to be dependent both as on doping level of HTSC, as CuO_2 layer number per elementary cell.

Two-dimensional Hubbard model allows one to study main phenomena proper for the superconductive cuprates. At half-filling energetic structure, it defines antiferromagnet Mott–Hubbard state. At introducing doping, the pseudogap arises, and *d*-wave pairing and state with charge stripes (stripe phase, see Sect. 2.2.1) display at low temperature. Various numerical methods have been used to investigate the Hubbard model and *t–J* model. The approach, connected with the Lanczos diagonalization of small clusters has been developed in [112, 325, 770, 1005, 1373, 1374], the density matrix by using the renormalization-group method (RGM) for *n*-leg "ladders"[7] has been studied in [326, 617, 1796], quantum

[7] In particular, note that Sr–Ca–Cu–O compounds with so-called "spin ladders" present themselves an analogs of HTSC, because of contain layers of almost one-dimensional "ladders" (that is number of "legs") since they contain layers of almost one-dimensional "ladders" from copper and oxygen atoms. Increasing transverse sizes of this "ladder" (namely, the number of "legs") leads to its gradual development into copper-oxygen plane, which as it is assumed answer for high-temperature superconductivity. It has been theoretically shown that shot-acting spin correlations lead to formation coupled pairs of holes in the "ladders". Superconductivity has been discovered in $(Sr, Ca)_{14}Cu_{24}O_{41}$, while at high pressure.

modeling by using Monte-Carlo method in two-dimensional lattices has been carried out in [49, 205, 1599]. In principle, these methods could be extrapolated to three-dimensional treatment or in the RGM case—to the infinite length of the "ladders". Other important numerical approaches are the extrapolation with zero dispersion based on the Monte-Carlo method [49, 1599], the variation cluster approximations [327, 1073, 1396], the renormalization-group flow techniques [702, 1473], and also high temperature series expansions [914, 1412].

Cluster generalizations of the mean-field dynamical theory [327, 1073, 1396] ensured new information on low-temperature properties of HTSC. Significant advantage of these numerical methods (for instance, in relation to the Hubbard model) consists in that what they are capable to study antiferromagnetism, charge stripes, pseudogap behavior and $d_{x^2-y^2}$-pairing. Moreover, the computational methods have shown as could be balanced these models between almost degenerate phases.

A doping breaking half-filling can displace the balance from antiferromagnetism to the charge state, at which the region bounds with half-filling divide the antiferromagnet regions with phase sheared on π. The stated balance also reveals in different results obtained by using various computational approaches for the same model. For example, RGM for density matrix for *t–J* "ladder" structures [1796], has stated existence of charge stripes in ground state. However, variation approach and calculations of Green function based on Monte-Carlo method for doped *t–J* lattice [1599], found ground states characterized by the $d_{x^2-y^2}$-superconductive order at presence of only weak signs of existence of the charge stripes.

The pointed computational methods state modern level of investigations. They can lead to different results, that is direct sequence of various boundary conditions or/and different other aspects of used computational lattices. Open boundary conditions in the RGM-computations of the "ladder" structures can favor to formation of stripe phase. On the contrary, finite sizes of cluster lattices and boundary conditions can prevent from its formation. It is also possible, that these differences reflect imperfection of different numerical methods. Nevertheless, these results, collected together, show that the stripe and $d_{x^2-y^2}$-superconductive phases are to be almost degenerate low-energetic states of doped system.

The quantum modeling by using the Monte-Carlo method [1401] also as different cluster calculations demonstrate that the Hubbard model for underdoped case shows pseudogap states. Tremendous likeness of this behavior with a set of phenomena observed in cuprates provides correct witness that what the Hubbard model and *t–J* model cover actually significant part of physical essence of given problem. The computational methods have been used to study interaction structure which is responsible for pairing in the Hubbard model. The dependence of superconductive gap on impulse (momentum) and oscillation frequency has been numerically studied in [1495]. The obtained results showed that the pairing interaction in the doped Hubbard model is repulsive into elementary cells, attractive between nearest cells and slowed in time at rearrangement of spin-fluctuation spectrum. It is important to note, that this spectrum includes a continuum of particles and holes. By this, it is necessary to answer on question, is this interaction actually mechanism responsible for pairing in HTSC, and how learn this from experiment?

Since this issue touches dependence of pairing interaction on impulse then investigations of superconductive gap, fulfilled by using the angle-resolved photoemission spectroscopy (ARPES) [328], together with the transport study [275], and also the phase structure study [1714] ensure strict witness of nodal *d*-wave character of this gap. At the same time, as it has been stated in [876, 1586], existing couples in YBCO lead to admixture of *s*-wave, and studied impulse regions in majority dispose along the Fermi surface. Therefore, based on the observed $\boldsymbol{k}$-dependence of superconductive gap (where $\boldsymbol{k}$ is the wave vector), it may be presented that actually pairing interaction is repulsive into cell and attractive for singlets formed between nearest cells.

Other characteristic of the interaction is its frequency dependence. In relation to it, there is lesser information, however, similar that the frequency dependence of superconductive gap and renormalization parameter can give important contribution in understanding the pairing mechanism for HTSC. It is known, that the frequency dependence of superconductive gap for conventional LTSC has provided identification of phonon exchange pairing mechanism, while some time existed doubts on this connection. In the case of HTSC, it was initially to be assumed that the *d*-wave impulse dependence of superconductive gap should ensure enough exact proof. However, this did not occur. For instance, it is known that exchange by B_{1g}-phonons favors to *d*-wave pairing while its total contribution in superconductive transition temperature T_c is small in accordance with standard theory.

On the other hand, two-level Cu–O-model assuming that fluctuations in circulating currents ensure the mechanism of *d*-wave coupling has been proposed in [1745]. Even, into framework of the Hubbard model, there are different viewpoints on the process dynamics. In the model [44] (see Sect. 2.2.2 in details), it is suggested that the dynamics take place in energetic scale associated with the Mott–Hubbard gap. However, the numerical results [1495] support a picture, in which dominating contribution is caused by excitations of "article-hole" type inside of relatively narrow energetic strip. While the spectrum of these excitations spreads down to zero energy, the main contribution is associated with wide spin-fluctuation continuum [291]. Thus, obviously again the dynamics will important in identification of the superconductivity mechanism.

In addition to traditional electron tunneling and infra-red conductivity, the ARPES measurements ensure important witnesses for frequency dependence of the renormalization parameter and superconductive gap. Therefore, it is important to extend numerical investigations of the Hubbard model in order to compare obtained results with corresponding test data. Finally, in addition to frequency and impulse dependence of pairing interaction, arises issue on its strength. Actual Cu–O structure has additional ways of exchange interaction, and as it has been shown [335] the Hubbard and *t*–*J* models taking into account interactions in cells can lead to stronger pairing correlations.

Nevertheless, the issue on strength of pairing interaction remains to be open. As it has been noted, different numerical methods applied on various grids can witness in one case on *d*-wave pairing, and in other case on existence of charge stripes. Thus, small changes of local parameters could change a correlation nature, and then issue arises on treatment of role of heterogeneities in cuprates. That theory of dynamical heterogeneity, introduced by pairing, has been developed in [882]. In this approach, the pairing caused by repulsive interactions arises as mesoscopic effect, and HTSC phenomena are considered as arising owing to existence of the mesoscopic structures.

The measurements of admixtures in BSCCO by using scanning tunneling microscopy ensured important new information on local modulation of pairing and its strength [1290]. So, development of different numerical methods together with experimentally-theoretical approaches in future will continue to create new knowledge on electronic properties of wide class of the strongly correlated materials.

2.3 High-Temperature Superconductive Oxypnictides

In February 2008, Hosono et al. [806] synthesized a new superconductor $LaO_{1-x}F_xFeAs$ with the temperature of superconductive transition $T_c = 26$ K, so has been discovered a new superconductive family—superconductive oxypnictides.[8] The critical temperature very rapidly rose up to 43 K in $SmO_{1-x}F_xFeAs$, then to 52 K in $PrO_{1-x}F_xFeAs$, and to 55 K in $SmO_{1-x}F_xFeAs$, synthesized under high pressure. The maximal critical temperature $T_c \sim 56$ K has been stated in *M*FFeAs (M = Ca, Sr, Ba, Eu) with superconductivity arising upon modification with Ln^{3+} ions [269, 1809]. That high critical temperature near the liquid nitrogen temperature (77 K) allows us to relate the superconductive oxypnictides to the high-temperature superconductors. Similar to the cuprate HTSC, the fourfold compound LaOFeAs has layered crystalline structure with periods of tetragonal elementary cell $a = 0.40355$ nm and $c = 0.87393$ nm.

As in the cuprate HTSC, where superconductivity occurs in CuO_2 planes, the conductive layers (FeAs) of superconductive oxypnictides are to be spatially divided with the charge reservoir layers (LaO). If in the first discovered superconductive oxypnictides the superconductivity arose at doping of original non-superconductive compound LaOFeAs by electrons (by the way of O substitution on $O_{1-x}F_x$), then later it was became to be clear, that hole doping (at the La substitution on $La_{1-x}Sr_x$) also leads to superconductivity. Thus, in this relation the high-temperature oxide superconductors without copper lead themselves similarly to their cuprate analogues.

[8] Pnictides are compounds containing elements of Mendeleev periodical table of chemical elements such as, nitrogen, phosphorus, arsenic, etc.

However, their "parent" states are to be different. If non-doped cuprates are to be dielectrics (antiferromagnetic or Mott ones), then underdoped oxypnictides present themselves either non-magnetic metals or antiferromagnetic semimetals. According to that, there are two principle differences of oxypnictides from cuprates, namely: (i) magnetism of parent non-superconductive phase in the first case is caused by de-localized electrons, but in the second case—by local magnetic moments, and (ii) the superconductivity of HTSC cuprates arises only due to "chemical doping" (non-isovalent substitution of some elements by others or change of oxygen concentration), at the same time the superconductivity of HTSC oxypnictides arises additionally under influence of high pressure. Thus, the doping and strong compression by the same way changes definite properties of the superconductive oxypnictides which are significant for their superconductivity.

The effects of doping (Ba $\rightarrow$ $Ba_{1-x}K_x$) and pressure ($T_c = 31$ K at the $p = 5.5$ GPa) on structure of $BaFe_2As_2$ have been studied by using the neutron diffraction [871]. In both cases, occurred shortening length of couple Fe–Fe and angle As–Fe–As, and also suppression of transition from tetragonal to orthorhombic phase. These changes of crystalline structure, in its turn led to change of electronic structure. The performed calculations showed that at the doping and pressure increasing, the shape of energetic zones changes in vicinity of the Fermi level, nesting the Fermi surface weakens and the spin-density wave (SDW) destabilizes. Therefore, it has been made a conclusion that the main factor, defining superconductivity of the oxypnictides, was connected (in difference of cuprates) no concentration of moved carriers, but a kind of electronic zone structure.

Information on superconductivity of undoped compound $SrFe_2As_2$ has been presented in [690]. First, the thin films of $SrFe_2As_2$, received by laser evaporation, as it could be expected to be non-superconductive. However, at the repeated measurements performed after some hours, they demonstrated superconductive transition at the $T_c = 26$ K. All this time the films be kept at air and the cause of the cardinal change of the electric properties was caused by water steams penetrated into the films from environment atmosphere. As it was elucidated, the water penetration into $SrFe_2As_2$ caused a convergence of atomic layers and arising superconductivity. On the one hand, explanation of this phenomenon can be a key to the superconductivity mechanism of without-copper HTSC, and on the other hand, can prompt a way to increase their critical temperature.

Investigations of structure and magnetic characteristics of LaOFeAs by using neutron diffraction method [346] showed that at the decreasing temperature, first at the $T = 155$ K occurred structural transition from tetragonal to monoclinic phase, and then at the $T = 137$ K, it was formed far antiferromagnetic order with SDW. The substitution of O on $O_{0.92}F_{0.08}$ suppressed both these transitions and led to superconductivity with $T_c = 26$ K. Thus, as in superconductive oxypnictides, as in cuprate HTSC, superconductive state adjacent with antiferromegnetic, that gives some arguments to speak on non-phonon mechanism of pairing carriers in the both superconductive families.

The studies of structure and magnetic phase diagram of $CeFeAsO_{1-x}F_x$ into coordinates "temperature—concentration of charge carriers" [1875] by using neutron spectroscopy method showed that similarly to HTSC, the temperature of structural transition in oxypnictides becomes zero at the carrier concentration corresponding to maximum of T_c. At the same time, there is important difference of the phase diagrams of cuprates and oxypnictides, namely: in latter value of T_c, attaining at doping itself maximum, then very slowly decreases but does not turn into zero, that what overdoped non-superconductive state in their phase diagram is absent.

The measurements of temperature dependence of high critical magnetic field H_{c2} in $LaO_{0.89}F_{0.11}FeAs$ [730] showed that the value of $H_{c2}(0) \sim 65$ T was significantly greater than it may be expected for single zone superconductor on the base of value of the derivative $dH_{c2}/dT \sim 2$ T/K near T_c. Based on this result, it has been made a conclusion on two-zone superconductivity of oxypnictides as in MgB_2. However, studies of value of the superconductive gap D in $SmO_{0.85}F_{0.15}$-FeAs by using Andreev spectroscopy [267] led to the contrast conclusion on single zone superconductivity. The calculated value of $2D/(k_BT_c) = 3.68$ (where k_B is the Boltzmann constant) was very closely to corresponding value (3.53), obtained on the base of the BCS-theory. More that, the temperature dependence, $D(T)$, into all range $0 < T < T_c$ very good agreed with the BCS-theory. In this case, it may speak on involvement to superconductivity some bosonic excitations.

It may be affirmed, that the parameter of superconducting order in oxypnictides has not zeros on the Fermi surface in difference of the cuprate HTSC. Moreover, any pseudogap features also even have not discovered. In spite of their layered crystalline structure, oxypnictides demonstrate practically isotropic superconductivity. On that behavior results witness which have been received at the study of monocrystals (Ba, K)Fe_2As_2 into magnetic field of approximately 60 T [1852]. It has been elucidated, that in a whole temperature range lower than T_c, the value of H_{c2} is independent almost of direction of magnetic field that sharply contrasts with all known layered superconductors. The probable cause of the H_{c2} isotropy may be isotropy of the Fermi surface, but test data on its topology remain to be contradictory.

The Fermi surface of (Ba, K)Fe_2As_2 has been studied by using ARPES [1860], where shown its anisotropy even into *ab*-plane, in which it had a shape of "propeller" that differed of zone calculations. These authors assumed that their results could be explained by presence in superconducting oxypnictides some electronic order co-existing with superconductivity and keeping up to 300 K. It is not usual charge-density wave (CDW), but something more complex reminding hypothetical "hidden order" in cuprates. Underdoped oxypnictides being either non-magnetic metals or antiferromagnetic semimetals demonstrate magnetic properties which are expressed far weakly than at cuprates. They are very sensitive to smallest changes of crystalline structure and chemical composition that strongly impedes an interpretation of test data.

In [1108], it has been expressed proposition supported by calculations and tests on that, what magnetism of the HTSC without copper is caused by new type of magnetic excitations that is "magnetic antiphasons", which are connected with antiphase boundaries between regions with local SDW-order. Fast fluctuations of these excitations could lead to that, what a result of any experiment depends on corresponding time scale. The conception of antiphasons allows one to give natural explanation the temperature difference of structural and magnetic transitions in oxypnictides.

In [1034], it has been studied an influence of isotopic substitution of $^{18}O/^{16}O$ and $^{56}Fe/^{54}Fe$ on T_c of two typical representatives of the HTSC without copper, namely $Ba_{1-x}K_xFe_2As_2$ with $x = 0.4$ and $SmFeAsO_{1-x}F_x$ with $x = 0.15$. For oxygen, the factor $a = -d(\ln T_c)/d(\ln M)$ (where M is the isotope mass), characterizing isotopic shear of T_c, turn out to be very small (about 0.05), at the same time for iron $a \approx 0.35$, that to be closely to "phonon standard" ($a = 0.5$). In these tests, it has been also studied as isotopic substitution influences on the transition temperature T_{SDW} of original compounds with $x = 0$ into state with spin-density wave. The corresponding value of $a_{SDW} = -d(\ln T_{SDW})/d(\ln M) \approx 0.05$ and ≈ 0.37 has been calculated for oxygen and iron, respectively.

In absence today, common theory of influence of the isotopic substitution on magnetic phase transitions, these results could be interpreted differently. From the one hand, the great value of a witnesses on that, what electron–phonon interaction plays definite role in superconductivity of oxypnictides, and from the other hand, it is obvious presence of strong phonon-magnon interaction. Therefore, it is quite possible, that the isotopic change of T_c is connected with change no phonon, but magnon spectrum of superconductive oxypnictides. Proper details of non-phonon coupling mechanism for account of carrier interaction with magnetic excitations are defined by microscopic nature of antiferromagnetic state. If magnetism in cuprates is caused by ordering local magnetic moments and well described by Heisenberg model, then the data on antiferromagnetic order in pnictides remain to be contradictory: some experiments witness on its local character, at the same time the others speak on definite role of collectivized electrons (as in metallic chrome).

An elastic scattering of neutrons on monocrystals of $CaFe_2As_2$ has been studied and it showed that from the one hand, dispersion of spin waves in a whole Brillouin zone of this "progenitor" of the superconductive oxypnictides is described by the Heisenberg model (in which are taken into account exchange interactions between nearest and following for the nearest neighbors) [1874]. From the other hand, very strong difference of interaction energies along a- and b-directions of Fe–As–Fe layers could not be explained, if account of all orbitals of Fe-atoms be localized. So, magnetism of pnictides is not to be purely local or purely collectivized, but presents itself a complex mix of the both types.

As it has been noted, the common structural elements for oxypnictides are to be conductive layers of FeAs. The layers from other atoms stabilize crystalline lattice and are to be sources of current carriers. Namely these layers (various for different pnictides) are responsible for a set of specific effects which have not relation to superconductivity, for example for phase delamination. They significantly complicate a picture and are difficult to an interpretation of test data. The idea to get rid of these layers led to study of iron chalcogenides FeX (X = S, Se, Te), related to pnictides, in especially iron selenides FeSe, a possibility of superconductivity in which has been predicted theoretically [1620]. It has been stated that FeSe, no containing "excess" layers, has very low T_c (about 8 K), and therefore a doubt arose on that, what it could be treated as a ground compound for study of superconductivity of oxypnictides [712].

However, in [1123], it has been shown, that FeSe becomes high-temperature superconductor at enough high pressure, p. In this compound, the value of T_c increases rapidly with growth of p (it is possible because of increasing carrier concentration in FeSe layers) and attains 37 K at the $p = 9$ GPa (it is almost record for binary superconductors). In difference of cuprate HTSC and compounds with heavy fermions, superconductivity of FeSe at the following increasing p disappears does not because of exceeding optimal carrier concentration, but because of structural transition into hexagonal phase. If this transition will be avoided, then maximal T_c will be significantly increased. It has been also stated absence in FeSe of static magnetism at all values of p (in majority of undoped pnictides takes place a static magnetic order, which changes by superconductivity at doping or increasing p). Therefore, the growth of T_c under pressure is not connected with suppression of competing magnetic order. However, it does not sign, that magnetism has nothing to do superconductivity of FeSe. Data of nuclear magnetic resonance, received in [743] witness on presence of spin fluctuations in superconductive state of FeSe, which in principle could be responsible for the Cooper pairing.

2.4 Future Studies of High-Temperature Superconductivity Mechanism

From "electronic view", the main problem of construction of unified model consists in that what electrons in HTSC cannot considered separately: each of them interacts with others that is the given system to be "strongly correlated". The general methods, describing these systems until are to be only at the stage of development. Nevertheless, several theorists are believed that the electron pairing in HTSC occurs (as in usual BCS-superconductors) for account of their interaction with oscillations of crystalline lattice (phonons). Others insist on "magnetic mechanism" of pairing. Thirds assumes that the key could be found in charge heterogeneities (stripes, etc.). It should not be excluded some combination of all above factors.

One of the HTSC "fathers", Nobel Laureate, 1987—Müller to be so far convinced in polaron mechanism of high-temperature. Nobel Laureate, 2003 in the region of superfluidity—Leggett assumes that nearest goal of experimental studies should be statement as the causes of the property universality of various superconductive families (for example, similar dependence of the critical temperature T_c on number of the CuO_2 layers per elementary cell), as the causes of their non-universality (for example, sharp T_c-difference of single-layered HTSC based on Hg and Tl), and also clarification of role of the high-energetic excitations in HTSC superconductivity. Now, he develops a scenario for cuprate superconductivity in which a major factor is the reduction, due to increased screening by the Cooper pairs, of the long-wavelength, mid-infrared-frequency part of the Coulomb interaction. In addition, independently of this scenario, he attempts to explain the *c*-axis transport properties of the cuprates and look at some problems associated with the "pseudogap" regime and with the peculiar features resulting from the existence of gap nodes. Nobel Laureate, 1977—Anderson is faithful his RVB theory and assumes that it (may be with some modifications) can explain an essence of HTSC phenomenon. Now he focuses on achieving understanding of the crucial experimental anomalies in HTSC behavior. In his opinion, the key to the problem is a new method of dealing with the constrained Hilbert space which follows from the necessity of the Gutzwiller projection. Rice and Lee have the same opinion, and the last hopes that it will be success (at least by using numerical calculations) to prove existence of superconductivity in Hubbard model with repulsion. Pines is also convinced in non-phonon mechanism of HTSC superconductivity. From his view, we are near coordinated microscopic theory of superconductivity of overdoped HTSC which should be based on accurate description of electron interaction with antiferromagnetic fluctuations. However, in this case, it is necessary to understand why at the doping other antiferromagnetic dielectrics (for example, $LaTiO_3$, V_2O_3, NiS_2, Sr_2VO_4), a superconductivity does not arise even at the very low temperatures (in this fact pays attention Imada). Zaanen considers that HTSCs differ not their complexity, but on the contrary—simplicity which is displayed, for instance in strictly linear temperature dependence of electric resistance of the optimally doped samples upon to the melting point. This cannot be a simple accidence. Therefore, it is necessary to shun any dogmas, for instance as dogma of Anderson–Hubbard on electronic mechanism of HTSC superconductivity. Obviously, phonons play an important role while highly unusual with positions of microscopic theory. It is required to search basic principle laying in the basis of the phenomena. The key could be given by experiment. Chakravarty sees this key in a "dome"-like dependence of T_c on number of the CuO_2 layers per elementary cell, but Senthil is based on proximity of superconductive, dielectric (Mott's) and pseudogap states. Randeira attaches a great importance to search in HTSC new elementary excitations with unusual quantum numbers and detail analysis competition between different ordered states.

Thus, in relation to HTSC theory can be selected two polar positions, namely: (i) numerous unusual HTSC properties are caused by various no connected between them effects, that what can speak, for example on the theory of transport properties and theory of the pairing mechanism as on two different theories, (ii) natural explanation of all features of HTSC could be given from united view into framework of new theory of condensed matter with strong correlations. Schmalian supports the second view and assumes that systematic phenomenology of several HTSC families speaks in its favor. However, even it is not completely to be clear, as suggests Vojta, do these character features as pseudogap, stripes and heterogeneities observed by scanning tunnel microscopy to be necessary ingredients or on the contrary obstacle for high-temperature superconductivity. Varma thinks that experiment supports predicted by him the breaking symmetry relatively a time reversal in the pseudogap phase for account of spontaneous arising closed contours with current in CuO_2 layers. Therefore, the pairing charge carriers and anomalous properties of HTSC above T_c are caused by strong fluctuations in vicinity of quantum critical point, where the pseudogap "closes". Davis assumes that for solution the HTSC problem completely, it is necessary in a whole to understand the HTSC phase diagram but not only to study the superconductivity mechanism.

The above wide scattering of opinions of the leading scientists in the region of HTSC demonstrates a reality of existing discords which must be removed for statement of complete mechanism of high-temperature superconductivity. Today, the high-temperature superconductivity could be considered not only as one of non-discovered mystery of modern physics, but as a some polygon for selection ideas and methods to describe strongly correlated electrons and a barrier separating researches from principally novel physical concepts.

2.5 Weak Links and Josephson Junctions

In 1962, Josephson published a theoretical paper [796] predicting the existence of two remarkable effects. One was supposed to find them in superconducting tunnel junctions. The basic idea of the *first effect* was that a tunnel transition should be able to sustain a superconducting (i.e., zero-voltage) current. The critical value of this current was predicted to depend on the external magnetic field in a very unusual way. If the current exceeds the critical value, which is a characteristic of a particular junction, the junction begins to generate high-frequency electromagnetic waves. This phenomenon was called *second Josephson effect.*

Soon after their discovery, both effects obtained experimental confirmation [1547, 1831]. Moreover, it soon became clear that the Josephson effects exist not only in tunnel junctions, but also in other kinds of the so-called weak links, in particular, short sections of superconducting circuits, where the critical current is substantially suppressed [1021, 1022, 1597]. So, initiating weak superconductivity is based on the quantum nature of the superconducting state that assumes the existence of condensate of the Cooper pairs. This means that all electron pairs in the superconducting state occupy the same quantum level and are described by a single wave function, common to all of them. Their behavior is mutually conditioned and they are coherent.

Consider two bulk superconductors having the same temperature and completely isolated from each other. The behavior of the superconducting electrons in each of them is subjected to own wave function, when they are in the superconducting state. Then, because the temperatures and the materials of the superconductors are identical, the amplitudes of the wave functions must also be the same. However, the phases, in this case, are different. This situation remains as long as the superconductors are isolated from each other. Let us establish a weak contact between them, that is, the contact which is weak enough so as not to change radically the electron states of the two superconductors, but sufficient to initiate a perturbation. In this case, a new wave function is formed that is general for the joint superconductor, which can be considered as a result of interference between the wave functions of both its pieces. Therefore, phase coherence is a direct result of establishing the weak link.

In first stage of R&D of the high-temperature superconductive Josephson junctions (HTSC JJs) it has been assumed that main difficulties of their processing are connected with technologies. Due to the very small coherence length (several angstroms) in HTSC-materials, in order to receive qualitative and reproduced on parameters of the Josephson junctions, were required to solve three technological problems: (i) to grow epitaxial HTSC-films and multilayered structures (ii) to form a necessary structure of the film, and (iii) to carry out a control of boundaries in atomic scale. With these aims the unique processing methods of strontium titanate, sapphire, and neodymium gallate have been developed. Namely on these substrates have been processed the record on their noise characteristics HTSC JJs. SQUIDs manufactured on these substrates have record parameters at 77 K: sensitivity on magnetic flux $2.5 \cdot 10^{-5}\ \Phi_0/\mathrm{Hz}^{1/2}$ (on frequencies upper of 30 Hz), cutoff frequency noise $1/f \sim 10$ Hz, depth of output signal modulation up to 120 μV.

However, the required homogeneity of parameters for relatively complex devices consisting of several tens of HTSC JJs has not been attained. During time, it became to be clear, that the problem of homogeneity and reproduction of parameters has ground physical nature being a sequence of anisotropic pairing which is intrinsic for HTSC (different of traditional type for LTSC) *d*-coupling. It is known that the *d*-coupling is very sensitive to electron scattering on admixtures and suppresses at the comparatively small their concentrations. However, interior of HTSC, defects localize in interplane positions, due to transport properties in superconducting CuO_2-planes correspond with well accuracy to condition of pure limit.

The situation changes cardinally in vicinity of intergrain boundaries or artificially created interfaces of JJs. As a result of electron scattering on boundaries and defects, localized in their vicinities, the partial or total suppression of the pairing in *d*-channel occurs, accompanied by arising *s*-type superconductivity. This takes place owing to the electron, which incidents on the boundary from the region with positive potential of superconductive coupling, could in a result of the scattering chances to find oneself in the region of space with negative potential. It is qualitatively clearly, that this replacement of sign suggests presence of region, in which density of Cooper pairs crosses zero, at least on these quasiparticle paths. The fraction of these paths, and hence a degree of suppression of the superconductive properties by the boundary depends on the angle between normal to the boundary and crystallographic axes situated into CuO-planes. If these two directions coincide, then the suppression is absent. However, if the angle between them equals to 45°, then the value of the coupling potential on the boundary is strictly equal to zero. This local suppression of superconductive properties is accompanied by arising anomalously great normal conductivity in the region of small stresses, generation of Andreev coupled states at finite energies that is influences not only on local value of critical current, but on parameters responsible for transport of normal electrons.

In developed, today types of HTSC JJs (the junctions on bi-crystalline substrates, edge junctions, junctions received in a result of modification of the HTSC material properties, caused by any influence), a control for local value of the angle between normal to boundary and crystallographic axes is to be practically impossible. As a result, there is heterogeneity of parameters responsible for transport both superconductive and normal components of current. Therefore, the problem solution in this case lies in creation of technology which is able to manipulate atomic-smooth boundaries.

2.5.1 Classification of Josephson Junctions

The weak links (Josephson junctions, JJs) can be classified in the following way:

(1) *Devices without concentration of current such as tunnel junctions of the "superconductor–insulator–superconductor" type* (S–I–S) (Fig. 2.12a). The thickness of an insulating layer, as a rule, is about 1–2 nm, and the critical current density is in the order of 10^4 A/cm^2, that is, much less than the critical current density of the bulk superconductor, in particular, exceeding 10^5 A/cm^2 at 77 K and 0 T in Y(*RE*)BCO melt-processed samples.[9]

(2) *Layered structure of the "superconductor–normal metal–superconductor" type* (S–N–S). It involves the normal layer with thickness of ~1 μm (Fig. 2.12b). The wave functions of the superconducting electrons penetrate the normal metal due to the proximity effect. In the region of their overlapping, the wave functions interfere, establishing phase coherence between bulk superconductors. If the amplitude of the superconducting wave function in the weak link is small, then the critical current is also small.

(3) *Layered structures where a normal layer between two superconductors is replaced by a doped semiconductor or another superconductor with a small critical current density.* For example, if a narrow superconducting film is covered by thin film of a normal metal (Fig. 2.12c), then the amplitude of the superconducting electron wave function in the film is reduced, where the film contacts with normal metal, due to proximity effect. This causes local decrease of the critical current density, that is, weak link formation.

(4) *Devices with concentration of current.* The critical current density in the weak link is the same as in the bulk, but the absolute value of the critical current is much lesser. A superconducting film with a short narrow constriction (Dayem bridge) falls into this category, provided the size of the constriction is of the order of the coherence length, ξ (see Fig. 2.12d). Another example is a bridge

---

[9] High-quality tunnel Josephson junctions, made of niobium with a barrier layer of aluminum oxide, attain the critical current density of ~10^4–10^5 A/cm^2.

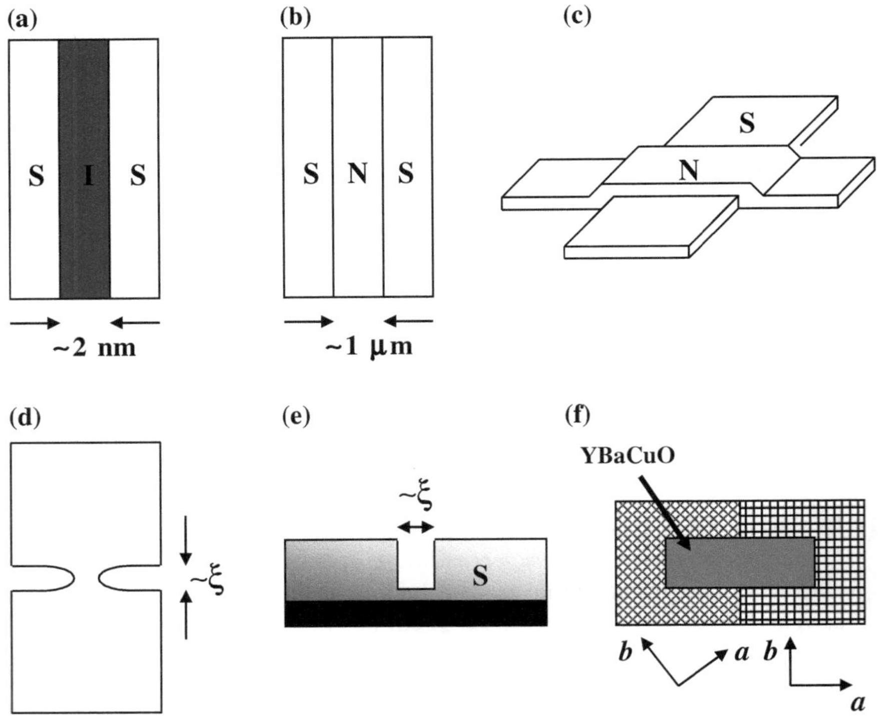

Fig. 2.12 Different types of weak links: (**a**) tunnel junction (S–I–S); (**b**) sandwich structure (S–N–S); (**c**) normal film (N) causes local suppression of the order parameter of superconducting film (S); (**d**) Dayem bridge; (**e**) bridge of variable thickness (longitudinal cross-section); (**f**) grain-boundary junction

of variable thickness, such that the thickness of the basic film is hundreds of nanometers, at the same time, the thickness of the bridge itself equals to only several tens of nanometers (Fig. 2.12e).

(5) *Grain boundary (or bicrystal) transition.* It is a typical one for high-temperature superconductors (Fig. 2.12f). Due to the extremely short coherence length in HTSC ($\xi \sim 1$ nm), defects in their crystalline structure can act as weak links. The best-controlled defects can be produced between two regions of an epitaxial high- temperature superconductive film with various crystal orientations (grain boundaries). The critical current density of such a weak link can be varied, changing the misorientation angle between two crystallites.

HTSC Josephson junctions can also be classified into the following three classes [583]:

(i) junctions without interfaces, in which the weak coupling is achieved by locally degrading the superconducting properties of a HTSC thin-film microbridges by focused electron or ion beam irradiation;

(ii) junctions with intrinsic barriers or interfaces formed, for example, using the intercrystalline boundaries of different crystallographic orientations;
(iii) junctions with extrinsic interfaces, in fabrication of which artificial barriers from normal metals and insulators are used.

From the standpoint of the damage and material strength problem, the last two classes of the Josephson junctions display most interest. HTSC JJs with extrinsic interfaces are prepared by using the film technology, and they themselves present multi-layered structures (hetero-structures) with rectilinear and inclined interfaces. There are different technological possibilities, which are discussed in detail in the following two chapters, for increasing superconducting properties of these systems. For example, an overdoping by calcium of intergranular boundaries in multi-layered structures and YBCO super-lattices permits to significantly increase J_c at all temperatures up to T_c, but also at magnetic fields up to 3 T [331].

2.5.2 Tunneling Junctions Based on High-Temperature Superconductors

In difference of conventional LTSC, clear difference between tunnel junctions and structures with directed conductivity does not preserve for HTSC JJs. Except several cases (for instance, by using scanning tunneling microscopy), a barrier occupies often intermediate site between tunnel and immediate contact. Below, we present main directions of R&D of the HTSC tunneling junctions.

2.5.2.1 Junctions with Single Crystals

First HTSC JJs could not be processed on high-qualitative films and used one of the concepts: (i) bulk materials as electrodes (ii) are manufactured at break junctions, in which bulk part could be divided into two parts at low temperatures (iii) junctions with point contact, in which one of electrodes to be sharp tip. The latter technology applied widely in the past years for conventional Josephson junctions. Creation of first break junctions and junctions with point contact generally was checked by quality of used single crystals and always characterized by bad reproduction of natural barrier in systems with intrinsic complex properties, for instance HTSC. Subsequent significant achievements became to be possible due to technologies of studying surfaces after precipitation (or deposition) artificial barrier and counter electrode (normal metal or LTSC) which allowed one to overcome critical stage of precipitation of the artificial barrier. These approaches have been successively applied to measurement of energetic gap [611] and investigation of order symmetry in YBCO [1805].

2.5.2.2 Grain-Boundary Junctions

This type of junctions [680] demonstrates advantage to significantly decrease critical current between two grains of different orientation which forms a weak couple and behavior of Josephson type between electrodes. Natural existing barrier allows one to overcome the problems which are intrinsic to artificial barriers. These contacts in spite of pointed below shortcomings could be assumed to be high-qualitative and have been formed in a set of experiments.

The critical current on the grain boundaries decreases exponentially with increasing their disorientation angle. This could be roughly interpreted as increasing thickness of the grain-boundary barrier with increasing the disorientation angle, θ. Majority existing data relate to YBCO-transitions however similar angle dependences of critical current density on the grain boundaries have been published for all other HTSCs including electron-doped materials. The terminology used for description of the grain-boundary junctions differs asymmetric case, at which one grain is aligned crystallographically with boundary (Fig. 2.13a) and symmetric case when crystallographic misorientations of two grains relatively of the grain boundary coincide (Fig. 2.13b). For example, the 45° symmetric boundary relates to the 22.5–22.5° misorientations, at the same time, the asymmetric boundary corresponds to the 0–45° misorientations. The 90° boundaries are to be proper partial case, which, for instance takes place in *a*-axis oriented thin films.

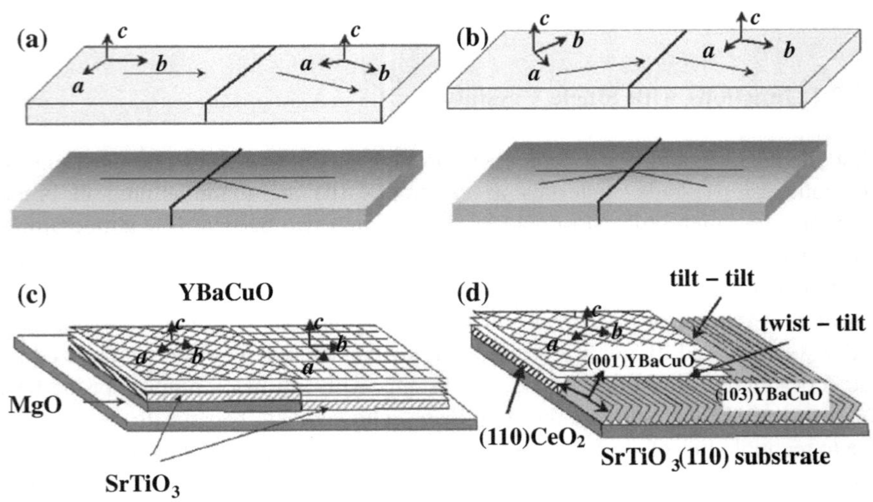

Fig. 2.13 Sketch of bicrystal junction in asymmetric (**a**) and symmetric (**b**) configuration, and of the grain boundary structures in classical biepitaxial (**c**) and CeO_2-based (**d**) out-of-plane biepitaxial junctions. In (**d**) the two limit configurations, tilt and twist, are indicated. In (**d**) the presence of the CeO_2 produces an additional 45° in-plane rotation of the YBCO axes with respect to the in-plane directions of the substrate [875]

Bicrystal technology used for processing the *bicrystal junctions* is based on unification of two substrates with different crystallographic orientations. This technology [1080] directly leads to creation of grain-boundary transitions. Epitaxial HTSC-films reproduce relative orientations of two substrates. Used turn of YBCO axes on relation to corresponding directions into substrate plane is a method of changing relative all-possible orientations of electrodes. Much less, the *c*-axis inclined grain boundaries are used. The bicrystal substrates have been processed on the base of numerous compounds, including $SrTiO_3$, doped $SrTiO_3$, MgO, zirconium, yttria-stabilized zirconia, YSZ), $NdGaO_3$, $LaAlO_3$, Si, etc. [584, 680, 762]. Other technologies (step, step-edge and biepitaxial) can overcome restrictions introduced by bi-, three- or tetracrystalline substrate in sites of the contact disposition. All other technologies apply means of lithography to state interfaces based on the grain boundaries. While they are more optimal for creation of junctions used in basic researches and more effective for designed electric circuits, however the grain boundaries based on these technologies are restricted by specific disorder angles and processing some of these junctions could be sufficiently complex.

Biepitaxial junctions use changes of orientation of HTSC-films introduced due to epitaxial growth on structured model layers. In original technology [257], model MgO layer on sapphire *r*-plane turned into plane on angle 45° to double layer of $SrTiO_3$/YBCO (Fig. 2.13c). Subsequent works used various material combinations in order to introduce alterations in orientation into plane. In the following years it has been modified biepitaxial technology for receiving new configurations [1641], in which one of the electrodes did not turn along *c*-axis. Specific feature of these structures is use (110)-oriented buffer layer of MgO or CeO_2 (Fig. 2.13d), deposited on (110) $SrTiO_3$-substrate. YBCO grew along [001]-direction on seeding layers of MgO and CeO_2.At the same time, it grew along [102]/[013]-direction on $SrTiO_3$-substrates. Presence of CeO_2 created additional 45°-turn of YBCO-axes into plane relatively of the substrate directions in the same plane. Due to this the grain boundaries were a product of two 45°-turns, first of which performed around *c*-axis, and second—around *b*-axis. This configuration led to required 45°-misorientation between two electrodes, which is necessary to increase *d*-wave effects of order parameter. Biepitaxial intergranular boundaries possess lesser probability of transport compared to other types of grain boundaries and to be nearer to the tunnel limit. Apparently, it was a key feature which led to first successive observation of angle dependence of critical current I_c in HTSC-junctions [1043], and first observation of macroscopic quantum tunneling and quantization of energetic levels in HTSC JJs [105, 106].

2.5.2.3 Step-Edge Junctions

The grain boundaries also form during growth of HTSC-film upper step duly formed inside substrate. The resulting structure depends on the step morphology. However, a boundary of one grain, as rule born on the step bottom, but other grain—on its upper part, and they couple electrically. The step-edge junctions could be disposed at any site of substrate that is found by lithography possibilities (in majority using amorphous carbon mask and ion-ray etching and reactive ion etching). The step height (200–300 nm) is usually greater than film thickness, at the same time both the step edge and substrate material plays decisive role. Better fulfillment is achieved for great angles of the step. The problem of reproduction is a serious restriction for using these junctions in various applications.

2.5.2.4 Electron Beam Junctions

As it has been noted above, the Josephson junctions also processed by using weakening superconductive properties in narrow microbrodges. With this aim, various sources of irradiation (in particular, electronic) are used. In HTSC, the irradiation creates displacement defects acting as scattering centers into Cu–O-planes and being mostly by oxygen defects for energies of order 100 keV. At the higher irradiation energy (>300 keV) could be created defects on copper. Decreasing concentration of charge carriers (due to, for instance removal of oxygen links at the diminished irradiation energies) decreases the doping level that leads to decreasing critical temperature, T_c. The irradiated barrier region during using this technology is no subjected never to air influence or fracture. Co-focused electron-beam irradiation is used also for modification of properties of the grain-boundary JJ. In this case, the situation becomes more complex due to presence of grain boundary acting as drain of defect migration defining kinetics of their accumulation. The irradiation by electrons changes volt-ampere characteristic, and apparently barrier also as the grain-boundary microstructure at the change of oxygen content in vicinity of these interfaces. These changes could be controlled altering electronic doze and partially restored by isothermal annealing of the junction. Other examples of the corrected barriers are connected with irradiation by protons and light which lead to decreasing I_c and increasing R_n similarly to the irradiation by electrons. Usually, the results are interpreted in terms of the barrier height of tunneling in dependence on physical influence.

2.5.2.5 Junctions with Artificial Barriers

While the grain-boundary junctions are only based on thin films, the junctions with artificial barriers have been processed as on the base of thin film as single crystals. It is obvious that the junctions based on single crystals have scientific interest only for basic experiments. For example, first junctions “YBCO single crystal-Pb (Nb) insulator” [611] have been rapidly replaced by second generation, based on thin films [832]. The barriers have been processed by using different methods, and as the barrier materials used noble metals and oxides. Various geometries and electrodes are used corresponding with selection of advantages and different features (greater coherence length, anisotropy, etc.). For instance, the ramp-edge junctions are manufactured by using well-developed c-axis HTSC thin-film technique allowing flow of transport current into ab-planes. The optimization of the interface electric resistance advanced a research activity directed to development of oxide barriers. The interface resistance could be caused by discrepancies of carrier density, lattice constants, thermal expansion and dimension. This has been moving force for development of strategies directed to decreasing the lattice discrepancy including agreement of the thermal expansion factors in direction of c-axis (PrBaCuO and Pr-doped YBCO), increasing the carrier density and displacement of YBCO into region of increased doping (Ca-doped YBCO), decreasing the carrier concentration at cation substitution of places in crystalline lattice far of CuO_2 planes ($YBa_2(Cu_{1-x}Co_x)_3O_{7-x}$) and substitution of Cu-atoms immediately into CuO_2 planes by using, for instance Zn or Ni [51].

2.5.2.6 Noble Metal Barriers

Gold and silver have been used as barriers for junctions based on single crystals and thin films because of their well compatibility with HTSC. Different counter-electrodes (as low-temperature superconductive as high-temperature superconductive) and configurations have been used together with HTSC as ground electrode. HTSC-junctions of S–N–S type (where S is the superconductor and N is the normal metal) in ramp-edge geometry have been processed by using special layers introduced with aim to possess proper division of superconductive electrodes and state a conductivity of the junction via barrier from gold. Focused ionic radiation has been applied to create narrow grooves, where occurred natural

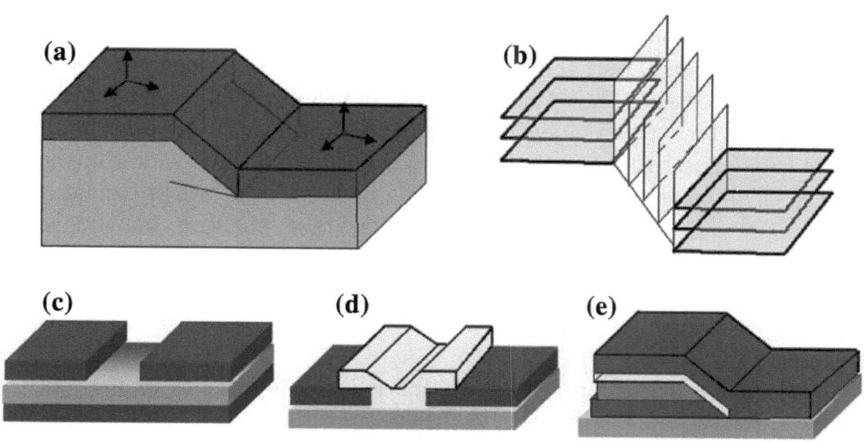

Fig. 2.14 (**a**) and (**b**) step junctions for high α angles (**c**) and (**d**) SNS coplanar junctions; in (**c**) the barrier is predeposited or occurs through a suitable substrate while in (**d**) the normal metal barrier is deposited in a narrow trench; (**e**) SNS ramp-edge junction: in improved versions a degraded layer of YBCO next to the Au is eliminated through suitable surface treatment before in situ deposition of the Au barrier and Nb counterelectrode [1585, 1587]

breaking YBCO films. The junctions with LTSC counter-electrodes are often simpler in processing than corresponding samples based on HTSC thin films however they have restricted range of work temperatures.

The processing technique of ramp-edge junctions [1587] (Fig. 2.14e) allowed one to use photolithography to process high-quality junctions. In this technique the ramp edges prepared on the base of [001]-oriented YBCO-films by using pulsed laser deposition (PLD) by using photolithography and ionic etching in argon. The devices have been etched and cleared, and then thin layers of YBCO and Au deposited by using PLD in situ. The junction processing has been finished by applying of Nb layer. This process eliminated the defective layer of YBCO sequent for Au improving the junction characteristics. This technique has been used for processing junctions with specific verges reproducing unusual magnetic interference pictures displaying in asymmetric 0–45° grain-boundary junctions and with aim to obtain great arrays of π-rings applied in π-SQUIDs.

2.5.2.7 Barriers Based on Perovskites and Layered Materials

Comparative investigations between cubic perovskite barrier materials ($CaRuO_3$, $SrRuO_3$, $La_{0.5}Sr_{0.5}CoO_3$) and layered materials ($Y_{0.7}Ca_{0.3}Ba_2Cu_3O_{7-x}$, $YBa_2Cu_{2.79}Co_{0.21}O_{7-x}$, $La_{1.4}Sr_{0.6}CuO_4$) have been carried out in junctions with ramp-edge geometry [52]. Deficiency/disorder of oxygen has been studied as a source of electric resistance of interface [256]. In the case of cubic perovskite barriers, characterized greater difference of the thermal expansion factors compared with YBCO, the greater values of resistance have been obtained in normal state of the material ($R_nA \sim 10^{-8}\ \Omega{\cdot}cm^2$) than corresponding values for layered materials ($R_nA \sim 10^{-10}\ \Omega\ cm^2$). For YBCO samples, doped by Co (Ca), being overdoped (underdoped) versions of YBCO with greater (lesser) carrier density, lesser critical temperature, T_c, and lesser (greater) anisotropy than YBCO, have been received comparative results. In [51, 354], it has been shown that the proximity effect had place for the both types of barriers. PrBaCuO and oxides based on Pr-doped YBCO ($Y_{0.3}Pr_{0.7}Ba_2Cu_3O_{7-x}$, $Y_{0.6}Pr_{0.4}Ba_2Cu_3O_{7-x}$) have been widely applied as barriers in different geometries also as Ga-doped samples which demonstrated $R_nA \sim 10^{-8} - 10^{-7}\ \Omega{\cdot}cm^2$. The barrier thickness changed into range of 6–30 nm, causing different values of critical current density, J_c. The value of I_cR_n changed linearly with the barrier thickness from 0.8 upon to 5 mV. At the same time, Ga-doped junctions demonstrated lesser sensitivity to changes of the barrier thickness.

Based on the barrier properties, the transport characteristics of the junction could be investigated. For instance, the behavior of PBCO bulk samples is defined by relatively high density of localized states, and PBCO-barriers allow ones to investigate effects of two localized states in inelastic process of tunneling [1814]. Moreover, *a*-axis YBCO–PBCO–YBCO junctions have been created on (100) $LaSrGaO_4$ substrates [1651] and on vicinal substrates (001) $LaAlO_3$–$SrAl_{0.5}Ta_{0.5}O_3$ [953] with spread of I_c and R_n equals to 11 and 8.8 %, respectively. It has been stated that Josephson behavior takes place for barriers with thickness of 80 nm and coherence length of order 20 nm.

2.5.2.8 Interface-Engineered Junctions

These junctions have thin barrier layer, as rule on ramped edges created due to damage of surface of the YBCO-ground electrode at ionic bombardment [1175]. During following deposition of counter-electrode occurs recrystallization of the surface. This technology has been applied in order to create three-layered *c*-axis YBCO-specimens [1176, 1846]. The studies of the processed high-quality interface engineered junctions with magnetic modulation of the critical current upper 80 % and critical current density $J_c = 10^2 - 10^6$ A/cm^2 at $T = 4.2$ K confirmed that they could be considered as an array of microscopic S–N–S junctions, introduced in isolating barrier of arbitrary orientation. This filamentary structure corresponded to special orientations and, in particular prevented to effects proper for *d*-wave paring symmetry.

2.5.2.9 Junctions Based on Other HTSC

HTSC which are different from YBCO family also used due to their special properties. For example, the compounds, based on Bi and Ta, with their great anisotropy are preferable for natural junctions. YBCO-junctions, doped by Ca, exploit a property of excessive doping of the sample, at the same time electron-doped compounds can display new physical aspects. Majority HTSC compounds demonstrate dominating *d*-wave symmetry of order parameter. Presence of additional sub-dominated components could be depended on material and geometry of interface.

2.5.2.10 $La_{1.85}Sr_{0.15}CuO_4$-Based Trilayer with One-Unit-Cell-Thick Barrier

Most significant step on way of processing HTSC trilayer structures with insulating barrier became creation of structure consisting of $La_{1.85}Sr_{0.15}CuO_4$ (LSCO) electrodes divided by barrier of La_2CuO_4 (LCO) with thickness of one elementary cell [167]. This achievement was to be a sequence of intensive studies of BiSrCaCuO (Bi-2212) structure in previous years [1755]. Then, it has been discovered "giant" proximity effect in LSCO-junctions [168], observing Josephson current in samples with LCO-barrier of thickness from 1 up to 15 elementary cells (upon to 20 nm), that is far greater than the coherence length and mean length of free run of electrons. This result could not be understood into framework of classical theory. The authors assumed that the superconductive current was averaged at resonance tunneling through a set of energetically ordered states inside barrier layer. These experiments used transformation of junction from S–N–S to S–I–S type due to low-temperature annealing in vacuum (that introduced an isolating state of LCO, practically preserving intactness of LSCO). Long-length or anomalous proximity effect has been also studied in another types of junctions and into framework of quantum phase transition between insulating antiferromagnet phase with low concentration of carriers and metallic or superconductive phase with high concentration of carriers [349].

2.5.2.11 Electron-Doped HTSC

Bicrystal junctions are most common type of junctions which includes electron-doped cuprate superconductors. As it has been stated in [115], the barriers in electron-doped compound $La_{1.85}Sr_{0.15}CuO_4$ possess in mean by lesser transmitted capability compared to YBCO-bicrystals demonstrating $J_c = 6{\cdot}10^3$ A/cm^2 at $T = 4.2$ K, $R_nA = 5{\cdot}10^{-8}$ Ω·cm^2 for misorientation angle of 24°, and $J_c = 3$ A/cm^2 at $T = 4.2$ K, $R_nA = 10^{-6}$ Ω·cm^2 for misorientation angle of 36°. The same technique which was used for YBCO-specimens allowed one to process zigzag-like junctions of ramp-edge type based on $Nd_{2-x}Ce_xCuO_{4-y}$ (NCCO). Then, the samples as optimally doped ($x = 0.15$) as with excessive doping ($x = 0.165$) have been prepared [55], which had double layer with thickness of 150 nm (001)-oriented NCCO and 35 nm $SrTiO_3$ (upper electrode was 160 nm Nb). As barrier have been used sub-layers of 12 nm NCCO and 12 nm Au. In the case of optimal doping were attained values of $J_c = 30$ A/cm^2 and $I_cR_n = 30$ μV at $T = 4.2$ K. Due to this the Josephson penetration depth $\lambda_J \approx 65$ μm turn out to be compared with the length of the zigzag face, and the resistance value was $R_n \approx 10^{-6}$ Ω·cm^2. By this, there were observed anomalous controlled magnetic images which witnessed on dominating *d*-wave symmetry without any alterations of *s*-wave at low temperatures.

2.5.2.12 Ca- and Co-doped YBCO and Overdoped Regime

As it has been shown in [627], an increasing critical current in bicrystal junctions was attained by using excessive doping of superconductor, in particular due to Ca- and Co-doping of YBCO-samples [331]. The latter samples of HTSC-junctions demonstrated best results in increasing J_c. The critical current has been studied in dependence on concentration of Ca demonstrating optimal doping at $x = 0.3$. Then, it has been defined $J_c = 7{\cdot}10^6$ A/cm^2 at $T = 4.2$ K for 24°-misorientation angle of grain boundary [1501], that is almost on order of magnitude greater than in the case of absence of the doping.

2.5.2.13 Ultra-Thin Films and Superlattices

Josephson junctions consisting only of several superconductive planes of CuO_2 (in particular, of 6 layers) have been processed on the base of ultra-thin $[Ba_{0.9}Nd_{0.1}CuO_{2+x}]_5/[CaCuO_2]_2/[Ba_{0.9}Nd_{0.1}CuO_{2+x}]_5/[CaCuO_2]_2/[Ba_{0.9}Nd_{0.1}CuO_{2+x}]_5$ (CBCO) structures 5/2/5/2/5). By this, the CBCO-film had thickness of only 8 nm. The Josephson effect has been measured even in junctions with wide of 5 mm [1640].

2.5.2.14 Intrinsic Stacked Junctions

Strongly anisotropic layered crystallite structure of cuprates allows creation of intrinsic stacked junctions, which are processed from bulk single crystals (see Fig. 2.15). Superconducting current in these junctions flows along of *c*-axis. Best results have been attained for single crystals and thin films of $Bi_2Sr_2CaCu_2O_8$ [887] and $Tl_2Ba_2Ca_2Cu_3O_{10}$ [888]. In *a*-axis oriented thin films of YBCO, it has been stated the Josephson pairing between double layers of CuO_2, and majority of materials led themselves similar to the stacked S–I–S Josephson junctions with effective barriers of order of the space thickness between double layers of CuO_2 (1.5 nm). By this typical value of critical current was 10^3 A/cm^2. Volt-ampere characteristics (*I*–*V*) demonstrated great hysteresis and numerous branches corresponding to subsequent link of high-qualitative junctions. Then, the practical recommendations on processing the intrinsic stacked junctions have been developed with aim to total elimination of heating effects [1859]. However, nevertheless it is necessary a carefulness by analyzing volt-ampere characteristics at high voltage because of possible keeping the heating effects. Finally, it has been confirmed that macroscopic quantum tunneling takes place in natural junctions of BSCCO [756, 786].

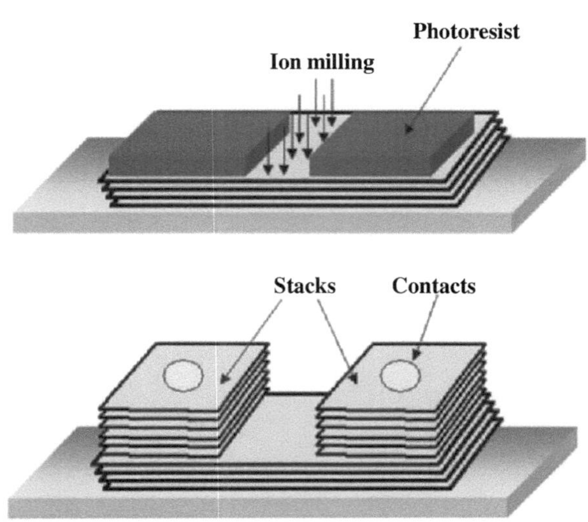

Fig. 2.15 An example of intrinsic stacked junctions. The mesas with *c*-axis transport are defined through mechanical etching [875]

2.5.2.15 Oxide Heterostructures

Development of technology of the growing multi-layer structures with atomic sharp boundary between the layers (for example by using molecular-beam epitaxy) allows one to create artificial material with advanced material demonstrating often very unusual properties. Especially great hopes are connected with oxides. For instance, while $SrTiO_3$ and $LaAlO_3$ are dielectrics, the interface boundaries in the $SrTiO_3/LaAlO_3/SrTiO_3$ heterostructure to be conductive, more that on the boundary of two insulating oxides has been discovered superconductivity [1427]. The characteristics of interfaces between cardinally different on their physical properties matters that as high-temperature superconductor $(Y,Ca)Ba_2Cu_3O_7$ and manganite $La_{2/3}Ca_{1/3}MnO_3$ with colossal magneto resistance have been studied experimentally and theoretically in [245]. It has been shown that due to the electric charge pass through interface, $d(3z^2 - r^2)$-orbitals of CuO_2 layers (which are totally filled in HTSC and do not take part in superconductivity) become to be filled only partially that is on them arise holes. This orbital reconstruction differs of usual electronic reconstruction which occurs on interfaces of other oxide compounds and consists in that what a distribution of electronic density near the interface differs of a volume distribution. This effect could be used to doping dielectric layers "no-chemical" method without effects of disordering, which are inevitably introduced non-isovalent admixture atoms. The practically important applications of this effect are connected, for example with magnetic tunnel junctions and "oxide electronics", one of directions of which to be R&D novel quasi-two-dimensional states (charge, spin, etc.), which in partially could be stabilized on the oxide interfaces.

Chapter 3
Composition Features and HTSC Preparation Techniques

Microstructure, strength and other HTSC properties are defined by the existence of numerous components. This circumstance supposes different chemical, physical and mechanical influences during numerous technological operations to prepare final sample from initial powders. Super-sensitiveness of HTSC final properties to the technical conditions, their manufacture and to composition features, and also numerous effects have determined various ways of the oxide superconductor preparation. HTSC samples in the forms of films at the mono- and polycrystalline substrates, coated conductors, tapes and superconducting bulks are the most interesting for applications. Their preparation techniques are considered in this chapter.

3.1 YBCO Films and Coated Conductors

In general, in order to synthesize HTSC films (as YBCO family as another ones) *in situ* and *ex situ* methods are used. In the first case, the film crystallization takes place directly during their deposition and an epitaxial growth occurs under corresponding conditions. In the second case, the films are deposited initially under low temperature, that is insufficient to form necessary crystalline structure, and then the films are sintered in O_2 atmosphere that leads to the crystallization of the necessary phase (e.g., this sintering temperature is equal to 900–950 °C for YBCO films). Most one-stage methods are realized under thermal treatment that is much worse compared with two-stage methods. The high-temperature annealing forms large crystallites and rough surfaces, defining small critical current density. Therefore, the *in situ* methods have advantages originally. According to the preparation methods and the deposition of HTSC components on a substrate, there are physical methods of deposition, including all possible evaporations and scatterings, and also the chemical methods of precipitation.

Methods of vacuum co-evaporation. These assume simultaneous or successive (layer-by-layer) co-precipitation of HTSC components evaporated from different sources by using, for example, electron beam guns or resistive evaporators.

I. A. Parinov, *Microstructure and Properties of High-Temperature Superconductors*,
DOI: 10.1007/978-3-642-34441-1_3,

The films, prepared by this technique, yield their superconductive properties to the samples, manufactured by methods of laser evaporation or magnetron scattering. Methods of vacuum co-evaporation are used in two-stage synthesis, when the structure of films, scattered in first stage has no principle consideration, as also the oxygen contents in them.

Laser evaporation. This is highly effective in the HTSC thin-film deposition. This method is simple in realization, demonstrates high rate of deposition and permits contact with small targets. Its main advantage is the evaporation, equally well, of all chemical elements contained in the target [833]. The films, having the same quality as the targets, may be prepared in the target evaporation under concrete conditions. The distance between target and substrate and also the oxygen pressure are the important technical parameters. Their right selection allows, on one hand, non- overheating of the growing film by energy of plasma, evaporated by laser, and accompanying formation of very big grains, but, on the other hand, to state an energetic regime that is necessary for film growth at perhaps very low temperatures of substrate. The high energy of the components deposited and existence in the laser flame of monatomic and ionized oxygen permit the preparation of HTSC thin films in one stage. In this case, the films are mono-crystalline or possess high texture with *c*-axis orientation (the *c*-axis is perpendicular to the substrate plane). The main disadvantages of the laser evaporation are the following: (i) small region in which stoichiometric films could be deposited, (ii) heterogeneity of their thickness and (iii) surface roughness. By using the radiation methods in film preparation, the interesting dependencies maybe stated, for example, between the degree of *a*-axis orientation of YBCO film,[1] the substrate temperature and the material, and also the deposition rate. These dependencies for the ion beam sputtering method in deposition of CeO_2 buffer layer are presented in Figs. 3.1 and 3.2. Obviously, by using the sapphire substrate in the preparation of the *a*-axis-oriented YBCO thin film higher rates of deposition are necessary. Moreover, film surfaces with smaller roughness are obtained on sapphire compared with $SrTiO_3$, defining its better superconducting properties.

Magnetron scattering. This permits to obtain in one stage YBCO films, not yielding their superconducting properties to the samples, deposited by the laser evaporation method. Moreover, they have more homogeneous thickness and higher smoothness of surface. As at the laser evaporation, the plasma formation at magnetron scattering creates high-energetic atoms and ions that permit to obtain HTSC films in one stage at not high temperatures. Here, the distance of "target–substrate" is also important. At small distance and insufficient pressure of environment, the substrate is subjected to intensive bombardment by the negative ions of oxygen that

[1] Due to the high anisotropy of HTSC, films with *c*-axis orientation only have good transport and screening properties. At the same time, films with *a*-axis orientation, possessing greater coherence length in direction that is perpendicular to the surface and distinguished by high smoothness, could be convenient to prepare qualitative HTSC Josephson junctions, consisting of successively deposited "HTSC-normal metal" (or "dielectric–HTSC") layers. Films demonstrating mixed orientation are not desirable in all cases.

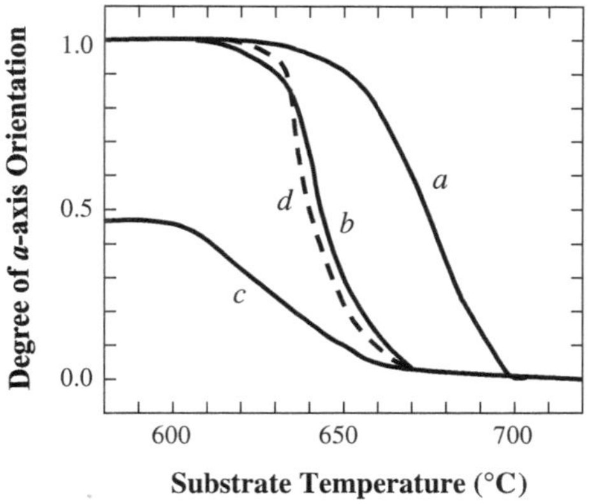

Fig. 3.1 Dependence of degree of *a*-axis orientation of YBCO film upon the substrate temperature. *Solid line* (**a**) Indicates deposition rate of 20 Å/min on sapphire; (**b**) For 10 Å/min; (**c**) For 5 Å/min. *Dashed line* (**d**) is for 5 Å/min on $SrTiO_3$ [780]

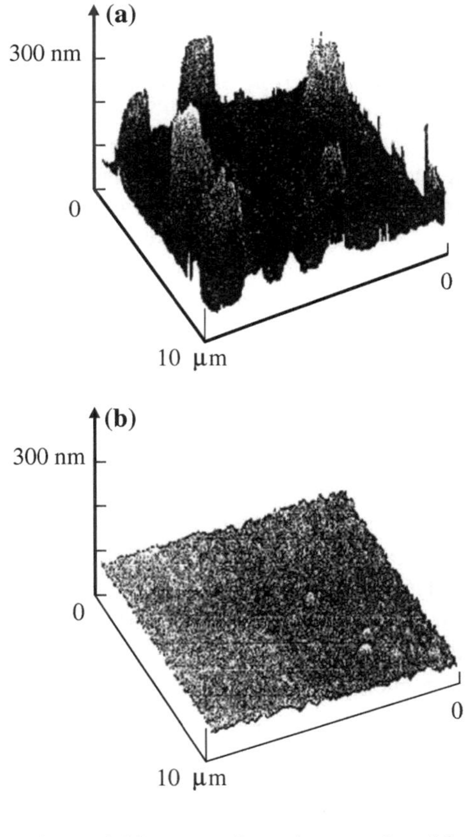

Fig. 3.2 Atomic force microscopy images of YBCO films deposited on (**a**) $SrTiO_3$and (**b**) sapphire substrate with CeO_2 buffer layers. Sputter conditions for the films are identical [558]

fracture the structure of growing film and its stoichiometry. In order to solve this problem, the set approaches are used [332], including the protection of the substrate from the bombardment of energetic ions and its location at optimum distance from gas-discharge plasma to ensure high rate of deposition and successive growth of the film at the maximum low temperatures.

In situ YBCO thin films prepared by out-axis magnetron scattering and possessing optimum electric properties have demonstrated $T_c = 92$ K and $J_c = 7 \times 10^6$ A/cm^2; and $T_c = 89$ K and $J_c = 2 \times 10^6$ A/cm^2 at optimum smoothness of surface [396, 397].[2]

Chemical precipitation from vaporous phase of metal-organic compounds. The core of the method is the transportation of metallic components in the form of steams of the volatile metal-organic compounds in a reactor, a mixing with gaseous oxidizer, decomposition of the steams and condensation of the oxide film on substrate. This method permits to obtain HTSC thin films possessing parameters, which are compared with samples prepared by the physical methods of deposition. The comparative advantages of the method considered to last ones are the following: (i) a possibility to obtain homogeneous films on details with non-plane configuration and possessing a great area; (ii) more high rates of precipitation under conditions of high quality; (iii) a flexibility of process at the preliminary stage of technological regime due to the smooth change of the vaporous phase composition.

Epitaxially grown thin films. Great progress has been achieved in *atomic-layer engineering* (ALE) of HTSC, that is in thin-film techniques in which one material by controlled way is deposited or precipitated above other material and by this at least one of the layers has thickness of several atoms. One of the main technique is the method of *molecular beam epitaxy* (MBE), consisting in simultaneous co-evaporation of several source of elements (by using thermal effusion cells or electron-beam guns) that allows one to obtain required atoms (for example, Y, Ba and Cu) in correct interrelation (for instance, 1:2:3). In this case, it is necessary a precise control method of relative rates of deposition for the elements used. In contrary, the atomic-layer engineering method is based on exact knowledge of absolute deposition rates which are much more difficult for measuring [166].

There are numerous methods based on epitaxy in interface engineering. The control of crystallographic properties in an epitaxial process is carried out by different methods. During c-axis growth an in-plane orientation could be changed, in particular by using different buffer layers (sub-layers). When lattice constants of film and substrate almost coincide (for example, in the case of $YBa_2Cu_3O_7$ on $SrTiO_3$ [780]), then it may be prevailing growth of cube-on-cube type. At the same time, in the case of strong mismatch of the lattice constants (for example, in the case of $YBa_2Cu_3O_7$ on MgO [1646]) the film axes are twisted in order to minimize free energy. On the other hand, the vicinal substrates are used, that is they are specially disoriented on small angle from ideal orientation of crystallographic axes. In particular, this allows one to obtain twin-less (single-domain) Bi-2212 films [166]. Other specific substrates are used for controlling film orientation out-of-plane that is compulsion HTSC-film growth with a-axes normal to the substrate

[2] The varieties of impulse laser deposition, used to obtain YBCO films and coated conductors with high texture on the different mono- and polycrystalline substrates with and without buffer layers, permit to reach the critical current density $J_c = 2.4$ MA/cm^2 at 77 K and zero magnetic field [463].

surface or inclined under definite angle [107]. There are attempts to use epitaxial strains [166, 436]. In the case of super-lattices (Nd, Ce)CuO_4/$YBa_2Cu_3O_7$ the strain of $YBa_2Cu_3O_7$ is tensile which accompanied by decreasing critical temperature, T_c. At the same time, in $La_{1-x}Sr_xCuO_4$ grown on $LaSrAlO_4$, there is a compression strain leading to increasing T_c.

It may be assumed, that film growth, at least partially is controlled by kinetics of adatoms. For ALE, ideal case corresponds to thermodynamic parameters (temperature and pressure) for which exists great surface mobility, that the atoms are well-ordered into strongly crystallite structures, and volume mobility is small restricting mixing layers. By acting into processing boundaries, there may be combined molecular layers of two or more stable compounds (for example Bi-2201 and Bi-2212) with formation of multilayer systems or superlattices. In ideal case, two material should be compatible chemically (this means, that there are not chemical reactions and interdiffusion of atoms), crystallographically (so, there exists good epitaxy) and electronically (so, there excludes electron accumulation or depletion which could eliminate superconductivity in HTSC-layer).

Record T_c has been attained in Hg-1223 under conditions of high pressure [1184]. It may be assumed that the same effect may be realized in very thin films subjected compressive epitaxial strain. This could be carried out by careful selection of substrate (or corresponding element of superlattice) with in-plane lattice constant which is lesser than for Hg-1223 [166].

The following techniques are as the most widespread ones:

- preparation of films and coated conductors (CC) with two-axes texture by using the *ion-beam-assisted-deposition* (IBAD) [740, 1811];
- preparation of films and coated conductors by using the *rolling-assisted biaxially textured substrates* (RABITS) [563, 564, 1249, 1848];
- preparation of two-axes ordered buffer layers by using the *inclined substrate-pulsed laser deposition* (ISPLD) [642];
- preparation of thin films by the *magnetron sputtering* [1653];
- preparation of buffer layers by the *electron beam evaporation* [1104], *laser ablation* [1051, 1444], *ion beam sputtering* [558,780] and *rf-sputtering* [277, 1024, 1141];
- preparation of thick epitaxially grown films by the *liquid phase epitaxy* (LPE) [1820];
- preparation of films by the *electrophoretic deposition* [1157];
- preparation of buffer layers by the *surface oxidation epitaxy* (SOE) [1102];
- preparation of coated conductors by using the *metal-organics decomposition* (MOD) [1567];
- preparation of thick films on the base of precursor films by the *metal-organic chemical vapor deposition* (MOCVD) [441].

As substrates in various techniques, the Ni-based alloys (Inconel, Hastelloy, Ni–Cr, Ni–V, etc.), metals (Ni, Ag, Zr), oxides (Al_2O_3, $SrTiO_3$, $NdGaO_3$, $LaAlO_3$, MgO, PrO_2) and non-metals (Si, glass) have been used. As buffer layers, YSZ, CeO_2, MgO, $SrTiO_3$, ZrO_2, $BaZrO_3$ and NiO have been used.

As it has been noted, the critical current could be enhanced due to the formation of artificial defects in the sample, playing role of the magnetic flux pinning centers. With this aim, the defects with linear sizes, l, which are near to the superconducting coherence length (e.g., $l = 2$–4 nm in Y-123 at $T < 77$ K), and density of order of $(H/2) \times 10^{11}$ cm^{-2}, where H is in Tesla, are the best of all. This high density of defects with nanometer sizes is reached with difficulty at the stage of sample fabrication; therefore the ionic irradiation is usually used for this goal. In [653], a method has been proposed for fabrication of Y-123 films, consisting of non-superconducting Y-211 particles with $l \sim 8$ nm, density of which was larger than 10^{11} cm^{-2}. This method consists of alternative deposition of the Y-123 and Y-211 layers, and further formation of the Y-211 "islands" (because of mismatches of the lattice periods of 2–7 % for Y-123 and Y-211 phases). In this case, the thickness of Y-123 layers and Y-211 layers was equal to $\sim$10 and $\sim$1 nm, respectively, but Y-211 nanoparticles were distributed uniformly into Y-123 matrix. These films have much more larger current density, compared to Y-123 films without defects, moreover, one decreases much slower at the enhancement of magnetic field.

From the view of material strength, the main aim is to select neighboring layers with like thermal and crystallographic properties, and also to prevent chemical reaction between them. It is necessary to note, that the critical current density of the film on polycrystalline substrate diminishes considerably at the existence of high-angle intergranular boundaries [375], causing the problem of weak links. Therefore, for tape applications, it is very important to prepare films with highly in-plane aligned crystalline structure, that is, to form the ordered structure not only along the c-axis, but along the a-direction. As a rule, thick YBCO-coated conductors are obtained layer-by-layer (metal substrate–buffer layer–superconducting layer), see Fig. 3.3. Every layer including metallic substrate plays a special role in the long length conductors. The candidate materials and the thickness of each layer are also presented in Fig. 3.3.[3]

In general, thick YBCO-coated conductors consist of epitaxially grown films on appropriate substrate with or without few buffer layers. Therefore, the substrate or the buffer layer must have a suitable textured surface structure to provide the required in-plane alignment growth of the HTSC crystalline films, resulting in the avoidance of weak links due to the existence of only low-angle grain boundaries. Typical problems, decided now at design and manufacture of coated conductors, are presented in the Table 3.1.

So, second generation tapes are based on the coated conductor methodology, where a multilayered structure is generated on top of a flexible thin metallic substrates. The methodology gives a long length coating free of high-angle grain boundaries allowing one to attain highest dissipation-less currents even at high temperature and under high magnetic fields. The most competitive techniques for

---

[3] For simplicity and clearness, a thin passivation layer for stabilization, insulation and encapsulation is excluded.

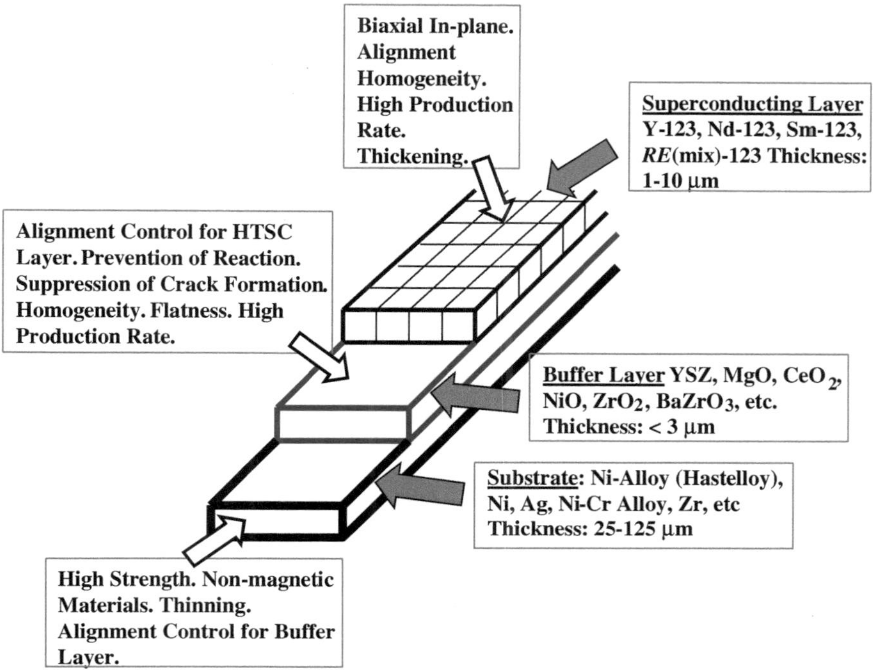

Fig. 3.3 The problems, solved in the design of high-qualitative-coated conductors and including the development of basic structure, estimation of key factors and use of candidate materials for each layer [1567]

the processing of textured metallic substrate for high current CCs are the following:

- A high strength polycrystalline substrate, such as stainless steel or Hastelloy, coated with a biaxially textured oxide by using IBAD. For the second generation tapes with a textured buffer layer processed by IBAD the values of critical current density from 200 up to 500 A/cm of width have been attained [1531].
- A softer metallic substrate such as NiW is textured by thermo-mechanical processing (RABIT). Second generation tapes with $J_c > 1$ MA/cm^2 and $I_c \sim 130$ A with the length of 8 m have been processed by RABIT [1531].

The processing of high performance and industrially competitive HTSC tapes requires the use of cost-effective deposition and growth techniques combined with the simplest conductor architectures possible for achieving the lowest costs and to accelerate implementation of a competitive processing. The CCs demonstrating now the highest performance are still based on very complex buffer layer architectures (3–5 various oxide layers). Radical advances in: (i) simplifying the architecture by using more robust buffer layers, and (ii) deposition methods, in particular via chemical methods have led to very significant advances in the development of IBAD and RABIT based CCs metallic substrates.

Table 3.1 Nearest goals of R&D for three different types of YBCO coated conductors [1567]

Type	Structure or technique features	Substrate	Buffer layer	HTSC layer	Key processes	Target
Textured substrate	Non-reactive high strength	Ni, Ag. Clad materials: Ni–Cr, Ni–V, Ni-base Alloys, etc.	None or NiO, ZrO_2, $BaZrO_3$, MgO, YSZ, Y_2O_3, CeO_2 etc.	Y-123 *RE*-123 etc.	Rolling/annealing. Surface polishing. Surface-oxidation epitaxy. Buffer layer. HTSC layer. Evaluation (J_c, etc.).	Length: 10–100 m Substrate thickness: ≤100 μm. $J_c \geq 10^5$–10^6 A/cm^2 (77 K)
Aligned buffer layer	ISPLD IBAD	Polycrys-talline hastelloy, Ni-base alloys, etc.	YSZ, MgO, CeO_2, etc.	Y-123 *RE*-123 etc.	Substrate polishing. IBAD process. ISPLD process. HTSC layer. Evaluation (J_c, etc.).	Length: 100–1000 m Substrate thickness: ≤100 μm $J_c \geq 10^4$–10^5 A/cm^2 (77 K) Production rate: >1 m/h
Rapidly grown HTSC layer	MOD LPE	Ni, Ag. Clad materials: Ni-base alloys, etc.	None or MgO, YSZ, NiO, $BaZrO_3$, CeO_2, etc.	Y-123 *RE*-123 etc.	Substrate polishing. Buffer layer. Homogeneous Seed film. MOD process. LPE process. Evaluation (J_c, etc.).	Length: 1–10 m Substrate thickness: ≤100 μm HTSC thickness: ≥5 μm. $J_c \geq 10^5$–10^6 A/cm^2 (77 K) Production rate: >1 m/h

The key critical parameters for electrotechnical applications of superconductors are the critical current density, J_c, the critical current, I_c (especially in the case of CCs because the thickness of the superconducting film only represents a small fraction of dimensions of the conductor) and the engineering critical current, j_e.

The main difficulty in processing high performance CCs consists in that J_c strongly decreases with film thickness, temperature and applied magnetic field due to Lorentz forces and thermal fluctuations. Moreover, the intrinsic crystalline anisotropy of YBCO needs to be surrounded through defect engineering to minimize the dependence of J_c on the magnetic field orientation. The best values attained up to now are 200–300 A/cm of width at 77 K and self-field in kilometer length, while in shorter lengths values as high as 1 kA/cm of width have been manufactured by using vacuum deposition techniques [1065].

Strong artificial pinning centers in addition to natural ones are required to control the magnetic flux dynamics. Different strategies for introducing artificial pinning centers have been studied, including substrate decoration [317], quasi-multilayers [653] and secondary phase distribution [1065]. These techniques are efficient for very thin films, but the critical current decreases sharply beyond a few hundred nanometers thickness. In the case of substrate decoration, Ag nano-dots were used for the first time for Tl-based superconducting films [314, 315, 317] then for YBCO [759], followed by Y_2O_3 [1101, 1124], Ce_2O_3 [1245], Ir [71], Au [1145], and $LaNiO_3$ [318].

Recent discovery of epitaxial nano-composites, where the secondary phase is either randomly distributed and generates nano-scale strain into HTSC matrix or is self-organized as nano-columns opened new possibilities for creation very effective magnetic pinning structures [613]. Very high vortex pinning properties have been demonstrated in these materials which smoothed the film thickness and field dependence of J_c and decreased the intrinsic anisotropy. At the same time, it has not be possible to sort out unambiguously the nano-structural source of the enhanced performance and thus the optimal compositions and required interfacial stresses are still to be unknown.

It maybe assumed that in thick films, pinning due to inclusions of secondary phases becomes less effective than in thin films because dislocations created by the inclusions cannot extend through the whole thickness of the films and may not reach the surface. Therefore, one of the possibilities of enhanced critical current densities in thick $YBa_2Cu_3O_{7-\delta}$ could be to grow multi-layer structures with defect centers in the interlayers. The nano-dot materials used to date in inter-layers include Y_2BaCuO_5 [653], Y_2O_3 [94, 217], yttria stabilized zirconia (YSZ) [218], $BaTMO_3$ (where transition metal TM = Ir, Ti, Zr, Hf) [78], various transition metal elements [632], noble metals (Au, Ag) [1144], Pd [1479], non-superconducting $YBa_2Cu_3O_x$ nano-dots [1478] and $PrBa_2Cu_3O_x$ [316].

So, the novel nano-dots/quasi-multilayer approach allows one to increase the thickness of films by retaining high J_c values and so significantly increasing I_c and j_e. It has been stated, that temperature, magnetic field intensity and orientation conditions under which significant improvements are obtained by using this approach strongly depend on the material type under consideration. This is due to the fact that different types, dimensionality and orientation of defects may be created in various compounds. For different electrotechnical devices, it is necessary to have high critical currents for specific field orientations, field intensities and/or temperatures. This led to necessity of an integrated nanotechnology of pinning centers that should combine several types of nano-dot materials. So, in future the nano-structured superconductive materials will contain 1D, 2D, and 3D defects with different orientations to design conductors for various applications. Into framework of the technique, combining the catalytic effect of noble metals nano-dots with nano-crystalline targets, containing different types of nano-inclusions, a 5 μm-thick Ag ($BaZrO_3$-doped YBCO) quasi-multilayer has demonstrated $J_c = 170$ A/cm of width at 77 K in 1T, and 60 A/cm of width at 77 K in 2T [1143].

The general problem of films prepared by the liquid phase epitaxy and electrophoretic deposition, demonstrating high rates of precipitation and sufficiently low cost, is the crack formation due to stress relaxation during sample cooling and different thermal properties of buffer layer and film. If we assume that fixed cracks are extended through the whole film thickness, we may follow a calculation for cracks in brittle films on elastic substrates [1674, 1677].

According to the model, the total energy is the sum of the strain energy in the cracked region and the energy necessary to create the cracks, that is, the new surface energy. The crack spacing, l, will be adjusted in order to minimize the total energy of the system to [1677].

$$l \approx 5.6\sqrt{K_{Ic}^2 h/(E\varepsilon)^2}, \tag{3.1}$$

where K_{Ic} is the fracture toughness of the mode I, h the film thickness, E the Young's modulus and ε the strain. The total energy of the cracked film cannot exceed the strain energy of the uncracked film. This result in minimum crack spacing and is consistent with the requirement that the energy release by cracking should be higher than the energy necessary to create the cracks.

This means that lower a critical thickness

$$h_c = 0.5K_{Ic}^2/(E\varepsilon)^2, \tag{3.2}$$

no cracks will appear.

Based on this approach, the predicted dependence of the average crack spacing l on the thickness h of a-axis-oriented YBCO films on (100)$SrPrGaO_4$ substrates is shown in Fig. 3.4. At the same time, the decreasing micron-sized particles to submicron-sized particles in colloids may diminish the film cracking at its electrophoretic deposition considerably (see Fig. 3.5).

In order to obtain YBCO samples with high properties, a preparation of qualitative precursor powders is also very important. A number of techniques for processing YBCO powders have been reported, including conventional solid-state reaction, precipitation [206], plasma spray [673], freeze drying [791], spray drying, combustion synthesis [927], sol–gel method [1185], acetate method [1114] and flame synthesis [1438].

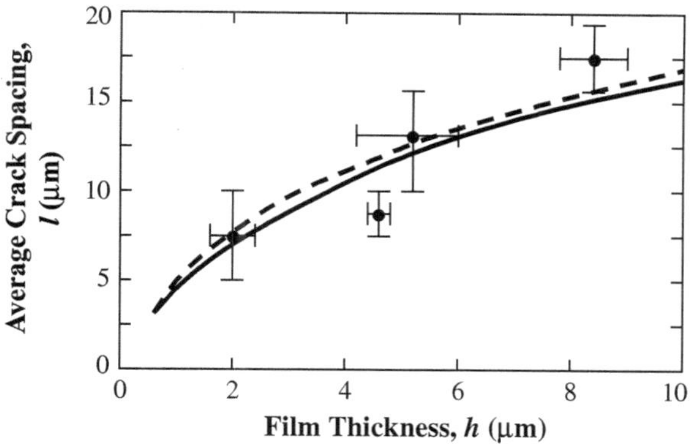

Fig. 3.4 Theoretical crack spacing (*solid curve*) as a function of the film thickness compared with the experimental data and a fit (*dashed curve*) to the data obtained from *a*-axis-oriented YBCO LPE films grown on (100) $SrPrGaO_4$ [6]

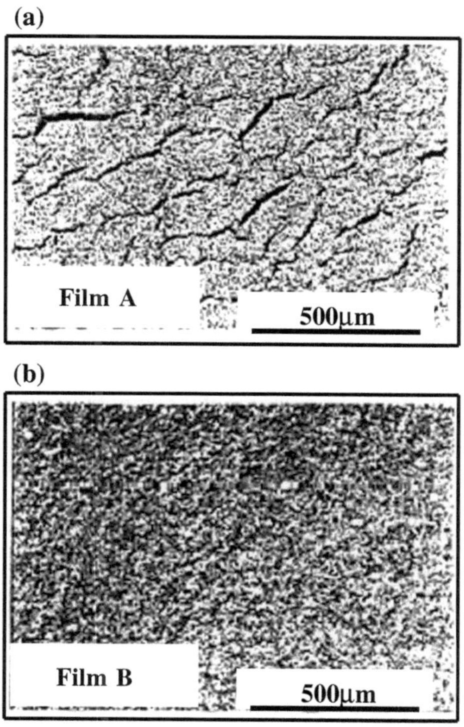

Fig. 3.5 Scanning electron microscopy images of sintered films. The images of (**a**) and (**b**) show films fabricated from colloids, consisting of micron-sized particles and sub-micron-sized particles, respectively. The heat treatment conditions for both are the same [1486]

The fabrication of a ceramic superconducting powder is presented in the flow chart of Fig. 3.6. First, the raw materials are weighed in a desired molar ratio and then mixed by conventional mixing/milling or liquid-solution mixing. The homogeneity, obtained by conventional mixing, is limited by the particle size of the powders, but the best mixing is generally obtained from powders with a particle size <1 μm. For ultra-fine powders (particle size much smaller than

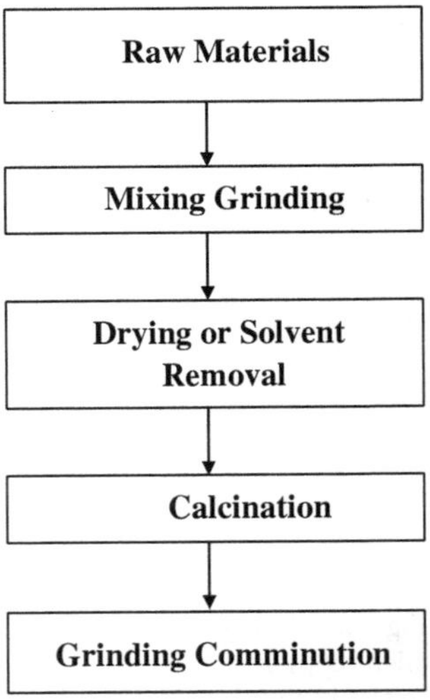

Fig. 3.6 A flow chart for the synthesis of ceramic superconducting powders

1 μm), the particles tend to segregate, thus resulting in poor mixing. This problem may be minimized by liquid-solution mixing, ensuring the precise compositional control and molecular-level chemical homogeneity. Moreover, this technique eliminates contamination from grinding and milling media that occurs during a conventional mix/milling process. For a multi-component system like HTSC, the mixing plays a key role in obtaining high phase purity. Better mixing also translates into faster reaction kinetics.

These powders can be calcined at lower temperatures and/or shorter time to achieve the desired phase purity. The next step is drying or solvent removal that is necessary to preserve the chemical homogeneity obtained by mixing. For multi-component systems, solvent removal by slow evaporation can lead to a very inhomogeneous residue due to the difference in solubility of various components. In order to minimize this effect, various techniques are used, including, in particular, the processes of filtration, sublimation, etc. [1538].

After drying, the powders are calcined in a controlled atmosphere for reaction until reaching a final composition and phase assemblage. The reaction path for HTSC systems depends on various processing parameters, such as calcination temperature and time, heating rate, atmosphere (oxygen partial pressure) and starting phases. The powders can also be synthesized directly from solution using pyrolysis techniques [1438] or an electro-deposition technique [132].

The phase diagram, presented for $YBa_2Cu_3O_x$ in Fig. 3.7, shows that even small fluctuations of compositions can lead to the formation of normal (non-superconducting) phases, namely Y_2BaCuO_5, CuO and $BaCuO_2$. An application of precursors with carbon content also complicates the formation of the $YBa_2Cu_3O_x$ phase and may lead to diminishing superconducting properties. In detail, the carbon problem will be considered in Chap. 5.

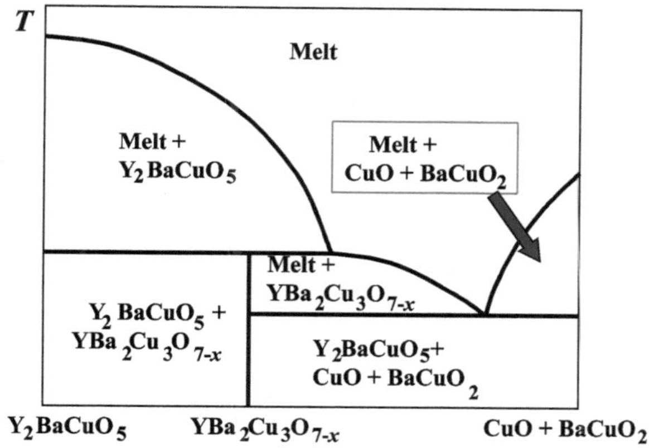

Fig. 3.7 A phase diagram of Y_2O_3-CuO-BaO system [545]

3.2 BSCCO Films, Tapes and Wires

Now for the preparation of Bi-2212 thick films on Ag and MgO substrates, the set of techniques are applied, namely melt-processing [662, 664], electrophoretic deposition [89], doctor-bladed [829], dip-coated [1684] and organic precursor [644] films. Last three procedures are presented in Fig. 3.8. It should be noted that the fabrication of Bi-2223, possessing higher superconducting properties, by using melt processing is impossible because of PbO loss in high-temperature process [441]. The powder for doctor-bladed and dip-coated films can be made using any of the processes listed in Table 3.2.

In the *doctor-bladed process*, one makes a "green" film (i.e., film before heat treatment) from the organic/powder mixture by pouring a pool of the slurry on a flat surface (e.g., a piece of glass), then leveling the slurry with a straight-edged blade, located above the flat surface at the desired film thickness (this allows to carry out corresponding control during its preparation). The film is dried, cut into strips that are placed on the silver foil and finally melt processed.

For the *dip-coated films*, the silver foil is passed through the organic/powder mixture, which adheres to the foil. The thickness of the film is controlled, changing the organic compounds, modifying their proportions in the mixture and adjusting the solids loading in the mixture. After passing through the organic/powder mixture, the coating is dried, the organics are burned out and the film is melt processed.

For the *organic precursor films*, a solution of organometallic compounds of Bi, Sr, Ca and Cu is deposited on the Ag foil, the solvent is burned out and the process is repeated until the desired layer thickness is built up. Finally, the film is melt processed. A Bi-2212 film can also be made, painting Bi-2212 powder onto a silver foil. Here, the powder is mixed with an organic liquid having a high vapor pressure (e.g., butanol). This slurry is brushed onto the foil and finally melt-processed.

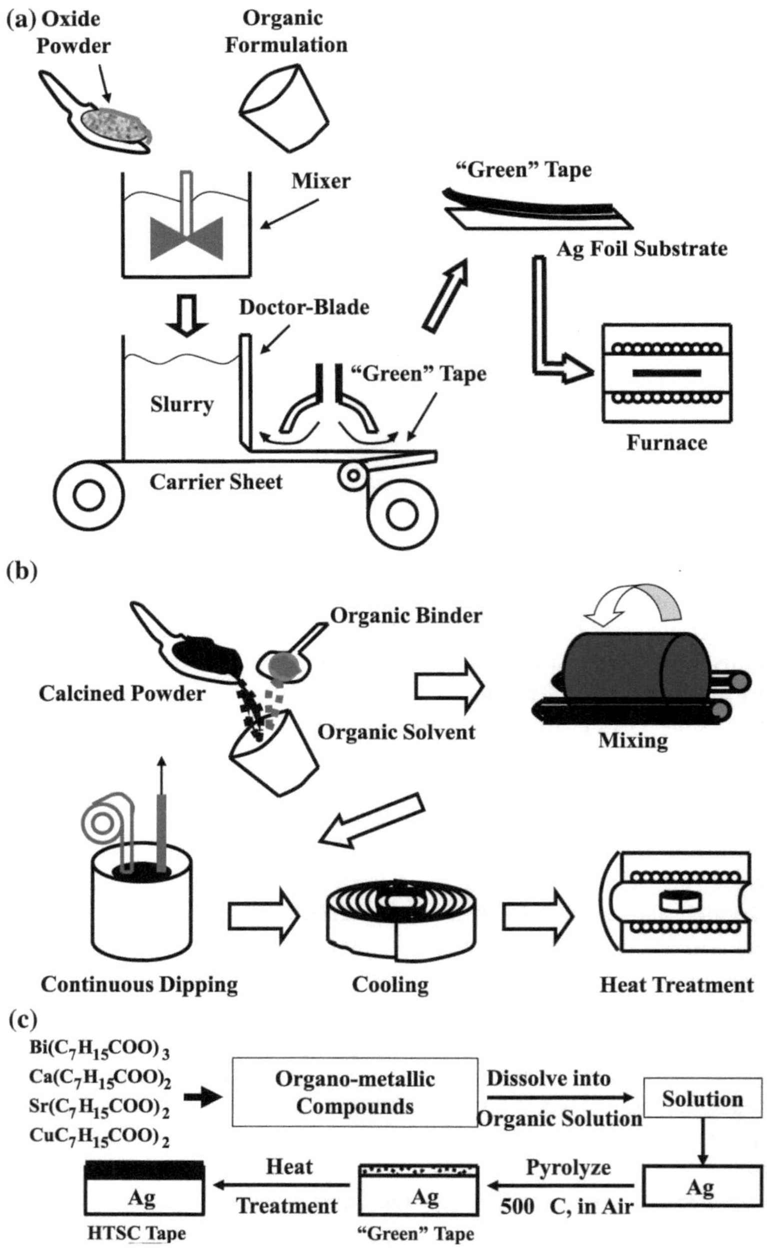

Fig. 3.8 Schematic diagram to make (**a**) doctor-bladed, (**b**) dip-coated and (**c**) organic precursor films [644]

Table 3.2 Synthesis techniques to make Bi-2212 and Bi-2223 powders

Method	Description	Advantages	Disadvantages
Solid state reaction [666]	Mix oxides, peroxides, carbonates or nitrates of Bi, (Pb), Sr, Ca and Cu. React at elevated temperature where no melting occurs. Grind sample and refire. Repeat until reaction is complete	Simple, inexpensive technique	Large grain size of reactants can cause slow reactions. Product can have large grain size. Impurities can introduce during grinding
Co-precipitation [666]	Dissolve Bi, (Pb), Sr, Ca and Cu compounds in acid. Add base to precipitate cations. Fire precipitate to yield the desired phase	Intimate mixing of cations	Not all cations may precipitate out at the same rate, causing segregation. Initial composition and precipitate composition may be different
Aerosol Spray Pyrolysis [666]	Make solution, containing cations. Produce fine mist of the solution and pass it through a hot furnace to form a powder of mixed oxides. Fire mixed powder to yield the desired phase	Intimate mixing of cations. Product has very fine grain size (1–2 μm). Product can have low carbon content (using nitrates)	Species can lose, partially Pb, during pyrolysis. Powder, formed in pyrolysis is not fully reacted to the desired phase
Burn technique [9]	Form nitrate solution of cations. Add organic species, such as sugar, to solution. Heat solution to remove water then heat powder at elevated temperature. The sugar (fuel) and nitrate ion oxidant react (i.e. burn) at elevated temperature, yielding a high temperature that forms mixed oxides. Fire this powder to yield the desired phase	Intimate mixing of cations. Powder can have fine grain size	Species can lose, partially Pb, during burn process. Powder formed in burn process is not fully reacted to the desired phase. Product may contain carbon

(continued)

Table 3.2 (continued)

Method	Description	Advantages	Disadvantages
Freeze drying [666]	Spray aqueous nitrate solution of Bi, (Pb), Sr, Ca and Cu into liquid nitrogen. Collect frozen droplets and freeze dry them to remove water. Fire dried powder to yield the desired phase	Intimate mixing of cations. Product can have low carbon content	Cations may demix during freeze drying, if the temperature is not carefully controlled. Nitrates, presented after freeze drying, may melt during firing, leading to large grains of non-superconducting phases
Liquid mix method [1345]	Form nitrate solution of cations then add glycol or citric acid. Heat to remove water and form polymerized gel, then heat to elevated temperature to yield the desired phase	Intimate mixing of cations. Powder can have fine grain size	Product may contain carbon
Micro-emulsion [949]	Form suspension of micro-droplets of aqueous nitrate solution of Bi, (Pb), Sr, Ca and Cu in oil. Add base to form precipitates. Separate precipitate from oil by washing in solvent. Fire precipitate to yield the desired phase	Intimate mixing of cations. Powder can have fine grain size	Product may contain carbon
Sol–gel [184]	Form alkoxide solution of cations. Add water or alcohol to cross link molecules, forming gel through polymerization and condensation reactions. Heat to elevated temperature to burn of the organics and yield the desired phase	Intimate mixing of cations. Powder can have fine grain size	Method is better suited to making films, than bulk powders

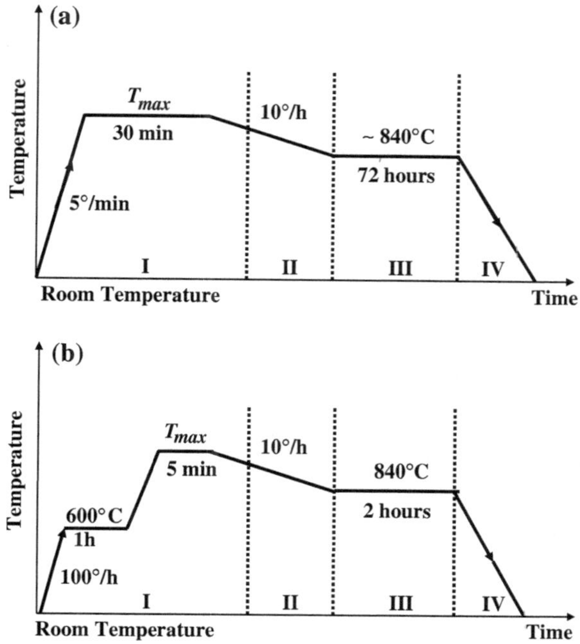

Fig. 3.9 Generic Bi-2212 melt-processing schedules used for (**a**) tapes and wires and (**b**) films. The schedules have been divided into four regions that are described in the text [666]

Since the pioneering work of Heine et al. [662] on Bi-2212 conductors, almost all Bi-2212 conductors have been melt-processed. This method is also named by *partial melt-processing* that reflects the fact that Bi-2212 melts incongruently forming liquid and crystalline phases (i.e., partial melt).[4] Numerous heating schedules that are currently used for the melt process of Bi-2212 can be generalized for tapes and wires (Fig. 3.9a), and also for films (Fig. 3.9a). Both schedules could be divided into four stages [666].

In Region I, the conductor is heated above the melting point of Bi-2212, held at this temperature for a short time and then cooled. In this stage, Bi-2212 powder melts incongruently, forming liquid and non-superconducting crystalline phase. At the end of this stage, Bi-2212 formation begins. At melting, the Bi-2212 powder releases oxygen, which is not a problem for films that are open to the environment, but can be a problem for tapes and wires, where it can cause the silver sheath to bubble. On the other hand, at elevated temperatures, used during melting, Bi evaporation from the melt is possible [1484]. The silver sheath of tapes and wires prevents this process, but this is actual danger for films. Also, it should be noted that critical condition to prepare homogeneous highly aligned microstructure is the minimum fraction of crystalline phases in the melt at the beginning of Bi-2212 phase formation.

[4] It should be noted that the term melt processing clearly distinguishes from a lower-temperature processing, where only a portion of the Bi-2212 is melted, which is called *liquid-assisted processing* [667].

In Region II, Bi-2212 forms from the melt, where growth and alignment of Bi-2212 grains occur. Since Bi-2212 melts incongruently, at cooling it should be formed by a reaction between liquid and crystalline phases. In this case, in order to form a homogeneous highly aligned microstructure, the melt must contain small grain of non-superconducting phases when Bi-2212 begins to form. The cooling rates, used for the melt process of Bi-2212 conductors should be sufficiently fast to intensify Bi-2212 phase formation and minimize fraction of normal phases, which are present in the final product always.

The problems with having non-superconducting phases in the fully processed conductor are that the phases are too large to pin flux, they block the supercurrent path and diminish useful properties of the conductor. In order to obtain high superconducting properties, a highly aligned grain structure is necessary. Plate-like Bi-2212 grains grow from the melt because growth of Bi-2212 is faster in the *ab*-plane than in the *c*-direction. This two-dimensional growth is critical for the alignment that develops during cooling. The experiments have shown [666] that the cooling rate and thickness of the oxide melt affect the alignment: slower cooling and thinner oxide yield higher alignment. The misorientation angle for a given grain size decreases with decreasing oxide thickness. The alignment mechanism requires that the large, properly oriented grains grow at the expense of the smaller misoriented grains. Films that are <20–25 μm thick align easily. In thicker films, the alignment is usually not uniform throughout, being higher close to the free surface than near the silver interface. It has been suggested that in the films the Bi-2212 growth and alignment begin at the free surface and proceed into the oxide layer. Since films are two-dimensional, one would expect a higher alignment in thinner films because of the smaller misorientation angle a grain could have and still grow to a given length. Moreover, as the aligned grains grow in films, they may rotate misaligned grains into alignment near the free surface more easily compared with the nearest neighborhood of the superconductor/metal interface. The free surface in films is also important from the point of general alignment of the film structure. Thick films (50–100 μm) can have a 20–25 μm thick layer of aligned Bi-2212 grains at the free surface with poorly aligned Bi-2212 below this layer [666]. In general, the grain alignment and J_c are higher in films than tapes.

In Region III, the formation and alignment of Bi-2212 is maximized. At the same time, in Region IV, where the sample cools to room temperature, the critical alteration of microstructure occurs. High J_c requires fast cooling rate for films (of order 1200 °C/h [1566]) and analogous rates for tapes and wires [1559]. However, at very high cooling rates the conductors may crack, leading to low J_c [1566]. At the same time, the Bi-2212 can decompose with slow cooling rates (at <300 °C/h) forming Bi-2201, possessing lower superconducting properties.

In order to minimize the number of intergranular weak links, that are proper for oxide superconductors, a high degree of crystallographic texture must be obtained. One possible route by which a strong crystallographic texture can be produced is to melt process the material under the effect of an elevated magnetic field [470]. In this case, the driving force for grain alignment is provided by the anisotropic

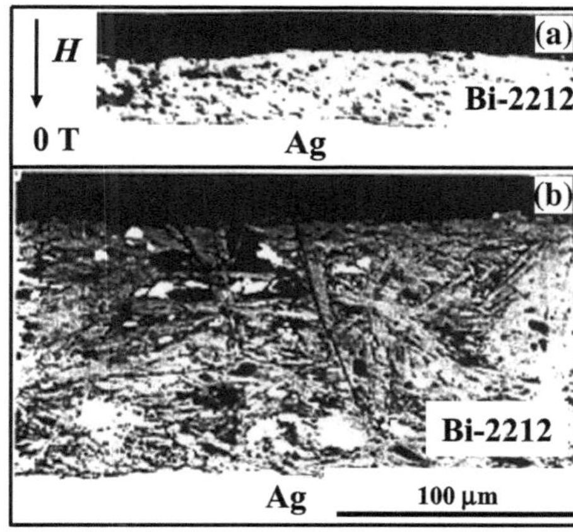

Fig. 3.10 SEM images of cross-sections of Bi-2212 films melt processed in the absence of magnetic field [470]

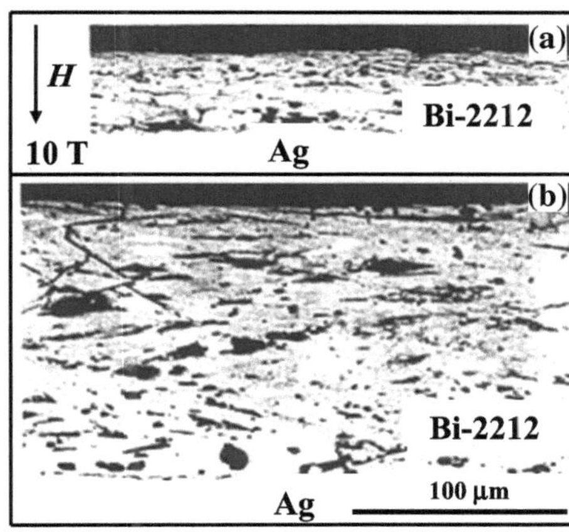

Fig. 3.11 SEM images of cross-sections of Bi-2212 films melt processed under 10 T magnetic field [470]

paramagnetic susceptibility, exhibited by the superconductor grains. When a superconductor grain is placed in a magnetic field, the axis of maximum susceptibility aligns with the magnetic field direction. As a result, in the case of superconductors, such as BSCCO and YBCO the grains should align with the *c*-axis parallel to the external magnetic field [97, 1731].

Scanning electron microscopy (SEM) backscattered images of polished cross-sections of Bi-2212 thick films with two different thicknesses, processed in zero field and at 10 T magnetic field, are shown in Figs. 3.10 and 3.11, respectively. In the first case, the degree of texture decreases with the increasing film thickness. In Fig. 3.12, the transport critical current densities of the films measured at 4.2 K are plotted as function of their thickness for films processed under zero and 10 T magnetic field. Analogous results of a higher degree of alignment at the higher magnetic field are also observed in Bi-2212 tapes (see Fig. 3.13).

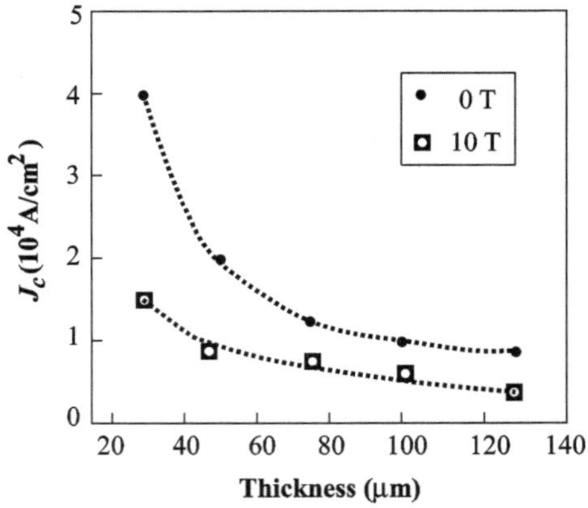

Fig. 3.12 Transport current densities at 4.2 K, zero field, as a function of thickness for films processed under 0 and 10 T magnetic fields [470]

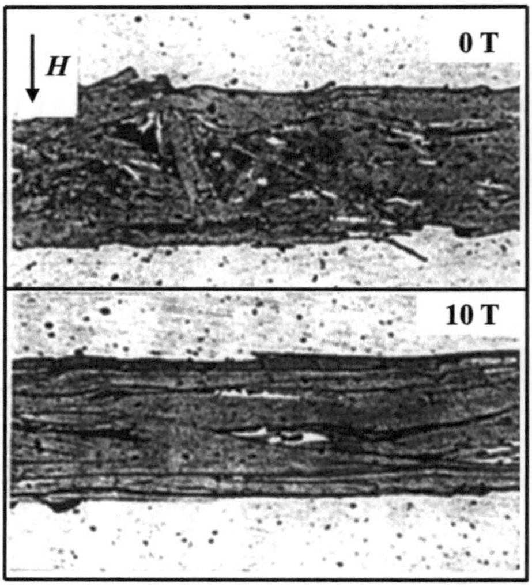

Fig. 3.13 SEM images of cross-sections of Bi-2212 tapes melt processed under 0 and 10 T magnetic fields [470]

The superconductors Bi-2212/Ag are also prepared by using *melt-solidification method* [643]. In its modification, *pre-annealing* and *intermediate rolling* are applied to improved the alignment of superconducting grains [1138]. At the same time, monocore and multifilament superconducting tapes and wires in silver sheath (Bi-2212/Ag and Bi-2223/Ag) are prepared most successively by using the *oxide-powder-in-tube* (OPIT) *method* [389, 1721, 1819]. Short Bi- 2223/Ag multifilament tapes have demonstrated $J_c > 80$ kA/cm^2 at 77 K using this technique [487]. However, this multistage method is characterized by numerous technical parameters and procedures which define (together with initial composition) useful properties of the final sample. In this case, it is very important to prepare superconducting Bi-2212 and Bi-2223 powders, possessing better phase composition, grain size and shape, and also chemical purity. Corresponding technical operations have been presented in Table 3.2. Today, the standard approaches to obtain

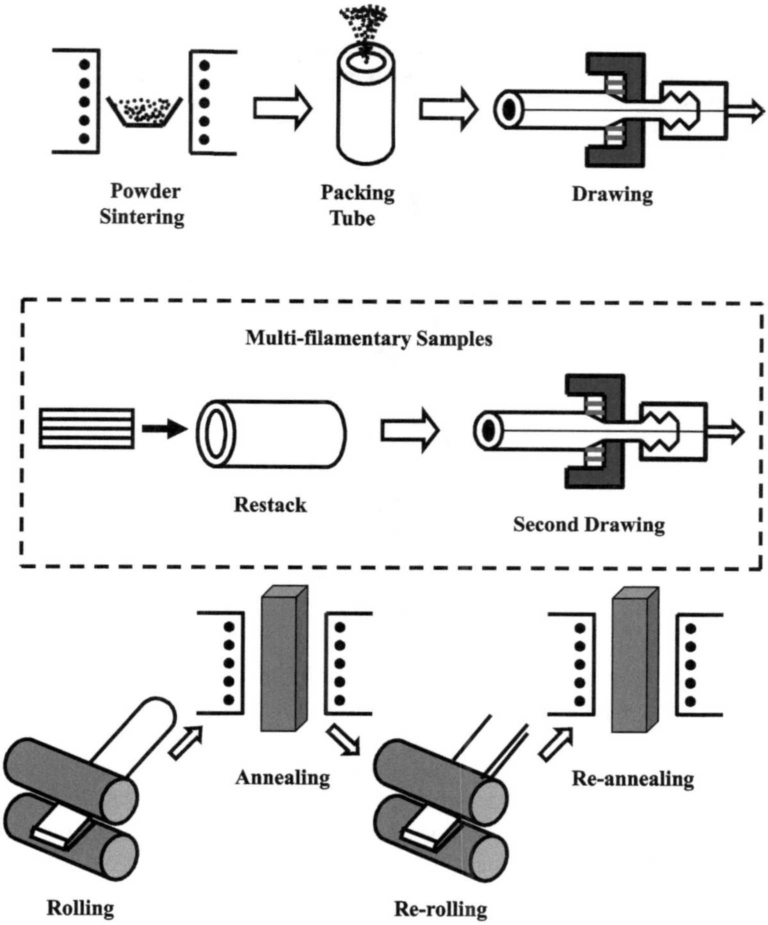

Fig. 3.14 Schematic diagram of the oxide-powder-in-tube method to make HTSC wires and tapes [666]

Bi-2223 precursor powders include the so-called methods of *one-powder synthesis* and *two-powder synthesis* [386]. In the first case, the precursor is prepared as the result of calcination of the oxides and carbonates mixture. In the second case, sintering of the mixture of two cuprate compositions is carried out.

The OPIT method for making conductors is shown schematically in Fig. 3.14 and includes the following procedures. After the preparation of precursor powder, it is packed in a tube from Ag or its alloy. Here, silver is used, in particular, because of its property to diffuse oxygen at high temperatures. This permits to control the oxygen pressure during superconductor fabrication [666]. The tube is sealed and mechanically worked into desired conductor form. Usually the Ag tube is drawn to a small diameter ($\sim$2 mm), using set of cone holes. Then, this wire is

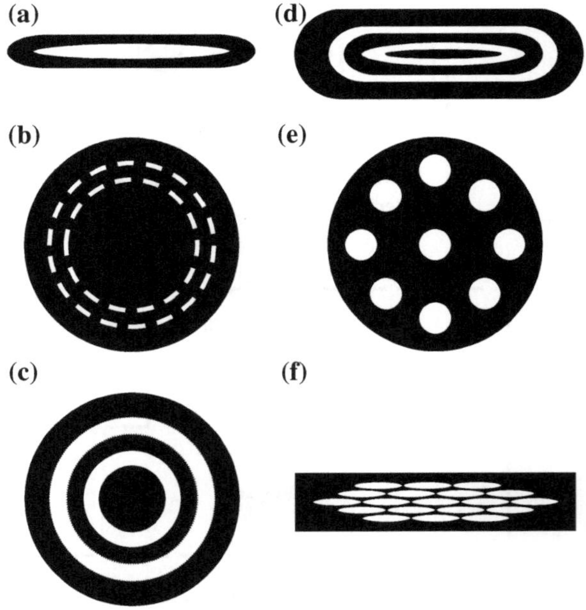

Fig. 3.15 Representative configurations for Ag-sheathed wires and tapes. *Black regions* are Ag and *white regions* are BSCCO: (**a**) Monocore tape, (**b**) Coaxial multifilament wire, (**c**) Wire with two BSCCO cores, (**d**) Tape rolled from the wire in (**c**), (**e**) Filament wire and (**f**) Filament tape

rolled into a flat tape with thickness ~0.1 mm. Multifilament superconductors are prepared, placing separate wires in a tube with a large diameter and carrying out analogous technical operations. At final stage, the samples are subjected to set of thermal treatments. During chemical reaction, necessary superconducting phases form at this stage. Mechanical deforming helps to align and construct the texture of crystalline structure in this technological link. Thus, the brittle superconducting oxide is encircled by the sheath from silver or its alloy which protects the superconducting core from chemical and thermodynamic influences. Figure 3.15 shows different OPIT conductor forms that have been made.

The use of silver as sheath material of BSCCO tapes and wires is caused by the compatibility of Ag with superconducting powders at high temperatures of sintering. However, for tapes with silver sheath, there is the problem of formation of an undulating oxide/Ag interface, which is known as *sausaging*. This occurs along the tape after its drawing, rolling and pressing (Fig. 3.16) [628, 1083], and also silver creeps at high temperatures [547]. Moreover, HTSC tape sheath, used in concrete devices or samples, must satisfy definite demands, namely (i) they must possess sufficient mechanical strength in order to bear strains, stated by high electric and magnetic fields, that may crack brittle superconducting core; (ii) they must possess low heat conduction in current conductor and high electric resistance to decrease current losses in cables [1843]. In order to solve these problems, the silver alloys with some metals, namely Au, Cu, Mg, Mn, Pd, Zr, (Ni, Y), (Zr, Al) and (Mg, Ni) are used [7, 25, 1528, 1575, 1680]. Thus, in total, there is a problem to obtain optimum microstructures of superconducting composite, taking into

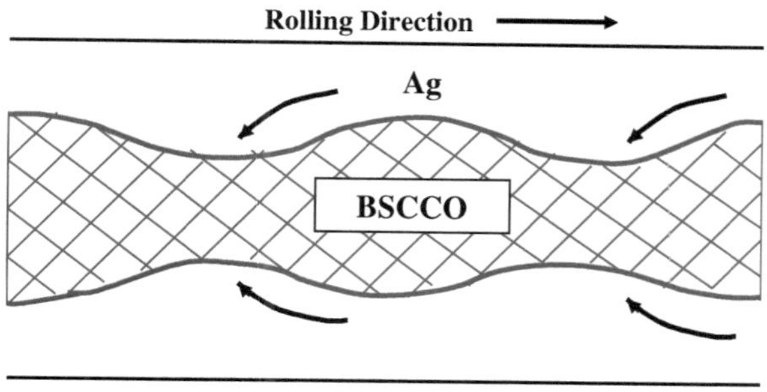

Fig. 3.16 Longitudinal cross-section of a BSCCO tape, showing undulations in the Ag/oxide interface. This is known as sausaging

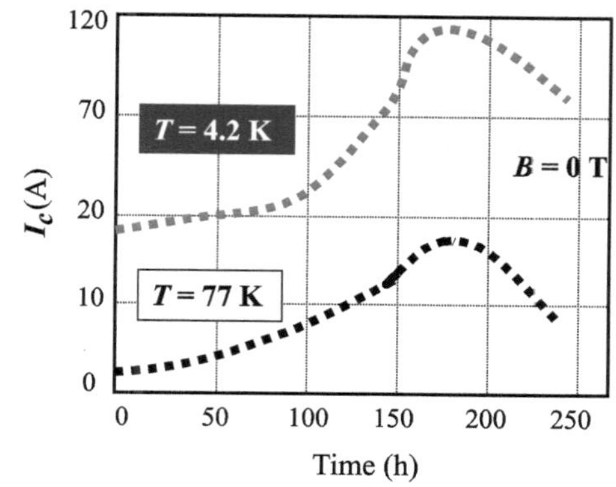

Fig. 3.17 The critical current versus sintering time ($B = 0$ T, $T = 4.2$ and 77 K)

account different behaviors of the silver (alloy) sheath and oxide core at thermal treatment [628, 1768].

Now, OPIT method is generally applied to prepare BSCCO superconductors. The advantage of this HTSC system is that the texturing and alignment of crystallites, influencing significantly superconducting properties of tapes, may be simply achieved at the final stage of the sample fabrication. In this case, the tape deformation leads to shear in BSCCO structure along the double layer Bi–O, acting as glass phase, due to the weak joints of the adjacent layers to each other. As a result, the grains form an aligned structure at the next heating. The YBCO family has a worse structure of grains and demonstrates higher isotropy compared with BSCCO and also possesses small intergranular transport currents that are caused by weak links. Therefore, all attempts to use OPIT method for this family were not successful [1056, 1621].

At long sintering of Bi-2223/Ag monocore tapes, a diminishing of magnetic flux pinning with corresponding decrease in critical current, I_c (Fig. 3.17 [5]), which could be caused by decreasing Pb content in the superconducting core at increasing calcination time [1091, 1092], is observed. However, we propose the more

[5] Obviously, the initial increase in critical current is connected with improvement in quality of intercrystalline boundaries. However, this factor has secondary value at longer sintering.

probable causes decreasing I_c to be inevitable processes of pore transformation at intergranular boundaries with pore displacement into growing grains, forming close porosity at prolonged sintering of the sample [1301, 1854].

In total, an increasing of low pinning, which is intrinsic to BSCCO family (compared to YBCO family), is one of the main goals of thermal treatment, applied to prepare BSCCO/Ag samples by OPIT technique [175, 479, 788]. There are other problems demanding their solution during tape processing by using OPIT technique, namely (i) an initiation of the local heterogeneities into powders during mechanical deformation (e.g., pressing) [1095]; (ii) the bubble formation into silver sheath due to isolation of gases [627, 918]; (iii) a possibility of the microstructure perturbations because of violation in crystallite alignment, initiated near secondary phases due to the formation of crack-like defects during inelastic strain of pressed powder [1336], and also due to various deviations from optimum-phase composition [1278]; (iv) difficulties in the texture formation of HTSC connected with the existence of secondary phases, heterogeneity and redistribution of material during thermal treatment [628, 1015]; (v) impossibility to totally heal the microcracks formed during mechanical deformation using high-temperature treatment [1382]. Finally, an increased length of tape initiates an action of scaling factor, which influences considerable decrease in the critical current density. Due to this, the doping additions of Ag (Ag_2O, $AgNO_3$) are added to HTSC for improvement of structure-sensitive properties [1471, 1593]. The complexity and multistage character of the OPIT technique, and also smaller more than order of magnitude the critical current density of tapes compared with films, define modifications of this standard technique of the BSCCO/Ag tape processing.

As examples, we point the following:

- using the hot extrusion (Fig. 3.18a) to obtain long-scale ($l > 150$ m) Bi-2223/Ag tapes [1095];
- intermediate mechanical deforming (one-axis pressing [568] (Fig. 3.18b) or cold rolling [1336]) of tapes after first sintering, that forms Bi-2223 phase in order to increase the density of superconducting core and corresponding J_c;
- application of deformation methods which are alternative to rolling, namely semi-continuous pressing [961], consecutive pressing [64] and periodic pressing [1093];
- modifications of the standard rolling: (i) out-center rolling with eccentricity, using two concentric rollers [912], (ii) sandwich rolling, in which a tape locates between two thick steel plates fastened with springs [1773], (iii) transversal (differing from standard or longitudinal) rolling [593], (iv) two-axis rolling, that replaces the wire-drawing [722], (v) groove rolling (Fig. 3.18c) [731];
- pressing Bi-2223/Ag tapes at cryogenic temperatures, based on relative increase in silver hardness compared with superconducting oxide at liquid nitrogen temperatures [309];
- application of hot isostatic pressing (Fig. 3.18d) to solve the problem of porosity and microcracks in Bi-2223/Ag tapes [162];

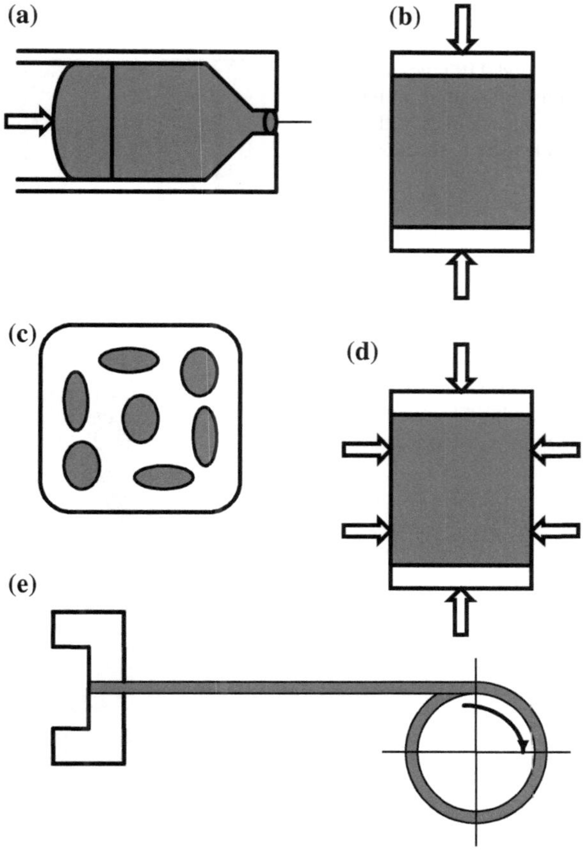

Fig. 3.18 Some additional technical operations used to improve quality of HTSC prepared by OPIT technique: (**a**) Extrusion, (**b**) One-axis pressing (**c**) Groove rolling (a cross-section of seven-filament wire is shown), (**d**) Isostatic pressing and (**e**) Continuous winding

- using an expensive pressure to pack a superconducting core at the temperatures of Bi-2223/Ag phase formation [1437];
- using the *wind-and-react* (W&R),*react-and-wind* (R&W) and *react-wind-sinter* (RWS) techniques of continuous winding at the final stage of Bi-2212/Ag tape processing, which decrease the damage of long-scale samples during their winding [163];
- using the *wind-react-and-tighten* (WRAT) technique, that combines the advantages of the W&R and R&W techniques and consists of free winding and reaction in the winding with final introduction of isolation after completion of thermal processes and tightness of the winding to its final size [1242].

Finally, the following techniques to prepare the composite tapes and wires are designed that use optimum geometry and placement of components or advantages of superconducting and ceramic materials, namely:

- *Tape-in-rectangular-tube* (TIRT) technique to prepare superconducting filaments with *c*-axis oriented in the direction that is not perpendicular to the tape surface (Fig. 3.19) [1369].

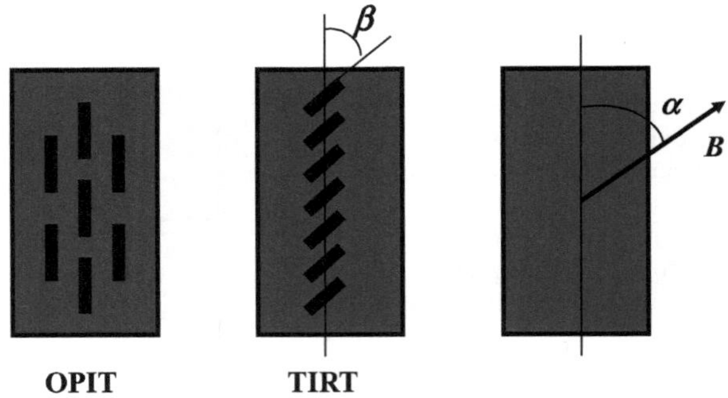

Fig. 3.19 Schematic view of the cross-sections of the OPIT and TIRT tapes. The mutual orientation of the filaments and that of the field with respect to the tape broad [1369]

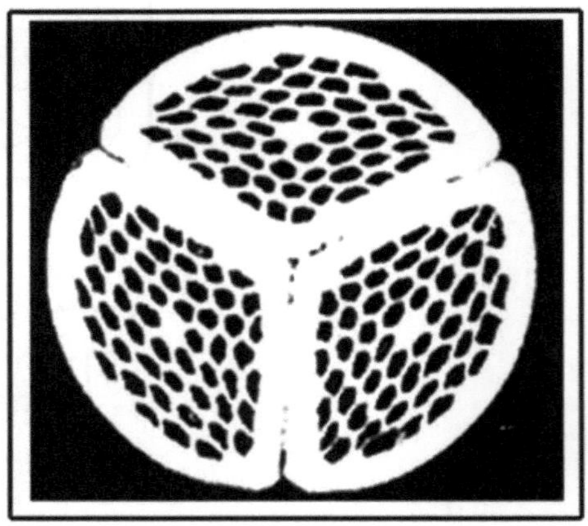

Fig. 3.20 Cross-section of a round wire fabricated by stranded-and-formed method [27]

- Processing round Bi-2212/Ag wires by using *stranded-and-formed method* (SAFM), in which first, filaments are bundled to consist of a segment, and face are expressed by the angles β and α [1369]then three segments are stranded, drawn and formed to a final round-shape wire. Round wires fabricated by SAFM are more tolerant against bending strain than the usual round wires and permit to reach high J_c (Fig. 3.20) [27].
- *Wrapping method* of BSCCO/Ag tapes around core to decrease the transport current losses (Fig. 3.21) [1627].
- *Rotation-symmetric-arranged tape-in-tube wire* (ROSATwire) technique with Bi-2212/Ag tape-shaped multifilaments, possessing triple rotation symmetry and having good crystal alignment in each filament. Since the present wire structure yields complete symmetrical arrangement of the tape-shaped filaments, it is no longer necessary to use a rolling, but permits to use a drawing (Fig. 3.22) [1264].
- Manufacture of Bi-2223/Ag one- and multifilament composite tapes with oxide barrier layers between filaments to diminish electric and magnetic losses (Fig. 3.23) [539, 721, 930].
- Development of multifilament tapes with central part, that consists of Bi-2223 filaments, ensuring transport current flow, and surrounded by the barrier ceramic

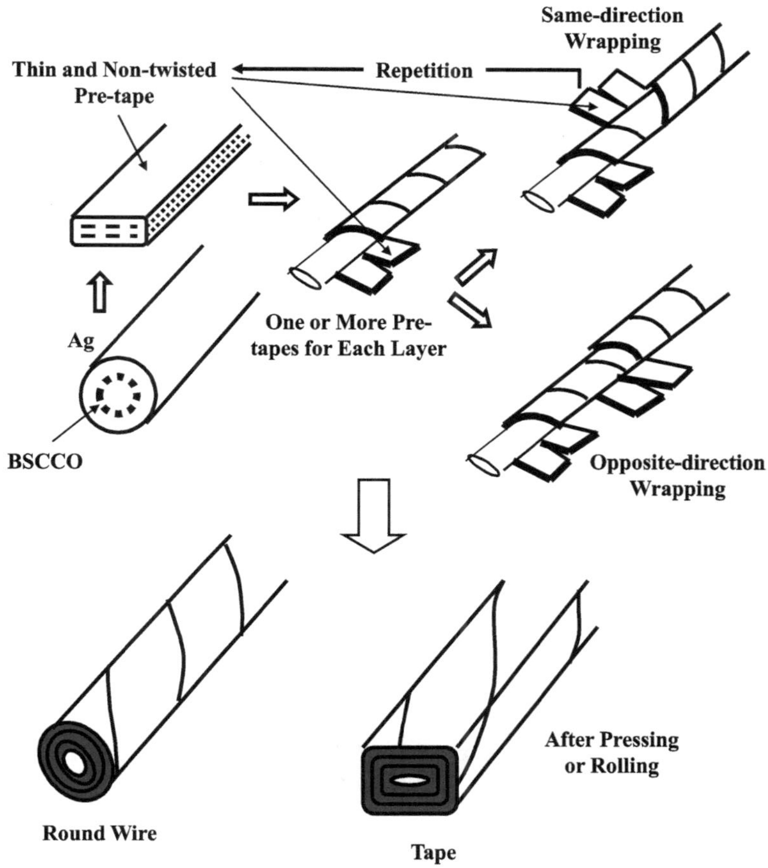

Fig. 3.21 Wrapping method to fabricate BSCCO/Ag wires and tapes [1627]

layer and YBCO thin film to screen external magnetic field and also to protect Bi-2223 filaments (Fig. 3.24) [85].

3.3 Films and Wires, Based on Thallium and Mercurial Cuprates

Processing techniques of the thallium thin films (Tl-1212, Tl-1223, Tl-2012, Tl-2212 and Tl-2223) coincide in many details with corresponding fabrication techniques of YBCO thin films. Best results have been obtained by using the following substrates: $LaAlO_3$ and $NdGaO_3$ [698, 986, 1221, 1222], and also buffer layers: YSZ and CeO_2 [1424] due to good agreement of their crystallographic properties. Moreover, MgO, $LaGaO_3$, $SrTiO_3$ have been used as substrates [986]. Tl-1223 thick films and wires with high superconducting properties have been

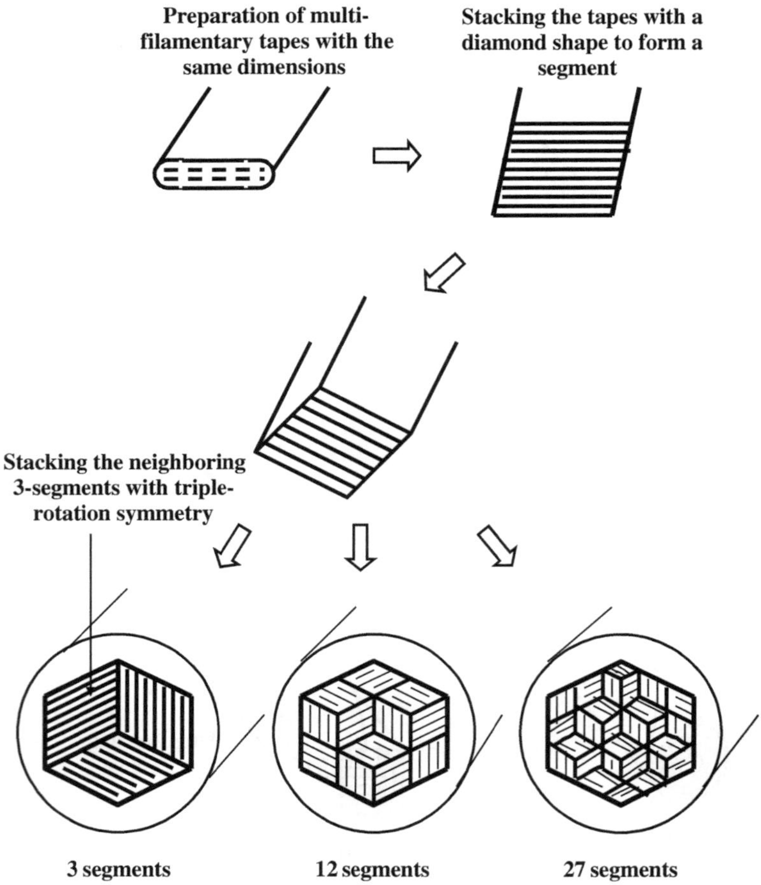

Fig. 3.22 Fabrication process of rotation-symmetric-arranged tape-in-tube wire with various Bi-2212/Ag segments [1264]

Fig. 3.23 Typical cross-section of Bi-2223/Ag/$BaZrO_3$/Ag monocore tape [930]

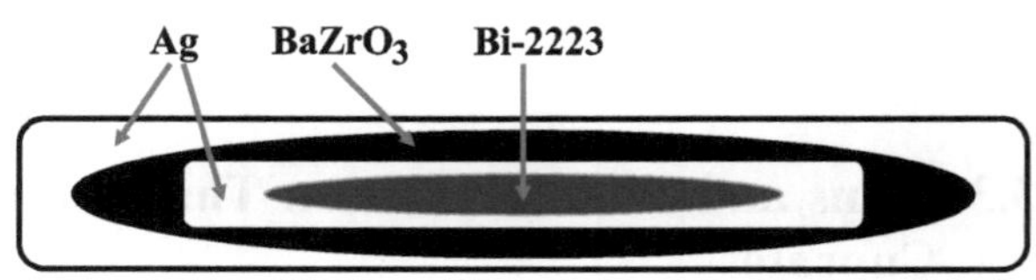

prepared on substrates from Ag and ZrO_2 [1187, 1188, 1625]. An expensive doping by fluorine has permitted to improve considerably the structure-sensitive properties of Tl-1223 samples [622]. Hg-1212 and Hg-1223 thick films demonstrating maximum critical temperatures, T_c, have been prepared in two stages [1186]: (1) a precursor layer, that consists of $Ba_2Ca_2Cu_3O_7$ and doping oxides, that is, PbO, Bi_2O_3 and ReO_2 has been deposited (2) superconducting phase has been formed by heating the precursor film at the partial pressure of mercury.

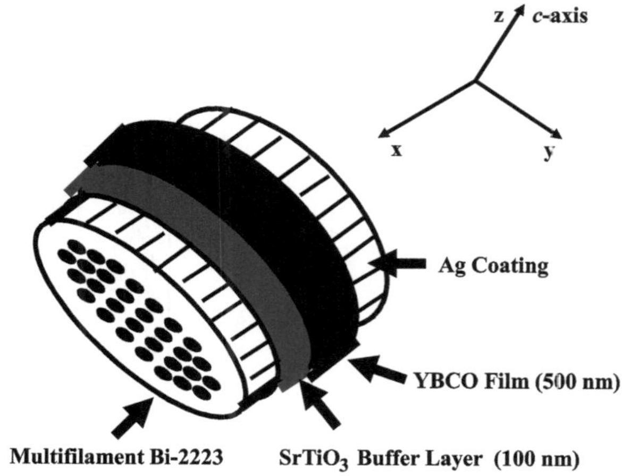

Fig. 3.24 Schematic diagram of Bi-2223 multi-filament tape with Y-123 screening layer [85]

The precursor layers could be obtained by using the precipitation processes including: (i) spray pyrolysis [1574]; (ii) pulverization of powder with solvent [1131]; (iii) sol–gel method [1706]; and (iv) application of powder-polymeric suspension [1844]. Majority of researches use mono-crystalline ceramic substrates (e.g., YSZ) [1574, 1706, 1844] Moreover, Ni-based substrates with Cr/Ag buffer layer have been used [1131]. The increased superconducting properties and improved formation of the mercurial conductor are reached by Re, Pb and F doping [459, 620, 621]. Optimum synthesis conditions permit to obtain good aligned (Fig. 3.25a) and phase purity (Fig. 3.25b) structures of thick films.

3.4 BSCCO Bulks

Bi-2212 superconducting coverings for tubes and rods are synthesized by the diffusion reaction between Sr-Ca-Cu oxide substrate, possessing high melting temperature, and Bi-Cu oxide coating layer [1638] (or mixing Bi-Cu and AgO_2 [1821, 1817, 1821]) with low melting temperature (Fig. 3.26). This technique permits to fabricate diffusive superconducting layer with thickness 150 μm (for roads of diameter 3 mm), demonstrating dense homogeneous structure of plate-like grains that are aligned almost perpendicular to the substrate surface (Fig. 3.27).

Bi-2212 ceramics are characterized by a rather high current carrying capacity in strong magnetic fields at 4.2 K ($J_c > 10^5$ A/cm^2, 10 T) [1139]. However, the rapid weakening of pinning properties and corresponding diminishing of the critical current density at above 20 K is intrinsic for Bi-2212 system [948]. There are mainly two factors responsible for this behavior: (i) the large anisotropy of the material [295, 510, 978, 1288] and (ii) the low density of defects [1136]. The practical requirement is to enhance this threshold that causes Pb doping of bulks in the *partial-melting technique* [10]. In order to obtain high-dense-textured Bi-2212

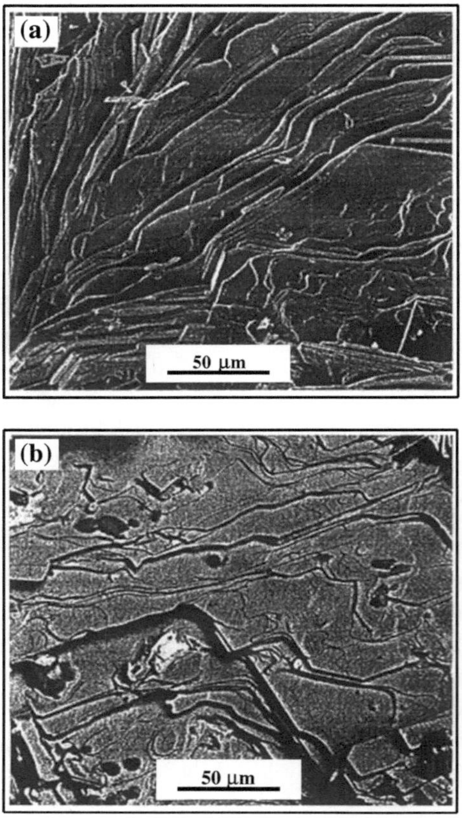

Fig. 3.25 (**a**) SEM micrograph of (Hg, RE)-1212 sample synthesized at 820 °C, (**b**) backscattered electron micrograph of the same sample [1483]

bulks, the *hot-forging technique* is used [551] and also the *solid-state reaction method* [1419]. The addition of PbO increases fraction, formation rate and stability of Bi-2223 phase [1647]. However, fabrication of Bi-2223 bulks with necessary for applications properties is the very difficult problem. *Hot-pressing technique*, used during processing of Bi-2223 bulks, promotes void elimination under pressure and high temperature. Moreover, it improves the crystallite alignment in the sample, causing an increasing of the critical current density [719]. At the same time, J_c can decrease at prolonged sintering of superconducting bulks due to corresponding decrease of the ceramic density. Hot-pressing causes high density near the theoretical level for Bi-2223 ($\sim$6.31 g/cm^3). The sufficiently long sintering before hot-pressing rises Bi-2223 phase content in the sintered samples. At the same time, in order to reach necessary properties of final superconducting ceramics, special control of processing parameters is required.[6] Such deformation

[6] This, in the first place, is explained by that Bi-2223 phase is stable only in very small temperature intervals (in difference from Bi-2212 phase) and kinetics of its formation is very slow to obtain mono-phase material.

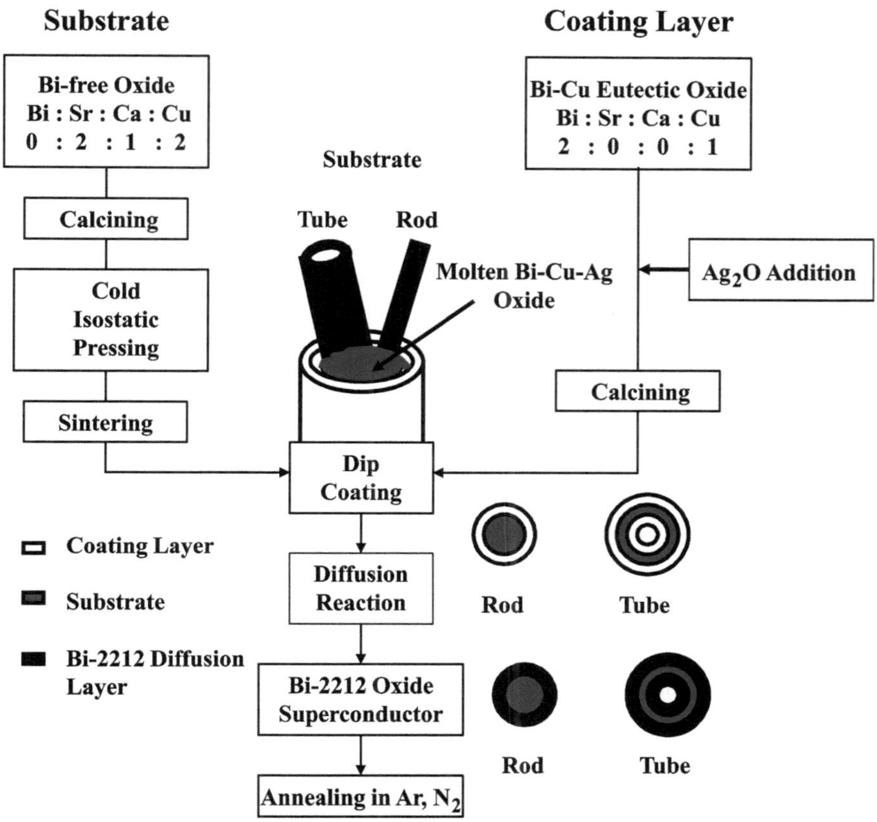

Fig. 3.26 Specimen preparation procedure [1821, 1821]

Fig. 3.27 SEM micrograph taken on the fractured cross-section of the Bi-2212 rod (Bi-2212 diffusion layer and substrate). The sample is obtained at 850 °C for 20 h on the (0213) substrate [1817]

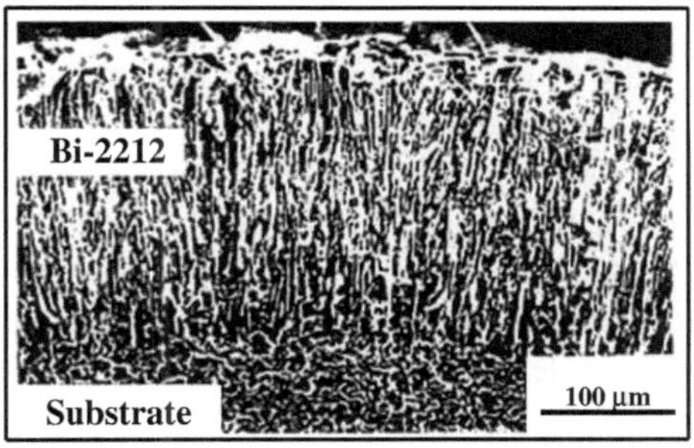

methods as hot isostatic pressing (HIP) [1147], HIP plus sinter-forging [288], or only sinter-forging [225, 523] are usually used with a deformation temperature range of 815–850 °C. However, since simple procedures of deformation based on uniaxial compression cannot provide high strain values, it is rather difficult to form strong texture together with a high density of dislocations.

Room temperature pressing is not effective because of high resistance to strains. By increasing the pressing temperature, this problem becomes less sharp. In this case, closed porosity is formed and the sample density increases. This increases the contact area between Bi-2212 and secondary phases. The formation of Bi-2223 phase is based on the epitaxial growth of the Bi-2223 crystallites into Bi-2212 matrix in accordance with the chemical reaction:

$$\mathrm{Bi}-2212+\ \text{secondary phases}\ \rightarrow\ \mathrm{Bi}-2223 \tag{3.3}$$

Therefore, the elevated contact area causes an acceleration of chemical reaction and Bi-2223 phase formation. The sample densification also increases a connectivity of superconducting grains (Bi-2223 or Bi-2212), which is found by a quality of intercrystalline boundaries. The increase of the grain connectivity causes a rise of the critical current in the sample. However, during the hot pressing, an excessive elimination of liquid phase, that is rich by Cu and Pb, is possible. This leads to considerable phase segregation in the sample, but a deficiency of the secondary phases prevents total transformation of Bi-2212 phase into Bi-2223 phase during hot pressing, decreasing T_c and J_c. Preliminary sintering causes a frame formation from Bi-2212 crystallites, preventing the secondary phase losses. After that, the hot-pressing redistributes secondary phases around Bi-2212 grains, accelerating the phase transformation considerably. Displacement of amorphous Bi-2212 phase and secondary phases at intergranular boundaries heals intercrystalline defects. At the same time, preliminary prolonged sintering can lead to the expense of larger part of secondary phases on the growth of arbitrary oriented Bi-2223 crystallites.

In this case, amorphous Bi-2212 phase has sufficient time to re-crystallize. Therefore, one could not be used to increase the grain connectivity during hot-pressing. Thus, preliminary super-long sintering before the hot-pressing, do not permit to fabricate bulks with high J_c. Moreover, expensive liquid phase, not dissolved into Bi-2223, forms non-superconducting phases and diminishes J_c [1278].

One-axis cold-pressing can lead to partial orientation of BSCCO plate-late grains in the pressing direction, taking into account great microstructure anisotropy [1537]. Therefore, the sample re-crystallization, presenting an orientated grain growth, provides maximum texture during next sintering, corresponding to external loading. There is a good correlation between the final angle distribution of BSCCO plates and sample *squashing*, namely a lower misorientation corresponds to higher loading [1250, 1452]. Then, the hot-forging followed by squashing of sample leads to improvement of alignment of Bi-2223 plates [1453]. These samples show $J_c = 8$ kA/cm^2 (zero magnetic field). This presents interest for current conductors and current limiters. Another multistage processing of Bi-2223 bulks includes one-axis cold pressing of superconducting powders, then annealing without pressure and finally the hot-forging [1654].

Taking into account the plate-late morphology of BSCCO grains, *magnetic melt-processing-texturing* (*MMPT*) technique has been developed, combining at high temperature two different parameters, namely an application of *in situ* magnetic field (8 T) and one-axis loading (60 MPa) [1251]. This technique has permitted to obtain materials with a good texture. At the same time, *hot-plastic deformation* technique permits the formation of sharp crystallographic texture and high density of defects in the crystalline lattice, which are effective pinning centers, in the BSCCO bulks [83, 1147, 1292, 1871]. As the result, the critical current density $J_c > 10^5$ A/cm^2 has been reached in Bi-2212 ceramic [329, 745, 746].

The use of complex loading procedures of deformation is more advanced for increasing strain values. One of such procedures is *torsion under pressure* [183] where quasi-hydrostatic pressure prevents destruction of the material and torsion leads to deformation up to very high strains [329, 748]. Recently it has been revealed that applying low quasi-hydrostatic pressure (slightly above 0.6 MPa) increases the onset temperature for melting of Bi-2212 by almost 60 °C, which essentially expands the useable temperature range for hot deformation [745]. An open question is the influence of the deformation performed near the melting temperature of Bi-2212 phase on structure and superconducting properties.

In order to form microstructure features—pinning centers—in BSCCO systems, following methods are used: (i) irradiation of samples after processing by protons [945], heavy ions [1669] or neutrons at doping HTSC by uranium [639], that forms super-thin amorphous cylindrical inclusions, the so-called "columnar defects" (Fig. 3.28); (ii) introducing defects, connected with doping additions, disperse particles and dislocations [5, 493, 1090]; and (iii) technique change, supposing an introduction of the mixed particles (e.g., Pb) into Bi-2212 phase [1074]. The enhanced pinning due to irradiation of superconductor by high-energetic ions, leads to significant increasing of the critical current density, J_c, sometimes in the order of magnitude and more. However, the critical temperature, T_c, in this case, as a rule decreases, according to the common law of the superconductivity suppression by structure. At small irradiation doses, small ($\sim$1 K) increasing of T_c (so-called "effect of small doses") is sometimes observed. Usually, the irradiation dose is selected so as to reach the maximal value of J_c at small decreasing of T_c. First, in [5], very significant (on 15 K) growth of T_c in Bi–Pb–Sr–Ca–Cu–O thick films after their irradiation by argon ions has been found. The original films consisted of mixing of Bi-2212 and Bi-2223 phases with the domination of the Bi-2212 phase (they has $T_c = 85$ K). The ionic irradiation stimulated the processes of local melting of the film and diffusion of atoms, resulting in the formation of high-T_c Bi-2223. Thus, the ionic irradiation can be an alternative technique to fabricate Bi-2223 phase, which is known for its great sensitivity to conditions of sample synthesis and annealing.

An intensive effect in improvement of superconducting properties is decreasing of cooling rate of the sample that increases J_c and T_c and forms more perfect microstructure of superconductor [1334]. In order to improve mechanical and superconducting properties of BSCCO during processing, the next additions are introduced into the samples: inclusions of Ag (Ag_2O, $AgNO_3$) [1591], Al_2O_3 strengthening filaments of Al_2O_3 [1103, 1806] and $ZrO_2 \bullet Y_2O_3$ [1103], whiskers of MgO [1854–1856], particle dispersion of MgO and additions of high-dense polyethylene [194], and also nanocrystalline MgO powder [744]. The additions of

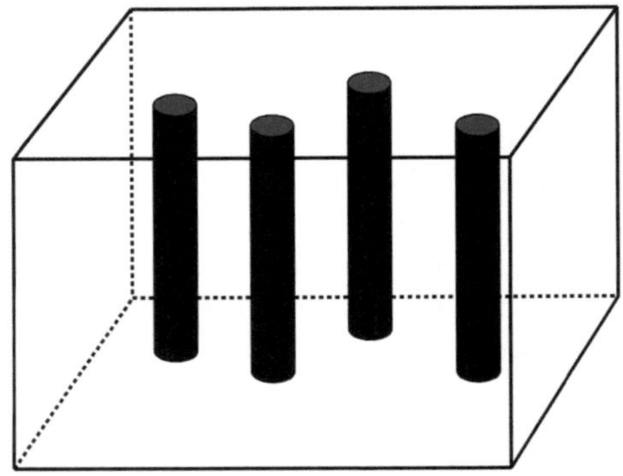

Fig. 3.28 Traces of heavy ions forming amorphous cylindrical regions in HTSC

Mo, W, Zr and Hf in Bi-2212 lead to formation within the temperature range 850–900 °C of submicron inclusions Sr_2CaMO_6 (M = Mo, W) and $SrAO_3$ (A = Zr, Hf), being chemically compatible with Bi-2212 [1075].

A schematic description of the solid-state processing method of the $(MgO)_w$/BPSCCO composite is shown in Fig. 3.29, and SEM micrographs taken from etched surfaces of polished monolithic and composite specimens are presented in Fig. 3.30. Moreover, in order to strengthen BSCCO bulks, the glass–epoxy tapes are used [1655]. At the same time, an application of doping additions of TiO_2 and ZrO_2, which leads to toughening of superconductor, can simultaneously cause considerable decomposition of Bi-2223 phase and diminish T_c [579, 1142].

If the size of these particles is much greater than the coherence length of the superconductor (actually it takes place almost always if the material was exposed to melt-texturing or hot deformation) they are not efficient for magnetic flux pinning. In this case the pinning is mostly conditioned by the structure features of the particle–matrix interface [1823] or such lattice defects as dislocations and stacking faults [999, 1770]. As a rule, the lattice defects appear at cooling since the particle and the matrix have different coefficients of thermal expansion. In this connection the efficiency of particles as vortex pinning centers can be improved if we try to increase the density of defects connected with them.

One of the methods for increasing the density of defects in the vicinity of the particles is plastic deformation. This well-known method is based on the fact that during deformation of two-phase materials the particles always promote trapping of dislocation tangles around themselves. These tangles are stable till very high temperatures [1094]. The plastic deformation of high-T_c superconducting composites Bi-2212/MgO has been studied and shown (at small strain value $\varepsilon < 10$ %) the increasing J_c as a result of the deformation [1344].

The particle size and their distribution mostly depend on deformation temperature [1094]. Moreover, the deformation temperature is one of the most important parameters effecting texture and dislocation density. That is why for more precise regulation of the size of particles and their distribution, as well as the density of dislocations and the texture sharpness the temperature range of the

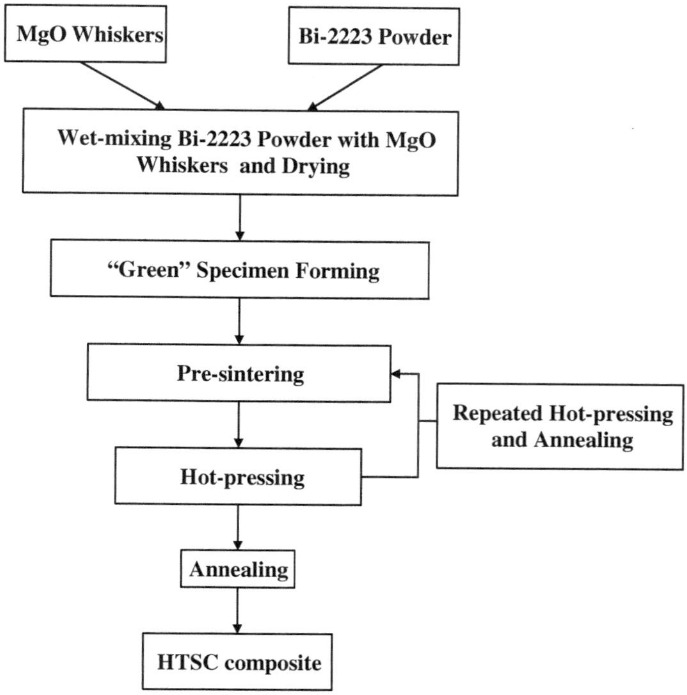

Fig. 3.29 Solid-state processing of $(MgO)_w$/BPSCCO composite [1854]

plastic deformation to be carried out should be as wide as possible. The Bi-2212-base composites are most interesting for such studies since Bi2212-phase is characterized by the evident lack of natural centers of flux pinning.

Unfortunately, the brittleness of Bi-2212 restricts the temperature range of its deformation when simple uniaxial compression is used. In order to expand the deformation temperature range (both to lower and higher temperatures) one should apply complex loading schemes—*torsion under quasi-hydrostatic pressure* [747]. On one hand, the quasi-hydrostatic pressure prevents initiation and growth of cracks and provides an opportunity to decrease the deformation temperature. On the other hand, it increases the melting temperature of the Bi-2212 [745] that provides increasing the deformation temperature essentially. Due to such a double effect from the quasi-hydrostatic pressure the deformation temperature range of the Bi-2212 was almost 150 °C (from 795 to 940 °C) [747]. Therefore, it is assumed that this deformation scheme is very prospective for studying the total influence of particles and lattice dislocations on superconducting properties of the composite.

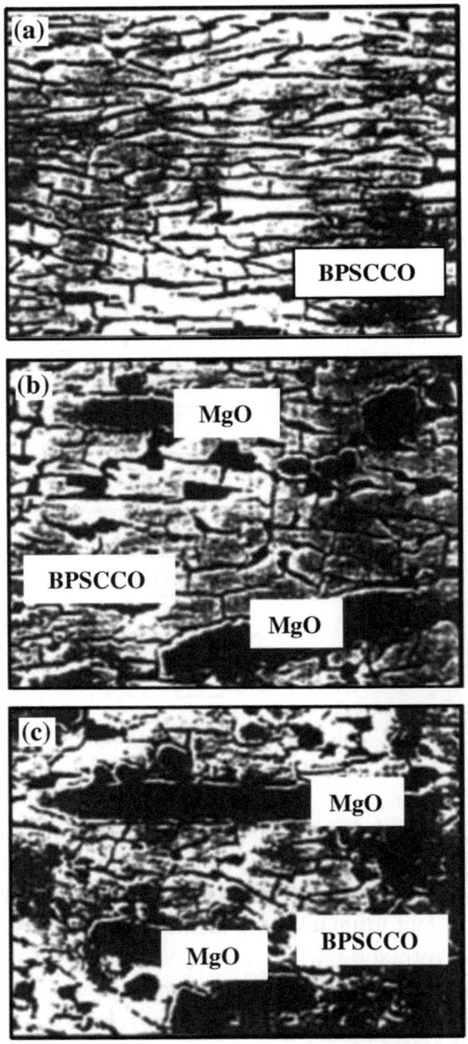

Fig. 3.30 Microstructure of BPSCCO phase in polished and etched cross-sections parallel to the hot-pressing direction of (**a**) a monolithic BPSCO, (**b**) and (**c**) A $(MgO)_w$/BPSCCO composite with 10 and 20 % concentration of whiskers, respectively (all specimens have undergone a three-cycle hot-pressing and annealing) [1855]

3.5 Y(*RE*)BCO Bulks

Intercrystalline boundaries can diminish considerably the transport properties of YBCO bulks and increase sensitivity to applied magnetic fields. Therefore, it is very expedient to create large grain specimens with maximum number of strongly connected intercrystalline boundaries. As a rule, $Y(RE)Ba_2Cu_3O_{7-x}$ crystallites (123 phase), prepared by the melt-processing technique, include $Y(RE)_2BaCuO_5$ particle dispersion (211 phase). The elevated critical current density in these samples can correlate with the concentration of the 211 inclusions [1539] and assumes that fine size of these particles causes directly elevated pinning. At the same time, the sufficiently large particles of 211 normal (non-superconducting) phase obviously diminish the superconducting properties. Therefore, in order to optimize the ceramic properties, it is necessary to control the concentration and the size of 211 particles in precursor powder used in YBCO processing [1041]. After fabrication of the precursor powder, it is pressed into pellets, which is then calcined and melted. The pellets, subjected to insignificant pressing, demonstrate a lost density compared with pellets subjected to cold isostatic pressing. Further sintering causes their densification next. However, the hard aggregates of grains, formed during the cold isostatic pressing, can prevent alignment of some crystallites, and, therefore limit an effective densification during the following sintering. Then, it could be assumed that the sintering is the optimum process to create high-dense HTSC samples. It should be noted that 211 particles grow into precursor bulk during sintering. It is not desired with the view of preservation of fine granularity of the 211 phase. Therefore, it is necessary to exclude this effect. Thus, the cold isostatic pressing in some cases may be more practical to prepare optimum precursor samples, which are then subjected to melting [1040].

All techniques, based on melt processing and used to prepare YBCO large–grain ceramics, are characterized by the peritectic reaction at $T_p = 1015\ ^\circ\mathrm{C}$ owing to the formation of 123 phase from 211 phase and liquid component [1040]:

$$\begin{array}{cccc} Y_2BaCuO_5 + & 3BaCuO_2 + 2CuO & \rightarrow & 2YBa_2Cu_3O_{6.5}, \\ \text{(211 phase)} & \text{(liquid phase)} & & \text{(123 phase)} \end{array} \tag{3.4}$$

or in another form [1038]:

$$\begin{array}{ccccc} Y_2BaCuO_5 + & Ba_3Cu_5O_{6.72} + & 0.42O_2 & \rightarrow & 2YBa_2Cu_3O_{6.28}. \\ \text{(211 phase)} & \text{(liquid)} & \text{(gas)} & & \text{(123 phase)} \end{array} \tag{3.5}$$

Pseudo-binary diagram, showing two peritectic transformations, is presented in Fig. 3.31. The 211 phase and liquid could be formed by fast heating a pre-sintered precursor sample to a temperature that is considerably above T_p. Then 123 phase forms very slowly during the cooling of YBCO partial melting through point T_p. In this case, 123 phase is added up to 30 wt% of the 211 particles before the melting process with the aim to create local pining centers and to prevent liquid loss during melting [1039]. Solidification processes, depicted by the formulas (3.4) and (3.5), put definite requirements on YBCO precursor samples [1040], namely (i) 211 particles must be sufficiently fine in the initial sample precursor in order to form fine-grain dispersion in the sample fabricated. (ii) The material should be able to retain the liquid phase, appeared in result of the peritectic reaction into samples at the temperatures, which are considerably above T_p. It is necessary in order to form 123 phase at cooling. Successive process is dependent on the precursor homogeneity and density and also on the size distribution of the 211 particles. (iii) The precursor sample should not contain foreign compositional and surface

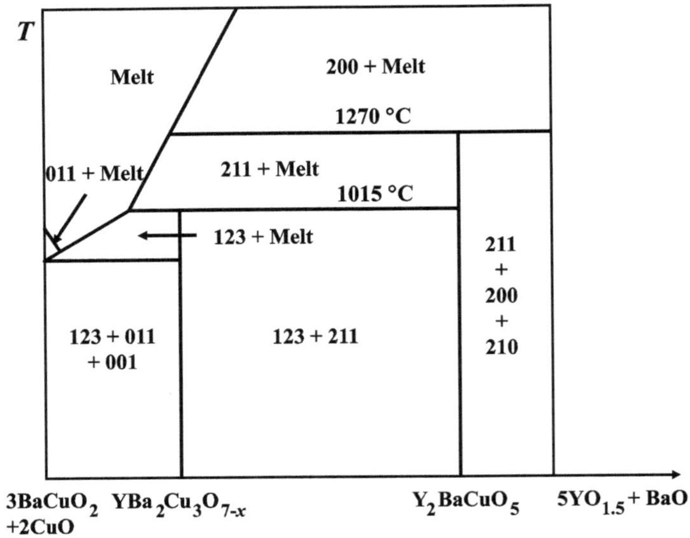

Fig. 3.31 Pseudo-binary diagram along the Y_2BaCuO_5-$YBa_2Cu_3O_{7-x}$ line showing two peritectic transformations [60]

admixtures, which form the heterogeneous nucleation sites of grains, and hence limit the grain sizes, that may be reached during the grain growth.

Obviously, it is very possible an enlargement of 211 fine-size inclusions during melting. As rule, the 211 enlargement occurs above peritectic temperature, solely. Therefore, the 211 inclusions with diameter up to 50 μm can form in the final ceramic sample. There are other problems, namely a very slow rate of solidification, a necessity to control thermal gradients during high-temperature treatment, a limited size of domains obtained, that is accompanied by their misorientation, a microcracking formation and heterogeneous composition. The progress could be reached by changing processing parameters and using doping additives.

Oxygen release increases sharply during melting. This forms voids, causing the corresponding rise of the final specimen volume. At the same time, a surface tension, connected with the melted state and decreased rate of the oxygen release, leads to specimen densification at increasing temperature. As it is shown in tests, the sample densification, controlled by the surface tension, is the dominating factor [1040]. Figure 3.32 demonstrates the YBCO cross-sections obtained at various heating rates. Even, using the same (in sizes) pellets does not preserve from considerable differences in geometry, distribution and structure of porosity in the melt-processed superconductors. Considerable changes of the sample geometry and homogeneity have been observed at the heating rate of 30 °C/h because of void formation with sizes above 1 mm (Fig. 3.32a). However, the void sizes diminished together with the heating rate (Fig. 3.32b and c) that caused their maximum homogeneity in the case of heating rate of 10 °C/h [1038]. At the same time, a void diminishing, caused by the following decrease of the heating rate, can

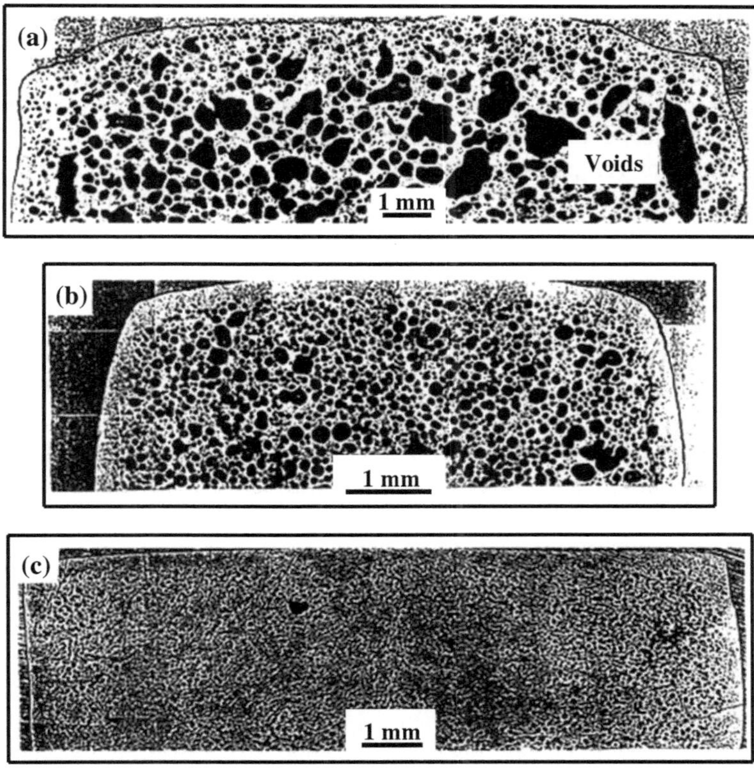

Fig. 3.32 Optical micrographs of the polished YBCO cross-sections obtained at various heating rates: (**a**) 30, (**b**) 20 and (**c**) 10 °C/h [1040]

lead to unfavorable consequences, in particular to elevated loss of liquid phase. This is not desirable to a high degree in view of stoichiometry and can lead to loss of control for composition and grain size during superconductor processing. Therefore, improved physical properties should be reached, using other methods, in particular by changing the specimen density and processing parameters of the peritectic solidification [1040].

HTSC applications require high-aligned grains and specimens with perfect texture that is reached more easily in smaller specimens in the presence of thermal gradients. The use of large thermal gradient, leading to better alignment, could cause microcracking of the sample. The latter spreads out as a consequence of the anisotropic thermal expansion of 123 phase and secondary phases ($BaCuO_2$, CuO), which precipitate at grain boundaries. The microcracks perpendicular to the *ab*-plane are the most detrimental for HTSC and can develop during cooling or re-oxygenation of the sample. Then, purification, associated with the crossing of the hot zone, can lead to chemical heterogeneities in long samples (which are always obtained using a melt-zone-type technique) owing to the different diffusion coefficients of the species. One observes in such a case a progressive decrease of the barium and copper amount, correlated with an increase of the amount of 211 phase during the process. Additives, such as Y_2O_3, which increase the viscosity of the melt, can drastically limit these phenomena. The more rapid heating rate of 123 phase causes the finer 211 precipitates. This means that the more rapid heating above the peritectic temperature, causes the higher temperature in which the decomposition of 123 phase begins. The lower the stability of the 123 phase, the more rapid is the precipitation of the 211 phase, that is, the higher the number of nucleation sites. In rapid heating, defining the fine-sized 211 phase a long plateau allows the particle coarsening by Ostwald ripening, due to the bigger grains consuming the smaller ones. In fact, as for the growth of the 123 phase, two mechanisms can be supposed to limit the growth of the 211 particles dispersed in a liquid, namely (i) diffusion of a solute in the liquid; and (b) reaction at the interface between 211 phase and liquid. Finally, the maximum temperature, reached above the peritectic temperature, controls the amount of liquid and its viscosity. So, one expects a large influence on the growth of the 211 phase. Taking into account these considerations, different *methods of melt crystallization* have been proposed during the development of melting techniques, namely:

- *melt-textured growth* (MTG) [785]. The textured growth of 123 superconducting phase from melting, in which the 123 ceramic is used as initial precursor.
- *liquid-phase processing* (LPP) [1468]. The liquid-phase technique based on decreasing of prolonged treatment of samples at maximum temperatures with the aim to prevent an undesired growth of the 211 particles, that occurs intensively above the peritectic temperature.
- *zone melting* (ZM) [1112]. This technique applies zone melting to obtain long samples using 123 ceramic as initial precursor.
- *quench-melt-growth* (QMG) [1213]. The growth of 123 superconducting phase from melting, based on the super-fast cooling, in which the 211 phase forms in results of rapid interaction Y_2O_3 phase with melt into high temperature region, where the 211 normal phase to be thermodynamically stable.
- *melt-powder-melt-growth* (MPMG) [502]. This process, using the same thermal schedule as the QMG process, introduces a drastic crushing of the quenched mixture after quenching in order to obtain a fine distribution of the Y_2O_3 phase, leading further to a finer 211 phase, which is uniformly distributed.
- *powder-melt process* (PMP) [1877]. The same as QMG process, starting from a $Y_2BaCuO_5 + 3BaCuO_2 + CuO$ mixture instead of pre-synthesized 123 phase. The result is equivalent to the QMG process, but without any overheating of the sample.
- *solid–liquid-melt-growth* (SLMG) [1558]. Using the same thermal profile as the PMP process, the SLMG process differs by the starting material: a mixture of $Y_2O_5 + BaCuO_2 + 2CuO$ is used here. These two techniques differ essentially in the morphology of the 211 phase: needle shape in the case of SLMG instead of quasi-spheres for PMP process.

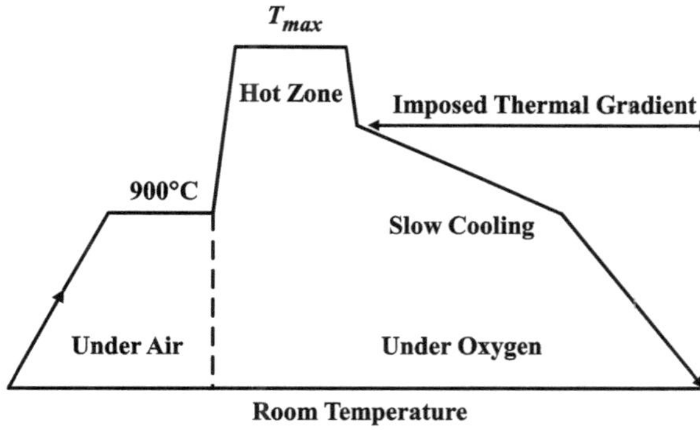

Fig. 3.33 Typical thermal cycle used in the MTG process [1764] (TSMG) technique [1193]

- *microwave-melt-texture-growth* (MMTG) [272]. This technique was developed to take into account the high thermal gradient for both increasing the solidification rate, and directing this process.
- *magnetic-melt-texturing* (MMT) [356]. This technique uses a static magnetic field in order to modify the microstructure of a material and to obtain samples with a good texture. This process exploits magnetic anisotropy of elementary cell that may be elevated by replacing Y ion with ion of rare-earth element (*RE*).
- doping YBCO by the RE ions in order to form local stresses caused by the deformation mismatch of the YBCO and *RE*BCO crystalline lattices, that is an important origin of the magnetic flux pinning and corresponding increase of J_c [1018, 1019].

The thermal cycle, used in a classical MTG process, is shown in Fig. 3.33. Here, the partial melting of a pre-sintered 123 sample is induced between 1100 and 1200 °C, preserving practically the geometry of the precursor sample. Finally, the sample is very slowly cooled (1–2 °C/h) under an oxygen atmosphere in a thermal gradient [1764].

With aim to improve magnetic flux pinning, the plastic deformation of high-T_c superconducting composites Y123/Y211 [1158, 1440, 1725, 1750, 1751, 1872] has been studied. At first the composites have been melt-textured and then hot-compressed by a small strain value ($\varepsilon < 10$ %). After deformation the density of dislocations and stacking faults in the vicinity of particles is indeed high and the increasing J_c as a result of the deformation has been observed in Refs. [1158, 1344, 1725, 1872]. At the same time, a degradation of the superconducting properties has been observed in [1751]. This degradation has been explained by the influence of the initial microstructure, in particular the size, concentration and distribution of the Y211 particles, which affect the deformation behavior and subsequent oxygenation process of the sample.

It is known that during deformation the density of dislocations trapped by the particles depends on the particle size but much more essentially it depends on the interparticle distance [1094]. The melt-texturing initiates the considerable growth of particles and increases the distance between them. Moreover, during melting can occur agglomeration of particles at inner interfaces [837]. So, it is desirable to exclude melting from processing of composite. For preserving a finer particle size and increasing the density of dislocations connected with the particles it is advisable to process textured composite by means of plastic deformation.

Introduction of silver [1576], resin impregnation [1053, 1686, 1687] and also using other additives improve the structure-sensitive properties of Y(*RE*)BCO. In general case, these additives are used to decrease the size and morphology modification of the 211 phase. They act in three ways, namely: (i) changing the 123/211 surface energy, (ii) changing kinetics of diffusion process, and (iii) forming nucleation sites for the 211 phase. In particular, an additional large dispersion of the 211 particles (up to 40 wt%) to the stoichiometric 123 phase is able to diminish the 211 particle size, formed during decomposition of the 123 precursor sample [1213]. To improve superconducting microstructure and properties, the doping additives used are: Ag (Ag_2O, Au/Ag) [1590, 1835], nano-size precipitates of Ag [1271], Pt (PtO_2) [808, 1257], Sn (SnO_2, $BaSnO_3$) [351, 1113, 1180], Zr (ZrO_2, $BaZrO_3$) [246, 549], Ce (CeO_2, $BaCeO_3$) [350, 352, 1082], Ca [637, 988, 989], SnO_2/CeO_2 [1082], Na_2CO_3 [1270], NaCl [1270],and $KClO_3$ [1270].

The formation of preferable orientation of grains is necessary for maximum use of anisotropic properties of HTSC in the specific products. The specimen texture could be formed and controlled by using the *top-seeded-melt-growth.* In this process, the crystallites-seeds from *RE* equivalents of the 123 phase are introduced.[7] They have the higher temperature of the peritectic decomposition. In this case, the 123 phase nucleates and grows in the specific direction (Fig. 3.34). Moreover, the weak links decrease considerably accompanied by proper increase of the grains. The peritectic solidification of YBCO by using seeds leads to growth morphology of faceted grains with symmetry, depending on nucleus crystallite, for example, SmBCO (Sm-123), NdBCO (Nd-123) or other *RE* barium cuprate. However, standard procedure by using seeding (e.g., SmBCO) proposes high overheating of the sample that is necessary to (i) avoid numerous formation of nuclei, (ii) increase liquid in the sample and (iii) eliminate an effect of precursor microstructure [1764].

In order to overcome the numerous nucleation and arbitrary growth of 123 crystals, beginning from substrate material at prolonged process of TSMG, as observed in test, the increasing of the formation temperatures for different *RE*BCO phases together with their ionic radiuses could be used [1192]. Then, a selection of corresponding *RE* composition permits to diminish an undercooling area $\Delta T = T_p - T_g$ (where T_p is the peritectic temperature and T_g is the temperature of grain growth in the sample under constant conditions) at slow cooling to homogeneous temperature (Fig. 3.34c). In this case, the grain growth rate decreases and unstable solidification eliminates. The kind of cooling process, that is, slow

[7] There are *hot seeding* and *cold seeding* [772]. In the first case, the seed is placed on YBCO sample at room temperature, in the second case the seed is placed at temperature T_{max} above the peritectic temperature (Fig. 3.34c).

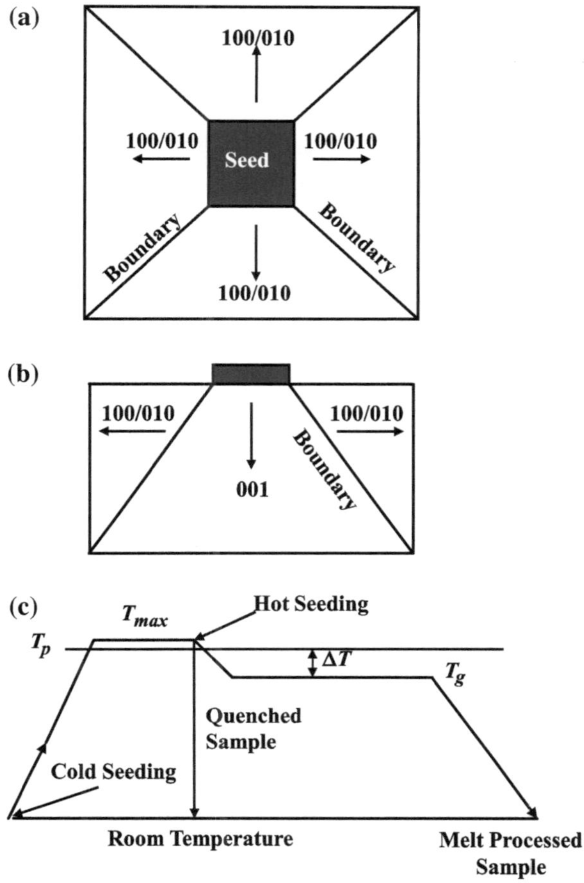

Fig. 3.34 Schematic diagram of the top-seeded melt growth process applied to form and control specimen texture (**a**) view from above, (**b**) View at the side and (**c**) applied heating schedule of the TSMG processing

continuous cooling or jump-like decreasing of temperature during one or some stages renders considerable effect on quality of the final crystallographic structure [771].

Carrying out the TSMG process in an air atmosphere with seeds from Nd, Sm, Eu and Gd forms *RE*BCO samples with decreased critical temperature T_c accompanied by broad superconducting transition. This is caused by the formation of hard precipitation of $RE_{1-x}Ba_{2-x}Cu_3O_{7-x}$ because of near cation radius between pointed *RE* and Ba. Carrying out the TSMG process in the controlled atmosphere with low partial pressure of oxygen (1 or 0.1 % of oxygen in Ar) (*oxygen-controlled-melt-growth-process* (OCMG)) [1217], it is possible to except formation this precipitation. In order to reduce the duration of TSMG process, the *multi-seeded-melt-growth* (MSMG) technique [1497] is used with corresponding thermal treatment. This process consists of some seeds on the YBCO sample. Now, *RE*BCO samples demonstrate high critical current density above 100 kA/cm^2 (77 K and 0 T), with corresponding control of composition and microstructure

[1461]. Considerable magnetic fields can be trapped in large-grain melt-processed HTSC (e.g., 10 T at 45 K [1423]), that is much more than in conventional magnets, and it is very important for various applications.

3.6 Superconductive Pnictides

There is a great diversity of chemical composition of the superconductive pnictides (see Appendix A.9) related with a possibilities to change into broad range the composition and structure of the charge reservoir layer and tolerance of the superconductive family to the substitution leading to significant variation of their physical properties. One of the most perspective lines of their processing is the synthesis of novel layered Fe-containing superconductors with intercalated organic cations and other organic moieties in the charge reservoir layer. Different available organic cations, neutral molecules and ligands cause various scenarios of arrangement of the interlayer space between Fe-containing layers. Moreover, the carefully controlled synthetic conditions allow achievement fine tuning of the doping level by governing the non-stoichiometry of the target compound with the incompleteness of the intercalation. All of the known processing methods for Fe-based superconductors require the use of high temperatures (800–1200 °C). Therefore, a creation of the hybrid superconductors with improved properties demands the use of soft synthetic approaches, including exchange reactions, low-temperature melts or ionic liquids, microwave heating, sonochemical treatment, supercritical fluids, etc.

The superconductive families *A*FeAs (A = Li, Na with $T_c = 21$ K) and $AEFe_2As_2$ (AE = K, Rb, Cs, Ca, Sr, Ba, Eu with $T_c = 38$ K), which contain no oxygen in above context to be most promising and convenient for the mentioned reactions. It is caused that what for these families the layer of the charge reservoir is formed by the cations of alkaline (*A*) and alkaline-earth (*AE*) metals. These metal ions demonstrate high mobility and chemical activity. For example, a study of synthesis and physical properties of NaFeAs and LiFeAs showed that the storage of the samples in the sealed quartz ampoules as well as in dynamic vacuum conditions at 300–400 °C led to decreasing the content of alkaline metal in the layer of the charge reservoir and significantly affected T_c and other physical characteristics of the superconductors [286]. The exposition of epitaxial $SrFe_2As_2$ film with a thickness of 100 nm during 4–6 h in air results in the appearance of superconductive properties of the specimen ($T_c = 25$ K) [690]. The effect has been explained by interaction of Sr with water vapor that led to the removal of a part of strontium from the crystal lattice of $SrFe_2As_2$ with the formation of vacancies.

Novel materials based on superconducting layered pnictides can be easily manufactured by the powder-in-tube method. For the wires made of $SmFeAsO_{1-x}F_x$ very high values of the upper critical field $H_{c2}(0) \geq 120$ T and critical current density $J_c = 3900$ A/cm^2 at 5 K have been demonstrated [520]. The transition temperature of the $SmFeAsO_{0.65}F_{0.35}$ wires was 52 K. It has been noted that J_c demonstrated very weak field dependence, indicating a good prospective for application this superconductor. The value of $J_c = 2 \times 10^5$ A/cm^2 at 5 K for the $SmFeAsO_{0.65}F_{0.35}$ powder with very weak field dependence showed that the $SmFeAsO_{0.65}F_{0.35}$ had a very high pinning force in grains. It has been shown that the oxygen-free superconductive pnictides *A*FeAs and *AE*Fe_2As_2 are more perspective than oxygen-containing family *Ln*OFeAs (*Ln* = La, Ce, Pr, Sm, Nd with $T_c = 56$ K), which are characterized by the higher anisotropy [1414]. Rather high upper critical fields over 100 T have been reported for iron oxypnictide superconductors. In particular a very large $H_{c2}(0) \geq 304$ T have been stated in $NdFeAsO_{0.82}F_{0.18}$ single crystal. Such high critical fields can be competitive with those of *A*15, MgB_2 and HTSC cuprates taking into account that iron pnictides have a great potential for high-field applications above 30 K, where conventional superconductors cannot play a role due to their low surface critical temperature, T_{cs} [1063].

Chapter 4
Experimental Investigations of HTSC

4.1 Experimental Methods of HTSC Investigations

4.1.1 Special Techniques

The complexity of HTSC structures and properties has resulted in a small number of directed observations and test dependencies of the type "structure–property." By this, *nuclear magnetic resonance* (NMR) and *electron-spin resonance* (ESR) are powerful experimental methods for investigation of condensed matter. The used techniques demand a doping superconductive cuprates by using small concentrations of Gd- or Mn-ions. The arising ESR-signal provides with information on electron spins of CuO_2-planes. Majority atoms in superconductive cuprates have isotopes which can lead to NMR-signals, namely: ^{89}Y, ^{135}Ba, ^{137}Ba, ^{63}Cu, ^{65}Cu, ^{17}O, ^{139}La, ^{207}Pb, ^{203}Tl, ^{205}Tl, ^{209}Bi, ^{151}Eu, ^{153}Eu. To regret, the abundant ^{16}O-isotope has not nuclear moment, therefore the oxygen studies require an enrichment of test specimen by ^{17}O-isotopes. NMR-nuclei are sensitive not only to magnetic effects, but if their spins are greater than ½, then electric and charge effects. As a result, NMR can differ the signals of definite nuclear isotope which presences in more than one crystalline cell. For example, NMR in YBCO can study atoms of Cu (or O) into CuO-planes separately of the atoms of Cu (or O) in corresponding links, and it can state a difference between oxygen atoms disposing into the same plane and forming Cu–O links in parallel and perpendicular directions of the links. Moreover, NMR can study both as static as dynamic effects in dependence on temperature and value of magnetic field. While NMR is a point probe, it can also state in some cases a dependence of wave vector. The studies of NMR intensity have discovered in some cases probable arising state of spin glass in HTSC at decreasing temperature [1584].

Among methods of three-dimensional observation the magnetic vortex structures in HTSC, the *small angle neutron scattering* and *spin precession of polarized muons* have an important application as the means of microscopic research of the local magnetic fields, but the *Bitter decoration* method is treated as the most often

I. A. Parinov, *Microstructure and Properties of High-Temperature Superconductors*, DOI: 10.1007/978-3-642-34441-1_4, used method of spatial resolution [172]. By using the neutron scattering method, it has been obtained direct proof for formation of wave of the spin (charge) density at the $T < T^*$, where T^* is the temperature of the pseudo-gap arising. The Bitter decoration method utilizes small ferromagnetic particles for decorating the magnetic domain structure. When these particles are sprinkled on a material, displaying at its surface an inhomogeneous distribution of magnetic flux density, the particles are attracted to the regions with the largest value of the local magnetic field. This method has demonstrated high effectiveness in visualization of the vortex structures localized at defects [381, 1752].

The great achievement in the magneto-optical characterization of the vortex structures consists of using the Bi-doped iron garnet thin films [751], which operate from below 4.2 K to above 500 K, that is, they very well fit the temperature window used in HTSC. These films come in two distinct varieties, one with magnetization vector perpendicular to the surface of the film (first investigations) and another with magnetization vector in-plane. As a result, a magnetic resolution of 10 μT has been reached. The spatial resolution can reach values less than the garnet film thickness, if the magnetic field gradients are sufficiently strong. The resolution of 0.4 μm has been reached for imaging obtained using films with thickness of 2 μm. Some achievements in using the *magnet-optical imaging* (MOI) technique for visualization of the vortex structures in different HTSC systems have been presented in overviews [902, 1389, 1756]. In combination with a digital camera and an image-processing system, time-resolved observations are possible. This method ensures the directed visualization of the superconducting currents and Abrikosov vortices, penetrating the sample. In this case, the total picture of defects is identified from voids and microcracks in polycrystal down to smallest heterogeneities in single crystallites. MOI technique has been used to research the structure of single crystals [382, 750, 902], bi-crystals [1338, 1391, 1717] and polycrystalline materials [750, 1337, 1790]. It should be noted that for the best understanding of HTSC intrinsic properties, it is very important to carry out comparative investigations of the “perfect” (i.e., monocrystalline) material and intergranular boundaries [835, 1766, 1767]. In particular, the discovery of the lamellar-like defects, which are weak links in Bi-2212 crystallites obtained by the directed solidification technique [1709, 1767], can be used to interpret comparative data on superconducting properties of Bi-2212 polycrystals and related materials.

The *microscopic Hall probe array* method is a promising novel technique to measure the component of the local magnetic field perpendicular to the sample surface [1864, 1865]. With this aim, two-dimensional Hall sensors for research of electronic gas are etched in a GaAs/Al GaAs hetero-structure. The active area of the Hall elements was initially $\geq 10 \times 10\ \mu m^2$ and was gradually reduced to $3 \times 3\ \mu m^2$. In the last case, the arrays have been fabricated, including up to 20 Hall elements with the magnetic field sensitivity better than 10 μT. In addition to a research of the vortex-lattice melting transition [1865], this method has been utilized for measurements of local magnetic relaxation in crystals of $YBa_2Cu_3O_{7-\delta}$ [1864] and $Nd_{1.85}Ce_{0.15}CuO_{4-\delta}$ [533].

The *scanning SQUID microscopy* provides highest sensitivity of vortex structures [874]. The acting microscope demonstrates a wide spectrum without-alternative applications based on high sensitivity of the device to weak magnetic fields into broad frequency range. It consists of a mechanical *xy*-scanning system combined with an integrated miniature SQUID-magnetometer. The SQUID is fabricated with 1 μm Nb–AlO_x–Nb junctions and octagonal pick-up loop with a diameter of 10 μm integrated on a silicon substrate and has a sharpened tip. The tip is brought in immediate contact with the sample during the scanning and serves for distance regulation. The spatial resolution (about 10 μm at the magnetic sensitivity 20 pT/$Hz^{1/2}$) is limited by the size of the pick-up loop. In 2002, *laser SQUID microscopy* replaced the SQUID microscopy ensuring significant expansion of the scanning area and the method sensitivity.

The rapid development of the *scanning probe microscopy* (SPM), following the pioneering work [143], has also contributed to vortex imagine research. The achievements in the investigation of HTSC systems by using this method are presented in overview [347]. The *scanning tunneling microscope* (STM) measures a tunnel current between conductive needle of the microscope and studied conductive surface, allowing observation of separate atoms and molecules with characterization and analysis of electronic nature around them. Moreover, there is a possibility to manipulate single atoms and molecules. At the same time, STM could be only used for study surfaces with definite electric conductivity. The atomic force microscope (AFM) has been developed for studying super-small forces (lesser than 1 μN), acting between the needle tip of AFM and specimen surface [141]. AFM is used for investigation of any surfaces (conductive or insulating) and normal forces in micro- and nano-scales. Then, AFM has been modified with aim to measure both as normal as lateral forces [133, 134, 136, 447, 504, 505, 1099, 1137]. Its modifications have been named lateral force microscope (LFM) and friction force microscope (FFM). Created noncontact atomic force microscope (NC-AFM) and based on it methods are powerful instruments of studying the surfaces in atomic-molecular scale due to NC-AFM possesses the following characteristics: (i) it has atomic-scale resolution, (ii) it can measure atomic forces (so-called, *atomic-force spectroscopy*), (iii) it can observe even insulators, and (iv) it can measure mechanical response (for instance, elastic strain).

At the same time, STM was applied to create features at nano-scale by means of local heating or introduction of chemical reactions by using STM needle and also for nano-scale mechanical processing. Moreover, STM and AFM have been used for extreme increasing (in 10^3–10^9 times) atomic-scale images up to macroimages with high resolution and with aim of their spectroscopy. While SPM techniques have been originally based on their displaying capability, the physical and chemical studies of interactions of the "probe-sample" type and the SPM-based quantitative analysis of the electronic, magnetic and chemical surfaces became especially popular in the last years. Nano-engineering and nano-technology use SPM often because this allows one to reduce research and manipulation by surfaces at atomic scale.

In a whole, the instrument families based on STM and AFM named scanning probe microscopes have been developed for different applications in scientific and industrial interests [305]. They include AFM, STM, FFM (or LFM), scanning electrostatic force microscopy (SEFM), scanning force acoustic microscopy (SFAM) or atomic force acoustic microscopy (AFAM), scanning magnetic microscopy (SMM) or magnetic force microscopy (MFM), scanning near-field optical microscopy (SNOM), scanning thermal microscopy (SThM), scanning electrochemical microscopy (SEcM), scanning Kelvin probe microscopy (SKPM), scanning chemical potential microscopy (SCPM), scanning ion conductance microscopy (SICM) and scanning capacitance microscopy (SCM). At the measurement of forces (as it occurs in AFM, FFM, SEFM, SFAM and SMM), it is also named scanning force microscopy (SFM). While these instruments achieve atomic resolution and be ideal for basic researches, they are used also for industrial applications not requiring atomic resolution.

Magnetic force microscopy has been used to imagine vortex systems and research their interaction with defects in the HTSC thin films [1851]. The imagines of Abrikosov- and Josephson-vortices trapped in thin-film YBCO washer *dc*-SQUIDs have been obtained, using low-temperature scanning electron microscopy at liquid nitrogen temperature [839]. During scanning of the sample surface by the electron beam, the signal is generated due to the beam-induced local displacement of the vortices, resulting in a change of the flux coupled from the trapped vortices into the SQUID loop. In this way, 1/*f*-noise sources in the device have been identified. The spatial resolution of this method is about 1 μm.

In SPM, an image is created by raster scanning of probe needle of the specimen surface and measurement of specific strongly localized parameters of interaction in dependence on disposition. SPM is based on several main types of interaction, including STM, which measures electronic tunnel current, AFM, measuring a force interaction, and SNOM, which studies local optical properties by using effects of the near field. These methods allow characterization of numerous properties (structural, mechanical, electronic, optical) in essence of any materials (in particular, superconductors) studied in any media (vacuum, liquid, air). Unique combination of resolution in nano-scale, achievements of electronic microscopy and wide introduction led to application of SPM practically in all regions of nano-engineering and nanotechnology. Several best technologies have been developed for SPM or borrowed from other techniques. Piezoelectric tube scans allow ones to state in sub-angstrom scale a disposition of microscope needle and/or sample in space. The systems of optical deviation and cantilevers created by using micro-technologies can estimate forces in AFM down to values of some piconewtons. The sensitive electronics can measure STM currents that are lesser than 1 pA. The filamentary optics of high resolution and sensitive photodetectors can manipulate and detect small optic signals of SNOM. The control of environment has been developed into ultraviolet at cryogenic temperatures and liquids. In this case, the vibration and displacement characteristics are controlled so that the probe needle can hold a single molecule during several hours at conducting observations [135].

By using scanning tunneling microscopy at low temperatures, the spatially varying quasiparticle density of states may be measured, yielding important information about electronic properties of vortices and vortex structures

[670, 671]. Experimental studies of superconductive gap by using STM have played an important role in development of understanding behavior of the low-temperature superconductors ensuring direct observation of gap in state density of superconductor, high-precise measurements of size, shape, temperature and field dependence of this gap, values of electron–phonon spectrum density, and also renormalized Coulomb's pseudo-potential. These measurements and calculations have ensured strict confirmation of electron–phonon mechanism for superconductivity of conventional superconductors. Tunneling Cooper couples between low-temperature superconductors has demonstrated macroscopic quantum coherence of superconductive state and also corresponding basic phenomena and applications. However, tunneling in superconductive cuprates is far more complex and varied [875]. The most important question from those which are addressed to scanning electron microscopy is the following one: what is the mechanism of Cooper pairing in cuprates? As it is known, HTSC mainly demonstrate $d_{x^2-y^2}$ pairing symmetry that is supported by definite STM observations. Dominant pairing symmetries can arise at surfaces and interfaces of non-conventional superconductors. Presence these dominant components could break the time of symmetry replacement leading to interesting tunnel effects. STM has shown that several HTSC specimens have significant spatial heterogeneities in their superconductive gap and in local tunnel density of states. The Fourie-imaging method has been developed for study of transport properties on the base of research data in real space by using scanning tunneling microscopy [1111].

AFM is most broadly used form of SPM because not requires electrically conducting specimen (as in STM) or optically transparent sample or substrate (as in SNOM). The basic AFM-modes measure the sample topography only with that restriction what the specimen is deposited onto plane surface and be enough rigid for displaying image. Since AFM can measure various forces including van der Waals forces, electrostatic, magnetic forces and also forces of adhesion and friction, then special AFM-modes can characterize electric, mechanical and chemical properties of specimen in addition to its topography.

Lorentz microscopy for vortex imaging uses a high-voltage field emission electron microscope, generating a coherent electron beam. The electron phase shift due to vortices in a tilted specimen is detected in an appropriately defocused image [634]. This method has been used to investigate the vortex lattice in Bi-2212 single crystals [635] and include the imaging vortex motion [1691].

Josephson tunneling presents very interesting experiments giving information not only on the magnitude but and phase of superconductive order parameter, that is on the gap [875]. "Tortuous" twining boundaries are observed by using *transmission electron microscopy*. At the same time, the information about junctions between parallel layers of CuO_2 in HTSC being intrinsic Josephson junctions could be obtained by analyzing volt-ampere characteristic.

Scanning tunneling spectroscopy in the mixed state of HTSC is difficult because of the problem of the sample surface quality and the strict requirements for highly reproducible tunneling conditions. The first successive experiments have been fulfilled on YBCO single crystals with vector ***B*** oriented along *c*-axis [1070]. The experiments on Bi-2212 single crystals brought an unexpected result; they showed that the spectra inside the vortices differ weakly from the zero field spectra, in contrast to the situation in YBCO [1425]. Therefore, the detection of the vortices in BSCCO is much more difficult than in the YBCO system.

Angle-resolved photoemission spectroscopy (ARPES) with ***k***-spatial resolution (where ***k*** is the wave vector) is used widely for investigation of HTSC transport properties [1878]. The methods of ARPES, angular magnetoresistance oscillations (AMRO) and Raman scattering create subsequent picture of strong anisotropy for the scattering rate (both as transport as quasiparticles) into basal plane in HTSC, which intensifies with decreasing doping.

Some other applications of the experimental methods for researching of HTSC structures are as follows:

- methods for measuring electromagnetic properties (J_c–H, E–H, magnetic sensitivity, etc.), used to separate components of the magnetic flux pinning and grain connectivity, causing J_c [419, 708, 1335];
- optical and scanning electron microscopy methods, used to study intercrystalline boundary structure, surface morphology and phase composition [518, 1186, 1187];
- atomic force microscopy, used to research surface morphology [1222];
- X-ray scattering to analyze HTSC epitaxial layers with very small twinned regions [475];
- X-ray diffraction methods, used to research texture and phase composition [1187, 1222];
- spectroscopy methods to study microstructure and phase composition [7, 1017];
- electron probe microanalysis to investigate chemical composition of surfaces and interfaces [1855];
- using transparent organic analogs to study peritectic reaction in YBCO [1505];
- using picture of the YBCO volume magnetization to identify internal defects [1166];
- investigation of individual filaments escaped from BSCCO tapes [221].

It should be noted that the broad spectrum of methods of the surface and interface reconstructions and also experimental approaches developed for semiconductor structures and nanomaterials [1179] can be used to study HTSC structures at nanosize scale.

4.1.2 ARPES

From date of HTSC discovery, the angle-resolved photoemission spectroscopy (ARPES) has ensured key experimental evidences allowing ones to state and study electronic structure of HTSC (see, for instance [1550, 1878]). The ARPES tests are instruments for observation and detalization of pairing behavior, caused by electron–phonon interaction, and also to register evolution of doping HTSC cuprates. Spectra, obtained from strongly coupled states, polaronic spectra observed in optimal-doped and underdoped cuprates relate to the spectra of Migdal-Eliashberg type in optimal-doped and overdoped cuprates. In addition to the pointed dependence on doping, cuprates demonstrate pronounced anisotropy in Brillouin zone, namely clear quasiparticles disposed along nodal direction which significantly run in anti-nodal region of the underdoped cuprates, anisotropic electron–phonon pairing displaying on specific modes and identified in optimal-doped and overdoped compounds and dominant scattering via two parallel pieces of Fermi surface in anti-nodal region for all levels of the doping. These phenomena also contribute into pseudo-gap behavior. Within limits, where it is applied Migdal-Eliashberg picture, the cuprate spectra are like with those, which have been stated in strongly coupled electron–phonon superconductors (for example, Pb). On other hand, cuprates can deviate from this conventional picture. In underdoped regime, the charge carriers could be considered as small polarons in antiferromagnet in background of high electronic correlation. At the same time, the doped compounds require anisotropic electron–phonon structure for detalization of prevailing pairing mode in superconductive state. A correction of electronic vortex structure to the electron–phonon pairing could increase and define for proper phonons the anisotropy of electron–phonon interaction. The subsequent picture arises for superconductive cuprates, which combines strong anisotropic electron–phonon pairing, specific phonon modes leading to this pairing, and electron–electron interaction modifying the electron–phonon structure. This combination together with subsequent experimental and theoretical efforts really could lead to understanding high temperature of superconductive transition with *d*-wave pairing in cuprate superconductors.

Among achievements of ARPES, it is necessary to point the following:

(i) most earlier identification of dispersion and great Fermi surface [1267];
(ii) investigation of anisotropic superconductive gap, suggesting *d*-wave order parameter, which has been observed in [1551];
(iii) observations of pseudo-gap in underdoped specimens [376, 1088].

After these observations, ARPES technique has been subjected to tremendous alterations connected with cardinal improvement its energetic and momentum resolution, that led to new discoveries which were impossible ten years ago. This revolution in the ARPES technique and its scientific value were a result of dramatic changes of four components, namely: (i) resolution and effectiveness of equipment, (ii) possibility to carry out different manipulations with samples, (iii) high quality of the specimens, and (iv) good statement of scientific problem. ARPES has ensured most important information on electronic structure of HTSC, including structure of energetic levels, Fermi surface, superconductive gap and pseudo-gap.

Among recent achievements of ARPES, it is necessary to name the following:

Energetic levels and Fermi surface. Two-layer splitting Fermi surface has been clearly stated in overdoped Bi-2212 [150, 290, 464]. It may be assumed that this phenomenon also exists in underdoped and optimal-doped Bi-2212 samples.

The measurements conducted in [913] have shown that there was a weak splitting along nodal direction (0, 0)–(π, π). Then, it has been stated [1878] that in four-layer $Ba_2Ca_3Cu_4O_8F_2$ were clearly identified at least two sheets of the Fermi surface.

Superconductive gap and pseudo-gap. After first identification of anisotropic superconductive gap [1551], the following measurements, fulfilled on different HTSCs: Bi-2212 [157, 376], Bi-2201 [1088, 1551], Bi-2223 [465, 1207, 1490], $YBa_2Cu_3O_{7-\delta}$ [1054], LSCO [755], stated universal character of behavior of the anisotropic superconductive gap in above superconductors doped by holes which agreed with *d*-wave pairing symmetry. At the same time, there is open question: is the gap shape simply *d*-wave or additionally should be included higher harmonics? The measurements, conducted in [57, 1489] on superconductors doped by electrons, also discovered a superconductive gap. Then, the next question arises: if the material has multi-sheet Fermi surface, then is the superconductive gap on various sheets of the Fermi surface the same, or not? This question returns to study of superconductive $SrTiO_3$ [140], for which it has been shown by using measurement of tunneling that different sheets of the Fermi surface can demonstrate various superconductive gaps. Significantly improved possibilities of ARPES have allowed ones to observe different superconductive gaps on various sheets of the Fermi surface in 2H-$NbSe_2$ [1838] and in MgB_2 [1600]. Among HTSC, Bi-2212 shows clearly two sheets of the Fermi surface, however in this case a difference between superconducting gaps is absent. At the same time, it has been stated [1878] that in $Ba_2Ca_3Cu_4O_8F_2$ observed clearly two sheets of the Fermi surface with different superconductive gaps.

Time breaking of symmetry replacement. In [1746], it has been proposed theoretically that utilization for ARPES the light with circular polarization could become a probe of the time breaking of symmetry replacement which associated with pseudo-gap statement in underdoped samples. First observation of this effect has been presented in [807]. However, this result has not been confirmed in [158], and additional data remain to be contradictory.

At the same time, today there is only weak consensus in relation to results obtained by using ARPES, that no allows their perfect comparison with existing transport properties. For instance, in [569] has been obtained linear dependence of the scattering rate on prevailing energy in normal state of HTSC on nodes, as it is followed from thickness of curves of the impulse distribution. At the same time, it has been stated [913] that this dependence is quadratic on energy in cuprates with optimal doping. In several cases this dependence demonstrated a kink related to pairing to some bosonic mode [976].

Influence of phonons on normal state of transport in HTSC is minimal, as it has been shown in [733]. However, this result significantly contradicts to assumptions of [976], that ARPES results demonstrate strong electron–phonon interaction in HTSC into a whole doping range. One of the ways to reconcile these contradictory results consists in remainder that ARPES measures longevity of quasiparticles, which is determined by all events of scattering, while the transport of electric charges is carried out by high-angle scattering. While electron–phonon scattering includes only weak transfer of charge at low energies, at the same time electron–electron (Umklapp) and spin-fluctuation scattering on their nature are far effective at perturbation of transport current. So, the electron–phonon interaction could be observed for ARPES, but in a whole be transparent for applied electric current.

Electronic Raman scattering is also a function of two-particle response, but in difference on *dc*- and optical conductivities possesses some $\boldsymbol{k}$-resolution. The regions in $\boldsymbol{k}$-space could be selected by using polarization changes of intrinsic and scattered photons. In this case, inclination of spectra at the circular frequency, $\omega = 0$, is equal to $\boldsymbol{k}$-resolved *dc*-conductivity. Actually, it has been again supported significant anisotropy of rate of the transport scattering, and that it decreased with increasing doping and in considerable part with rising T [616].

Tests show clearly that interlayer transfer in all cuprates with optimal doping (including $YBa_2Cu_3O_{6.95}$) is incoherent and real coherence of c-axis motion initiates only into region of phase diagram corresponding to optimal doping. ARPES measurements [1732], conducted on other oxides, confirm that in two-dimensional case a loss of interlayer coherence leads to development of incoherent response into plane.

Obviously, the underdoped case is most unusual region of phase diagram of the superconductive cuprates. In this case, one of the most basic problems for models based on the Fermi liquid is breaking Luttinger rule of sums, assumed from measurements of Hall effect. This significantly contradicts to situation of optimal doping, for which the Luttinger theorem is fulfilled. At the same time, novel technique of averaging, used in ARPES [1879], has allowed one to discover a whole Fermi surface in the underdoped cuprates with region which agreed with the Luttinger theorem.

4.1.3 Measuring Methods of Mechanical and Strength Properties

The measuring methods of mechanical and strength properties of HTSC repeat in many aspects the tests, which have been applied broadly for various ceramics and composites. In particular, the so-called R-curves are used to study fracture toughness of volume samples, changing with crack propagation [1835]. The fracture toughness has also been estimated by using the method of Vickers indentation [1002, 1404]. In this case, there are specific features connected, for example, with the cryogenic temperature effects on superconductors or with very small characteristic sizes of superconducting components. The concept of effective volume and Weibull's distribution function are used for test estimations of strength characteristics [1461].

In order to estimate elastic and acoustic properties, ultra-sound methods are used [550, 1419]; anisotropic stresses at sample surface are computed by application of X-ray diffraction [623]; mechanical behavior, strength and fracture toughness of

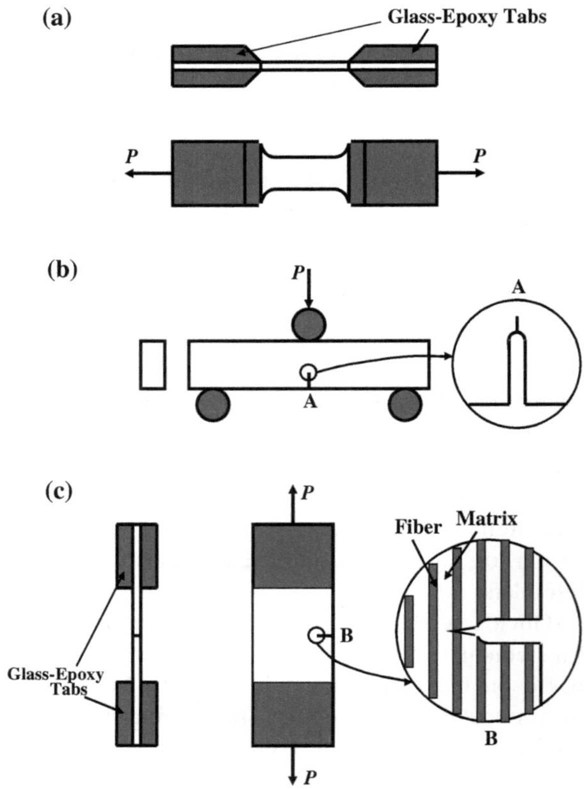

Fig. 4.1 Specimens for evaluation of the mechanical behavior of monolithic BPSCCO and unidirectional Al_2O_3/BPSCCO composite in: (**a**) tensile test, (**b**) three-point bending notched specimen and (**c**) single-edge notched fracture test [1172]

sample bulks are computed during tests fulfilled at room and cryogenic (77 K) temperatures, using tension/compression [1856], schemes of the three-point bending [1172] and four-point bending [551], rapture [1172] and bending [549] of notched specimens, creep at compression [1454] and also investigations of thermal cycling with cycles between room and cryogenic temperatures [1687].

In order to study bulk sample density the methods of immersion [1549], picnometry [1284] and Archimedes [1250] are applied. The methods of system analysis are developed, for example, for design of material properties and critical behavior of HTSC materials and composites [781, 1304, 1813]. In order to estimate stress–strain state of superconducting bulks, special devices are designed, in particular with integrated gauges, which are able to measure strains with required precision under different thermal and magnetic fields, taking into account anisotropy of properties in *ab*-plane and in the *c*-axis direction [1170, 1171]. Several methods for the mechanical tests of BPSCCO bulks are shown in Fig. 4.1. Figure 4.2 shows two schematic illustrations of how the strain gauges and Hall probe for the measurements of the thermal and magnetic properties are positioned on the YBCO surface.

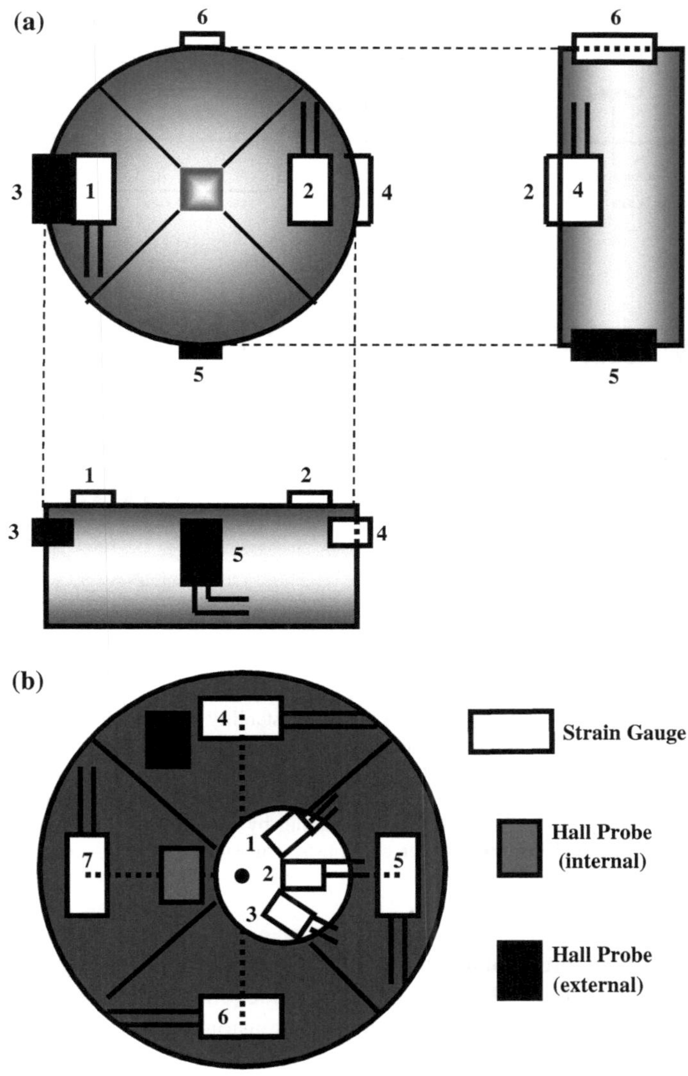

Fig. 4.2 (**a**) Schematic illustration showing how the strain gauges are mounted on the YBCO surface (gauges *1* and *2*, positioned on the top surface, are used for the measurements of the strain along the *ab*-plane, and gauges *3–6*, fixed on the side, are used to measure the strain along the *ab*-axis on the *ac*-plane (gauges *3* and *4*) and along the *c*-axis (gauges *5* and *6*) [1171]; (**b**) schematic illustration of how the strain gauges and Hall probe are attached to the YBCO sample for the measurement of the mechanical properties and trapped magnetic field (Nos. *1–3* are internal gauges of strain, Nos. *4–7* are external gauges of strain) [1170]

BSCCO/Ag tape superconductors are tested at room and cryogenic temperatures: on tension [1471], longitudinal compression (stability loss) [1753], transversal compression [425], compression and tension by using U-shaped spring dais

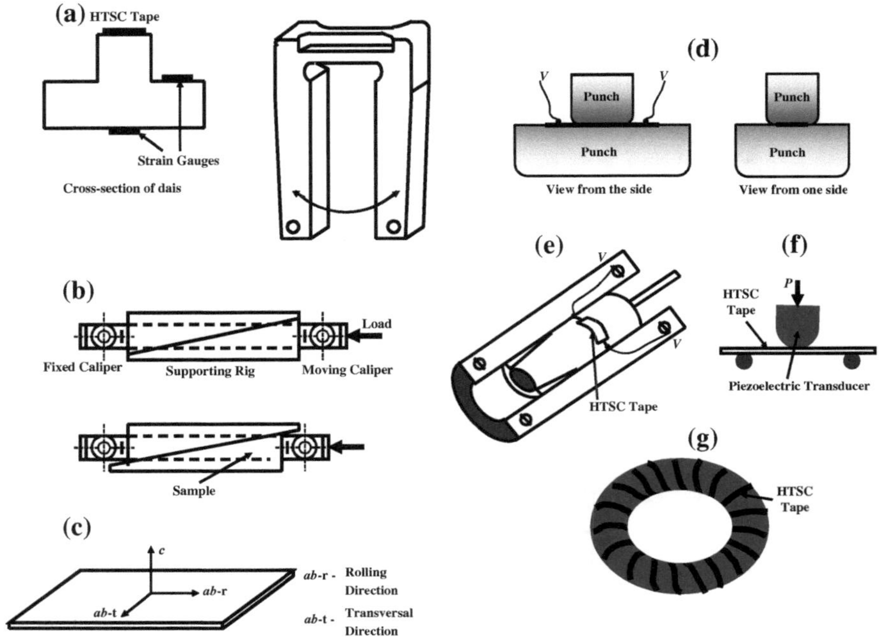

Fig. 4.3 Test methods for investigation of HTSC tapes: (**a**) on compression and tension by using U-shaped spring dais with soldered sample [878], (**b**) on stability loss at longitudinal compression [1753], (**c**) on tension or cyclic fatigue along *c*-axis [699], (**d**) on transversal compression by using two punches with evaluation volt–ampere characteristic [425], (**e**) on bending by using measuring devices of critical current depending on continuously decreasing (or increasing) diameter of bending [1664], (**f**) on three-point bending by using a device in which loading is carried out by the piezoelectric transducer of acoustic emission [1325], (**g**) on bending around helical surface [1247]

[1664], bend under three-point bending [1256], by hand [929] or by using special devices [931], around cylindrical surfaces (one side [931] and two side [1382]) and helical surfaces (an investigation of winding on tension and bending) [1247]. They are subjected to thermal cycling including change of temperature from room to cryogenic value and back [1754], to fatigue tests, including cycles of the tape bending and straightening [1076], to cyclic loading and unloading [880] or cyclic changing of external tension [1623], low-cyclic [1663] and multi-cyclic (up to 10^7 cycles) [700] fatigue under longitudinal tension, to tension in the *c*-axis direction and multi-cyclic (up to 10^5 cycles) fatigue [699]. Several test methods for HTSC tapes are shown in Fig. 4.3.

In order to evaluate superconducting core density of tapes, the measuring methods of microhardness, using Vickers diamond pyramid [1819] and Knoop non-symmetric indenter [1332], are applied. Relative changing of density is estimated for wire-drawing by using registration of its lengthening in dependence on decreasing of the cross-section diameter [627, 918]. The microhardness profiles

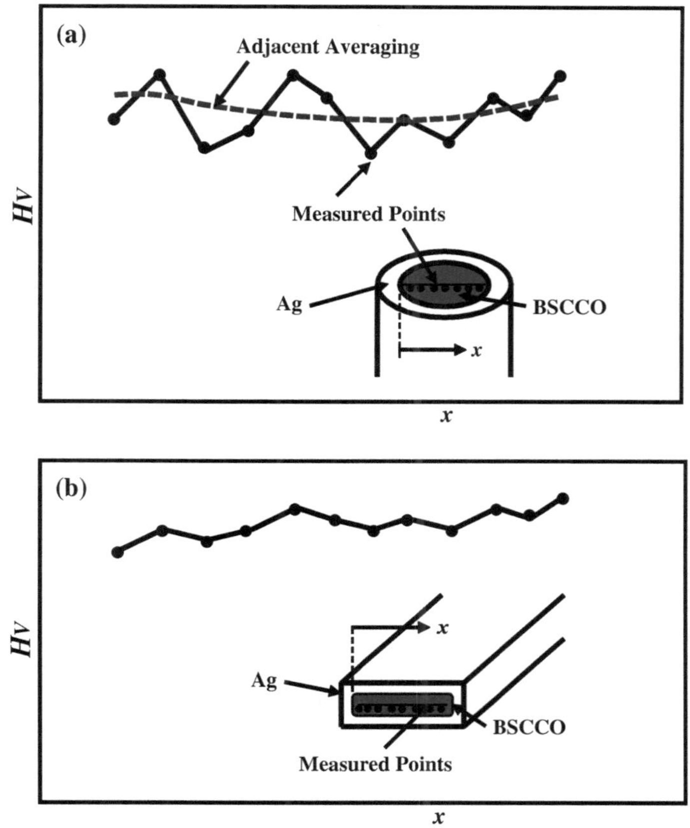

Fig. 4.4 Proper microhardness profiles of HTSC wires and tapes after: (**a**) drawing and/or extrusion, (**b**) rolling and/or pressing [732]

(i.e., monitoring of transversal cross-sections in different points of their diameters or main axes) are applied to estimate superconducting core density of tapes and wires after deformation (drawing, extrusion, rolling and pressing; see Fig. 4.4), [732] or single filaments of multifilament samples [731]. The ultrasound vibration method is used to evaluate the microcracking density [35, 1581]. The method for investigation of mechanical aging that model mechanical effects, developing during a conductor stranding, was proposed in [1098]. This test models bending, twisting and tensile forces, experienced by the HTSC wires during conductor stranding (Fig. 4.5). Moreover, to predict thermal cracking, the wire aging is studied under conditions of elevated temperatures [1098].

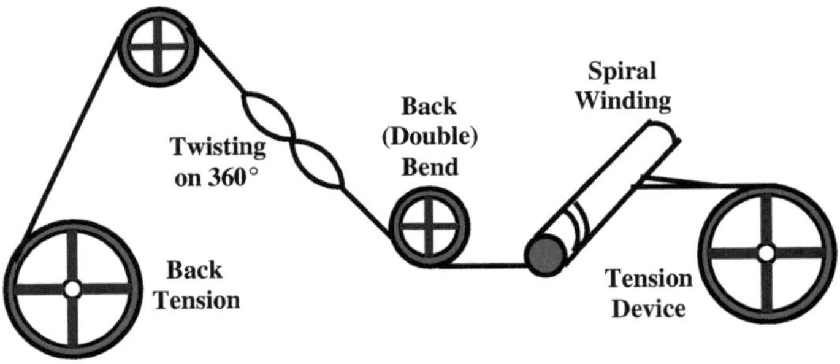

Fig. 4.5 Schematic of mechanical aging test, designed to simulate the conductor stranding process [1098]

4.1.4 Acoustic Emission Method

Acoustic emission (AE) is the physical phenomenon connected with irradiation of elastic waves by a solid under loading due to dynamic local reformation of the material's internal structure. This method has been applied to research different physical and mechanical properties of HTSC, accompanying microstructure and phase transformations during fabrication and loading of the superconductors at various internal and external loading. Intensive AE has been displayed at $YBa_2Cu_3O_{7-x}$ sintering [404] and thermal treatment [1608]. It is observed due to microcracking, caused by anisotropic compression of crystalline lattice in narrow temperature range at cooling after sintering. Acoustic emission has been studied in YBCO ceramic at the sample heating from liquid nitrogen temperature [1615]. The observed AE has been correlated with the microstress relaxation at the grain boundaries due to the thermal expansion anisotropy of grains. As possible mechanisms of stress relaxation, the dislocation displacement and microcracking initiation at most unfavorably oriented grain boundaries have been assumed. The oxidation kinetics of YBCO ceramic has also been studied by using AE method [405]. The oxidation has submitted to exponential law. The results showed that a narrow surface layer oxidized initially, but the oxygen solubility into material was controlled by the rate of the volume diffusion. YBCO ceramic has been investigated in the 25–700 °C range [403]; AE irradiation has been registered in the 260–300 °C; and it has been shown that this was caused by the structure phase transition of 90°-phase into 30° one during redistribution of oxygen into material. AE method has been used to register the structure transformations in HTSC during thermal cycling [1541]. The effect of the cyclic change of current in YBCO ceramic has been studied by the AE method at 77 K [407]. In this case, a penetration of magnetic flux lines into sample under the action of own magnetic field has been stated. A pick of acoustic emission in $Tl_2Ba_2CuO_{6+x}$ sample at critical temperature T_c [1283] has been observed. The relaxation anomalies stated have been correlated with a change of charge state of linear defects and parametric changes of the cuprate layer structure.

Fig. 4.6 Bi-2223/Ag monocore and multi-filamentary tapes

The processes of microcracking and microplasticity of $GdBa_2Cu_3O_y$ and $Bi_2Ca_2Sr_2Cu_3O_y$ monocrystals and also of YBCO ceramic have been studied by the local indentation method accompanied by registration of AE signals [165]. Comparison of AE data and fracture pictures permitted to estimate a character of inelastic strain of HTSC. Then, it may be concluded that AE method can operatively control mechanical properties of HTSC. In the case of $GdBa_2Cu_3O_y$ monocrystals, the AE signals have been caused by formation of characteristic microcracks at the angles of the indenter imprint. In the process of local loading of $Bi_2Ca_2Sr_2Cu_3O_y$ crystals, formation of wedge-shaped defects, accompanied by initiation of low-frequency component of AE signals, was observed. It may be assumed that these defects are result of joint processes of microcracking and delamination in BSCCO. The AE method was also applied to research inelastic strains of HTSC ceramics during macroscopic mechanical tests [164] and microindentation [567]. AE was used to investigate secondary sintering of Bi-2223/Ag tape during heating after primary sintering and rolling [406]. In this case, it has been assumed that the process of liquid-phase healing of cracks, formed during rolling, has appeared as a possible source of AE. This result leads to the conclusion on the possibility of application of the AE method as the non-destructive control method in the processing of Bi-2223/Ag tapes.

In order to answer the issue of the possibility to apply the AE method for estimation of microdamage initiation and propagation during bending monocore and multi-filamentary Bi-2223/Ag tapes (see, Fig. 4.6), a test device has been designed, which is presented in Fig. 4.7 [1325].[1] In the tests carried out at room temperature, the following parameters were estimated: (i) sample displacement under indenter (Δl), (ii) the applied loading (P_m), (iii) the number of AE impulses

[1] The modified device for studying damage of HTSC-tapes under three-point bending by using AE and optic interferometry methods together with structure scheme of optic device measuring small displacements are shown in Fig. 4.8.

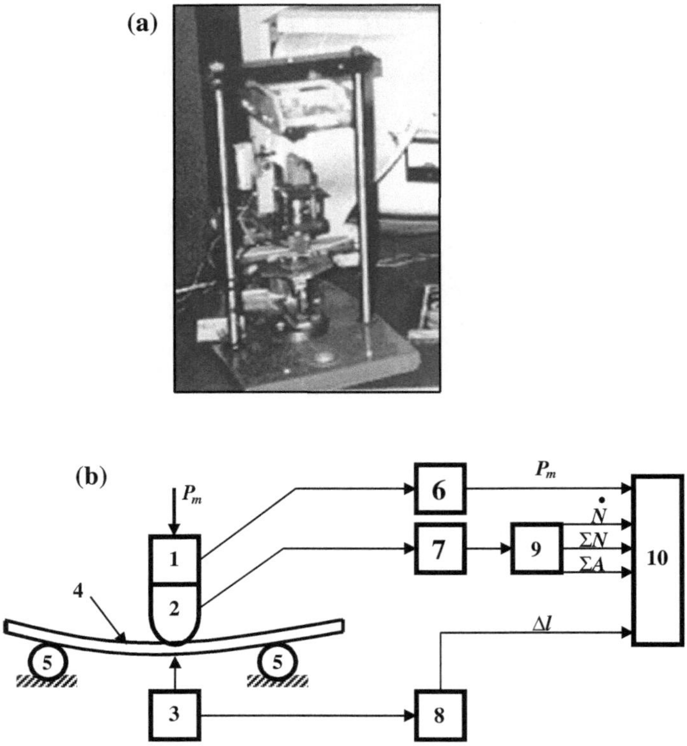

Fig. 4.7 Test device (**a**), and its schematics (**b**). Dinamometer (*1*), cylindrical indenter of 20 mm diameter with AE transducer (*2*), measurer of displacements (*3*), Bi-2223/Ag tape (*4*), supports (*5*) with 10 mm diameters and 35 mm spacing, tension-metric devices of agreement (*6, 8*), preliminary amplifier (*7*), system block of AE signal treatment (*9*) and registering device (*10*)

(ΣN), (iv) a sum of AE signal amplitudes (ΣA), (v) AE signal activity ($\dot{N}$), (vi) the loading time (t).

Deformation was estimated as [1256] $\varepsilon = 6\ d\Delta l/L^2$ and deformation rate as $\dot{\varepsilon} = 6dV_n/L^2$, where $V_n = \Delta l/t$ is the loading speed, d is the thickness tape, L is the distance between supports. A test specimen is loaded consistently in several sites. The AE signal activity was discontinuous in the form of single bundles, characterizing non-uniformity of the fracture process or defect initiation under loading, which was proper for brittle fracture.

First, a contribution in the AE activity from deformation of a silver sheath was stated. For this, the tests of pure silver tapes with the same sizes and conditions of loading, as for the tested Bi-2223/Ag tapes, have been carried out. Moreover, it has been proved that loading device did not cause acoustic noises. In order to confirm a character of fracture and correlation of the AE activity with the sample damage during loading, a comparison was carried out between the obtained test data and the experimental results (obtained by Polyanskii [1387] by MOI method [1390]) of

(a)

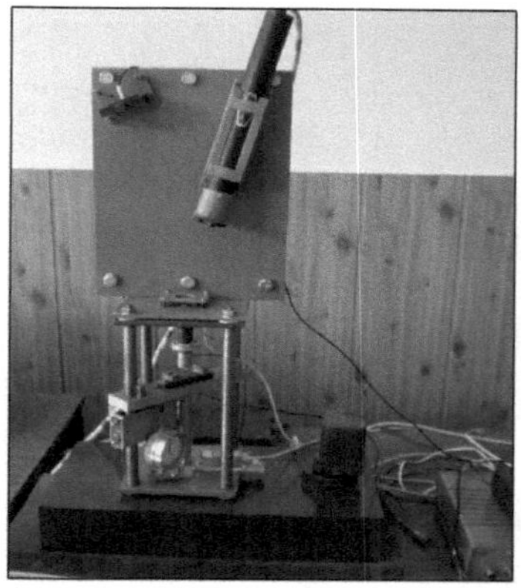

(b)

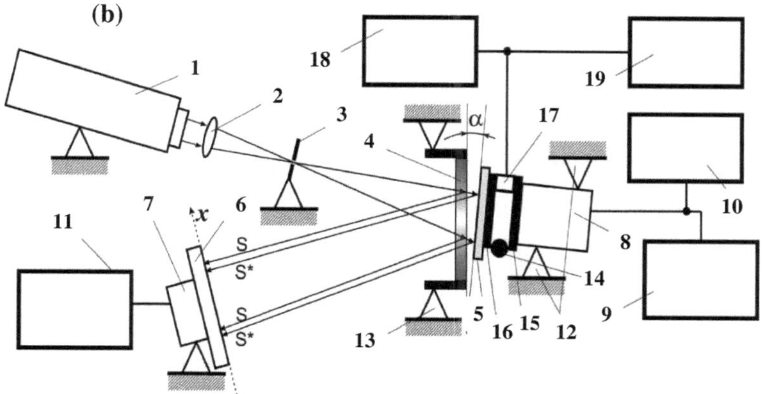

Fig. 4.8 Modified device for studying damage of HTSC-tapes under three-point bending by using AE and optic interferometry methods (**a**), and structure scheme of optic device measuring small displacements (**b**) [1159]. Laser (*1*), optic system (lens *2* and diaphragm of spatial filter *3*), beam splitter (*4*), reflector (*5*), screen (*6*), digital photo-receipt device (*7*), piezoelement (*8*) with source (*9*) and digital voltmeter (*10*), PC (*11*), attachment mechanism (*12*) of the piezoelement 8, device (*13*) for fastening the beam splitter *4*, plates (*15*) and (*16*) which are jointed by pivot (*14*) from one edge, but rigidly connected with the piezoelement *8* from other edge, piezoelement (*17*) with source (*18*) and digital voltmeter (*19*)

critical current I_c and critical current density J_c in magnetic fields directed along c-axis of the tapes. Magneto-optical images demonstrated pictures of the magnetic flux arrangements directly in regions loaded for two regimes: zero field cooled (ZFC) and field cooled (FC). In ZFC, the sample was cooled below T_c in the absence of a field, and then a field was applied. In the resulting image (Figs. 4.9 and 4.10) the sample was shielding the magnetic field. For the FC regime, the sample was cooled below T_c in the presence of a magnetic field, and then the field was turned off, so the sample was trapping magnetic flux (Fig. 4.11).

The test results for monocore tapes #1 and #3 and also for multifilamentary tape #2 by using the AE and MOI methods are presented in Tables 4.1 and 4.2 and

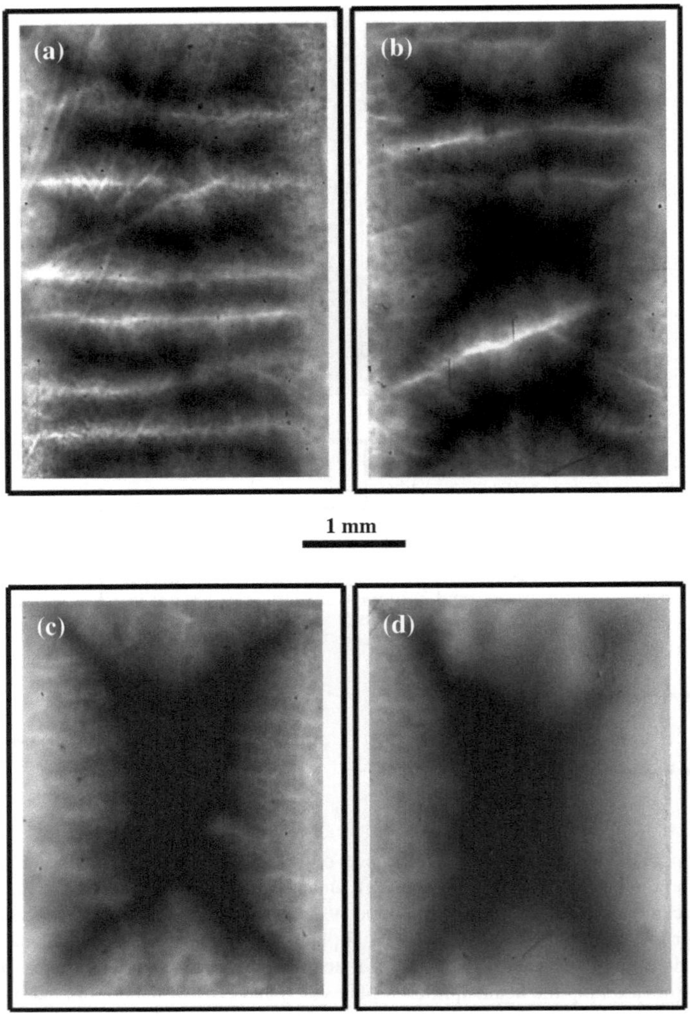

Fig. 4.9 Magneto-optical images of the monocore tape #1, ZFC regime ($T = 13$ K, $H = 600$ Oe): (**a**) site 1 (two-side bending), six cracks are seen; (**b**) site 2, 3 cracks are seen; (**c**) site 3, no cracks; and **d** site 4, weak defects [1325]

Figs. 4.9, 4.10, 4.11. As it follows from Table 4.1 and Figs. 4.9, 4.10, in a whole, there is a good correlation between results obtained by the above two methods. Moreover, for sample #2, a tendency of the damage increasing with increasing strain and deformation rate is obvious. The correlation of the results in the case of sample #3 (see Table 4.2 and Fig. 4.11) is less obvious. Based on the AE tests of this sample, the following conclusions are possible:

(1) At sites 2 and 3 with the smallest critical current, AE signals were also observed. The site 3 (with the smallest critical current) was loaded in the tests

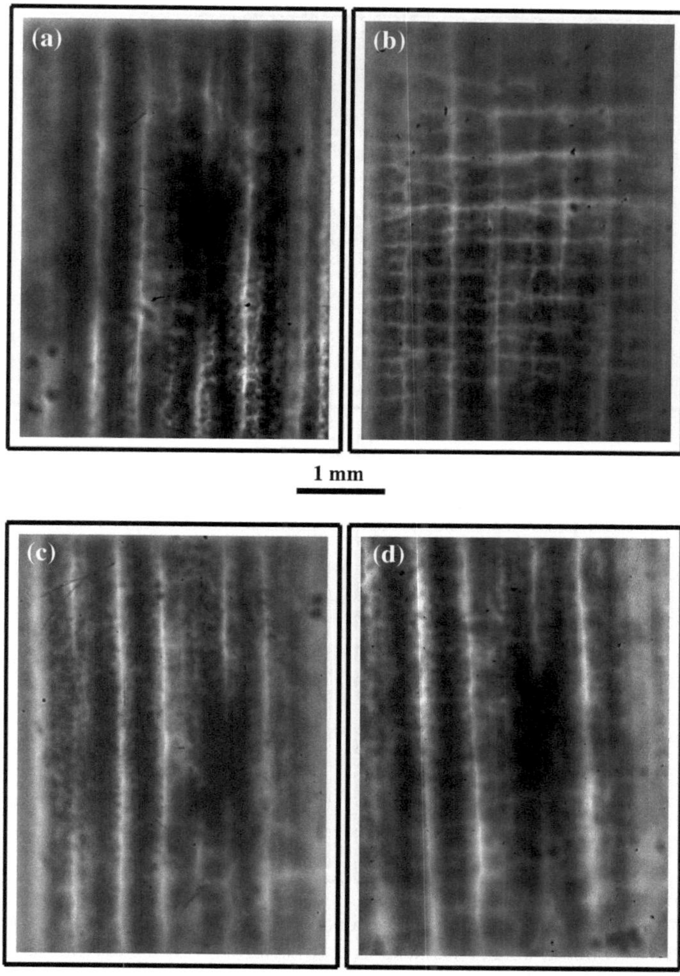

Fig. 4.10 Magneto-optical images of the multi-filamentary tape #2, ZFC regime ($T = 13$ K, $H = 1200$ Oe): (**a**) initial unstrained tape; (**b**) site 1, 10–12 great cracks are seen; (**c**) site 2, no cracks; (**d**) site 3, some cracks are possible [1325]

twice. It may be assumed that during first loading (when electric hindrances were observed), a sample cracking was initiated that was forced at repeated loading.

(2) An absence of AE signals at the sites 1 and 4 (at the existence of cracks, revealed by the MOI method) is related with the possible absence of a good acoustic contact of the sample with the transducer or/and due to the cracks that could be formed at the following accompaniment of the sample (i.e., at marking of the sample, its straightening or/and in the following measurements).

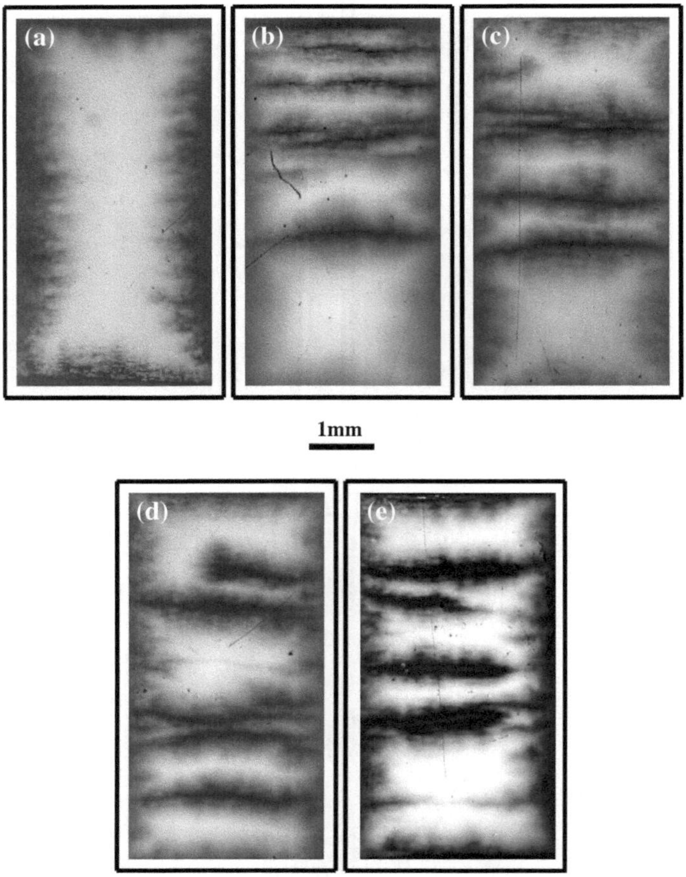

Fig. 4.11 Magneto-optical images of the monocore tape #3, FC regime at $H = 600$ Oe ($T = 12$ K, $H = 0$ Oe): (**a**) initial unstrained tape; (**b**) site 1, 4–5 great cracks are seen; (**c**) site 2, 3–4 great cracks are seen; (**d**) site 3, 5 great cracks are seen; (**e**) site 4, 4–5 great cracks are seen [1325]

Additionally, the AE tests of the Bi-2223/Ag monocore tapes were carried out with the following measurement of I_c (at 77 K). At the testing of three samples with different strain rates (one test for every sample) there is correlation between changes of the strain rate and critical current. We obtained the critical currents $I_c = 31$, 24 and 18 A for deformable samples (for initial unstrained sample $I_c = 39$ A). They corresponded to strain rates: $\dot{\varepsilon}/d = 0.00030$, 0.00049 and 0.00054 (s mm)$^{-1}$, normalized in the thickness, at visible absence of the dependence on strain. Corresponding correlation was also observed for acoustic emission.

Thus, an increase of the strain rate caused a forcing of the AE activity, accompanying greater microdamage and corresponding diminution of critical current.

Table 4.1 Test results for the samples #1 and #2 obtained by the AE and MOI methods

No. of site	Δl (mm)	V_n (mm/s)	ε/d (mm^{-1})	$\dot{\varepsilon}/d$ (s·mm)$^{-1}$	$\dot{N}$ (imp/s)	ΣN (imp)	ΣA, (m) ($\times 10^{-7}$)	P_m (g)	MOI data
#1, site 1	3	0.1	0.0147	0.00049	–	–	–	5	–
#1, site 1[a]	1.5	0.06	0.0073	0.00029	80 imp/s in 6 bundles	800	0.2	5	6 cracks
#1, site 2	2.4	0.15	0.0118	0.00074	130 imp/s in 3 bundles	400	–	5	3 cracks
#1, site 3	2.9	0.2	0.0142	0.00098	–	–	–	5	No cracks
#1, site 4	2.9	0.26	0.0142	0.00127	1500 imp/s in 1 bundle	3000	0.6	5	Weak defects
#2, site 1	3	0.2	0.0147	0.00098	$\approx$1300 imp/s in 1 bundle	4000	3	80	10–12 great cracks
#2, site 2	1.4	0.06	0.0069	0.00029	–	–	–	75	No cracks
#2, site 3	2.9	0.08	0.0142	0.00039	150 imp/s in 4 bundles	600	0.2	80	It is possible some cracks

[a] At second loading of considered site from the opposite side

Table 4.2 Test results of AE and critical current for the sample #3

No. of site / Property	#3, site 1	#3, site 2	#3, site 3	#3, site 3[a]	#3, site 4
Δl (mm)	6	6.5	4.3	6.5	6.5
V_n (mm/s)	0.33	0.23	0.3	0.3	0.26
ε/d (mm^{-1})	0.0294	0.0319	0.0211	0.0319	0.0319
$\dot{\varepsilon}/d$ (s·mm)$^{-1}$	0.00162	0.00113	0.00148	0.00148	0.00127
$\dot{N}$ (imp/s)	–	300 imp/s in 9 bundles	AE with electric hindrances	400 imp/s in 1 bundle	–
ΣN (imp)	–	1000		400	–
ΣA (m) ($\times 10^{-7}$)	–	1		0.2	–
P_m (g)	14	18	18	20	22
I_c, (A)	23	18	–	14	21
J_c (kA/cm^2)	15.6	12.2	–	9.5	14.2
AE data	No AE	There are active defects	AE is not distinguished	Weak AE	No AE

For original no deformed edge of the sample placed at the outer side of supports at the first loading, the measured critical current had the next values: $I_c = 45$ A and $J_c = 30.6$ kA/cm^2 (at $T = 77$ K). Maximum thickness of Bi-2223 core is equal to 60 μm, and $S = 1.47 \times 10^{-3}$ cm^2 is the cross-section area

[a] At second loading of considered site from the same side

Hence, it is necessary to take into account the sample strain rate at evaluation of the superconducting tape damage, additionally to the sample deformation.[2]

Based on the above tests, it could be concluded that AE method is suitable to estimate the microdamage formation and propagation during Bi-2223/Ag tape bending. An improvement of the test results may be related to using the designed interference method [1322] and the created optic devices for estimation of small linear and angle displacements [1159–1162, 1323].

4.1.5 Optical Interferometry

The difficulties of structure defects analysis in HTSC tapes are caused by small applied loads, specimen strains and displacements, intrinsic distortions of the specimen surfaces and simultaneous requirements to high accuracy of damage estimation. In this paragraph, I present, following [1163], some original techniques and technological solutions (which has been patented in Russia) directed to improvement of measuring means for technical control of thin composites and

[2] Usually, the sample strain is only taken into account at mechanical treatment of the deformation behavior of structure-sensitive properties.

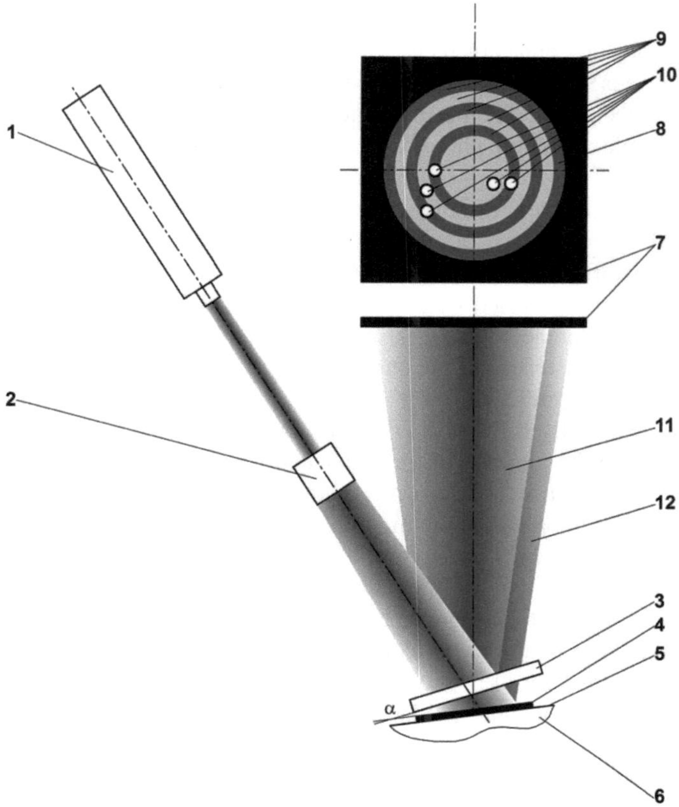

Fig. 4.12 Scheme of prototype optical interference meter of displacements (the designations are explained into text)

goods (in particular HTSC tapes). The obtained solutions ensure (i) modification of composition and measuring schemes leading to raising quality of test results, (ii) expansion of functional possibilities due to coupling in measuring device linear and angle displacements and also increasing range of measured displacements, (iii) raising accuracy of test results based on the complex correction of the results directly during the test fulfillment and by increasing hindrance protection of the test means from vibrations and various perturbations on the base of own functional possibilities of the device. As proper optic methods there are used the following methods: (i) under-shining method of control object surface by laser interferometer (LI), (ii) measuring method of small linear and angle displacements by LI, (iii) complex correction method of test results by LI.

It is well known [104] an optical interference meter of displacements (see Fig. 4.12) based on two-way laser interferometer with combined branches providing a possibility of direct application for solution above problems. The meter consists of optically coupled and consecutively situated (i) source 1 of coherent optical radiation, (ii) optical system 2, (iii) beam-splitter 3, (iv) reflector 4 fastened at surface 5 of a control object 6, and (v) screen 7 on which an interference picture 8 (presenting itself a set of rings 9 with various intensities) is projected. The photo-detectors 10 are also stated on the meter. The beam-splitter 3 and reflector 4 are situated one from other under angle α, and the photo-detectors 10 located into rings 9 of interference picture 8.

The action principle of the displacement meter consists in the following. Radiation of the source 1 of coherent optical radiation after passing optical scheme 2 transforms in diverging light beam and catches the beam-splitter 3, where its splitting is carried out, namely first part reflects from surface of the beam-splitter 3 (reference beam 11), and second part reflects from the reflector 4 (object beam 12), fastened on the surface 5 of the control object 6. Obtained in combination of the reference beam 11 and the object beam 12 the interference picture 8 presenting itself the set of rings 9 with various intensities is projected on the screen 7 with the photo-detectors 10 stated on it. At the displacement of the surface 5 of the control object 6, the photo-detectors 10 measure intensity of optical field of the interference picture in place their location directly coupled with the pointed displacement.

The intensity alteration of the interference picture rings from minimal to maximal value occurs at displacement of the control object surface on value of $\lambda/4$, where λ is the wave length of optical radiation (for example, for optical radiation of helium–neon gas laser of red color $\lambda = 0.63\ \mu\text{m}$). This meter provides without-contact measurement of small displacements of control object surface and has high precise characteristics (at scale of working standard). It has been proposed to use volumetric Fourie hologram as the beam-splitter 3 [1322, 1323].

At the same time, significant shortage of the devices [1322, 1323] is caused by enough low quality of measurements. Under quality of measurements, it is understood [1542] a set of properties leading to result with required precise characteristics in necessary form and stated time, and the quality of measurements is characterized by the proper factors, namely accuracy, correctness and reliability. The pointed shortage is caused by that interference pictures reproductive by devices [1322, 1323] have strongly pronounced features of distribution of the interference picture optical field in the screen plane. These peculiarities are revealed in that in the some regions of interference pictures a change of intensity of the optical field occurs heterogeneously even into bounds of the ring of first order at the displacement of the control object surface. Therefore, an arbitrary disposition of the photo-detectors into interference picture rings in the screen plane (as it has been realized in [1322, 1323]) leads to significant errors in measuring results diminishing their quality.

Technical result of the device for the displacement measuring [1160, 1163] is connected with qualitatively improved measurements of small displacements of the control object surfaces due to introduction in the device construction of new constructive element, namely a new-shape matrix of photo-detectors. This new geometry of the matrix must coincide on shape and sizes with most informative region of the interference picture and dispose in the pointed area of the

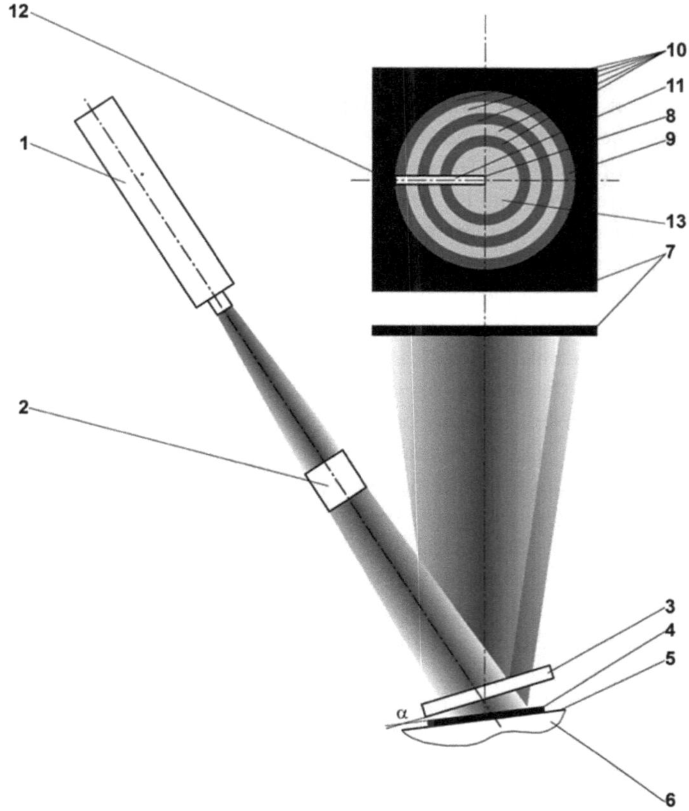

Fig. 4.13 Scheme of modified optical interference meter of displacements (the designations are explained into text) [1163]

interference picture. The data obtained in experimental studies of distributive features of the optical fields in interference pictures are obvious proof of achieving the new technical result [1164].

The essence of the device [1160, 1163] is found by the scheme presented in Fig. 4.13. The optical device for measurement of displacements consists of optically coupled and consecutively situated (i) source 1 of coherent optical radiation, (ii) optical system 2, (iii) beam-splitter 3, (iv) reflector 4 fastened at surface 5 of control object 6, and (v) screen 7 with photo-detector 8 stated on its. The beam-splitter 3 and reflector 4 are situated one from other under angle of α. By combining the reference and object beams the interference picture 9 (presenting itself a set of rings 10 with various intensities) is projected on screen 7.

The photo-detector device 8 presents itself rectangular matrix of photo-detectors 11 and is stated in the screen plane 7 in radial direction to the rings 10 of interference picture 9 at distance from edge 12 of the interference picture 9 opposite to the angle tip of α between the beam-splitter 3 and the reflector 4 before center of the interference picture 9. The photo-detector 8 is connected with device for registration and image of measuring results (it is not shown in Fig. 4.13).

The device works in the following scheme. Radiation of the source 1 of coherent optical radiation after passing the optical scheme 2 transforms in diverging light beam and catches the beam-splitter 3, where its splitting is carried out, namely first part reflects from surface of the beam-splitter 3 (reference beam), and second part reflects from the reflector 4 (object beam), fastened on the surface 5 of the control object 6. Obtained in combination of the reference and object beams the interference picture 9 presenting itself set of the rings 10 with various intensities is projected on the screen 7 with the photo-detector device 8 stated on it. At the displacement of the surface 5 of the control object 6, the photo-detector device 8 measures intensity of optical field of the interference picture in place their location that is in the region of interference picture, demonstrating maximal functional characteristics (great contrast ratio, wide dynamic range, etc.). The measurement results pass from the photo-detectors 8 to device (no shown) for registration and reflection of the measuring results.

In this case, possible variants of the intensity registration could be registration of summary intensity of the optical field on surface of the photo-detector device 8, registration of intensity of the optical field in the ring of first order, in the rings of first and second orders, etc. The proposed device provides realization practically of all known variants which could be required and defined as aims and problems of concrete experimental research.

The device [1160, 1163] by keeping positive qualities of devices [1322, 1323] differs from them by increased quality at measuring the small displacements of control object surfaces, and could be applied for high-precise measurements of small linear and angle displacements of the control object surfaces by fulfilling experimental investigations of prospective goods, estimating and controlling their technical state, studying AE processes in solids, damage in HTSC tapes, wave processes in layered goods and goods from anisotropic heterogeneous materials used in various technical regions, etc.

Without-contact optical methods of laser interferometry are most perspective methods of measuring nano-displacements. They provide high sensitivity ($<10^{-3}$ nm) to displacement of object, linearity in wide dynamical range, high locality, absolute character of calibration of the measurements in fractions of wave length of the laser irradiation, possibility to automatize definition and treatment of measurement information.

Below, it is presented a modified device for testing thin composite samples (in particular, HTSC tapes) on bending [1161, 1163]. Essence of the modification of the test device concludes in that the device for testing samples consists of a base with the loading device fastened on it which provides displacement of moved plate. Supports for disposition of test sample are stated on the plate. The test device also contains test sample freely stated on the supports and no moved plate with central support fastened on it. The base has a cavity in which it is disposed a liner fulfilled from damping material for statement of device creating loading. Moreover, the supports stated on the moved plate are supplied by rigidly fastened

П-shaped restrictors of longitudinal and transverse displacements of the sample. The no moved plate has central cavity and is stated by using racks on additional base. In the central support, a cavity is made in which it is disposed piezo-transducer of system for registration of AE signals in the sample during tests. Tension-metric meter of force, fastened on no moved plate, is carried out in the form of hollow cylinder with flanges. Upper flange is fastened with no moved plate and has central cavity co-axial to cavity in the no moved plate. Lower flange is joined with the central support from external side and rigidly fastened from internal side with one end of cylindrical rod passing via central cavity in the upper flange of the tension-metric meter of force and central cavity in the lower flange. At the contrary end of the tension-metric meter is stated reflector for providing without-contact couple with the optical meter of displacements. The proposed technical solution is explained by Figs. 4.14 and 4.15. Figure 4.14 depicts schematically the modified device for testing specimens, and Fig. 4.15 presents increased image of the support on the moved plate with П-shaped restrictor of longitudinal and transverse displacements of the sample.

The device for testing the sample on three-point bending contains (i) base 1 with cavity 2 in which is disposed liner 3 fulfilled from damping material and device 4 for creating load with stock providing displacement of moved plate 6, (ii) supports 7 and 8 stated on the moved plate 6 supplied, respectively by П-shaped restrictors of longitudinal 9 and transverse 10 displacements of test sample 11 freely disposed on supports 7 and 8, (iii) no moved plate 12 stated on additional base 13 by using racks 14 and having central cavity 15, central support 16 with cavity 17 in which piezotransducer 18 is disposed, (iv) system for registration during tests of AE signals in the sample 11 (is not shown in Fig. 4.14), (v) tension-metric meter 19 of force fulfilled in the form of hollow cylinder 20 with upper 21 and lower 22 flanges. The upper flange 21 is fastened with the no moved plate 12 and has central cavity 23 co-axial to the cavity 15 in the no moved plate 12. The lower flange 22 is joined with the central support 16 from external side and rigidly fastened from internal side with one end 24 of cylindrical rod 25 passing via the central cavity 23 in the upper flange 21 and the central cavity 15 in the no moved support 12. At the other end 26 of the rod 25 is stated reflector 27 for providing without-contact couple with the optical meter of displacements (is not shown in Fig. 4.14), for example with triangulation meter of displacements, interference meter of displacements, etc.

The test device works by following way. Preliminary the test sample 11 is freely disposed on supports 7 and 8 between the П-shaped restrictors 9 and 10 excluding its longitudinal and transverse displacements. Due to command of the control system by tests the device of loading 4 guarantees, by using the stock 5, movement of the moved plate 6 with the supports 7 and 8, the restrictors 9 and 10, and the tested sample 11 stated on it in direction to the central support 16. The loading of the tested sample 11 is carried out with a moment of contact its surface with the central support 16 and continues up to moment definite by the aims and problems of the test, for instance up to moment of achievement of the given flexure in the place of contact with the central support 16.

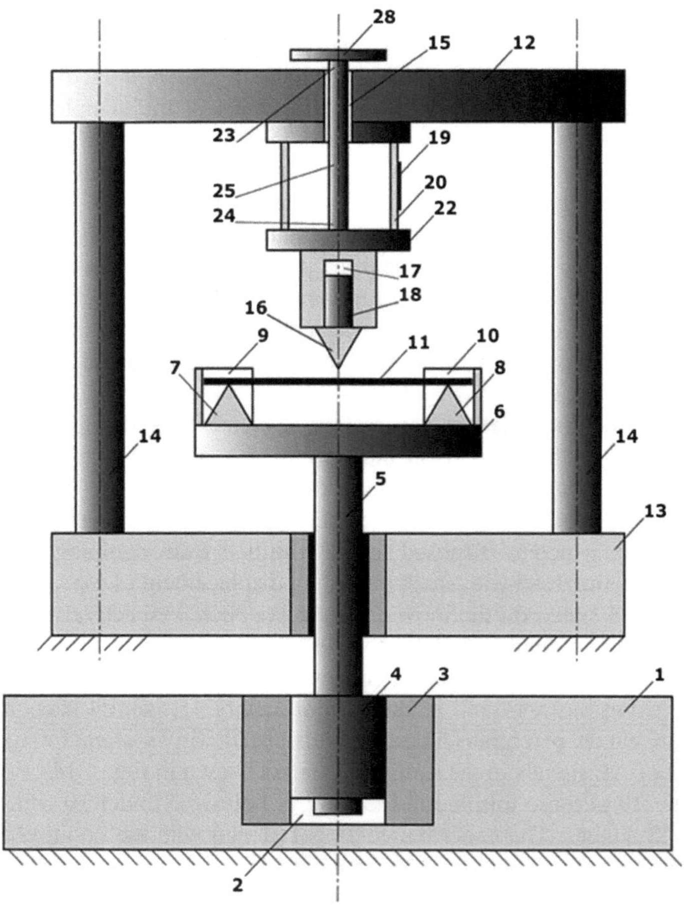

Fig. 4.14 Scheme of the modified device for testing specimens (the designations are explained into text) [1163]

During loading process at the deformation of the sample 11 it is carried out registration of AE signals characterizing the defect-formation processes in the sample 11 by using the piezoelement 18, axis force by using the tension-metric meter 19, and axial displacement of the lower flange 22 of the meter 19 by using coupled with it the element 25, the reflector 28 and without-contact optical meter of displacements (is not shown in Fig. 4.14). In realization of this meter, several technical solutions patented in Russia [12, 13, 1160, 1162] have been used which allowed one to distinguish it among typical laser interferometers.

In some cases, for example at minimal deviation of the sample physical and mechanical characteristics, it may be fulfilled a control of the loading parameters by using measurement results of without-contact optical meter of displacements. The technical result of the modification of the test device [1159] consists in

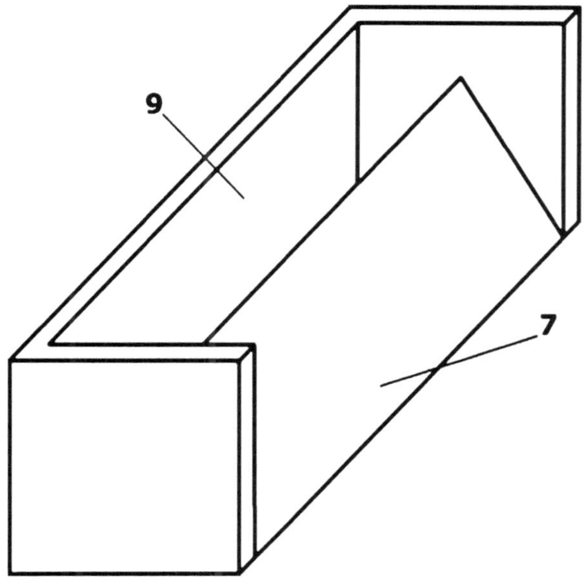

Fig. 4.15 Increased image of support on moved plate with П-shaped restrictor of longitudinal and transverse displacements of the sample (the designations are explained into text) [1163]

extension of functional possibilities for account of conducting investigations of the damage processes in thin specimens from advanced materials at testing on bending. The preliminary experiments have been fulfilled by using the test device. They supported a correctness of technical solutions used in its construction and achievement of the pointed technical result. External views and schemes of the developed devices and their parts for testing thin samples on bending are presented in Figs. 4.16, 4.17, and 4.18.

So, the modified test device by keeping positive qualities of device [1159] and existing analogues extends their functional possibilities and could be applied, for example to study the defect-formation processes in thin HTSC tapes used in various areas of industry.

4.2 Intergranular Boundaries in HTSC

The properties of high-angle grain boundaries are believed to control the macroscopic $J_c(H)$ characteristics of all polycrystalline HTSC. This control occurs because most high-angle grain boundaries act like barriers to the current and have electromagnetic properties, such as Josephson junction-like properties [374]. On the other hand, in the melt-processed polycrystalline YBCO, actual currents can penetrate through intergranular boundaries misoriented up to 30° and more. So, HTSC properties are connected closely with distribution of the intergranular boundary misorientations [374, 1056].

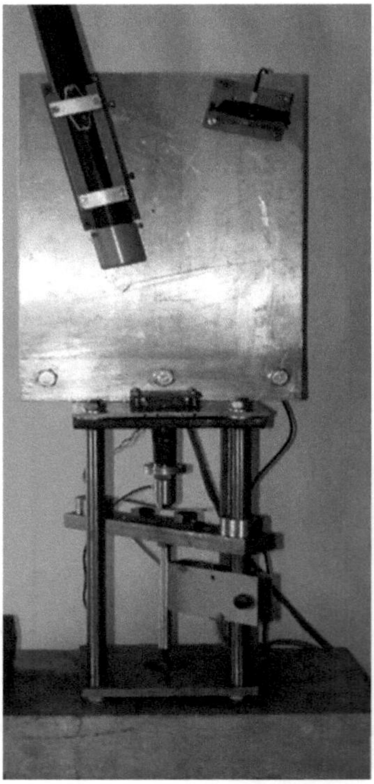

Fig. 4.16 External view of device for testing thin samples on bending [1163]

From both the high-field flux-pinning viewpoint and low-field, Josephson junction-based electronics viewpoint, there is strong motivation to develop a detailed picture of the grain boundary structure and microstructure and to describe their effects on the electromagnetic properties of the grain boundaries. The superconducting coherence length, ξ, defines the grain boundary thickness that may be penetrated by the supercurrent. In this case, barriers with thickness up to a few coherence lengths can still show superconducting coupling, albeit of reduced strength. At the same time, the grain boundaries in HTSC are defects with thickness in the 0.5–1.0 nm range, indeed, approaching ξ [74, 941]. Super-short coherence length and great values of n in the power dependence of E–J (where E is the electric field intensity and J is the current density) proper for HTSC [605, 607], sign, that the defects with size of some nanometers can prevent supercurrent and create obstacles with effective size, that is considerably higher than nominal size of defect. There are other planar defects, causing magnetic flux pinning and supercurrent percolation in HTSC, namely (i) twinning in YBCO [488, 962, 1011], (ii) stacking faults [467, 1100], (iii) colonies of low-angle c-axis intercrystalline boundaries [467, 1100, 1727], (iv) twist intergranular boundaries of the "brick wall" type [203] and low-angle a-, b-axis boundaries of the "railway switch" type [1855] in BSCCO, (v) overgrowth of Bi-2223 phase into Bi-2212 phase [902], (vi) overgrowth of superconducting phase into silver sheath [1436], (vii) amorphous

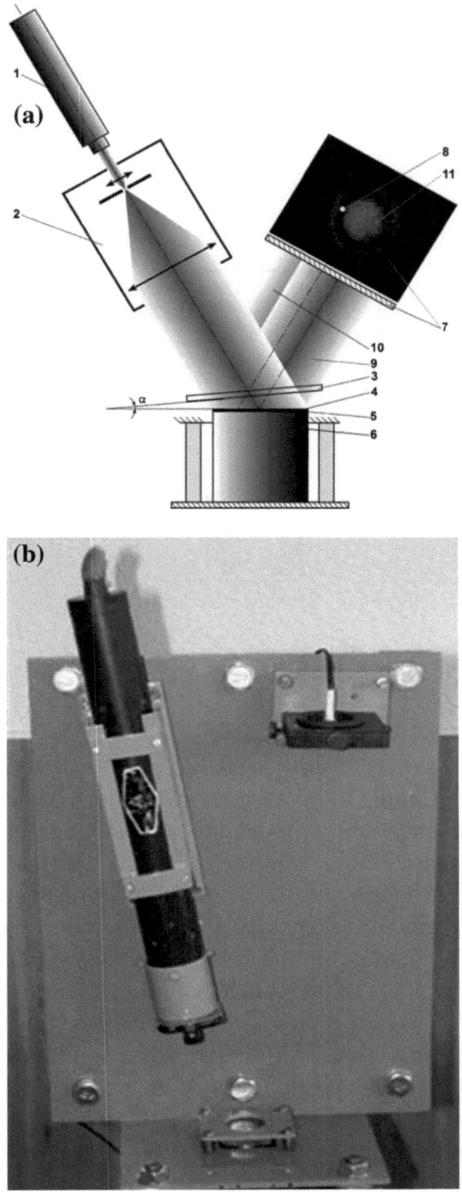

Fig. 4.17 Scheme (**a**) and external view (**b**) of device for measuring displacements. At the scheme are shown the next designations: *1* source of coherent optical irradiation, *2* optical system, *3* beam-splitter, *4* reflector, *5* control surface, *6* rod, *7* screen, *8* photo-detector device, *9* object beam, *10* reference beam, *11* interference picture [1163]

and normal (non-superconducting) phases [489], (viii) voids, microcracks and other crack-like defects, which are proper for all oxide superconductors [659, 1211], (ix) macrodefects into Josephson junctions [1611], and also various dislocation networks [1028], discussed in detail below. Different types of planar defects causing the structure-sensitive properties of HTSC are shown in Fig. 4.19.

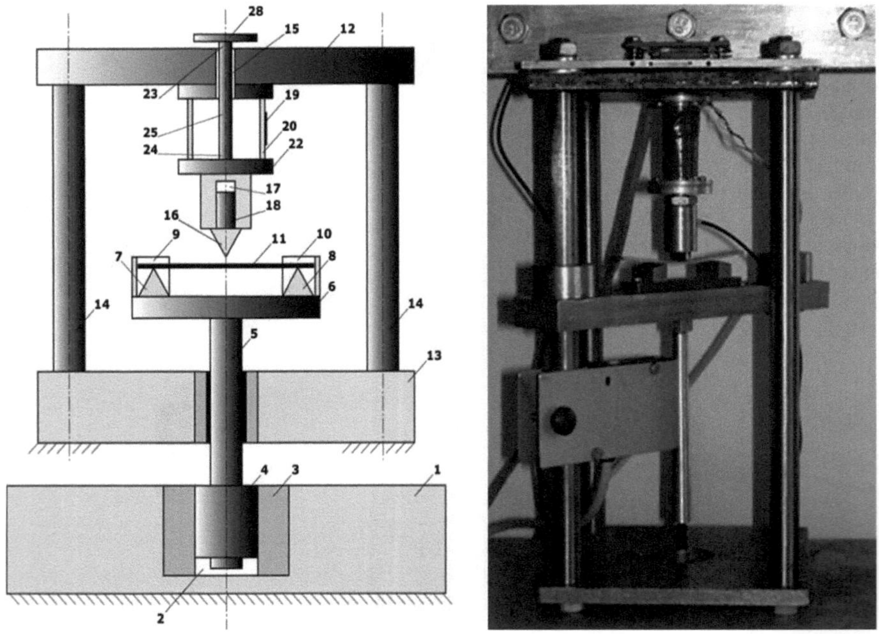

Fig. 4.18 Scheme and external view of device for testing thin samples on bending [1163]

Polycrystalline YBCO samples can be divided into two broad classes, according to microstructure features, namely (i) specimens heated above peritectic temperature with the aim of obtaining oblong well-orientated grains [20] and (ii) samples sintered with grain structure that is near to equal-axes one [1557]. The experimental studies relate to investigation of both individual isolated grain boundaries and polycrystalline samples with averaged effects of great number of the intercrystalline boundaries. Both test types often supplement each other. So, the study of high-angle boundaries in YBCO thin epitaxial films onto bicrystalline $SrTiO_3$ substrates [374] has confirmed the behavior of intercrystalline boundaries of Josephson type. Earlier, this result has been predicted by using tests on bulks [1355]. On the other hand, the tests on oriented YBCO bi-crystals [72] showed that some grain boundaries, including high-angle ones, demonstrate under high magnetic fields much more higher transport properties compared with behavior that is proper for weak links subjected to Josephson effects. Analogous phenomena take place in bulks, in which there is often small intergranular current component that does not depend on field [427].

The general results are usually based on the bi-crystal investigation of three main types, namely (i) epitaxial thin films that are grown on bi-crystalline [374, 581, 762] or otherwise specially prepared [551, 1008, 1759] substrates with predetermined misorientation relationships and boundary planes (on a macroscopic scale); (ii) very large grain (millimeters) melt-processed bi-crystals (MPB) that

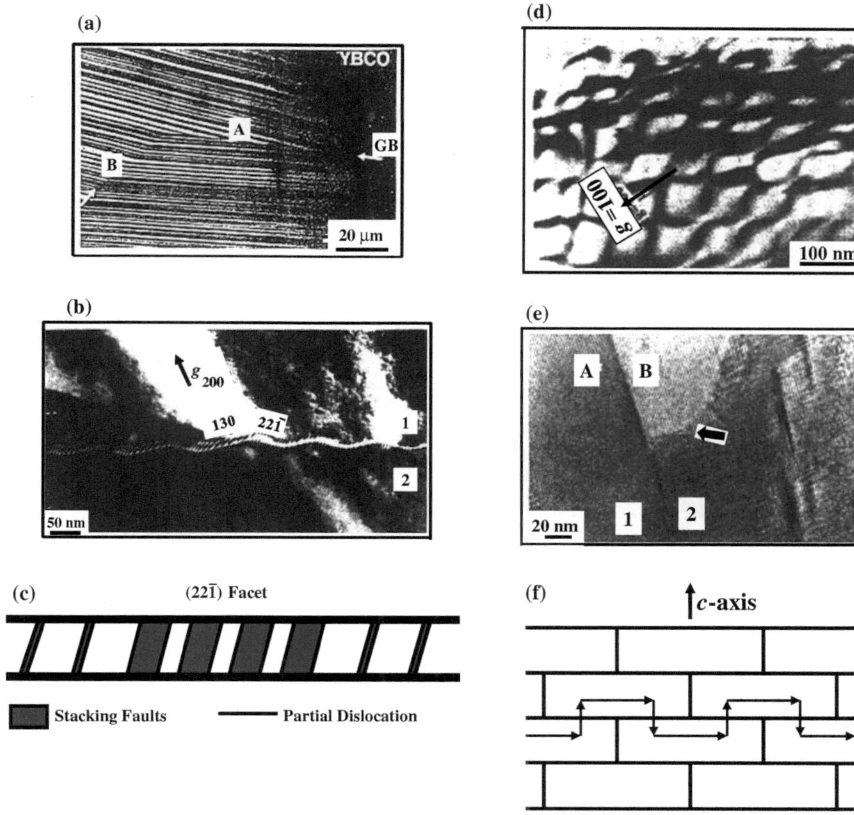

Fig. 4.19 Different types of planar defects in HTSC: (**a**) {110} twinning boundaries in the 10°-bicrystal of YBCO (*A* and *B* are the macroscopic facets) [1710]; (**b**) diffraction contrast image of the spatial dislocation network configuration in grain boundary (g = [200] is the diffraction vector), the *dark* and *bright* areas are associated with long-range strain contrast in facet junctions [1707]; (**c**) schematic diagram, showing the arrangement of dislocations in the (22$\bar{1}$) facet [1707], (**d**) screw dislocation in twist intergranular boundary [1769]; (**e**) lattice fringe image, showing examples of a twist grain boundary (grains *A* and *B*), a low-angle *c*-axis tilt colony boundary (colonies *1* and *2*) and a low-angle *ab*-axis tilt colony boundary (*arrow*) [418]; (**f**) twist intergranular boundaries of the "brick wall" type [1855], (**g**) different models of the grain boundaries of the "railway switch" type [1855]; (**h**) overgrowth of Bi-2223 phase into Bi-2212 phase [489], (**i**) longitudinal cracks in BPSCCO superconducting core [1285], (**j**) scheme of macrodefect (*black area* in *gray* intermediate layer) in Josephson junction [1611]

contain considerable volume fractions of secondary phases and have random misorientation relationships and macroscopically meandering boundary planes [476, 1295]; and (iii) flux-grown bicrystals (FGB) with dimensions of the order of a few hundred microns, misorientation relationships of the θ [001] type and comparatively straight, but uncontrolled boundary planes with large tilt sites [72, 73, 979].

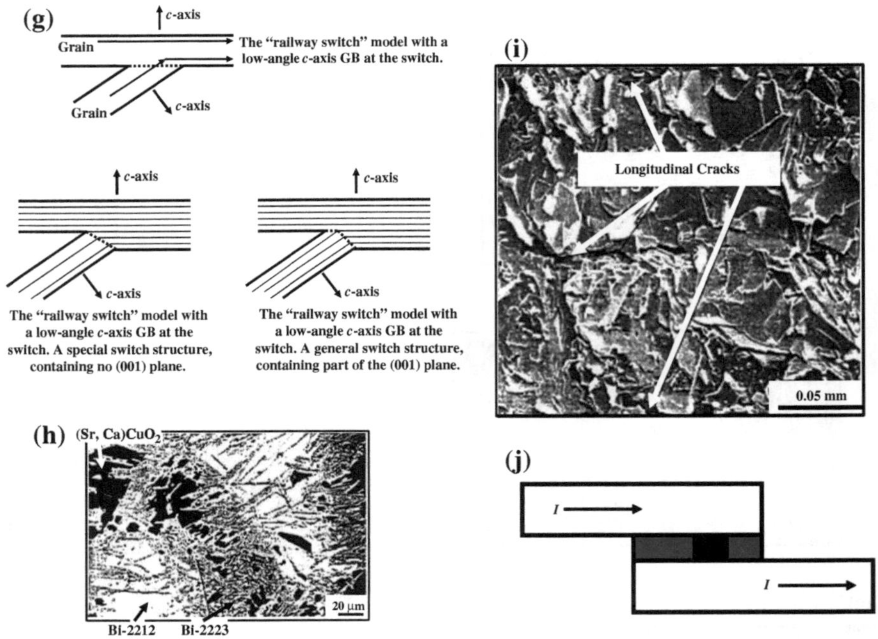

Fig. 4.19 (continued)

The study of electric resistance of the [001] symmetrical tilt boundaries at the normal state of sample revealed common dependence of electromagnetic properties on the misorientation angle θ. The value of $J_c(gb)/J_c(g)$ (where $J_c(gb)$ and $J_c(g)$ are the critical current densities of the boundaries and neighboring grain interiors, respectively) decreased rapidly with a $1/\theta$-like dependence for θ values up to about 20°. For $\theta > 20°$, an approximately θ-independent, uniformly low value of $J_c(gb)/J_c(g) \approx 0.01$ was observed [374, 375]. The same behavior of grain boundaries with low/high angles of misorientation is observed in all HTSC. In particular, epitaxial bi-crystal experiments using thin films of Bi-2212 [29, 1107, 1767], Tl-1223 [1223], Tl-2212 [229] and Tl-2223 [1481, 1767], all showed a qualitatively similar dependence of $J_c(gb)$ on θ. The $1/\theta$ dependence/plateau behavior of $J_c(gb)/J_c(g)$ versus θ led researchers [229, 374, 1222] to postulate that the dislocations, covering the structure of intercrystalline boundaries, determine their electromagnetic properties. As the misorientation angle of the boundary increases, barriers with weak link properties progressively pinch off the supercurrent. At some critical transition angle, the barrier regions completely control the boundary properties and a weak link forms. The dependence on θ suggests that the primary grain boundary dislocations (PGDBs), constituting the boundary structure, cause the weak link behavior [374, 581]. In this model, the boundary is frequently described as a uniform array of like edge dislocations with spacing, which are characteristics of symmetrical tilt boundaries as specified by Frank's formula [1420].

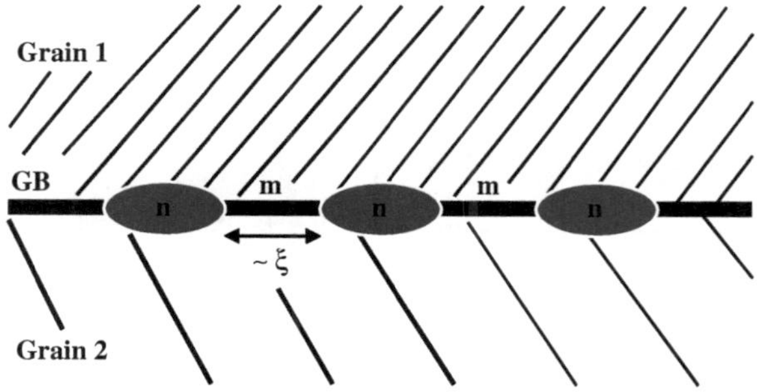

Fig. 4.20 Schematic illustration of the heterogeneity, proposed for weak-linked grain boundaries in YBCO. The current flows through microbridges (labeled by *m*) that exist between non-superconducting regions (labeled *n*). The electrical character of the microbridges is assumed to be normal, superconducting or insulating depending on the model used

There are many attributes of dislocation network, including the weak link behavior, namely (i) overlapping of dislocation cores with cation stoichiometry [1222]; (ii) overlapping of elastic strain fields, caused by PGDBs [1436]; (iii) agreement of lattice parameters through stoichiometry on oxygen for elimination of lattice mismatch at intercrystalline boundary [1883, 1884]; (iv) local alterations of T_c and chemical potentials due to elastic strains [608]. The test data show the transition from strong to weak coupling at $\theta \approx 10°$,[3] which is most evident in thin-film YBCO bicrystals [374, 581]. This transition could be explained by narrowing of superconducting connecting dislocation cores [374] or overlapping of their strain fields [279]. This model assumes absence of superconductivity of dislocation cores and limitation of the supercurrent paths by channels, connecting these cores, and also by neighboring boundary regions. The regions of perfect crystalline structure into space between dislocations or symmetric facets in the absence of the void-like defects can be potential sources of current-carrying paths defined by strong links [190]. The orthorhombic crystalline structures of cuprate superconductor could considerably complicate investigation of their intercrystalline boundaries. In addition to different lattice parameters *a* and *b*, YBCO family possesses a good developed twinning structure [572], but BSCCO family demonstrates non-regulated modulation along *b*-axis [1753]. Both the lattice parameters and YBCO superconducting properties change with oxygen concentration. In particular, the lattice parameters *a* and *c* increase with decreasing oxygen in the structure, causing an expansion of unit cell. Therefore, the deficiency of oxygen and corresponding weakening of superconducting properties may be intrinsic to intercrystalline boundaries in YBCO.

The electromagnetic properties of weakly coupled boundaries vary substantially and perhaps systematically with position along the grain boundary. The patches of "better" material are separated by weak or non-superconducting regions, as shown

[3] Strong link behavior can be observed in melt-processed YBCO bi-crystals [477, 1295] and in flux-grown YBCO bicrystals [77] at angle θ up to 20°.

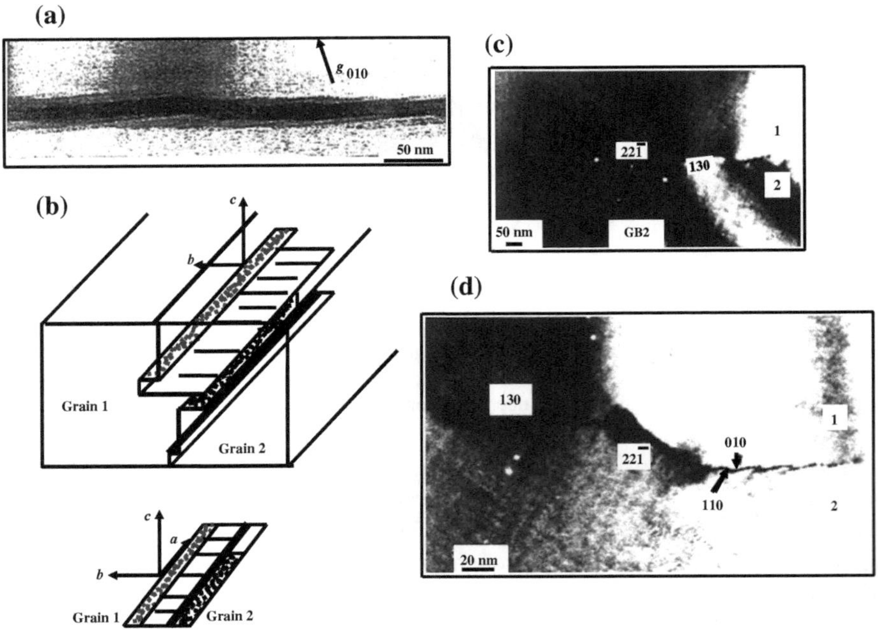

Fig. 4.21 Faceted structures in HTSC: (**a**) a ***g***|| [010] diffraction-contrast TEM image of nominally pure [001] tilt boundary in 10°[001] bi-crystal of Bi-2212 [1710]; (**b**) perspective and [001] plan view schematics of the "stepped" boundary topography, composed of pure symmetric tilt facets and pure twist facets, that was observed in many sections of the [001] tilt 2212 bi-crystals [1710], (**c**) diffraction-contrast TEM image showing saw-tooth faceting onto (130) and ($22\bar{1}$) plane in GB2 [1707]; (**d**) enlargement of a portion of (**c**), showing that the (130) facets further facet onto (010) and (110) sub-facets; the dot-like strain contrast along the sub-facets is produced by the grain boundary dislocations [1707]

schematically in Fig. 4.20. One of the main causes of these structure and composition alterations could be the local oxygen depletion and oxygen disorder in YBCO structure [190, 391, 1882]. Other possible microstructure sources of heterogeneity include: (i) cation composition modulation, existing within the boundary plane [607, 1739]; (ii) "wavy" boundaries, facets with arbitrary configurations and facet junctions, observed in epitaxial thin film bicrystals of YBCO and Bi-2212 [11, 681, 1701]; (iii) strain fields, caused by intrinsic and extrinsic intercrystalline dislocations and also by regular distribution of facets [1707, 1708]; (iv) the intersections of twin planes with the intercrystalline boundary plane in the case of YBCO [77]; and (v) oscillatory changes in misorientation of the neighboring grains due to twinning in YBCO [77, 1707, 1708]. Different faceted structures are shown in Fig. 4.21.

The superconductor crystals that nucleate on one substrate crystal can often grow past the substrate boundary and over the "other" crystal for appreciable distances before impinging on a crystal growing in the "correct" orientation. This overgrowth, which may be more prevalent for certain misorientation angles, results in "wavy" boundary topography, causing boundary morphology and properties [802, 990, 1708]. On the other hand, the twinning in each YBCO single crystal alone results in discrete, systematic changes in the local misorientation relationship along the boundary, which should produce concomitant oscillations of PGDBs, as observed for a high-angle boundary in sintered material [517].

There are numerous examples of PGDBs at the low- and high-angle YBCO intercrystalline boundaries. Some high-angle boundaries include regularly positioned amorphous regions [1133, 1436]. Nevertheless, usual observations show high degree of the intercrystalline structure localization, ranging only on 1–2 elementary cells to neighboring grains [279, 517, 990]. These observations agreed with the width of dislocation cores, which are observed on representative intercrystalline boundaries in ceramics [1133].

For the framework of description of the high-angle boundary structure in terms of PGDB [75], concentrated picture of the "good/bad" regions may be used, according to defect distribution. Intercrystalline dislocations are localized sufficiently in order to create periodic field of the elastic physically signed strains. In this case, the accommodation dislocations of the lattice mismatch at [100] boundary are partial dislocations and can dispose sufficiently far from the boundary into one from grains [232, 517, 1222]. These partial dislocations are joined with boundary by stacking faults, which rather introduce copper excess in a boundary region. Nevertheless, this boundary can act as strong link despite its extended structure and possible stoichiometry [444, 947, 1008].

Nano-facets are observed at different intercrystalline boundaries [239, 1133]. In this case, the facet planes are often dictated by a crystalline structure. Facets are observed on boundaries, possessing relatively simple crystallography and low-coincidence index (Σ) values, where Σ is the fraction of lattice sites of one crystal that are coincident with the other [1222], and also on boundaries with complex geometry [990]. Facets aspire to align parallel to planes with lowest index of one from crystals and are observed, as rule, in (100), (010) and (001) planes. The facet formation introduces considerable structure heterogeneity in the boundary. Due to limited size of facets, macroscopic boundaries with different planes may be considered as mixture of some types of the boundary facets of fixed structure. The intersection lines of facets possess their own atomic structure, and secondary dislocations are often associated with them [77]. In this structure level, heterogeneities, which are important for superconducting properties, include saw-tooth structure of facets, additional faceting of facets, initially appearing plane, intensive strains at facet junctions, in homogeneously distributed dislocations at boundaries in the range of individual facets and favorable conditions for dislocation-division with Burgers vector $\boldsymbol{b} = \langle 100 \rangle$ in partial dislocations at the center of some facets (see Fig. 4.21c, d) [1707].

The facet junctions can influence stoichiometry of intercrystalline boundaries; in particular, it is possible for the concentrated excess of copper to be connected with facet distribution in volume bi-crystals with a low misorientation. Some more oscillations of stoichiometry on copper with period of about 100–200 nm are observed at high-angle boundaries [76]. Contrast strains take place in the regions

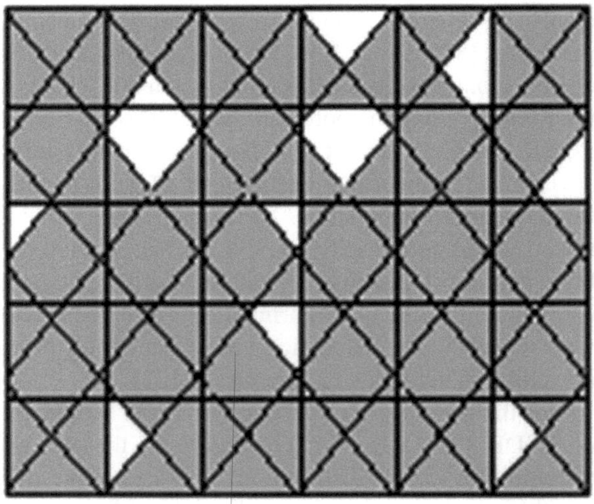

Fig. 4.22 Schematic presentation of possible structures for intercrystalline boundaries in general case. The *horizontals, verticals* and *diagonals* show dislocation network, *grey* regions present barriers for supercurrent, *white sites* are the channels of strong links. Contrast to the model of boundaries directed along [001] direction (see Fig. 4.20), the strongly connected channels at these boundaries to rather point, than linear contact

with copper excess, but the facet morphology and the strain origin have not exactly been stated. These copper redistributions together with facet topography and/or strains in facet junctions can cause intercrystalline electromagnetic properties.

Dissociation of PGDBs, forming pairs of partial dislocations, can considerably influence the value of the critical angle at transition from weak links to strong ones due to the accompanying decreasing of dislocation spacing. During PGDBs dissociation, superconducting channels narrow and can lead analogously to microbridges, demonstrating Josephson effect, that is, weak link behavior [1021]. This transition could be expected by decreasing lesser than the coherence length (about 1 nm in *ab*-plane). These narrow superconducting channels, caused by the dislocation dissociation in partial ones, form weak links at intercrystalline boundary (one could be strongly connected, if consists of PGDBs only). Thus, a transition to lower energy structure of boundaries (at dislocation dissociation in partial ones) will suppress an intercrystalline supercurrent owing to mechanism of the superconducting channel narrowing.

Strain fields near tilt boundaries must considerably differ from fields near twist boundaries due to differences between parallel rows of edge dislocations and cross-shaped lattices of screw dislocations, covering their structures, respectively. While the boundaries along [001] direction have, as a rule, one row of dislocation, in the case of mixed orientation of grains, there are at least two intersecting rows of dislocations. This alters a form of channel with strong coupling owing to inclusion of both rectangular sites, located between parallel rows of dislocations, and almost point contacts of high-angle boundaries with mixed orientation (Fig. 4.22). These boundaries retain a good connected system of flaw pattern of the supercurrent (see test results [477]), despite far smaller number of "strong" channels predicted by the PGDBs model [374, 375].

Faceted structures in YBCO and Bi-2212 differ in the type of facets, as well as in regularity of their distribution. At nanoscale in Bi-2212, there is non-regular distribution of symmetric tilt and twist [001] facets. At the same time, very regular saw-tooth distributions of tilt facets dominate in YBCO [1707, 1739]. In this case, the strains observed in YBCO bi-crystals are always in accordance with saw-tooth facet intercrystalline structure [1707, 1708]. Elevated strains could be caused by incomplete elimination of long-range component because of small number of dislocations that are contained at the facets with limited sizes.

Despite of differences in the dislocation topography and distribution at the boundaries in Bi-2212 and YBCO, there is one common feature concerning the Burgers vectors of these dislocations, which may correlate with superconductivity of boundaries. It is known that existence of the partial dislocations is common for intercrystalline boundaries in Bi-2212 and YBCO [1707, 1708, 1739]. Moreover, in both materials, Burgers vector value and direction correspond to distance between two nearest atoms of oxygen in the CuO_2 planes of unit cells. Then, the strain caused by these dislocations can initiate local superconductivity. In particular, the shear strain in $\{110\}$ planes associates with the beginning of superconductivity in YBCO on the basis of analysis of the anomaly in thermal expansion factors in a-axis and b-axis directions [211, 212, 707].

It is known that a layer of material with different superconducting properties exists at YBCO grain boundaries. This layer appears to be hole-deficient YBCO [73, 190, 1883]. However, key questions about the nature of this layer remain unanswered: (i) What is the average width of the hole-depleted zone, and does it depend on the form of the material (bulk or thin film) or processing method? (ii) Are the magnitude of depletion and width of the depleted zone uniform along the boundary? (iii) What is the microstructural origin of the holes' depletion?

There is correlation between the value of hole depletion and initiation of effects, demonstrating the weak link behavior [73]. The hole-concentration profile studies are based on the concept that the hole depletion zone width exceeds both the superconducting coherence length for YBCO and the apparent structural width of the boundary. The widths reported vary from one material form to another: ~ 8 nm wide affected zone at polycrystalline thin film boundary [190]; the width of the depletion zone in FGBS is ~ 60 nm [73] and, in sintered materials, a width the order of 10–20 nm [1883]. All of these values exceed the width of the structurally disordered region (~ 1 nm) and the metal cation non-stoichiometry (~ 5 nm) [76], and also the coherence length ξ. Unfortunately, it is not possible to associate the reduced density of holes with a specific atomic defect such as oxygen non-stoichiometry or oxygen disorder, these being the most likely candidates. In this case, there maybe important differences in the structure, composition and electronic properties of grain boundaries in thin films as compared to their bulk counterparts. Therefore, it is expected that hole density variations might occur on a number of different length scales in both strongly and weakly coupled boundaries, and also in superconductors with different forms and sizes.

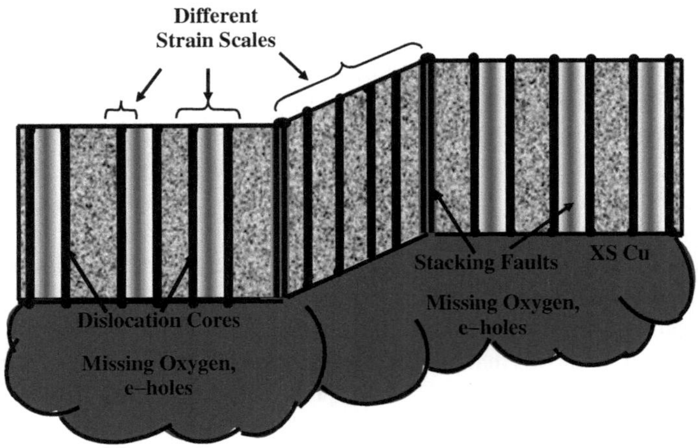

Fig. 4.23 Schematic illustration of various types and scales of structural and compositional heterogeneities at [001] tilt boundaries in YBCO and in other HTSC

Thus, the main features of grain boundaries in the considered HTSC systems (Fig. 4.23 illustrates some of them) include:

- the cores of the grain boundary dislocations, where the nearest neighbor configurations are disrupted and the strains are very large;
- the ribbons of stacking fault, disposing between dislocations that dissociate to form pairs of partial dislocations;
- the variety of changes in boundary plane that occur as facets form, for example, due to twinning;
- the different dislocation distributions in the various facets and the uneven arrangement of dislocations that occur in smaller facets;
- the compositional and structural changes, including heterogeneous but periodic copper excess (XS Cu), that are caused by local diminution and disorder of oxygen, and also cation modulations;
- the electron–hole depletion that occurs at boundaries and also appears to be non-uniform;
- stress–strain states of different length scales that arise as a result of the dislocation network, dislocation dissociation, facet junctions, phase transitions, twinning, initiation and growth of pores and microcracks.

These features comprise a boundary that is heterogeneous on a variety of length scales with respect to atomic distributions, electronic structure, composition and stress–strain state. The above heterogeneities influence all structure-sensitive properties of bi-crystals and polycrystalline materials.

4.3 Superconducting Composites, Based on BSCCO

4.3.1 BSCCO/Ag Tapes

Investigations of BSCCO/Ag tapes show that the *c*-axis alignment reaches maximum near the "superconductor–silver sheath" interface, but is often observed in the tape center [468, 980]. This microstructure assumes that an oxide core of Bi-2223 tapes consists of the plate-late grain colonies, in which the grains are divided by twist boundaries. The boundaries of colonies could be divided into four types, namely [20, 669, 1033] (i) the boundary in the *c*-axis direction approximately parallel to the *c*-axis of colony; (ii) the colony boundary adjacent to basis plane at one side of the boundary; (iii) the colony boundary, neighboring to basis planes at both sides of the boundary; and (iv) three-point junctions of colonies. Generally, in the above cases, the twist boundaries are low-angle ones existing at the screw dislocations. Despite high texture, high-angle boundaries are often observed in the tapes. Majority of colony boundaries in the *c*-axis direction are the mixed type, including tilt and twist components and also dislocation network. The value of J_c could be controlled by residual layers of Bi-2212 phase at the twist [001] boundaries [1727, 1728] and also by Bi-2201 inclusions [1772], if their content is sufficiently great.

Amorphous buffer layers can exist at all four types of boundaries and considerably influence conductive properties, because their thickness usually exceeds considerably the superconducting coherence length [20, 1855]. The triple point junctions of the colonies in the form of triangles (Fig. 4.24a) are usually formed in tapes owing to plate-late growth of the colonies and misorientation between them. In these triple point junctions, the amorphous phase and admixtures are often observed in the tapes even with high J_c. Obviously, the triple point junctions of colonies, exceeding on some orders of magnitude the coherence length, render negative effect on the transport current in the tape. Secondary phases, consisting of amorphous structure and admixtures, form into Bi-2223 colonies separated particles of $(Ca, Sr)_2CuO_3$ and other oxides (see Fig. 4.24b) [1333]. Large particles of secondary phase lead to arbitrary orientation of Bi-2223 colonies and serious distortion of crystalline lattice, creating numerous defects in these regions. Mutual intergrowths of Bi-2212 and Bi-2223 phases that can be usually observed in colonies in the oxide core of the tape eliminate near silver sheath [1728]. At specific thermal treatment, the intergrowth of superconducting phase is possible in silver [1436]. The propagation of these intergrowths increases Ag/BSCCO interface and assumes an energetic advantage of their growth along concrete crystallographic planes of Ag. Then, the silver texture is able to effectively influence superconducting grain alignment.

There are other microstructure features of Bi-2223/Ag tapes, namely [001] edge dislocations, high density of stacking faults, bending and interrupting of (001) planes in local small regions, which can destroy perfection and continuity of CuO_2 plane, decreasing J_c. At the same time, strain around these defects can act as effective magnetic flux-pinning centers, improving superconducting properties

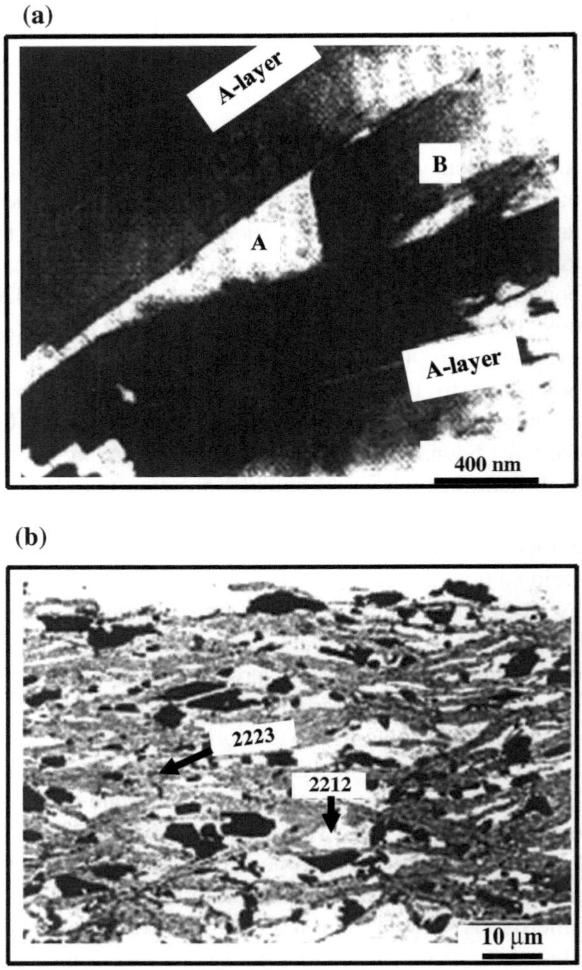

Fig. 4.24 (**a**) Thin amorphous layer at the boundaries of colonies with two neighboring basis planes (*B*) and amorphous structure (*A*) in triple-point junction of colonies [20]; (**b**) inclusion of admixed non-superconducting phase of (Ca, Sr)$_2$CuO$_3$, shown by using more *dark color* [1333]

[1293]. The dislocation density in monolithic tapes near Ag/Bi-2223 interface can exceed on one order of magnitude the corresponding value in the core center [1753]. A more higher plastic deformation near the interface in the tape processing is explained. An addition of silver dispersion permits to form relatively uniform density of core. In this case, an increased dislocation density leads to higher and homogeneous residual stresses in the core, allowing the preparation of superconductor with better properties. Moreover, an increasing of the irreversible strain ε_{irr} (the strain level, above which J_c decreases irreversibly) is caused by fact that the silver inclusions deflect crack and pin their surfaces in the mechanism of crack bridging, preventing the growth of crack into superconducting core [1753]. In this case, $AgNO_3$ additives render better effect on the formation of Bi-2223 phase and the value of J_c compared with Ag or Ag_2O dispersion [1592]. Fabrication of layered tape compositions of Bi-2212/Ag with intermediate buffer layer of silver provides the improved superconducting properties at bending compared with single tapes, having the same cross-sections of superconducting component [742]. The level of irreversible strain also changes in dependence on the number of superconducting filaments. So, the bending strain above 0.1 % at the surface of monocore tape leads to decreasing of critical current (I_c). At the same time, the tape with 1296 filaments can sustain strains of 0.7 % without actual diminution of I_c [1485].

The test data (Fig. 4.25a) of tension of the 61 filamentary Bi-2223/Ag tapes under $T = 20$ K and $B = 0$–8 T show small changes (5 % decreasing of I_{c0} at $\varepsilon = 0$) in the range of small strain (upon to $\varepsilon_{irr} = 0.4$ %) [1415]. Upper this level, there is intensive fracture of superconducting filaments, destroying current paths and leading to sharp diminution of critical current. This behavior is observed for all measured magnetic fields. In this case, corresponding normalized values of I_c/I_{c0} (see Fig. 4.25b) state generalized curve for all considered values of the magnetic field, causing independence of mechanical properties from field. Moreover, the moment of irreversible decreasing of I_c by straining does not depend on external magnetic field, implying ε_{irr} to be an intrinsic parameter of superconductor, found by its material properties. This result is confirmed by comparison of corresponding curves $I_c = I_c(\varepsilon)$, obtained at various temperatures (see Fig. 4.25c). Small differences between values of ε_{irr}, observed in broad temperature range, are explained by material differences of test samples, rather than by some other factors. It should also be noted that a strain increasing much more than ε_{irr} can demonstrate saturation at decreasing of I_c. In many cases, the critical current makes up 20 % from initial value even after a strain of 0.8 % [425].

The most important effects on superconducting properties are rendered by voids and microcracks, forming in the BSCCO core during multi-stage thermal treatments [1336, 1870]. The distribution of microcracks in Bi-2223/Ag tapes is introduced by mechanical deformation during superconductor processing. Comparative studies of longitudinal-rolled and transversal-rolled "green" (i.e., without thermal treatment) tapes, and also of tapes prepared by using one-axial pressing, give the following results [593]. The longitudinal-rolled and transversal-rolled tapes demonstrate smooth interfaces with a small number of short cracks, respectively, in the direction of the tape width and along its length. At the same time, the one-axial pressed tapes show wavy interfaces and long cracks directed to the sample length. These results are explained by the concept that in the case of rolling, maximum shear stress will develop along plane with normal vector, having component coinciding with the rolling direction, promoting the formation of transversal cracks that block transport currents. In the case of pressing, the stress state twists at 90° about the normal to the tape plane, causing cracking along current direction (Fig. 4.26). It should be noted that the contact length (of roller with material deformed) during rolling in the rolling direction is far lower than in perpendicular direction (along the axis of rollers) and agrees with acting friction force. Then, the strains $\varepsilon_x \gg \varepsilon_z$ develop in the central part of oxide core under conditions of restraining of the surrounding material. However, the situation is

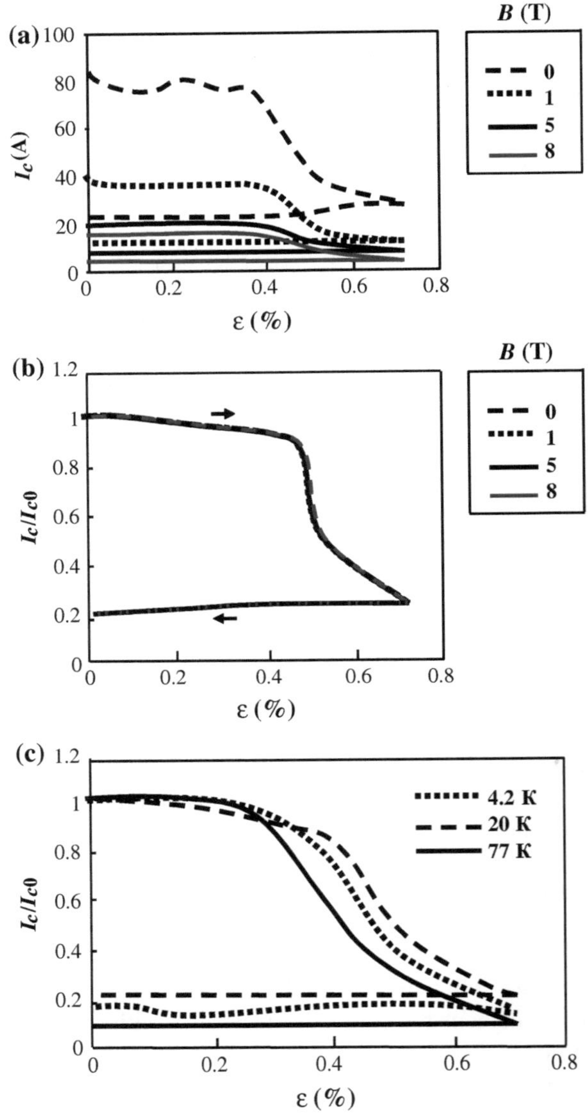

Fig. 4.25 The test data of the Bi-2223/Ag tape tension: (**a**) critical current and (**b**) its normalized value, obtained at $T = 20$ K and $B = 0$–8 T; (**c**) comparison results for different temperatures [1415]

different at the tape edges because of absence of the restraining in the direction outside of the free side surface. It leads to side extension of the tape, strains $\varepsilon_z > \varepsilon_x$ and possible longitudinal cracking of the sample (Fig. 4.26b).

In the case of one-axial pressing, the situation is complicated because of high friction between tape and punches, causing considerable heterogeneity of stress–strain state, which grows with decreasing of the ratio h_{Ag}/d_{Ag}, where h_{Ag} and d_{Ag} are the thickness and width of silver layer in the tape, respectively. Therefore, the

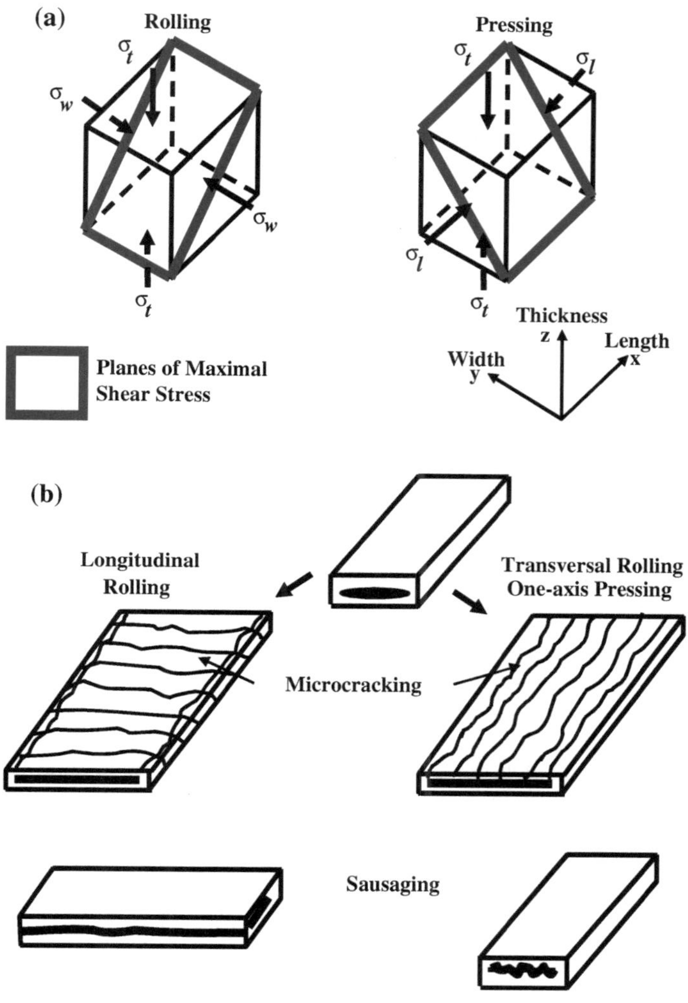

Fig. 4.26 (**a**) Stress state at rolling and pressing of "green" tapes; (**b**) corresponding behavior of microcracking and sausaging

pressing effect on the core density will be directly found by the pressing parameters (i.e., pressure and friction) and also by the tape composition and sizes.

Decreasing of microcracks and porosity may be reached by using modification of standard rolling [722, 908, 1773], alternative methods of pressing [64, 961, 1093], intermediate deforming [568, 1336], regulating the rate of the tape thickness decreasing under mechanical loading [1032], changing radii of rollers in rolling [1580], applying excessive pressure at temperature of Bi-2223 phase formation [1437], controlling the process of the Bi-2223 phase decomposition-restoration at high temperatures [1333], sintering samples after their deforming [118, 1388] and optimizing the cooling rate [708, 947, 1334]. In these cases, a hot deformation in BSCCO systems improves weak links, but a cold deformation increases a magnetic flux pinning [1031].

Pressing of Bi-2223/Ag tapes under cryogenic temperatures permits to transfer more mechanical deformation energy to the oxide core, due to a relative increasing in the hardness of silver to oxide in this thermal treatment. Then, the fracturing of grains during deformation can produce multiple new surfaces, promoting better final alignment of superconducting grains [309]. In this sense, it may be assumed that the mechanism of the grain alignment under pressing would be expected to involve at least two steps:

(1) Initial fracturing of the grains to produce multiple surfaces. The degree of fracture depends on the deformation introduced into the grains.
(2) Rotation and/or movement of the fractured grains into (00*l*) alignment. It is suggested that powder flow is the main mechanism for the rotation and/or movement of the granular bulk. Powder flow also exposes the multiple surfaces created by the fracturing step.

The room temperature process (RTP) of deformation includes the following steps. A small amount of force is applied and a small degree of fracture occurs. Further force is applied and the silver sheath begins to deform plastically. As a result, there are both powder flow and fracturing of the grains. At this step only few new surfaces are created and a ridge formation starts to occur (the ridge formation mechanism in the RTP tapes is shown in Fig. 4.27). The tape density increasing in the final step reduces the amount of powder flow, and the fractured grains are further broken into small pieces.

The cryogenic process (CP) of deformation (77 K) includes the following steps. First, a force is applied that fractures the platelets. The silver sheath is harder and powder flow does not occur at the same amount of applied force as for the RTP tapes. This causes the platelets to fracture, but with very small powder flow. This step creates many new surfaces in the superconductor, but the surface density of the tape increases, as there is no lateral powder flow. As the force increases, the silver sheath becomes to significantly be deformed and both powder flow and fracturing occur. As the force continues to increase, the silver continues to plastically deform, and the tape density increases, reducing the amount of powder flow. The fractured grains are further broken into small pieces. The RTP and CP are demonstrated in Fig. 4.28. In conclusion, note that the increased silver sheath hardness at cryogenic deformation increases rate of deformation in the Bi-2223/Ag processing compared to usual conditions. As a result, an optimum process of tape fabrication improves connectivity and texture of grains, and also J_c [901]. Nevertheless, a densification of superconducting core may be accompanied by aggravation of Bi-2223 texture and diminishing of J_c [1678]. Then, the texture of Bi-2223 superconducting grains could be improved as a result of directed growth of the Bi-2212 grains due to residual stresses during tape [1876] or during reaction of Bi-2212 with secondary phases [1132].

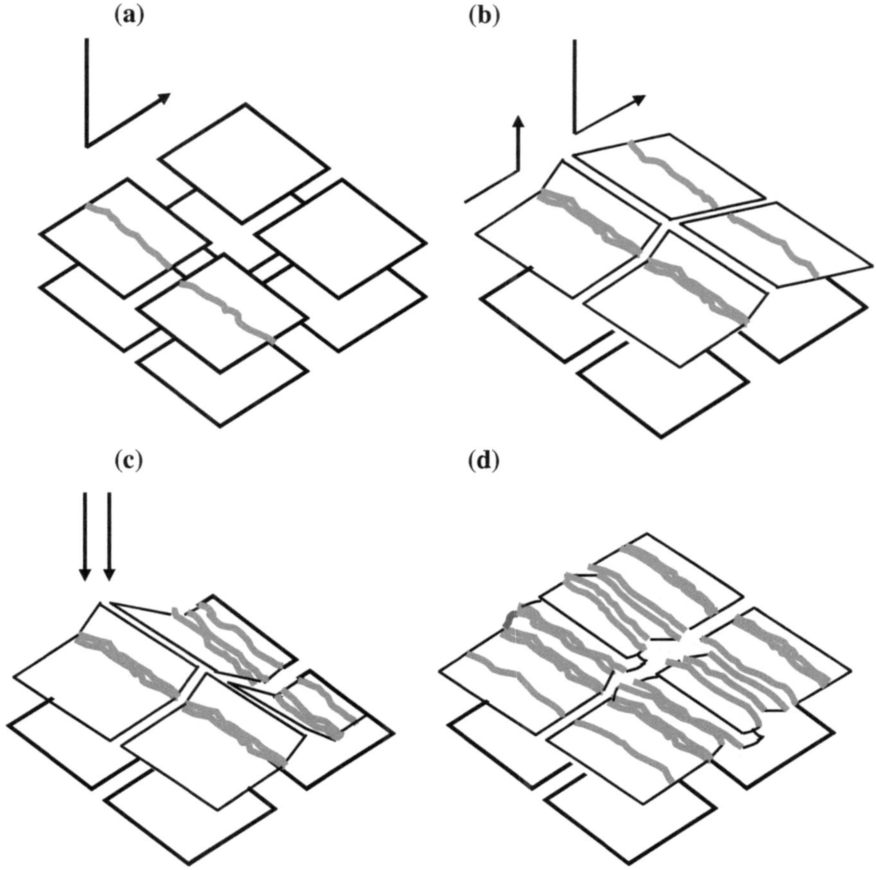

Fig. 4.27 A schematic representation of the steps, involved in the ridge formation in the RTP tapes [309]

In the case of substitution of silver sheath by a silver alloy, all alloying additions form in the sheath particles of oxides as a result of the internal oxidation and promote formation of Ag_2O particles, serving as seeds. Moreover, they reduce grain sizes of the sheath and also decrease pre-stress of oxide core, arising from differences in thermal expansion coefficients of sheath and ceramics that is revealed at tape cooling from sintering temperature to cryogenic one of HTSC application [25]. All these factors result in microhardness enhancement of the alloyed silver sheaths, as well as an increase of the ultimate strength and the yield strength of a composite as a whole. It is accompanied by the growth of J_c stability relative to thermal cycling. Proper test data (tension and bending) demonstrating improved mechanical properties of tapes with sheath from silver alloy in comparison to tapes possessing silver sheath are shown in Fig. 4.29.

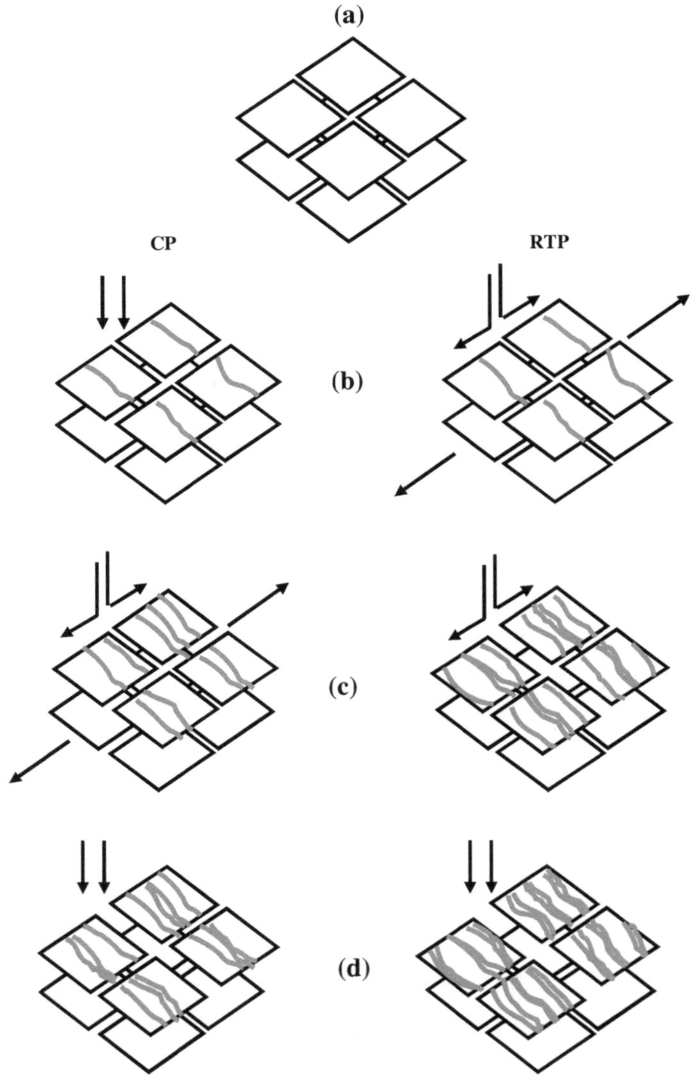

Fig. 4.28 A schematic representation of the steps, involved in the grain alignment mechanisms for both cryogenically pressed (CP) and room temperature pressed (RTP) tapes [309]

At the same time, the doped Ag-sheath leads to some decreasing of critical current density, amount of Bi-2223 phase and T_c, and also to growth of secondary phase and the core thickness variability (sausaging) [1331]. Besides, there is little silver in the superconducting core in the case of the alloyed Ag-sheath. At the same time, Ag intensively penetrates from the sheath into ceramics of an unalloyed sheath composite during annealing and forms particles of pure silver there. These

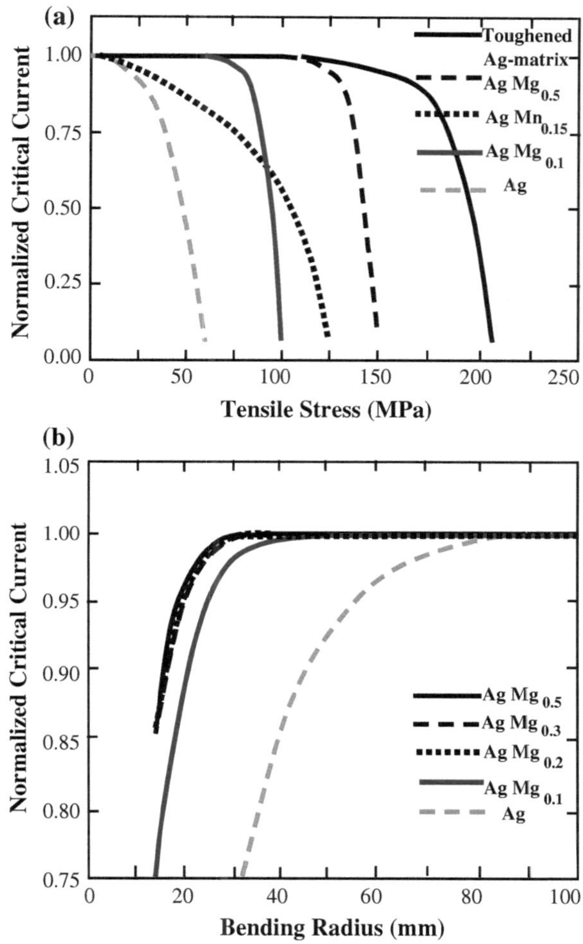

Fig. 4.29 Improved mechanical properties of the alloyed Ag-sheath Bi-2223 tapes with 55 filaments in comparison with pure Ag-sheath Bi-2223 tapes: (**a**) tension and (**b**) bending [479]

particles provide an enhancement of J_c. The above-mentioned decreasing of Bi-2223 phase amount and T_c is the result of the internal oxidation of the alloyed Ag-sheath and accompanying decrease of oxygen flow to the ceramics [1017]. It should be noted that the critical current density in multi-filamentary tapes is very sensitive to technical parameters and microstructure properties. In the case of silver sheath, the value of J_c in the central part of the tape is found to be considerably higher relative to the side sites [490] owing to heterogeneous strain caused by rolling. The silver alloys with improved mechanical properties decrease heterogeneous strains during rolling and increase J_c in the side sites [1867]. Further decreasing of sausaging and differences in the values of critical current density in the central and side sites causes fabrication of the alloyed Ag-sheath multi-filamentary tapes, possessing higher J_c in comparison with corresponding counterparts, having pure silver sheath (Fig. 4.30).

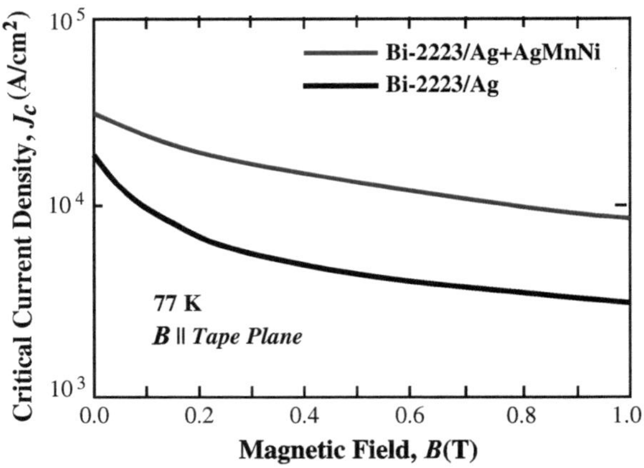

Fig. 4.30 Critical current density versus magnetic field at 77 K [1867]

Investigations of the after-heat-treatment of BSCCO-core Vickers microhardness, assuming directed conformity between Vickers microhardness (H_V) and volume density of the material [1819], state a correlation between high density of superconducting phase and high values of J_c [732, 1331]. These measurements, carried out at different stages of the oxide-powder-in-tube process, estimate an effectiveness of different thermal treatments. In particular, a pressing leads to a greater increasing of J_c in comparison with rolling due to a greater oxide core density after first processing, than after the second one. Moreover, the degree of transformation from Bi-2212 to Bi-2223 phase is always found to be greater in the pressed tapes [732, 1331]. Corresponding results for different thermal treatment (i.e., cycles of "heat–strain") are shown for rolled and pressed Bi-2223/Ag monocore tapes in Fig. 4.31. It is also noted that the sausaging increases with the increasing of superconductor thickness in the process of intermediate mechanical deformation between heat cycles. At this thermal mechanical regime, the rolling leads to greater sausaging in comparison with the pressing (see Fig. 4.31c). Moreover, an increase of microhardness in the case of prolonged heating correlates almost linearly with the growth of the critical current density (see Fig. 4.31a), [1331]. The measurement of the Knoop microhardness (H_K) estimates difference of the J_c behavior in *ab*-plane (that is parallel to the rolling plane) and along *c*-axis direction (that is perpendicular to the rolling plane). So, in the first case, the microhardness H_K remains approximately constant, at the same time; in the second case, it increases with the annealing time (Fig. 4.32) [918]. Destroying of the microhardness growth with the density increasing permitted to state heterogeneities, linked with formation of more dense blocks of grains, separated by cracks [828] that was caused by dominating mechanisms of the core deformation, namely sliding and fracture of grains.

Key factors, stating critical strains and stresses for superconducting tapes, are the following: (i) connectivity and alignment of grains [468, 980], (ii) uniformity and high texture of ceramic core [628, 828], (iii) interface regions in Ag/BSCCO and sheath material [1017], (iv) temperature during loading [699]. For multi-filamentary

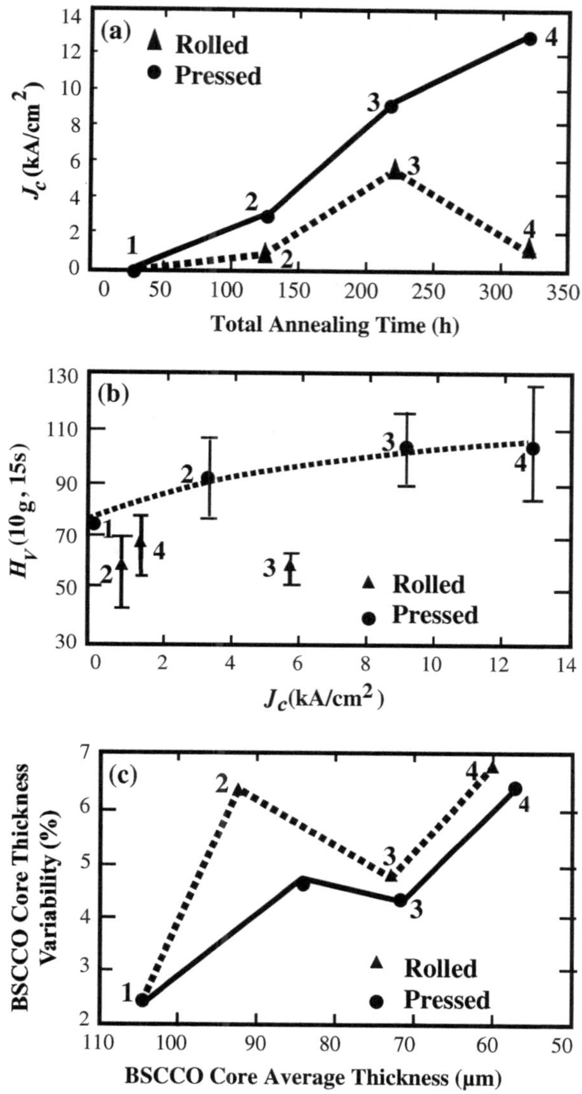

Fig. 4.31 Comparative results of intermediate rolling and pressing between heat cycles for Bi-2223/Ag tapes: (**a**) J_c (0 T, 77 K) versus total heat treatment time; (**b**) Vickers hardness versus critical current density; (**c**) BSCCO core thickness variability as a function of average core thickness. The *numbers* point to the number of heat treatments, which the tape is subjected [1331]

tapes, number and distribution uniformity of filaments play considerable role in the estimation of mechanical properties [1382, 1415]. Critical mechanical characteristics can be considerably improved, adding Ag (Ag_2O, $AgNO_3$) dispersion in the superconducting core. For example, the value of ε_{irr} increases more than two times compared to the monolithic tape in the case of addition into monocore tape of 7 wt.% $AgNO_3$. In the case of multi-filamentary tape, this value increases more than three times in comparison with the monolithic monocore tape [1593]. This growth of mechanical properties compensates well a small decreasing of J_c, observed in the case of $AgNO_3$ additives [64].

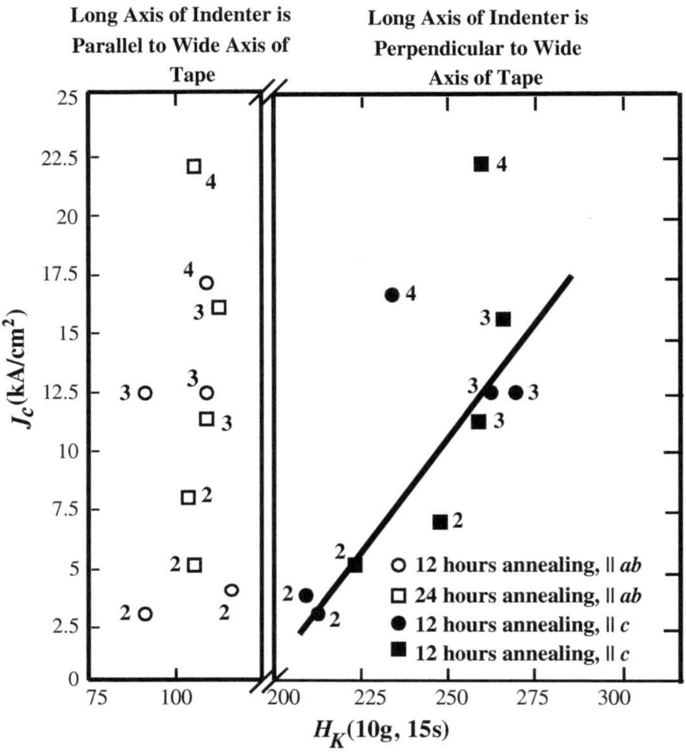

Fig. 4.32 J_c (0 T, 77 K) as a function of Knoop microhardness of BSCCO transverse cross-sections. The data to the *left* of the axis is for H_K parallel to the rolling plane (i.e., approximately || *ab*-plane), and the data to the *right* is for H_K perpendicular to the rolling plane (i.e., approximately || *c*-axis). The *numbers* point to the number of heat treatments, which the tape is subjected [918]

Based on MOI method, intensive researches of BSCCO/Ag samples under bending and at different values of strain permitted to observe the stage of microcrack formation and also to understand how magnetic flux penetrated superconductor [901, 1383]. It is confirmed that the defects of the void type or non-superconducting phase type are the sites of crack initiation [426, 1342]. The bending, also as another strain, changes the microstructure and critical current in two directions: (i) as a result, the intergranular contacts become worse, decreasing critical current density, and (ii) preliminary existing defects are developed that diminish local critical current due to decreasing of effective superconducting cross-section [723, 1383].

Intergranular contact, destroyed at bending, could be restored after straightening of tape and the tape cooling down to cryogenic temperature [929]. The observed effect is explained by the core compression transferred by silver sheath on cooling. The compression causes a sliding of BSCCO grains, restoring the broken contacts.

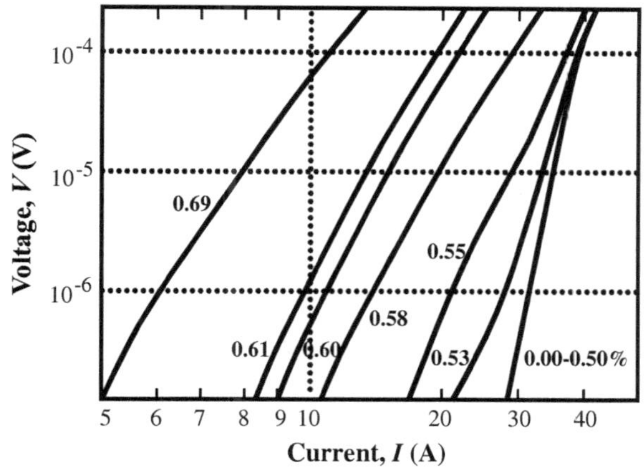

Fig. 4.33 Characteristics of transport current *V–I* versus strain in multi-filamentary Bi-2223/Ag [878]

Obviously, this effect will depend on the core density and texture. In this case, the lattice of transversal microcracks, formed during the longitudinal rolling, possesses elevated sensitivity to the tape bending [1382]. In bending, most tensile and compressing loading exist at external surfaces of the tape and render effect mainly on the metal (alloy)/ceramic interfaces. Fatigue tests of the tape bending/straightening type identify the mechanism of microcracking formation in these interfaces and the microcrack growth in ceramic. It is found that tensile loading favors formation of intercrystalline cracks. At the same time, the compressing stresses form transcrystalline cracks, depending on the orientation of the *ab*-plane [1076]. When tape is subjected to cyclic strains, which are lower than ε_{irr}, the cracks are transcrystalline type irrespective of stress kind.

The tensile tests of Bi-2223/Ag tapes demonstrate three typical stages, namely (i) very narrow region of elastic behavior, (ii) sufficiently broad stage of microcrack initiation and growth and (iii) a macroscopic flow, accompanied by multiple cracking and macrocrack formation [1281, 1282]. Fatigue tests of monocore and multi-filamentary tapes show that the microcracks do not reach the macrofracture threshold in the second stage [880, 1623]. An effect of tension on the sharpness of transition, depicted in curves of the "transport electric field–current density" (*E–J*) dependence (that is very important for devices, based on HTSC), may also be estimated experimentally [878]. Typical *V–I* curves dependent on strain for multi-filamentary Bi-2223/Ag tapes are shown in Fig. 4.33. As shown in the figure, before sharp decreasing of I_c its degradation takes place initially in the region of lower electric field. The test curves also estimate change of value *n*, defining the dependence $V–I^n$.

Joint effect of strain and external magnetic field defines considerable decreasing of I_c even in the smallest growth of the field [723]. Two current carrying paths co-exist in the tapes: one is through the Josephson junction network, consisting of weakly linked grains, and the other is through the strongly linked grains. Investigation of these paths shows that in addition to material cracking, strain deteriorates the grain connectivity, leading to the easier suppression of I_c by

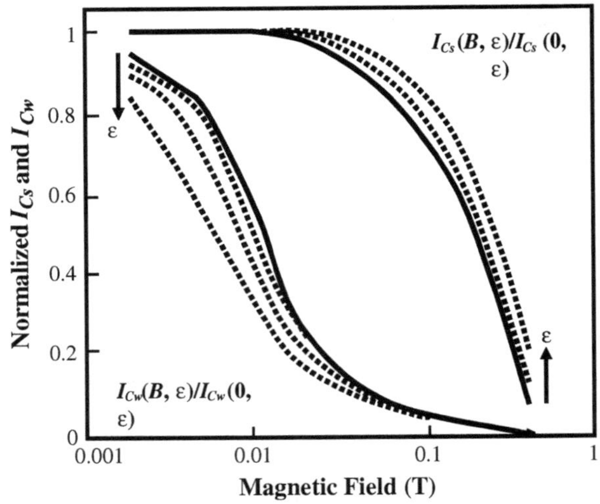

Fig. 4.34 Normalized I_{cw} and I_{cs} as a function of magnetic field, showing the strain effects on both the strong-link and weak-link current-carrying paths [723]

magnetic field in the low field region. For the well-connected current-carrying path, strain on one hand damages well-connected grain boundaries through cracking or sliding. On the other hand, strain can introduce new defects in grains, increasing intercrystalline components of the magnetic flux pinning. Typical dependencies of transport currents, flowing through weak-linked boundaries (i.e., the Josephson junction network) I_{cw} and through strong-linked grains I_{cs} (the resulting critical current $I_c = I_{cw} + I_{cs}$), on applied magnetic field and strain are shown in Fig. 4.34 [723].

4.3.2 Irreversibility Lines for BSCCO

As has been noted in Sect. 2.2.6, the current-carrying capability of superconductor was found by the irreversibility line, presented by the "magnetic field–temperature" non-linear dependence. Above this line, there is a region with the condition that direct current decreases rapidly, making superconductor to be useless for many applications. This property is connected with the material anisotropy. Due to the greatest anisotropy of Bi family in comparison with other HTSC, its irreversibility line is the lowest among high-temperature superconducting families (Fig. 4.35). Therefore, the irreversibility line is one of the main problems of BSCCO techniques. Comparative test data, obtained for single Bi-2212 crystals, Bi-2212/Ag and Bi-2223/Ag tapes [387], show that the irreversibility line for Bi-2223/Ag tapes occupies a region of higher temperatures than the irreversibility line for Bi-2212/Ag tapes, which is higher than the one for Bi-2212 crystals (Fig. 4.36). Defects introduced, in particular by using mechanical loading, act as pinning centers and are responsible for trapping of vortex lines.

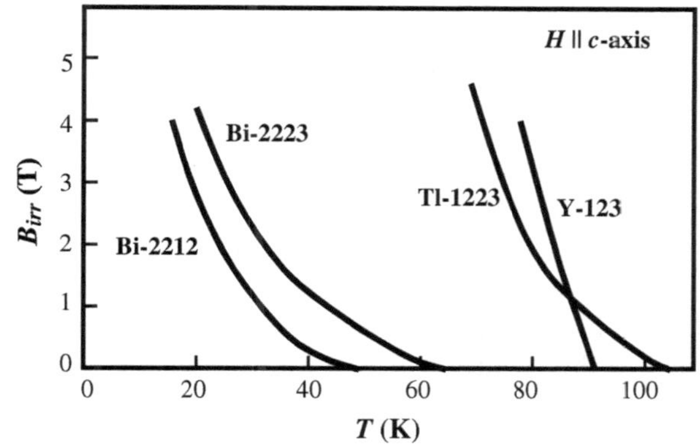

Fig. 4.35 Irreversibility lines for Bi-2212, Bi-2223, Y-123 [283] and Tl-1223 [1682]

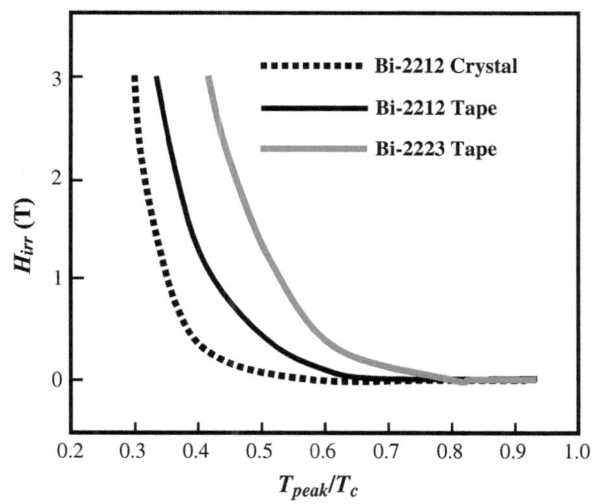

Fig. 4.36 Irreversibility lines for single Bi-2212 crystals, Bi-2212/Ag and Bi-2223/Ag tapes possessing comparative sizes [387]

The irreversibility lines are a measure of connectivity of the CuO_2 superconducting planes and the vortices trapping (i.e., pinning strength) [387, 758]. All three samples (Bi-2212 crystals, Bi-2212/Ag and Bi-2223/Ag tapes) are superconductors of the same family and possess identical spacing between CuO_2 planes. Moreover, the samples investigated in [387] had comparative sizes. Hence, it may be assumed that shear of the irreversibility line location characterizes directly the pinning strength and is thus maximum for Bi-2223/Ag tapes, followed by Bi-2212/Ag tapes and smallest for Bi-2212 crystals. Elevated magnetic flux pinning could be due to higher dislocation density in tape in comparison with crystal. At the same time, the difference between Bi-2212/Ag and Bi-2223/Ag tapes, in this case, is rather caused by different processing conditions. So, in order to obtain the Bi-2212/Ag tapes, the melt-texturing technique has been used, different from Bi-2223/Ag tapes prepared by using the solid-phase reaction techniques and partial application of liquid-phase sintering. In this case, the defects, introduced during mechanical strain in Bi-2212/Ag tapes, are healed partially on melting. At the same time, the solid-phase reaction in the Bi-2223/Ag tapes does not lead to considerable decreasing of material damage [387]. The experiments show [758] that the dislocation line density in Bi-2223/Ag tapes is approximately greater by an order of magnitude, than in Bi-2212/Ag tapes and single Bi-2212 crystals.

The comparison of irreversibility lines for 27 filamentary tapes, processed by using different techniques with application of hot-pressing, rolling and cold-pressing, has been carried out in [628]. All three tapes had the same sizes and have been subjected to the same thermal and mechanical treatments before final sintering. Therefore, in this case also, shear of irreversibility line must be attributed only to the characteristic of strength pinning. The cold strain must create more defects than hot deformation because in the latter case, the grains possess elevated mobility. Then, it may be expected that dislocation density in the tapes, obtained by using cold-pressing technique, is more, than in the case of hot-pressing technique. This states a higher location of the irreversibility line in the tapes, obtained by cold-pressing method, in comparison with the rolled tapes. The latter, in its turn, exceeds corresponding data for tapes, fabricated by hot-pressing technique (Fig. 4.37).

Then, the dependence J_c–H for the hot-pressed sample (Fig. 4.38) shows improved behavior in the region of low magnetic field in comparison with other two samples. The fast diminution of J_c at low external field is proper for Josephson weak links at intercrystalline boundaries. In this case, the rolled tapes demonstrate maximum diminution of critical current under the low field. Under high magnetic field, there is a plateau due to magnetic flux pinning in single grains. The dependence J_c–H for the hot-pressed and cold-pressed tapes is almost parallel to one another, when applied field is parallel to the tape surface. In the case of perpendicular field to the tape surface, prepared by using cold pressing, the critical current density decreases slowly in external field, exceeding 100 mT, in comparison with the hot-pressed and rolled tapes. This result proves that hot-pressing increases grain connectivity and therefore improves weak link behavior in low fields. At the same time, cold isostatic pressing forms a great number of defects, forcing the magnetic flux pinning [628].

4.3.3 BSCCO Bulks

As it has been pointed in previous chapters, the high-temperature deformation by torsion under pressure allowed one to process the melted Bi-2212 ceramics with high superconductive properties [744, 747]. By this, for all deformation conditions a microstructure (typical for Bi-2212), consisting of colonies of plate-like grains, is present in Fig. 4.39. In this processing technology, the deformation leads to formation of a distinct metallographic texture, with the [001] axis of the colonies approximately parallel to the axis of compression (see Fig. 4.39). The dependencies of the Lotgering factor *F*, characterizing material texture [1052], together with the mean length, *L*, and the mean thickness, *H*, of the colonies on deformation

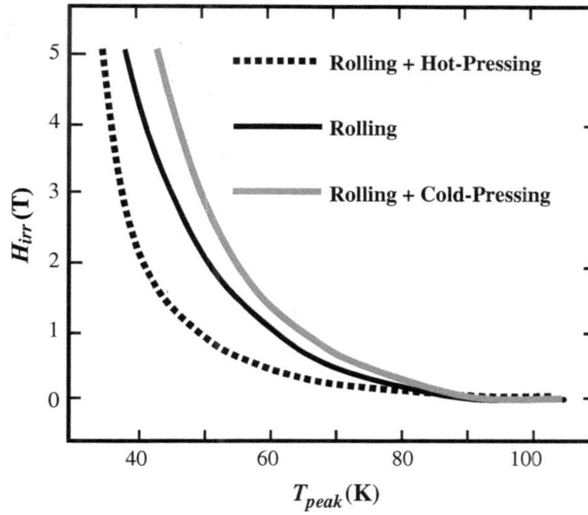

Fig. 4.37 Irreversibility lines for the cold-pressed, rolled and hot-pressed 27-filamentary Bi-2223/Ag tapes, possessing the same sizes [628]

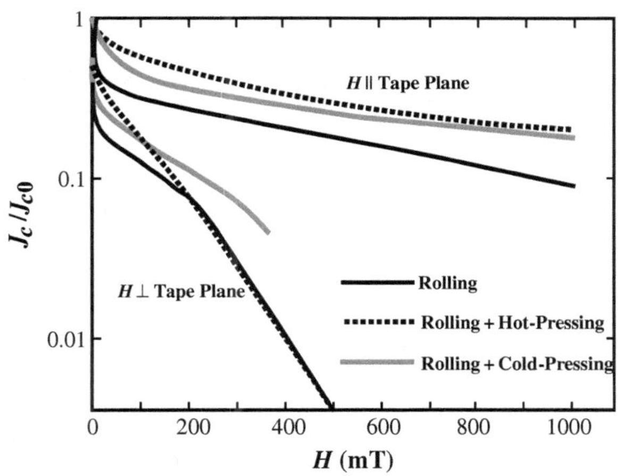

Fig. 4.38 Dependencies J_c–H for the cold-pressed, rolled and hot-pressed 27 filamentary Bi-2223/Ag tapes at 77 K in magnetic fields that are parallel and perpendicular to the surface tape [628]

temperature, T_d, are shown in Fig. 4.40. The dependences of the current densities on deformation temperature for 4.2 and 30 K at 1.5 T are depicted in Fig. 4.41.

By analyzing the deformation mechanisms one should take into account intercolony sliding (ICS). Consequently the total deformation (e_Σ) of Bi-2212 can be represented as a combination of intercolony sliding and intracolonial deformation (e_{ID}) [1375]:

$$e_\Sigma = e_{\mathrm{ICS}} + e_{\mathrm{ID}}. \tag{4.1}$$

Since Bi-2212 does not form twins during deformation, three mechanisms might contribute to e_{ID}, namely intracolonial dislocation slip (IDS), grain boundary sliding and diffusion creep [1375].

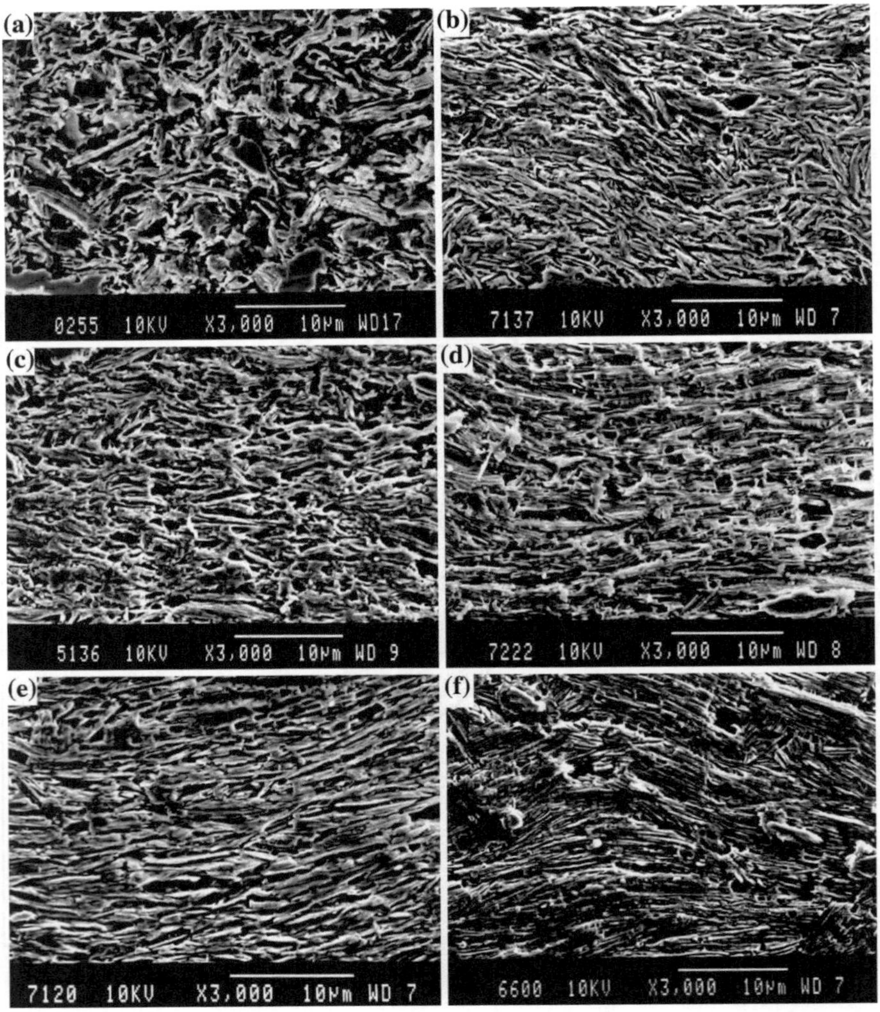

Fig. 4.39 SEM micrographs of the Bi2212 samples (**a**) as-sintered at 855 °C and deformed under standard conditions at $T_d = 795$ °C (**b**), 815 °C (**c**), 895 °C (**d**), 915 °C (**e**) and 940 °C (**f**). The axis of compression is vertical [744]

As it has been concluded in [744], the intracolonial deformation is mainly due to IDS, that assumption was supported by the observed bending of colonies which connected with the presence of lattice dislocations and with prevailing of one sign edge dislocations in the bent sections. At the same time, due to only two independent systems of dislocation slip [100](001) and [010](001) operate in Bi-2212 [1598], only ICS can provide the possibility to preserve the compatibility of deformation of colonies and is responsible for the high plasticity of Bi-2212 ceramic. This conclusion is confirmed by that what as it is well-known the basal

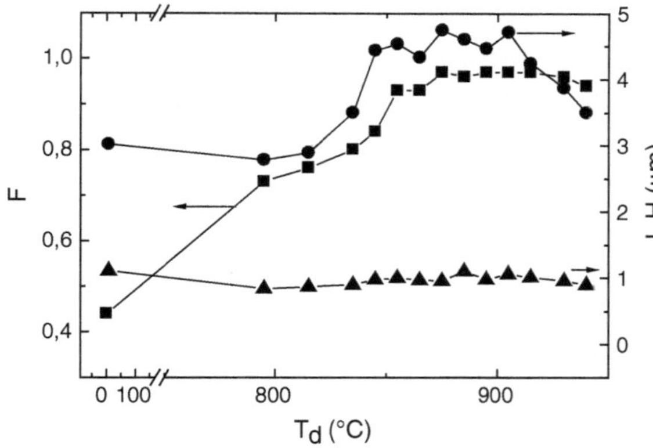

Fig. 4.40 Dependence of Lotgering factor F (*square*), mean colony length L (*circle*) and thickness H (*triangle*) on deformation temperature T_d under standard conditions [744]

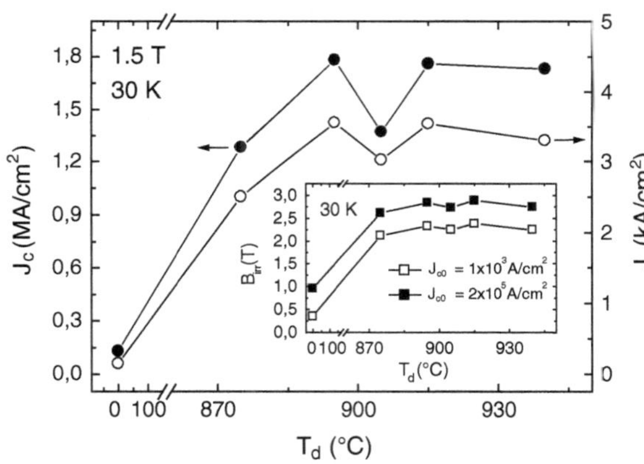

Fig. 4.41 Dependence of J_c at 1.5 T and 30 K on deformation temperature T_d under standard conditions. Inset: Dependence of irreversibility field at 30 K determined for J_c criterion of $1 \cdot 10^3$ and $2 \cdot 10^5$ A/cm^2 on deformation temperature T_d. Full (open) symbols: J_c interpreted as intra- (inter-) granular current density [744]

texture can result from three different mechanisms: IDS [1779], grain boundary sliding [748, 749, 911] and oriented growth of grains/colonies due to applied stress [911, 1876]. IDS by two slip systems cannot provide rotation of colonies as a whole relative to other colonies. It can only initiate the appearance of an additional texture component with an [*hk*0]-axis oriented along the sample radius. Therefore, IDS does not become an important deformation mechanism and cannot be considered as main mechanism for basal texture formation.

Moreover, the colony growth during deformation was a negligible process in the formation of texture in the melted Bi-2212 ceramics [744]. Thus, it has been concluded that the main mechanism of deformation and formation of basal texture is ICS, due to which the plate-like colonies are rotated and oriented in such a way that their [001] direction is parallel to the compression axis. It has been also stated, that the colony length L depended on three processes, namely: (i) thermal activated growth of colonies, (ii) formation of transverse sub-boundaries in the colonies, and (iii) the retarding action of other phase particles and pores. Depending on deformation conditions the contributions of these processes vary and, as a result, either a decrease or increase of L is observed. Since particles of secondary phase $(Sr, Ca)_{4-y}Bi_2O_z$, presented in the initial sintered state, precipitate mainly at colony boundaries the contribution of IDS to the total deformation is increased and that of ICS is decreased. The action of IDS contributes to the increase of the amount of sub-boundaries, thus decreasing L. The increased amount of secondary phases also results in a reduced texture.

Concerning the superconductive properties, it is well known that low-angle boundaries can be strong centers of magnetic flux pinning [362, 1228, 1292]. That is why the increase in J_c and B_{irr} is attributed not only to the decrease in the amount of weak links at intercolony boundaries, but also to increase the mean effective pinning energy. Both factors are caused by the increasing number of low-angle colony boundaries. The observed complex dependence of superconducting properties on deformation temperature (see, Fig. 4.41) has been explained on the base of two simplified assumptions [744]: (i) the deformation temperature influence the density of defects acting as pinning centers of magnetic flux, instead of the pinning energy on the corresponding defects, and (ii) there exist three main types of pinning centers—intracolonial lattice defects (point defects, dislocations and stacking faults), low-angle colony boundaries and particles of secondary phases. In particular, in the vicinity of 895 °C low-angle colony boundaries and intracolonial lattice defects (point defects, dislocations and stacking faults) have been pointed as dominant pinning centers. At higher deformation temperatures both the low-angle colony boundaries and those due to the decomposition of the Bi-2212 phase near the melting point causing an increasing number of nonsuperconducting particles, led to a further increase in the critical current density.

An addition of nano-size MgO particles to starting precursor for composite Bi2212/$Mg_{1-x}Cu_xO$[4] also leads to improvement of superconductive properties of the material. Figure 4.42 shows the various samples of the composites processed by using the high-temperature deformation by torsion under pressure, but Figs. 4.43, 4.44 present dependencies of superconductive properties on deformation temperature.

As it has been shown [744], in the undeformed composite the $Mg_{1-x}Cu_xO$ particle size was rather small. However, the energy of pinning exceeded the corresponding value for the undoped Bi-2212 sample only by 1.5 meV (19.5 against 18 meV) [747]. Neighbor values were obtained in the melted composites Bi-2212 with submicron inclusions $SrZrO_3$, $SrHfO_3$, Sr_2CaMoO_6 and Sr_2CaWO_6, in which due to doping the value of pinning energy $\langle E \rangle$ is increased by no more than 3 meV [390]. Such a small increase in $\langle E \rangle$ is likely to be typical for undeformed Bi-2212 base composites with submicron impurities of secondary phases.

When deformation was applied the $Mg_{1-x}Cu_xO$ particles being obstacles for sliding contributed to accumulation of high density of lattice defects (dislocations and stacking faults) in their vicinity. That is why the essential growth of $\langle E \rangle$ (up to 26 meV) and the rather high J_c in the sample deformed at 815 °C were attributed to the joint influence of particles $Mg_{1-x}Cu_xO$ and lattice defects on flux pinning.

---

[4] Formation of the $Mg_{1-x}Cu_xO$ solid solution is initiated by the interaction of the nanocrystalline MgO powder with Bi-2212 at 850 °C [838].

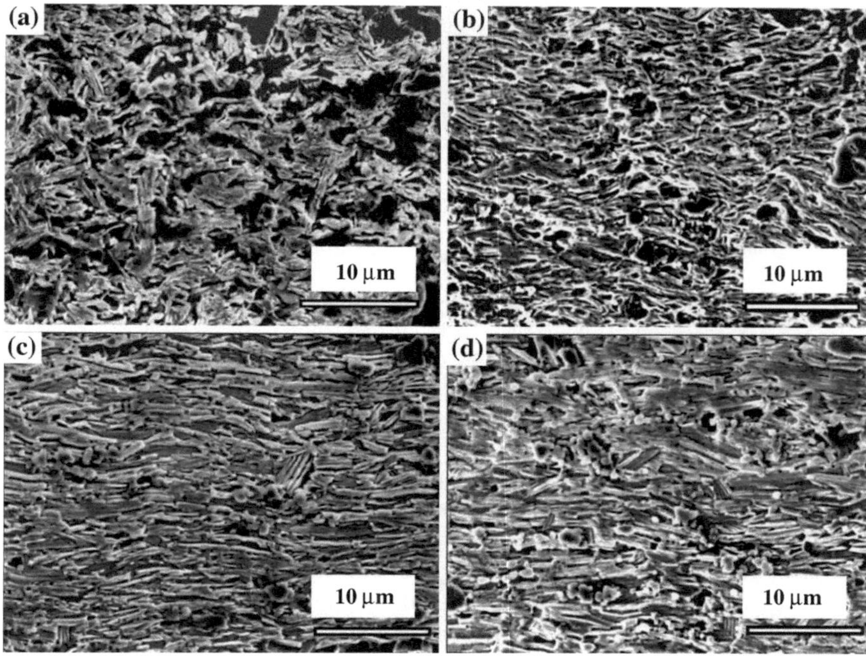

Fig. 4.42 SEM micrographs of the Bi-2212/$Mg_{1-x}Cu_xO$ samples, (**a**) as-sintered at 855 °C and deformed by torsion under pressure at (**b**) 815 °C, (**c**) 895 °C and (**d**) 915 °C. The axis of compression is vertical. *Grey* submicron particles protruding on the surface of the metallographic section are the $Mg_{1-x}Cu_xO$ [744]

Fig. 4.43 Dependence of J_c at 1.5 T and 30 K on deformation temperature T_d of composite Bi-2212/$Mg_{1-x}Cu_xO$, J_c is interpreted as both intra-(*full square*) and inter-(*full circle*) granular current density, unfilled symbols correspond to undeformed sample [744]

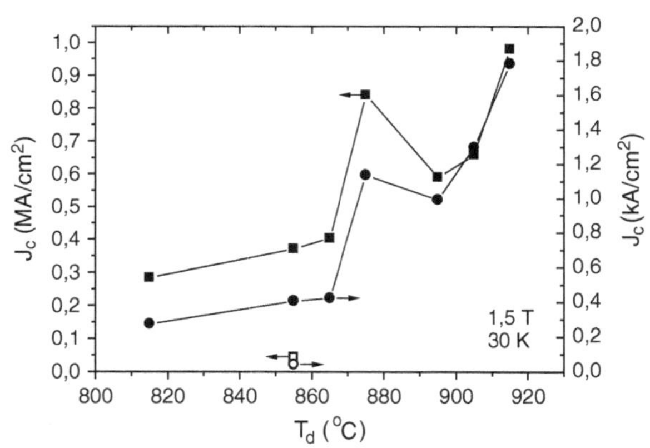

Note, that deformation of the undoped Bi-2212 sample at 815 °C did not lead to such strong growth of the pinning energy, where $\langle E \rangle = 19$–21 meV [747].

With increasing deformation temperature, took place an increase in the particle size and the distance between particles. It is known that the size of the stable dislocation substructure formed around the particles is conditioned by the geometry

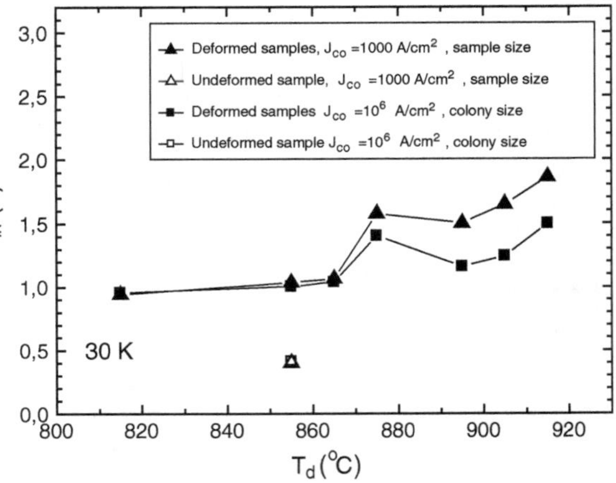

Fig. 4.44 Dependence of irreversibility field B_{irr} at 30 K determined for J_c criterion of $1 \cdot 10^3$ and $1 \cdot 10^6$ A/cm^2 on deformation temperature (T_d) of composite Bi-2212/$Mg_{1-x}Cu_xO$, J_c is interpreted as intra- (*square*) and inter- (*triangle*) granular current density [744]

of disperse particles (mainly by the distance between the particles) [1094]. That is why the increase of the distance between the particles decreased the density of dislocations connected with them. Moreover, regardless of the presence of particles, took place an acceleration of recovery processes within the grain interior, grain boundaries and intercolonial boundaries with increasing deformation temperature.

The studies of microstructure and superconducting properties of the deformed undoped Bi-2212 ceramics have allowed establishing that there are two enhancements of superconducting properties: at 895 and 940 °C [747]. The first peak was explained by the joint action of low-angle colony boundaries and intracolonial lattice defects (point defects, dislocations and stacking faults), while the second one by the joint action of low-angle boundaries and fine particles of the non-superconducting phases occurring in the process of decomposition of the Bi-2212 phase near the melting temperature.

The results presented in Figs. 4.43, 4.44 can be explained in terms of the scheme based on the simplified assumption that there are four main types of pinning centers of the magnetic flux in the Bi-2212/$Mg_{1-x}Cu_xO$ composite [744]: (i) intracolonial lattice defects, (ii) low-angle colony boundaries, (iii) non-superconducting phase particles occurring at decomposition of the Bi-2212 phase, and (iv) $Mg_{1-x}Cu_xO$ particles. In particular, at low temperature (30 K) the composite displayed the best superconducting properties in two samples obtained at 875 and 915 °C, respectively. The main pinning centers in the sample deformed at 875 °C were low-angle colony boundaries and intracolonial lattice defects (point defects, dislocations and stacking faults). The low-angle colony boundaries and particles of non-superconducting phases arising from decomposition of the Bi2212 phase near the melting point dominated as a flux pinning centers in the sample deformed at 915 °C.

At the same time, it has been stated a duality effect of the $Mg_{1-x}Cu_xO$ particles and the lattice defects connected with them on superconducting properties of the composite. On the one hand, they essentially increased the pinning energy and critical current density at low-temperature (30 K). On the other hand, during deformation the particles located at colony boundaries became sources of

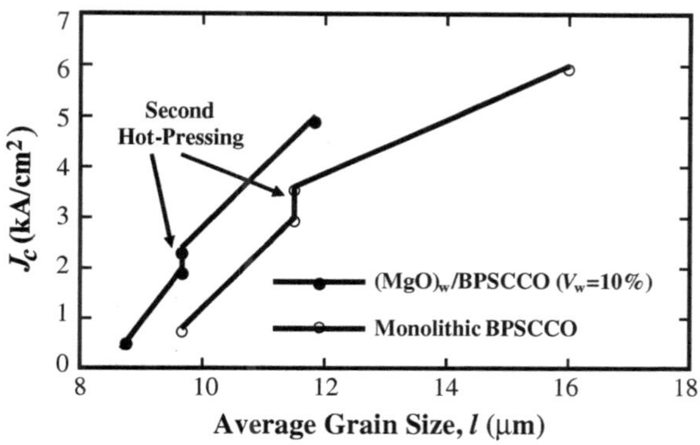

Fig. 4.45 Effects of average grain size (l) on critical current density (J_c) in monolithic BPSCCO and $(MgO)_w$/BPSCCO composite with $V_w = 10$ % [1855]

uncompensated stresses, distorted oxygen stoichiometry near the boundaries, weakened intercolonial J_c and increased ΔT_c. The highest value of ΔT_c was observed in the sample deformed at 815 °C. As the deformation temperature increased, the value ΔT_c decreased while J_c at 60 K and 77 K increased, respectively.

An addition to BPSCCO (Bi-2223) bulks of thermo-mechanical and chemical compatible reinforcement also can improve the HTSC microstructure, superconducting and mechanical properties. In particular, an addition of MgO whiskers oriented parallel to *ab*-plane of superconducting grains [1855], together with carrying out cyclic procedures of hot-pressing and annealing, increase the average grain size both in the *ab*-plane and on the sample thickness. At the same time, when a whisker concentration (V_w) is 20 % and more, there is considerable constraint of the BPSCCO phase grain growth, especially in the *ab*-plane, which is most favorable for improvement of superconducting properties. By introducing whiskers, the current-carrying capability of superconductor increases in comparison with analogous monolithic sample at the same grain size (Fig. 4.45).[5] Additional pinning centers, for example, microdamages, defects of crystalline lattice or low chemical substitutions, can form at the MgO/BPSCCO interfaces. These interfaces remain clear without chemical reaction zone even after prolonged annealing (Fig. 4.46), which is an additional factor of J_c increasing. Long thermal treatment can also contribute to obtain denser BPSCCO microstructure with lower content of admixtures, causing better superconducting properties [1853].

In order to reach desired phases, microstructures and structure-sensitive properties of superconductor, it is important to control technological parameters. In the case of hot-pressing technique, these parameters include temperature, pressure and process duration. Based on the tests, the optimum temperature of hot-pressing for BPSCCO bulk is equal to 800–855 °C [1618]. An initial hot-pressing leads to J_c

[5] An addition of Al_2O_3 filaments is less advantageous due to physical and chemical incompatibility of BPSCCO with Al_2O_3 leading to chemical reactions on interfaces at elevated temperatures of processing, decreasing J_c [1806].

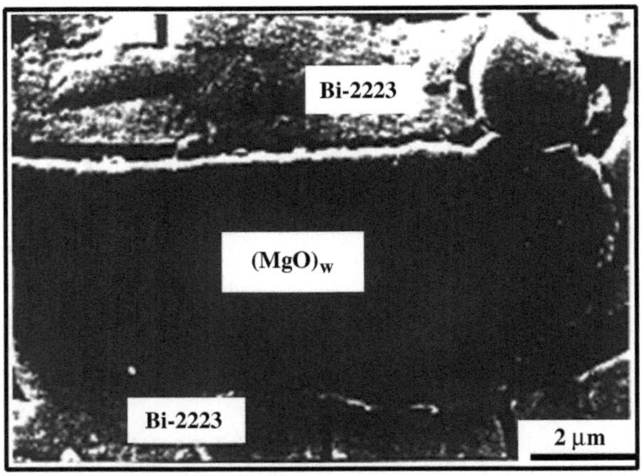

Fig. 4.46 High-magnification SEM micrograph of the interface region in a polished and etched cross-section of a $(MgO)_w$/BPSCCO composite with $V_w = 10\ \%$ [1855]

increasing in HTSC directly after this procedure, as well as following annealing (Fig. 4.47). However, as has been discussed in Sect. 3.4 single hot-pressing alone cannot create a dense and alignment microstructure with large grains. Moreover, it is impossible to reach high purity of the Bi-2223 phase. Sometimes, a diminution of its content is observed because of partial decomposition, caused by combination of high temperature and high pressure [1854]. Due to the above causes, it is necessary to carry out additive annealing or cyclic procedures of "annealing–pressing" to increase J_c (Fig. 4.48). Next, annealing improves superconductor properties, but it should be noted that exceeding its definite duration (this mark is equal to 160 in Fig. 4.47) causes a diminution of J_c. There are other causes for decreasing of the critical current density: (i) stopping of densification in the absence of additional pressure [1301, 1854] and (ii) definite alterations of the superconductor chemical composition because of a possible evaporation of Bi and Pb during prolonged annealing [1091, 1092].

High texture along *c*-axis due to one-axis cold pressing could be formed in the samples with MgO dispersion and in the superconductors with polymeric matrix (BPSCCO/polyethylene) [194]. An addition of silver in different forms into the bulk leads to improved superconducting and mechanical properties. In this case, $AgNO_3$ additives show greater effect on formation of superconducting phase and J_c in comparison with Ag and Ag_2O particles [1591]. Introducing Al_2O_3 filaments into BPSCCO matrix does not lead to considerable decrease in superconducting properties [1806], but provokes substantial toughening of superconductor simultaneously, caused by non-regular and rough surfaces of fracture [1172]. In the case of room temperature (293 K), macroscopically plane surfaces of crack are observed corresponding to brittle fracture with small de-laminations (Fig. 4.49a) and microscopically short pulled-out fibers (see Fig. 4.50a). At the same time, at cryogenic temperature (77 K), there are intensive de-laminations at the fracture surface (Fig. 4.49b) and pulled-out long fibers (Fig. 4.50b). The pulled-out fibers are covered in both cases by BPSCCO grains oriented preferably along fibers (Fig. 4.51), demonstrating strong interfaces between fibers and matrix. Due to brittleness of the

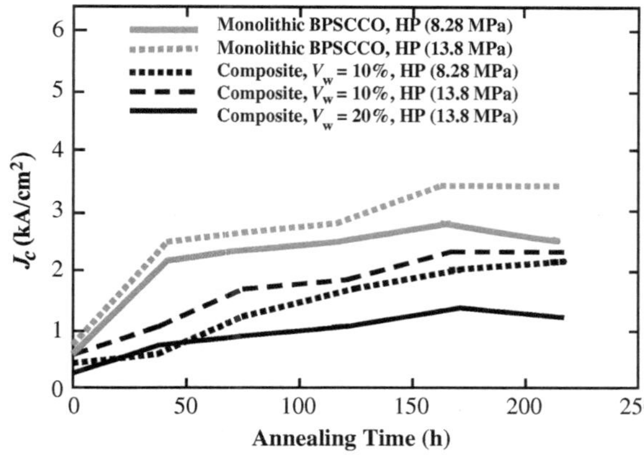

Fig. 4.47 Effect of annealing time on J_c (77 K and zero field) for monolithic BPSCCO and $(MgO)_w$/BPSCCO composite (first, hot-pressed at 820 °C with different pressures and then continuously annealed at 832 °C in 8 % O_2) [1854]

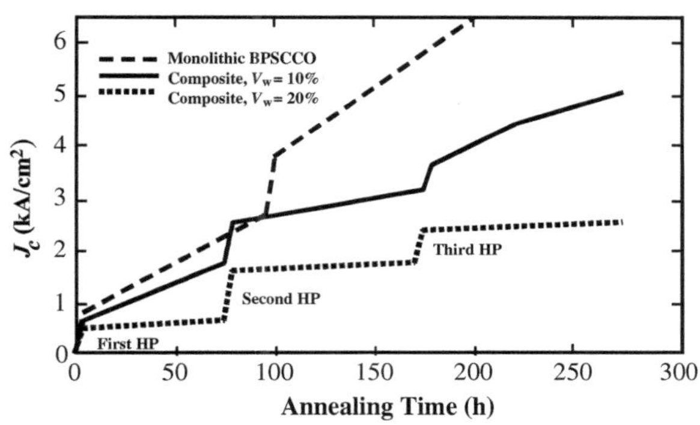

Fig. 4.48 J_c (77 K and zero field) of monolithic BPSCCO and $(MgO)_w$/BPSCCO composite with $V_w = 10$ % and $V_w = 20$ % in 2-cycle and 3-cycle hot-pressing and annealing processes [1854]

BPSCCO monolithic sample at 77 K and as its fracture is controlled by microstructure defects, a considerable weakening of Al_2O_3/BPSCCO composite takes place at 77 K because of intensive cracking of matrix, caused by thermal stresses. Corresponding weakening of interface region between matrix and fibers causes greater increasing of fracture toughness at 77 K in comparison with room temperature. The toughening mechanisms, acting at cryogenic temperature, include the macrocrack deflection, branching and blunting and also crack propagation along the fiber direction. Thus, the differences of the fracture toughness and fracture modes at room and cryogenic temperatures may be explained by thermal stresses, forming in the composite during cooling down to cryogenic temperature. At the same time, strength of the Al_2O_3/BPSCCO composite may decrease in comparison with monolithic superconductor, if the interfaces between matrix and fibers are weak [1103]. In this case, the fibers act as pores and also as stress concentrations.

Best thermodynamic compatibility of MgO [388] assumes application as toughening elements of high-strength MgO whiskers that ensure significant increasing stiffness and fracture toughness of the monolithic superconductor [1856]. The toughening mechanisms of this composite are related to whisker fracture, whisker/matrix interface de-bonding, pushing of whiskers by crack surfaces and also

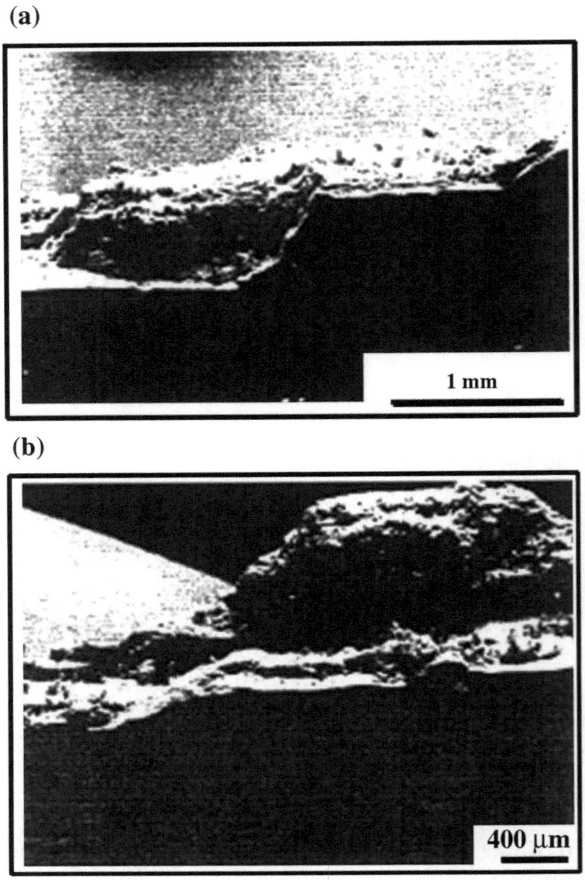

Fig. 4.49 Macroscopic failure modes in Al_2O_3/BPSCCO composite at: (**a**) 293 K and (**b**) 77 K [1172]

the fracture modes caused by the crack deflection (Fig. 4.52). An appreciable increasing of plane-strain fracture toughness (K_{Ic}) is observed both for monolithic BPSCCO specimens and $(MgO)_w$/BPSCCO composite after cooling from room temperature (293 K) to cryogenic (77 K) (Fig. 4.53). For monolithic ceramic, it can be related to greater surface energy (γ) at lower temperature by using Griffith's formula [576] as

$$K_{Ic}(T) = \left[\frac{2E\gamma(T)}{1-\nu^2}\right]^{1/2}, \tag{4.2}$$

where E and ν are Young's modulus and Poisson's ratio, respectively.

A formation of small residual thermal stresses may be expected in the $(MgO)_w$/BPSCCO composite that introduce minimum number of defects at interfaces during cooling to cryogenic temperature because of proximity of the thermal expansion factors for whiskers and superconducting matrix and also because of their small sizes. Hence, an elevated fracture toughness of composite is caused in

Fig. 4.50 SEM micrographs of fracture surfaces of Al_2O_3/BPSCCO composite at: (**a**) 293 K and (**b**) 77 K [1172]

(a)

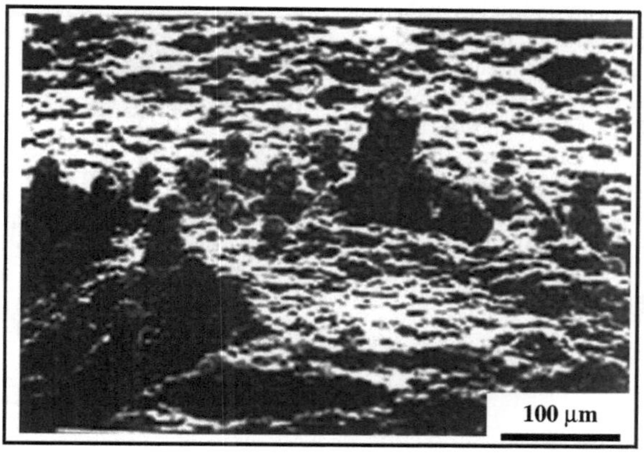

(b)

Fig. 4.51 Surface morphology of a pulled-out fiber observed in fracture surface of Al_2O_3/BPSCCO composite at 77 K [1172]

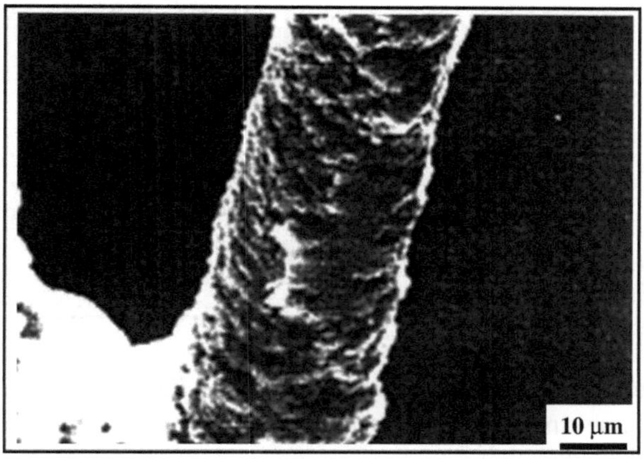

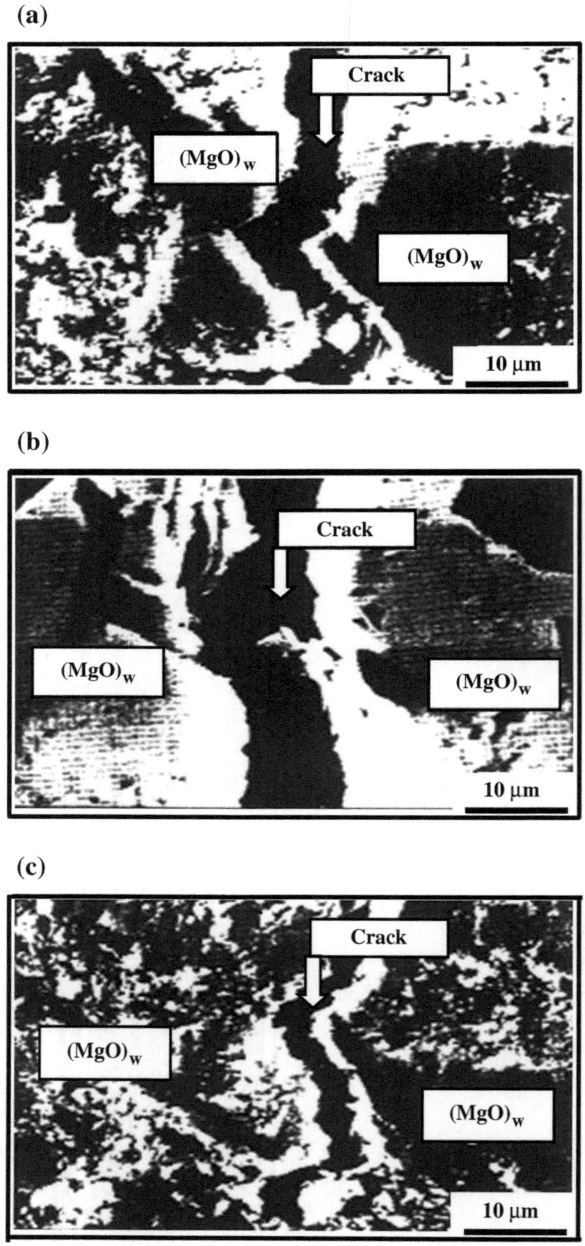

Fig. 4.52 SEM micrographs, showing multiple toughening mechanisms in fractured specimens (three-point notched bending) of $(MgO)_w$/BPSCCO composite, tested at room temperature: (**a**) whisker breakage and whisker/matrix interface debonding, (**b**) whisker pulled out and (**c**) crack deflection [1856]

total by the surface energies of BPSCCO matrix and MgO whiskers [1856]. At the same time, an initial increasing of strength with volume fraction of whiskers may be replaced by its decrease at growth of the whisker fraction up to 20 % and more. It could be explained by microdefect formation, including grain boundaries and

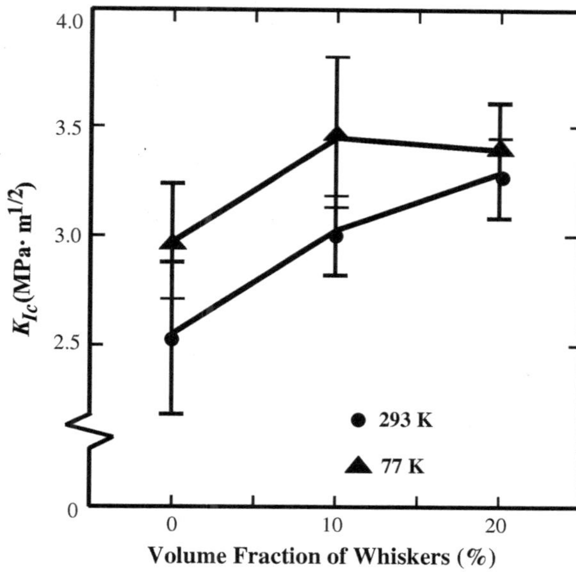

Fig. 4.53 Fracture toughness of monolithic BPSCCO and $(MgO)_w$/BPSCCO composite at room temperature and 77 K [1856]

BPSCCO/MgO interfaces, that creates significant stress concentrations. An increasing of the whisker content leads to the same effect on change of elastic module. By introducing MgO particles in BPSCCO matrix, the bending strength increases. In this case, a decreasing of compliance in comparison with monolithic sample is possible. However, an application of polymeric addition increases the composite compliance.

4.4 Melt-Processed Y(*RE*)BCO

4.4.1 Microstructure Features

Now, the melt-processing techniques, presented in Sect. 3.5, are considered the main perspective for preparation of large-grain YBCO samples with high values of J_c. Therefore, the main attention will be directed to investigation of the melt-processed Y(*RE*)BCO samples. The normal phases, chemical and physical heterogeneities at boundaries of a Y-123 (or 123) phase cause weak links, leading to decreasing J_c in YBCO. In the case of melt-processed multi-domain samples, a liquid is often observed at boundaries of these regions as a result of incomplete peritectic reaction (so-called defects caused by grain growth) and is one of the causes of small J_c [364]. Moreover, a non-symmetric HTSC structure, leading to anisotropy of transport current, is also proper for small critical current density. So, J_c in *ab*-plane is much greater than the corresponding value along *c*-axis [615]. The value of J_c in the melt-processed YBCO strongly depends on the microstructure of superconducting grains

Fig. 4.54 Scanning electron micrograph, showing $BaCuO_2$ lamellae along the *ab*-plane of a Y-123 domain [851]

[856, 1212, 1775]. Compared to the conventionally sintered samples, many more defects are included in the interior of melt-processed YBCO grains. The defect density is a function of fabrication conditions including peritectic heat treatment and subsequent oxygenation treatment [856] and also a special concentration of the *ab*-planar defects (microcracks or lamellae boundaries) around trapped Y-211 (or 211) particles [364, 1775]. In this case, these defects do not considerably influence the transport current parallel to the *c*-axis [1775]. The Y-211/Y-123 interface [1212] and the structure defects [1775] are proposed as flux-pinning centers of a Y-123 phase. Other advantages, associated with the Y-211 additives, are the following: the improved fracture toughness [503], homogeneous microstructure [857] and suppression of the microcrack formation [1212].

The microstructure of melt-processed YBCO, as a rule, consists of Y-123 lamellae (pseudo-grains), oriented almost parallel to the *ab*-planes and possessing common *c*-axes (Fig. 4.54). However, the formation process of this structure is not yet to be understood. For example, it is argued in [1037] that lamellae formation occurs during peritectic solidification and is associated with the irregular growth morphology of a Y-123 growth front in the presence of a build-up of Y-211 particles. On the other hand, it is suggested that the lamellae structure forms during the tetragonal to orthorhombic phase transition (TOPT), which occurs at 400–500 °C, in subsequent sample oxidization rather than at high temperature during peritectic solidification [1775]. The latter point of view can be supported by the absence of lamellae in a tetragonal, unoxygenated Y-123 grain, at simultaneous existence of a number of lamellae in the oxygenated Y-123 grain [851]. Therefore, it may be suggested that the driving force for lamellae formation is the stress, induced during the TOPT, and the subsequent instability of the Y-123 phase.

Besides twins and microcracks, another defect frequently observed in the melt-textured microstructure is a CuO stacking fault. The stacking fault presents itself an additional CuO layer between two BaO layers [1775]. These additional CuO layers exist together with partial dislocations both in a Y-123 matrix and around trapped

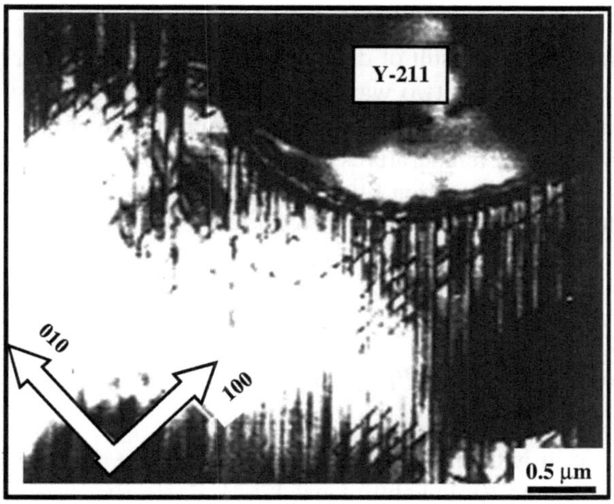

Fig. 4.55 Stacking faults and partial dislocations, developed around a trapped Y-211 particle [851]

Y-211 particles (Fig. 4.55). In [14], it was proposed that CuO stacking faults formed during high-temperature peritectic reaction. However, in [862], it was found that the stacking faults were formed during oxygen annealing in the same manner as that of the lamellae, because Y-123 was not stable at oxygen annealing and decomposed into other stable secondary phases, following the reaction [1801]:

$$4YBa_2Cu_3O_{7-x}+(1/2-3/2\delta+2x)O_2 \rightarrow 2Y_2BaCuO_5+\ 3Ba_2Cu_3O_{6-\delta}+\ CuO. \quad (4.3)$$

The Y-123 decomposition seems not to be associated with the formation of a Y-211 phase, because yttrium-enriched phase regions are observed around the trapped Y-211 particles [1775]. Similar to the $BaCuO_2$ lamellae, a driving force for the formation of CuO stacking faults may be stress induced by the TOPT and also stress caused by the difference in thermal expansion between Y-211 and Y-123 phases. The stresses caused by the TOPT in the melt-processed YBCO are much more than corresponding values of sintering ceramic due to comparatively greater Y-123 grain size. It also can provoke decomposition of Y-123 phase around trapped Y-211 particles. In the case of application of the melting processing (PMP), stacking faults form in the sample because of incomplete diffusion during peritectic reaction between Y-211 particles and the liquid ($BaCuO_2$ and CuO). The amount of defects could be decreased using an additional annealing of the sample under elevated temperature. The considerable decreasing of J_c, observed after this annealing, proves that stacking faults are proper magnetic flux-pinning centers [1868].

There are two main sources of void formation in Y(*RE*)BCO. (i) Inert gas in technological atmosphere remains in the form of voids in sample after its fabrication. In this case, the void number increases with the partial pressure of inert gas, which cannot diffuse from the sample [1329]. (ii) A great amount of oxygen realises as a result of decomposition of Y-123 phase, forming voids [1462]. In this case, there are two ways for oxygen evaporation: usual diffusion and O_2 bubbles, moving through liquid. In low-temperature region, oxygen diffusion is dominated, but at high temperatures, bubble formation occurs, which may be accompanied by sample deformation [365, 1462]. A melting in pure oxygen renders to be effective for decreasing of Sm-123 porosity; however, in this case, a decreasing of superconducting properties is possible [741].

4.4.2 Growth Processes in Seeded Sample

The growth of Y-123 crystal occurs in the form of a parallelepiped with (100), (010) and (001) habit planes [764]. Basically, three modes of crystal growth from the seed are observed (Fig. 4.56), namely:

(1) Epitaxial growth of single Y-123 crystal from the seed with its *c*-axis parallel to the sample axis (the case of using Nd-123 plate-shaped seed and direct seeding at the beginning of solidification) [365]. In this case, there are five growth sectors (GSs), namely: four *a*-GSs with habits perpendicular to the [100], [$\bar{1}$00], [010] and [0$\bar{1}$0] directions and a *c*-GS with habit perpendicular to the [00$\bar{1}$] direction. The top angle of the *c*-GS depends on the ratio of the growth rates in the *c*- and *a*-directions.
(2) A cubic Nd-123 seed leads to five domain samples with *c*-GSc dominating in each grain [1683]. In this case, four narrow *a*-GSs develop along both sides of the 90°-boundaries in each grain. These 90°-high-angle grain boundaries between grains are strongly coupled and *ab*-microcracks do not disturb current in *ab*-planes.
(3) Seed from MgO single crystal does not dissolve in the partially melted sample and it can be placed on a cooled green sample before melt processing. Due to the misfit of the MgO and 123 lattices, it often happens that samples with the *ab*-plane parallel to the sample axis are produced. They consist of two *c*-GSc and three *a*-GSs [695].

In Y-123 crystals, grown from partially melted bulks, subgrains (SGs) can form, which present crystal regions, divided by low-angle grain boundaries [365]. In general, these boundaries are not to act as weak links; however, some low-angle grain boundaries can act as weak links in high-field regions [368]. Therefore, for high-field applications, it is necessary to control the low-angle grain boundaries, that is, it is important to study the formation mechanism of the subgrains. Five different types of subgrain can be classified, according to growth directions active in their formation [370–372], namely: (i) *a*-subgrains (*a*-SGs) have subgrain boundaries (SGBs) parallel to the *a*-axis; (ii) *a-a* subgrains (*a-a*-SGs) have SGBs tilted from the *a*-direction (Fig. 4.57a); (iii) *a-c* subgrains (*a-c*-SGs) have SGBs

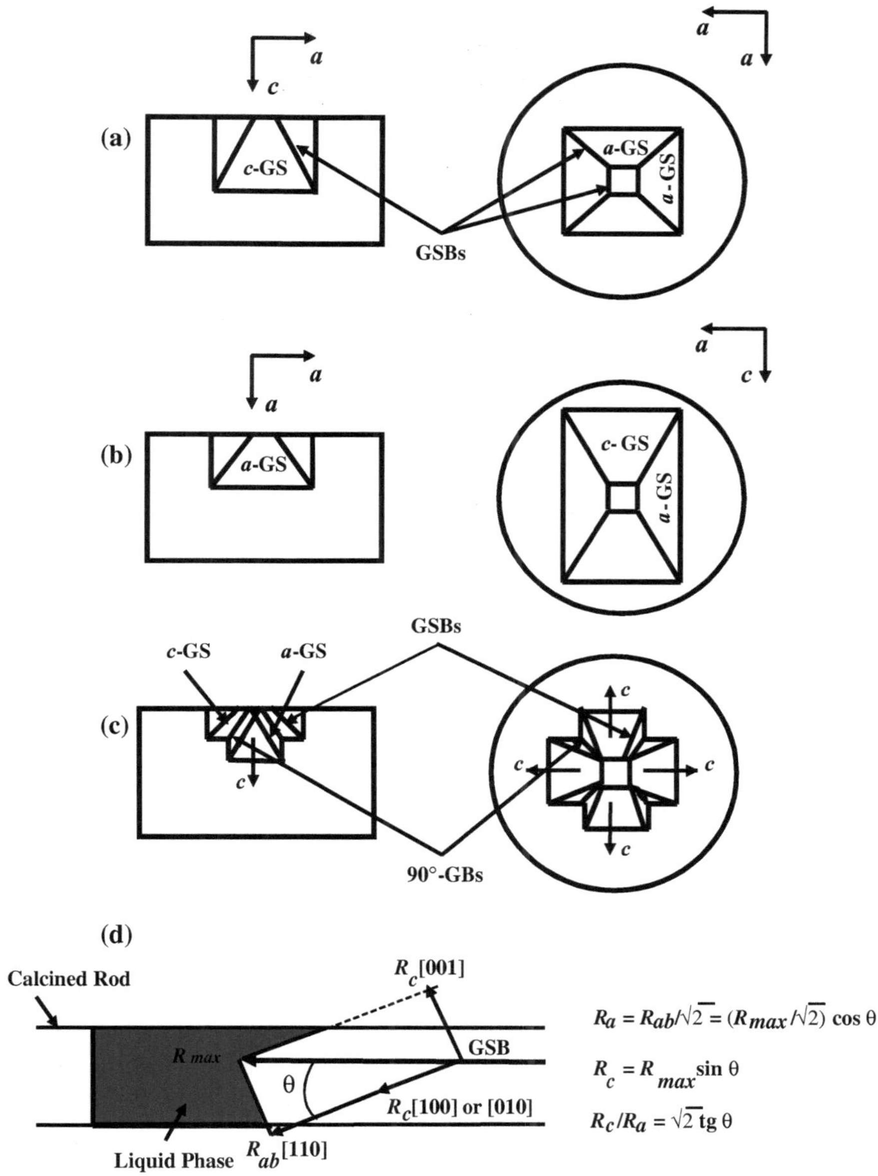

Fig. 4.56 Schematic illustrations of GSs, divided by GS boundaries (GSBs), formed in the melt-grown 123 bulk: TSMG-processed single-grain bulks with the *c*-axis parallel (**a**) or perpendicular (**b**) to the sample axes and in the five-grain bulk (**c**). Schematic illustration of longitudinal interface for typical sample, prepared by directional solidification (**d**) [365]

parallel to the *c*-axis and develop at the stepped planar *a* growth front; (iv) *c*-subgrains (*c*-SGs) have SGBs parallel to the *c*-axis; and (v) *c*-*a* subgrains (*c*-*a*-SGs) have SGBs parallel to the *a*-axis and develop at the stepped planar *c* growth

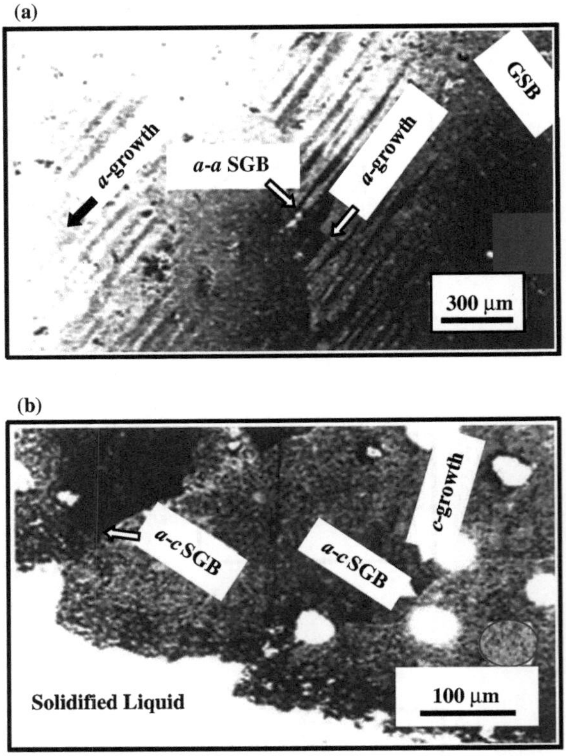

Fig. 4.57 Polarized optical micrographs of TSMG-processed Y-123/Y-211: (**a**) a higher macrostep at the growth front along *a*-axis, forming *a*–*a*-SGB; (**b**) the cross-section both parallel to the *c*-axis and the growth direction, *c*-*a*-SGBs, which are connected with the inner corners of the steps, form on the stepped growth front [365]

front. The first three SGs form in the *a*-GS and the last two SGs form in the *c*-GS. It is interesting that the *c*-SGs are always larger than the *a*-SGs in cross-section. The arrangement of subgrains in the single-grain melt-grown bulk is demonstrated in Fig. 4.58.

In particular, subgrains are manifested by the changes in the direction of microcracks parallel to the *ab*-plane (Fig. 4.59) [370]. In this case, the SGBs are clean and not cracked. The growth-related subgrains, observed in melt-grown Y-123, are supposed to be formed by the dislocation arrangement into dislocation walls during the crystal growth [769]. The subgrain formation is assisted by the edge dislocations with Burgers vectors parallel to the growth front. In the case of melt-grown 123 grains, the dislocation formation due to Y-211 particle seems to be the most plausible mechanism because Y-123/Y-211 interface is incoherent [371]. The density of dislocations should be proportional to the density of 211 particles, which is supported by the fact that the subgrain size increases with an increase in the average spacing between 211 particles [367]. Moreover, it is observed that the subgrain size is reduced with an increased cooling rate [1622]. The fact that subgrains are not observed for single crystals supports the idea that the incorporation of the 211 particles into a 123 matrix is responsible for the subgrain formation.

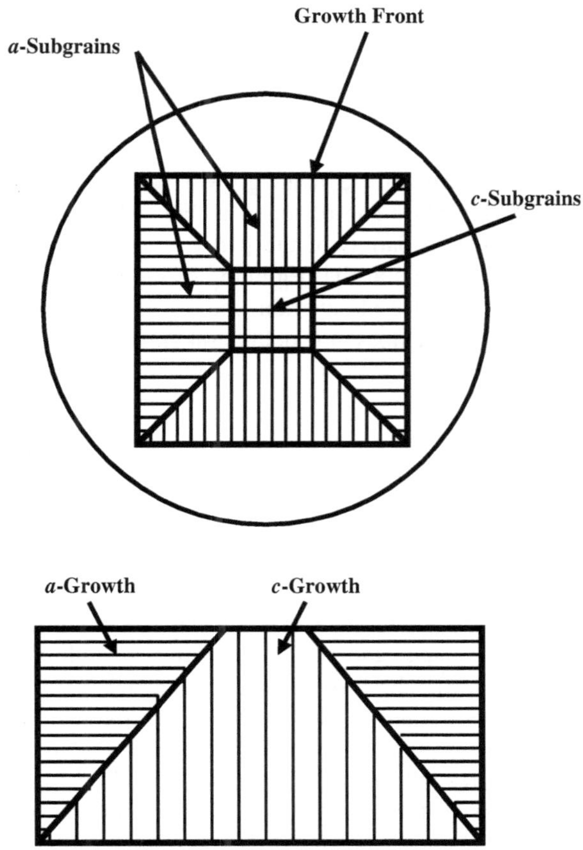

Fig. 4.58 The subgrain arrangement in the single-grain melt-grown bulk [365]

Fig. 4.59 The tilting of *ab*-microcracks at SGBs in a section parallel to the (100) plane [365]

The consequence of the proposed model of subgrain formation is the existence of an incubation period free of subgrains (e.g., subgrain-free region) at the beginning of growth, when dislocation density is not high enough. Such regions are observed at the seed and at the GSBs (Fig. 4.60). The regions free from subgrains can also form along SGBs. In such regions, the grain growth occurs perpendicularly to the main growth front, forming a step at the growth front (Fig. 4.61 [365]). The region *B* in Fig. 4.61 is the barrier for GSBs to continue from part *A* to part *C* of the [010] *a*-GS. Hence, the incubation period free of subgrains appears at the beginning of the *C* part of the [010] *a*-GS. At the same time, the presence of subgrain-free regions along the GSBs and SGBs shows that cellular growth does not define subgrain formation in melt-grown Y(*RE*)BCO bulks [365].

4.4.3 Behavior of 211(422) Disperse Phase

A uniformity of Y-211 phase distribution within a Y-123 matrix is necessary to increase superconducting and mechanical properties of a Y-123 phase. For this, two problems should be solved, namely (i) the spherical Y-211-free regions [853] and (ii) 211 segregation along specific crystallographic orientations of a Y-123 phase[6] [1738]. The formation of spherical Y-211-free regions is attributed to the formation of spherical pores due to gas (oxygen) evolution during incongruent melting of a Y-123 phase [853]. Figure 4.62 demonstrates spherical pores in liquid, pores filled by liquid and Y-211-free regions. When a Y-123 powder compact is heated above a peritectic temperature, oxygen gas is released, forming spherical pores in the liquid. If oxygen gas diffuses out of the pores, they will disappear by the liquid filling process. Compared to the liquid motion to pores, the mobility of solid 211 particles is relatively slow. It makes a non-uniform Y-211 distribution around the liquid pockets (these regions are shown by circles in Fig. 4.62b). During peritectic reaction, liquid pockets turn into spherical Y-211-free regions. At the same time, due to the lower Y-211 density around the liquid pockets, the reaction to form a Y-123 phase is not easy. Therefore, unreacted liquid phase ($BaCuO_2$ and CuO) is often observed in the center of Y-211-free regions [851]. The size and amount of pores are dependent on the heating rate to a peritectic temperature [1040]: larger pores are developed at higher heating rate. Prolonged holding at the partial melting state can eliminate spherical pores by providing enough time for diffusion of oxygen gas out of the sample [1604]. However, the prolongation of this process leads to significant coarsening of Y-211 particles and decreasing of J_c [1040].

[6] Heterogeneity of 211 particle distribution can also be classified in the following forms [365]: (i) an increasing of the 211 phase density along *c*-axis direction, (ii) a development of particle heterogeneity at 123 grain and subgrain boundaries and (iii) oscillations of 211 concentration perpendicular to the *c*-axis.

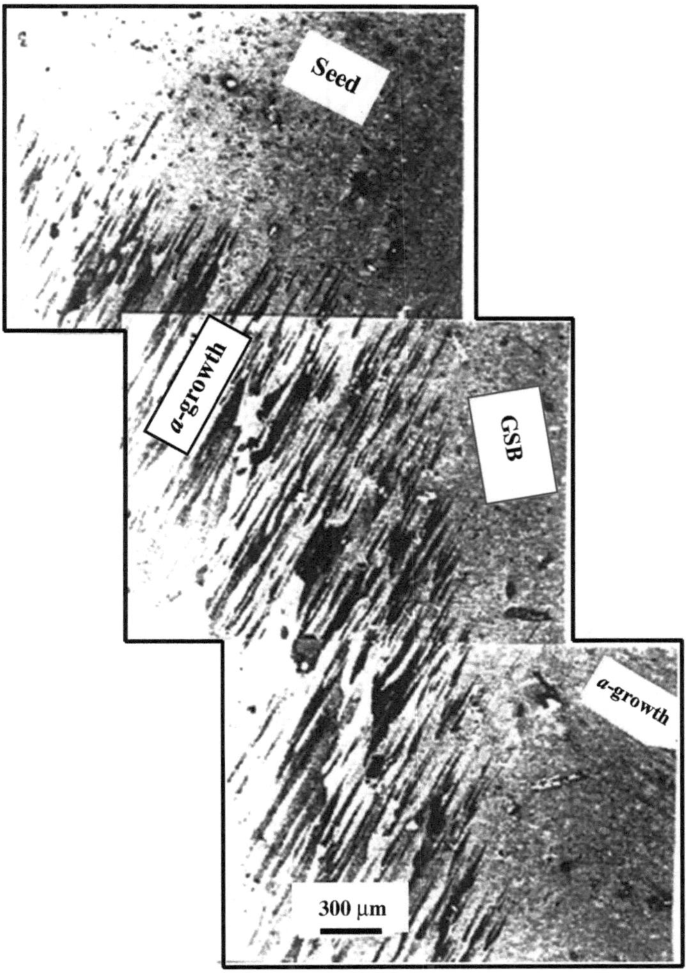

Fig. 4.60 Polarized optical micrograph of TSMG-processed Y-123/Y-211 [365]

When stoichiometric Y-123 powder is used as a starting material for the melt processing, Y-211 particles are formed due to incongruent melting. Most of them are consumed completely to form a Y-123 phase during the peritectic reaction. However, some of them often remain unreacted in the form of trapped particles within Y-123 domains. In this case, the trapped Y-211 particles form X-like linear Y-211 tracks (Fig. 4.63) along the diagonal directions of Y-123 domains [853, 855, 1166, 1738]. The shape of the Y-211 track pattern is dependent on the crystallographic orientation of the polished surfaces of the Y-123 domains, but generally, the Y-211 tracks meet the corners of Y-123 domains. The boundary planes to produce the Y-211 tracks are {110} planes of a Y-123 matrix [851]. Other secondary phases ($BaCeO_3$, $BaSnO_3$, etc.) that are formed by impurity

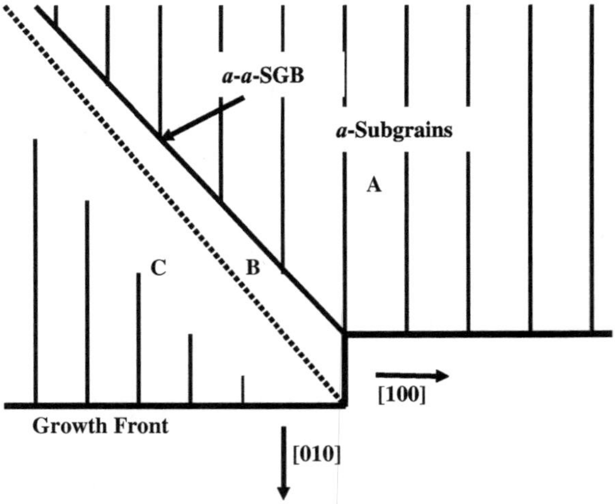

Fig. 4.61 Schematic view of the *a*–*a*-SGB, formed by the growth in the [100] direction at the step on the growth front of the [010] *a*-GS. The layer B, grown by [100] *a*-growth, is a barrier for dislocation walls (*a*-SGBs) to travel from part A to part C of the [010] *a*-GS

additions do not make such a pattern. When the particle density is low, most of them are segregated in liquid at the Y-123 domain boundary [853]. When their density is high, they are trapped within the Y-123 domains in normal direction to the growth fronts in the form of agglomerates [855].

In the case of using Y-211 excess powder, instead of linear Y-211 tracks, a planar segregation mode is developed. Specific crystallographic parts of Y-123 domains are filled with Y-211 particles, but other parts are free from Y-211 particles. The planar segregation of Y-211 particles appears as a symmetrical butterfly-like pattern presented by both (100) and (010) growth interfaces on a polished surface [855]. The formation of planar Y-211-free regions is dependent on the size of Y-211 particles and the growth rate of Y-123 domains. Large-sized Y-211 particles are trapped randomly; at the same time, small-sized Y-211 particles form a segregation pattern, containing Y-211-free regions. The Y-211 segregation pattern is preferentially developed when the cooling rate is low [851].

The Y-211 density in the liquid of the Y-211 excess system is relatively higher than that of the stoichiometric Y-123 system. The presence of many particles at the advancing solid interface decreases the critical velocity for particle trapping by increasing the viscosity of the liquid (η) given in the form [1612]:

$$\eta * = \eta\left(1 + 2.5f_p\right), \tag{4.4}$$

where $\eta*$ is the effective viscosity and f_p is a volume fraction of particles ahead of the growing interface.

Even at a constant Y-211 content, the distribution of Y-211 particles can be changed with growth rate of Y-123 fronts. As can be seen in Fig. 4.64, Y-211 particles are trapped randomly at a cooling rate of 20 °C/h through a peritectic temperature, while they make planar Y-211 segregation patterns at a rate of 5 °C/h.

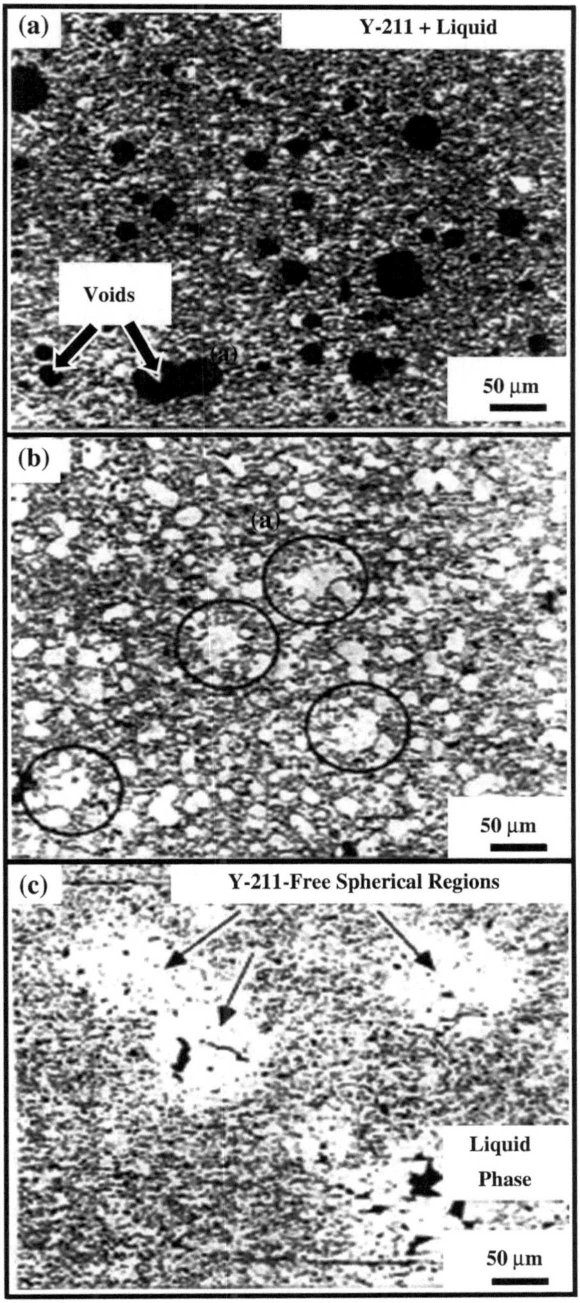

Fig. 4.62 (**a**) Spherical pores, developed in liquid, (**b**) liquid filling into pores and (**c**) spherical Y-211-free regions in a Y-123 domain [851]

This implies that the formation of Y-211 segregation pattern is a function of the growth rate of Y-123 interfaces [851].

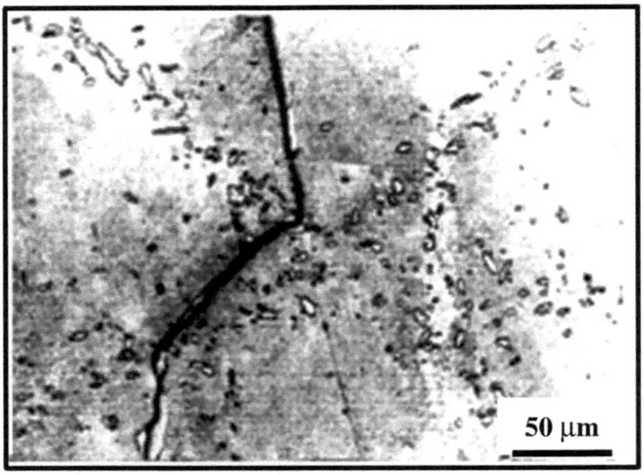

Fig. 4.63 X-like Y-211 pattern [851]

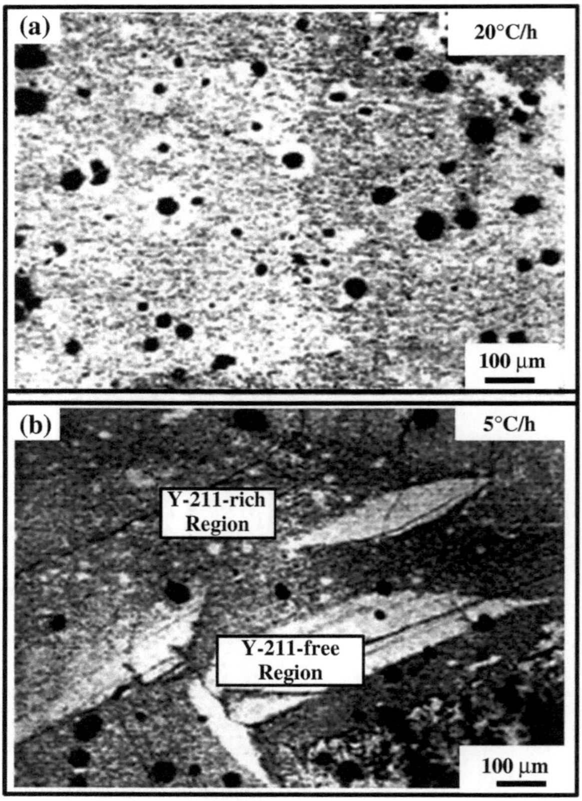

Fig. 4.64 Y-211 distribution within Y-123 domains of samples melt-processed at cooling rates of (**a**) 20 °C/h and (**b**) 5 °C/h [851]

The spatial distributions of Y-211 particles have been studied in the samples obtained by using *RE*-seeds with application of the undercooling technique. A macrosegregation of Y-211 particles occurs during growth process from seeded

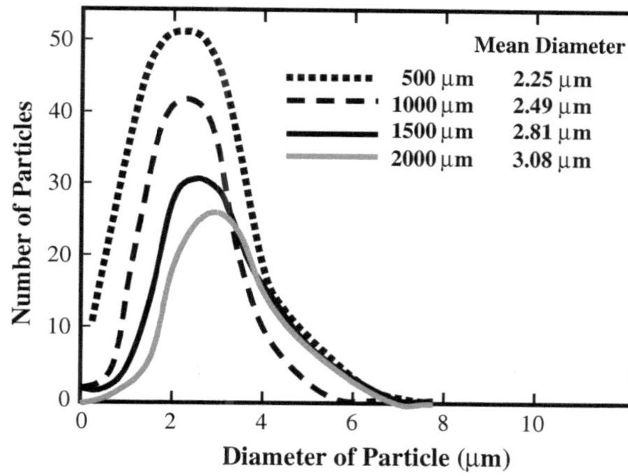

Fig. 4.65 Size distribution for Nd-422 particles into Nd-123 matrix as function of distance from NdBCO seed. The sample is grown at decreased pressure of O_2, 1030 °C, during 50 h [230]

liquid. It depends on the interface direction and growth rate, and also introduced undercooling of the sample [439, 440]. The Y-211 particle density into Y-123 matrix is higher at lower degree of the undercooling sample below peritectic temperature ($\Delta T = T_p–T_g$) [439], showing that the particle trapping is caused by the growth rate of Y-123 interface. However, the single relationship between the growth rate and undercooling (ΔT) is not sufficient to explain the macrosegregation. The interaction of the advancing solid–liquid interfaces and the particles in the melt must be taken into account. By considering the particle radius, the growth rate and the number of particles per unit volume in the vicinity of the growth front, two critical growth ranges may be stated [353]. When 15 °C $\leq \Delta T \leq$ 25 °C both growth rate and number of particles per unit volume exhibit low values. Therefore, only the largest particles can be trapped by Y-123/Y-211 interface. When $\Delta T \geq$ 25 °C, the undercooling is increased, leading to an increase of the growth rate that causes the smallest particle trapping also. The Y-211 particle density near seed can be smaller than at the sample edge. It may be explained by the processes of their pushing, trapping, ripening and coalescence, occurring in liquid, and also by microcracks, observed between 123 lamellae and spherical pores near seed, which cause lower density of this sample part.

When using the hot-seeding technique by Nd-123 crystals at lowered oxygen gas pressure, the Nd-422 particle sizes increase considerably with distance from seed (Fig. 4.65) that correlate with a time of crystallization [230]. However, in this case, an increase of density of the small trapped Nd-422 particles is not observed, that is, processes of small particle pushing and large particle trapping are realized at the definite growth conditions of NdBCO grain. Generally, the growth rate of Nd-123 grain is greater in the *c*-axis direction, than in the *ab*-plane at any values of ΔT (Fig. 4.66). The growth rate increases considerably with growth of ΔT and becomes non-linear in Nd-123 samples even at ΔT = 5 °C, causing the processes of unstable solidification [230]. The observed gradual decreasing of growth rate together with duration of the sample fabrication time at lowered content of oxygen

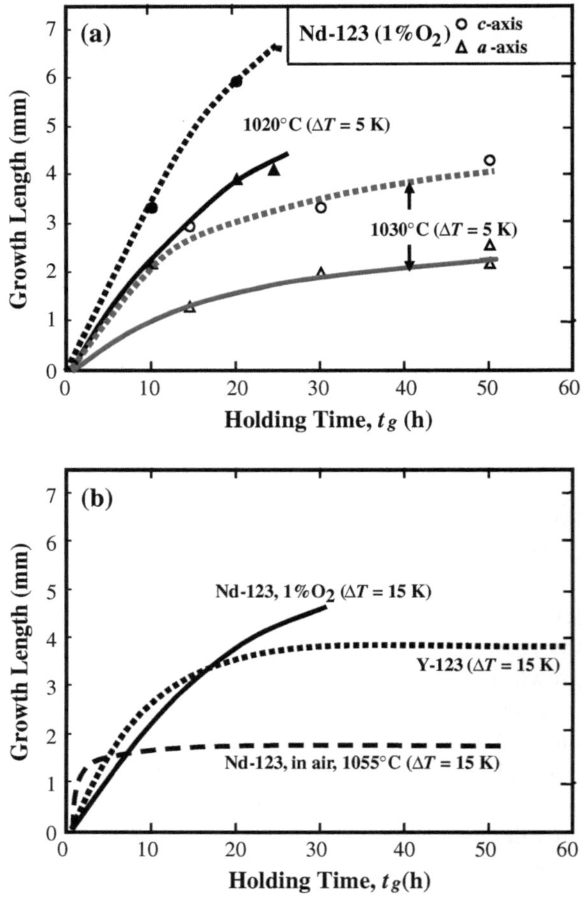

Fig. 4.66 Growth rate of NdBCO grain: (**a**) as function of undercooling ΔT and (**b**) at different processing conditions [230]

gas can be explained by increased volume fraction of Nd-422 particles into liquid, which, in this case restrains diffusive flux directed from growing front to melt.

As ΔT increases, many undesirable subsidiary nuclei begin to form. The growth mode of Y-123 crystal can be classified into four groups of morphology (Fig. 4.67), namely: planar facet mode ($\Delta T \leq 30$ °C), cellular growth mode (30 °C $< \Delta T \leq$ 40 °C), cellular growth with undesirable nuclei (40 °C $< \Delta T \leq$ 45 °C) and formation of the polycrystalline sample by nuclei with random orientation ($\Delta T >$ 45 °C). Two-step undercooling technique, in which the Y-123 nucleus is stabilized at the first undercooling step, where the planar growth mode is maintained, and then crystal growth is accelerated at a second undercooling step, permits to reduce considerably the required processing time for single crystal fabrication compared to the conventional technique without degradation of the sample magnetic properties and texture [771]. The dependencies of the Y-123 phase growth rate at {100} and {001} facets on undercooling degree are subjected to parabolic law.

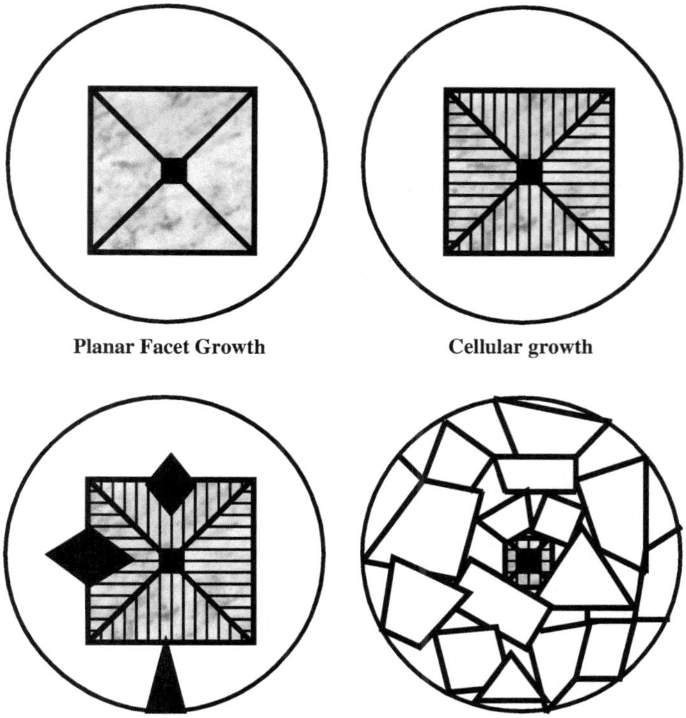

Fig. 4.67 Schematic illustrations of the growth morphology of Y-123 depending on the degree of undercooling, ΔT [771]

Apparently, it changes a direction of crystalline growth and combines orientation of the *ab*-plane with direction of sample solidification [439].

By using some seeds (MSMG technique), accelerating a processing of melted Y(*RE*)BCO samples, a decreasing of levitation force due to residual products of melt at intergranular boundaries [1497] is possible. In order to study this problem samples were prepared by using two seeds with different crystallography [852]. The obtained superconductors with junctions (100)/(100) and (110)/(110) demonstrated better properties (i.e., levitation force and trapped magnetic field) than the samples with junctions (100)/(001) and (110)/(001).

In melted YBCO samples with and without PtO_2 additives, 211 particles are disposed so that 211 longer axes are parallel or perpendicular to the *ab*-direction or the *c*-axis of crystalline lattice of 123 phase [231]. The orientation relations between 123 matrix and trapped 211 particles ([001]123 || [001]211and [110]123 || [111]211) are stated in [70]. In contrast to these results, investigations with application scanning electron microscopy [1666] show absence of definite relationships between crystallographic orientations of 123 matrix and 211 inclusions. Due to own magnetic properties, the 211 particles align along an applied magnetic field, when sample is prepared by the melt-processing technique in the presence of the magnetic field [411]. However, it may be observed that Nd-422 particles [1821] and Sm-211 particles [1626] are arranged parallel to growth direction in one-direction solidified YBCO system even without magnetic field. The 211 particles have polygonal form with faceted surfaces, possessing interface anisotropy. The particle anisotropy increases with addition of PtO_2 [1612] and CeO_2 [856] that are used to control the 211 particle sizes. Additives form 211 needle-shaped particles in liquid. In order to diminish a free energy of interface, 211 particles should be reoriented in liquid to dispose with the least surface energy before the moment of their trapping by the growing 123-front. An alignment of 211 particles also depends on the growth rate of 123 interfaces, size and shape of 211 particles [851].

4.4.4 Effects of Doping Additives

To improve the structure-sensitive properties of HTSC, metal oxide additions are used. Among the impurity elements, PtO_2 [1257], and CeO_2 [855] are known to be effective materials to increase J_c through the refinement of Y-211 particles and/or possible substitution in Y-123 lattices. The impurity elements not only reduce the Y-211 size effectively [1257], but also change its morphology [1738]. During heating of Y-123 powder with doping additives to a peritectic temperature, the additives react with a Y-123 phase to form $BaCeO_3$ [858] and $Ba_4CuPt_2O_9$ [1847]:

$$YBa_2Cu_3O_{7-x} + CeO_2 \rightarrow BaCeO_3; \tag{4.5}$$

$$YBa_2Cu_3O_{7-x} + PtO_2 \rightarrow Ba_4CuPt_2O_9. \tag{4.6}$$

This causes Ba deficiency in the composition of the remainder and formation of the free CuO phase. The reaction of the latter with Y-123 phase leads to the pseudo-peritectic reaction, forming a Y-211 phase prior to peritectic melting [851]:

$$YBa_2Cu_3O_{7-x} + CuO \rightarrow Y_2BaCuO_5 + \text{liquid}. \tag{4.7}$$

These Y-211 particles, formed before melting, serve as seeds for the Y-211 phase during peritectic reaction, increasing the number of peritectic Y-211 particles [577].

One of the ways to reduce the size of Y-211 particles is to provide heterogeneous nucleation sites for peritectic Y-211 particles. The $Ba_4CuPt_2O_9$ compound, which is formed as result of PtO_2 addition in accordance with reaction (4.6), can act in this role due to the similar lattice parameters between $Ba_4CuPt_2O_9$ and a Y-211 phase [1257]. A similar heterogeneous nucleation mechanism acts in the case of CeO_2 additives. Y_2O_3 particles, forming as a result of the reaction between CeO_2 and a Y-211 phase:

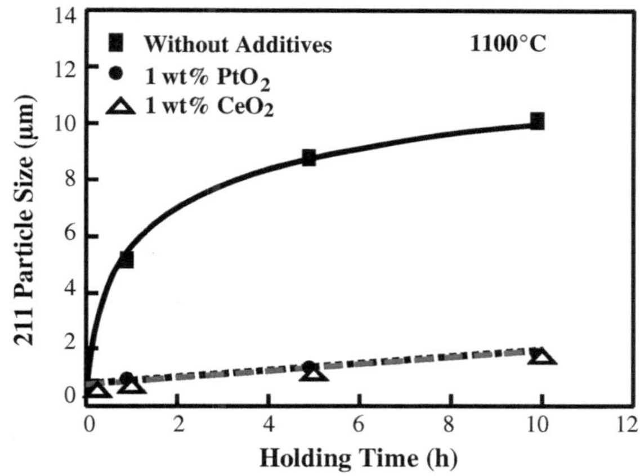

Fig. 4.68 Variation of Y-211 particle size against holding time at 1100 °C in samples with and without PtO_2/CeO_2 addition [851]

$$Y_2BaCuO_5 + CeO_2 \rightarrow BaCeO_3 + Y_2O_3 + CuO, \quad (4.8)$$

can be similar nucleation sites for the peritectic Y-211 phase [1749].

Y-211 particles are coarsen easily in liquid by the Ostwald ripening process [131]. However, it is difficult to estimate the degree of contribution of Y-211 coarsening to the Y-211 refinement because of the nucleation and growth of Y-211 particles, taking place simultaneously. The attempt to separate one process from the other was concluded in a melt infiltration into Y-211 compacts, where no Y-211 nucleation was involved [854]. The technique was applied to Y-211 powder compacts with and without CeO_2 and PtO_2 additions [854, 860]. The results show that the addition of either CeO_2 or PtO_2 significantly suppresses the Y-211 growth in liquid (Fig. 4.68).

Shape control of Y-211 particles is required to yield fine Y-211 particles because the dissolution behavior of Y-211 particles in liquid during the peritectic reaction depends on the Y-211 shape [859]. It is a function not only of the initial powder composition, but also of the type of precursor powder [1181]. In general, the shape of peritectic Y-211 particles of a stoichiometric Y-123 system is a prism-like one, while the shape of a Y-211 excess system is an almost equiaxed granular with facets [851]. During a peritectic reaction the faceted Y-211 particles are dissolved in liquid and their shape is changed to an irregular form with round surfaces. Sometimes, a longer Y-211 particle is divided into several parts, demonstrating that Y-211 particles with a high aspect ratio are better in reducing size than block-like or equiaxed Y-211 particles. Introducing CeO_2 and PtO_2 additions changes the Y-211 shape in liquid to more anisotropic, that is, the needle-like one (Fig. 4.69). These transformations can be a result of change by the additives of the Y-211/liquid interfacial energy.

The melt-processed YBCO samples with $BaSnO_3$ addition could be grown with higher rate than in the case of addition absence. This implies that SnO_2 doping improves effectively yttrium diffusive factor [351]. However, in this case, ripening

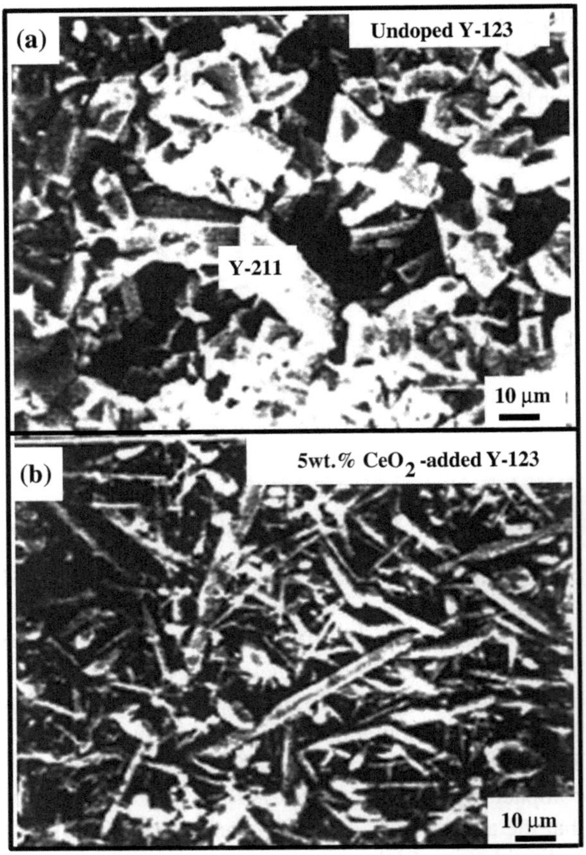

Fig. 4.69 Y-211 particles in liquid of (**a**) undoped Y-123 sample and (**b**) 5 wt.% CeO_2-added Y-123 sample [851]

of 211 particles is often observed. CeO_2/SnO_2 additives (at greater amount of CeO_2) permit to increase growth rate of 123 oxide and to reach submicron size of secondary phase particles, demonstrating square shape and distributed uniformity in the matrix [1082]. Partial substitution of yttrium by calcium in large-grain YBCO introduces additional hole-carriers in CuO_2 planes, ensuring a regime of excessive doping [624, 625]. These additives permit to improve the structure and to reach sub-micron size of 211 particles without using Pt [637]. Then, it has been shown on the test data for monocrystals [896] that the substitution Y/Ca occurred without strain. However, the substitution Ba/Ca or Cu/Ca, which competes with the substitution Y/Ca, becomes predominant at the compression or tension strain above ~6 %. This is found due to Y and Ca ionic radii being approximately the same. At the same time, Ba and Cu ionic radii on 20 % are greater and smaller than Ca ionic radius, respectively. Because the regions are highly deformed near intergranular boundaries in polycrystallite samples, Ca substitutes mainly Ba and Cu, but not Y, decreasing the local strain. In this case, the formation energy of oxygen vacancies in CuO_2 planes and Cu–O links decreases at the deformation.

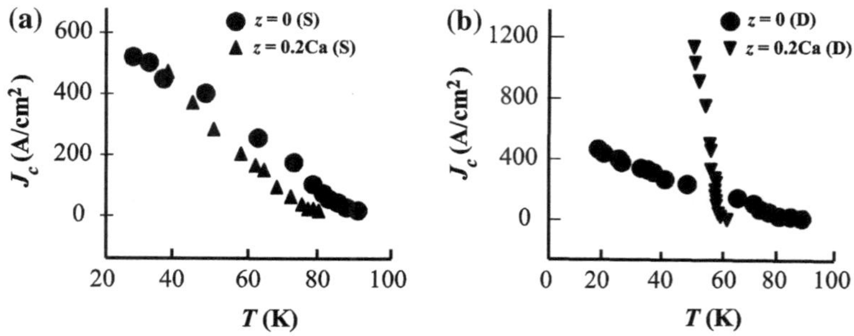

Fig. 4.70 Temperature dependencies of critical current density for S (**a**) and D (**b**) samples [988]

Therefore, there is segregation of oxygen vacancies in the undoped, by Ca of the $YBa_2Cu_3O_{7-\delta}$ samples, near intergranular boundaries, and superconducting properties of the YBCO diminish because of the decreasing of hole carriers. The Ca additives decrease strain and make the formation of oxygen vacancies disadvantageous energetically. The chemical composition of intergranular boundaries and adjoining regions remains to be relative to stoichiometrical one, and J_c increases compared to the value in the samples without Ca.

In Refs. [988, 989], it has been studied the effect of Ca-doping on transport properties of Dy-123. In this case Ca was introduced by using two ways: (I-type) Ca was added as potential substitution of Dy in quantity of Dy deficit that is processed ceramics of nominal compositions $Dy_{1-z}Ca_zBa_2Ca_3O_{7-\delta}$ ($z = 0.2$) (S-samples) and $Dy_{1-z}Ca_zBa_2Ca_{2.8}O_{7-\delta}$ ($z = 0.2$) (D-samples) [988], (II-type) Ca was added during the ceramic processing as excessive element to stoichiometrical relationship Dy:Ba:Cu = 1:2:3 for main cations $DyBa_2Ca_3O_{7-\delta}/Ca_z$ [Dy-123/Ca_z ($0 \leq z \leq 0.2$)] [989]. Ca-doping on I-type decreased significantly T_c, but its diminishing was different for S and D samples ($\Delta T_c \approx 10$ K and 30 K, respectively) (see Fig. 4.70).

The substitution in the I-type ceramics was governed by formation of the microstructure under non-equilibrium conditions, in which local stress (or strain) can have important value and differ significantly for coarse-grain (S) and fine-grain (D) samples. In S-samples at decreasing temperature the grains with higher T_c more rapidly transformed in superconductive state ensuring formation of percolation lattice. It explained their higher $T_c \approx 81$ K. The prevailing substitution by Ca of Ba-sites (S-ceramic) in grain boundaries determining critical current unchanged practically J_c (Fig. 4.70a). At the same time, this substitution of Dy-sites (D-ceramic) caused great increase (in 5–7 times at $T \approx 0.8T_c$) of critical current (Fig. 4.70b). So, the difference in valence (but not in content) of Ca and the substituted cation in grain boundaries was the main factor of improving intergrain critical current at the Ca-doping.

In the II-type ceramics Dy-123/Ca_z ($0 \leq z \leq 0.2$), concentration of Ca in 1.5–3.5 times exceeded the Ca-concentration in grains forming the boundary (the width of Ca-segregation zone at the grain boundary was approximately ±5 nm).

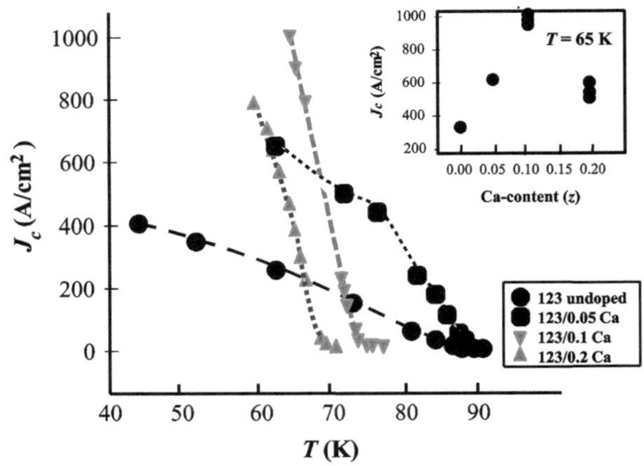

Fig. 4.71 Temperature dependencies of critical current density for ceramics Dy-123/Ca$_z$. Insert: dependence of critical current density on Ca-content for system Dy-123/Ca$_z$ at 65 K [989]

The prevailing substitution by Ca of Dy in the grain boundaries ensured great increasing of critical current ($J_c > 10^3$ A/cm^2 at $T \leq 0.8\ T_c$) (Fig. 4.71) similar to the case of the I-type ceramic ($Dy_{1-z}Ca_zBa_2Ca_{2.8}O_{7-\delta}$). By this, at the same nominal content of Ca, the value of T_c was approximately on 10 K greater in the II-type samples for account of lesser Ca-concentration in grains. The prevailing substitution by Ca of Dy-sites in the grain boundaries led to change of weak state of the grain-boundary lattice from S–I–S to S–N–S type at the both types of Ca-doping.[7] At the substitution of Ba-sites (S-samples, I-type), S–I–S behavior remained.

Addition of NaCl and $KClO_3$ to oxides of main elements at the processing Dy-123 ceramic can also significantly improve superconductive properties [1270]. In this case, Na and K were introduced on type of nominal substitution of Dy or added as excessive to 1:2:3 stoichiometry of main components. By this, there was observed, increasing J_c in 3–4 times at $T = 77$ K and change of character of the grain-boundary weak links from S–I–S to S–N–S type. This is caused in the both cases by segregation of Cl at the 'clean' grain boundaries. By this, the segregation profile (a enriched layer around boundary) width of Cl varied in the limits from ±5 up to ±15 nm for different boundaries ensuring a pinning of magnetic flux. Moreover, Cl was absent into grains that prevented decreasing T_c.

Higher temperature incongruent melting of Nd-123 and Sm-123 compared to Y-123 [1193, 1845] and greater solubility of Nd and Sm in liquid compared to Y [1833] increase both a ripening Sm-211/Nd-422 and a growth of Sm-123/Nd-123 regions. Addition of PtO_2 does not significantly influence decreasing of particle size compared to the effect on Y-211 particles [1194]. A precursor with small Nd-422 particles defines dispersion of small Nd-422 particles in the final sample [1214]. The improved Sm-211/Nd-422 structure can be reached by combining

[7] The character of weak-connected behavior is defined by the dependence $I_c = \text{const}\,(1-T/T_c)^{\alpha}$, where $\alpha = 1$ for S–I–S and $\alpha = 2$ for S–N–S junctions.

CeO_2 additives with grinding of precursor powder [863]. In this case, an anisotropy of Sm-211/Nd-422 particles into melt due to change of energy of the *RE*-211(422)/liquid interface is increased. Doping of precursor by Au/Ag additives diminishes Nd-422 particles down to sub-micron level that is approximately three times smaller than that in the absence of additives and decreases ripening of Nd-422 particles in peritectic region. In this case the doped grains consist of spherical Nd-422 inclusions compared to needle ones that are observed usually in monolithic sample [771].

Additives of Pt [1272, 1275] and ZrO_2 [1273] in YBCO ceramics improve J_c only insignificantly (on 20–30 %) at $H = 0$ and stabilize J_c in magnetic fields $H < 40$ Oe. This correlates with microstructure features caused by the doping such as nano-size (~20 nm) twins coming out on grain boundaries (Pt-doping) and nano-size (2–5 nm) facets on grain boundaries (Zr-doping) which can ensure effective pinning of magnetic vortices due to comparability with coherence length, ξ.

An increase of critical current together with the amount of Ag has discrepant character [543, 1263]. Nevertheless, the presence of silver (up to 20 wt.%) in YBCO samples, which demonstrate initially weak links, can improve them significantly [1469]. This is manifested in the form of corresponding increasing of sample density, orientation and growth of superconducting grains. As a result, an increasing of superconducting properties together with silver amount (Fig. 4.72) is possible. The increase of J_c for small amounts of Ag ($\leq$20 wt.%) maybe due to the acceleration of densification and grain growth. At the same time, the decrease of J_c for large amounts of Ag (>20 wt.%) maybe due to the excessive increasing of the normal (non-superconducting) phase; 20 wt.% of Ag_2O can increase critical current density in the *ab*-plane (J_c^{ab}) by ~200 % and critical current density along *c*-axis (J_c^c) by ~150 % at 5 K. These results are correlated with a reduction of the microcrack density parallel to the *ab*-planes in YBCO/Ag_2O composite. The proper increase of superconducting properties is depicted in Figs. 4.73 and 4.74. Silver additives can additionally improve the flux pinning that also leads to increasing of superconducting properties [1130].

Sol–gel technique has allowed processing high-homogeneous YBCO/Ag ceramics, original components of which were mixed at atomic scale [1085, 1271]. In this case, Ag dissolved very weakly in YBCO-lattice and yielded as Ag-precipitates with diameter of 1–2 μm. Silver segregated in majority of 'clean' boundaries in limits of very narrow region with width of ~1 nm and concentration in two times greater than its solubility limit into grain lattice. This led to change of the grain-boundary weak links from S–I–S to S–N–S type. Significant increasing J_c more than three times at $T = 77$ K was stated only for ceramic with great Ag-doping ($x = 0.4$). In this case, it was observed a microstructure feature, namely segregation of nano-size (2–5 nm) precipitates in 'clean' boundaries disposed on mean distance of ~30 nm one of other. These precipitates are compared with coherence length, ξ, and therefore could be effective pinning centers significantly increasing J_c.

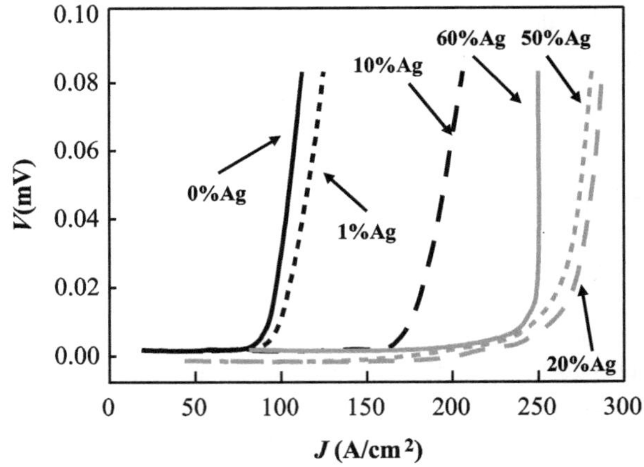

Fig. 4.72 *V–J* diagrams for different Ag-doping of YBCO [1469]

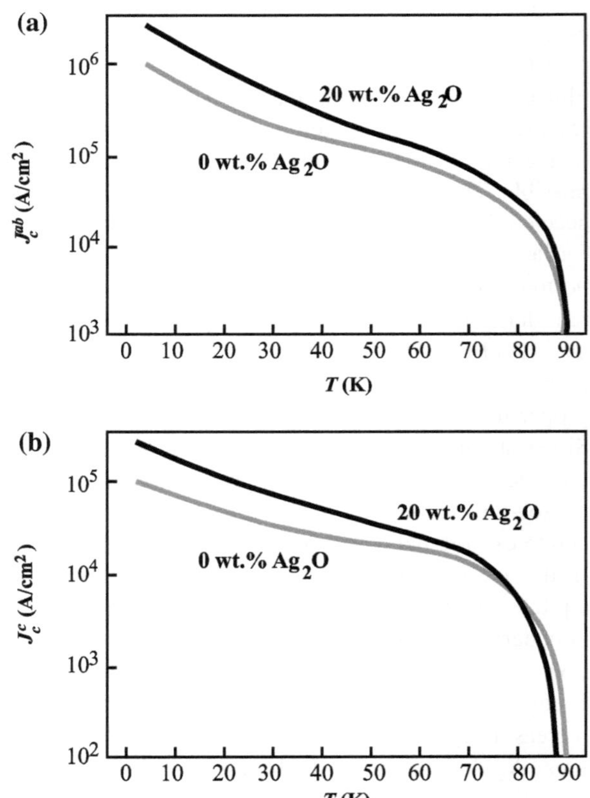

Fig. 4.73 Temperature dependencies of the self-field critical current density: (**a**) at $H \parallel c$ and (**b**) at $H \parallel ab$ for 0 wt.% and 20 wt.% Ag_2O [1130]

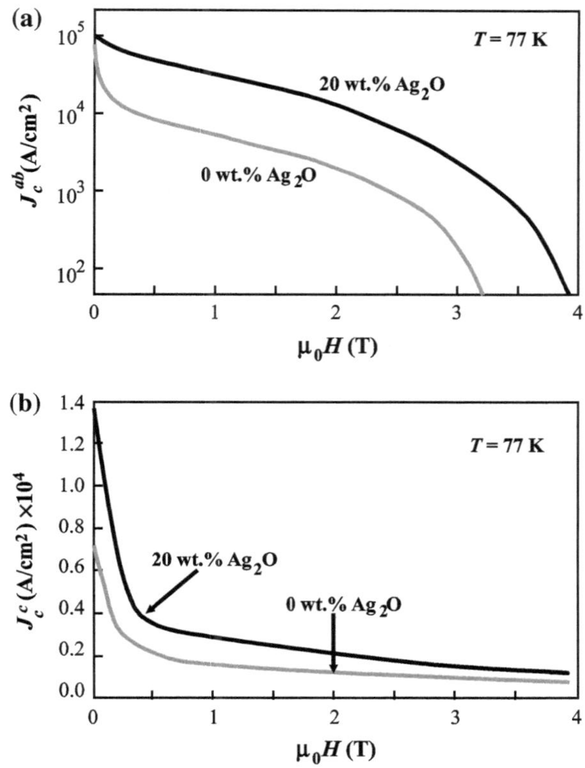

Fig. 4.74 Magnetic field dependencies of the critical current density (at $T = 77$ K): (**a**) at $H \parallel c$ and (**b**) at $H \parallel ab$ for 0 wt.% and 20 wt.% Ag_2O [1130]

4.4.5 Mechanical Loading Effects

As it has been shown by tests on one-axis compression of different Y(*RE*)BCO samples, absolute change of critical current, ΔI_c (or its density ΔJ_c) increased in first approximation together with value of J_c. At the same time, relative change $\Delta J_c/J_c$ and factor of sensitivity of the critical current to stress, $k_c = \Delta J_c/(J_c \Delta\sigma)$, increased with decreasing J_c (see, Fig. 4.75a). Influence of the loading on volt-ampere characteristic (VAC) of the HTSC ceramics has place only at the temperatures lower than T_c. The loading of the samples in normal state both as at room temperature as the temperature of 95 K demonstrated total absence of the loading effect.

In granular HTSC systems, the value of critical current and its change under loading is restricted by weak links (that is grain-boundary junctions) taking part in formation percolation lattice for critical current. An inclination of VAC (dV/dJ) at its linear part characterizes differential resistivity of intergranular junction lattice taking part in transfer of transport current. A different influence of mechanical loading on T_c and J_c is connected with that what in the structure with weak links the loading acts simultaneously on various structural components, namely on

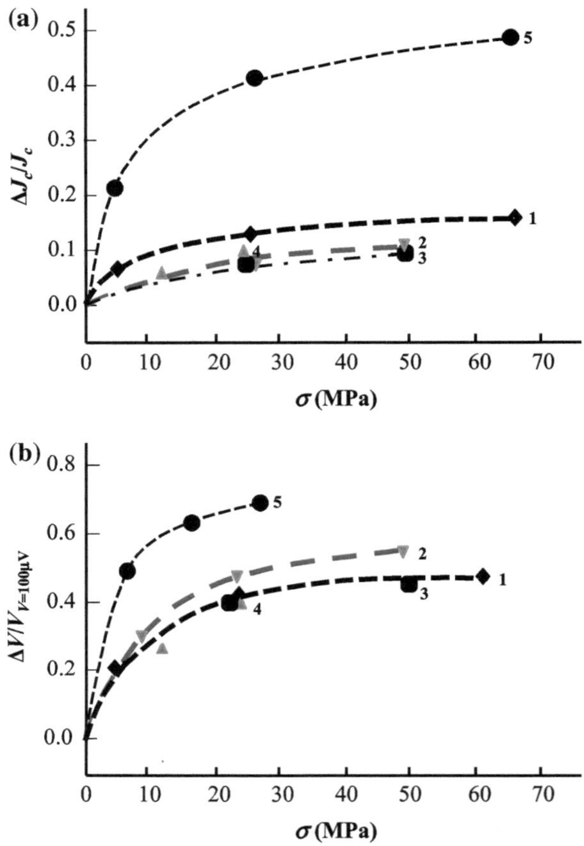

Fig. 4.75 Relative change of critical current density $\Delta J_c/J_c$ (**a**) and relative shear of VAC $\Delta V/V$ at point $V = 100$ μV of linear region of VAC at I = const (**b**) under one-axis compression in dependence of stress σ for $YBa_2Cu_3O_{7-\delta}$ ceramics with various values of J_c (A/cm^2): 40 (1), 86 (2), 110 (3) [1274, 1276], $Y_{0.75}Er_{0.25}Ba_2Cu_3O_{7-\delta}$ ceramic with $J_c = 80$ A/cm^2 (4) [330], and hydrogen-treated $YBa_2Cu_3O_{6.7}$ with $J_c = 25$ A/cm^2 (5) [1588, 1589]

superconductive phase (grains) in mainly defining T_c and on grain-boundary weak links of S–I–S or S–N–S type, characterizing J_c. The change of critical current density $\Delta J_c/J_c$ (Fig. 4.75 a) and the relative shear of $\Delta V/V$ at the linear region of VAC at I = const > I_c (see, Fig. 4.75 b), defining a change of resistance, $\Delta R/R$, increase non-linearly with applied stress, σ, approaching a saturation at σ = 30–50 MPa for all presented ceramics.

The Y(*RE*)BCO samples with different structural features, namely: with deviation from stoichiometry in cation composition, decreased oxygen content due to special treatment with hydrogen, and also with Ag-additives have shown that the loading effect is the structure-sensitive characteristic. In particular, a grain-boundary decoupling due to the treatment with hydrogen lead to strengthening the loading effect on J_c in 3–4 times and on VAC shear in 2–3 times (see Fig. 4.75 a, b). By introducing silver, the loading effect is suppressed independently of the doping method that, in particular could be connected with metallization of current-currying grain boundaries in narrow (~1 nm) area around grain boundary and leads to prevailing S–N–S type of grain-boundary weak links [1271]. In this

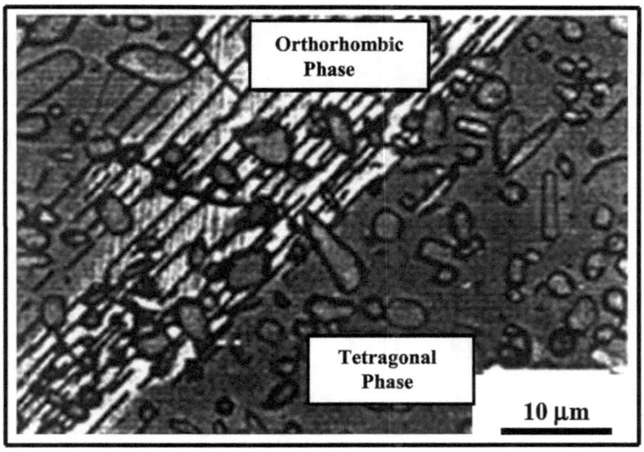

Fig. 4.76 Dense *ab*-microcracks in the orthorhombic Nd-123 phase and no *ab*-microcracking in the tetragonal Nd-123 phase (small 422 particles) [369]

narrow metalized junctions ($d_N/\xi_N < 1$, where d_N is the thickness of normal metal layer, and ξ_N is the coherence length in the normal metal layer, the order parameter, Δ_G at the junction boundary will be suppressed owing to the proximity effect, and reaction of J_c on compression will be practically absent because the loading influence insignificantly on ξ_N in normal metal. Moreover, Ag-additives in ceramics demonstrate strong micro-plastic deformability leading to more uniform distribution of stresses on current-currying boundaries, decreasing local stresses on weal links, that is the loading effect.

4.4.6 Mechanical and Strength Properties

Microcracks form in the *ab*-plane of Nd-123 orthorhombic phase during cooling even at small numbers of Nd-422 particles due to the difference of thermal expansion between matrix and inclusions. These planar defects are absent both in the Y-123 tetragonal phase [371, 856] and in the Nd-123 tetragonal phase with small Nd-422 particles (Fig. 4.76) [369], that is, they are not defined by crystallization process [615]. In the samples with much larger Nd-422 particles, the *ab*-microcracks also develop in the tetragonal parts of the sample (Fig. 4.77). The crack spacing in the orthorhombic Nd-123 is much smaller than in the tetragonal part. This suggests that for both the orthorhombic and tetragonal 123 phases there is a critical size of 422 (211) particles, below which *ab*-microcracking associated with 422 (211) particles does not occur. The higher critical particle size for the tetragonal 123 phase is associated with the absence of additional shortening of the *c*-axis due to oxygen uptake or/and influenced by the higher fracture toughness of the tetragonal 123. Besides the *ab*-microcracks, the *c*-microcracks are also observed in melt-grown YBCO. These *c*-cracks stop at the boundary between

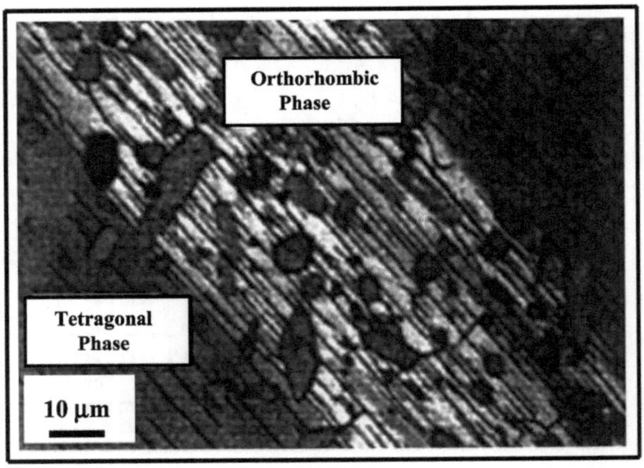

Fig. 4.77 Higher *ab*-microcrack density in orthorhombic than in tetragonal Nd-123 phase (large 422 particles) [369]

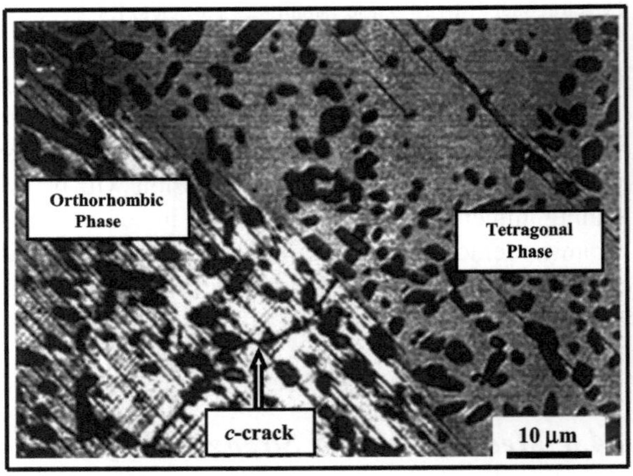

Fig. 4.78 *c*-crack stop at the boundary between orthorhombic and tetragonal phase [366]

orthorhombic and tetragonal phase (see Fig. 4.78), demonstrating formation of the *c*-cracking during tetragonal-orthorhombic phase transition [366].

Figure 4.79 shows interesting change of the five-domain sample shape. An originally circular-shaped sample can deform during solidification. The *c*-growth parts of the sample are bulged and the *a*-growth parts have shortened radii when the sample is observed from above.[8] Due to this, there is specific macrocracking of the sample.

A deficiency of the 211 particles can induce tensile residual stresses in the *c* direction in the central *c*-GS and also tensile tangential stress in the *ab*-plane at the sample surface (in the *a*-GS). These stresses can cause *ab*-microcracking and detrimental radial macrocracking [365]. Other reasons for macrocracking are the

[8] Higher 211 concentration at 90°-boundaries confirms existence of the *a*-GSs along 90°-boundaries [365].

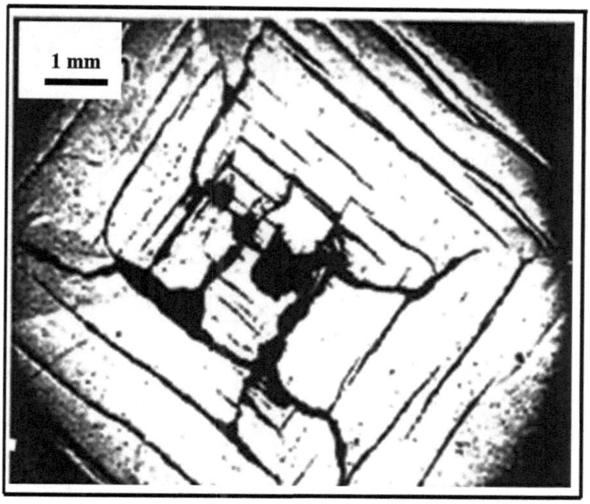

Fig. 4.79 Polarized-light photomicrograph, revealing the higher density of 211 particles (*gray diagonal* regions) in *a*-GSs along 90°-grain boundaries and also bulging of *c*-GSs and macrocracks parallel to *ab*-plane [365]

great thermal gradients at sample cooling [1463] and magneto-elasticity effects connected with magnetic flux pinning, causing residual stresses during the sample magnetization [1422]. As a result, the cracks grow along the (100)/(010) plane and decrease a trapped magnetic field [1169]. At increasing oxidization duration of the melt-textured YBCO samples, an aging of mechanical and superconducting properties [1476] may be observed. It is accompanied by increasing dislocation density, formation of great chippings at rather low-angle grain boundaries and around 211 inclusions and also by intensive degradation of Y-123 matrix near microcracks and the 211 particles. The studies of textured YBCO ceramics at helium temperatures and high magnetic fields show that the latter can cause the dynamics of twinning dislocations, leading to cleavage stresses that can fracture the sample along the planes parallel to the basic *ab*-plane [414]. 211 particles deflect crack, increasing its path. In this case, the crack branching at the 211 inclusions [1002] and de-lamination of 211/123 interface [365] may be observed. As the 211 particle has a higher Young's modulus compared to the 123 matrix, they can also act as grains-bridges, pinning down the crack opening and decreasing tensile stresses in the crack tip [565].

Silver dispersion (Ag_2O, Au/Ag) decreases considerably cracking and porosity both in the *ab*-plane and along *c*-axis, increasing J_c significantly [615, 1130, 1590]. Figure 4.80 shows the magnetic field distributions in YBCO bulk (0 wt.% Ag) at 9.0 and 8.5 T in the decreasing field process from 10 T. Magnetic field distributions show clearly the sample fracture during the decreasing field process from 9.0 to 8.5 T. At the same time, the additives of 10 and 20 wt.% Ag prevent the sample fracture at the same decreasing field process even after total removal of external field [1169]. The observed fracture of the sample into two pieces by macrocrack occurs along (001) or (010) planes, showing that those are the cleavage planes in YBCO. This is followed from the concept that twin planes are {110}, but the

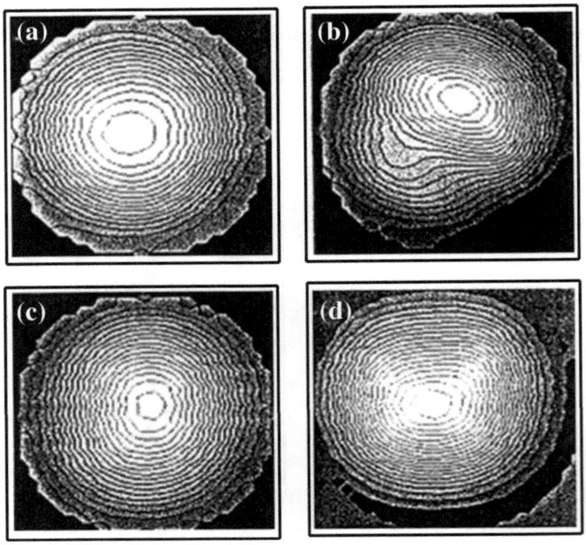

Fig. 4.80 Magnetic field distribution in YBCO samples during decreasing process from 10 T down to: (**a**) 9.0 T, 0 wt.% Ag; (**b**) 8.5 T, 0 wt.% Ag; (**c**) 0 T, 10 wt.% Ag; (**d**) 0 T, 20 wt.% Ag [1169]

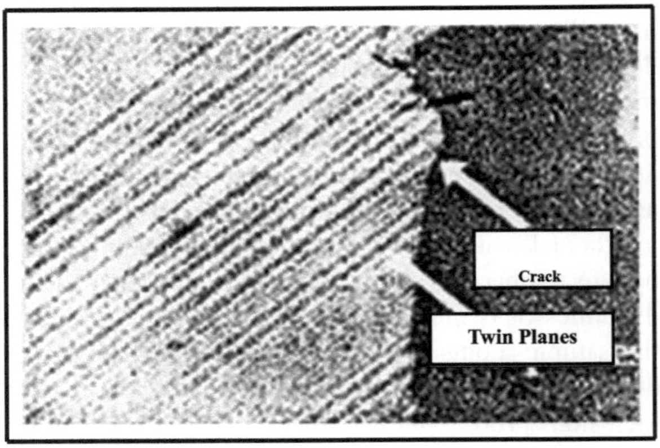

Fig. 4.81 Light micrograph of YBCO sample (without Ag) near the crack edge with higher magnification observed under polarized light. Twin planes can clearly be seen on the sample surface. Note that the crack crosses twin planes at 45°, showing that the cleavage surface is (001) or (010) plane [1169]

fracture plane crosses at 45° from twin planes (Fig. 4.81). Then, it may be concluded that the cleavage occurs at (001) or (010) planes. The Ag addition (Ag_2O) decreases considerably the *ab*-microcracking, introduced in TOPT by micro-stresses, caused by 211 particles (Fig. 4.82). In this case, significant increase of spacing between both *ab*-cracks and *c*-cracks (Table 4.3) [366] is observed.

The crack, initiated due to Vickers micro-indentation, deflects when one reaches the gap between two silver inclusions (Fig. 4.83). After the crack passes through the gap, it continues in the same direction of the initial growth. Thus, a toughening is demonstrated, increasing intrinsic fracture toughness of YBCO sample [1066]. Silver increases both maximum size and incline of *R*-curve (Fig. 4.84). This increase of fracture toughness is caused by reinforcing, developing into so-called, *wake zone*, where a plastic strain of Ag inclusions-bridges,

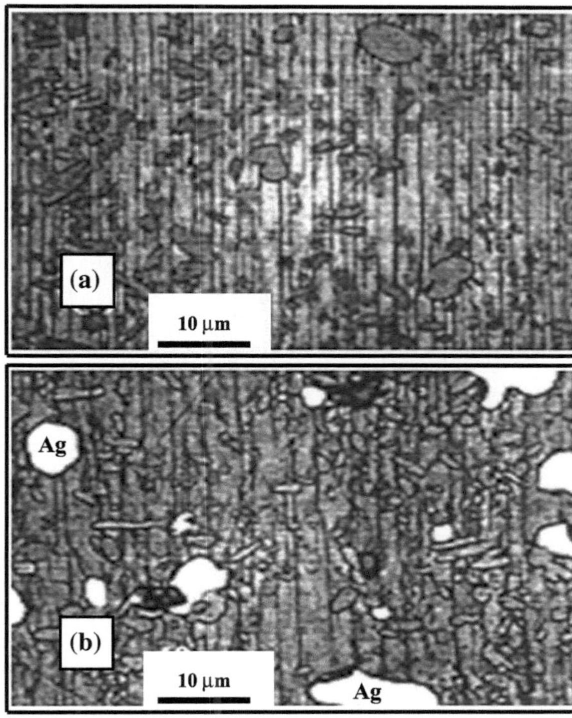

Fig. 4.82 Microcracks in YBCO/Ag_2O samples after etching: (**a**) 0 wt.% Ag_2O and (**b**) 12 wt.% Ag_2O [366]

Table 4.3 Crack spacing [366]

Sample	$\lambda_{ab} \pm \sigma$ (μm)	$\lambda_c \pm \sigma$ (μm)	d_{211} (μm)	V_{211} (wt.%)	d_{Ag} (μm)	V_{Ag} (wt.%)
0 wt.% Ag_2O	1.8 ± 0.1	34.2 ± 17.0	1.95	21.1	–	–
12 wt.% Ag_2O	3.7 ± 0.4	133.5 ± 43.2	2.02	20.9	5.2	6.1

λ_{ab} is the *ab*-microcrack spacing, λ_c is the *c*-microcrack spacing, d_{211} is the 211 particle size, d_{Ag} is the Ag particle size, V_{211} is the 211 volume percent, V_{Ag} is the Ag volume percent, σ is the standard deviation

pinning the crack surfaces, states most mechanism of toughening and fracture resistance in the absence of material transformations [1835].

An addition of tetragonal ZrO_2 particles can significantly reinforce monolithic YBCO samples due to the martensitic (tetragonal-monoclinic) phase transition [549]. In particular, when 10 mol.% ZrO_2 particles, coated with the Y-211 phase by using a sol–gel process, were added to Y-123, only modest improvement of fracture toughness (K_{Ic}) was observed. However, 20 mol.% ZrO_2 improved the K_{Ic} by nearly 50 %. In this case, for the samples with the same density, the 10 mol.% ZrO_2 additions also increased K_{Ic} by nearly 50 %. At the same time, the average strength of the Y-123 and Y-123/Y-211/ZrO_2 remained nearly constant despite the differences in K_{Ic}. On the contrary, a critical concentration of ZrO_2 additions (~15 wt.%) [1287] was stated, at which the maximum superconductor strength

Fig. 4.83 Crack deflection between silver particles, dispersed in YBCO superconducting matrix [1066]

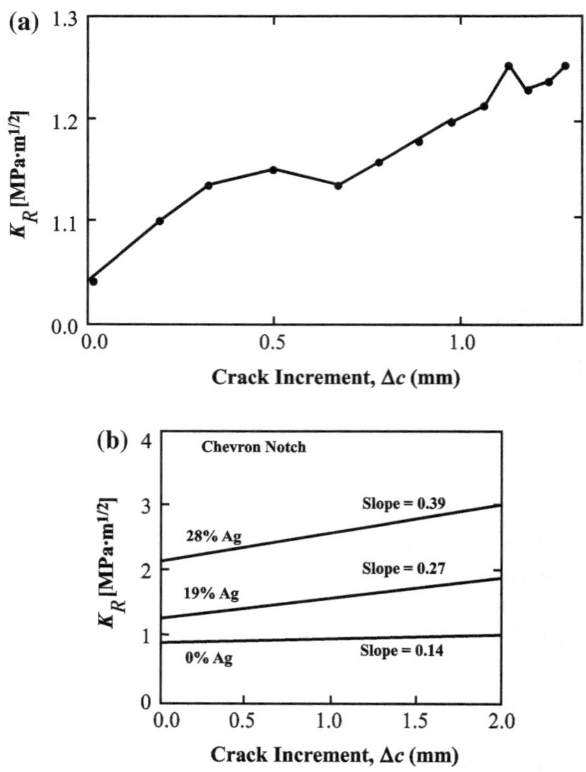

Fig. 4.84 *R*-curves obtained for different YBCO/Ag composites in samples with single edge notch: (**a**) 123 superconductor; (**b**) three composites with various amount of silver. Ag increasing improves both the level and incline of the fracture resistance curves, showing cumulative character of the toughening with the crack growth [1835]

and plasticity were reached. In this case, the plastic deformation occurs in range of high temperatures in mechanism of intercrystallite sliding on nanophase with Zr-contents.

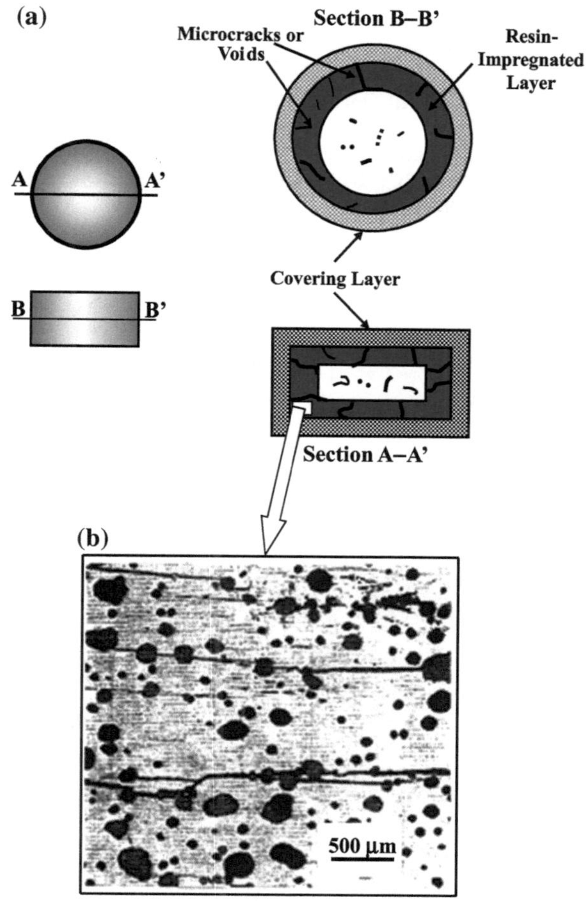

Fig. 4.85 (**a**) Structure of superconducting bulk reinforced by resin impregnation [1687]; (**b**) optical micrograph for cross-section of Sm-123 bulk after the resin impregnation [1690]

The reinforcement by surrounding the Y(*RE*)BCO bulk with a metal ring is also effective in increasing mechanical properties [1191]. In this case, the bulks can be pre-strained in compression, since the thermal expansion coefficient of a metal ring is much smaller than that of the Y(*RE*)BCO bulks. The epoxy resin impregnation into *RE*BCO bulk effectively improves mechanical properties [1053, 1686, 1687]. In this case, almost all defects (cracks, pores and scratches) can be filled by the epoxy resin within a distance of about 2 mm from the sample surface (for superconducting pellets of 3 cm in diameter and 2 cm in thickness) (Fig. 4.85), [1686]. These bulks do not show a decreasing of the trapped magnetic field even after intensive thermal cycling (cycles 77–293 K) [1687] and demonstrate an

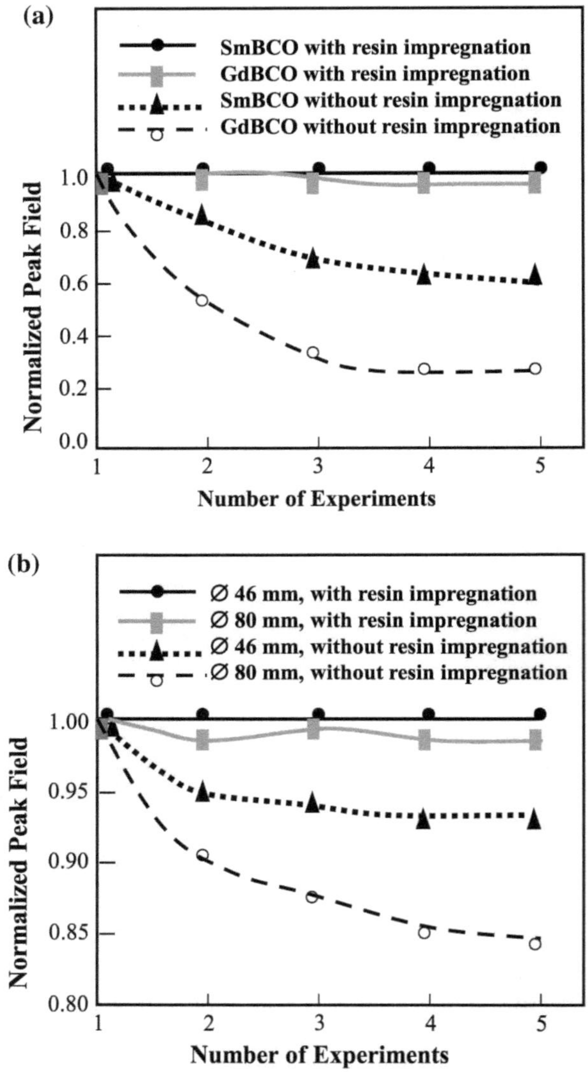

Fig. 4.86 Trapped field values versus the number of thermal cycles (77–293 K): (**a**) SmBCO and GdBCO with and without resin impregnation [1685]; (**b**) the same results for YBCO bulks with different diameters of pellets [1687]

increasing of levitation force [1689]. Compared to usual samples, the characteristics of reinforced specimens are presented in Fig. 4.86. Moreover, a large current of 1000 A could be passed via the *RE*BCO bulk without the transition of the superconductor in normal state [1690]. The epoxy resin could be successfully reinforced by dispersing quartz fillers which also promote a decreasing of thermal expansion difference between polymer and superconductor [1687]. Finally an additional high pressure at high temperature permits to obtain during short time non-porous practically *RE*BCO bulks with improved mechanical properties at simultaneous preservation or improvement of superconducting properties [1404].

The mechanical properties of disc-shaped $YBa_2Cu_3O_7$ magnets have been reinforced significantly due to an epoxy resin impregnation and introducing of carbon filaments [1688]. Moreover, an aperture has been drilled in the sample center, which has been filled by Bi–Pb–Sn-Cd alloy with aluminum wire. As a result, the static trapped field has been increased up to 17.24 T at magnet cooling with the absence of mechanical damage.

Chapter 5
Carbon Problem

5.1 YBCO System

During all stages of HTSC preparation from calcining of precursor powder up to fabrication of final product, there is a problem of carbon segregation into sample deteriorating structure-sensitive properties. So, during calcining in preparation of YBCO, the $BaCO_3$, Y_2O_3 and CuO powders form 123 phase reacting according to the following [1536]:

$$BaCO_3(\text{solid}) + CuO(\text{solid}) \rightarrow CO_2(\text{gas}) + BaCuO_2(\text{solid}) \tag{5.1}$$

$$4BaCuO_2(\text{solid}) + Y_2O_3(\text{solid}) + 2CuO(\text{solid}) \rightarrow 2YBa_2Cu_3O_x(\text{solid}) \tag{5.2}$$

When $BaCO_3$ decomposes, according to reaction (5.1), CO_2 is released, and its localized concentration quickly reaches equilibrium value. The localized CO_2 pressure, depending on the temperature and other thermodynamic conditions, can cause other decomposition reactions, forming undesired phases, which in turn contribute to reduce the superconducting properties of the final product. It has been reported [1330, 1729] that in the superconductors carbon remains in an amount of 0.1–1.0 % or more, even if the powder is fired at high temperatures. In particular, an existence of residual carbon in oxide superconductors, synthesized by a sol–gel method [946, 1116, 1726], is possible. However, the residual carbon content of the powder, synthesized by the conventional solid-state reaction method, can be about three times as much as that of the powder, prepared by the sol–gel method for a similar condition of heat (Fig. 5.1).

The powders, synthesized by the sol–gel method, demonstrate superconductivity, when they are heated at temperatures higher than 900 °C and the total carbon content is less than 0.4 %. Dependencies of the $T_{c,onset}$ property and the magnetic susceptibility of YBCO powders on total carbon contents are shown in Fig. 5.2. Obviously, diamagnetism volumes are increased and $T_{c,onset}$ is improved as the carbon contents are reduced.

I. A. Parinov, *Microstructure and Properties of High-Temperature Superconductors*,
DOI: 10.1007/978-3-642-34441-1_5, © Springer-Verlag Berlin Heidelberg 2012

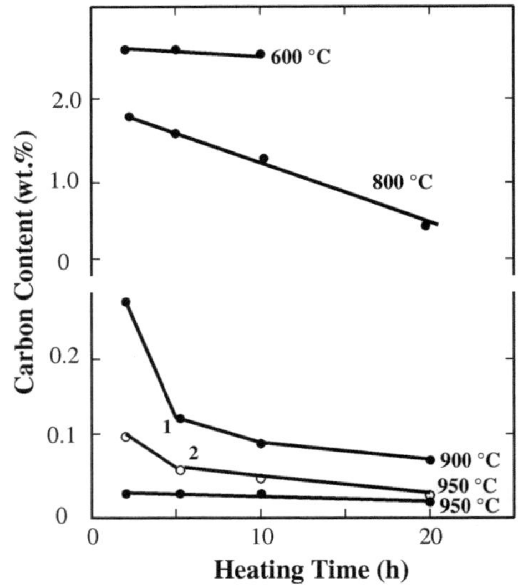

Fig. 5.1 Change of carbon content with heating time for various firing temperatures. (•) in O_2 flow, (o) in air flow, (◆1) and (◆2): samples synthesized by the conventional method [1097]

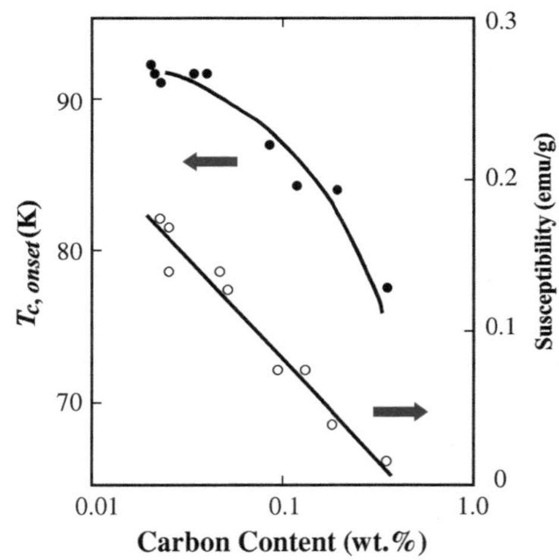

Fig. 5.2 Dependencies of $T_{c,onset}$ and magnetic susceptibility on carbon contents. The susceptibility is measured at 10 K [1097]

The next hot isostatic pressing (HIP) can lead to the following improvement of the powder's superconducting properties. So, the powder, heated at 500 °C with HIP, has a higher $T_{c,onset}$ and a larger volume of diamagnetism than the powder without HIP treatment (Fig. 5.3). In this case, the orthorhombic distortion increases when the HIP treatment is performed at 500 °C, but the distortion decreases when the temperatures of the HIP treatment are raised to 800 and 900 °C (Fig. 5.4).

It should be noted that the carbon content of the powder, prepared by the sol–gel method, decreases more effectively than that of the powder by the conventional solid-state reaction, in spite of the presence of many more carbonaceous materials

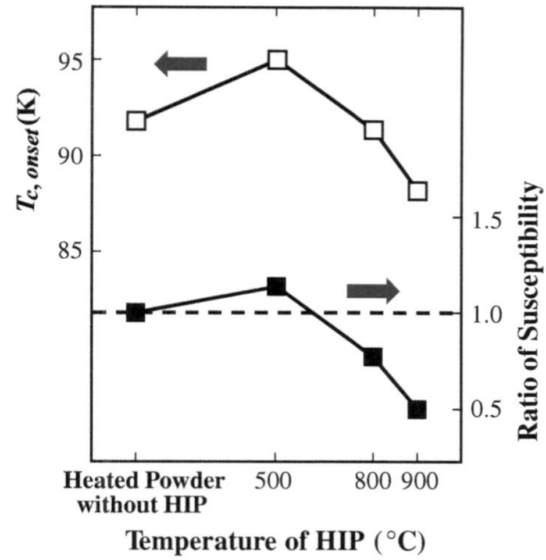

Fig. 5.3 Influence of HIP treatment in O_2 on $T_{c,onset}$ and magnetic susceptibility at 77 K. The HIP was performed with 100 MPa and the duration time 10 h [1097]

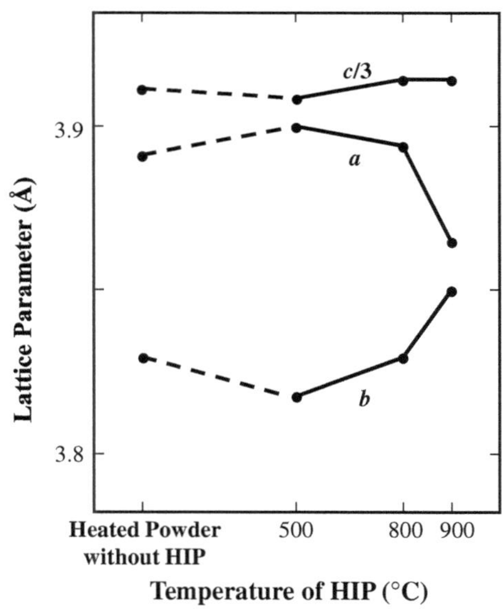

Fig. 5.4 Change of the lattice parameters of the YBCO powder by HIP treatment in O_2 [1097]

in the starting materials used in the sol–gel method. The reason for this behavior maybe explained by assuming that $BaCO_3$, introduced in the powder synthesized by the sol–gel method, would be decomposed rapidly during heating at higher temperatures than 900 °C due to the very small size (some dozens of nanometers) of $BaCO_3$ particles. It is possible that such fine particles maybe highly reactive.

When the HIP treatment is performed at 900 °C, a degradation of superconducting properties is observed. It is assumed that the degradation is caused by an increase of the tetragonal 123 phase over the orthorhombic 123 phase. This agrees with test observation that orthorhombic distortion of the lattice parameters decreases as HIP temperatures increase from 500 upto 900 °C.

YBCO superconductor reacts during sintering at different temperatures and conditions of gas mixtures. Its reaction with CO_2 maybe carried out in two stages. At 815 °C, it is given as [519]

$$2YBa_2Cu_3O_{7-x}(\text{solid}) + 4CO_2(\text{gas}) \rightarrow 4BaCO_3(\text{solid}) + Y_2Cu_2O_5(\text{solid}) + 4CuO(\text{solid}) + (0.5 - x)O_2(\text{gas}). \quad (5.3)$$

The complete reaction of YBCO with CO_2 at 950 °C is [519]

$$2YBa_2Cu_3O_{7-x}(\text{solid}) + 3CO_2(\text{gas}) \rightarrow 3BaCO_3(\text{solid}) + Y_2BaCuO_5(\text{solid}) + 5CuO(\text{solid}) + (0.5 - x)O_2(\text{gas}). \quad (5.4)$$

For both cases, the decomposition of YBCO is only partially completed and starts at grain boundaries. In the sample annealed at 815 °C, the reaction products consist of two phases: $BaCO_3$ and $Y_2Cu_2O_5$. The primary reaction products are all insulators, which essentially coat all grain boundaries. It can make the material loses its superconductivity even if the major phase is still superconducting. In the case of sintering in a 5 % CO_2/O_2 mixture, the reaction products maybe nucleated at grain boundaries as separate precipitates rather than as a thin layers which coat the grain boundaries. In the case of these individual precipitates at grain boundaries, the J_c value may not decrease, as long as most boundaries remain to be superconducting. On the contrary, since defects, such as twin boundaries and tiny secondary phases, can act as flux pinning centers, the J_c value may in fact increase due to the presence of small precipitates at grain boundaries.

The interaction between YBCO and CO_2 at a total pressure of 1 atm. (0.999 atm. of O_2 and 0.001 atm. of CO_2) leads to two reaction mechanisms [485]:

$$YBa_2Cu_3O_{6+}x(\text{solid}) + 2CO_2(\text{gas}) \rightarrow 2BaCO_3(\text{solid}) + 0.5Y_2O_3(\text{solid}) + 3CuO(\text{solid}) + (2x - 1)/\ 4O_2(\text{gas}), \quad (5.5)$$

$$YBa_2Cu_3O_{6+x}(\text{solid}) + 2CO_2(\text{gas}) \rightarrow 2BaCO_3(\text{solid}) + 0.5Y_2Cu_2O_5(\text{solid}) + 2CuO(\text{solid}) + (2x - 1)/\ 4O_2(\text{gas}). \quad (5.6)$$

Both of these reactions lead to the formation of non-superconducting phases. The partial pressure of CO_2 can affect the partial pressure of O_2, which in turn affects the oxygen content of the YBCO compound formed. It is well known [794, 795, 958] that for $x \approx 1$, $YBa_2Cu_3O_{6+x}$ is in the orthorhombic phase and exhibits superconductivity at 90 K. As the oxygen content, which is dependent on both the

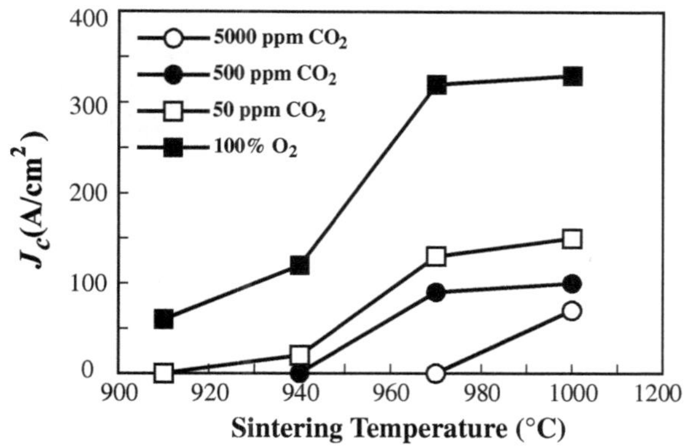

Fig. 5.5 The effect of sintering temperature on J_c [1536]

temperature and oxygen partial pressure, decreases, the transition temperature of the orthorhombic phase decreases to 60 K, at $0.6 < x < 0.7$. At $x \approx 0.4$, the material becomes tetragonal, and superconductivity is destroyed. The critical current density is affected by both sintering temperature and CO_2 content in the sintering atmosphere (Fig. 5.5). J_c values decrease with decreasing sintering temperature and increasing CO_2 content.

The crystals of the sintered samples in CO_2 atmospheres with 50 or 500 ppm CO_2 can demonstrate twins, whereas those sintered in 5 % CO_2 do not show any twinning [1536]. The twin structure in YBCO determines its orthorhombic superconducting properties [1637]. Hence, the samples sintered in 5 % CO_2 are not orthorhombic, but the extent of the orthorhombic structure decreases with increasing CO_2 in the sintering atmosphere. Moreover, increasing CO_2 contents in the sintering atmosphere increases the electrical resistivity of the sample [1536].

We do not know test evidence of secondary phases at grain boundaries in samples sintered in a 100 % O_2 atmosphere. At the same time, two distinct types of grain boundaries can be observed in the samples sintered in 0.5 % CO_2 + 99.5 % O_2 [1536]. Approximately 10 % of the grain boundaries are found to be wet by a thin layer of a second phase, as shown in Fig. 5.6, and the remaining boundaries are sharp grain boundaries and free from the presence of secondary phases. The local change of YBCO stoichiometry leads to the eutectic reaction, in accordance with the phase diagram, and formation of $BaCuO_2$ and CuO at the grain boundaries. The decomposition of YBCO phase at the grain boundaries weakens the path for the current, thereby decreasing the overall J_c value. The wetting of the boundaries with non-superconducting phases decreases the effective contact area between the superconducting grains and decreases J_c. From Fig. 5.6, the width of the secondary phase region along the grain boundaries is determined to be approximately 50 nm, which is much larger than the coherence length in this material. Therefore, these secondary phases block the passage of currents.

There are two potential sources of CO_2, when YBCO are processed by solid-state sintering, namely (i) the CO_2, contained in the oxygen gas, used during sintering and/or annealing and (ii) the CO_2, derived from the decomposition of $BaCO_3$ during

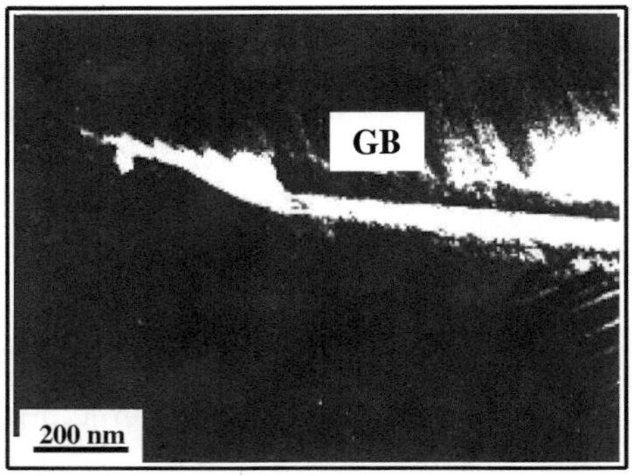

Fig. 5.6 The sample sintered at 970 °C in 0.5 % CO_2 + 99.5 %O_2. TEM micrograph for the grain boundary (GB) which is wet by a layer of $BaCuO_2$ second phase [1536]

the calcining step. Each source affects the quality of the final product in different ways and can lead to a drop in the critical current density. An additional potential source is the organic binders that are used in the YBCO processing. The CO_2, formed by the interaction of the organic binders and the oxygen from the gas atmosphere, could have a deleterious effect on the performance capabilities of the final product. By using thermodynamics, it may by shown [1536] that for pressure values, $p_{CO_2} <$ 1 atm., sintering temperature must increase as p_{CO_2} increases. Therefore, increasing the temperature results in higher value of p_{CO_2}. The sintering temperatures can be lowered if p_{CO_2} can be lowered, and one of expedient means of doing this is to remove the CO_2 contained in the O_2 gas.

Carbon also affects negatively the temperature and broadness of superconducting transition [1637]. It is known that YBCO samples, sintered to closed porosity, exhibit broad and suppressed superconducting transitions (Fig. 5.7). The observations suggested that during sintering of the closed porosity material, "impurities" that suppress the superconducting transition become trapped in the center of the sample, while in the open-porosity sample and near the surface of the closed-porosity material, volatilization of the impurity can occur during sintering.

The highest carbon content of ~1.2 wt.% is detected in the grains of the central part from the closed-porosity material. Significantly lower carbon contents are detected in specimens, cut from both the exterior of the closed-porosity material and the central part of the open-porosity material (~0.6 wt.%). Carbon removal occurs more readily in the open-porosity material and near the surface of the denser material, consistent with the suggestion that carbon removal is limited by the slow diffusion of carbon out of the YBCO grains.

Under appropriate conditions, carbon has an appreciable solubility in YBCO bulk. The depression of the transition temperature, detected by magnetic susceptibility measurements, correlates well with carbon content. This correlation suggests that incorporation of carbon in the structure is the dominant cause of the lowering of the transition temperature in the high-dense superconductors.

In the interior of the closed-porosity material, CO_2 or CO gas may be trapped in the pores. As a result, a pressure of CO_2 or CO gas (in the range 0.01–0.1 atm.) can

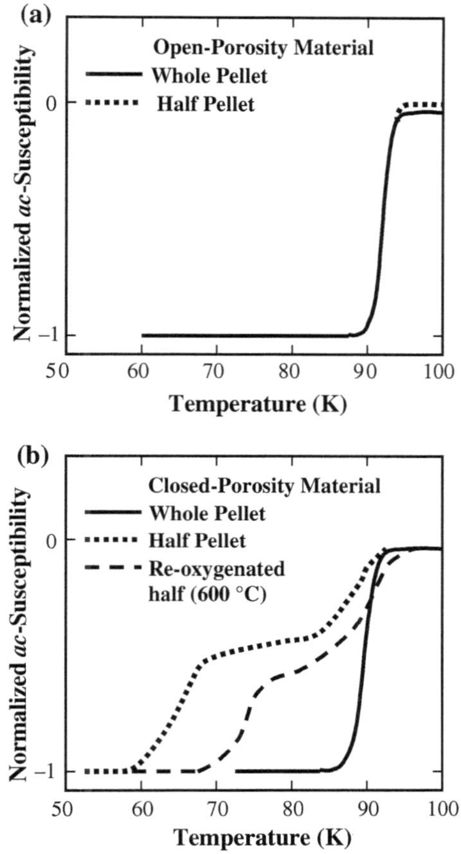

Fig. 5.7 Change in ac-susceptibility with temperature for (**a**) an open-porosity material ($\rho = 87$ %) and (**b**) a closed-porosity material ($\rho = 92$ %). In the open porosity material both whole and halved samples exhibit sharp transitions, whereas in the closed-porosity material a stepped transition is seen in the halved sample even after an additional oxygenation treatment [1637]

develop in the pores, that is enough to slow down sintering processes. Then, several steps may be taken to lower the amount of carbon, trapped in a dense material, namely:

(i) The external partial pressure of CO_2 should be as low as possible in order to favor decomposition of carbonate in the material.
(ii) If the densification process can be delayed (to allow a longer period, in which the sample contains interconnected porosity), more decomposition and gas evolution may occur before the gas becomes to be trapped.
(iii) A successful approach to carbon removal is to heat more slowly to the sintering temperature.

Carbon in the crystal structure of YBCO interacts with Ba and O, reducing the Ba–O coordination [1637]. The carbon connected with Ba ion could be incorporated in the structure either substitutionally or interstitially. In the first case, a direct substitution of oxygen by carbon ions is possible (it is not supported in tests). An alternative possibility is that the carbon lies in an interstitial site adjacented to the barium ion. In the absence of a detailed theoretical understanding of the origin of the superconducting transition in YBCO, the origin of the depression of T_c, produced by the incorporation of carbon, may be speculated. It is known that there is a strong correlation between T_c and the concentration of holes in the material [1545]. The presence of the carbon causes localization of some of the holes and thus reduces the mobile holes' concentration in the sample. This suggestion is consistent with the location of the carbon in an interstitial site adjacent to the CuO_2 planes. Because the carbon concentration can approach one atom per unit cell, the location of the carbon in such a site could be expected to cause appreciable local distortions of the CuO_2 planes and thereby lower T_c. It may be shown [1064] that carbon is released from the matrix at temperatures higher than that of the peritectic decomposition. Nevertheless, there is no doubt that even in a melt-processed sample, depression of T_c may still occur. Consider proper technological procedures, leading to decreasing of carbon content in the melt-processed YBCO samples by using as example the results of [1765]. The melt-texturing was performed by a melt-texture growth (MTG) process, using a temperature gradient furnace (temperature gradient about 25 K/cm). As a result, two samples were synthesized, namely, (i) the sample Sr5a, containing large grains (100 μm) with a very dense, closed porosity structure, and (ii) the sample Sr5b, containing small grains (<10 μm) with open porosity. Their superconducting transition curves are shown in Fig. 5.8. The $T_{c,onset}$ of sample Sr5a is depressed to 80 K, with a broad transition. Sample Sr5b has $T_{c,onset}$ at 92 K, which is the optimum value for the Y-123 system. The analysis shows rather a high degree of carbon retention in the two samples: 2600 ppm for sample Sr5a and 1200 ppm for sample Sr5b. Thus, almost all carbon of the initial powder (3000 ppm) is retained in sample Sr5a, and its low $T_{c,onset}$ is in accordance with the correlation stated above.

In sample Sr5a, three experimental features can explain the carbon content in this sample as follows:

(1) The obtained lattice parameters (a = 3.8288 Å, b = 3.889 Å and c = 11.667 Å) showed that the decrease of the c parameter, which was sensitive to the carbon content, was considerably less than that observed for samples prepared under a CO_2 atmosphere (c = 11.585 Å) [161].
(2) The characteristic features of the YBCO carbonated phases is the existence of oriented domains, where the c-axes are perpendicular; the formation of these

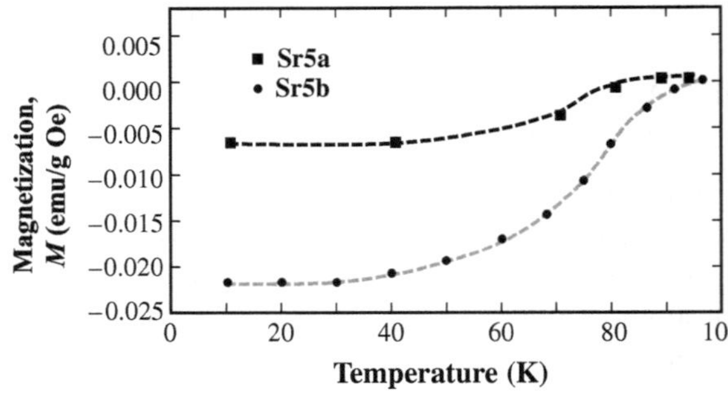

Fig. 5.8 The temperature dependence of magnetization of Y-123 ceramics sintered at different temperatures: Sr5a (sintered at 950 °C in oxygen flow) and Sr5b (sintered at 920 °C in oxygen flow). Average particle size of precursor (2.5 μm) is the same for both ceramics [1765]

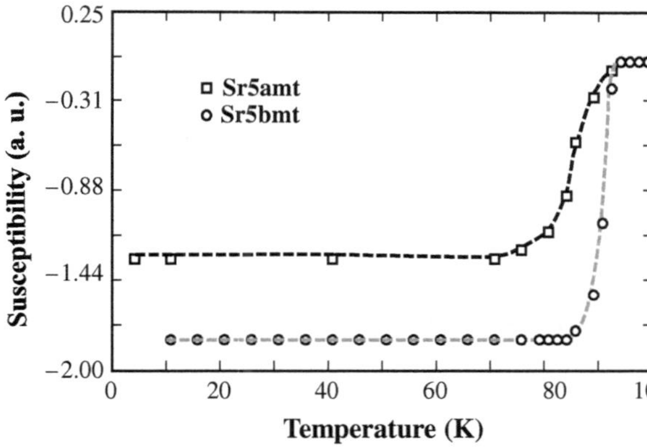

Fig. 5.9 The dependence of magnetic susceptibility on temperature of melt-processed samples. Sr5amt (or Sr5bmt) is an MTG sample, using Sr5a (or Sr5b) as a pre-sintered precursor [1765]

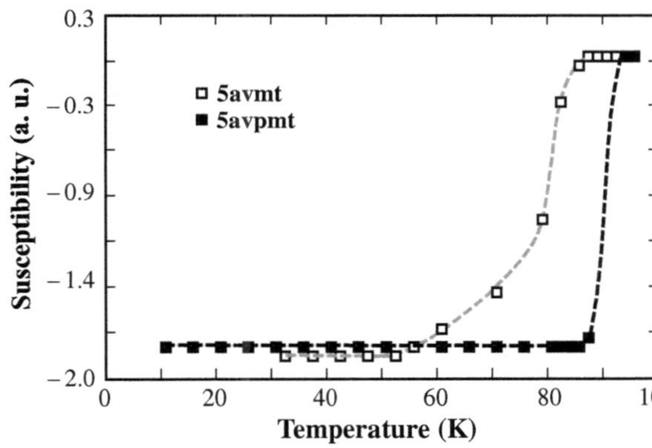

Fig. 5.10 The temperature dependence of the susceptibility curve for the samples processed by an MTG method with starting composition $Y_{1.4}Ba_{2.2}Cu_{3.2}O_x$, with (5avpmt) and without (5avmt) addition of Pt [1765]

domains is correlated not only to the decrease of the *a/b* and *a/c* ratios, but also to the presence of carbonate groups, which are preferentially located at the level of the domain boundaries.

(3) In carbon-rich materials, the existence of distributions of local superstructures, leading to the formation of streaks along *a*-direction and to short-range aligned areas, is clearly visible on the images.

The above two sintered samples were further processed by MTG. The results presented in Fig. 5.9 show that this procedure allowed further decrease the carbon content. However, an excessive amount of carbon can still be retained, if the carbon content of the initial precursor is too high, as shown for sample Sr5amt. Therefore, the absolute necessity of controlling the quality of the precursor material before melt texturing is obvious. Then, in order to examine the influence of the 211 fine-disperse additives and Pt doping on carbon retention, two sets of samples were prepared, one without Pt addition (5avmt) and another with the addition of 0.5 wt.% PtO_2 (5avpmt). Both sets had nominal composition $Y_{1.4}Ba_{2.2}Cu_{3.2}O_x$. Addition of Y-211 does not significantly decrease carbon retention in the Y-123. In contrast, as it follows from Fig. 5.10, for sample 5avpmt with added Pt, $T_{c,onset}$ is enhanced up to the optimum value of 92 K, and a remarkably sharp transition ($\Delta T_c = 2$ K) is obtained. Analysis shows that the carbon content in this sample is as low as 200 ppm. At the same time, the sample 5avmt contains about 870 ppm of carbon. Although the exact mechanism by which the platinum acts to decrease the carbon content is unknown, it may be suggested that Pt acts as a catalyst for de-carbonation.

5.2 BSCCO Systems

The behavior of BSCCO samples due to carbon effect somewhat differs from the behavior of YBCO superconductors. The dependencies of sintering temperature and density of Bi-2223 precursor samples on carbon content are shown in Fig. 5.11. Obviously the higher sintering temperature and increasing density of the Bi-2223 samples cause the lower carbon content.

Oxide superconducting powder can absorb atmospheric moisture and then react with CO_2, worsening superconducting properties. Therefore, a preparation process, proposed for Bi-2223 superconductor [1491], in the first place restricts absorption of moisture in order to suppress the deterioration of the calcined powder. The preparation process includes the processes of dissolving, co-precipitation,

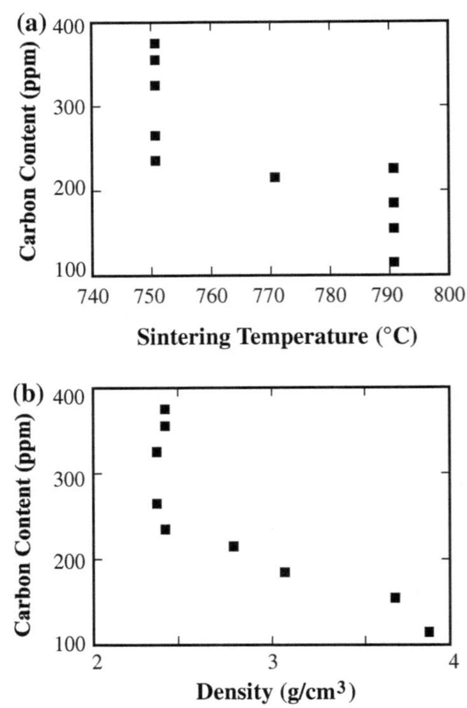

Fig. 5.11 The dependencies of (**a**) sintering temperature and (**b**) density of Bi-2223 precursor samples on carbon content [1826]

Table 5.1 Moisture content and carbon content of calcined powders [1491]

Sample	Process	Moisture content (wt.%)	Carbon content (wt.%)
A	After drying in air	1.01	0.610
	2nd-calcined powder at 800 °C	0.13	0.041
B	After drying in vacuum	0.65	0.486
	2nd-calcined powder at 800 °C	0.11	0.028
C	After drying in vacuum	0.64	0.482
	2nd-calcined powder at 815 °C	0.12	0.025

filtering, primary calcining ($T = 800$ °C), wet pulverizing to obtain a fine powder, drying and secondary calcining ($T = 800$–815 °C). Then, in order to fabricate bulk sample the calcined powder is molded by using a coaxial pressing technique. The sample is sintered at 850 °C some times by using intermediate cold isostatic pressing (CIP). The results obtained for three various samples are presented in Table 5.1. The intermediate pressing leads to increasing of the sample density (Fig. 5.12a) and the corresponding rise of critical current density (Fig. 5.12b). The saturation of the results occurs after third intermediate pressing. The results obtained confirm that superconducting properties of the bulk superconductor are found by content of moisture and carbon in the calcined powder. Carbon and admixture phases segregate at the grain boundaries, forming weak links and decreasing J_c.

The cause of the porosity formation and broken uniformity of Bi-2212/Ag tapes maybe as a result of the gas release, such as CO_2, when Bi-2212 phase reforms from the liquid phase [1870]. Carbon exists in the sample in the carbonate form and can transfer into CO_2 gas in proper condition during the heat treatment, such as at high temperature by the reaction:

$$SrCO_3 \rightarrow SrO + CO_2 \quad (5.7)$$

The CO_2 gas maybe released and causes severe bubble formation and porosity, depending on the carbon content. It may also dissolve into the liquid phase and define its behavior. The carbon content effects on the melting behavior of the Bi-2212 phase [1869] state that increasing the carbon content of the powder can greatly decrease the melting and reformation temperatures of the Bi-2212 phase in the OPIT tapes.

During cooling, when Bi-2212 phase reforms from liquid, [C] will be released in accordance with the reaction:

$$[C] + O_2 \rightarrow CO_2, \quad (5.8)$$

where [C] is the carbon, which is dissolved in the liquid phase.

The volume change, caused by the formation of CO_2 gas from the solid-state carbon, can be tremendous. A rough calculation indicates that 200 ppm of carbon can cause about 36 % porosity in the core, if all carbon forms CO_2 at high temperature [1870].

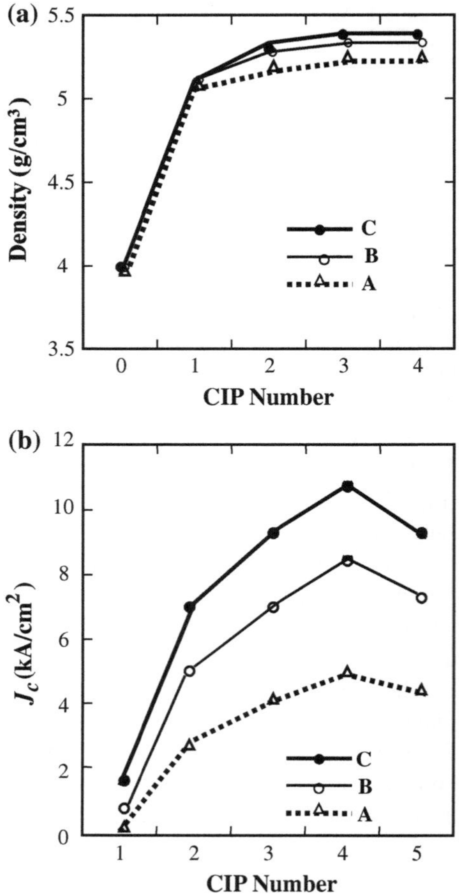

Fig. 5.12 Increase of (**a**) sample density and (**b**) J_c (77 K, 0 T) due to intermediate pressing (CIP) and sintering [1491]

The effect of the introduction of carbon into YBCO and Bi-2212 bulks, obtained by using a partial melting (both samples were heated some above the peritectic temperature), was studied in [1540]. The presence of carbon has strong effect on the YBCO melting, but is not essential for Bi-2212. In this case, the exposed carbon excess can cause an increasing of the magnetic flux pinning. At slow cooling of the YBCO and Bi-2212 from the temperature of partial melting, it is necessary to strictly control temperature and gas atmosphere in order to obtain high superconducting properties. Most important factors are the oxygen gas release [266, 665, 1210, 1539] and the presence of carbon and CO_2 in the sample volume [1733, 1869]. Thus, due to the reaction of YBCO and BSCCO with CO_2 atmosphere, the carbon trapped in the sample can render unfavorable effect on the sample density [1870], intergranular boundaries [519, 1536], critical currents [1548], and critical current density [1733]. Besides atmospheric sources, carbon maybe introduced due to carbonized precursors and organic binders during sample processing [385] and also by solvents [516]. Finally, the carbon can be precipitated into superconductor due to carbonized gases and liquids. Existing techniques can form carbon particles with size of some nanometers, which can be the centers of magnetic flux pinning into superconducting grains. Then, on one hand, carbon can render useful effect, in particular, using Pb and some rare-earth elements [1210], and, on the other hand, it can deteriorate a structure of the intergranular boundaries and structure-sensitive properties of HTSC.

5.3 Carbon Embrittlement and Fracture of YBCO Superconductor

Carbonates, formed in result of the chemical reactions (5.3)–(5.6), are the brittle phase, leading to embrittlement of superconductor. It may lead to delayed carbonate cracking, a subcritical crack growth mechanism with the next formation and fracture of carbonate. The carbon-induced embrittlement is a complicated mechanism, which results from the simultaneous operation of several coupled processes, namely (i) carbon diffusion, (ii) carbonate precipitation, (iii) non-mechanical energy flow and (iv) material deformation. In this case, carbon diffusion occurs due to gradients of chemical potential and temperature [355, 1556]. Similar to hydride formation in metals [144, 1009, 1348, 1413], it may be proposed that the carbon chemical potential depends on stress, and therefore the carbon diffusion is coupled with material deformation and non-mechanical energy flow. The final carbon dissolution into material depends on the thermal stresses due to carbonate expansion at precipitation defining coupling between carbonate precipitation, material deformation and non-mechanical energy flow. The material deformation depends on all other processes due to material expansion, which is caused by carbon dissolution, carbonate formation and temperature increase. Microscopic models of the HTSC intergranular cracking during carbon segregation, depicting slow, fast and steady-state crack growth, screened by linear dislocation array [1317] will be considered in Sect. 5.4. In this section, the governing equations for carbon embrittlement and fracture of HTSC are derived, taking into account thermal stresses in the framework of the thermodynamic theory of irreversible processes [355], based on the thermal diffusion of carbon (Soret's effect). In this case, the delayed fracture of carbonate is modeled by using the de-cohesion model, taking into account de-cohesion energy changed during time due to the time-dependent carbonate precipitation process [1308, 1309].

5.3.1 Mathematical Model for Carbonate Precipitation and Fracture

The model is developed for a superconductor, forming carbonates. The presence of the carbonates is described by the carbonate volume fraction. Energy flow and diffusion of mass are coupled processes. A temperature gradient leads to flow of matter, and therefore to a concentration gradient. Conversely, a diffusion process gives rise to a small temperature difference. A detailed discussion for thermodynamic treatment of energy flow/diffusion is based on Onsager's principle of microscopic reversibility [355]. In the following, the theory is applied to the process of carbonate diffusion and energy flow, occurring in YBCO interaction with CO_2, forming carbonates and cuprates under stress and temperature gradient.

According to the empirical law of Fourier, heat flux is linearly related to the temperature gradient, which is the thermodynamic force, driving heat flow. In an isothermal system, the flux of a diffusing substance is proportional to the gradient of its chemical potential, that is, chemical potential gradient is the thermodynamic force, driving diffusion under isothermal conditions. When the above-pointed processes operate simultaneously, coupling is taken into account by assuming that the non-mechanical energy and carbon fluxes are linearly related to both thermodynamic forces:

$$J_i^E = L^E X_i^E + L^{EC} X_i^C, J_i^C = L^{CE} X_i^E + L^C X_i^C, L^{EC} = L^{CE}, \tag{5.9}$$

J_i^E and J_i^C are the components of the non-mechanical energy flux and the carbon flux, respectively. Note that the energy flux includes conducted heat, described by Fourier's law, and the energy transported by the diffusing carbon; X_i^E and X_i^C are the thermodynamic forces, driving non-mechanical energy flow and carbon diffusion, respectively; L^E, L^C, L^{EC} and L^{CE} are phenomenological coefficients. Third relation (5.9) is valid due to Onsager's reciprocity relation.

The thermodynamic forces are related to the gradients of the absolute temperature T and chemical potential of carbon in the carbonate μ^C as

$$X_i^E = -\frac{1}{T}\frac{\partial T}{\partial x_i}, X_i^C = -T\frac{\partial}{\partial x_i}\left(\frac{\mu^C}{T}\right) \tag{5.10}$$

When carbon and superconductor form carbonate, the carbon flux satisfies the following relation [1556]:

$$J_k^C = -\frac{D^C C^C}{RT}\left(\frac{\partial \mu^C}{\partial x_k} + \frac{Q^C}{T}\frac{\partial T}{\partial x_k}\right), \tag{5.11}$$

where R is the gas constant; C^C, D^C and Q^C are the concentration, diffusion coefficient and heat flux due to transport of carbon in the carbonate, respectively. The concentration of carbon as well as other components or phases is given in moles per unit volume. Relation (5.11) is a special case of (5.9) and (5.10), and therefore satisfies for a HTSC under stress. Because of the absence of carbon diffusion in the cuprate, the total carbon flux in cuprate/carbonate composite is given as

$$J_k^{CT} = fJ_k^C, \tag{5.12}$$

where f is the volume fraction of the carbonate in the material. Mass conservation requires that the rate of total carbon concentration, C^{CT}, inside a volume V, is equal to the rate of carbon flowing through the boundary S:

$$\frac{d}{dt}\int_V C^{CT} dV + \int_S J_k^{CT} n_k dS = 0. \tag{5.13}$$

Relation (5.13) is valid for an arbitrary volume. Then, one may derive the respective differential equation by using divergence theorem:

$$\frac{dC^{CT}}{dt} = -\frac{\partial J_k^{CT}}{\partial x_k}. \tag{5.14}$$

Note that the total carbon concentration, C^{CT}, is only related to the concentration of carbon in the carbonate, C^C, because of the absence of carbon in cuprate:

$$C^{CT} = fC^C. \tag{5.15}$$

This is equal to the carbon terminal solid solubility, C^{TS}, when $f \neq 0$. Value of C^C is defined with respect to the volume occupied by the carbonate, that is, fV. Note that in the carbon diffusion model, the effect of carbon trapping in the solid by dislocations and voids is neglected, since the bulk of the material is assumed to behave elastically. However, the effect of traps is more important at low temperatures, where the lattice solubility (i.e., ability to substitute atoms and formation of atom interstices) is relatively small [1556]. Corresponding model of carbon trapping by dislocations and crack-like defects can be developed similar to the hydrogen precipitation in metals [692].

Then, the governing equation for non-mechanical energy flow is determined. With this aim, the energy flux is found, that is coefficients L^E and L^{EC} are calculated. Comparing (5.9)–(5.11), we have

$$L^C = \frac{D^C C^C}{RT}, L^{CE} = L^{EC} = \frac{D^C C^C}{RT}(Q^C + \mu^C). \tag{5.16}$$

The remaining coefficient L^E can be determined by taking into account Fourier's law for heat conduction:

$$J_i^E = -k\frac{\partial T}{\partial x_i}, \tag{5.17}$$

where k is the thermal conductivity of the superconductor. Note that (5.17) is valid when there is no carbon diffusion. Then, by stating $J_k^C = 0$, we have from (5.11)

$$\frac{\partial \mu^C}{\partial x_i} = -\frac{Q^C}{T}\frac{\partial T}{\partial x_i}. \tag{5.18}$$

Substituting (5.10, 5.17) and (5.18) in (5.9), one may derive L^E:

$$L^E = kT + \frac{D^C C^C}{RT}\left(Q^C + \mu^C\right)^2. \tag{5.19}$$

Thus, all the coefficients of the phenomenological equations (5.9) have been determined. Consequently, one may derive the expression for energy flux in a carbonate, substituting (5.16) and (5.19) into (5.9) and taking into account the relation (5.11) for carbon flux:

$$J_i^E = (Q^C + \mu^C)J_i^C - k\frac{\partial T}{\partial x_i}. \tag{5.20}$$

The first term of the right-hand side is the energy flux, which is produced by the diffusion of carbon atoms, and the second term is the conducted heat. According to (5.20), the heat of carbon transport is the heat flux per unit flux of carbon in the absence of temperature gradient. Due to the absence of carbon diffusion in the cuprate, the energy flow in the cuprate is only thanks to heat conduction. It is assumed that the thermal conductivity of the carbonate equals the thermal conductivity of the cuprate. Then by using (5.17), the next relation provides the total energy flux in the cuprate/carbonate composite:

$$J_i^{ET} = fJ_i^E - (1-f)k\frac{\partial T}{\partial x_i}, \tag{5.21}$$

which, when combined with (5.12) and (5.20), leads to

$$J_i^{ET} = (Q^C + \mu^C)J_i^{CT} - k\frac{\partial T}{\partial x_i}. \tag{5.22}$$

In other to completely describe the process of energy flow, the conservation of energy should be enforced. It requires that the internal energy rate equals the energy input rate due to the external stress power and gradient of the non-mechanical energy flow [1078]:

$$\rho\frac{du}{dt} = \sigma_{ij}\frac{d\varepsilon_{ij}}{dt} - \frac{\partial J_k^{ET}}{\partial x_k}, \tag{5.23}$$

where ρ, u, σ_{ij} and ε_{ij} are the mass density of material, the internal energy per unit mass, the stress tensor and the strain tensor, respectively. The minus sign in (5.23) is due to convention that the energy flux is positive when it leaves the body. According to the discussion of continuum thermodynamics, based on a caloric equation of state, the rate of internal energy is related with the specific entropy, stress–strain state and total carbon concentration rates:

$$\rho \frac{du}{dt} = \rho T \frac{ds}{dt} + \sigma_{ij} \frac{d\varepsilon_{ij}}{dt} + \mu^C \frac{dC^{CT}}{dt}. \tag{5.24}$$

Equations (5.14) and (5.22–5.24) lead to the following differential equation:

$$\rho T \frac{ds}{dt} = \frac{\partial}{\partial x_i}\left(k \frac{\partial T}{\partial x_i}\right) - Q^C \frac{\partial J_m^{CT}}{\partial x_m} - J_n^{CT} \frac{\partial \mu^C}{\partial x_n}. \tag{5.25}$$

Entropy rate is also related to the rates of temperature, carbonate volume fraction and total carbon concentration. This relation is derived taking into account the dependence of entropy on all thermodynamic variables (i.e., temperature, stress as well as carbonate, cuprate and carbon concentrations) [1743]:

$$\rho T \frac{ds}{dt} = \rho c_p \frac{dT}{dt} + \frac{\Delta \overline{H}^{car}}{\overline{V}^{car}} \frac{df}{dt} + Q^C \frac{dC^{CT}}{dt}, \tag{5.26}$$

where c_p is the specific heat of the superconductor at constant pressure; $\Delta \overline{H}^{car}$ is the enthalpy, associated with the formation of a mole of carbonate and $\overline{V}^{car}$ is the carbonate molal volume. In the determination of (5.26), the following conditions are taken into account: (i) the total number of HTSC moles in the cuprate and carbonate remains constant; (ii) the partial derivative of entropy with respect to temperature is related to the specific heat of the cuprate/carbonate composite under constant stress, $c_\sigma = T(\partial s/\partial T)_\sigma$ (in the present analysis, $c_\sigma = c_p$ is assumed); (iii) by neglecting thermo-elastic coupling, the partial derivative of entropy with respect to stress is taken equal to zero[1]; (iv) the change of entropy due to carbonate formation is equal to $\Delta \overline{H}^{car}/T$; (v) the change of entropy due to the addition of a mole of carbon in the carbonate is equal to Q^C/T.

Substitution of (5.26) into (5.25), taking into account (5.14), gives the governing equation for the flow of non-mechanical energy:

$$\rho c_p \frac{dT}{dt} + \frac{\Delta \overline{H}^{car}}{\overline{V}^{car}} \frac{df}{dt} = \frac{\partial}{\partial x_i}\left(k \frac{\partial T}{\partial x_i}\right) - J_n^C \frac{\partial \mu^C}{\partial x_n}. \tag{5.27}$$

Therefore, the variation of the heat content in the cuprate/carbonate composite depends on conducted heat, heat generated during carbon diffusion and heat released during carbonate formation.

According to the above mathematical formulation for carbon diffusion and energy flow, the knowledge of carbon chemical potential and terminal solid solubility is necessary. Both quantities depend on applied stress and are derived below.

The chemical potentials of mobile and immobile components in stressed solids have been derived [1009]. The chemical potential of a component B is given as

[1] Note that, in ceramics, thermo-elastic coupling effects are quite small [152].

$$\mu^B = \mu_0^B + \frac{\partial w}{\partial N^B} - W^B, \tag{5.28}$$

where μ_0^B is the chemical potential of component B under stress-free conditions, for the same concentration as that under stress; w is the strain energy of the solid and N^B is the number of B moles. Therefore, the second term in the right-hand side of (5.28), $\partial w/\partial N^B$, represents the strain energy of the solid per mole of component B; in the determination of $\partial w/\partial N^B$, temperature and stress are held to be constant. Finally, W^B is the work performed by applied stresses, σ_{ij}, per mole of addition of component B. For immobile components, since the addition or removal of the component takes place at an external surface or an interface, chemical potential is considered as a surface property. However, this is not the case for mobile components.

Equation (5.28) is applied to barium carbonate under stress. For material particle of volume, V, under uniform stress, we have

$$\begin{aligned} \frac{\partial w}{\partial N^C} &= \frac{\partial}{\partial N^C}\left(\int_0^{\varepsilon_{mn}} V\sigma_{ij} d\varepsilon_{ij}\right) = \frac{\partial}{\partial N^C}\left(\int_0^{\sigma_{mn}} V\sigma_{ij} M_{ijkl} d\sigma_{kl}\right) \\ &= \left(\int_0^{\sigma_{mn}} \frac{\partial V}{\partial N^C} M_{ijkl}\sigma_{ij} + \frac{\partial M_{rskl}}{\partial \chi^C}\frac{\partial \chi^C}{\partial N^C} V\sigma_{rs}\right) d\sigma_{kl}, \end{aligned} \tag{5.29}$$

where M_{ijkl} is the elastic compliance tensor of the superconductor and χ^C is the mole fraction of carbon in the carbonate. The work performed by the applied stresses per mole of carbon addition in carbonate is given as

$$W^C = V\sigma_{ij}\frac{\partial \varepsilon_{ij}}{\partial N^C} = V\sigma_{kk}\frac{\overline{V}^C}{3\,V} = \frac{\sigma_{kk}}{3}\overline{V}^C. \tag{5.30}$$

Substituting (5.29) and (5.30) in (5.28), the final expression is found for the chemical potential of carbon, being in carbonate under stress:

$$\mu^C = \mu_0^C + \overline{V}^C\left(\frac{1}{2}M_{ijkl}\sigma_{ij}\sigma_{kl} - \frac{1}{3}\sigma_{mm}\right). \tag{5.31}$$

In determination of (5.31), it has been taken into account that the derivative of volume with respect to carbon moles equals the partial molal volume of carbon $\overline{V}^C$. Moreover, it has been assumed in the first approximation that there is no effect of carbon on the elastic modules of the material (this assumption is discussed in detail in Sect. 5.3.2). Consequently, the derivative of the elastic compliance with respect to the mole fraction of carbon is equal to zero. Note that the first term in parenthesis in (5.31) is of the order of σ^2/E, where E is Young's modulus of the superconductor. The second term is of the order of σ and, therefore, is significantly larger than the first term. If the first term in parenthesis of (5.31) is neglected, a relation is derived, which is more often used in the literature.

The obtained relations for the chemical potential of mobile and immobile components are used for derivation of carbon terminal solubility in a superconductor under stress. According to (5.28), carbonate chemical potential in a stressed material is given by

$$\mu^{car} = \mu_0^{car} + \frac{\partial w}{\partial N^{car}} - W^{car}; \frac{\partial w}{\partial N^{car}} = \overline{w}_{acc} + \overline{w}_{\text{int}} + \overline{w}_{af}; \overline{w}_{acc} = -\frac{1}{2}\int\limits_{\overline{V}^{car}} \sigma_{ij}^{I}\varepsilon_{ij}^{T}dV;$$

$$\overline{w}_{\text{int}} = -\int\limits_{\overline{V}^{car}} \sigma_{ij}\varepsilon_{ij}^{T}dV; \overline{w}_{af} = \frac{1}{2}\int\limits_{\overline{V}^{car}} \sigma_{ij}\varepsilon_{ij}dV, W^{car} = \sigma_n \overline{V}^{car}. \tag{5.32}$$

In determination of relations (5.32) it was taken into account that carbonate formation is accompanied by a deformation, ε_{ij}^{T}, which is mainly a volume expansion. Under no external loading, the above deformation leads to the development of stresses, σ_{ij}^{I}, in the carbonate, defining strain energy of material per mole of precipitating carbonate $\overline{w}_{acc}$. Under externally applied stress, σ_{ij}, the interaction energy, $\overline{w}_{\text{int}}$, as well as the strain energy of the applied field, $\overline{w}_{af}$, should also be taken into account [448]. In the last equation of relations (5.32), σ_n is the normal stress at the location of cuprate/carbonate interface, where the chemical potential is considered.

The chemical potential of the stressed Ba–O component of barium carbonate is defined as

$$\mu^{Ba-O} = \mu_0^{Ba-O} + \frac{\partial w}{\partial N^{Ba-O}} - W^{Ba-O}; \frac{\partial w}{\partial N^{Ba-O}} = \frac{1}{2}\int\limits_{(\overline{V}^{car}-\overline{V}^{C})} \sigma_{ij}\varepsilon_{ij}dV;$$

$$W^{Ba-O} = \sigma_n(\overline{V}^{car} - \overline{V}^{C}). \tag{5.33}$$

The carbonate is assumed to be in equilibrium with carbon and cuprate either under stress or under stress-free conditions. Then, taking into account the chemical formula of barium carbonate $BaCO_3$, we have:

$$\mu^{car} = \mu^{Ba-O} + \mu^{C}(C^{TS}); \mu_0^{car} = \mu_0^{Ba-O} + \mu_0^{C}(C_0^{TS}), \tag{5.34}$$

where C^{TS} and C_0^{TS} are the values of carbon terminal solid solubility under applied stress and stress-free conditions, respectively. By substitution of relations (5.32) and (5.33) in (5.34), one may derive

$$\mu^{C}(C^{TS}) - \mu_0^{C}(C_0^{TS}) = \overline{w}_{acc} + \overline{w}_{\text{int}} + \frac{1}{2}\int\limits_{\overline{V}^{C}} \sigma_{ij}\varepsilon_{ij}dv - \sigma_n\overline{V}^{C} \tag{5.35}$$

It has been implied that carbonate and Ba–O equilibrium concentrations do not change significantly with stress. Because of material continuity, the molal volume of the carbonate equals that of the cuprate at the cuprate/carbonate interface; therefore $\overline{V}^{\,C} \approx 0$ and the last two terms in the right side of (5.35) can be assumed to be zero. Moreover, in ideal or dilute solutions (Raoult's law), the stress-free carbon chemical potential satisfies the following well-known relation:

$$\mu_0^C = \mu_{RS}^C + RT\ln(C^{CT}\overline{V}), \tag{5.36}$$

where μ_{RS}^C is carbon chemical potential in the "standard" (i.e., reference) state and $\overline{V}$ is the molal volume of composite. Then, by invoking (5.15) and (5.36) and substituting (5.31) in (5.35), one may derive the terminal solid solubility of carbon in the composite under stress:

$$C^{TS} = C_0^{TS}\exp\left(\frac{\overline{w}_{acc} + \overline{w}_{\text{int}}}{RT}\right)\exp\left[\frac{\overline{V}^{\,C}}{RT}\left(\frac{\sigma_{mm}}{3} - \frac{1}{2}M_{ijkl}\sigma_{ij}\sigma_{kl}\right)\right]. \tag{5.37}$$

The above derivation is based on similar arguments of [1009] for the formation of cementite in ferrite. When the compliance term is neglected, (5.37) is similar to the relation for the hydrogen terminal solid solubility [1413]. Using (5.37) or its simplified version implies that chemical equilibrium occurs under local thermal stress conditions.

Then, it is assumed that all material phases are elastic, and the elastic properties of carbonate and cuprate are identical in the first approximation and do not depend on carbon concentration. The material strain is coupled with carbon diffusion and energy flow due to the strains, caused by carbon dissolution, carbonate formation and thermal expansion. Similar to the processes of hydride formation in metals [1743] and taking into account carbon absence in cuprate, the following equation may be obtained:

$$\begin{aligned}
&\frac{d\sigma_{ij}}{dt} = M_{ijkl}^{-1}\left(\frac{d\varepsilon_{kl}}{dt} - \frac{d\varepsilon_{kl}^C}{dt} - \frac{d\varepsilon_{kl}^E}{dt}\right); M_{ijkl}^{-1} = \lambda\delta_{ij}\delta_{kl} + \mu(\delta_{ik}\delta_{jl} + \delta_{il}\delta_{jk});\\
&\frac{d\varepsilon_{ij}^C}{dt} = \frac{1}{3}\delta_{ij}\frac{d}{dt}(f\theta^{car}); \frac{d\varepsilon_{ij}^E}{dt} = \alpha\delta_{ij}\frac{dT}{dt},
\end{aligned} \tag{5.38}$$

where λ and μ are Lame constants of the HTSC; $\theta^{car} = \varepsilon_{kk}^T$ is carbonate expansion, occurring during its precipitation, and α is the thermal expansion factor of the composite, assumed to be equal to that of the carbonate.

Simulation of fracture by considering cohesive tractions [92, 402] assumes that ahead of a crack tip there is a fracture process zone, where the material deteriorates in a ductile (void growth and coalescence) and/or brittle (carbonate cleavage) mode. According to de-cohesion model, de-cohesion layer with the thickness as fracture process zone is taken off the material, along the crack path. Along the boundaries, created by the cut, the cohesive traction is applied. All information on the damage is contained in the distribution of the cohesive traction, which depends on boundary displacements of the de-cohesion layer. The shape of the traction – displacement function depends on the failure process. In the case of tensile separation, the most important features are the maximum cohesive traction, $\sigma_{\max}$, and the energy of de-cohesion, ϕ_0:

$$\phi_0 = \int_0^{\delta_c} \sigma_n d\delta_n, \tag{5.39}$$

where σ_n is the normal cohesive traction and δ_n is the respective normal displacement, which equals the sum of the displacements on both sides of the de-cohesion layer. Moreover, δ_c is the normal displacement, which corresponds to complete fracture and consequently to zero normal cohesive traction. Details of the model for crack growth under plane strain conditions have been presented in [1740, 1741]. The model has been used for the solution of several fracture problems [1237, 1718, 1744].

Here, cohesive traction is assumed to vary, according to the next relation:

$$\sigma_n = \begin{cases} E_i \frac{\delta_n}{\delta_0}, & \delta_n \le \delta_l; \\ \sigma_{\max}, & \delta_l \le \delta_n \le \delta_f; \\ \sigma_{\max} - E_f \frac{\delta_n - \delta_f}{\delta_0}, & \delta_f \le \delta_n \le \delta_c; \\ 0, & \delta_c \le \delta_n, \end{cases} \tag{5.40}$$

where δ_0 is a constant length of the order of carbonate thickness; E_i and E_f are the de-cohesion modules, which are assumed to be constant; δ_l is the normal displacement at initiation of damage, at which maximum cohesive traction is reached. Unloading starts when normal displacement exceeds δ_f. Note that δ_f depends on σ_{max} and ϕ_0, according to the following relation:

$$\delta_f = \left[\phi_0 + \frac{1}{2}\sigma_{\max}^2 \delta_0 \left(\frac{1}{E_i} - \frac{1}{E_f}\right)\right] \sigma_{\max}^{-1}. \tag{5.41}$$

As shown in [1430], the energy of de-cohesion, related to a cohesive zone ahead of a crack tip, in elastic material, is equal to the critical value, J_c, of J-integral, when fracture is imminent:

$$\phi_0 = J_c = \frac{1 - v^2}{E} K_{Ic}^2, \tag{5.42}$$

where $K_{\mathrm{I}c}$ is the critical value of stress intensity factor under plane strain conditions; E and v are, respectively, Young's modulus and Poisson's ratio of the material, which ahead of the crack is a composite made of brittle carbonate and relatively tough cuprate. Therefore, the fracture toughness of the material, expressed by the energy of de-cohesion, depends on carbonate volume fraction, f, along the crack plane. In accordance with a mixture rule, we have

$$\phi_0 = f\phi_0^{car} + (1-f)\phi_0^{cup}, \tag{5.43}$$

where ϕ_0^{car} and ϕ_0^{cup} are the de-cohesion energies of the material, which relate with the critical values of the stress intensity factor for carbonate, K_{Ic}^{car} and cuprate K_{Ic}^{cup}, by using a relationship similar to (5.42). Note that the carbonate is surrounded by cuprate, and its delayed cracking is assumed. The experimental values of the threshold stress intensity factor include the energy required for the generation of the new surface due to crack growth, as well as any plastic dissipation in the cuprate matrix, which surrounds the crack tip carbonate. It is necessary to take these into account by using K_{Ic}^{car}. The cuprate can sustain different maximum cohesive tractions from that of the carbonate. Therefore, the maximum cohesive traction of the composite material along the crack plane depends on the carbonate volume fraction. Here, the maximum cohesive traction is given as

$$\sigma_{\max} = \sqrt{f\sigma_{car}^2 + (1-f)\sigma_{cup}^2}, \tag{5.44}$$

which is derived by assuming that the part of de-cohesion energy during loading satisfies a relation similar to (5.43). In (5.44), σ_{car} (σ_{cup}) is the maximum cohesive traction, sustained by the material during crack growth in the presence of the carbonate (cuprate) along the crack plane over a distance from the crack tip $x \gg \delta_0$. According to theoretical studies for the crack tip field in elastic plastic materials [399, 735, 1431], the maximum hoop stress, along the crack plane, is nearly equal to three times the yield stress of the material. Assuming that σ_{car} is equal to the fracture strength of carbonate, relations (5.43) and (5.44) provide the average properties of the de-cohesion layer over its thickness. Due to carbon diffusion and carbonate formation, carbonate volume fraction changes locally with time, causing a corresponding change of the de-cohesion properties. Maximum cohesive traction is reached when δ_n corresponds to a normal traction, satisfying relation (5.44). As time increases, the de-cohesion energy changes, according to (5.43), but unloading starts when relation (5.41) is satisfied. It is assumed that the de-cohesion energy does not change during unloading. The part of de-cohesion energy during unloading is minimized by choosing the largest value of E_f for which quasi-static unloading exists [1744].

5.3.2 Discussion of Results

The developed mathematical model of the carbon embrittlement and fracture of YBCO takes into account the coupling of the operating physical processes, namely (i) carbon diffusion, (ii) carbonate precipitation, (iii) non-mechanical energy flow and (iv) cuprate/carbonate composite deformation. Material damage and crack growth are simulated by using the de-cohesion model. Governing equations are obtained for superconductor/carbon system, in which brittle carbonate may precipitate and be accommodated elastically, forming cuprate/carbonate composite. It is assumed that carbon does not affect the elastic modules of the composite. The elastic and thermal properties of the carbonate and composite are taken to be identical in the development of the governing equations. The model can be extended to single or multiphase alloys without any changes, if the additional elements have a very small concentration and do not affect operating processes.

The consideration of elastic behavior in the bulk of the body, causing elastic carbonate accommodation, leads to re-dissolution of the crack tip carbonates, after their fracture, and the reduction of the hydrostatic stresses. In the case of greater plasticity of superconductor, on carbonate precipitation, cuprate matrix should be yielded similar to the hydride behavior in metal matrix [1743]. As a consequence, the crack tip carbonates are more stable and may re-dissolve only partially after fracture. An extension of the present model, which would take into account elastic–plastic carbonate accommodation is useful for simulation of processes beyond crack growth initiation as well as for consideration of all parameters, which play a role in fracture resistance of material. Then, the model of carbon embrittlement and sub-critical fracture of YBCO may be added to earlier developed simulation of toughening mechanisms, acting in HTSC [1300, 1304, 1317, 1319, 1326, 1327]. Another direction for further development of the present model is the consideration of different mechanical and thermal properties of cuprate and carbonate. In this case, the approaches, which are used in composite materials [1841], could be adopted for developing the governing equations. Along this direction is also the consideration of the carbon effect on the elastic modules of carbonate, defining carbon chemical potential in carbonate [1009], and therefore on carbonate terminal solid solubility.

Generally, similar to precipitation of hydrogen in metal and intermetallic compounds [1004], the carbon effect on the elastic constants of superconductor can be characterized by a factor:

$$r = \frac{(C_C - C_0)/C_0}{C^C/C^{YBCO}} \tag{5.45}$$

where C_C and C_0 are the elastic moduli of material with and without carbon, respectively; C^C/C^{YBCO} is the concentration ratio of carbon to superconductor. However, the factor r should be regarded only as a rough approximation of the carbon effect, because the effect is not linear in C^C/C^{YBCO} in some cases. The parameter (5.45) should be estimated in experiment and depends on the test temperature and crystallographic properties of HTSC. In particular, the absorption of carbon by a superconductor with free boundaries almost invariably leads to expansion of crystalline lattice and corresponding change of elastic properties.

Similar to the hydrogen effect in intermetallic compounds [1004], an existence of the carbon influence on elastic properties (in particular, shear modulus) may be assumed due to an electronic effect. The addition of carbon and the contribution of electrons at the Fermi level moves the Fermi level and results in a corresponding change of temperature dependence of the elastic properties. Carbon contributes electrons in the conduction band and thereby changes the concentration of the conductivity electrons [1634] that also changes electronic contribution to the elastic constants. The sign and magnitude of the effect depend on the electronic band structure and the density of states at the Fermi level. In addition to these long-range effects of the carbon, there are other, more local effects: (i) the direct carbon—superconductor ion potential contributes to the elastic constants, and the interstitial carbon may also affect the superconductor ion—superconductor ion potential [1004]; (ii) direct influence on elastic constant of the optical phonons due to the carbon vibrations [527]; and (iii) mechanical relaxation (Snoek effect) of the interstitial carbon in response to strain [1069]. However, note that for actual statement of the above carbon effects on elastic properties of HTSC, it is necessary to carry out intensive test investigations.

Then, in the present model simple mixture rules (5.43) and (5.44) have been used for the derivation of the energy of de-cohesion and the maximum cohesive tractions. However, note that the maximum hydrostatic stress, which is expected ahead of the crack tip in an elastic–plastic material before the precipitation of near-tip carbonates, is recovered after their fracture. Consequently, the strong effect of hydrostatic stress distribution on carbon diffusion and carbonate precipitation near the crack tip has been taken into account. At the same time, if perfect cohesion is considered, carbon precipitates at the tip, and no ductile ligament is formed during fracture similar to its occurrence for hydrides in metals [1742]. Therefore, the de-cohesion model improves the performance of the carbon embrittlement model, which is based on elastic behavior of the material. Moreover, the de-cohesion model takes into account the time variation of the de-cohesion energy due to the time-dependent process of carbonate precipitation.

Finally, note that a similar mathematical model and governing equations may be obtained for other HTSC under carbon embrittlement. Finite-element scheme for numerical realization of the governing equations is presented in Appendix B.

5.4 Modeling of Carbon Segregation and Fracture Processes of HTSC

It could be proposed that carbon can segregate not only to grain boundaries, but also to crack surfaces and dislocations, where lattices are distorted. Therefore, two microcracking processes are possible: continuously slow crack growth and discretely rapid crack growth, associated with high amounts of acoustic emissions. Then, the carbon segregation processes can be studied by using the microscopic models of the equilibrium slow and fast crack propagation and also a steady-state crack growth, which are screened by dislocation field [1317].

5.4.1 Equilibrium Slow and Fast Crack Growth

Consider an intergranular crack of length, $2a$, in a carbonated HTSC (Fig. 5.13). The crack lies along x-axis in an elastic–plastic isotropic body with shear modulus, G, Poisson ratio, ν, yield strength, σ_y, and work hardening factor, n. The body is loaded by a remote stress, σ_a parallel to the y-axis at a constant temperature, T. At the x-axis, two linear dislocation arrays with the length, r_y, are located at the distance, d, from the crack tips. This model proposes that an intergranular crack tip maintains an atomistic sharpness and a local equilibrium condition in the presence of screening dislocations. It is assumed that all geometrical parameters of the crack tip, presented in Fig. 5.14 (in particular, the size of the arc-shaped crack tips, q, and a crack tip displacement, δ_c), remain constant during plastic deformation. The condition of the local equilibrium at the crack tip is that the crack must be screened by dislocation field and maintains a dislocation-free zone with the size d. The loaded system "crack–dislocation arrays" maintains a local stress, σ_d, in the dislocation-free zone and produces the next stress intensity in the screening dislocation zone, given by Hutchinson–Rice–Rosengren model as [1430]

$$\sigma_{yy} = \sigma_d, \text{at} \quad a < |x| < \quad a + d; \tag{5.46}$$

$$\sigma_{yy} = \beta\sigma_y(K_a/\sigma_y)^{2n/(n+1)}(|x| \; - a)^{-n/(n+1)}, \text{at} \quad a + d < |x| < a + d + r_y, \tag{5.47}$$

where K_a is the applied stress intensity and β is the factor, which depends on the elastic and plastic deformation properties (see Fig. 5.15).

The carbon segregation process is found by the crack tip profile and by the stress field ahead of the crack tip. The chemical potentials of carbon and superconductor can be stated, following [804] in various grain boundary and crack surface zones, namely, I is the zone not affected by the stress intensity ($|x| > a + d + r_y$); II is the zone of screening dislocations ($a + d < |x| < a +$ - $d + r_y$); III is the dislocation-free zone ahead of the crack tip ($a < |x| < a + d$);

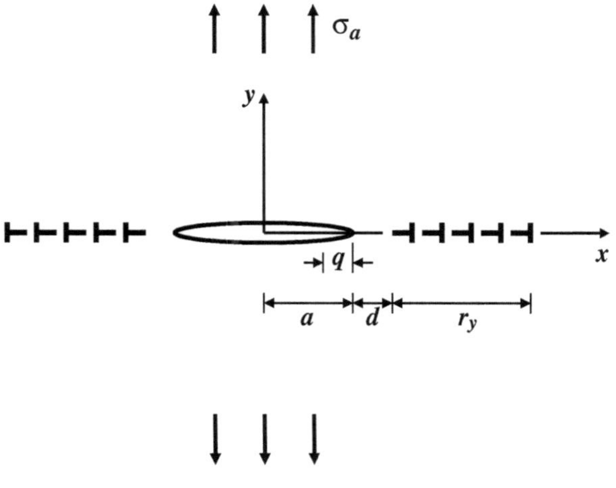

Fig. 5.13 Schematic representation of intergranular crack, screened by linear dislocation array

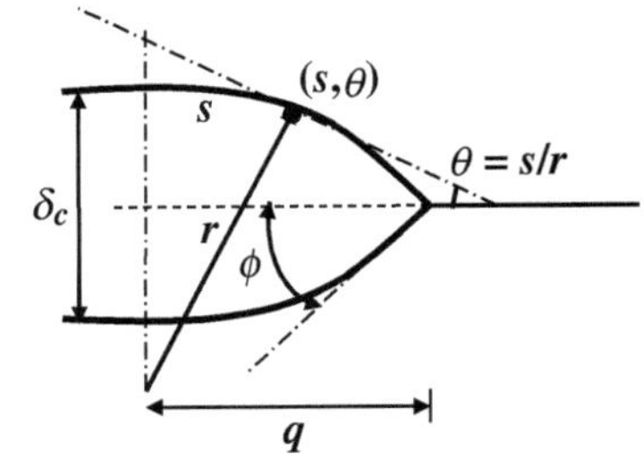

Fig. 5.14 Illustration of crack tip profile, indicating crack tip angle, 2ϕ, and polar coordinates (s, θ)

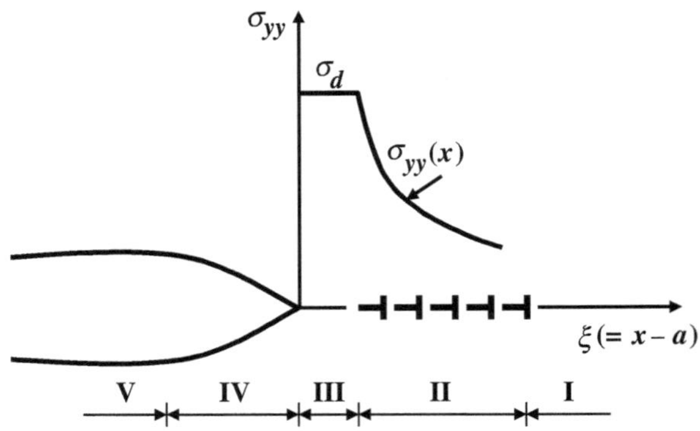

Fig. 5.15 Illustration of local stress and stress intensity ahead of the dislocation-screened crack tip and definition of various regions, where different chemical potentials are given

IV is the arc-shaped crack tip zone ($a - q < |x| < a$); and V is the parallel flat crack surface zone ($|x| < a - q$).

At equilibrium, the chemical potentials of carbon and superconductor must be the same in all the regions, respectively. So, the equilibrium carbon segregation depends on the binding energies and crack tip conditions. The binding energies of carbon at grain boundaries and crack surfaces $(H_B)_b$ and $(H_B)_s$, respectively, are found through the standard chemical potentials of C and HTSC as

$$(H_B)_b = (\mu_{m0})_C - (\mu_{b0})_C - (\mu_{m0})_{HTSC} + (\mu_{b0})_{HTSC}; \tag{5.48}$$

$$(H_B)_s = (\mu_{m0})_C - (\mu_{s0})_C - (\mu_{m0})_{HTSC} + (\mu_{s0})_{HTSC}. \tag{5.49}$$

Here and further the subscript m is the matrix, b is the grain boundary and s is the crack surface.

The basic assumption of the model is that the embrittlement occurs as a reduction of the surface and grain boundary energies due to the carbon segregation. Moreover, it is taken into account that slow fracture occurs when the segregation is sufficiently rapid to maintain the same chemical potential between the grain boundary and the crack surface. Fast fracture occurs when the carbon concentration remains the same at the grain boundary and the crack surface. Then, from the thermodynamic theory proposed by Seah, Rice and Hirth, the ideal works, expended in slow (γ^s) and fast (γ^f) fracture, can be obtained as [804]

$$\gamma^s = \gamma_0 - RT(2C_V/\Omega_s - C^s_{III}/\Omega_b); \tag{5.50}$$

$$\gamma^f = \gamma_0 - (C^f_{III}/\Omega_b)\Delta\mu. \tag{5.51}$$

Here, the equilibrium carbon concentrations in zones III and V have forms:

$$C_{III} = \frac{C_m \exp\left\{\left[(H_B)_b + \sigma_d V_h - \frac{\sigma_d^2 V_h}{4G(1+v)}\right]/(RT)\right\}}{1 - C_m + C_m \exp\left\{\left[(H_B)_b + \sigma_d V_h - \frac{\sigma_d^2 V_h}{4G(1+v)}\right]/(RT)\right\}}; \tag{5.52}$$

$$C_V = \frac{C_m \exp[(H_B)_s/(RT)]}{1 - C_m + C_m \exp[(H_B)_s/(RT)]}, \tag{5.53}$$

where C_m is the bulk carbon concentration; V_h is the molar volume of carbon; R is the gas constant; $1/\Omega_i$ is the carbon coverage at interfaces; C^s_{III} and C^f_{III} are the critical values of carbon concentration in zone III, required for slow and fast fractures, respectively; γ_0 is the ideal work of intergranular fracture in the absence of carbon ($= 2\gamma_{s0} - \gamma_{b0}$; γ_{s0} is the surface fracture energy and γ_{b0} is the fracture energy of grain boundary); and $\Delta\mu = RT\ln(2C_V/C^f_{III})$ is the chemical potential difference between the crack surface and the stressed grain boundary.

The equations for constant carbon concentrations can also be found in zones I and IV. In this case, C_{I} coincides with C_{V}, in which $(H_{\mathrm{B}})_{\mathrm{s}}$ is replaced by $(H_{\mathrm{B}})_{\mathrm{b}}$, and for C_{IV} we have

$$\exp\left(\frac{C_{IV}V_h}{\Omega_s r}\right) + \frac{C_{IV}(1 - C_m)}{(1 - C_{IV})C_m} = \exp\{[(H_B)_s + \gamma_{s0}V_h/\Omega_s]/(RT)\}, \tag{5.54}$$

where r is the curvature radius of the arc-shaped crack tip (Fig. 5.14). At the same time, due to the variable stress distribution (5.46) and (5.47), the carbon content in zone II is not constant. The relationship between the critical stress intensity, required to propagate the crack (for slow, fast or steady-state fracture) and to change the ideal work due to the carbon segregation, is stated by using the local energy balance condition as [804]

$$-(1 - v)K_d^2/2G + \gamma^c \leq 0, \tag{5.55}$$

where the superscript c corresponds to certain fracture state, K_d is the local stress intensity factor connected with the dislocation-free zone size ahead of the crack tip (d) and local stress (σ_d) in this zone by the equation, approximately derived from the load balance condition between a crack with a linear stress intensity and that with local stress [804]: $\pi d = (K_d/\sigma_d)^2$. Moreover, a relation between σ_d, K_d and δ_c follows from the condition that the elastic energy release rate is the same as the J-integral, i.e., $\sigma_d = K_d[2(1 - v^2)/\delta_c]^{1/2}$. Then the threshold apparent stress intensity, K^c_{th}, is given by the relationships (5.47), and (5.55):

$$K_{th}^{c} = K_c(\gamma^c/\gamma_0)^{(n+1)/4n}(\delta_{c0}/\delta_c^c)^{(1-n)/4n}, \tag{5.56}$$

where K_c is the fracture toughness, δ_c^c is the critical crack opening displacement (CCOD), required for various fracture processes (superscript c), and δ_{c0} is the CCOD in the absence of carbon, defined as

$$\delta_{c0} = \frac{[4G(1+v)\gamma_0]^{(n+1)/(1-n)}}{[2\pi(1-v^2)]^{2n/(1-n)}\beta^{2(n+1)/(1-n)}\sigma_y^2 K_c^{4n/(1-n)}}. \tag{5.57}$$

Note that for the crack to maintain the dislocation-free zone during the growth, besides the inequality (5.55), it is necessary to satisfy an additional condition, namely the total energy balance criterion in the form [804]:

$$-(1-v)K_a^2/2G + \gamma^c + \gamma_p \leq 0, \tag{5.58}$$

where γ_p is the plastic work due to the generation and motion of screening dislocations, which could be found numerically, for example, in the case of a linear dislocation array [804, 1430].

5.4.2 Steady-State Crack Growth

Assume that the carbon diffusion along stressed boundaries and crack surfaces is the mechanism which controls the intergranular embrittlement and affects the crack growth rate. In this case, the bulk diffusion effects on carbon-induced intergranular cracking (CIIC) are neglected. Under the geometrical and loading conditions of the equilibrium crack growth problem, the steady state case indicates subcritical intergranular crack growth with constant velocity, v_c (Fig. 5.16). Taking into account the grain boundary and crack surface zones (II–V), the fluxes of carbon in these regions, J_i^j, can be stated as

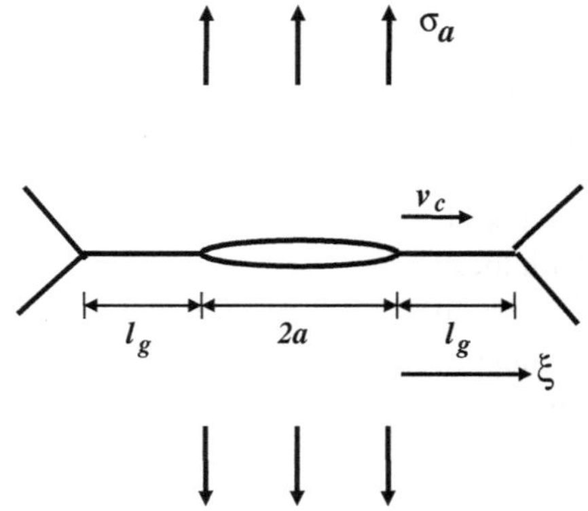

Fig. 5.16 Schematic representation of steady-state growth of intergranular crack and boundary condition

$$J_i^j = -\frac{D_i C_i^j}{RT} \frac{d\mu_i^j}{d(x \text{ or } s)}, \tag{5.59}$$

where D_i is the diffusivity of carbon, $C_i^j$is the carbon concentration, i is the subscript indicating b or s and j is the superscript indicating various interface zones; μ_i^j are the corresponding chemical potentials. The differentiation with respect to s is carried out only in zone IV; in this case s is the variable arc length in the corresponding part of the arc-shaped crack tip (see Fig. 5.14). The continuity equation of fluxes is

$$\frac{dC_i^j}{dt} + \frac{dJ_i^j}{d(x \text{ or } s)} = 0, \tag{5.60}$$

where t is the time. Based on (5.46, 5.47, 5.59) and (5.60) and also the relationships between the interface energies, γ_i^j, and the amounts of carbon, C_i^j, in the various zones, derived from Gibbs theory and dilute solute approximation as $\gamma_i^j = \gamma_{i0} - (RT/\Omega_i)C_i^j$, the second-order differential equations controlling the carbon diffusion in the intergranular cracking regions can be obtained, similar to [805]:

Zone II:

$$J_b^{II} = -D_b \frac{dC_b^{II}}{dx} - \left(\frac{D_b V_h}{RT}\right) \frac{n}{(n+1)} \beta \sigma_y^{(1-n)/(n+1)} K_a^{2n/(n+1)} (x-a)^{(-2n-1)/(n+1)} C_b^{II}; \tag{5.61}$$

$$\frac{d^2 C_b^{II}}{dx^2} + \left[\left(\frac{v_c}{D_b}\right) + \left(\frac{V_h}{RT}\right) \frac{n}{(n+1)} \beta \sigma_y^{(1-n)/(n+1)} K_a^{2n/(n+1)} (x-a)^{(-2n-1)/(n+1)}\right] \frac{dC_b^{II}}{dx} - \left(\frac{V_h}{RT}\right) \frac{n(2n+1)}{(n+1)^2} \beta \sigma_y^{(1-n)/(n+1)} K_a^{2n/(n+1)} (x-a)^{(-3n-2)/(n+1)} C_b^{II} = 0. \tag{5.62}$$

Zone III:

$$J_b^{III} = -D_b \frac{dC_b^{III}}{dx}; \tag{5.63}$$

$$\frac{d^2 C_b^{III}}{dx^2} + \left(\frac{v_c}{D_b}\right) \frac{dC_b^{III}}{dx} = 0. \tag{5.64}$$

Zone IV:

$$J_s^{IV} = -D_s \frac{dC_s^{IV}}{ds} - \left(\frac{D_s V_h}{r\Omega_s}\right) \frac{dC_s^{IV}}{ds} C_s^{IV}; \tag{5.65}$$

$$\left[1+\left(\frac{V_h C_s^{IV}}{r\Omega_s}\right)\right]\frac{d^2C_s^{IV}}{ds^2}+\left(\frac{V_h}{r\Omega_s}\right)\left(\frac{dC_s^{IV}}{ds}\right)^2+\sec\phi\left(\frac{v_c}{D_s}\right)\frac{dC_s^{IV}}{ds}=0. \quad (5.66)$$

Zone V:

$$J_s^v=-D_s\frac{dC_s^v}{dx}; \quad (5.67)$$

$$\frac{d^2C_s^v}{dx^2}+\left(\frac{v_c}{D_s}\right)\frac{dC_s^v}{dx}=0. \quad (5.68)$$

The boundary values of carbon concentration, $C_V^{\bullet\bullet}$(at $x=0$) and $C_b^{\bullet\bullet}$(at $x=a+l_g$) used in the solution of the second-order differential equations, have forms:

$$C_v^{\bullet\bullet}=\frac{C_m\exp\left[(H_B)_s/(RT)\right]}{1-C_m+C_m\exp\left[(H_B)_s/(RT)\right]}; \quad (5.69)$$

$$C_b^{\bullet\bullet}=\frac{C_m\exp\left\{\left[(H_B)_b+\sigma_{yb}V_h\right]/(RT)\right\}}{1-C_m+C_m\exp\left\{\left[(H_B)_b+\sigma_{yb}V_h\right]/(RT)\right\}}, \quad (5.70)$$

where σ_{yb} is the local stress in a triple point of intergranular boundary.

The conditions at the boundaries between zones are given as

Zones II–III:

$$\left(C_b^{II}\right)_{x=a+d}=\left(C_b^{III}\right)_{x=a+d}; \quad (5.71)$$

$$\left(\frac{dC_b^{II}}{dx}\right)_{x=a+d}+\left(\frac{V_h}{RT}\right)\frac{n}{(n+1)}\beta\sigma_y^{(1-n)/(n+1)}K_a^{2n/(n+1)}d^{(-2n-1)/(n+1)}\left(C_b^{II}\right)_{x=a+d}=\left(\frac{dC_b^{III}}{dx}\right)_{x=a+d}. \quad (5.72)$$

Zones III-IV (at the crack tip):

$$\mu_{b0}+RT\ln\left(C_b^{III}\right)_{x=a}-\sigma_d^{\bullet}V_h+\frac{\sigma_d^{\bullet 2}V_h}{4G(1+v)}=\mu_{s0}+RT\ln\left(C_s^{III}\right)_{s=r\phi}-(V_h/r)\left\{\gamma_{s0}-\left[RT\left(C_s^{IV}\right)_{s=r\phi}/\Omega_s\right]\right\}; \quad (5.73)$$

$$D_b\left(\frac{dC_b^{III}}{dx}\right)_{x=a}=2D_s\left[1+\left(\frac{V_h\left(C_s^{IV}\right)_{s=r\phi}}{r\Omega_s}\right)\right]\left(\frac{dC_s^{IV}}{ds}\right)_{s=r\phi}. \quad (5.74)$$

Zones IV-V:

$$RT\ln\left(C_s^{IV}\right)_{s=0} - (V_h/r)\left\{\gamma_{s0} - \left[RT\left(C_s^{IV}\right)_{s=0}/\Omega_s\right]\right\} = RT\ln\left(C_s^{V}\right)_{x=a-q}; \quad (5.75)$$

$$\left[1 + \left(\frac{V_h\left(C_s^{IV}\right)_{s=0}}{r\Omega_s}\right)\right]\left(\frac{dC_s^{IV}}{ds}\right)_{s=0} = \left(\frac{dC_s^{V}}{dx}\right)_{x=a-q}. \quad (5.76)$$

It is assumed that the steady-state crack growth maintains the equilibrium values at the crack center and at the triple point of grain boundaries ahead of the crack. It should be noted that the present boundary condition is a first order approximation because the equilibrium content of carbon at the triple grain junction is difficult to attain, especially at sufficiently high velocity of the crack. The interface conditions show that the chemical potentials and fluxes of carbon must be the same at each interface in order to maintain the continuity of the carbon flux. So, the boundary-value problem is stated for the solution of which some relationships, defined in the equilibrium crack growth, are to be used, namely: the local equilibrium condition at the crack tip, the geometrical crack tip conditions, and also the crack tip condition, derived from the local energy criterion (5.55). The carbon diffusivity effect is determined by the ideal work of steady-state fracture as

$$\gamma^{\bullet} = \gamma_0 - RT\left(2C_V^{\bullet}/\Omega_s - C_{III}^{\bullet}/\Omega_b\right) + \left(C_V^{\bullet}/\Omega_s + C_{III}^{\bullet}/2\Omega_b\right)\Delta\mu^{\bullet}, \quad (5.77)$$

where

$$C_V^{\bullet} = \left(C_s^{V}\right)_{x=a-q}, C_{III}^{\bullet} = \left(C_b^{III}\right)_{x=a}; \quad (5.78)$$

$$\Delta\mu^{\bullet} = \mu_{b0} - \mu_{s0} + RT\ln\left(C_{III}^{\bullet}/C_V^{\bullet}\right) - \sigma_d^{\bullet}V_h + \frac{\left(\sigma_d^{\bullet}\right)^2 V_h}{4G(1+\nu)}. \quad (5.79)$$

The superscript • indicates the steady-state fracture. The boundary-value problem can be solved numerically, for example, by using the Runge–Kutta method. The boundary conditions at the triple points permit to study the effect of grain sizes on the kinetics of CIIC. At the same time, the size effects cannot be estimated in the cases of the equilibrium slow and fast cracks.

5.4.3 Some Numerical Results

Numerical results can be obtained in the case of equilibrium crack growth for different values of the bulk carbon concentration, C_m. Equating the right parts of the (5.46), and (5.47) at $|x| = a + d$ and $K_a = K_c$ and considering the relations (5.50, 5.52, 5.53) and (5.55) for slow crack and (5.51–5.53) and (5.55) for fast crack, the problem is reduced to numerical solution of transcendental algebraic equations. These equations state the relationships γ^c and the critical tip parameters

Table 5.2 Numerical results for slow crack

Parameters	$C_m =$ 50 ppm	$C_m =$ 100 ppm	$C_m =$ 150 ppm	$C_m =$ 200 ppm	$C_m =$ 250 ppm	$C_m =$ 300 ppm
d^s, μm	185	176	167	159	150	142
σ_d^s/σ_y	1.438	1.444	1.451	1.458	1.465	1.472
K_d^s/K_c	0.347	0.340	0.332	0.326	0.318	0.311
γ^s/γ_0	0.963	0.925	0.882	0.850	0.809	0.774
δ_{c0}/δ_c^s	1.050	1.104	1.164	1.222	1.295	1.368
K_{th}^s/K_c	1.007	1.007	0.996	1.004	0.999	0.999

Table 5.3 Numerical results for fast crack

Parameters	$C_m =$ 50 ppm	$C_m =$ 100 ppm	$C_m =$ 150 ppm	$C_m =$ 200 ppm	$C_m =$ 250 ppm	$C_m =$ 300 ppm
d^f, μm	188	182	176	169	163	157
σ_d^f/σ_y	1.436	1.440	1.444	1.450	1.454	1.459
K_d^f/K_c	0.349	0.344	0.340	0.334	0.329	0.324
γ^f/γ_0	0.974	0.947	0.925	0.892	0.866	0.840
δ_{c0}/δ_c^f	1.034	1.067	1.104	1.150	1.192	1.238
K_{th}^f/K_c	1.002	0.997	1.007	1.000	1.000	1.000

(σ_d^c, δ_c^c) on C_m. Then, the values of K_{th}^c are determined from (5.56), using the calculated values of γ^c and δ_c^c.

Based on test data, the necessary parameters for numerical calculations are $\gamma_0 = 1$ J/m^2, $1/\Omega_b = 1/\Omega_s = 8.1 \times 10^{-5}$ mol/m^2, $(H_B)_s = 50$ kJ/mol, $(H_B)_b = 10$ kJ/mol, $n = 0.1$, $\sigma_y = 10$ MPa, $V_h = 2 \times 10^{-6}$ m^3/mol, $R = 8.316$ J/mol·K, $v = 0.2$, $K_c = 1$ MPa·m$^{1/2}$, $G = 50$ GPa, $T = 1110$ K. The numerical results are presented in Tables 5.2 (slow crack) and 5.3 (fast crack) [1317].

As shown by the numerical results, all auxiliary parameters $(d^c, \sigma_d^c, K_d^c, \gamma^c, \delta_c^c)$ change monotonously with C_m for both slow and fast fractures. In particular, the normalized parameter, γ_c/γ_0, decreases with an increase of C_m. At the same time, the normalized parameter, δ_{c0}/δ_c^c, increases together with C_m. These alternative contributions to K_{th}^c/K_c cause its non-monotonous behavior in the dependence on C_m. In this case, the strengthening effect on K_{th}^c (i.e., when $K_{th}^c/K_c > 1$) occurs when the segregation of carbon in the crack regions strongly affects the crack tip condition (i.e., reduction of δ_c^c), but does not produce a substantial reduction in γ^c (see (5.56)). More evident change of all auxiliary parameters in the case of slow crack compared with the fast crack at the considered range of C_m proposes the greater susceptibility of slow growth on CIIC increase. Then, it is apparent that under the condition of a dislocation-screened crack, carbon segregation induces slow fracture more readily than fast fracture. The weak change of K_{th}^c/K_c on C_m in both cases of slow and fast cracks is apparently caused by the small range of C_m (while this is a real bulk carbon concentration in HTSC systems). The presented numerical example should be specified with more accurate selection of key parameters for certain HTSC.

For the used local energy criterion, which is controlled by CIIC, the dependence of K_{th}^{c} on the ideal work of fracture, the crack tip conditions and the plastic deformation properties (see (5.56)) is somewhat similar to that obtained in [1781]. The difference is that the present analysis explicitly includes not only the embrittlement effect of carbon, but also the crack tip conditions, affected by the carbon segregation. It should be noted that the presence of a dislocation-screened crack and the microscopic behavior of plastic deformation, associated with CIIC, have not yet been experimentally verified in HTSC compositions. However, as has been known, an intergranular crack remains atomistically sharp when an energy barrier for the nucleation of a dislocation loop at a crack tip is present. This barrier is produced due to a low level of stress intensity at the crack tip in the presence of screening dislocations, stated by the dislocation sources (e.g., such as intergranular boundaries, particles, defects, etc.) [1028]. Thus, the relative strength of energy barriers for dislocation nucleation, produced by the crack tip or/and the other dislocation sources, states whether a crack tip maintains an atomistic sharpness or emits dislocation loops. Obviously, the carbon segregation in HTSC causes a complex effect on the dislocation behavior. Then, the occurrence of the dislocation-screened crack tips is possible in the presence of carbon, depending on how carbon affects the generation of dislocation at the crack tips and the other dislocation sources. Besides the crack tip conditions, the strength of carbon binding at crack surface and intergranular boundaries also controls the carbon embrittlement of the grain boundaries. It is known that the value of $(H_B)_s$ is much higher than that of $(H_B)_b$, and they are found by the conditions at interfaces such as carbon coverage, structure and roughness. Moreover, a high degree of lattice incoherence (e.g., at the interphase boundary [1028]) can also state a higher susceptibility to carbon embrittlement of HTSC. Obviously, there are also other factors controlling the strength of carbon binding at interfaces. So, the values of carbon binding need an accurate experimental foundation for the considered HTSC systems.

The detailed analysis of numerical results for steady-state crack outruns the book. Note, only the parameters corresponding to the obtained ones for slow and fast cracks are found to be dependent on the crack velocity and the carbon diffusivities at grain boundaries and crack surfaces. In particular, the numerical results, obtained for steady-state crack, indicate that the crack growth rate is higher for smaller grain size that is as the grain size decreases the susceptibility to CIIC increases. In total, the above solutions and numerical results can be used in the finite element formulations and other numerical codes by which the stress—strain state distributions, kinetics and parameters of intergranular defects during HTSC bulk manufacture can be predicted.

Chapter 6
General Aspects of HTSC Modeling

The problem of fabrication of the oxide superconductors with high structure sensitive properties suggests as one from primary tasks a design and creation of property monitoring of the HTSC ceramics and composites.[1] This includes (i) observation and modeling of microstructure transformations, causing formation (during processing) and change (during loading) of the material fracture resistance that renders defining influence on superconducting properties; (ii) estimation of change of the microstructure, strength and conducting properties under different loading; and (iii) property prognosis of final product, depending on the sample composition, parameters and features of processing technique. An approximate scheme of numerical monitoring is presented in Fig. 6.1 and includes: (i) specific stages of the superconductor preparation and loading, accompanied by microstructure transformations; (ii) the stages of modeling strength, superconducting and other structure-sensitive properties; (iii) and proper processes, determining the final sample parameters. Note that Fig. 6.1d–f shows modeling in square lattice. However, corresponding realization of the computational scheme in other lattices (i.e., triangular, hexagonal, overlapping elements, etc.) does not present principal difficulties. Table 6.1 shows possibilities of the monitoring and spectrum of superconductor properties, which may be defined at the monitoring realization.

As it has been pointed in the previous chapters, the formation of microstructure defects and weak links during material compaction and sintering renders a significant effect on structure-sensitive properties of HTSC. In this chapter, two important problems are considered, which relate to the technique optimization of HTSC systems, namely the yield criteria are stated, which can describe both displacement into press-powder bulk and its consolidation during compaction [1305, 1306], and the void formation and transformation due to diffusion processes during sintering are studied [1301, 1305]. The proposed yield criteria are based on

[1] For non-cubic Al_2O_3 ceramic, a numerical scheme of the monitoring has been realized in [1303].

I. A. Parinov, *Microstructure and Properties of High-Temperature Superconductors*,
DOI: 10.1007/978-3-642-34441-1_6, © Springer-Verlag Berlin Heidelberg 2012

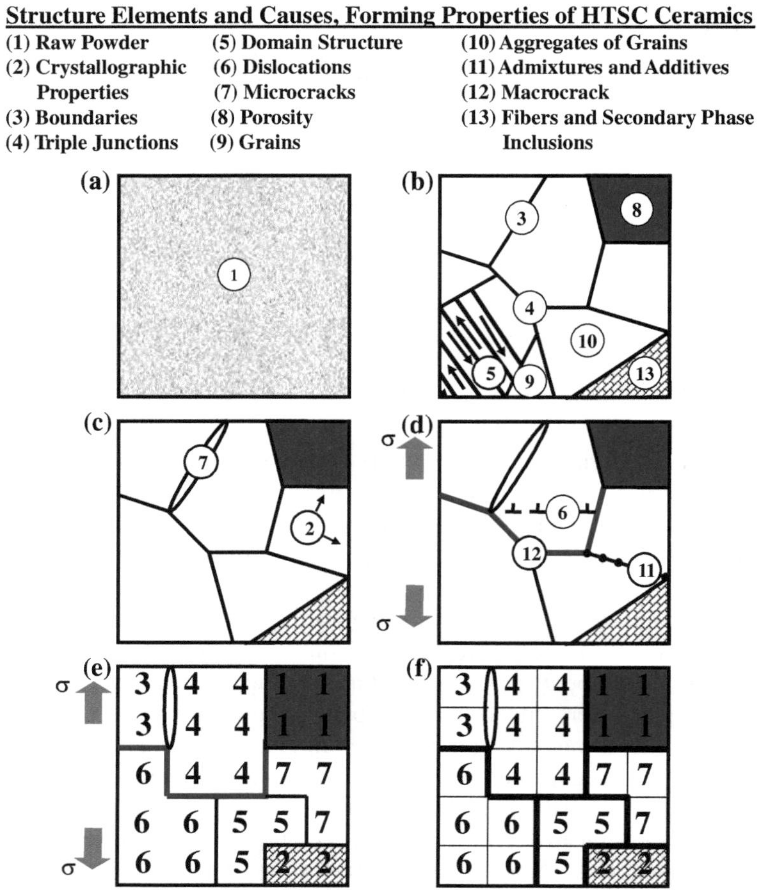

Fig. 6.1 General scheme of computational monitoring of HTSC structure-sensitive properties: (**a**) initial powder (**b**) sintering (**c**) cooling (**d**) macrocrack propagation (**e**) presentation of structure fragment in PC (**f**) model structure for study of percolation properties

the associated and non-associated flow rules. The microstructure transformations of porosity are investigated in the framework of phenomenological models of the shrinkage, coarsening and coalescence of intergranular voids, and also their separation from grain boundaries and displacement to interior of grains is possible. Moreover, there is presented modeling epitaxial films. Second part of the chapter is devoted to modeling micro- and macrostructure processes during processing and fracture of HTSC-ceramic, taking into account heating, shrinkage and cooling of material, grain growth and sample microcracking.

Table 6.1 Possibilities of computer monitoring and spectrum of estimated properties

Investigated processes and effects	Estimated and accounted properties and characteristics
(1) Thermal and mechanical loading (2) Solidification (3) Shrinkage (4) Secondary re-crystallization (grain growth) (5) Phase transitions (6) Superconducting transition (7) Non-superconducting fibers, inclusions and additives into superconducting matrix (8) Superconducting inclusions into non-superconducting matrix (9) Defect structures, in particular carbon effect (10) Dislocation structures (11) Percolation (12) Structure effect on magnetic flux pinning	(1) Elastic anisotropy (2) Thermal expansion anisotropy (3) Deformation and temperature mismatch of phases and components (4) Residual (internal) stresses (5) External physical and mechanical loading (6) Fracture conditions and types (7) Microcracking (8) Phase transformations (9) Twinning and domain re-orientation (10) Crack deflection and twisting (11) Crack branching (12) Crack interaction with structure heterogeneity (pores, dislocations, microcracks, inclusions, etc.) (13) Crack bridging (14) *R*- (or *T*-) curve behavior (15) Pushing of grains (or fibers) by crack surfaces (16) Crack coalescence (17) Crack healing (18) Crack cohesion (19) Scaling factor (20) Effective superconducting properties (21) Percolation properties (22) Magnetic flux pinning

6.1 Yield Criteria and Flow Rules for HTSC Powders Compaction

Obviously, the optimization of the process of the HTSC powders compaction plays important role in the preparation of superconductors with optimum structure-sensitive properties. As it has been demonstrated in the previous chapters, the numerous techniques can be used to prepare oxide HTSC materials and propose an application of the complex thermal and mechanical treatments of powdered precursors (e.g., cold one-axis pressing [28, 1370, 1654], cold isostatic pressing [479, 771, 1370, 1817, 1821], hot isostatic pressing [244], hot forging [1654], etc.). Final aim of the pointed techniques consists in the preparation of strong coupled and aligned structures of grains in the final product. Nevertheless, a formation of microstructure defects and weak links is unavoidable due to intrinsic brittleness of oxide superconductors. In this case, it could be noted that selection of thermal and mechanical loading as rule is badly caused. Therefore, in order to solve the above problem, understanding densification mechanisms and features of HTSC powder deformation is required. The critical mechanical behavior during HTSC powder compaction may be described by using yield criterion and flow rule, taking into account microstructure properties, material parameters, applied loading, etc.

Below, consider the problem of improved compaction technique for powdered HTSC precursors, coupled with selection of optimum yield criterion with the associated (or non-associated) flow rule, which can describe both displacement into bulk and consolidation of the powder during compaction process. Moreover, a validity of the non-associated flow rules for description of HTSC powder compaction is discussed, taking into account the dissipation energy due to particle re-arrangement and friction.

6.1.1 HTSC Compaction and Yield Criterion

Effective application of finite element method and other numerical methods to modeling of microstructure transformations is needed in detail information about material properties, critical mechanical characteristics and applied loading and densification mechanisms. In particular, density distribution, damage initiation and growth depend on the stress–strain state during HTSC powder compaction. In this case, the densification mechanism consists of two processes, namely displacement into powdered volume and plastic deformation of particles. Because the aggregate includes multiple particles, one may be considered as uniform continuum. Then, it is necessary to state a critical condition or a yielding criterion in order to describe the powder strain and densification. Successive selection of the yielding criterion and flow rule are capable of helping in creation of optimum HTSC composition and microstructure. The deformation, caused by sliding of grains, leads to increase in volume during compaction; at the same time it leads to decrease in volume during consolidation of grains. At the so-called critical state (transition point), there is no tendency for volume change. Different yield criteria for free powder filling, porous and granular specimens have been discussed (see, e.g., [736]). Theories of yielding with an increase in density have been developed in [1446]. On the other hand, theories of plasticity with a decrease or no change in density have been discussed in [1604] but stated numerous issues. However, very often, the yield criteria do not take into account the density effect on the sample strain. Moreover, due to volume changes during sample compaction, an effect of hydrostatic pressure should be included into yield criterion. At the same time, the limiting envelopes stated by the Mohr–Coulomb's criterion and similar criteria suggest an infinitely large shear stress to cause slip at compressive stresses. This does not apply to the powder compaction. Therefore, the yield criterion for powder aggregate should be in the form of closed curve (e.g., an ellipse) with asymmetric conditions for compressive and tensile stresses, because the powders are not able to sustain significant tensile stresses. Considering the isotropic case of HTSC powder compaction, a 3D yield criterion has the form:

$$f = \alpha(I_1 + s)^2 + J_2 = \beta Y^2, \tag{6.1}$$

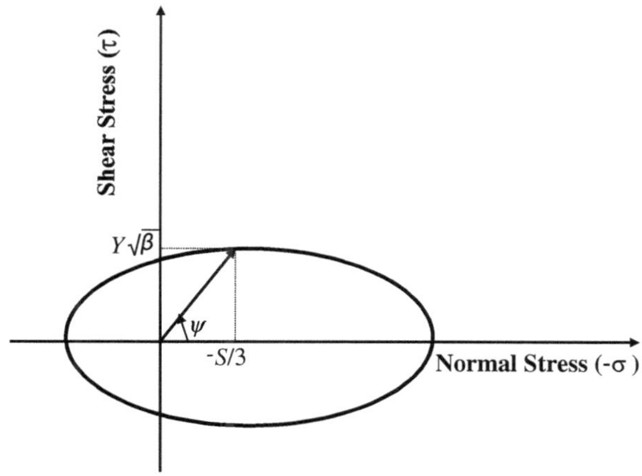

Fig. 6.2 The proposed yield locus for a shear test

where α, β and s are functions of relative density, $I_1 = \sigma_1 + \sigma_2 + \sigma_3$ is the first invariant of the stress tensor, $J_2 = \frac{1}{6}\left[(\sigma_1 - \sigma_2)^2 + (\sigma_2 - \sigma_3)^2 + (\sigma_3 - \sigma_1)^2\right]$ is the second invariant of the deviatoric stress tensor (σ_k are the principle stresses) and Y is the yield stress of the solid material.

Using the yield function, f, as the plastic potential, the associated flow rule is obtained by imposing the normality condition between the plastic strain rate and the yield surface:

$$\dot{e}_{ij}^p = \lambda \frac{\partial f(\sigma_{ij})}{\partial \sigma_{ij}} = \lambda\left[2\alpha(\sigma_{kk} + s)\delta_{ij} + s_{ij}\right], \tag{6.2}$$

where δ_{ij} is Kronecker's delta function, σ_{kk} and s_{ij} are the hydrostatic and deviatoric stress, respectively. A dot above a symbol implies the material time derivative and upper superscript p signs plastic component.

Considering the principal strain rates, $\dot{e}_k^p$, and determining the volumetric strain rate, $\dot{e}^p = \dot{e}_1^p + \dot{e}_2^p + \dot{e}_3^p$, we obtain from (6.1) and (6.2) the positive constant λ, defining the strain amount at a given point, as

$$\lambda = \left[(\dot{e}^p)^2/(18\alpha) + \dot{d}_{ij}^p \dot{d}_{ij}^p\right]^{1/2} \Big/ \left(\sqrt{2\beta}Y\right), \tag{6.3}$$

where the deviatoric deformation rate has the form: $\dot{d}_{ij}^p = \dot{e}_{ij}^p - \frac{1}{3}\dot{e}^p\delta_{ij}$. Then (6.1) can be presented by using normal stress, σ, and shear stress, τ:

$$\frac{(\sigma + s/3)^2}{\left(\sqrt{(1 + 12\alpha)/9\alpha}\sqrt{\beta}Y\right)^2} + \frac{\tau^2}{\left(\sqrt{\beta}Y\right)^2} = 1. \tag{6.4}$$

The material constants α, β and s can be determined by using shear tests [1715]. For this, the yield locus (Fig. 6.2) is described in the σ–τ plane by an ellipse represented by (6.4). Due to the assumption of an isotropic deformation in a shear test, the transition point almost coincides with the apex of the minor axis of the ellipse. Then, the inclined angle, ψ, of the critical line against the abscissa, σ, and the ratio, R, of the major axis to the minor axis of the ellipse can be considered as material constants, and by using (6.4), they have the forms:

$$tg\psi = \sqrt{\beta}Y \Big/ (s/3); \tag{6.5}$$

$$R = \sqrt{(1 + 12\alpha)/9\alpha}. \tag{6.6}$$

Finally, we select the relation between the density of a powder compact and pressure, required to achieve that density, as remaining third equation [658]:

$$KP = \ln[(1 - \rho_0)/1 - \rho], \tag{6.7}$$

where P is the applied pressure, ρ_0 and ρ are the average densities of the loose powder and the compressed powder, respectively, and K is the test constant.

Then taking into account the relation

$$P = s/3 + R\sqrt{\beta}Y, \tag{6.8}$$

the test constants ψ, R and K can be used to define the material constants α, β and s:

$$\begin{aligned} \alpha &= \frac{1}{(9R^2 - 12)}; \; s = \frac{3}{K(1 + R\mathrm{tg}\psi)} \ln \frac{1 - \rho_0}{1 - \rho}; \\ \beta &= \frac{1}{Y^2} \left[\frac{\mathrm{tg}\psi}{K(1 + R\mathrm{tg}\psi)} \ln \frac{1 - \rho_0}{1 - \rho} \right]^2. \end{aligned} \tag{6.9}$$

A validity of the proposed yield criterion for description of the HTSC powder compaction can be verified by using, for example, a 3D compaction test (Fig. 6.3a), in which three pairs of anvils compress the powder in three mutually perpendicular directions. However, note that tests of granular materials (e.g., soils) or cemented materials (e.g., concrete or rock), presented in Fig. 6.3b and *c*, have shown that associated flow rule cannot depict test data satisfactorily [1748]. At the same time, satisfactoriness of associated (or non-associated) plasticity has not been stated for compacted HTSC powders. Therefore, it is very important to additionally consider non-associated flow rules for which the plastic strain rate is not orthogonal to the yield surface.

6.1.2 Non-Associated Plasticity of HTSC Powders

Base hypothesis of normality [398], which forms a basis of associated plasticity and is very successful in the description of metallic compositions, can be broken for non-metallic granular materials in free filling, in particular for HTSC powders.

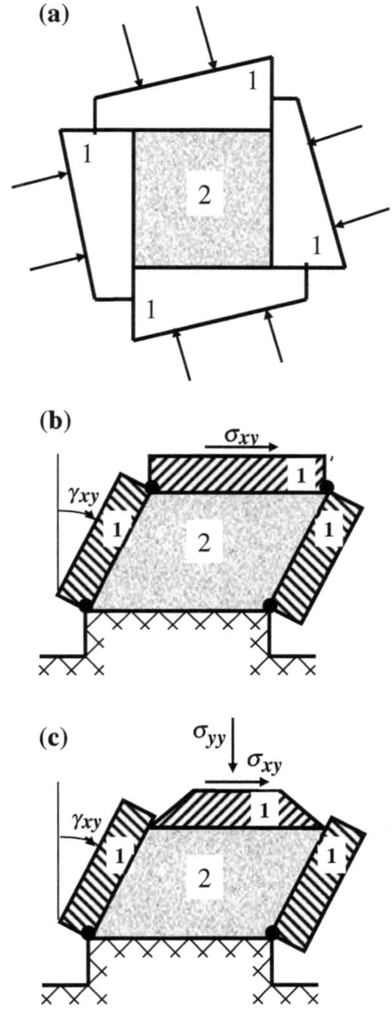

Fig. 6.3 Different schemes of loading for triaxial compression. Third pair of anvils compresses the powder in the direction perpendicular to the figure plane. The numbers sign anvils (*1*) and powder (*2*). *Arrows* show applied loads

The key circumstance is that the associated plasticity cannot validly describe shear dilatancy of granular material (i.e., the change in volume that is associated with shear distortion of an element in the material, consisting of multiple particles as microelements). As has been shown [1748], the material, retaining constant volume under plasticity, reacts otherwise on loading, demonstrating plastic expansion. The differences are related to both the "load–deformation" curve and the value of critical loading. In order to characterize a dilatant material, dilatancy angle Ψ is introduced, presenting the ratio of plastic volume change to plastic shear strain.

An ideal triaxial test should permit independent control of all three principal stresses so that general states of stress could be examined. Typical results in a standard triaxial test of granular material in free filling are shown in Fig. 6.4.

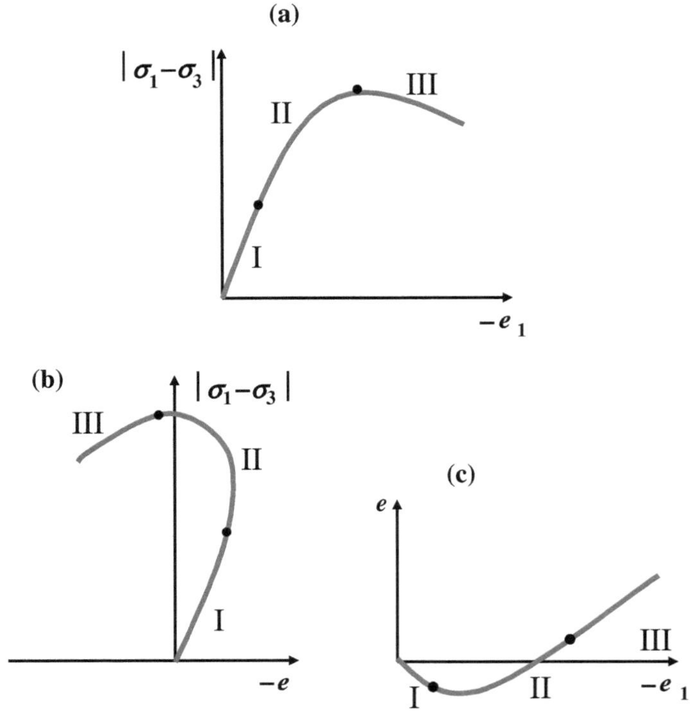

Fig. 6.4 Typical triaxial test results for loose granular material

Fig. 6.5 Sliding between groups of particles, leading to dilatation

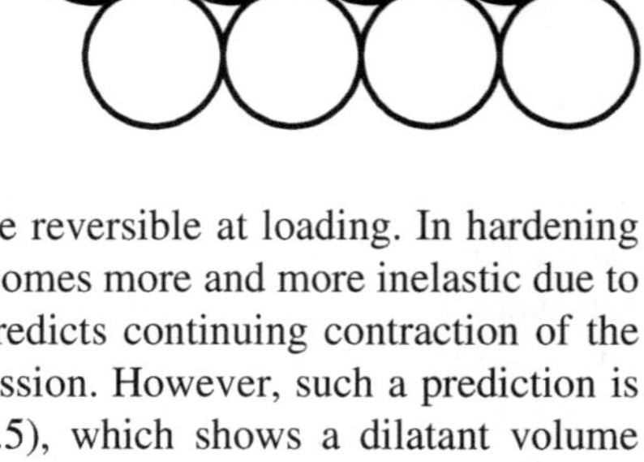

In elastic region (I), as usually, the strains are reversible at loading. In hardening regime (II), the strain of granular material becomes more and more inelastic due to particle sliding. Here, non-linear elasticity predicts continuing contraction of the specimen under continued loading in compression. However, such a prediction is disproved by experimental evidence (Fig. 6.5), which shows a dilatant volume increase at subsequent loading [1748]. This phenomenon takes place due to frictional sliding along particles. The elastic strain rate in the hardening regime is almost zero. Moreover, there exists a linear relation between the volume change and the change of the axial strain near the end of the hardening regime (II) and in the softening regime (III) (see Fig. 6.4c). Then, following [1748], the constant dilatancy angle, Ψ, may be introduced as

$$\sin \Psi = \frac{\dot{e}^p}{-2\dot{e}_1^p + \dot{e}^p}. \tag{6.10}$$

Equation (6.10) defines constant rate of dilatation, and it is valid under conditions of triaxial compression. For broad row of granular materials, the dilatancy angle, Ψ, approximately equals 0–20°; at the same time, the internal friction angle, $\Phi = 15$–$45°$ [1748]. Because the dilatancy angle can be considerably smaller than internal friction angle, it is necessary to apply for considered materials the non-associated flow rules. The function of plastic potential, g, used to study plasticity of granular media, coincides nearly with the yielding function, f, applied to separate plastic and elastic states. Difference between these functions consists in that the internal friction angle, Φ, in the equation for f is replaced by the dilatancy angle, Ψ, in relation to g. Thus, in order to state a validity of the associated or non-associated flow rules for HTSC powder compaction, it is necessary to define the dilatancy angle, Ψ, from (6.10) and to compare its value with the friction angle, Φ.

After a pick value in the "stress–strain" curve (see Fig. 6.4a) is reached, unstable behavior of material arises in the softening regime (III), in particular caused by thin shear bands, which separate the specimen in two more or less rigid bodies. For such macroscopic non-uniform deformations, connected with bulging of specimen in triaxial compression test, the strain increment is not measured correctly. At the same time, the strain rate ratio is not so strongly affected by the localization into a shear band. The axial strain–volumetric strain curve of Fig. 6.4c is much more informative, but the dilatancy angle, Ψ, can be measured with acceptable accuracy.

Violation of the normality law, caused by actual resistance to deformation, comes from two sources: (i) the dilatation of the bulk material during shear enhances the yield stress under conditions of compression—due to this it is necessary to do a work against the applied pressure and (ii) the frictional dissipation of energy at the contact patches between the granules also enhances the yield stress under applied loading. Below, we will not take into account the work executed during material hardening and softening and also the localization processes at definition of yield surface. Then, assuming that the material is the strain rate independent of small deformation, the strain rate is split up into elastic and plastic components. Each of these strain rates can be also split up into volumetric, $\dot{e}$, and deviatoric, $\dot{d}_{ij}$, components. Consider a typical dilatancy rule for rigid particles caused by the material volume expansion at granule rearrangement [252, 254]:

$$\dot{e}^p = \nu \left(\dot{d}_{ij}^p \dot{d}_{ij}^p \right)^{1/2}, \tag{6.11}$$

where the constant of proportionality, ν, is a generalization of the dilatancy angle, Ψ. For granules of finite strength, some deformation occurs at the contact patches, so restricting this expansion. Hence, it may be assumed that energy is dissipated during this damage at the rate:

$$\dot{D}_1 = l\left[v\left(\dot{d}^p_{ij}\dot{d}^p_{ij}\right)^{1/2} - \dot{e}^p\right], \tag{6.12}$$

where parameter l is proportional to the strength of the granules and the size of the contact patches between granules. Then, the deformation of granular material caused by the granules rolling and sliding defines the energy dissipated due to friction at the contact patches. The rate of this energy can be approximated by

$$\dot{D}_2 = -\mu\sigma\left(\dot{d}^p_{ij}\dot{d}^p_{ij}\right)^{1/2}, \tag{6.13}$$

where μ is some measure related to the friction factor. Note that both energy dissipation rates are obtained taking into account that (i) there can never be negative dissipation function for all non-zero values of strain increment and (ii) the resistance to deformation is rate-independent.

During densification, there are deformation and/or rearrangement of particles depending on the ratio of shear stress to compression. Bearing both mechanisms in mind, a suitable dissipation function might be

$$\dot{D} = \left\{\mu^2\sigma^2\dot{d}^p_{ij}\dot{d}^p_{ij} + l^2\left[v\left(\dot{d}^p_{ij}\dot{d}^p_{ij}\right)^{1/2} - \dot{e}^p\right]^2\right\}^{1/2}. \tag{6.14}$$

The rate working per unit volume, $\dot{W}$, by the boundary tractions is reduced by Gauss's theorem and the condition of equilibrium to

$$\dot{W} = \sigma_{ij}\dot{e}_{ij}. \tag{6.15}$$

This energy input can be stored as an increase in elastic energy rate, $\dot{U}$, or dissipated, $\dot{D}$. Energy conservation gives $\dot{W} = \dot{U} + \dot{D}$. Splitting the deformation rate into elastic and plastic components, we obtain

$$\sigma_{ij}\dot{e}^p_{ij} - \dot{D} = 0. \tag{6.16}$$

From Euler's theorem on homogeneous functions, it is clear that

$$\sigma_{ij} = \frac{\partial\dot{D}}{\partial\dot{e}^p_{ij}}. \tag{6.17}$$

Equations (6.14) and (6.17) provide an implicit flow rule, which would satisfy the energy balance:

$$s_{ij} = \frac{\dot{d}^p_{ij}}{\dot{D}}\left\{\mu^2\sigma^2 + \frac{vl^2}{\left(\dot{d}^p_{ij}\dot{d}^p_{ij}\right)^{1/2}}\left[v\left(\dot{d}^p_{ij}\dot{d}^p_{ij}\right)^{1/2} - \dot{e}^p\right]\right\}; \tag{6.18}$$

$$\sigma = \frac{l^2}{\dot{D}}\left[\dot{e}^p - v\left(\dot{d}^p_{ij}\dot{d}^p_{ij}\right)^{1/2}\right]. \tag{6.19}$$

The yield surface can be found in this case by eliminating the strain increments in the energy balance relation (6.16) using (6.18) and (6.19). We obtain

$$\frac{\left[\left(s_{ij}s_{ij}\right)^{1/2}+\sigma v\right]^2}{\sigma^2\mu^2} + \frac{\sigma^2}{l^2} - 1 = 0. \tag{6.20}$$

Again, using (6.18) and (6.19), the flow rule can be derived in terms of the components of stress and the overall rate of energy dissipation:

$$\dot{d}^p_{ij} = \dot{D}s_{ij}\frac{\left[\left(s_{ij}s_{ij}\right)^{1/2} + v\sigma\right]}{\left(s_{ij}s_{ij}\right)^{1/2}\mu^2\sigma^2}; \tag{6.21}$$

$$\dot{e}^p = \dot{D}\sigma\left\{\frac{1}{l^2} + \frac{v\left[\left(s_{ij}s_{ij}\right)^{1/2}+v\sigma\right]}{\mu^2\sigma^3}\right\}. \tag{6.22}$$

Equations (6.21) and (6.22) predict that at small σ/l the granules behave in an almost rigid manner, but as $\sigma/l \approx 1$, compaction takes place. This is in accordance with experiments for granular materials [960]. From (6.21), it follows that these flow rules give sensible results as long as

$$\left(s_{ij}s_{ij}\right)^{1/2} + v\sigma > 0. \tag{6.23}$$

Successive development and application of constitutive models for HTSC powder compaction should be based on the three-level hierarchical structure of the models. At first step, the non-hardening perfectly plastic model under consideration provides an excellent introduction to the modeling of the HTSC powders under compression, taking into account the particle cohesion and friction. Second step includes consideration of the isotropic hardening models together with study of isotropic softening and estimation of damage variables, based on ideal plasticity, taking into account non-associated behavior. Finally, in the third step, anisotropic hardening and softening should be studied together with modeling of material instability that leads today to numerous difficulties in the application of finite element codes and other computational methods for numerical solution of corresponding boundary-value problems.

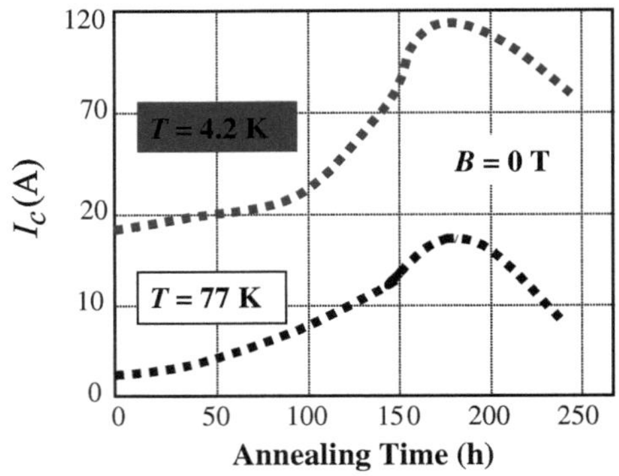

Fig. 6.6 Annealing time dependence of the self-field I_c at 4.2 K and 77 K [1091, 1092]

6.2 Void Transformations during Sintering of Sample

The experimental studies of the sintering duration effect on superconducting properties of monocore Bi-2223/Ag tapes [1091, 1092] have shown that critical current, at first, increases together with annealing time, reaches maximum value and then decreases with reaction time (Fig. 6.6). Monotonous enhancement of weight losses of the sample with simultaneous diminution of Pb content and increase of the Bi-2223 phase homogeneity (Figs. 6.7 and 6.8) at remaining constant average size of superconducting grains about 25 μm has been observed, which has not depended on the annealing duration (in the range of 40–200 h). This behavior of critical current has been related by the authors to acting during calcining competing mechanisms, which initially improve the quality of grain boundaries (or weak links),[2] and then decrease the pinning properties of the superconductor. A decrease in critical current has been explained by lead expelling during annealing and by the corresponding degradation of the pinning strength. At the same time, as rule, during BSCCO/Ag processing, there are significant processes of the void formation and other microstructure transformations due to CO_2 release, in particular leading to bubbles arising and the critical current decrease. As it has been noted in Sect. 5.2, 200 ppm of carbon can cause about 36 % porosity in the core of Bi-2212/Ag tapes if all carbon forms CO_2 at high temperature [1870].

Though an average grain size observed in monocore Bi-2223/Ag tapes has remained constant during various duration of annealing, the *grain size distribution* could be altered during reaction. This is supported by considerable lead expelling monotonously increasing with annealing time. But then these additions heterogeneously distributed into the material can inhibit only a local grain growth in the Bi-2223 core. Moreover, as it has been shown in the modeling of fracture

[2] At relatively badly coupled structure of grains, magnetic flux can penetrate intergranular boundaries and trap pores or secondary phases. This non-superconducting volume will be effectively screened in the sample with better intercrystalline properties, and corresponding fraction of superconducting volume will be enhanced.

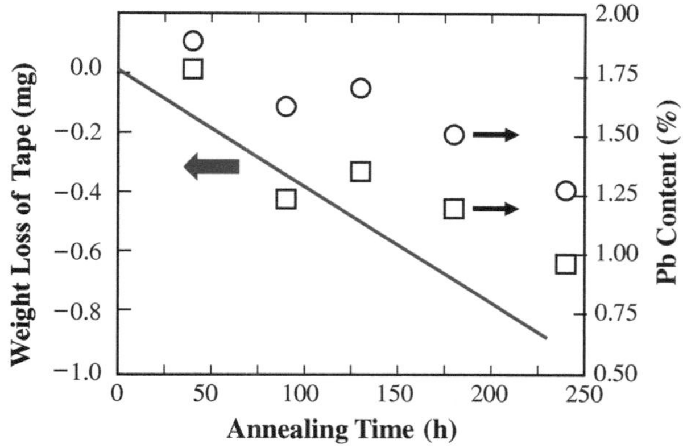

Fig. 6.7 Weight loss of total sample and the lead concentration (in atomic percent) versus annealing time. The circulars sign values averaged on several single grains and squares show measurements on the large region of about 400 grains [1091, 1092]

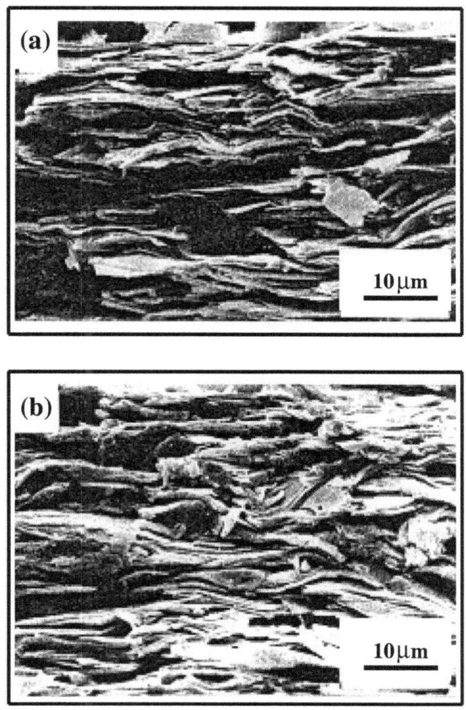

Fig. 6.8 SEM micrographs of a typical transverse fracture surface after (**a**) 40 h and (**b**) 240 h annealing time [1092]

processes in YBCO, *the size distributions of microstructure elements* play more important role to compare with average sizes [1326]. Hence, it may be assumed that the observed preservation of the *average* grain size in different annealing times cannot assert an absence of grain growth and other microstructure alterations. Therefore, another mechanism of diminution of critical current during prolonged annealing is possible, namely inevitable transformation of pores attached to intergranular boundaries, which can lead to the formation of significant closed porosity. Consider some regimes of change of the void space: their displacement, shrinkage, coarsening, coalescence and possible separation from grain boundaries to the interior of the grains that directly define useful properties of superconductor.

6.2.1 Void Separation from Intergranular Boundary

In discussion of pore breakaway processes, first, it should be noted that pores attached to grain boundaries decrease due to boundary-grain diffusion. However, when a pore separates from grain boundary and locates into the grain, one can shrink only thanks to much more slow diffusion process of the crystalline lattice. Total separation of pores occurs after their displacement at the boundaries between two grains (Fig. 6.9). Therefore, preliminary displacements of pores from triple junctions of grain boundaries to interfaces of two grains must precede to total pore breakaway.

The phenomenological analysis is essentially based on solutions for the interaction of a boundary with a rigid second-phase particle. Specifically, the interaction between a pore and a moving grain boundary assumes a spherical pore moving in isotropic, homogeneous material with a rate determined by the surface diffusion coefficient (into the framework of surface diffusion mechanism,

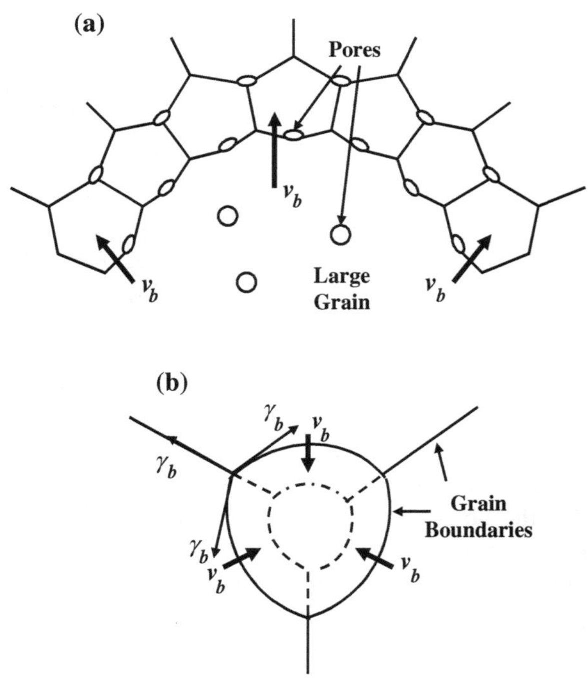

Fig. 6.9 (**a**) A schematic of the grain disappearance process involved in grain growth of large grain at the expense of five-sided small grains located in its periphery; (**b**) the three-sided configuration associated with ultimate grain disappearance

appropriating to small pores). Then, unique pore mobility is derived, retaining the spherical symmetry of the pore (i.e., neglecting the changes in pore shape needed to maintain the atom flux over the pore surface). The corresponding pore mobility is given by [1555]

$$M_p = \frac{D_s \delta_s \Omega}{kT\pi a_0^4}, \tag{6.24}$$

where $D_s\delta_s$ is the surface diffusion parameter, a_0 is the pore radius and Ω is the atomic volume. The force F exerted by the grain boundary on the pore (which eventually dictates separation) is derived from the assumption that the contact line between the boundary and the pore can move freely over the pore surface. The force has a maximum value:

$$F_{max} = \pi a_0 \gamma_b. \tag{6.25}$$

Equations (6.24) and (6.25) yield a peak pore rate as

$$v_p^0 = \frac{\Omega D_s \delta_s \gamma_b}{kTa_0^3}. \tag{6.26}$$

The pore breakaway is considered to occur when the grain boundary rate exceeds this peak pore rate.

Following [716], consider the motion of pores with grain boundaries, taking into account a flux of atoms from the leading to the trailing surface of the pore (see Fig. 6.10). The driving force for the atom flux is caused by the existence of a gradient in the curvature of the pore surface, that is, pore distortion is a necessary consequence of pore motion. The configuration selected is an axisymmetric pore, moving due to surface diffusion. Initially, steady-state motion (all points of the surface move with the same rate) is considered. The axisymmetric pore exhibits two curvatures (see Fig. 6.11): an in-plane curvature, k_1, and an axisymmetric curvature, k_2. These curvatures are related to the coordinates of the problem (x, y) by

$$k_1 = \left(d^2y/dx^2\right)\left[1 + (dy/dx)^2\right]^{-3/2}; \tag{6.27}$$

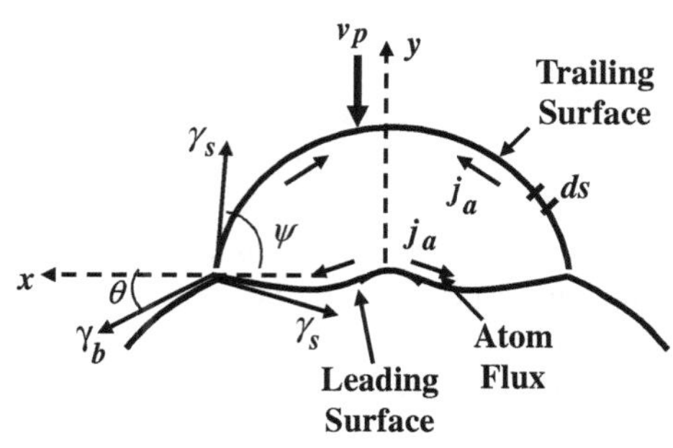

Fig. 6.10 A schematic of a moving pore, indicating the atom flux and the inclination of the grain boundary, θ

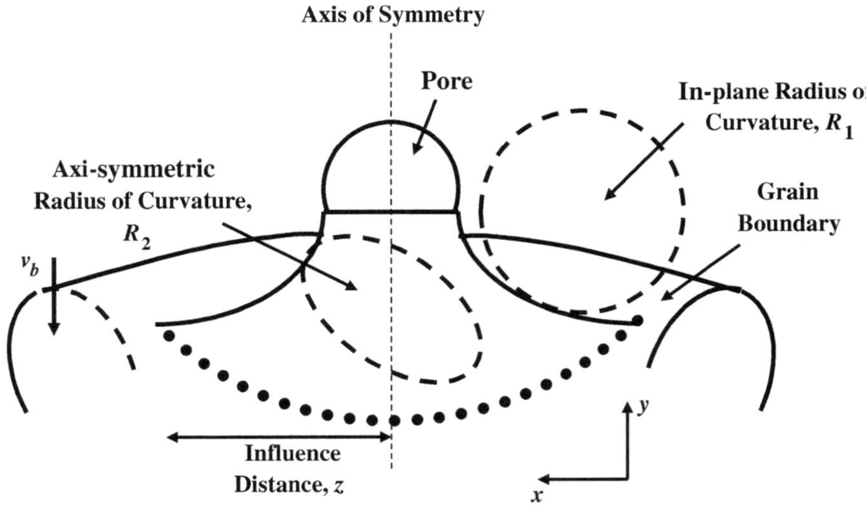

Fig. 6.11 The axisymmetric configuration associated with pore drag, illustrating the important curvatures, the influence distance and the pore and grain boundary rates

$$k_2 = (1/x)(dy/dx)\left[1 + (dy/dx)^2\right]^{-1/2}. \tag{6.28}$$

The flux J_s of atoms in the presence of these curvatures is given by

$$J_s = -\left(\frac{D_s\delta_s\gamma_s}{kT}\right)\frac{d(k_1 + k_2)}{ds}, \tag{6.29}$$

where ds is an element of pore surface in the flow direction (see Fig. 6.10), γ_s is the surface energy (assumed isotropic). For a pore moving with a rate, v_p, in the y-direction, conservation of matter requires that

$$2\pi x J_s = \pm\pi x^2 v_p/\Omega, \tag{6.30}$$

where the positive sign refers to the leading surface and the negative sign to the trailing surface.

By using the symbols, p and q ($\equiv dy/dx$) for the slopes of the trailing and the leading surfaces, respectively (6.27–6.30) lead to two differential equations of second order. The motion of the pore is subject to the requirement that the total dihedral angle, ψ, between the grain boundary and the pore surfaces (Fig. 6.10) be invariant. Moreover, it is necessary that the chemical potential be continuous at the intersection of the leading and trailing surfaces. Finally, the atom flux and surface slope must be zero at the axis of symmetry. The solution of the differential equations at the pointed conditions is a non-linear problem. A solution is obtained by linearizing a trial solution and then using a finite difference scheme [716].

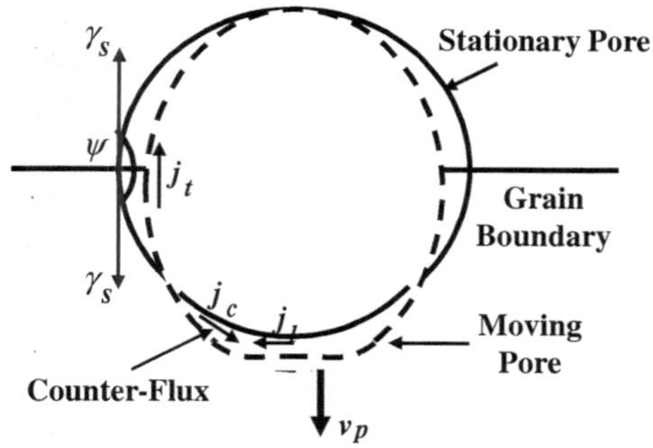

Fig. 6.12 A schematic illustrating the development of a counter flux with $\psi = \pi$ is distorted to achieve a net atom flux and hence pore motion with the grain boundary

The inclination, θ, of the grain boundary tangent to the plane of contact between the grain boundary and the pore emerges from the analysis as a unique function of the dihedral angle, ψ, and normalized pore: $v_p = (v_p/v_p^0)(\gamma_b/\gamma_s)$. The radius of contact, a, between the boundary and the pore can also be deduced by requiring that the pore volume be independent of pore rate in order to permit a unique comparison between the dimensions of the stationary and moving pores.

Convergent pore shape solutions are found to exist over a limited range of pore rates, implying the existence of a steady-state rate maximum, v_p^m. Its existence is associated with an ability to simultaneously satisfy the requirements that (i) the dihedral angle is specified (ii) the curvature is continuous and finite, and (iii) the pore rate is uniform. The consequences of increasing the rate greater than v_p^m leading to violation of one of these imposed conditions, can be presented when the dihedral angle $\psi \approx \pi$ (Fig. 6.12). Shape changes, which induce a continuous atom flux in the requisite direction for pore motion (i.e., a continuous gradient in surface curvature), cannot be constructed due to an inevitable counter-flux of atoms. Thus, steady-state motion by surface diffusion of a pore with $\psi \sim \pi$ is impossible.[3] The steady-state rate maximum may be required as the equivalent of the peak rate, v_p^0, given by (6.26), and can be expressed as [716]

$$v_p^m = v_p^0(17.9 - 6.2\psi)/2\cos(\psi/2). \tag{6.31}$$

In this case, for all reasonable choices of the dihedral angle, the peak steady-state rate exceeds the value anticipated by the phenomenological model.

Non-steady-state solutions, in which the rate varies over the pore surface (with a maximum at the axis of the leading surface), can be found for net rates in excess of v_p^m. These solutions coincide with a marked change in pore shape and a decreasing in the grain boundary contact radius, a, as shown by the shape depicted in Fig. 6.13. It is presumed that, in this case, the contact radius a will rapidly diminish to zero,

[3] The inability of a pore with $\psi \sim \pi$ to exhibit steady-state motion implies that such pores will *always* detach from grain boundaries. This is intuitively clear because when $\psi = \pi$, the grain boundary energy is zero (surface energies are always finite) and there is no preference for pores to locate on grain boundaries.

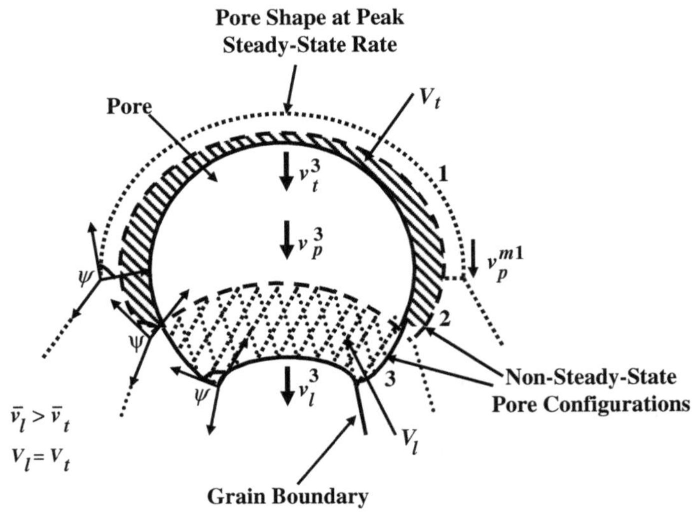

Fig. 6.13 Schematic illustration of the pore shapes under non-steady-state conditions. For the pore volume to remain constant, the matter removed from the leading surface, V_l, must equal the matter deposited on the trailing surface, V_t. Moreover, under non-steady state conditions, the average rate of the leading surface, $\langle v_l \rangle$, must exceed that for the trailing surface, $\langle v_t \rangle$. In order to satisfy these requirements, the contact radius, a, must decrease rapidly with increase in net pore rate

causing the grain boundary to converge onto the pore axis and to initiate breakaway. Hence, the upper bound steady-state pore rate can be used as a rate, which, if exceeded, will inevitably result in non-steady-state pore motion and breakaway.

It may be shown that the grain boundary, subjecting to pore drag, exhibits a rate component normal to the axis of symmetry, indicative of a tendency toward instability [716]. Therefore, when the grain boundary rate, v_b, exceeds v_p, the axisymmetric radius of curvature, R_2 (see Fig. 6.11), decreases during the motion of the pore–grain boundary configuration, especially within the immediate vicinity of the pore. The decrease in R_2 exceeds the change in R_1, and therefore, the configuration is intrinsically metastable, but steady state of the complete pore/grain boundary ensemble is impossible. The pore drag will increasingly distort the boundary and eventually induce separation. It is interesting to note that grain boundary coalescence is likely to result in a dislocation or sub-grain boundary attached to the pore (especially in the presence of some symmetry) [365, 851].

The specific condition that dictates the separation of pores from grain boundaries depends on the grain configuration to which the pore is attached. Consider two configurations that occur during grain growth,[4] namely (i) a pore located on

[4] Other configurations, such as several pores on single grain boundary, will provide different separation conditions. However, these separation events usually occur after the first separations have been induced and are, probably, less critical. For example, a modified phenomenological

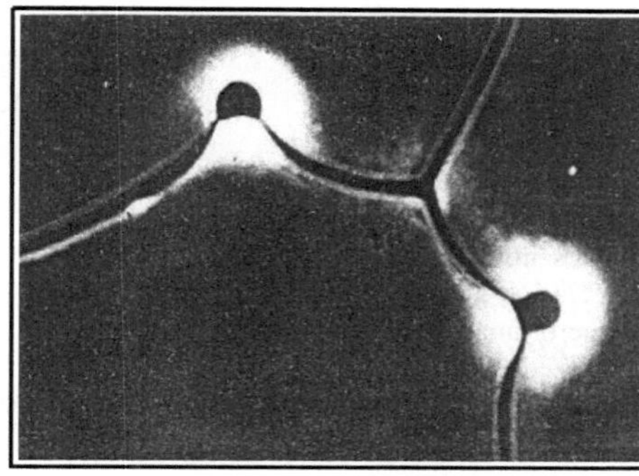

Fig. 6.14 Pores in MgO subjected to motion with the grain boundary [716]

one boundary of a three-sided grain (Fig. 6.9b) (the configuration of grain that invariably precedes grain disappearance) and (ii) a pore located on one boundary of a five-sided grain at the perimeter of a large grain subjected to exaggerated grain growth (Fig. 6.9a). Pore drag observations (Fig. 6.14) indicate that the pore perturbs grain boundary motion over a certain influence distance, z (see Fig. 6.11). Therefore, it is appropriate to examine the motion of the grain boundary outside this influence distance relative to that of the pore (obviously, it is assumed that the influence distance is less than the grain radius). The rate of the grain boundary outside the influence zone in the direction of pore (Fig. 6.11) is [716]

$$v_b = \sqrt{3}\gamma_b M_b \Omega^{2/3}/R. \tag{6.32}$$

By allowing this rate to exceed the peak steady-state pore rate, v_p^m, separation should be inevitable, provided a grain boundary displacement (i.e., grain size), sufficient to create boundary convergence at the dragging pore, is available. This condition yields a pore size for separation given by

$$a_0^2 \geq \left(\frac{R}{a_0}\right)\left(\frac{\Omega^{1/3}D_s\delta_s\gamma_s}{kTM_b\gamma_b}\right)\frac{(17.9 - 6.2\psi)}{\sqrt{3}}. \tag{6.33}$$

For the five-sided grain, R remains essentially constant (Fig. 6.15). However, the most stringent condition for breakaway exists when R attains its *smallest* value, which corresponds to the three-sided grain configuration, preceding grain disappearance (Fig. 6.9b). In this case, separation is averted if the pore converges onto the prospective three-grain junction ($R \sim 2a_0$) before (6.33) can be satisfied. Thus, the critical condition becomes

(Footnote 4 continued)

analysis with a pore $\sim R$ gives the two limiting cases [716]: $R \leq 2M_b kT\gamma_b a_0^3 \big/ D_s\delta_s\gamma_s\Omega^{1/3}$ $(17.9 - 6.2\psi)$, and $R \geq \delta_s\gamma_s\Omega^{1/3}(17.9 - 6.2\psi)/2\gamma_b a_0$.

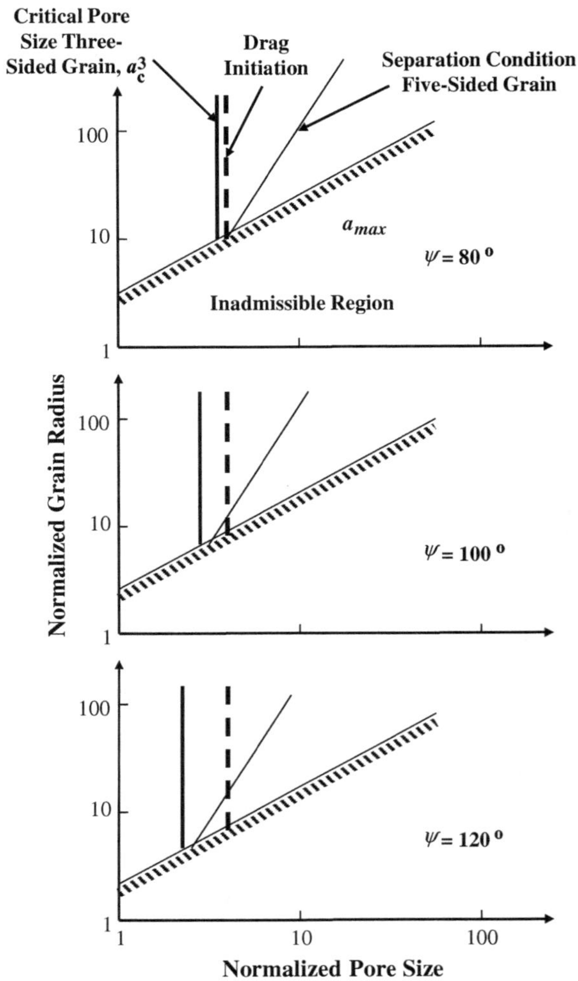

Fig. 6.15 The various separation and admissibility conditions, identified in the analysis of [716], are plotted for three values of the dihedral angle. The abscissa axis presents normalized pore size $(kTM_b\gamma_b/\Omega^{1/3}D_s\delta_s\gamma_s)^{1/2}a_0$; the ordinate axis shows normalized grain size $(kTM_b\gamma_b/\Omega^{1/3}D_s\delta_s\gamma_s)^{1/2}R$

$$\left(a_0^2\right)_c = \left(\frac{\Omega^{1/3}D_s\delta_s\gamma_s}{kTM_b\gamma_b}\right)\frac{2(17.9 - 6.2\psi)}{\sqrt{3}}, \tag{6.34}$$

as shown in Fig. 6.15. This critical pore size represents a lower bound for pore separation at all reasonable values of a_0/R. Obviously that obtained critical value increases with increasing of surface diffusion or with increasing of grain boundary mobility or dihedral angle.

HTSC materials demonstrate a spectrum of dihedral angles connected, in the first place, with various grain boundaries formed during grain growth. Then, there is a range of the pore critical sizes that the very fine pores can localize at low angle or special boundaries, in the sites of coincidence of the crystalline lattices. Similarly, a distribution of the surface diffusion parameters and grain boundary mobility due to inhomogeneity of admixture distribution should exist. Hence, different spectrum of the pore critical sizes will correspond to various superconducting systems and should be taken into account in consideration with the pore breakaway. The microstructure changes, suppressing the pore separation, should be estimated in comparison with the pore critical size, with the trajectories of alteration of the pore and grain sizes at the final stage of sintering. These processes will be examined in detail in the next section.

6.2.2 Size Trajectories in the Pore/Grain Boundary System During Sintering

The sintering process is generally accompanied by grain growth, pore shrinkage and pore coalescence [300, 1827]. These concomitant processes result in pore/grain size trajectories that typically entail pore size enlargement during the intermediate stage and pore size reduction prior to final densification Moreover, exaggerated grain growth may initiate as a result of the pore separations from grain boundaries [233, 234]. Therefore, it is very important to study the pore/grain size trajectory, especially with reference to the critical size for pore breakaway. In this section, following [1601], concurrent grain growth, pore shrinkage and coalescence are analyzed and compared to the maximum pore size achieved during densification, with the critical pore size at pore–grain boundary separation. Mutual dependencies between the grain and pore sizes, caused by concomitant grain growth, pore coalescence and pore shrinkage, can be used together with expressions for the grain boundary rate and the pore shrinkage rate to establish features of the pore/grain size trajectories during sintering.

It is assumed that pores exist at each three-grain junction (see Fig. 6.16). Subject to this condition, pores coalescence during the disappearance of small grains results in a simple magnification of the microstructure, while concomitant

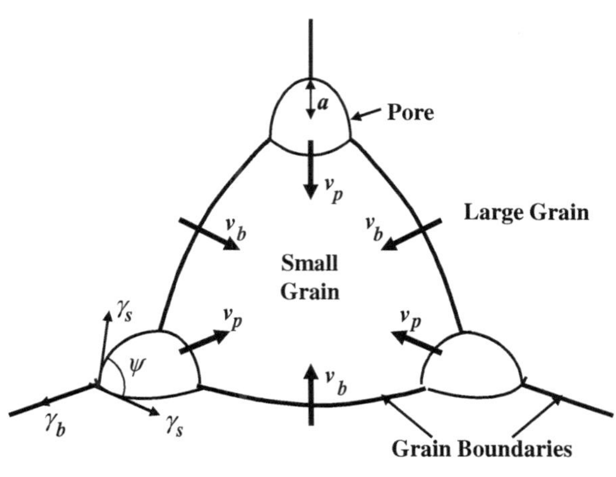

Fig. 6.16 A schematic indicating the grain and pore configurations used to calculate pore drag effects

pore shrinkage reduces the size of the pores relative to the grain size. Thus, the relative change in pore volume comprises an increase due to coarsening and a decrease due to shrinkage given directly by

$$\frac{dV_p}{V_p} = \frac{dV_g}{V_g} - \frac{\left|dV_p^s\right|}{V_p}, \tag{6.35}$$

where dV_p is the resultant volume change of the average pore in a small time interval, dt; dV_g is the corresponding grain volume change; dV_p^s is the absolute pore volume change due to shrinkage; and V_p and V_g are the average pore and grain volume, respectively. Dividing (6.35) throughout by the time interval dt, and rearranging, we obtain

$$\frac{dV_p}{dV_g} = \frac{V_p}{V_g} - \frac{\left|\dot{V}_p^s\right|}{\dot{V}_g}, \tag{6.36}$$

where a dot above a symbol signs the time derivation. Herein, it will be assumed that the pore coarsening will be balanced by the shrinkage, that is, $dV_p/dV_g = 0$. In particular, this condition defines the pore volume, V_p^+, at which a transition from pore coarsening to pore shrinkage occurs during normal sintering (i.e., when the average number of pores per grain remains). This is given by

$$V_p^+ = \frac{V_g\left|\dot{V}_p^s\right|}{\dot{V}_g}. \tag{6.37}$$

Taking into account the relations:

$$V_g \propto l^3;\ \dot{V}_g \propto 3l^2\dot{l};\ V_p \propto a^2 l;\ \dot{V}_p \propto a^2\dot{l} + 2l\dot{a}a;\ \dot{l} = \lambda v_b, \tag{6.38}$$

where l is the grain facet length, a is the pore size, v_b is the grain boundary rate and λ is a grain shape coefficient $\left(1/3\sqrt{3}<\lambda<1/\sqrt{3}\right)$ [685, 1827]. Substitution of (6.38) into (6.36) and (6.37) yields a coarsening pore rate as

$$da/dl = a/l - |\dot{a}^s|/\lambda v_b, \tag{6.39}$$

and a transition pore size as

$$a^+/l = |\dot{a}^s|/\lambda v_b. \tag{6.40}$$

6.2.2.1 Pore Rate

A moving pore, located at a three-grain junction, must exhibit a distortion with general characteristics (depicted in Fig. 6.17) contingent upon the next

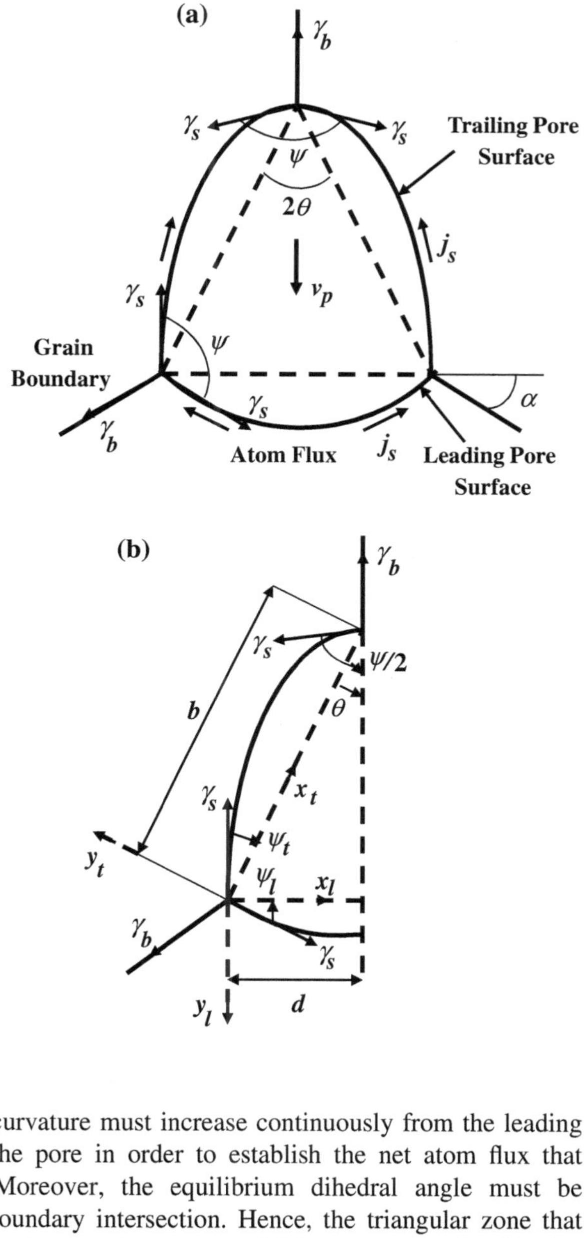

Fig. 6.17 Schematics of the geometry of moving pores at the grain junctions: (**a**) the pore distortion (**b**) the coordinate system (x, y) used for analysis of the distortion of the leading and trailing surfaces

requirements. The surface curvature must increase continuously from the leading to the trailing surface of the pore in order to establish the net atom flux that constitutes pore motion. Moreover, the equilibrium dihedral angle must be maintained at each grain boundary intersection. Hence, the triangular zone that connects the grain boundary intersections must exhibit an apex angle, 2θ, that diminishes as the pore rate increases. This angular change influences the orientation of the grain boundary tangent at each pore intersection, increasing its curvature and reducing the permissible grain boundary rate (Fig. 6.18).

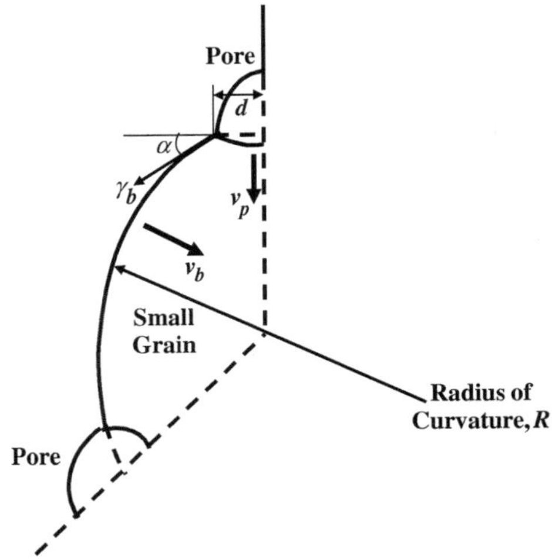

Fig. 6.18 The grain boundary parameters used to determine the grain boundary rate

A linearized analysis of surface diffusion (see Fig. 6.17b) provides an adequate solution as the deviations of the surface tangent from the ordinate are sufficiently small that the following approximations apply:

$$ds \approx dx,\ 1/r \approx d^2y/dx^2, \tag{6.41}$$

where ds is an element of surface and r is the radius of curvature of the surface. Subject to this linearization, the atom flux can be expressed as [289]

$$j_s = -\frac{D_s\delta_s}{kT}\left[\frac{d}{dx}(\gamma_s/r)\right] = -\frac{D_s\delta_s\gamma_s}{kT}\left(\frac{d^3y}{dx^3}\right), \tag{6.42}$$

and the atom flux gradient (based on matter conservation) becomes [289]

$$\frac{dj_s}{dx} = \frac{v_n}{\Omega}, \tag{6.43}$$

where $D_s\delta_s$ is the surface diffusivity, γ_s is the surface energy, Ω is the atomic volume and v_n is the component of the pore rate in the y-direction. Combining (6.42) and (6.43), the governing differential equation is given by

$$\frac{d^4y}{dx^4} = -\frac{v_n kT}{D_s\delta_s\gamma_s\Omega}. \tag{6.44}$$

This equation can be solved for the leading and trailing pore surfaces subject to the next boundary conditions (see Fig. 6.17b):

The leading surface: $j_s = 0$ and $dy/dx = 0$, at $x_l = d$, where upon

$$\frac{y_l}{d} = \frac{\psi_l}{2}\left[1 - \left(\frac{x_l}{d}\right)^2\right] - \frac{v_p(d/a_0)^3}{24}\left[\left(\frac{x_l}{d}\right)^4 - 2\left(\frac{x_l}{d}\right)^2 + 1\right], \tag{6.45}$$

where ψ_l is the pore surface inclination at the boundary intersection, subscript l refers to the leading surface and $v_p = v_p kTa_0^3/D_s\delta_s\gamma_s\Omega$ is the dimensionless pore rate identified in Sect. 6.2.1 (a_0 is the initial, zero rate, pore dimension).

The trailing surface: $j_s = 0$ and $dy/dx = \theta - \psi/2$, at $x_t = b$, giving

$$\frac{y_t}{b} = \frac{v_p \sin\theta(b/a_0)^3}{24}\left[\left(\frac{x_t}{b}\right)^4 - 4\left(\frac{x_t}{b}\right)^3 + 5\left(\frac{x_t}{b}\right)^2 - 2\left(\frac{x_t}{b}\right)\right] + \left(\frac{\psi}{2} - \theta\right)\left[\left(\frac{x_t}{b}\right) - \left(\frac{x_t}{b}\right)^2\right]. \tag{6.46}$$

Continuity of chemical potential requires that the curvature of the leading and trailing pore surfaces be the same at the boundary intersection. Incorporation of this condition permits the apex angle θ to be related to the pore rate and to the pore dimension, b:

$$v_p = \frac{3[2\sin\theta(\psi/2 - \theta) - 2\theta - (\psi - \pi)/2](a_0/b)^3}{\sin^3\theta + (5/4)\sin^2\theta + (1/4)\sin\theta}. \tag{6.47}$$

Finally, requiring that the volume V of the pore be independent of the rate:

$$V/l = \int_0^d y_1 dx + \int_0^b y_2 dx + \left(b^2 \sin\theta\cos\theta\right)/2, \tag{6.48}$$

the dimensionless pore rate can be expressed in terms of the apex angle θ and the dihedral angle ψ. Simple functional relations that adequately describe the dependence of θ, α and d on the normalized pore rate in the important application angular range ($\pi/3 < \psi < 2\pi/3$ and $\pi/12 < \theta < \pi/6$) are given by [1601]

$$\theta = (\pi/6)\exp\left[-v_p/(0.07 + 0.4\psi)\right];\ \alpha \approx \pi/6 + 0.013 v_p(1 + 3\psi);$$
$$d/a_0 \approx 1/2 - 0.013 v_p(1 + \psi). \tag{6.49}$$

6.2.2.2 Grain Boundary Rate

The grain boundary rate, v_b, is found by the grain boundary radii of curvature R_1 and R_2 and also by the boundary mobility, M_b, that is regarded as independent of the driving force:

$$v_b = \gamma_b \Omega^{2/3}\left(\frac{1}{R_1} + \frac{1}{R_2}\right)M_b. \tag{6.50}$$

The dependence of the radii of curvature on the extent of pore distortion, through the pore drag angle, α (Fig. 6.18), establishes a unique link between the

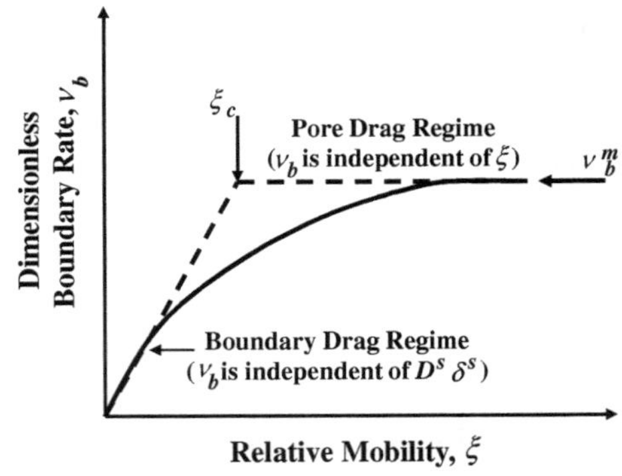

Fig. 6.19 A schematic indicating the pore drag and boundary drag regimes [1601]

boundary rate and the pore rate (with the intermediate determination of α). For a boundary of uniform curvature, the in-plane radius of curvature, R_1, is given by

$$R_1 = \frac{\sqrt{3}l - d}{\sqrt{3}\cos\alpha - \sin\alpha}. \tag{6.51}$$

We have from (6.49) and (6.51):

$$R_1/l \approx \frac{\sqrt{3} + [0.013(1+\psi)v_p - 1/2]a/l}{[2.5 - 0.023v_p(1+3\psi)]}. \tag{6.52}$$

Thus, the radius of curvature is dictated by the relative pore size, *a/l*, the dihedral angle, ψ, and the dimensionless pore rate, v_p. The out-of-plane curvature is assumed to be of secondary significance because one is independent of the pore movement. Hence, the radius R_2 may be neglected. Then, the grain boundary rate can be obtained from (6.50) and (6.52) as

$$\frac{v_b l}{M_b \gamma_b \Omega^{2/3}} \approx \frac{2.5 - 0.023v_p(1+3\psi)}{\sqrt{3} + [0.013(1+\psi)v_p - 1/2]a_0/l}. \tag{6.53}$$

Since the pore and boundary remain attached, it is required that

$$v_b \cos\alpha = v_p. \tag{6.54}$$

A dimensionless grain boundary rate $v_b = v_b kTa_0^3/D_s\delta_s\gamma_s\Omega$ can be determined directly from (6.49, 6.53) and (6.54) in terms of a dimensionless parameter, $\xi = M_b kTa_0^2\gamma_b/D_s\delta_s\gamma_s\Omega^{1/3}$, that reflects the relative mobility of pores and grain boundaries. The boundary rate can be characterized by two regimes (see Fig. 6.19). For small values of ξ, the motion of the pore/grain boundary ensemble is limited by the motion of the grain boundaries. Within this boundary mobility limited regime, grain boundary motion is independent on the presence of the pores. Therefore, the boundary rate increases linearly with ξ and has form:

$$v_b = M_b \gamma_b \Omega^{2/3} \Big/ l. \tag{6.55}$$

For large values of ξ, the motion of the ensemble is caused by the mobility of the pores (i.e., by the rate of surface diffusion) and the rate of the boundary becomes independent of ξ (Fig. 6.19), causing $\alpha \to \pi/3$, and the boundary rate approaches the pore drag limit:

$$v_b^m = 8/(1 + 3\psi). \tag{6.56}$$

The transition between these two regimes occurs over a range of ξ, associated with a critical value, ξ_c (Fig. 6.19). The grain boundary mobility dominated grain growth is considered to obtain when $\xi < \xi_c$, that is, when

$$a_0^2(a_0/l) < 8D_s\delta_s\gamma_s\Omega^{1/3}/M_b kT\gamma_b(1 + 3\psi). \tag{6.57}$$

6.2.2.3 Pore Shrinkage

The shrinkage of pores at three-grain junctions in a material, consisting of grains of equal size, is given for grain boundary diffusion dominance by [713]

$$\dot{a}a^2(1 - a/l)^2 = -(8/3)F(\psi)(\Omega D_b\delta_b\gamma_s/kTl), \tag{6.58}$$

where

$$F(\psi) = \sin(\psi/2 - \pi/6)\left\{1 + \frac{\sqrt{3}[\psi - \pi/3 - \sin(\psi - \pi/3)]}{2\sin^2(\psi/2 - \pi/6)}\right\}^{-1}, \tag{6.59}$$

$D_b\delta_b$ is the boundary diffusivity. Evidently, the pore shrinkage rate, $\dot{a}$, increases as the grain facet length, l, diminishes. It might be anticipated that the shrinkage rate in the three-sided grain configuration would increase as the grain boundaries converge. However, the tendency toward enhanced pore shrinkage results in the development of residual tensile stresses. These residual stresses constrain the local shrinkage, such that the shrinkage rate is approximately that of circumventing material, as given by (6.56), in which l is the mean facet length. Similarly, a tendency toward enhanced shrinkage, induced by the change in curvature at the distorted moving pore, is suppressed by the matrix constraint.

6.2.2.4 Coarsening Trajectories

The pore and grain coarsening that occur during normal sintering may be deduced, inserting (6.58) and (6.59) into (6.39) and (6.40), in the case of the constrained pore shrinkage rate, and inserting (6.55) and (6.56) into (6.39) and (6.40), in the case of the boundary rate. We obtain in the boundary mobility regime, $\xi < \xi_c$, inserting (6.55, 6.58) and (6.59) into (6.39) and (6.40):

$$\frac{da}{dl} = \frac{a}{l} - \frac{8F(\psi)\omega}{3\lambda(1 - a/l)^2 a^2}; \tag{6.60}$$

$$\frac{a^+}{l}\left(1 - \frac{a^+}{l}\right)^2 = \frac{8F(\psi)\omega}{3\lambda(a^+)^2}, \tag{6.61}$$

where $\omega = D_b\delta_b\gamma_s\Omega^{1/3}/M_bkT\gamma_b$ is a dimensionless parameter that reflects the ratio of the grain boundary mobility to the pore shrinkage rate by the grain diffusion. The trends in a^+/l indicate that solutions exist over a limited range of ω. For values of ω above a critical value ω_c, the absence of a solution indicates that the *shrinkage rate always exceeds the coarsening rate*, that is, the maximum pore size coincides with initial value, a_0. Therefore, pore coarsening is excluded below a critical pore size, a^*, when [1601]

$$a < a^* \approx \left(7D_b\delta_b\gamma_s\Omega^{1/3}\big/M_bkT\gamma_b\right)^{1/2}. \tag{6.62}$$

Thus, low boundary mobility and large boundary diffusivity are the most desirable conditions for averting pore–boundary separation.

The densification process is accompanied by initial coarsening that causes a decrease in a/l and an increase in pore size. Therefore, if a/l decreases at a sufficiently rapid rate that the transition line $\widehat{a}/l$ is attained, further pore coarsening is prohibited and the pore shrinkage commences.

When the grain boundary rate is limited by pore drag, the trajectory is obtained from (6.39, 6.40, 6.56, 6.58) and (6.59) as

$$\frac{da}{dl} = \frac{a}{l}\left[1 - \frac{F(\psi)(1 + 3\psi)}{3\lambda\Delta(1 - a/l)^2}\right]; \tag{6.63}$$

$$\frac{a^+}{l} = 1 - \left[\frac{F(\psi)(1 + 3\psi)}{3\lambda\Delta}\right]^{1/2}, \tag{6.64}$$

where $\Delta = D_s\delta_s/D_b\delta_b$. The transition pore size again shows pore coarsening and shrinkage regions. Coarsening is invariably excluded when the diffusivity ratio and dihedral angle satisfy the condition:

$$D_b\delta_b/D_s\delta_s > 3\lambda/F(\psi)(1 + 3\psi). \tag{6.65}$$

Otherwise, there is no maximum, and pore coarsening inevitably initiates whenever a/l becomes smaller than a^+/l. Both considered regimes of grain growth can be generalized as it is shown in Fig. 6.20. Recognizing that the transition pore sizes pertinent to the individual grain growth regimes will typically intersect, this intersection may be found by using inequalities (6.62) and (6.65). When both

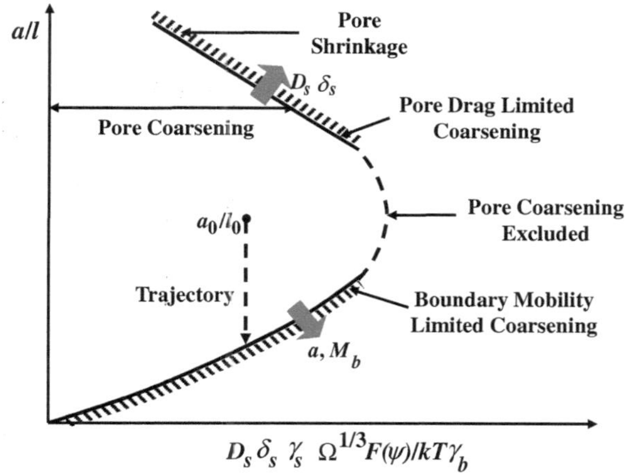

Fig. 6.20 A schematic illustration of the general tendencies in the transition pore size

inequalities are closely approached, the intersection occurs within an intermediate range of *a/l* (~0.2–0.3), and exclusion of pore coarsening is contingent upon less stringent conditions than suggested by either inequality (6.62) or inequality (6.65) (Fig. 6.20). Otherwise, the coarsening behavior is dominated either by pore drag or by the grain boundary mobility. The intersection is caused by the initial pore size, the grain boundary mobility and the surface diffusivity. In particular, small values of a_0, $D_s\delta_s$ and M_b simultaneously induce intersection behavior, narrowing the region of pore coarsening.

The pore coarsening value outside the exclusion region can be deduced, in principle, by superimposing onto Fig. 6.20 the line that indicates the transition from pore drag to grain boundary mobility limited grain growth and evaluating the coarsening that occurs in each region. An upper bound pore size, when it exists, has a functional form [1601]:

$$\widehat{a}/l = F_1(D_s\delta_s/D_b\delta_b)F_2\left(M_b a_0^2 kT/D_b\delta_b\right)F_3\left(\psi^{-1}\right), \tag{6.66}$$

where F_n are increasing functions of the pertinent variables.

6.2.3 Estimation of Pore Separation Effects for HTSC

The coarsening parameter that exerts a major influence on the final microstructure is the peak pore size, $\widehat{a}$ (see Fig. 6.21). Its comparison with the critical pore size, a_c, defined by (6.34), states its breakaway from grain boundary. The pore-breaking behavior is caused primarily due to the pores in triple junctions (and pores with this morphology dominate the peak pore size, $\widehat{a}$). At the same time, only those pores at two-grain interfaces are amenable to separation from grain boundaries [716] (at pore sizes $>a_c$). Thus, in the comparison of $\widehat{a}$ and a_c, it is necessary to

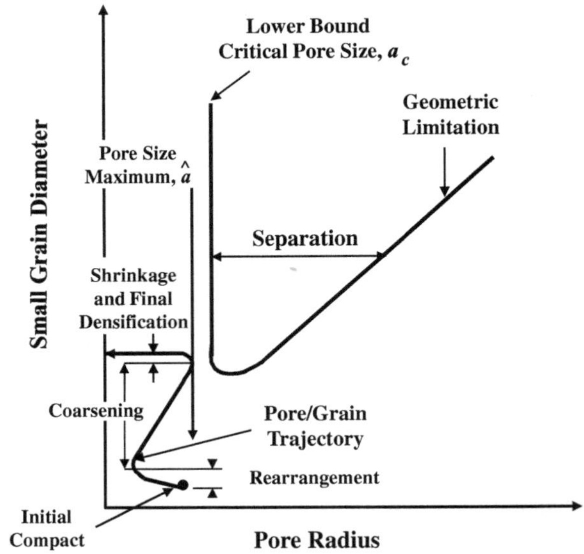

Fig. 6.21 The relation between critical pore size for separation, a_c, and peak pore size, $\hat{a}$

assume the concurrent existence of pores of similar dimensions on both two- and three-grain interfaces. Non-separating pores on two-grain interfaces become inevitably attached to three-grain corners during small grain disappearance [716] that leads to subsequent enlargement by pore coalescence. A part of the coarsened pores may subsequently translate again onto two-grain interfaces causing the process to repeat until either the peak size is reached or the pores on two-grain interfaces become large enough to separate.

Comparing (6.3) and (6.66), the understanding for the possibility of averting pore separation in small microstructures is provided, for which pore shrinkage and pore motion are dominated by grain boundary diffusion and surface diffusion, respectively. Note that requirements of small grain-boundary mobility and large grain-boundary diffusivity are evident. However, these conditions can only be simultaneously satisfied in the presence of appreciable solute (or precipitate) drag throughout the grain disappearance process. Hence, a vital influence of drag-inducing solutes upon the attainment of optimum microstructures is apparent.

The trends connected with the surface diffusion can be stated for two extremes, namely (i) when $a_c \gg a_0 \gg a^*$ pore breakaway can be avoided by increasing the ratio of the boundary to surface diffusivity, consistent with condition of $a_0 < a_c$, until (6.65) is satisfied. This trend is entirely compatible with that needed to achieve densification during the initial stage of sintering [233, 672]. Conversely, in the case (ii) when $a_0 \sim a^*$, large values of the surface diffusivity are needed to ensure that $a_c > a$ during subsequent pore coarsening [672]. This requirement is distinctly different from the initial stage of densification. Hence, different temperatures are undoubtedly desirable during the initial and final stages of sintering. Also note that the explicit statement of different thermal treatments depends on the values and temperature dependencies of both the surface and boundary diffusivities and also the grain boundary mobility.

The presented analysis of pore transformation as a result of grain growth, based on the surface diffusivity mechanism, is well satisfied for fine pores. Transformations of larger pores, when their motion is caused by the difference of the leading and trailing surface curvatures, can be also studied, taking into account corresponding changes of gas pressure in different points of the pore surface [714]. This case is less interesting for HTSC and therefore is not considered here.

Finally (6.34) is used for quantitative analysis of possible pore sizes, which break away from grain boundaries during sintering of monocore Bi-2223/Ag tapes. Selecting required parameters as follows [1091, 1092, 1395]: $\Omega = 2.2 \times 10^{-30}$ m^3, $D_s\delta_s = 2.5 \times 10^{-21}$ m^3/c, $\gamma_c = 2\gamma_b$, $\psi_{max} = \pi/2$, $T = 1110$ K, $2R = 25$ μm, $k = 1.38 \times 10^{-23}$ J/K (here the some parameters are selected for Al_2O_3, because corresponding data are absent for Bi-2223), we obtain a very high value of $M_b \approx 4 \times 10^4$ m/(N · s) even for $a_c = 100$ nm. For smaller pores separated from grain boundary, longer grain mobility is required. Obviously, the size of pores which can be separated from grain boundaries during prolonged annealing on some orders of magnitude is longer than the coherence length ($\sim$1 nm) in Bi-2223. Therefore, these separated pores cannot serve effective pinning centers and because of percolation features must considerably diminish the critical current. Apparently, in prolonged annealing, this effect is more important than deteriorating pinning strength due to lead expelling [1091, 1092]. We think that namely numerous pore separations and their movement into grain inside have found the critical current decrease in longer calcining after observed I_c maximum at about 180 hrs-annealing of monocore Bi-2223/Ag tapes [1091, 1092]. Thus, the lead expelling causes a decreasing of critical current in long reaction, but rather due to pore transformations, occurred during annealing (because Pb can inhibit only a local grain growth in this case), than thanks to decrease of its pinning efficiency in the grains.

6.3 Modeling Epitaxial Growth of Thin Films

Growth of thin crystallite films includes physical transport of material toward surface, to which bring chemical components of film. Atomistic processes in nanoscale, in particular attachment of single atoms to step or defect could in significant measure to cause an evolution of common surface morphology in larger scales on some orders of magnitude [90]. Moreover, in dependence of epitaxial conditions, the mechanisms of exact transport and kinetics should be changed. The growth modes could be varied and correspond to growth which forms steps or islands, and also serial growth of layers with possible transition from one type to other. The model of epitaxial growth should correspond to three conditions: (i) describe spatial and time scales on several orders of magnitude (ii) ensure direct extension on subsequent areas of transport and kinetic values, and (iii) be capable to take into account topological changes, originating at transition from one growth mode to other. Numerical approaches based on modeling behavior of single particles, in particular kinetic algorithms of Monte-Carlo method could be applied to modeling atomistic processes owing to their ability to numerical realization of atomistic kinetic processes taking into account actual rates of the particle deposition [1418].

However, modeling thin-film systems based on above approach is impossible because the modeling is found by spatial and time scales relate to single atoms and rates of adatoms. In order to fulfill above-mentioned conditions (i–iii) and model required parts of surface, the continuum models ensure both as rapid calculation as introduction of quantitative kinetic information in atomic scale.

6.3.1 Sharp Interface Formulation

The model of two-side sharp interface for epitaxial growth is described by the following equations [436]:

$$\frac{\partial \rho}{\partial t} = D\nabla^2 \rho - \frac{\rho}{\tau^{ev}} + F; \tag{6.67}$$

$$\pm D\left(\frac{\partial \rho^{\pm}}{\partial n}\right)_{\Gamma_{ij}} = k^{\pm}\left(\rho^{\pm} - \rho^{eq}\right)_{\Gamma_{ij}}; \tag{6.68}$$

$$v_n = \Omega D\left(\frac{\partial \rho^{+}}{\partial n} - \frac{\partial \rho^{-}}{\partial n}\right)_{\Gamma_{ij}}. \tag{6.69}$$

Here ρ is the adatom density over surface (ρ^{eq} is the equilibrium density), which to be transport variable in transport equation (6.67) in two regions Ω_1 and Ω_2, divided by boundary Γ_{ij}; v_n is the normal rate at interface. The adatom dynamics on surface is the process of "diffusion – relaxation" type at external delivery of atoms to surface defined by parameter F. The adatoms can evaporate after time τ^{ev}. Equations (6.68) and (6.69) are boundary conditions defining, respectively conservation of energy and mass on boundary Γ_{ij}; D is the diffusion constant included into the process; Ω is the atomic square of the treated atoms and $k^{\pm}$ are different factors determining a kinetics of atom attachment to boundary corresponding to cases of attainment of the boundary above (−) and below (+). Thus, the physical picture described by the obtained equations includes transport interior volume (see (6.67)) and the kinetics of atom attachment to interface (see (6.68, 6.69)).

It is evident, that the boundaries can have very different geometries in dependence on the epitaxial growth mode of the substrate. Figure 6.22 demonstrates boundaries during step flow and island growth. Note, that it may be observed also the epitaxial growth anisotropy [435, 438].

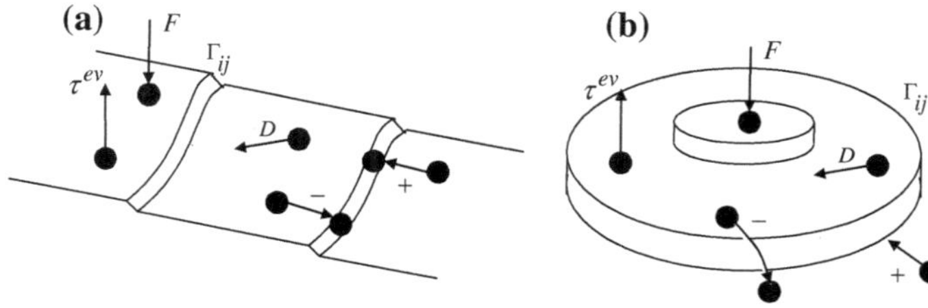

Fig. 6.22 Schematics of kinetics and dynamics during epitaxial growth. Adatoms reach the surface with a velocity which is proportional to F. They diffuse on the surface and may be desorbed from it according to desorption parameter, τ^{ev}. Reaching a boundary Γ_{ij} (e.g. a step **(a)** or island **(b)**) they are incorporated with kinetic attachment velocities depending on their positions with respect to the surface that is from up or down. This leads to growth, i. e. the boundary movement [436]

Fig. 6.23 The phase-field function, ϕ, changing over some layers of an epitaxial substrate

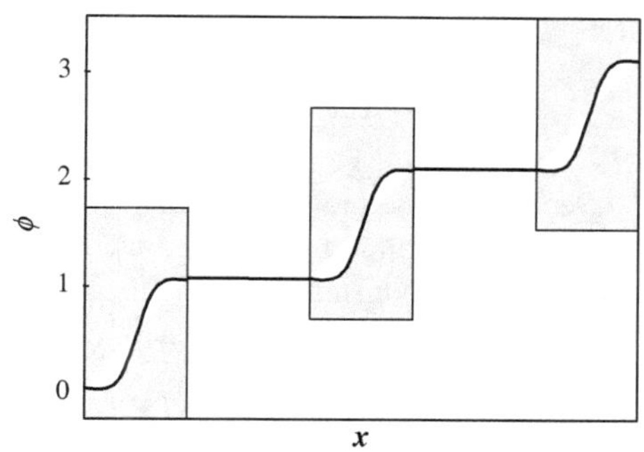

In accordance with modeling phase region of epitaxial growth, the boundary conditions (6.68, 6.69) of the formulation of the sharp interface could be replaced by differential equation for order parameter, ϕ, which smoothly changes at transition through boundary Γ_{ij} such that various layers of the substrate are labeled by different integer numbers (see Fig. 6.23).[5] This convention has been used, in particular in [820], where equal attachment kinetics at the interface has been assumed. The approach allowed one to describe dendritic growth with equal diffusion constants in both phases.

Below, following [436] we present an extended model which (i) handles the case of unequal attachment kinetic from both sides of interface basing on the conservation laws of energy and entropy, and (ii) introduces a “continuous attachment function” which changes smoothly in the interfacial region. With this aim write the conservation laws of energy and entropy as

[5] The variation takes place on a scale much less than the structuring of the substrate.

$$\frac{d\varepsilon}{dt} + \int\limits_{\Gamma_{ij}} J_E \cdot \boldsymbol{n}\, dA = 0; \tag{6.70}$$

$$\frac{dS}{dt} + \int\limits_{\Gamma_{ij}} J_S \cdot \boldsymbol{n}\, dA = \int\limits_V \dot{s}_P dV, \tag{6.71}$$

where $\boldsymbol{J}_E$ and $\boldsymbol{J}_S$ are the fluxes of energy and entropy through boundary Γ_{ij}, which has external unit normal $\boldsymbol{n}$; $\dot{\boldsymbol{s}}_P$ presents entropy production.

For functional of internal energy, ε, and entropy, *S*, there may be also obtained relationships [436]:

$$\varepsilon = \int\limits_V \left[e(\phi, \rho) + \frac{\xi}{2} |\nabla \phi|^2 \right] dV; \tag{6.72}$$

$$S = \int\limits_V \left[s(\phi, \rho) + \frac{\xi}{2} |\nabla \phi|^2 \right] dV. \tag{6.73}$$

Second terms under integrals in relationships (6.72) and (6.73) define a diffusion contribution on interface, where ξ is the gradient value which could be accounted for constant, *e* and *s* are the densities of internal energy and entropy, respectively. They are determined via density of free energy, $f(\phi, \rho)$, as

$$s(\phi, \rho) = -\frac{\partial f}{\partial \rho}, \; e(\phi, \rho) = f(\phi, \rho) + \rho s(\phi, \rho). \tag{6.74}$$

The thermodynamic consistency demands a positive entropy production. The constitutive equations providing positive production of local energy correspond to the following selection:

$$J_E = -k \nabla \rho - \xi \frac{\partial \phi}{\partial t} \nabla \phi; \tag{6.75}$$

$$J_S = -\frac{k}{\rho} \nabla \rho - \xi \frac{\partial \phi}{\partial t} \nabla \phi, \tag{6.76}$$

where *k* is a function of the vector of spatial coordinates $\boldsymbol{x}$ and second terms in the right parts of Eqs. (6.75, 6.76) define a flux connected with interface region. Then, the entropy production is determined as

$$\rho \dot{s}_P = \frac{\xi}{\rho} |\nabla \rho|^2 + \frac{\partial \phi}{\partial t} \left[\xi (1 + \rho) \nabla^2 \phi - \frac{\partial f}{\partial \phi} \right]. \tag{6.77}$$

The corresponding equation for phase field is selected by that that the right part of Eq. (6.77) has the form: $\frac{\xi}{\rho}|\nabla\rho|^2 + \left(\frac{\partial\phi}{\partial T}\right)^2\tau$, where $\tau > 0$. Then, by considering the free energy in the form:

$$f(\phi,\rho) = \frac{V_E + \rho V_S}{2}d(\phi) + \frac{(\rho - \rho^{eq})}{\rho^{eq}}k(\phi) - \rho\ln\left(\frac{\rho}{\rho^{eq}}\right) + \rho \tag{6.78}$$

we obtain the following system of the constitutive equations for case of epitaxial growth:

$$\begin{aligned}\tau\frac{\partial\phi}{\partial t} &= \xi(1 + \rho)\nabla^2\phi - \frac{\partial f}{\partial\phi}\\ &= \xi(1 + \rho)\nabla^2\phi - \frac{V_E + \rho V_S}{2}\frac{\partial d}{\partial\phi} - \frac{(\rho - \rho^{eq})}{\rho^{eq}}\frac{\partial k}{\partial\phi};\end{aligned} \tag{6.79}$$

$$\frac{\partial\rho}{\partial t} = D\Omega\nabla^2\rho + G(\phi)\frac{\partial\phi}{\partial t} + F + \frac{\rho}{\tau^{ev}}. \tag{6.80}$$

Here $1/\tau > 0$ is the mobility factor; $d(\phi) = \phi^2(1 - \phi)^2$ is the dowble-well potential function with barrier height defined as $V_E + \rho V_S$; $\xi(1 + \rho) > 0$ is the factor of energy gradient and $k(\phi) = \phi^2(3 - 2\phi)$. The term of atom source, $G(\phi)\frac{\partial\phi}{\partial t}$, describes influence of movement of the boundary Γ_{ij} on evolution of transport field, ρ, where $G(\phi) = \xi\nabla^2\phi - \frac{\partial d}{\partial\phi}\frac{\xi}{2} + 1$.

In order to simplify the asymptotic analysis of sharp interface Eqs. (6.67–6.69), we introduce "function of continuous attachment", ξ_*, which changes smoothly in interface region and is found as [436]

$$\xi_* = -\frac{D}{k^-}\frac{12\chi}{l}(1 - \phi) + \frac{D}{k^+}\frac{12\chi}{l}\phi. \tag{6.81}$$

Equations (6.79, 6.80) could be made non-dimensional at the transition to units of macroscopic length scale, χ, time scale, χ/D, and scale of governing field, ρ. Moreover, l defines interface width, that is a scale in which the phase field, ϕ, changes from one integer number to other; l_c is the capillary length proportional surface tension, γ, that is

$$l_c = \frac{\gamma\Omega}{k_B T}, \tag{6.82}$$

where k_B is Boltzmann constant, and T is the temperature. By introducing designations: $V^{eq} = V_E + \rho^{eq}V_S$ and $\xi^{eq} = \xi + \rho^{eq}\xi$, we define the next non-dimensional parameters:

$$\varepsilon = \frac{l}{\chi};\ \psi = \frac{\rho^{eq}D\tau}{6ll_c};\ \lambda = \frac{l}{6l_c};\ \nu = \frac{V_s}{V^{eq}};\ \gamma = \frac{\xi}{\xi^{eq}}. \tag{6.83}$$

Based on these non-dimensional parameters, Eqs. (6.79, 6.80) could be written as

$$\varepsilon^2 \psi \frac{\partial \phi}{\partial t} = \varepsilon^2 (1 + \gamma c) \nabla^2 \phi - \frac{(1 + \nu c)}{2} \frac{\partial d}{\partial \phi} - \lambda c \frac{\partial k}{\partial \phi}; \tag{6.84}$$

$$\frac{\partial c}{\partial t} = D_* \Omega_* \nabla^2 c + G_*(\phi) \frac{\partial \phi}{\partial t} - c, \tag{6.85}$$

where re-scaled function, $G_*(\phi) = \varepsilon^2 \xi_* \nabla^2 \phi - \frac{\mu}{2} \frac{\partial d}{\partial \phi} + 1$

In order to satisfy the model equations, we use method of "matched asymptotic expansions" (MAE), in which the consistency conditions are applied to calculate integrating constants arising in internal expansion. In this case, external solution which is satisfactory far from interface consists with internal solution which fulfils in the interface region. By this, the following internal expansions are used for the solution with the lowest order terms [436]:

$$\nabla^2 = \frac{1}{\varepsilon^2} \frac{\partial^2}{\partial \zeta^2} + \frac{\langle \kappa \rangle}{\varepsilon} \frac{\partial}{\partial \zeta} + O(1); \ \frac{\partial}{\partial t} = -\frac{v_{\text{int}}}{\varepsilon} \frac{\partial}{\partial \zeta} + +O(1). \tag{6.86}$$

Here non-dimensional mean curvature, $\langle \kappa \rangle$, and non-dimensional interface rate, v_{int}, are assumed to be order of 1; ζ is the stretched coordinate, through of which is defined internal expansion $\boldsymbol{x} = \boldsymbol{x}_0 + \varepsilon \zeta \boldsymbol{n}$ near point $\boldsymbol{x}_0$, where $\boldsymbol{n}$ is the local unit normal vector to the interface. In the internal region, the phase and transport fields are expanded on degrees of ε:

$$\Phi = \Phi^0 + \varepsilon \Phi^1 + O(\varepsilon^2); \ c = c^0 + \varepsilon c^1 + O(\varepsilon^2). \tag{6.87}$$

The resulting equations are solved subsequently order by order in ε together with boundary conditions in far zone, which are obtained by alignment with external solution. The external solution for the phase field has the form: $\Phi(\boldsymbol{x}) = 0$ or $\Phi(\boldsymbol{x}) = 1$ for all orders of ε and on any side of boundary Γ_{ij}. The transport field is represented by expansion $C(\boldsymbol{x}) = C^{(0)}(\boldsymbol{x}) + \varepsilon C^{(1)}(\boldsymbol{x}) + O(\varepsilon^2)$, but limit behavior near $\boldsymbol{x}_0$ is given as

$$C(\boldsymbol{x}_0 + \varepsilon \zeta \boldsymbol{n}) = C_\pm^{(0)}(\boldsymbol{x}_0) + \varepsilon \left[C_\pm^{(1)}(\boldsymbol{x}_0) + \zeta \frac{\partial C_\pm^{(0)}}{\partial n}(\boldsymbol{x}_0) \right] + O(\varepsilon^2), \tag{6.88}$$

where $C_+^{(0)}$ and $C_-^{(0)}$ are limits of C, respectively at $\zeta \to 0_+$ and $\zeta \to 0_-$. Internal solution should be matched with this behavior at $\zeta \to \pm\infty$. By using the approach of asymptotic matching [436], it may be shown that model Eqs. (6.70, 6.71) are equivalent formulation of closed system of equations stating formulation of sharp interface (6.67–6.69).

Application of MAE-method to non-dimensional Eqs. (6.84, 6.85) leads to the leading-order equation for the phase region [436]:

$$\frac{\partial^2}{\partial \zeta^2}\phi^{(0)} - \frac{1}{2}\frac{\partial d(\phi^{(0)})}{\partial \phi} = 0. \tag{6.89}$$

The leading-order equation for transport current has form:

$$D_* \Omega_* \frac{\partial^2 c^{(0)}}{\partial \zeta^2} = 0. \tag{6.90}$$

By double integrating this equation and matching with external solution we obtain that $c^{(0)} = C_+^{(0)}(\boldsymbol{x}_0) = C_-^{(0)}(\boldsymbol{x}_0)$ is a constant. The first-order phase-field equation is found as [436]

$$\frac{\partial^2 \phi^{(1)}}{\partial \zeta^2} - \frac{1}{2}\frac{\partial^2 g(\phi^{(0)}\phi^{(1)})}{\partial \phi^2} = (\psi v_{\text{int}} + \langle \kappa \rangle)\frac{\partial \phi^{(0)}}{\partial \zeta} + \Lambda c^{(0)}\frac{\partial k(\phi^{(0)})}{\partial \zeta}, \tag{6.91}$$

where g is a negative double-well function and the condition of solubility for this equation has the form:

$$c^{(0)} = d_0 \kappa + \beta_* v_{int}/v_0, \tag{6.92}$$

where v_0 corresponds to velocity of a straight step. Condition (6.92) could be determined as modified boundary Gibbs–Thomson condition. At $\beta_* \to 0$, it defines equilibrium concentration corresponding to ρ^{eq} in Eqs. (6.68). So, in this leading-order expansion of concentration at interface, parameters $\rho^{\pm}$ in (6.68) would simply be ρ^{eq}. The difference in this equilibrium value is origin of kinetics and restores in expansion of following order.

The first-order equation has the form:

$$D_* \Omega_* \frac{\partial^2 c^{(1)}}{\partial \zeta^2} = v_{\text{int}} \frac{\partial \phi^{(0)}}{\partial \zeta}. \tag{6.93}$$

By integrating this equation into interval $-\infty < \zeta < \infty$ and using the matching condition, the mass conservation condition of leading order is presented in non-dimensional form as

$$D_* \Omega_* \frac{\partial C_+^{(0)}}{\partial n} - D_* \Omega_* \frac{\partial C_-^{(0)}}{\partial n} = v_{\text{int}}. \tag{6.94}$$

Thus, the boundary condition (6.69) restores into framework of the sharp interface formulation. It still remains to be proven that the boundary conditions (6.68) can be obtained for this order of the expansion.

Following procedure [436], the phase-field leading-order and first-order equations are obtained in the form:

$$\frac{\partial^2 \phi^{(0)}}{\partial \zeta^2} - \frac{1}{2}\frac{\partial d(\phi^{(0)})}{\partial \phi} = 0; \tag{6.95}$$

$$\frac{\partial^2 \phi^{(1)}}{\partial \zeta^2} - \frac{1}{2}\frac{\partial^2 d(\phi^{(0)})\phi^{(1)}}{\partial \phi^2} = r_1, \tag{6.96}$$

where

$$r_1 = -\gamma c^{(1)} \frac{\partial^2 \phi^{(0)}}{\partial \zeta^2} + \frac{v}{2} c^{(1)} \frac{\partial d(\phi^{(0)})}{\partial \phi} - (\tau v_{\text{int}} + \langle \kappa \rangle) \frac{\partial \phi^{(0)}}{\partial \zeta} + \lambda c^{(1)} \frac{\partial k(\phi^{(0)})}{\partial \phi}. \tag{6.97}$$

Taking into account Eq. (6.95), the first two terms in right part of (6.97) give

$$-\gamma c^{(1)} \frac{\partial^2 \phi^{(0)}}{\partial \zeta^2} + \frac{v}{2} c^{(1)} \frac{\partial d(\phi^{(0)})}{\partial \phi} = \frac{v_1}{2} c^{(1)} \frac{\partial g(\phi^{(0)})}{\partial \phi} \tag{6.98}$$

where $v_1 = v + \gamma$. Then the solubility condition for Eq. (6.96) is written as

$$\begin{aligned} 0 &= \int_{-\infty}^{\infty} \frac{\partial \phi^{(0)}}{\partial \zeta} r_1 d\zeta \\ &= \lambda \int_{-\infty}^{\infty} c^{(1)} \frac{\partial k(\phi^{(0)})}{\partial \phi} \frac{\partial \phi^{(0)}}{\partial \zeta} d\zeta + \frac{v_1}{2} \int_{-\infty}^{\infty} c^{(1)} \frac{\partial d(\phi^{(0)})}{\partial \phi} \frac{\partial \phi^{(0)}}{\partial \zeta} d\zeta - \frac{1}{6} (\psi v_{\text{int}} \\ &\quad + \langle \kappa \rangle) \end{aligned} \tag{6.99}$$

and includes first-order solution for transport field, $c^{(1)}$.

At $c^{(1)} = 0$, the equation for transport field becomes identical to leading-order equation. First-order transport equation has the form:

$$D_* \Omega_* \frac{\partial^2 c^{(1)}}{\partial \zeta^2} - \xi_1 v_{\text{int}} \frac{\partial \phi^{(0)}}{\partial \zeta} \frac{\partial^2 \phi^{(0)}}{\partial \zeta^2} = v_{\text{int}} \phi^{(0)} \frac{\partial \phi^{(0)}}{\partial \zeta} - \frac{\mu}{2} v_{\text{int}} \frac{\partial d(\phi^{(0)})}{\partial \phi} \frac{\partial \phi^{(0)}}{\partial \zeta}, \tag{6.100}$$

which when is simplified by means of (6.95) can be integrated reducing to the equation:

$$D_* \Omega_* \frac{\partial c^{(1)}}{\partial \zeta} - v_{\text{int}} \phi^{(0)} - \frac{\xi_{1*} v_{\text{int}}}{2} \left(\frac{\partial \phi^{(0)}}{\partial \zeta} \right)^2 = D_* \Omega_* \frac{\partial C_-^{(0)}}{\partial n} = D_* \Omega_* \frac{\partial C_+^{(0)}}{\partial n} - v_{\text{int}}, \tag{6.101}$$

where $\xi_{1*} = \xi_* - 1$. The integrating constant has been estimated by two methods, namely at transition to limits at $\zeta \to \pm \infty$, and by using the matching condition (6.88). Then, the solution can be presented into two equivalent forms:

$$c^{(1)}(\zeta) = C_+^{(1)}(x) + \zeta\frac{\partial C_+^{(0)}}{\partial n} - \int_{\zeta}^{\infty}\frac{1}{\Omega_* D_*}\left[v_{\text{int}}\phi^{(0)} + \frac{\xi_{1*}v_{\text{int}}}{2}\left(\frac{\partial\phi^{(0)}}{\partial\zeta}\right)^2\right]d\eta; \tag{6.102}$$

$$c^{(1)}(\zeta) = C_-^{(1)}(x) + \zeta\frac{\partial C_-^{(0)}}{\partial n} - \int_{-\infty}^{\zeta}\frac{1}{\Omega_* D_*}\left[v_{\text{int}}\left(1 - \phi^{(0)}\right) + \frac{\xi_{1*}v_{\text{int}}}{2}\left(\frac{\partial\phi^{(0)}}{\partial\zeta}\right)^2\right]d\eta. \tag{6.103}$$

The jump in the adatom concentration at transition through boundary Γ_{ij} now is equal to

$$C_+^{(1)} - C_-^{(1)} = \frac{\xi_{1*}v_{\text{int}}}{12}. \tag{6.104}$$

The estimations for frontal and back sides of interface Γ_{ij} and substitution into full expansion for C up to first order define, respectively the relationships:

$$\left(C_{\pm} - C_{eq}\right)\frac{12\chi}{\xi_{1*}l} = \pm\frac{\partial C_{\pm}}{\partial n}, \tag{6.105}$$

which restore equivalents of conditions (6.68). This shows that the model Eqs. (6.70, 6.71) are equivalent to full equation system in formulation of sharp interface (6.67–6.69).

6.3.2 *Epitaxial Growth Involving Elasticity*

In the case when epitaxial surface consists of deformed layers, the free energy of system includes corresponding energy of elastic strain. Then, Eq. (6.78) takes the form:

$$f\left(\phi, \rho, \varepsilon_{ij}\right) = \frac{V_E + \rho V_S}{2}d(\phi) + \frac{\left(\rho - \rho^{eq}\right)}{\rho^{eq}}k(\phi) - \rho\ln\left(\frac{\rho}{\rho^{eq}}\right) + \rho + f_{el}(\phi). \tag{6.106}$$

Here $f_{el}(\phi)$ is the free energy density taking into account global elastic energy of both phases. A suitable start form of f_{el} for the phase-field approach is based on the Hooke's law presenting the dependence between components of tensors of the stress, σ_{ij}, and strain, ε_{ij}:

$$f = \mu\varepsilon_{ij}\varepsilon_{ij} + \frac{\lambda}{2}\varepsilon_{ii}^2, \tag{6.107}$$

where summation over double subscripts is used; λ and μ are the Lamé constants; f is the free energy per unit volume of a general system, of which only energy contribution is elasticity.

Then, the stress tensor has the form:

$$\sigma_{ij} = 2\mu\varepsilon_{ij} + \lambda\varepsilon_{kk}\delta_{ij} = 2\mu\left(\varepsilon_{ij} - \varepsilon_{kk}\frac{\delta_{ij}}{d}\right) + K\varepsilon_{kk}\delta_{ij}, \tag{6.108}$$

where $K = \lambda + 2\mu/d$ is the bulk modulus, d is the dimension of the problem considered (that is $d = 2$ or 3) and δ_{ij} is the Kronecker delta-function, An existing mismatch on substrate surface causes arising an additional mismatch strain. Following [1605], this additional strain may be determined by additive contribution, σ_{ij}^m:

$$\sigma_{ij}^m = -2\mu \in^m \frac{(1+v)}{(1-2v)}\delta_{ij}, \tag{6.109}$$

where $\in^m$ defines mismatch strain, v is the Poisson ratio. This allows one to formulate directly required dependence of f on ϕ:

$$f_{el}(\phi, \{\varepsilon_{ij}\}) = k(\phi)\left[\mu\varepsilon_{ij}\varepsilon_{ij} + \frac{\lambda}{2}\varepsilon_{ii}^2 - 2\mu \in^m \frac{(1+v)}{(1-2v)}\right] + [1 - k(\phi)]\frac{K_{ns}}{2}\varepsilon_{ii}^2. \tag{6.110}$$

Here K_{ns} is the bulk modulus of liquid or gaseous phase in dependence on implied epitaxial technology. Then, the governing equations in adiabatic approximation (i.e. at $\delta F_{el}/\delta u_i = 0$) have the form:

$$\begin{aligned}\tau\frac{\partial\phi}{\partial t} &= \xi(1+\rho)\nabla^2\phi - \frac{V_E + \rho V_S}{2}\frac{\partial d(\phi)}{\partial\phi} - \frac{(\rho - \rho^{eq})}{\rho^{eq}}\frac{\partial k(\phi)}{\partial\phi} \\ &+ \frac{\partial k(\phi)}{\partial\phi}\left[\mu\varepsilon_{ij}\varepsilon_{ij} + \frac{\lambda - K_{ns}}{2}(\varepsilon_{ii})^2 - 2\mu \in^m \frac{(1+v)}{(1-2v)}\right];\end{aligned} \tag{6.111}$$

$$\frac{\partial\rho}{\partial t} = D\Omega\nabla^2\rho + G(\phi)\frac{\partial\phi}{\partial t} + F + \frac{\rho}{\tau}. \tag{6.112}$$

They are solved together with the elasticity equations in which strains ε_{ij} (where $i, j = 1\ldots d$) are dependent values, that also defines dependence of variational derivatives, $\delta F/\delta\varepsilon_{ij}$. At the same time, the displacement components u_i ($i = 1\ldots d$) are independent. By assuming that proper time scales of the problem are well comparable with time of the sound propagation, the variational derivatives $\delta F_{el}/\delta u_i$ of the elastic free energy may be equated to zero which is exactly adiabatic case. Then we have

$$0 = \frac{\delta F_{el}}{\delta u_i} = \frac{\partial}{\partial x_j}\frac{\delta F_{el}}{\delta\varepsilon_{ij}} = \frac{\partial}{\partial x_j}\{k(\phi)\sigma_{ij} - [1 - k(\phi)]\lambda\varepsilon_{kk}\delta_{ij}\}. \tag{6.113}$$

The second term could be related to the system pressure:

$$p = p_{0l} - K_{ns}\varepsilon_{kk},, \tag{6.114}$$

where p_{0l} is the equilibrium pressure in liquid phase. By assuming that on liquid side, the derivatives of this equilibrium pressure are small, Eq. (6.113) reduces to the form:

$$0 = \frac{\partial}{\partial x_j}\{k(\phi)\sigma_{ij}\}. \tag{6.115}$$

Equations (6.111, 6.112) and (6.115) form closed system at replacement of σ_{ij} and ε_{ij} by components of displacement, u_i, due to the expression for strain tensor:

$$\varepsilon_{ij} = \frac{1}{2}\left(\frac{\partial u_i}{\partial x_j} + \frac{\partial u_j}{\partial x_i}\right) \tag{6.116}$$

and Hooke's law. Again, by returning to the sharp interface model determined by relationships (6.67–6.69) note that account for elastic effects demands a coupling to the following evolution equations for displacements of the solid phase:

$$(1 - 2v)\frac{\partial^2 u_i}{\partial x_k^2} + \frac{\partial u_k}{\partial x_i}\frac{\partial u_k}{\partial x_k} = 0. \tag{6.117}$$

In order to estimate these equations in each time moment, the boundary conditions in displacements on moving boundary of the phase interface should be stated in each calculation step. Physically, the correct boundary condition presents a force balance, acting on film surface and pressure in liquid (or gaseous) phase, that is

$$\sigma_{ij} \cdot n_j = p^{l/v}\big|_{\Gamma_{ij}}. \tag{6.118}$$

Obviously, in this case implementation of the boundary conditions for strains, ε_{ij}, based on (6.118) presents greater difficulties at numerical realization compared to equations for phase field (6.111–6.116). A more detailed discussion of this point could be found for special case of an assumed zero-pressure condition above the substrate in [1609].

It should be noted, that the diffuse interface approach successively applied to study solidification morphologies [870, 1410, 1411, 1771, 1795]. Into framework of the thin interface limit [818] the side branching effects in dendritic growth have been studied [818, 819]. The diffuse interface approach could be extended to study alloy solidification that is a growth problem with coupled transport fields. Corresponding problems have been formulated in [259, 260, 1534, 1792, 1793]. It may be applied to investigate the phenomenon of solute trapping [258, 304, 305], dependence of surface tension on composition [258], isothermal dendritic growth and microsegregation [306, 1777], Ostwald ripening and coalescence [1778], recolescence during dendritic solidification [148], cell to front phase transitions during directional solidification [149], and eutectic alloy solidification [394, 428, 429, 585, 816, 1794]. The diffuse interface modeling coupled with hydrodynamics has been developed in [116, 215, 1692] and used to some extended computations [322, 437]. The model could be also applied to simulate wetting phenomena [1208]. Thermodynamically consistent diffuse interface models including hydrodynamics have been formulated as for pure materials as for binary alloys [30, 31]. The model including convection and anisotropy for a pure material [32, 33], has been extended to multicomponent alloy solidification. In the area of solid–solid transformation a diffuse interface field model with multiple order parameters has been used to explore order–disorder phase boundaries and transformations in alloys [179, 180]. A phase-field model involving two order parameters has been used for modeling polycrystalline materials with grain boundaries [899]. Finally, it has been solved the problem of two-sided dendritic growth [817].

6.4 HTSC Microstructure Formation During Sintering

For investigating the processes of HTSC ceramic preparation, plane sample of superconducting powder compact in gradient furnace is considered. Two-scale modeling, consisting of the macroscopic study of the precursor powder sintering and microstructure formation into the region of the heat front propagation is carried out. For this, the considered rectangular region $[a, b]$ is divided into square lattice with characteristic size of elementary cell, δ, which corresponds to either particle or pore. The sample moves into gradient furnace with constant rate, v (which can be correlated with the temperature change rate of the sample surface). Moreover, it is assumed that temperature distribution T into furnace depends on one coordinate x and consists of sites with constant temperature and linear dependence on this coordinate. In order to solve an initial-boundary problem, the method of summary approximation (MSA) is used (see Appendix C.1).

Microstructure modeling begins from the pore generation in the initial sample by using Monte-Carlo procedure. In this case, it is suggested that the pores are distributed in accordance with the normal distribution, and original pore concentration (i.e., porosity), C_p^0, is given in different variants of the computation. This procedure can be carried out, for example, filling the cells of the initial lattice by arbitrary numbers, using the generator of arbitrary numbers (GAN) and then selecting a number of minimum values that corresponds to the pore number. In order to obtain the statistically reliable results, the computations are accompanied by averaging distributions of the pores and crystallites in the sample microstructure.

The model includes the following main stages, namely (i) a heat front displacement and definition of the material sintering region; where a temperature above the sintering temperature, u_s; (ii) a press-powder re-crystallization into the corresponding region; and (iii) a shrinkage of the microstructure formed. The first from pointed steps relates to the macroscopic modeling and the other two to the microscopic modeling.

The computation of effective heat conduction (see Appendix C.2) for non-sintered part of the sample consists of the following stages. First, a coordination number, N_c, and the sizes of element with averaged parameters (y_1, y_2) using (C.2.12) (C.2.22) and (C.2.28) is defined. In order to calculate the heat conduction of gas into gaps between particles, δ_s, λ_{sr}, and λ_s from (C.2.37) (C.2.35) and (C.2.34) are computed successively. Then, the heat conduction of frame is computed depending on the porosity of the considered region, using (C.2.43). The porosity of the second-order structure is computed on the basis of (C.2.17) and then, using (C.2.5) and (C.2.6), $c_2 = c$ is calculated. In this case, the heat conduction of gas into pores of the second-order structure, λ_{22}, is defined, using (C.2.20). Finally, the effective heat conduction of the non-sintered part of the sample is calculated using (C.2.13).

In order to study a displacement of the thermal front, the first main problem for quasi-linear equation of heat conduction with a variable $u = T - T_0$ is considered, where T_0 is the environment temperature:

$$\frac{\partial u}{\partial t} = \frac{\partial}{\partial x}[k(u, C_p)\frac{\partial u}{\partial x}] + \frac{\partial}{\partial y}[k(u, C_p)\frac{\partial u}{\partial y}], \tag{6.119}$$

with initial condition

$$u(0, x, y) = 0, \tag{6.120}$$

and boundary conditions

$$u(t, 0, y) = u_1(t);\ u(t, a, y) = u_2(t);\ u(t, x, 0) = u(t, x, b) = u_3(x, t). \tag{6.121}$$

Here, $k(u,C_p)$ is the temperature conductivity factor; and C_p is the pore concentration or porosity. The boundary conditions are shown schematically in Fig. 6.24 for different time, t. In this case, the function $u_3(x, t)$ has the form:

$$\text{(a)}\quad u_3(x, t) = \begin{cases} A_0 t^{\frac{(vt-x)}{vt}}, & x \leq vt; \\ 0, & x > vt; \end{cases} \tag{6.122}$$

$$\text{(b)}\quad u_3(x,t) = A_0 t\frac{(a - x)}{a};\quad \text{(c)}\quad u_3(x,t) = u_s + \frac{(u_{\max} - u_s)(a - x)}{a}, \tag{6.123}$$

where A_0 is the given constant. The calculation is finished in the case of reaching in the whole region the temperature, $u(x, y) \geq u_s$.

The boundary conditions (6.121–6.123) correspond to a scheme of the ceramic gradient sintering [825]. In the case of hot-pressing, these boundary conditions are replaced by the following ones [1328]:

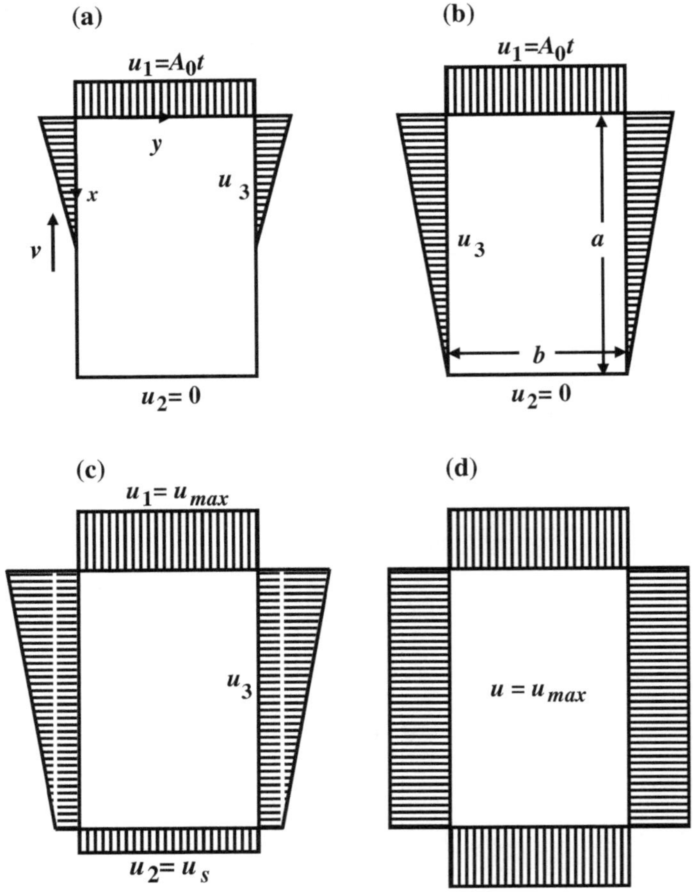

Fig. 6.24 Change of boundary conditions during heating of the sample

$$u(t,\ 0,\ y) = u(t,\ a,\ y) = u(t,\ x,\ 0) = u(t,\ x,\ b) = A_0 t. \tag{6.124}$$

In the beginning of the modeling, preliminary, the start porosity, C_p^0, the sintering temperature, $u_s < u_{max}$ (where u_{max} is the maximum temperature into gradient furnace for considered site of sintering) and the sample movement rate, v, are assumed to be known. The temperature conductivity factor at the first stage is found by constant values of heat capacity, c_V, and material density, ρ, and also by heat conduction factor, λ, depending on press-powder porosity. As λ is selected a value of the heat conduction factor, λ_{ef}, is calculated on the basis of the generalized conductivity principle (Appendix C.2). Hence, $k\ (u,\ C_p) = \lambda_{ef}/(c_V \rho)$. A solution of the problems (6.119–6.123) is obtained on the basis of the method of summary approximation (MSA) by using pure non-evident local one-dimensional scheme (LOS) (Appendix C.1). For this aim, a finite-difference counterpart of the initial-boundary problems (6.119–6.123) is written. The finite-difference equation is solved, using the run method. In order to obtain the required temperature field at any step of the considered process, there is a non-linear equation solved by iterations. After definition of temperature distribution into region, where $u \geq u_s$ a modeling of the material re-crystallization and shrinkage is carried out. Note that the grain formed at any stage can penetrate into earlier "sintered" region if it is allowed by the microstructure porosity. After that, the heat conduction factor is altered and corresponding value of the temperature conductivity factor is calculated. At the every stage of the microstructure modeling, λ_{ef} will be defined by the heat conduction of sintered and non-sintered regions. The first region possesses heat conduction of fabricated material. At the same time, this parameter is calculated for non-sintered region taking into account an existing porosity by using the generalized conductivity principle (Appendix C.2). Therefore, λ_{ef} in the whole considered region is defined by concentration of both components and by the values of their heat conduction by using the rule of mutual-penetrating components (see (C.2.13)). Then, a transition to the next time interval is carried out. The process finishes after microstructure formation in the whole considered region, which, as a result, consists of pores and grains.

Separately, the microstructural mechanisms of the superconducting press-powder re-crystallization and shrinkage for the formed structure in the temperature region, $u \geq u_s$, are modeled. The process of the crystallite nucleation into precursor press-powder is assumed to be the thermal activated process. Therefore, an arbitrary number corresponds to each remaining (after modeling of porosity) cell of the considered region. This arbitrary number characterizes the initiation time of a single crystallite, and these numbers are obtained by using GAN from the law of exponential distribution [1594]: $P_{ij}\ (t) = 1 - \exp(-t/\tau_{ij})$, where $\tau_{ij} \sim \exp(U/k_B u_{ij})$ is the mean expectation time of the crystallite nucleation in the node of i-line and j-column; U is the activation energy of crystallite nucleation; k_B is Boltzmann constant; u_{ij} is the temperature in the lattice cell with coordinates (i, j). It is suggested that a crystallite with minimum nucleation time, t^*_{ij}, first nucleates, and its nearest neighbors with coordinates (k, l) obtain priority. According to this, the values of t_{kl} are decreased. At the next step, a minimum nucleation time, t^*_{ij}, is again found among of all remaining lattice cells, and a new crystallite nucleates in the corresponding cell. This cell is either the nearest cell to the earlier-nucleated crystallite or far from it. In the first case, there is a grain growth in the re-crystallization process. Generally, the nucleation time of crystallite, t^c_{kl}, near with earlier-nucleated one is defined as

$$t^c_{kl} = \left| t_{kl} + \frac{t^*_{ij} - t_{kl}}{S \exp(1 - S)} \right|, \tag{6.125}$$

where S is the grain square (which is defined by the corresponding number of cells) near of that it is possible an nucleation of a new crystallite. Note that relation (6.125) causes a decrease of the nucleation time of neighbor cells, until S is sufficiently small, and the time increases together with the grain area due to pushing of secondary phases at the intergranular boundaries during grain growth.

After completion of the crystallite system formation into region of order of the sintering front width, shrinkage of the sample is modeled. One includes pushing of gaseous component from the sample and decreasing of closed porosity, thanks to grain movement. Computational algorithm of the shrinkage provides successive alternating displacements of grains along two orthogonal directions upon total exhaustion of possibility of their movement. In the displacement process, it is assumed that grains preserve own volume, shape and spatial orientation. This is achieved due to successive definition for each lattice layer of possibility of the grain movement (this considered grain consists of single cells) and its displacement as a single whole. In the calculation, it is assumed that the shrinkage process occurs instantaneously.

The completion of the grain structure formation during the re-crystallization and shrinkage is accompanied by beginning secondary re-crystallization, that is, abnormal grain growth due to existence of the pores and admixture phases. This grain growth occurs at maximum pressure and a linear character of the temperature change ($0 \leq T \leq T_{max}$, where T_{max} is the maximum temperature into furnace) [1642]. The modeling of the abnormal grain growth is carried out on the basis of the Wagner–Zlyosov–Hillert's dynamic growth models [1, 2, 423].

Then briefly, we discuss influence of the crystallite package kind. While in the book, the microstructure processes are considered at square greed, however, it is expected that modeling by using other right lattices will lead to similar results. We are only restricted by example of consideration of the secondary recrystallization (grain growth). In [423], it has been shown, that at enough high temperature, the grain growth kinetics and microstructure morphology at triangle and square lattices have not obvious differences. Similar results have been also obtained for hexagonal structure. In relation of selection of the specific number for cells of the calculation mesh, during computation process in each specific case preliminary has been realized variant with double number of the cells compared to the used one. The calculation results of estimated parameters witnessed on insignificant errors into range of 2 %. A simulation veracity of computational algorithm of the microstructure shrinkage has been found by direct verification on test example. Block-scheme for one realization of computational process in HTSC microstructure formation at heating is presented in Fig. D.1.

6.5 Microcracking of Intergranular Boundaries at Sample Cooling

A plane sample of sintered ceramic into gradient furnace is considered. After sintering, a uniform temperature, $u = u_{max}$, is stated in the whole sample. Due to a shrinkage during the sample cooling, the rectangular site of the front with sizes

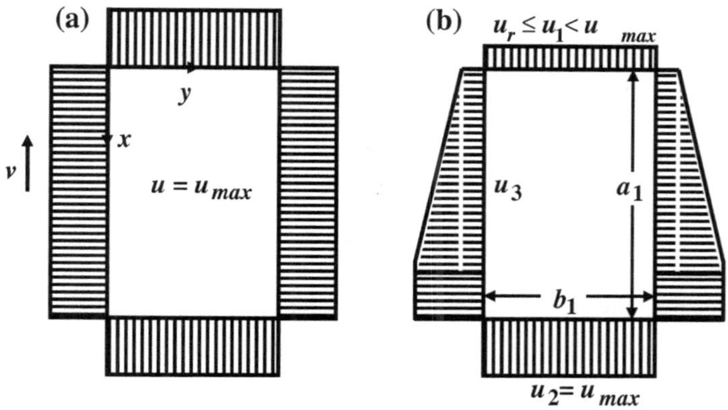

Fig. 6.25 Boundary conditions for problem of sample cooling

$a_1 \le a$, and $b_1 \le b$ is considered, which excludes the layers, including only elements of open porosity. At macroscopic modeling of the cooling, it is assumed that heat conduction factor is found by the thermal conduction of the sintered material and depends only on temperature, u. The modeling of the heat front movement continues down to attainment of room temperature, u_r, at one of the sample boundaries. Because intergranular boundaries (the main stress concentrators in this case) are subject to cracking at cooling, the required values (i.e., temperatures and stresses) are calculated in the lattice nodes but do not in the lattice cells, as in the case of sintering.

A modeling of microcracking consists of the following stages: (i) a displacement of heat front and definition of temperature field (ii) a calculation of normal thermal stresses in the lattice nodes (iii) a division of all intergranular boundaries into separate sections (iv) a computation of mean normal stress, acting onto given section, and (v) a satisfaction to microcracking condition.[6] Similar to the problem of ceramic sintering, it is assumed that the sample displaces with constant rate, v (which can be also correlated with the temperature change rate of the sample surface), from gradient furnace. The temperature changes in linear law into furnace. In the calculation of temperature field, the initial-boundary problem (6.119–6.121) is considered, where initial condition reduces to $u(0,\ x,\ y) = u_{max}$, and boundary conditions are depicted in Fig. 6.25. In this case, function $u_3(x,\ t)$ is given as

$$\text{(a)}\quad u_3(x,\ t) = u_{max}; \tag{6.126}$$

$$\text{(b)}\quad u_3(x,\ t) = \begin{cases} u_{\max} + \frac{(u_1 - u_{\max})}{vt}(vt - x); & x \le vt; \\ u_{\max}; & vt \le x \le a_1. \end{cases} \tag{6.127}$$

---

[6] Similar numerical algorithm may be applied to modeling of microcracking processes during cooling of oxide superconductor from room temperature down to cryogenic one its application.

As a result of the solution of this thermal conduction problem, a temperature distribution, causing corresponding thermal stresses, is calculated.

In order to define 2D stress state, the method of finite differences [1081] is applied. For framework of the thermal stresses problem, stress state is calculated through Airy function, φ

$$\sigma_x = \frac{\partial^2 \varphi}{\partial y^2}; \ \sigma_y = \frac{\partial^2 \varphi}{\partial x^2}; \ \sigma_{xy} = -\frac{\partial^2 \varphi}{\partial x \partial y}. \tag{6.128}$$

In this case, the function φ satisfies the differential equation in partial derivations:

$$\Delta^2 \varphi + E\alpha \Delta u = 0, \tag{6.129}$$

where $\Delta = \frac{\partial^2}{\partial x^2} + \frac{\partial^2}{\partial y^2}$; $\Delta^2 = \frac{\partial^4}{\partial x^4} + 2\frac{\partial^4}{\partial x^2 \partial y^2} + \frac{\partial^4}{\partial y^4}$; $E\alpha = \text{const}$; E is Young's modulus; α is the thermal expansion factor and $u = u(x, y)$ is the temperature, calculated from a some value, in which the thermal stresses are absent.

Replacing the partial derivations in (6.129) by finite differences, we obtain for arbitrary point 0 of the considered region (see Fig. 6.26):

$$\begin{aligned} &20\varphi_0 - 8(\varphi_1 + \varphi_2 + \varphi_3 + \varphi_4) + 2(\varphi_6 + \varphi_8 + \varphi_{10} + \varphi_{12}) \\ &+(\varphi_5 + \varphi_7 + \varphi_9 + \varphi_{11}) + E\alpha\delta^2(u_1 + u_2 + u_3 + u_4 - 4u_0) = 0. \end{aligned} \tag{6.130}$$

In the case of simply connected region and absence of external loading, the boundary conditions can be presented as [1679]

$$\varphi = \frac{\partial \varphi}{\partial y} = 0, \text{ at } x = 0; \ x = a_1; \tag{6.131}$$

$$\varphi = \frac{\partial \varphi}{\partial x} = 0, \text{ at } y = 0; \ y = b_1. \tag{6.132}$$

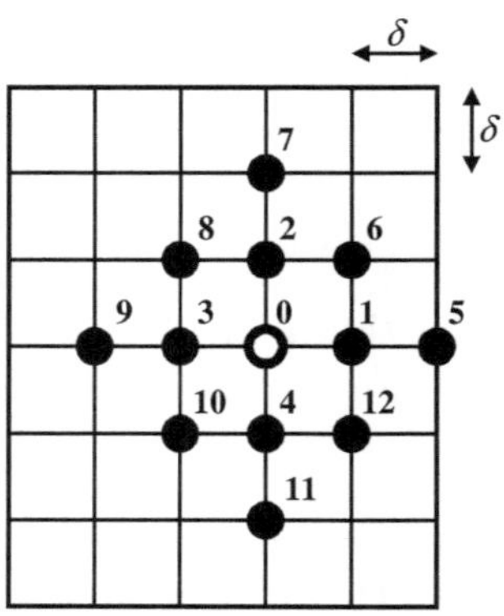

Fig. 6.26 Illustration to the definition of thermal stresses at point 0

The conditions (6.131) and (6.132) sign a definition of zero values of principal vector and principal moment of the acting forces at the region boundaries. Then, by using the function φ the normal stresses are calculated as

$$(\sigma_x)_0 = \frac{1}{\delta^2}(\varphi_2 - 2\varphi_0 + \varphi_4); \ (\sigma_y)_0 = \frac{1}{\delta^2}(\varphi_1 - 2\varphi_0 + \varphi_3). \tag{6.133}$$

After computation of the thermal stresses, the length of each section at intergranular boundary is estimated. The mean value of normal stress, acting on the given section of intergranular boundary, permits to verify the condition of the boundary microcracking:

$$\overline{\sigma}_n \sqrt{\pi l} \geq K_{Ic}^0, \tag{6.134}$$

where $\overline{\sigma}_n$ is the mean value of the normal stress on the intergranular boundary section with length l; K_{Ic}^0 is the fracture toughness of ceramic. When for the section, the condition (6.134) is satisfied, it is replaced by a microcrack. Obviously, in the first place, microcracks form on the sufficiently long and stressed boundaries. The microcracking modeling is repeated cyclically together with solution of heat conduction problem and calculation of thermal stresses. Block-scheme for one realization of computational process of the intergranular boundary microcracking during cooling is presented in Fig. D.2.

A simpler scheme of the microcracking modeling on intergranular boundaries may be the next. Based on the test data, a critical grain size during spontaneous cracking, D_c^S, for considered composition of superconducting ceramic is selected. Assuming, that the grain sizes and facet sizes per grain are subject to the normal distribution, we obtain [336] $D_c^S \approx 2l_c^S$, where l_c^S is the critical facet size. Taking into account misorientation of grains, a criterion of microcrack formation at the boundary l is stated [499] as

$$l/l_c^S \geq 2/[1 + \cos(2\Theta_1 - 2\Theta_2)], \tag{6.135}$$

where $\Theta_i (i = 1, 2)$ is the angle between axis of maximum compression in i-grain and the boundary plane of the grain. These angles are calculated by using Monte-Carlo procedure [336]. Computational algorithm takes into account that a microcrack grows from triple junctions (i.e., junctions of three or four grains at square lattice) [499, 928] or is initiated by pore and stops at the nearest facet node because the facets are usually subject to internal compressing stresses [499].

6.6 Study of Statistical Properties of the Model Structures

The computational approach presented in Sects. 6.4 and 6.5 is the good base to investigate the size, quantitative, topological parameters and morphological features and also to define structure-sensitive properties of polycrystalline superconductor. Typical description of ceramic microstructure includes definition of the size, shape and spatial orientation of the pores and grains. There are different classifications of materials in the porosity parameters. Another approach to microstructure description is based on the assumption about its non-regularity. Either microstructure is estimated by own constant statistical properties. The void space is correlated with the solid surfaces and forms by using the grain facets and edges. Architecture of the void space is caused by the shape, size and mutual disposition of crystallites. The description of solid matrix characterizes the material space as united whole, including the void space. Similar approach does not exclude an identification of single spatial structure elements (e.g., pores, grains, etc.) [1294].

HTSC microstructure may be described by integral (i.e., global) parameters, defining joint physical properties. An application of integral characteristics permits to abstain from simplified selection of the grain and void shapes and to characterize the material space by using summary values of volume, surface, curvature, etc., obtained from stereological measurements. Latter estimate only global metrics, that is, volume fraction, square of surface, length of linear elements, curvature of surface, and also their combined parameters [822].

The method to define the volume structure characteristics by using measurements in the observation plane has been named by the *statistical reconstruction* [271]. It is based on two statistical principles, namely (i) the structure sample should have representative volume, and (ii) a statistical correlation of the depicted structure characteristics in the observation plane with actual structure is necessary. Generally, a calculation of necessary number of the measurements to obtain unbiased estimation of any stereological characteristic is carried out [271]:

$$n = (200/y)(\sigma_x/\overline{x}), \tag{6.136}$$

where y is the accuracy level (%); σ_x is the average quadratic deviation; $\overline{x}$ is the mean value of the stereological characteristic. In the case of given accuracy, the number of measurements depends on the variation factor, $\sigma_x/\overline{x}$, that characterizes the uniformity of analyzed structure element. The procedure for the calculation of necessary measurement number includes the following steps:

(1) the mean value of the stereological characteristic, $\overline{x}$, and variance, σ_x, are estimated for some random sampling;
(2) it is given a necessary accuracy level (y, %) for the mean value of the measured magnitude;
(3) the number of measurements, n, is calculated from (6.136) which ensures the necessary accuracy level.

Note that (6.136) is obtained by using central limit theorem of statistics, according to which average values of characteristic, obtained for representative samplings, are distributed in the normal law relative to the mean value for corresponding universe. This approach to define the required number of

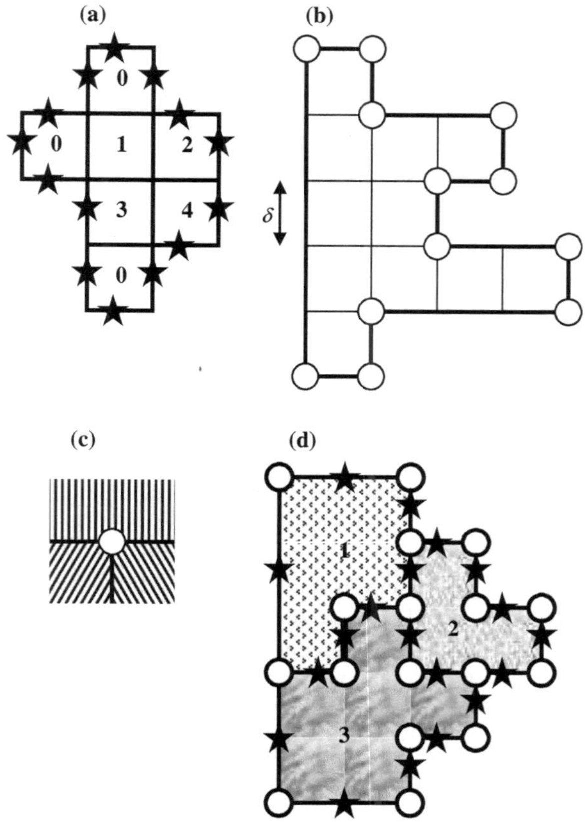

Fig. 6.27 Examples of calculation of some geometrical and topological parameters: (**a**) normalized perimeter of grain, L/δ (6.137), $n = 7$, each from two elements (2, 4) have two neighbors, element (3) has three neighbors and element (1) has four neighbors: $L/\delta = 3 \times 7 - (2 \times 1 + 1 \times 2 + 1 \times 3) = 14$; (**b**) number of grain tip (side) $N_l = 12$; (**c**) triple junction (point of intersection of three grain boundaries in square lattice); (**d**) Euler's relation for polyhedron [271]: $G - E + V = 1$, where G is the grain number (3), E is the side number (19), V is the tip number (17)

measurements is used in the next chapters to study microstructure, strength and current-carrying parameters of model superconductors.

A square of single grain or pore is usually estimated by using a detailed depiction of microstructure. Here, the realized algorithm defines a disposition of considered grain and computes the total number of its cells. Then, based on these squares, the proper grain sizes (or radii) are calculated. The grain or pore perimeter, L, in 2D case of square lattice may be found (normalized to the elementary cell size, δ) as

$$L/\delta = \begin{cases} 2(n+1); & n \le 2; \\ 3n - \sum_{k=2}^{4} l_k(k-1); & n > 2, \end{cases} \tag{6.137}$$

where n is the cell number of considered grain (or pore), k is the possible number of nearest neighbors for each cell in 2D case and l_k is the cell number of grain (or pore) with k number of neighbors. An example of (6.137) application is presented in Fig. 6.27a.

6.7 Modeling of Macrocracks

High-temperature superconductors (YBCO and BSCCO) are the brittle materials. Cracks, forming in these materials, as rule, are of I Mode, that is, fracture (or critical) loading is perpendicular to the crack plane. The cleavage planes in HTSC usually correspond to one of the direction of [100] type. The algorithm for the definition of arbitrary angles, Θ, formed by the normal to cleavage plane with direction of tensile stresses, based on the method of arbitrary twisting of cube, is presented in monograph [1307]. This algorithm models a transgranular cracking of HTSC ceramic, using a critical stress condition. In this case, it is suggested that definite grain fraction is fractured, when the stress normal to cleavage plane $\sigma_n = \sigma \cos^2 \Theta$ attains critical value of σ_n^* (where σ is the tensile stress). At the $\sigma = \sigma_n^*$, the microcrack formation begins in the grains, where the cleavage planes are perpendicular to the tensile axis ($\Theta = 0°$). The following increase of the stress leads to the crack propagation in the grains, for which $0 < \Theta < \Theta^*$, where Θ^* is determined from the relation: $\sigma \cos^2 \Theta^* = \sigma_n^*$. When applied stress attains the value of σ_{max}, a grain cracking into cleavage plane occurs. Then, the angle $\Theta^*_{\max}$ between the normal to the cleavage plane and tensile direction is obtained as

$$\Theta^*_{\max} = \arccos(\sigma_n^*/\sigma_{\max})^{1/2}. \tag{6.138}$$

Experimentally observed crack growth in oxide superconductor can take place as in grain boundaries (intergranular crack growth), as by cleavage through grain volume (transgranular crack growth). Additionally, there is a mixed character of macrocrack propagation, combining above two fracture types. Moreover, an existence of pores and microcracks into the material structure renders significant influence on crack growth.

Consider a model sample, including layers of rectangular form, $N \times M$ into units of mean size of the elementary cell, δ. As a first mechanism, the transgranular fracture through grain volume is considered. It is known, that in the case of crack growth arbitrarily oriented to tensile direction, the fracture toughness, K_{Ic}, can be found as [270]

$$K_{Ic} = \frac{K_{Ic}^0}{\cos^2 \Theta}, \tag{6.139}$$

where $K_{Ic}^0 = \sqrt{E\gamma_0}$ corresponds to the normal-opening crack, E is Young's modulus, γ_0 is the specific surface fracture energy and Θ is the arbitrary angle between the normal to the crack plane and tension direction.

Assume that there is cleavage crack, crossing crystallites in the planes of $\{100\}$-type. Then, the value of $K_{Ic}^{(j)}$ for crack path along j-line of the coordinate lattice with length, h_j, is found as

$$K_{Ic}^{(j)} = \sum_i \frac{K_{Ic}^0}{\cos^2 \Theta_{ij}} \sqrt{\frac{d_{ij}}{h_j}}, \tag{6.140}$$

where d_{ij} is the length of i-grain in the j-line and Θ_{ij} is the corresponding arbitrary angle formed by normal to cleavage plane and tension direction. Mean fracture toughness of the considered region may be estimated as

$$K_{Ic} = \frac{1}{N_l} \sum_{j=1}^{N_l} K_{Ic}^{(j)}, \tag{6.141}$$

where N_l is the number of lines in the selected rectangular region without the lines, including only elements of open porosity.

Then, consider a macrocrack growth along intergranular boundaries without microcracks (Fig. 6.28a) and in the existence of microcracks (Fig. 6.28b) onto the part of the intergranular boundaries. The material fracture toughness, K_{Ic}, as function of the specific surface fracture energy, γ_0, and Young's modulus, E, has the form:

$$K_{Ic} = \sqrt{E\gamma_0}. \tag{6.142}$$

In the case of the rectilinear intergranular boundary, $\gamma_0 = \gamma_b$, where γ_b is the specific grain boundary energy. In the actual sample, either crack trajectory is the arbitrary broken line. The value of γ_0 can be found through the ratio of the crack path length, L, to the sample width, h, as

$$\gamma_0 = L\gamma_b/h. \tag{6.143}$$

In order to compute L, we use a presentation about crack path as on the graph branch, joining the points at the opposite sides of the model layer. In this case, the graph branches are constructed, taking into account the intergranular boundary lattice. The origin of the graph tree coincides with the intersection point of one from intergranular boundaries with the left side of the sample (see Fig. 6.28). This selection of the crack growth origin corresponds to test data, showing that usually a fracture starts from sample surface. It is clear that a set of the graph tree branches may be selected, connecting a given point at one side of the layer with arbitrary point at the opposite side. From the energy minimum condition, a minimum

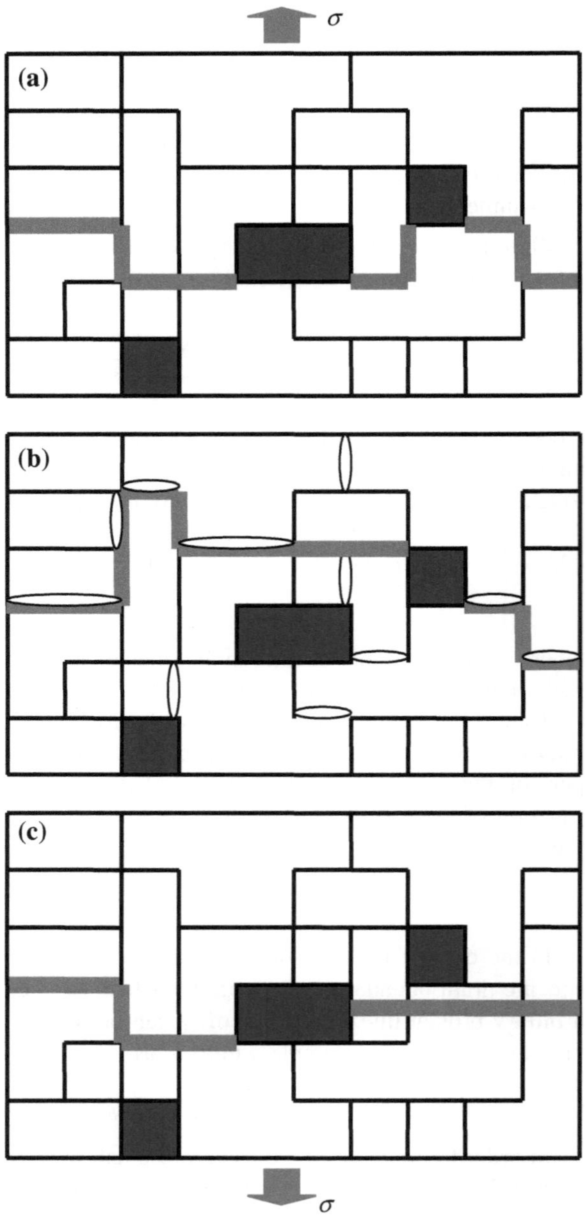

Fig. 6.28 Examples of macrocrack growth, shown by *gray* line, along intergranular boundaries in the case of (**a**) microcracking absence (**b**) microcracking presence onto part of intergranular boundaries and (**c**) mixed mechanism of macrocrack growth

trajectory corresponds to actual crack path. In order to define this minimum path, the Bellman-Kalaba's algorithm is used [905]. In this case, a problem of minimization of the numerical $(n + 1)$-order graph with tips, x_i, is reduced to solution of the next equation set:

$$\begin{cases} V_i = \min\left(V_j + C_{ij}\right);\ i = 0,\ 1\ldots,\ n - 1;\ j = 0,\ 1\ldots,\ n;\ i \neq j; \\ V_n = 0, \end{cases} \quad (6.144)$$

where V_i is the length of optimum path between points of x_n and x_i; $C_{ij} \geq 0$ is the value corresponding to the graph arc (x_i, x_j). The Bellman-Kalaba's algorithm suggests an iteration method for the solution of the minimization problem (6.144). By supposing $V_i^{(0)} = C_{in}$; $i = 0,\ 1\ldots,\ n - 1$; $V_n^{(0)} = 0$, successively compute

$$\begin{aligned} V_i^{(1)} &= \min\left(V_j^{(0)} + C_{ij}\right);\quad i = 0,\ 1\ldots,\ n - 1;\quad j = 0,\ 1\ldots,\ n;\quad i \neq j; \\ &\ldots\ldots\ldots\ldots\ldots\ldots\ldots\ldots\ldots\ldots\ldots\ldots \\ V_i^{(k)} &= \min\left(V_j^{(k-1)} + C_{ij}\right);\quad i = 0,\ 1\ldots,\ n - 1;\quad j = 0,\ 1\ldots,\ n;\quad i \neq j; \\ &\ldots\ldots\ldots\ldots\ldots\ldots\ldots\ldots\ldots\ldots\ldots\ldots \\ & V_n^{(k)} = 0, \end{aligned} \quad (6.145)$$

up to carrying out of the equalities, $V_i^{(k)} = V_i^{(k-1)}$; $i = 0,\ 1\ldots,\ n - 1$. In this case, the values of $V_i^{(k)}$ are the minimum values, which define an optimum branch of the graph tree, and $(n - 1)$ iterations are sufficient for its determination.

In the considered case, the graph tips coincide with the lattice nodes disposed at intergranular boundaries. In order to accelerate a sorting out of possible graph branches, Viterbi's algorithm is used [147]. Only the graph branches with length, which is not longer than a given inter-nodal distance, l_j (or length of the crack unit jump), are considered. If these paths are to be more than one, a priority is given to the path with the final tip disposed from the initial one at maximum distance along the coordinate perpendicular to tensile direction. The final graph tip, found in this way, corresponds to the crack tip after its jump; at the same time, this tip is assumed to be initial one for the next graph tree, and so on. In order to optimize the selection process, it is expedient to limit a number of considered graph tips in accordance with the length of the macrocrack unit jump. The region of possible crack unit jumps is shown in Fig. 6.29a.

In order to define a crack path, taking into account ceramic porosity, it is natural to assume that in the case of the crack hit in a pore, the crack tip blunts, causing a diminution of stress concentration. In our model, the pointed drag is depicted by adding the pore boundary length, crossed by the crack, to the crack length. The normalized boundary length of the pore, L_p/δ, is calculated in 2D case by using (6.137).

Now, the values of C_{ij} used in (6.145) for calculation of the crack path can be found:

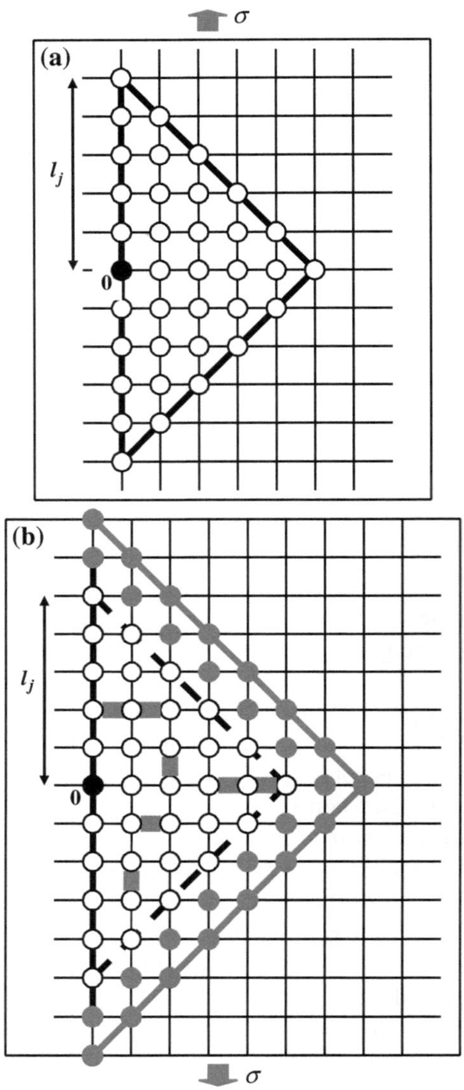

Fig. 6.29 Possible region of macrocrack unit jump during intergranular fracture (where l_j is the length of the crack unit jump): (**a**) boundary microcracking is absent; (**b**) part of intergranular boundaries is replaced by microcracks (*gray*); *circles* show considered nodes; additional nodes, caused by the intergranular microcracking, are shown by *gray color*

$$C_{ij} = \begin{cases} 0; & i = j; \\ L_b/\delta; & i,\ j \text{ are two nearest nodes at the grain boundary;} \\ L_p/\delta; & i,\ j \text{ belong to grain boundary of the same pore;} \\ \infty; & \text{in other cases,} \end{cases} \tag{6.146}$$

where $C_{ij} = C_{ji}$ and $L_b = 2\delta$ is the double length of elementary cell side.

Relationships (6.144–6.146) define the whole crack trajectory. The calculation of fracture toughness for the model sample is carried out as

$$K_{Ic} = \frac{1}{N}\sum_{i=1}^{N} K_{Ic}^{(i)}, \tag{6.147}$$

where $K_{Ic}^{(i)} = \sqrt{E\gamma_0^{(i)}}$; $\gamma_0^{(i)} = \frac{L_i}{h_i}\gamma_b$; and N is the number of considered crack paths by which an averaging of the fracture toughness is carried out.

Then, consider a model sample of superconducting ceramic, taking into account microcracks, substituting a part of intergranular boundaries. As it has been demonstrated, these microdefects can nucleate, in particular due to residual thermal stresses, caused by the thermal expansion anisotropy of grains and/or by the deformation mismatch of phases.

It is natural to assume that a microcrack, crossing the macrocrack path, renders a drag effect if the macrocrack crosses the microcrack at a junction of two sections of the microcrack (it is not important, with same direction or disposed at angle). In other cases (i.e., at the macrocrack hit in one of the microcrack tips), an acceleration of the macrocrack growth occurred. In the first case, the length of the microcrack surfaces is added to the macrocrack length. At the same time, in the second case, it is suggested that the macrocrack propagates on the whole microcrack length. Hence, relation (6.146) should be modified as in the following equation:

$$C_{ij} = \begin{cases} 0; & i = j, \text{ or at favorable disposition of microcrack;} \\ L_b/\delta; & i, j \text{ are two nearest nodes at intergranular boundary;} \\ L_p/\delta; & i, j \text{ belong to the boundary of the same pore;} \\ L_m/\delta; & \text{at unfavorable disposition of microcrack;} \\ \infty; & \text{in other case.} \end{cases} \tag{6.148}$$

In this case, $C_{ij} \neq C_{ji}$ and L_m/δ is the normalized length of the microcrack boundary. The microcrack existence leads to a change of the crack unit jump compared to the case of absence of the grain boundary microcracking (see Fig. 6.29b). When there are points for favorable growth of the crack, initially considered region (Fig. 6.29a) should be expanded. The number of additional layers is found by the maximum summary number of rectilinear sections of microcracks disposed in one line (or column) of the initial region (see Fig. 6.29b).

Third mechanism of crack growth is mixed. It implies a possibility of transition from one fracture mechanism to the other at definite conditions (see Fig. 6.28c). Here, at either stage, three possible variants of crack growth are considered (either two intergranular and one transgranular cracks or two transgranular and one intergranular cracks) depending on character of the crack growth at the previous stage. For either from the three competing cracks, the value of K_{Ic} is calculated by using (6.140, 6.141) and (6.147). Then, a minimum value of K_{Ic} states a path of the crack growth at considered stage. Again, a final crack tip becomes an initial one at the next stage of the crack growth. The process is continued up to attainment by the crack of opposite side of the sample or open porosity. The consideration of mixed mechanism is caused by the replacement of the crack growth character at different sections that is supported by existing test results for different ceramics [1368]. Block-scheme for one realization of computational process of the macrocrack growth in model structure is presented in Fig. D.3. A simulation veracity of computational algorithm of the macrocrack propagation taking into account microstructure microstructure has been found by direct verification on test example.

A fabrication of optimum HTSC microstructures implies an inclusion in the consideration of proper crack shielding (amplification) mechanisms and the material fracture resistance. These mechanisms for YBCO and BSCCO ceramics and composites will be considered in Chaps. 10 and 11.

Chapter 7
Modeling of BSCCO Systems and Composites

7.1 Transformation of Bi-2212 to Bi-2223 Phase

Understanding the mechanism and kinetics of the Bi-2212-to-Bi-2223 phase transformation is very important to fabricate Bi-2223 superconductors with high structure-sensitive properties. The most widely used method of processing Bi-2223 materials is to mix Bi-2212, $CaPbO_3$ and CuO powders and to anneal the mixture in a sealed tube at about 830 °C during 10–100 h [224]. All models proposed for phase transformation of Bi-2212 into Bi-2223 suggest a diffusion-controlled, two-dimensional transformation due to different mechanisms of chemical reactions [65, 581, 592, 1030, 1058, 1189, 1880]. In order to analyze the kinetics of Bi-2223 formation, the Avrami equation is usually used [592, 1058, 1880]:

$$\ln\left(\frac{1}{1-C}\right) = K_0 \exp(-U/RT)t^{\alpha}, \tag{7.1}$$

where C is the fraction of Bi-2223 phase transformed at time t, T is the temperature, U is the activation energy, $K_0 = 1.71 \times 10^{-22}$ is the rate constant, R is the universal gas constant and α is the Avrami exponent. The Avrami exponent α obtained from test data can provide some insight into the reaction mechanism. Almost all of the above-mentioned authors conclude that their data support a diffusion-controlled, two-dimensional transformation of Bi-2212 into Bi-2223 phase. The values of the Avrami exponent, which they obtained, varied greatly, ranging from 0.5 [1880] to 1–1.5 [592, 1058]. This seemed to be more consistent with one-dimensional diffusion-controlled transformation mechanism with a varying nucleation rate [726]. At the same time, transmission electron microscopy (TEM) investigations, using electron diffraction and lattice imaging, show that during the annealing process, the Bi-2212/Bi-2223 system consists of fast-growing intercalating Ca/CuO_2 bilayers instead of compact Bi-2223 domains [137]. Unlike the conventional nucleation-and-growth mechanism, where reactant diffuses from grain boundaries into the interior of the bulk material, leading to a compactly propagating "front" of the product, the transformation from Bi-2212 to Bi-2223

I. A. Parinov, *Microstructure and Properties of High-Temperature Superconductors*,
DOI: 10.1007/978-3-642-34441-1_7,

phase appears to be accomplished via the layer-by-layer intercalation of the extra Ca/CuO_2 planes into the Bi-2212 matrix. In this case, the Ca/CuO_2 bi-layer insertion into the matrix during transformation leads to formation of an edge dislocation [223]. Hence, it becomes obvious that the fast intercalation of individual Ca/CuO_2 plane into the Bi-2212 matrix is more probable as the path for Bi-2212 to Bi-2223 transformation than nucleation and growth of the Bi-2223 compact region.

The growth of the Ca/CuO_2 planes occurs much faster than the bulk cation diffusion. Then, it may be assumed that a distinct diffusion mechanism acts in the Bi-2212/Bi-2223 system, in which cation diffusion takes place through the cylindrical cavity created by the edge dislocation, which accompanies the insertion of a Ca/CuO_2 plane. These cavities are located at the interfaces between Bi-2212 and Bi-2223 phases, where the reaction occurs. As the transformation progresses, these pores move with the Bi-2212/Bi-2223 interface; thus, the reaction progresses uninhibitedly. Based on the above assumption and the layer-rigidity model [222, 1671] modified to the Bi-2212/Bi-2223 system, we consider the cation diffusion mechanism and calculate the size of the pointed void [223, 224].

7.1.1 Edge Dislocations as Channels for Fast Ion Diffusion

The transformation of Bi-2212 phase to Bi-2223 one can be regarded as chemical reaction between the Bi-2212 precursor and the secondary phases such as $CaPbO_3$ and CuO, to provide the required surplus of Ca and Cu. As shown in Fig. 7.1, the only structural difference between Bi-2212 and Bi-2223 is a pair of extra Ca/CuO_2 planes (in the case of Bi-2223) at the corresponding lattice expansion. Therefore, it is convenient to divide the Bi-2212-layered structure into two parts, namely (i) the block layers (called "host layers" below), which consist of BiO, SrO and the Ca/CuO_2 plane in Bi-2212 and (ii) the "gallery layers" into which the extra Ca/CuO_2 planes are inserted in the case of Bi-2223. Thus, we can consider the Bi-2223 as a stacking fault of Bi-2212. Therefore, the interfaces in a transforming system between Bi-2212 and Bi-2223 layers can be considered as an edge dislocation.

The approximately cylindrical pore created at the additional half plane of this dislocation is a line of vacancies, which can be an easy path for the additional Ca, Cu and oxygen ions to diffuse from the surface (or grain boundary) into the bulk material, provided the size of the pore is large enough. The size of the pores depends on the rigidity of the block layer as well as the compressibility of the Ca/CuO_2 plane along the *c*-axis. Obviously, the more rigid the layer causes the larger the pore. In the limit of infinite layer rigidity, the size of the pore will be infinite. At the annealing temperature about 830 °C, both $CaPbO_3$ and CuO are liquid. The grinding and mixing prior to the annealing make sure that this liquid phase is evenly coated around each Bi-2212 grain. As the Ca/CuO_2 plane nucleates near the surface of the Bi-2212 grain (or grain boundary), the pore created by the partially inserted Ca/CuO_2 plane opens up a channel for the reactant ions to diffuse into the

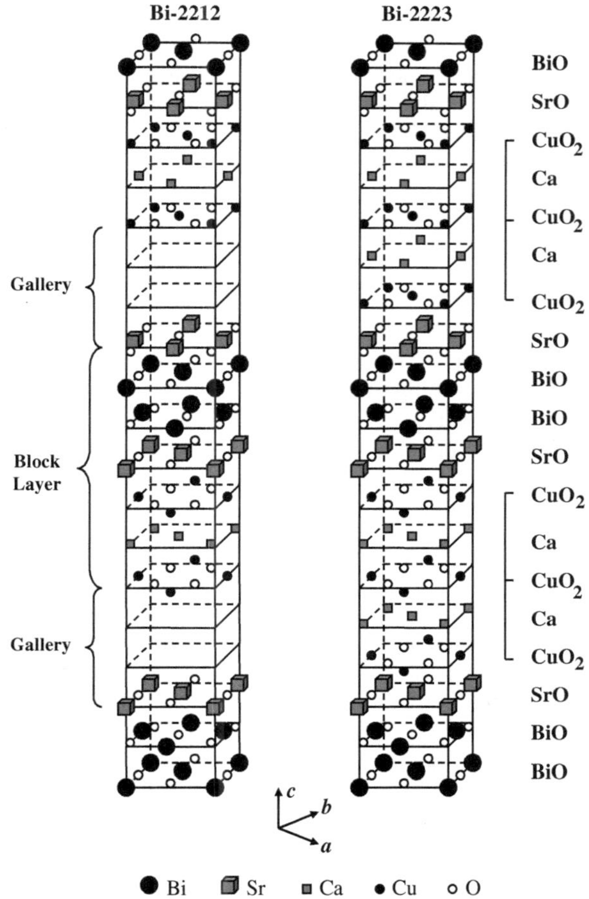

Fig. 7.1 The unit cells of Bi-2212 and Bi-2223. The layered structure of Bi-2212 can be divided into block layers and galleries for intercalation of the additional Ca/CuO_2 plane to form Bi-2223. The "gallery height" in the Bi-2212 structure is shown here expanded from its normal value for clarity of comparison of the two structures [224]

bulk. Thus, it permits the edge dislocation to climb and the reaction to proceed rapidly at this location. Therefore, it can be assumed that the transformation from Bi-2212 phase to Bi-2223 is limited by the nucleation rate of the Ca/CuO_2 plane near the Bi-2212 grain boundary and the diffusion of reactant ions along the moving dislocation lines.

7.1.2 The Layer-Rigidity Model

The volume expansion at the cores of the edge dislocations (caused by additional atomic half-planes) formed during the Bi-2212 to Bi-2223 transformation is connected with the large anisotropy in physical properties of these layered systems. It is reasonable to assume that the major expansion takes place along the direction perpendicular to the layers, denoted as the *c*-axis. As it has been shown by diffraction tests the lattice parameters along the *a*- or *b*-axis change very small during Bi-2212/Bi-2223 transformation. At the same time, the lattice parameter along the *c*-axis increases from 30.9 up to 37.8 Å [224].

Basing on the structural difference between Bi-2212 and Bi-2223 phases, it is regarded as an insertion of Ca/CuO_2 in the interior of Bi-2212 matrix. Then, the Bi-2212/Bi-2223 system can be considered as a type of intercalation compound in the form of $A_{1-x}B_xL$ with $0 \le x \le 1$, where B is the intercalant (additional Ca/CuO_2 layer in Bi-2223 phase), A is a vacant layer in the Bi-2212 phase (see Fig. 7.1), which will be considered as an intercalant of a smaller size, and L denotes the host layer which represents the rest of the structure. The single phases Bi-2212 (AL) and Bi-2223 (BL) present the limiting cases for $x = 0$ and $x = 1$, respectively. The dislocation at the Bi-2212/Bi-2223 interface can be modeled as an intercalation compound with B, occupying the semi-infinite plane, as shown in Fig. 7.2.

Two different models have been proposed to study the *c*-axis expansion of intercalation compounds [222, 1671]. In both models, it is assumed that the host layers have finite transverse as well as bending rigidity. At the same time, the intercalants have finite compressibility and different sizes. In the first model (*bi-layer model*) [1671], it is assumed that compressibility of the host layer is much smaller than that of the intercalant, that is, the correlation between various galleries can be ignored. In the second model (*multilayer model*) [222], it is assumed that the compressibility of the host layer is much larger than that of the intercalant, that is, the correlation between various galleries can be mapped into an Ising-type model. It may be suggested that the first type model is more suitable for Bi-2212/Bi-2223 compound. Let the compressibility of the host layer be about 10 % of that of the intercalant, because the compressibility of the host layer is inversely proportional to its thickness, but this thickness is much larger than the mean

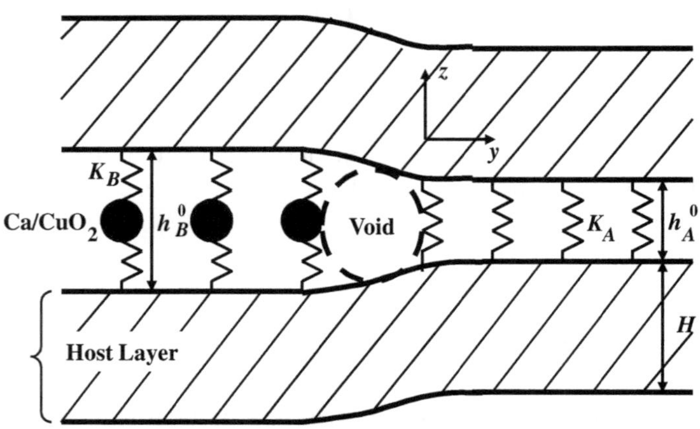

Fig. 7.2 Side view of the intercalants in the transforming Bi-2212/Bi-2223 system. The additional Ca/CuO_2 plane is shown schematically as large intercalant "atoms". The rest of the structural features are shown schematically as the "host layer." The separation between the host layers is given by the local gallery height, h_i [224]

gallery height in the Bi-2212/Bi-2223 compound. Here, a spring model that describes both the layer rigidity and the size and stiffness of the intercalant is considered (see Fig. 7.2).

Consider a layered system with composition $A_{1-x}B_xL$, where L represents the host layer of thickness H, which consists of BiO, SrO and Ca/CuO_2 planes in the Bi-2212 phase; A and B are two various types of intercalants, which occupy a set of well-defined lattice sites. The total energy of the system can be presented by a sum of two major contributions: one is associated with the interaction between the intercalants and the host atoms and the other - between host atoms themselves. It is assumed that the intercalants are "frozen" into the align structure (i.e., the nucleation time of the Bi-2223 phase is much shorter than the time for diffusion of the Ca, Cu and O ions from the grain boundary in the bulk). Thus, the direct interaction energy between the intercalants does not play any role in layer distortion. Because the compressibility of the host layer is much smaller than the intercalant, the interlayer correlation is mostly related to the various thickness of the host layer instead of the correlation between intercalants of various galleries. As the variation of the host layer thickness, ΔH, is very small compared to H, it can be neglected by the interlayer correlation and be considered a single gallery, bounded by two host layers with intercalants, occupying the lattice sites inside the gallery.

Following [224], the total energy of the host layer-intercalant system, presented in Fig. 7.2, can be written as

$$E = \frac{1}{2}\sum_i K_i(h_i - h_i^0)^2 + \frac{1}{2}K_T\sum_{i,\delta}(h_i - h_{i+\delta})^2 + \frac{1}{2}K_F\sum_i\left[\sum_\delta(h_i - h_{i+\delta})\right]^2, \tag{7.2}$$

where h_i is the local gallery height at site i, where an intercalant (either A or B) sits; h_i^0 is the gallery height for the pure system AL or $BL(h_A^0 = 2.21$ Å for Bi-2212 (AL) and $h_B^0 = 6.18$ Å for Bi-2223 (BL) from the diffraction measurements); K_i is the spring constant, representing the compressibility of the local intercalant i. The terms, involving the spring constants K_T and K_F, describe, respectively, transverse and bending rigidity of the host layers [1671]. They are related to the elastic constants of the Bi-2212 compound by the following relations:

$$K_T = \frac{H}{a_0}c_{44}; \tag{7.3}$$

$$K_F = \frac{H^3}{12a_0^3}\left[\frac{(c_{11} + c_{33})c_{33} - 2c_{13}}{c_{33}}\right], \tag{7.4}$$

where $a_0 = 3.8$ Å is the lattice constant in the ab-plane and $H = 14.35$ Å is the thickness of the host layer. The required elastic constants for Bi-2212 have the next values [224]: $c_{11} = 24.08 \times 10^{11}$ dyn/cm^2; $c_{13} = 6.71 \times 10^{11}$ dyn/cm^2; $c_{33} = 14.59 \times 10^{11}$ dyn/cm^2; $c_{11} = 6.18 \times 10^{11}$ dyn/cm^2.

The compressibility $K_A = K_i\ (A)$ of the intercalant A (the vacancy) can be obtained from the elastic constants as

$$K_A = \frac{a_0}{h_A^0} c_{33}. \tag{7.5}$$

The compressibility $K_B = K_i(B)$ of the intercalant B (the Ca/CuO_2 plane) cannot be calculated due to the lack of data for the elastic properties of Bi-2223. However, it can be assumed that $K_A < K_B$. Then, we shall calculate the gallery height for the case of $K_A = K_B$, estimating thus the minimum size of the pore. Minimizing E in (7.2) with respect to h_i, we find

$$\mathbf{Mh} = \mathbf{\Phi}, \tag{7.6}$$

where $\mathbf{M}$ is a tridiagonal matrix with elements

$$M_{ii} = K_i + 2K_T + 6K_F; \quad M_{i,i\pm 1} = -K_T - 4K_F; \quad M_{i,i\pm 1} = K_F, \tag{7.7}$$

but $\mathbf{h}$ and Φ are two column vectors:

$$\mathbf{h} = \begin{pmatrix} h_1 \\ h_2 \\ \cdots \\ h_N \end{pmatrix}; \quad \mathbf{\Phi} = \begin{pmatrix} K_1 h_1^0 \\ K_2 h_2^0 \\ \cdots \\ K_N h_N^0 \end{pmatrix}. \tag{7.8}$$

By diagonalizing matrix $\mathbf{M}$, h_i can be obtained. In the case of $K_A = K_B = K$, the dispersion relation for the algebraic system is given as [224]

$$\lambda_q = K + 2K_T[1 - \cos(qa_0)] + 4K_F[1 - \cos(qa_0)]^2, \tag{7.9}$$

and the expanded form of $(\mathbf{M}^{-1})_{nm}$ can be expressed as [224]

$$\left(\mathbf{M}^{-1}\right)_{nm} = \frac{1}{N}\sum_q \frac{\exp[iqa_0(n - m)]}{\lambda_q}, \tag{7.10}$$

with $q = 2\pi r/Na_0$; $r = 1, 2\ldots N$ and λ_q being the eigenvalues of the matrix $\mathbf{M}$. Then we have

$$h_n = \frac{1}{N}\sum_m \left(\mathbf{M}^{-1}\right)_{nm}\Phi_m = \frac{1}{N}\sum_q \frac{K}{\lambda_q}\sum_m e^{iqa_0(n-m)} h_m^0. \tag{7.11}$$

Considering a system with half of the yz-plane occupied by intercalant B (i.e., Bi-2223 phase), we assume the dislocation disposes at $y = z = 0$, and the equilibrium height h_n^0 is defined as

$$h_n^0 = \begin{cases} h_B^0, & \text{for } n = -\frac{N}{2} + 1, -\frac{N}{2} + 2, \ldots, -1; \\ h_A^0, & \text{for } n = 1, \ldots, \frac{N}{2} - 1, \frac{N}{2}. \end{cases} \tag{7.12}$$

Taking the limit of $N \to \infty$ and changing the summation over q by integration, we find that

$$h(y) = \frac{h_A^0 + h_B^0}{2} + \frac{h_A^0 - h_B^0}{4\pi} \int_0^{\pi} \frac{K\sin[\theta(y/a_0 - 1/2)]}{\lambda_\theta \sin(\theta/2)} d\theta, \qquad (7.13)$$

where $\theta = qa_0$, and $\lambda_\theta = K + 2K_T[1 - \cos(\theta)] + 4K_F[1 - \cos(\theta)]^2$. The solid line in Fig. 7.3 shows the profile of h_n for $K_A = K_B$ and the values of parameters defined in (7.3–7.5). The obtained results can be compared with the model of the stacking fault, presenting an additional CuO plane in the $YBa_2Cu_3O_7$ matrix. For this, the theoretical profile of h_n is calculated, using the elastic constants, obtained from the atomistic simulation of $YBa_2Cu_3O_7$ [82], and one is shown by dashed line in Fig. 7.3.

Of the three types of reactant ions (Ca^{2+}, Cu^{2+}, and O^{2-}), the largest is O^{2-}, which has a diameter of about 2.9 Å. As it is shown in Fig. 7.3, the lateral size of the pore is about twice as large as diameter of the oxygen ion, if we consider the pore to be the region, where $h_n \geq 3$ Å for $y \geq 0$. Hence, the dislocation line is indeed likely to be an easy path for reactant ions to diffuse from the liquid phase at the Bi-2212 grain boundary into the bulk, causing the edge dislocation (the additional Ca/CuO_2 plane) to rapidly climb into Bi-2212.

7.1.3 Dynamics of Bi-2223 Phase Growth

The Avrami exponent in the Bi-2212-to-Bi-2223 transformation using (7.1) and presented layer-rigidity model is estimated. Because the reactant always contacts with the Bi-2212/Bi-2223 interface, the reaction rate of Bi-2212 to Bi-2223 phase perpendicular to the dislocation line is independent of the volume fraction of

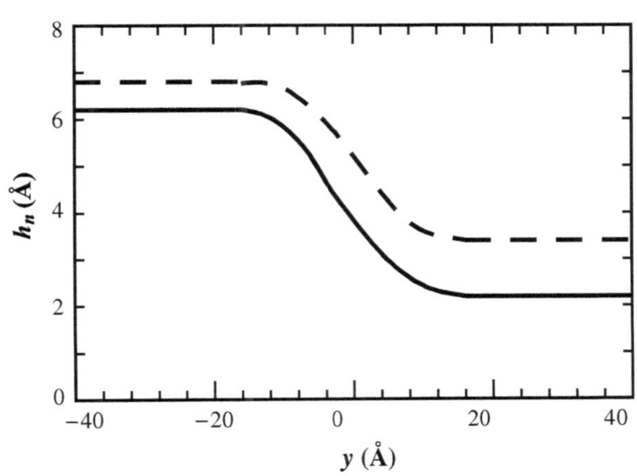

Fig. 7.3 The spatial profile of the gallery h_n. *Solid line* presents Bi-2212/Bi-2223 system, but *dashed line* corresponds to $YBa_2Cu_3O_7$/$YBa_2Cu_4O_8$ system (at $K_A = K_B = K$)

Bi-2223 phase, and the growth in this direction is very fast. Therefore, it is necessary to regard only the diffusion along the dislocation line and assume that the reactants are consumed immediately upon reaching the Bi-2212/Bi-2223 interface, because the diffusion is probably a much slower process than any local rearrangement of atoms, which might be required. Then, the growth rate of the product layer along the dislocation line is given as [224]

$$dl/dt = D/l, \tag{7.14}$$

where l is the distance from the grain boundary measured along the dislocation line and D is the effective diffusion factor of the reactant ions. Because the diffusion factor and the pore size (see the previous section) are independent of the volume fraction of Bi-2223, dividing variables and integrating Eq. (7.14), we obtain

$$l = \sqrt{2Dt}. \tag{7.15}$$

The volume of the reactant ions, consumed by the formation of the product phase, is given as

$$R(t) = S\sqrt{2D}t^{1/2}, \tag{7.16}$$

where S is the size of the pore. For a Bi-2212 grain with volume V, the number of Ca/CuO_2 planes nucleated at the grain boundary in the time range between t' and $(t' + dt')$, is $I(t')Vdt'$, where $I(t')$ is the nucleation rate per unit area of the grain boundary. Hence, the total volume of reactant ions, consumed at time t (assuming nucleation can occur everywhere at the grain boundary, including the transformed region), is given as

$$V_e^C = A\int_0^t I(t')R(t - t')dt' = S\sqrt{2D}A\int_0^t I(t')(t - t')^{1/2}dt', \tag{7.17}$$

where A is the total area of the grain boundary. Note that V_e^C differ from the actual volume of reactant consumed because it is assumed that nucleation can take place in the transformed region. All regions are considered, including those continuing growth irrespective of other regions. In order to correct this problem, we sign the volume of the transformed region through V^C and state the relation between V^C and V_e^C. Consider a small region, of which a fraction $[1 - (V^C/V)]$ remains non-transformed (where V is the total volume of the grain). During the time range, dt, taking into account the total transformed volume, $d\,V_e^C$, a fraction $[1 - (V^C/V)]$ on the average will be in previously non-reacted material and thus contributes to dV^C, while the reminder of the $d\,V_e^C$ will be in the already transformed region. Then, we obtain

$$dV^C = \left(1 - \frac{V^C}{V}\right)dV_e^C. \tag{7.18}$$

Dividing variables and integrating, we have

$$V_e^C = -V\ln\left(1 - \frac{V^C}{V}\right). \tag{7.19}$$

Let $[1 - (V^C/V)] = 1 - C$, where C is the volume fraction of Bi-2223. Substituting (7.19) into (7.17), we obtain

$$\ln\left(\frac{1}{1-C}\right) = \frac{S\sqrt{2D}A}{V}\int_0^t I(t')(t - t')^{1/2}dt'. \tag{7.20}$$

The volume fraction of Bi-2223 depends directly on the nucleation rate $I(t)$. It is often assumed that the time dependence of the number of nucleation sites is a classical first-order rate process, that is, [224]

$$dN(t)/dt = -fN(t), \tag{7.21}$$

where f is the frequency of an empty gallery in Bi-2212 phase turns into a nucleation site for Bi-2223 phase and $N(t)$ is the number of such nucleation sites at time t. Integrating the above equation yields

$$I(t) = fN(t) = fN_0\exp(-ft), \tag{7.22}$$

where N_0 is the number of empty galleries in the Bi-2212 grain at $t = 0$. Substituting (7.22) into (7.20) and integrating yields

$$\ln\left(\frac{1}{1-C}\right) = \frac{S\sqrt{2D}N_0A}{V}\left[t^{1/2} + \frac{i}{2}\sqrt{\frac{\pi}{f}}\exp(-ft)\mathrm{erf}(i\sqrt{ft})\right], \tag{7.23}$$

where erf(z) is the error function with complex variable z, which can be expanded as [915]

$$\mathrm{erf}(z) = \frac{2}{\sqrt{\pi}}\int_0^z e^{-\zeta^2}d\zeta = \frac{2}{\sqrt{\pi}}\left(z - \frac{z^3}{3} + \frac{1}{2!}\frac{z^5}{5} - \frac{1}{3!}\frac{z^7}{7} + \ldots\right). \tag{7.24}$$

Now, we can estimate the Avrami exponent α for various values of f.

In the limiting case, when ft is very small, that is, the time to produce Bi-2223 nuclei is much longer than the time to grow the Ca/CuO_2 plane through the sample, we can expand (7.24) in terms of ft. Taking into account (7.24) and the expansion [915]

$$e^z = 1 + z + \frac{z^2}{2!} + \frac{z^3}{3!} + \cdots, \tag{7.25}$$

we obtain

$$\ln\left(\frac{1}{1-C}\right) = \frac{S\sqrt{2D}N_0A}{V}\left(\frac{2}{3}ft^{3/2} - \frac{4}{15}f^2t^{5/2} + \cdots\right). \quad (7.26)$$

Comparing (7.26) with (7.1), we have $\alpha \approx 1.5$, which agrees well with the test results for the preheated sample [592, 1058]. Note that the assumption of this limiting case is valid only if reactants (Ca, Cu and O) are evenly coated around the Bi-2212 grains and the Bi-2212 grains are small enough. Therefore, the model is more consistent with tests with small precursor powder, which are heat-treated before annealing to ensure the even distribution of the reactants.

On the other hand, if ft very large, that is, the time to produce a Bi-2223 nuclei is much shorter than the time to grow the Ca/CuO_2 planes through the sample, we can consider that all the empty galleries in Bi-2212 have turned into nucleation sites for Bi-2223 before the Ca/CuO_2 planes begin to grow. In this case, we find from (7.23)

$$\ln\left(\frac{1}{1-C}\right) = \frac{S\sqrt{2D}N_0A}{V}t^{1/2}. \quad (7.27)$$

Hence, $\alpha \approx 0.5$. This will be the case if there is local shortage of reactant ions, which will lead to the slow growth of Ca/CuO_2 plane. Tests in [1880] apparently fall into this situation. The pointed study has stated an increase of the Avrami exponent α together with the temperature from 0.5 at 840 °C to 0.79 at 870 °C. The small value of α in [1880] is due to either the large Bi-2212 grain size as indicated by the high porosity ($\sim$40 %) or the uneven distribution of the reactant ions. When the temperature rises, the growth of the Ca/CuO_2 plane accelerates due to the increase of ion diffusion constant, *D*. In this case, the reactant ions distribute more evenly around the Bi-2212 grain, thanks to the decreased viscosity of the ionic liquid of Ca, Cu and O.

Thus, the Avrami exponent depends directly on the nucleation rate of the Bi-2223 phase and growth of the additional Ca/CuO_2 plane in the Bi-2212 matrix changing between 0.5 and 1.5.

7.1.4 Formation Energy of Bi-2223 Phase

As it has been shown in experiments, less than 30 min is to be enough for intercalation of one Ca/CuO_2 plane into Bi-2212 matrix across 2 μm sample [223]. Moreover, due to the presented model, the ion diffusion should be faster in the pore than that in the bulk. Then, we can estimate the formation energy of Bi-2223 from Bi-2212 phase.

The intercalation of the Ca/CuO_2 bi-layer into the Bi-2212 matrix can be presented as a single jog on the otherwise straight dislocation (see Fig. 7.4a). The climb motion of the dislocation can be accomplished by jog nucleation and propagation along the dislocation line. When the reactant ions (Ca^{2+}, Cu^{2+} and

O^{2-}) are deposited at the jog, one propagates and the dislocation climbs upward by a distance, Δx, as shown in Fig. 7.4b. Both these processes require net mass transport, whose rate is controlled by the diffusion of the reactant ions. These ions, moving to a preexisting jog, simply translate the jog. This translation does not change core energy and core structure of the dislocation, as shown in Fig. 7.4c.

Considering the reactant ions as particles, suffering random force of their surroundings, the equation of the jog motion can be written as

$$\eta v = \boldsymbol{F}_c + \boldsymbol{F}_{el}, \tag{7.28}$$

where η is the viscosity of the surrounding matrix, v is the jog rate, $\boldsymbol{F}_c$ is the force caused by the reduction of chemical energy at Bi-2212/Bi-2223 transformation and $\boldsymbol{F}_{el}$ is the elastic force suffered by the jog.

Assuming the rate of jog motion to be constant, the time t which is necessary to insert a single Ca/CuO_2 bi-layer in Bi-2212 matrix of a $L_1 \times L_2$ rectangular sample is found as

$$t = \frac{L_1 L_2}{\Delta x v}, \tag{7.29}$$

and we have from Fig. 7.4c:

$$|\boldsymbol{F}_{el}| = \sigma_{33}\varepsilon_{33}b_z l, \tag{7.30}$$

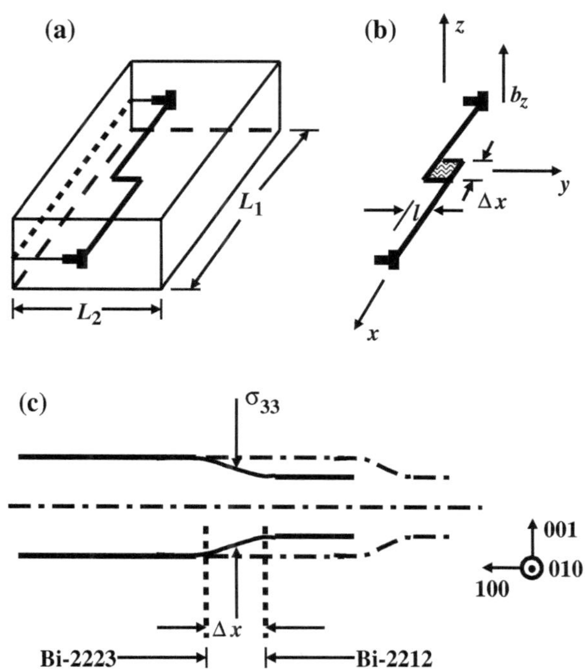

Fig. 7.4 Schematic drawing of the dislocation model for intercalation of Ca/CuO_2 bi-layer into Bi-2212 matrix: **(a)** The insertion of a Ca/CuO_2 bi-layer, described as climb motion of a dislocation in a grain with rectangular platelet shape; **(b)** jog configuration of the dislocation, the shading area shows the transition region; **(c)** core configuration of the dislocation view along the [010] direction, the channel of diffusion for the reactants is shown [223]

where $\sigma_{33} = c_{33}\varepsilon_{33}$ is the stress, ε_{33} is the strain, $c_{33} = 14.59 \times 10^{11}$ dyn/cm^2 is the elastic constant of the Bi-2212 matrix and $b_z = 4.4$ Å is Burgers vector of the dislocation [223]. If we regard the intercalation of a single Ca/CuO$_2$ bi-layer for each Bi-2212 unit-cell, that is, a local Bi-4435 structure, which is often observed in the early stage of the transformation, then the strain $\varepsilon_{33} = 0.142$ [137]. Moreover, we take into account that $l \approx \Delta x \approx b_z$ [224].

The bulk diffusion factor of O^{2-} in the ab-plane of Bi-2212 phase at 850 °C is approximately 1.1×10^{-9} cm^2/s [223]. The diffusion constant for the reactant ions in the pore depicted in Fig. 7.4c is at least one order of magnitude larger than that. Using the relation between the diffusion constant, D, and the viscosity constant, η,

$$\eta = \frac{k_B T}{D}, \tag{7.31}$$

where k_B is Boltzmann constant and $T = 850$ °C is the temperature we obtain value of η. Finally, we estimate the formation energy per Ca/CuO$_2$ bi-layer E_c from (7.28) and (7.30) as

$$E_c = -\boldsymbol{F}_c \cdot \Delta\, \boldsymbol{x} \leq 3.5\ \text{eV}. \tag{7.32}$$

7.1.5 Effect of Deformation on Bi-2212/Bi-2223 Transformation

Obviously, it may be assumed that deformation processes render direct influence on Bi-2212/Bi-2223 phase transformation. In the case of fabrication of the BSCCO tapes by using oxide-powder-in-tube method, the mechanical deformation effects have been investigated at rolling the wires into tapes with different thickness reduction [1016]. Each rolling step produced an approximate 20 % thickness reduction. The deformation ratio R is defined as the thickness reduction of the tape:

$$R = (t_0 - t_f)/t_0, \tag{7.33}$$

where t_0 and t_f are the original and final thicknesses of the superconductor core, respectively. The obtained test dependencies of Bi-2223 phase concentration on deformation and duration of final annealing at $T = 833$ °C are shown in Fig. 7.5. These results demonstrate that the mechanical deformation strongly affects the phase transformation of Bi-2212 to Bi-2223 in the early stages of annealing, implying that different mechanisms may dominate the transformation kinetics upper and lower definite value of the deformation ratio (here 60 %). This unusual behavior, demonstrating minimum value of transformation kinetics at $R = 60$ %, states an existence of two major effects for BSCCO tapes during the mechanical preparation, namely (i) a part of deformation energy is absorbed by the oxide core as fracture energy, splitting the Bi-2212 crystal and producing new interfaces; (ii) a part of the energy is stored in the crystals as the result of lattice distortion of the Bi-2212 crystals. The first effect takes place when the mechanical deformation ratio, R, is relatively low, while the second effect occurs when higher mechanical deformation is applied to the BSCCO tapes [1016]. Both effects have different contributions to the Bi-2212/Bi-2223 phase transformation process. Obviously, the area of the fractured surfaces increases as the mechanical deformation ratio enhances, increasing the surface energy. This surface energy causes re-crystallization of the fractured Bi-2212 crystals, reducing the energy of the system during final annealing. At the same time, the lattice distortion increases the internal energy and the phase balance, facilitating phase transformation, forming Bi-2223 instead of the re-crystallized Bi-2212 crystals. Hence, there are two possible types of annealing behavior for the Bi-2212 phase, either the Bi-2212 re-crystallization to minimize the surface area or reaction leading to the phase balance and forming the Bi-2223 phase. Hence, the annealing behavior of the deformed Bi-2212 crystals is determined either by the surface energy or by the chemical potential and internal energy of the Bi-2212 crystals. When the surface energy is higher than the chemical potential and the internal energy, re-crystallization of the Bi-2212 crystals occurs. At the same time, when the chemical potential and internal energy dominate the system, the chemical reaction takes place. Thus, it could be assumed in this case that re-crystallization is probably a favorable process at $R < 60\ \%$ and the trend increases with the deformation ratio. The surface area rapidly increased with the increasing of R, until the surface area reached a "saturation" value. This slowed down the Bi-2223 formation with increasing mechanical deformation. However, at $R > 60\ \%$, the lattice distortion of the Bi-2212 crystals increased rapidly, causing an increase of the internal energy with increasing mechanical deformation.

Figure 7.6 is a schematic illustration, showing that the driving force for the Bi-2223 phase formation is enhanced by the internal energy. Annealing the sample at temperatures above the equilibrium temperature of the Bi-2212 and Bi-2223 phases, T_e ($T > T_e$), allows the Bi-2223 phase to form because of the lower free energy. At the same time, the formation of the Bi-2212 phase occurs at lower annealing temperatures ($T < T_e$). As the internal energy of the Bi-2212 phase is increased by the mechanical deformation, the free energy of Bi-2212 phase becomes $G'_{\text{Bi}-2212} = G_{\text{Bi}-2212} + E_{\text{int}}$. One is presented by the curve parallel to the original curve, defining free energy of the Bi-2212 phase ($G_{\text{Bi-2212}}$) without the mechanical deformation, but is shifted up to $G'_{\text{Bi}-2212}$ (see Fig. 7.6). Obviously, it

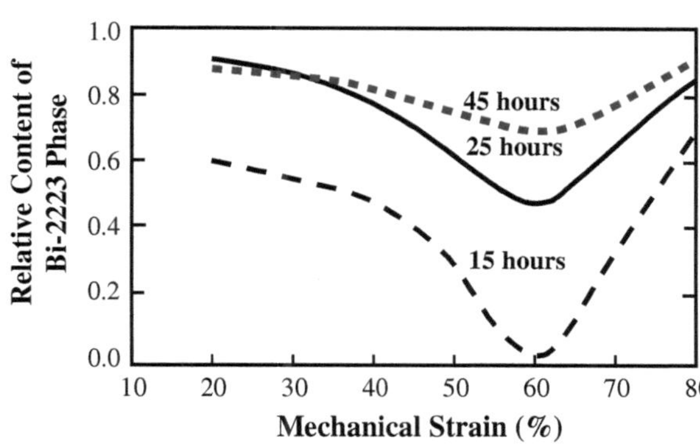

Fig. 7.5 Bi-2223 phase content versus the mechanical deformation ratios [1016]

results in a decrease in the equilibrium temperature of the Bi-2212 and Bi-2223 phases ($T_e' < T_e$). As a result of the reduced equilibrium temperature, the driving force for the Bi-2223 phase formation from the overheating (833 °C – T_e') increases from ΔG to $\Delta G'$, as shown in Fig. 7.6. In this case, the chemical potential and the internal energy replace the surface energy as the dominant energies of the system. Therefore, after annealing, the content of the Bi-2223 phase increased together with the deformation ratio, when $R > 60\ \%$ [1016]. Moreover, apart from the pointed effects of mechanical deformation, there are other parameters defining the Bi-2212/Bi-2223 phase transformation, namely the effective diffusivity, concentration gradient and grain size. For example, the fine grain size leads to an increase in the grain boundary (interface) area. In this case, the effect of the surface diffusion would be stronger than that of the lattice diffusion during the phase transformation and grain growth processes. This implies that the formation of Bi-2223 phase should increase together with the parameter of R. In addition, the mechanical deformation might also influence the number of nuclei and the nucleation rate. Generally, the nucleation rate is independent of the nucleation sites. The dependence between the number (or volume) of nuclei, N, the nucleation rate, v, the area of the nucleation sites, A, and the annealing time, t, can be expressed as [1016]

$$N = vAt. \tag{7.34}$$

Obviously, for a certain nucleation rate, higher mechanical deformation exposes more surfaces, providing a large amount of Bi-2223 nuclei that facilitate the phase transformation from Bi-2212 to Bi-2223. However, the various mechanical deformation processes may result in different grain boundary areas, defining value of R, corresponding to the minimum formation of Bi-2223 phase. Thus, mechanical deformation causes both the thermodynamics and kinetics of the BSCCO phase formation.

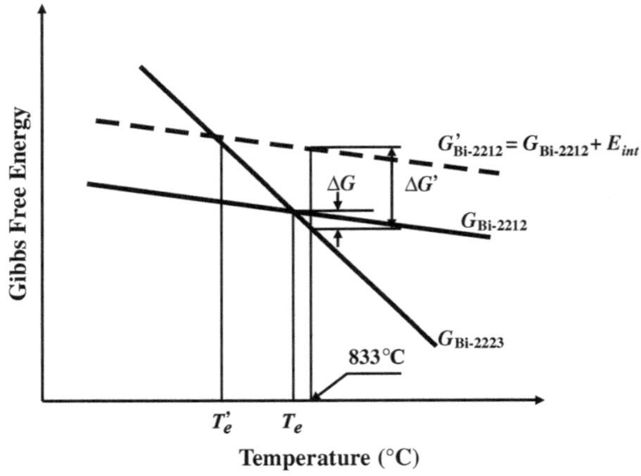

Fig. 7.6 A schematic illustration shows the mechanism, in that the driving force of the Bi-2223 phase formation is enhanced by the internal energy [1016]

7.2 Modeling of Preparation Processes for BSCCO/Ag Tapes

As it has been noted in the previous chapters, HTSC tapes are subjected to multi-staged and complex thermal, mechanical and magnetic treatments during preparation. Below, we consider several model and computational approaches directed to the process optimization of BSCCO/Ag tape fabrication and attainment of the improved structure-sensitive and superconducting properties.

7.2.1 Sample Texturing by External Magnetic Field

When superconductor is placed in a magnetic field, the axis of maximum susceptibility for each grain aligns with the magnetic field direction, that is, the grains should align with the *c*-axis parallel to the external magnetic field. For quantitative estimation of the material texturing degree under magnetic field, consider the rotation of superconductor grains into liquid in the early stages of their growth (i.e., when the particles can rotate without interacting) [470]. We assume in this model that the magnetic field does not influence on the processes of grain nucleation and growth into liquid.

Let an anisotropic grain with volume V be placed in a magnetic field $\boldsymbol{H}$. Then, the change in magnetic energy of the grain with a change in magnetic field can be written as

$$dE_m = -\boldsymbol{M}Vd\boldsymbol{H} = -(M_c\cos\theta + M_{ab}\sin\theta)VdH, \tag{7.35}$$

where $\boldsymbol{M}$ is the magnetic moment per unit volume, which can be resolved in the two directions c and ab and θ is the angle between the magnetic field and the c-axis of the grain. For HTSC in their normal state, the magnetic moments M_c and M_{ab} are paramagnetic moments. Then, (7.35) can be reduced to the form:

$$dE_m = -(\chi_c\cos^2\theta + \chi_{ab}\sin^2\theta)VHdH, \tag{7.36}$$

where χ_c and χ_{ab} are the paramagnetic susceptibilities along the c-axis and in the ab-plane, respectively. Integrating (7.36), we obtain the expression for the magnetic energy of a grain:

$$E_m = \int_0^H dE_m = -(\chi_c\cos^2\theta + \chi_{ab}\sin^2\theta)VH^2/2, \tag{7.37}$$

or

$$E_m(\theta, H) = -(\chi_{ab} + \Delta\chi\cos^2\theta)VH^2/2, \tag{7.38}$$

where $\Delta\chi$ is the difference in the volume susceptibilities of the grain.

Subsequent to the nucleation event, nuclei will start their growth under the influence of a magnetic field. In the early stages of growth, the grains are completely surrounded by a liquid phase and thus they can be regarded as small particles, rotating in a free medium without intergrain interactions. Consider the probability, $f(\theta)$, that a grain has an orientation with angle θ under the influence of a magnetic field. It can be expressed, according to a classic Boltzmann statistics, as

$$f(\theta)d\theta = \frac{\exp[-E(T,H,\theta)/k_BT]d\theta}{\int_0^{\pi/2} \exp[-E(T,H,\theta)/k_BT]d\theta}. \tag{7.39}$$

Then, for total number of grains, n, the mean number of grains with an orientation between θ and $\theta + d\theta$ can be given as

$$n(\theta)d\theta = nf(\theta)d\theta = n\frac{\exp[-E_m(T,H,\theta)/k_BT]d\theta}{\int_0^{\pi/2} \exp[-E_m(T,H,\theta)/k_BT]d\theta}. \tag{7.40}$$

Hence, the distribution, $n(\theta)$, can be related to an alignment parameter, which is used to quantify the degree of texture in melt-processed superconductors under the influence of a magnetic field. This alignment parameter, F, such that $F = 1$ for a completely alignment structure and $F = 0$ for a totally random structure, can be found as

$$F = 1 - \frac{s^2}{s^2_{H=0}}, \tag{7.41}$$

where s^2 is the variance of the grain distribution for a particular processing condition and $s^2_{H=0}$ is the variance of the distribution in the absence of a magnetic field. It is known that the anisotropy of molar magnetic susceptibility $\Delta\chi^{molar} \approx 22.5 \times 10^{-5}$ cm^3/mol [97, 1731]. Then, the anisotropy of volume magnetic susceptibility, $\Delta\chi$, used in (7.38), is 1.5×10^{-6}, if we assume a density 5.5 g/cm^3 [1393] for the superconductor. Figure 7.7 shows the dependence of the parameter F on external magnetic field for various grain sizes at the temperature, $T = 875$ °C, which is approximately 5 °C smaller than the melting temperature of Bi-2212 phase.

As it is followed from Fig. 7.7, when the magnetic field increases, there is a trend for the texture to increase, except in cases where the grain size is small. Therefore, a high degree of alignment can be obtained by increasing the magnetic field and the grain size. In the case of larger grain sizes, the magnetic field tends to saturate and thus increasing the magnetic field has only a negligible effect on the degree of texture.

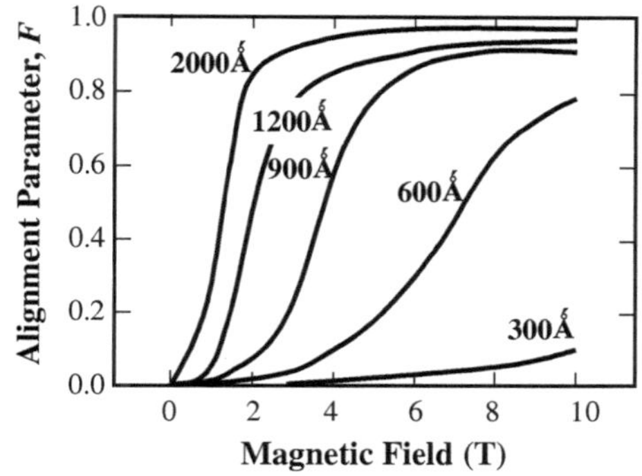

Fig. 7.7 Alignment parameter as a function of magnetic field for various grain sizes. The processing temperature is 875 °C [470]

7.2.2 Deformation at Tape Cooling

Deformations caused by compressing stresses, forming into ceramic core during sample cooling are defined. Known estimations of mechanical properties of the silver matrix and oxide superconducting filaments at all considered temperature intervals are used.

Young's modulus. Following [119], we assume a linear dependence of Young's modulus on temperature, $E^{Ag}(T)$; for BSCCO use data obtained in [548].

Plastic deformation. The linear relation between yielding stress, σ_y^{Ag}, and temperature, suggested in [119], uses for adaptation to the room-temperature results [641] and to value $\sigma_y^{Ag} = 0$, corresponding to the melting temperature. The yielding strain is found from Hooke's law. The behavior of the pressed multi-phase ceramic core is suggested to be elastic.

Thermal expansion. The behavior of parameters $\alpha^{Ag}(T)$ and $\alpha^{BSCCO}(T)$ at low temperatures is stated from results of [1265] and is fitted to the value 19×10^{-6} K^{-1} for Ag and to the room-temperature data for BSCCO [548]. At higher temperatures, we suggest a linear dependence up to 22×10^{-6} K^{-1} for Ag and 16.5×10^{-6} K^{-1} for BSCCO (at 1110 K).

At high temperatures, a different thermal compression leads to elastic deformations, both matrix and core. In this case, the deformations at temperature change on the value of ΔT are estimated, using the law "action—counteraction" for two-component tape, consisting of filaments and matrix. From Hooke's law,

$$\Delta\varepsilon^{BSCCO}(T) = \left[1 + \frac{FE^{BSCCO}(T)}{(1 - F)E^{Ag}(T)}\right]^{-1} (\alpha^{Ag} - \alpha^{BSCCO})\Delta T; \qquad (7.42)$$

$$\Delta\varepsilon^{Ag}(T) = \left[1 + \frac{(1 - F)E^{Ag}(T)}{FE^{BSCCO}(T)}\right]^{-1} (\alpha^{BSCCO} - \alpha^{Ag})\Delta T, \qquad (7.43)$$

where $F = \%BSCCO/(\%BSCCO + \%Ag)$ is the filling factor. The values of $\varepsilon^{BSCCO}(T)$ and $\varepsilon^{Ag}(T)$ are defined, integrating (7.42) and (7.43):

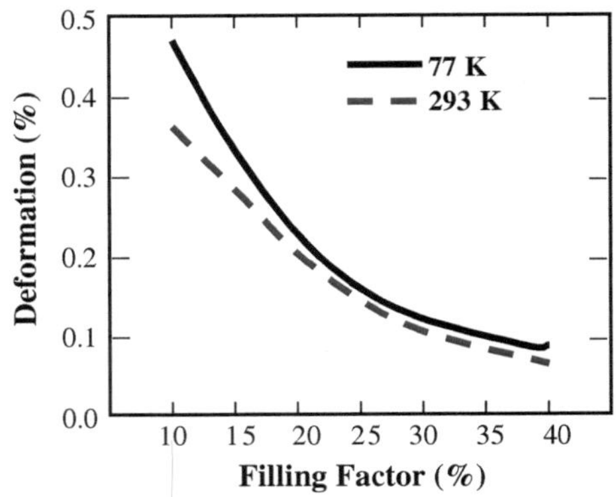

Fig. 7.8 Deformation of BSCCO core as a function of filling factor for tapes, cooled down to room temperature and cryogenic temperature (77 K)

$$\varepsilon^{BSCCO}(T) = \int_{T_m}^{T} \frac{\Delta\varepsilon^{BSCCO}}{\Delta T} dt; \; \varepsilon^{Ag}(T) = \int_{T_m}^{T} \frac{\Delta\varepsilon^{Ag}}{\Delta T} dt, \tag{7.44}$$

where T_m is the temperature of the tape annealing. A transition from elastic to plastic regime (i.e., from linear to non-linear dependence) occurs, when $\varepsilon^{Ag}(T)$ attains the yielding strain, $\varepsilon_y^{Ag}(T)$. Lower than the corresponding transition temperature, the silver matrix deforms plastically (non-linearly), leading to a loading of the superconductor by constant stress, σ_y^{Ag} [1341]. Finally, the required core deformation is found as

$$\varepsilon^{BSCCO}(T) = \varepsilon_y^{Ag}(T) \frac{(1 - F)E^{Ag}(T)}{FE^{BSCCO}(T)}. \tag{7.45}$$

The obtained numerical results are presented in Fig. 7.8 [1341].

7.2.3 Effects of Mechanical Loading

The introduction of "freedom parameter," Δf, estimates qualitatively the influence of different constraint factors on the mass-flow behavior and also presents the models of re-distribution of mass and flux in the powder compact due to plastic deformation of composite [628].

Freedom Parameter

Consider the compression of a two-dimensional slab of an ideal plastic solid between two flat anvils as shown in Fig. 7.9. If there is friction between the slab and the anvils, "barreling" occurs during the process (see Fig. 7.9b). In this case,

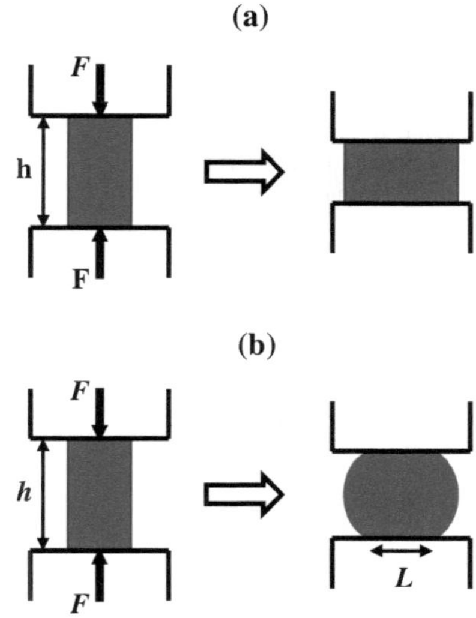

Fig. 7.9 Processes of two-dimensional compression: (**a**) two sides are free, and there is no friction between the slab and the anvils; (**b**) two sides are free, but there is substantial friction between the slab and the anvils

there is inhomogeneous stress–strain state of material. The pressure at the interface between the anvil and the slab needed to induce plastic deformation is given as [709]

$$P = 2k \exp\left(\frac{2\mu x}{h}\right), 0 \leq x \leq \frac{L}{2}. \tag{7.46}$$

The mean pressure is

$$P_c \approx 2k\left(1 + \frac{\mu L}{2h}\right), \tag{7.47}$$

where $2k$ is the flow yield, coinciding with the mean pressure at $\mu = 0$ and/or $L \ll h$; μ is the sliding friction factor; h is the sample height and L is the contact size. If there is sticking friction at the interface, then the pressure is [709]

$$P = 2k\left(1 + \frac{x}{h}\right), 0 \leq x \leq \frac{L}{2}, \tag{7.48}$$

and the mean pressure is

$$P_c = 2k\left(1 + \frac{L}{4h}\right). \tag{7.49}$$

The increasing pressure, or "friction hill," is caused by friction and reaches a maximum in the center (see Fig. 7.10). The higher the friction, the higher the pressure needed to cause mass flow. Obviously, the pressure strongly depends on the ratio between the height and the contact size, *h/L*. If the ratio *h/L* is small, then the required pressure to induce plastic flow in the material is high, and vice versa.

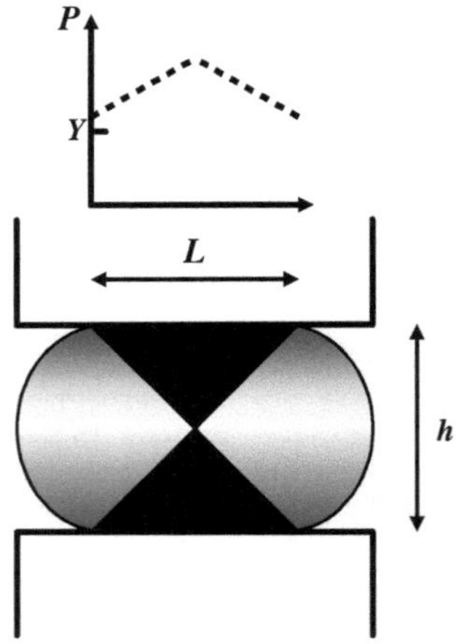

Fig. 7.10 Inhomogeneous deformation under compression (Y is the yield value of the sample). The *shaded regions* have experienced less deformation than the remainder of the slab

This means that the material has a large freedom to flow, which is found by the freedom parameter, $\Delta f = h/L$.

Pressing. The freedom parameter, Δf, can be used to describe how freely the material expands. During plane pressing, the larger the Δf, the larger the expansion. Therefore, the width increase of a narrow thick tape is larger than that of a wide thin tape under the same pressure. The values of Δf will be different if the contact sizes are not the same in the x- and y-directions. For example, if the contact size in the x-direction is shorter than the contact size in the y-direction, then the expansion in the x-direction will be larger than that in the y-direction.

Rolling. Figure 7.11 presents the flat rolling process. Due to the friction at the interface between the rollers and sample, a friction hill is built up during rolling. Lubrication can reduce the constraint and therefore also reduce the stress in the sample, when other rolling parameters are fixed. If the sample height is much smaller than the roller diameter, (7.49) could be used as a rough estimation with the mean height h:

$$h = (h_0 + h_f)/2, \tag{7.50}$$

where h_0 and h_f are the original and final heights of the sample, respectively.

The contact size in the sample length direction, L_L, is given as

$$L_L \approx \sqrt{R(h_0 - h_f)} = \sqrt{R\delta h}, \tag{7.51}$$

where R is the roller radius and δh is the height reduction of the sample [709]. If a tensile stress, σ_t, is applied to the sample during rolling, then (7.49) gives

$$P_c = (2k - \sigma_t)\left(1 + \frac{L}{4h}\right). \tag{7.52}$$

Fig. 7.11 Illustration of the flat rolling process

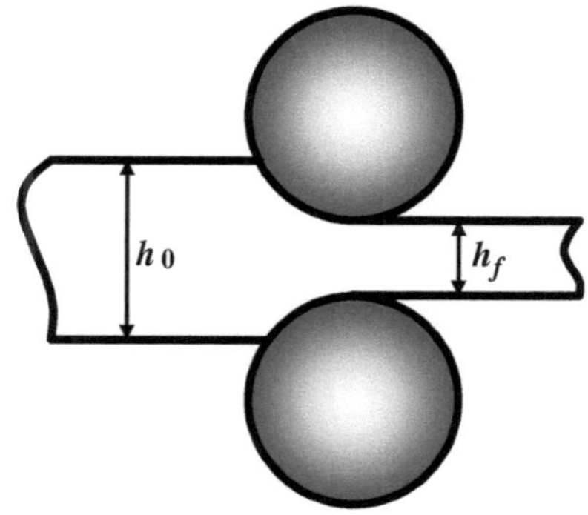

This means that the rolling pressure and therefore the stress in the sample could be decreased, applying a tensile stress during rolling.

The freedom parameter in the sample length direction, $\Delta_{f,L} = h/L_L = h/\sqrt{R\delta h}$, depends on the roller radius, sample height and the height decreasing. For example, the rolling of a thick tape, using larger rollers, could be similar to that of a thin tape, using small rollers. The freedom parameter in the sample width direction is given by $\Delta_{f,w} = h/L_w$, where L_w is the width of the sample. The ratio between the freedom parameters, $\Delta_{f,L}/\Delta_{f,w} = L_w/\sqrt{R\delta h}$, is useful to determine the direction in which the material is most apt to deform. In particular, large rollers, a large decreasing and a small tape width could lead to a significant width increase during rolling and *vice versa*.

Drawing and extrusion are axis-symmetric deformation processes (see Fig. 7.12), which are in principle similar. There are two main differences between them, namely (i) the compressive stress in the deformation zone during extrusion is larger than that of drawing, especially because the area decreasing during extrusion is often much larger than the decrease in a drawing process[1]; (ii) there is a tensile stress in the exiting wire during drawing, which limits the maximum drawing force and the maximum decrease by the strength of the material. The fact that it is possible to obtain very large decreasing ratios in the extrusion process makes it an efficient process for large-scale production. However, it is rather difficult to obtain homogeneous Bi-2223/Ag composites by extrusion [732].

The freedom parameter, $\Delta_f = h/L$, could be determined using (7.50) and the contact size, L, as

$$L = \frac{h_0 - h_f}{2\sin\alpha} = \frac{\delta h}{2\sin\alpha}, \tag{7.53}$$

[1] This means that the powder density in a BSCCO/Ag composite is larger after extrusion than after drawing.

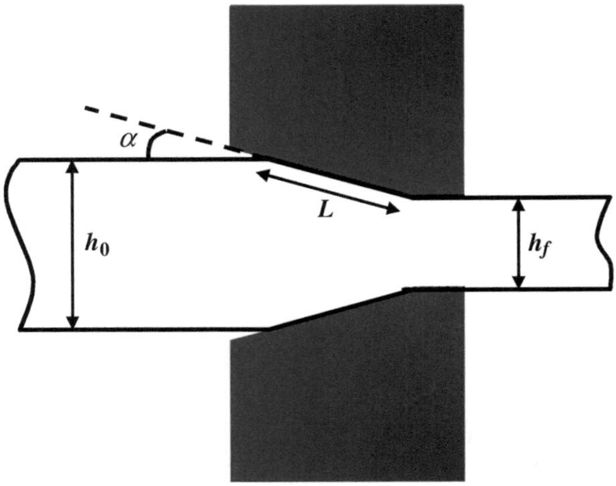

Fig. 7.12 Illustration of the drawing and extrusion process

where h_0 and h_f is the original and final diameters of the sample, respectively, δh is the diameter decreasing and α is the half-angle of the die.[2]

Thus, the different constraint factors have a strong influence on the deformation process. By introducing the freedom parameter, Δ_f, it is possible to estimate these effects using only one parameter. Table 7.1 gives a summary of the expressions of the freedom parameters for various deformation processes.

The presented description of the mechanical deformation process is only for single-element samples. The duplex systems, such as BSCCO/Ag composites, are needed in additional consideration. The difference of the mechanical properties of each component leads to different behavior of the mass flow of each component during deformation, resulting in a "mass re-distribution." As shown in Fig. 7.13 the deformation of a weaker component will be larger than that of a stronger one. A weak component is defined by a lower yield stress or a different flow behavior. This leads to the symmetry change of the component mass distribution for the composites and could be the main reason of inserted heterogeneity during mechanical deformation of BSCCO/Ag samples. A lightened deformation of a weak component leads to convey a high pressure to the strong component under certain constraint conditions (see Fig. 7.13b). Due to the friction at the interfaces between the anvils and the sample, between weak and strong components, a high stress could be built up in the center region of the weak parts, especially when the height of the sample is much smaller than the width of the sample, that is, the freedom parameter is small. This leads to transfer of pressure through the weak region to the strong part of the sandwich. At the edges of the sandwich, the weak component is less constrained and can transfer only insignificant pressure to the

[2] It should be noted that the condition of drawing or extrusion is usually far from that of plane pressing. However, the concept of the freedom parameter could be useful for qualitative estimations. For example, the stress in the sample will be large if the die angle is small or the diameter decreasing is large or the diameter of the sample is small.

Table 7.1 Freedom parameters for various deformation processes

Pressing	Rolling	Drawing or extrusion
$\Delta_f = h/L$	$\Delta_{f,L} = h/\sqrt{R\delta h}$ $\Delta_{f,w} = h/L_w$	$\Delta_f = \frac{2h}{\delta h}\sin\alpha$

strong part. However, the shear stresses might still be considerable, defining deformation at the edges of the strong part. Obviously, constraint conditions or the freedom parameter have a strong influence on the mass re-distribution process.

If the density of the superconductor is not very high, then the BSCCO/Ag core can be considered as the weaker part of the composite, resulting in a more significant deformation of the core compared to the silver sheath. On the other hand, if the core density is very high, then the Ag sheath could be the weaker component unable to convey enough high pressure to the core for further densification of the core. However, by lowering the freedom parameter of the process ($\Delta_{f,L} = h/\sqrt{R\delta h}$), the mass-flow behavior could be changed so that the Ag sheath could again transfer an enough high pressure to increase the density of the superconductor, although the silver part is still the weaker component of the composite. Obviously, when the roller diameter is larger than the wire diameter, the freedom parameter is small, defining larger mass re-distribution.

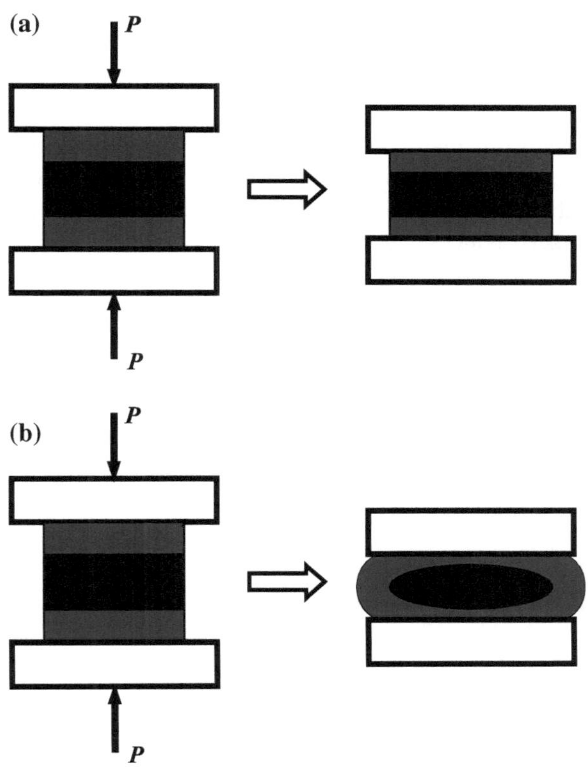

Fig. 7.13 Mass re-distribution during pressing: **(a)** homogeneous deformation at equivalent mechanical properties of both components, and **(b)** inhomogeneous deformation, causing the mass re-distribution (the *dark area* represents a stronger component)

The freedom parameter also has influence on the sausaging. If the freedom parameter is large, the mass flow of the Ag sheath is large, and the sausaging (at least the sausaging frequency) as well as the density of the core will be decreased. The freedom parameter at rolling ($\Delta_{f,L} = h/\sqrt{R\delta h}$) assumes that sausaging could be decreased, using small rollers, which has been supported by test results [918]. This also implies that sausaging will be more pronounced when the thickness of the tape becomes small; this has also been demonstrated [918]. It should be pointed that the smaller factor of thickness decreasing causes the larger freedom of silver sheath in the direction of the wire length. Therefore, it is not surprising that the BSCCO core density, obtained in this way, is smaller than the core density obtained, using a larger decreasing factor.

Compatibility Line

The stresses developed in a material due to an external loading can be estimated using the so-called Mohr circles [1178]. The intrinsic curve, giving the behavior of a material for a given deformation ε, is obtained by drawing the envelope of these Mohr circles. In particular, Fig. 7.14a shows the intrinsic curves for Bi-2223 and Ag at a deformation $\varepsilon = 25$ %. The existence of this intersection (P_c) is called the *compatibility point*. Its existence shows that in spite of quite different mechanical behavior at atmospheric pressure of the matrix and filaments, there are external mechanical stresses defining the same ductile behaviors of silver and superconductor. By drawing such intrinsic curves for different deformation rates, one can obtain the line along which P_c moves, called the *compatibility line*. One corresponds to conditions where the superconducting powder and the sheath both have the same mechanical behavior, especially in the ductile domain. Then, instead of fabricating wires at atmospheric pressure, that is, out of the plastic domain for the superconducting powder (S_p limit, see Fig. 7.14b) and out of the ductile zone common to the powder and to the sheath (S_c limit, see Fig. 7.14b), it is much more expedient to use an analysis of triaxial stress state. This approach helps to minimize Ag/BSCCO interface sausaging, using yielding criterion, defining the deformation compatibility of silver matrix and superconducting core [356].

7.2.4 Mathematical Modeling of Cold Drawing of Filamentary Composites

Problem Statement

The cold drawing is one of the effective method of plastic deformation and strengthening of metallic samples. It allows one to process filamentary composites with high degree of structural heterogeneity. Sokolovsky [1596] was one of the

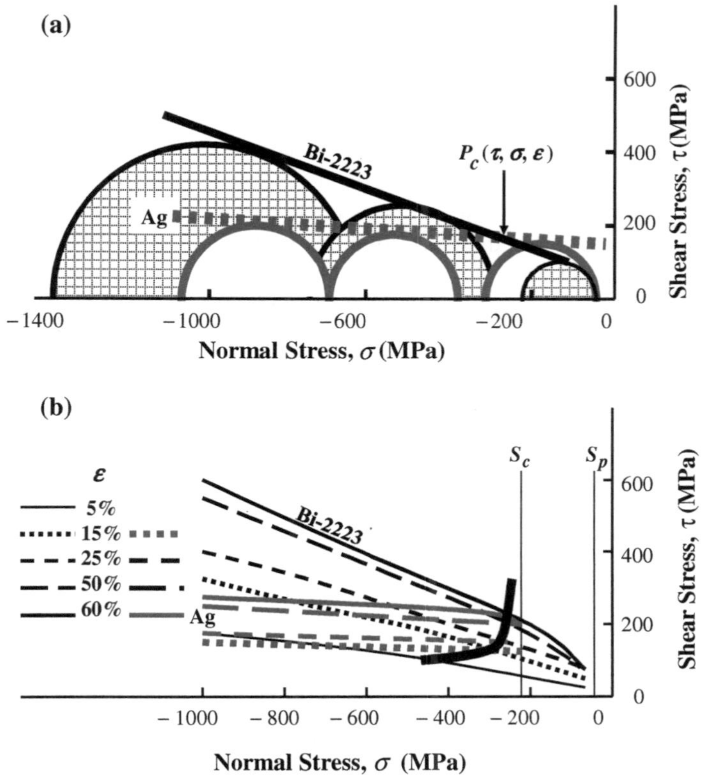

Fig. 7.14 (**a**) Determination of the compatibility point P_c (σ, τ, $\varepsilon = 25$ %) at the intersection of the intrinsic curves for a Bi-2223 ceramic and metallic Ag; (**b**) the compatibility line between Bi-2223 and Ag: trace of the compatibility points P_c as ε varies from 5 to 60 %. S_p corresponds to the plasticity threshold of the Bi-2223 ceramic, and S_c to the compatibility strain threshold defining initiation of the brittle fracture between the Bi-2223 ceramic and metallic Ag [1243]

first who proposed a model of plastic flowing of metal through cone hole (draw plate or die). In the following this problem has been studied by many researchers [68, 696, 1670]. By using finite-element modeling, complex regimes of flow have been calculated [273, 379]. The modern investigations are connected with analysis of effects of various yield criteria [18], modeling non-stationary processes [1434], optimization of geometry of draw plate [276, 1036]. Plastic flowing bi-material compositions has been in the center of attention of the works [19, 276, 333]. Mutual interaction of the components causes a local stress concentration at interfaces. In this case, the interface properties and its carrying capability are the key factors defining the flow regime and characteristics of the composite.

Following to [154] we consider axi-symmetrical plastic flowing of composite wire consisting of matrix $\Omega^{(1)}$ and filament $\Omega^{(2)}$ through cone die (see Fig. 7.15). In the dependence on the drawing angle, γ, strain degree $Z = 1-(z_1/z_0)^2$ and value of friction at the die boundary, the plastic deformation may be of various types.

We shall treat only *radial regime of flowing*, being steady at the small values of γ, presenting a significant interest from practical view-point. In this case all current lines are directed along radial direction of the spherical coordinate system $\{r, \theta, \varphi\}$ (see Fig. 7.15). Cross-sections of die and matrix decrease proportionally due to volume fractions of the components are not change. The zone of plastic strain is restricted by boundaries $\partial\Omega_0$ and $\partial\Omega_1$ at the die entrance and exit, respectively. For the treated radial regime, these surfaces could be approximated by planes [68, 379]. We also propose that the discontinuities in the rate field are localized at the boundaries $\partial\Omega_0$ and $\partial\Omega_1$.

During the cold drawing, low strain rate and corresponding grease provide absence of significant thermal effects. We shall also neglect inertial terms in equilibrium equations due to consider stationary process of flowing. Owing to the plastic strains exceed significantly elastic one, we also neglect the elastic strains. Moreover, it is assumed that filament and matrix are ideal elastic–plastic materials. Based on these hypothesizes the constitutive equations have the following form.

Equilibrium equations:

$$\frac{\partial\sigma_r^{(a)}}{\partial r} + \frac{1}{r}\frac{\partial\sigma_{r\theta}^{(a)}}{\partial\theta} + \frac{2(\sigma_r^{(a)} - \sigma_\theta^{(a)}) + \sigma_{r\theta}^{(a)}\operatorname{ctg}\theta}{r} = 0; \tag{7.54}$$

$$\frac{\partial\sigma_{r\theta}^{(a)}}{\partial r} + \frac{1}{r}\frac{\partial\sigma_\theta^{(a)}}{\partial\theta} + \frac{3\sigma_{r\theta}^{(a)}}{r} = 0, \tag{7.55}$$

where $\sigma_r^{(a)}$, $\sigma_\theta^{(a)} = \sigma_\varphi^{(a)}$, $\sigma_{r\theta}^{(a)}$ are the stress components; $\sigma_{r\varphi}^{(a)} = \sigma_{\theta\varphi}^{(a)} = 0$.

Von Mises yield condition:

$$(\sigma_r^{(a)} - \sigma_\theta^{(a)})^2 + 3(\sigma_r^{(a)})^2 = (\sigma_y^{(a)})^2, \tag{7.56}$$

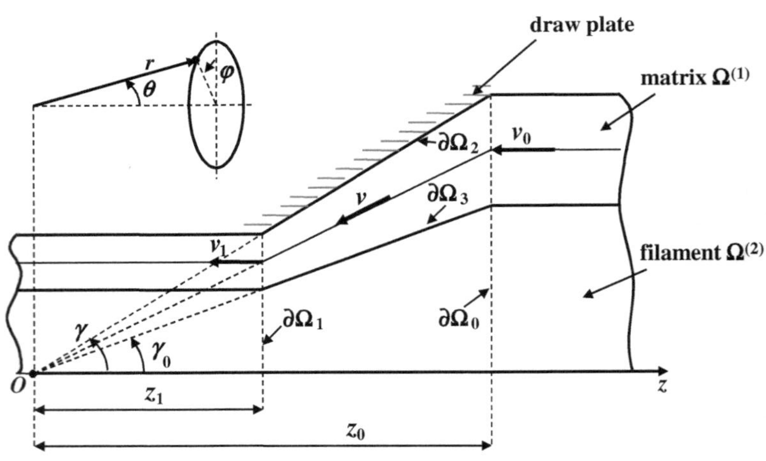

Fig. 7.15 Flowing of composite wire through cone die [154]

where $\sigma_y^{(a)}$ is the yield limit at tension, and $\sigma_y^{(a)}/\sqrt{3}$ is the yield limit at shear.

Flaw law:

$$\frac{\sigma_{r\theta}^{(a)}}{\sigma_r^{(a)} - \sigma_\theta^{(a)}} = \frac{\dot{\varepsilon}_{r\theta}^{(a)}}{\dot{\varepsilon}_r^{(a)} - \dot{\varepsilon}_\theta^{(a)}}, \tag{7.57}$$

where $\dot{\varepsilon}_r^{(a)}$, $\dot{\varepsilon}_\theta^{(a)}$, $\dot{\varepsilon}_{r\theta}^{(a)}$ are the strain rates. As it is usually, the dot over variable designs derivative on time.

Incompressibility condition:

$$\dot{\varepsilon}_r^{(a)} + 2\dot{\varepsilon}_\theta^{(a)} = 0. \tag{7.58}$$

To the constitutive equations it is necessary to add the following boundary conditions:

(i) *Boundary* $\partial\Omega_2$. Tangential stresses on this boundary are proportional to the yield limit at shear, that is

$$\sigma_{r\theta}^{(1)} = m\sigma_y^{(1)}/\sqrt{3}, \text{at} \quad \theta = \gamma, \tag{7.59}$$

where m is the friction factor.

(ii) *Boundary* $\partial\Omega_3$. Ideal contact matrix with filament defines the equality of rates $v^{(a)}$:

$$v^{(1)} = v^{(2)}, \text{ at } \quad \theta = \gamma_0 \tag{7.60}$$

and stresses $\sigma_\theta^{(a)}$, $\sigma_{r\theta}^{(a)}$:

$$\sigma_\theta^{(1)} = \sigma_\theta^{(2)}, \text{at} \quad \theta = \gamma_0; \tag{7.61}$$

$$\sigma_{r\theta}^{(1)} = \sigma_{r\theta}^{(2)}, \text{at} \quad \theta = \gamma_0. \tag{7.62}$$

(iii) *Central axis*. From axial symmetry of the problem, it is followed that the shear stresses at the filament center are equal to zero:

$$\sigma_{r\theta}^{(2)} = 0, \text{at} \quad \theta = 0. \tag{7.63}$$

(iv) *Boundary* $\partial\Omega_0$. In the case of absence of friction between matrix and filament ($m = 0$), the longitudinal component of rate (on absolute value) is equal to initial rate of the sample, v_0, before its entrance in the plastic zone:

$$v^{(a)}\cos\theta = -v_0, \text{ at } \quad z = r\cos\theta = z_0. \tag{7.64}$$

Current lines break in radial direction at the boundary $\partial\Omega_0$. The averaged longitudinal stress σ_0 at $\partial\Omega_0$ should be enough in order to ensure corresponding discontinuity in the rate field. We shall calculate σ_0 by the next way.

The rate of energy dissipation at the surface $\partial\Omega_0$ is defined as

$$\dot{E}_0 = \frac{\sigma_y^{(a)}}{\sqrt{3}} \iint\limits_{\partial\Omega_0} \int |\Delta v_0| dS_0 = \frac{2\sqrt{3}}{9} \pi z_0^2 v_0 [\sigma_y^{(2)} \tan^3 \gamma_0 + \sigma_y^{(1)} (\tan^3 \gamma - \tan^3 \gamma_0)], \tag{7.65}$$

where $dS_0 = z_0^2 \frac{\tan\theta}{\cos^2\theta} d\theta d\varphi$, $|\Delta v_0| = v_0 \tan\theta$ is the value of discontinuity in the rate field along tangential direction (parallel to plane $\partial\Omega_0$).

The rate of energy change of drawing at the entrance in the plastic zone is defined as

$$\dot{E}_{0d} = \sigma_0 v_0 S_0 = \sigma_0 \pi z_0^2 v_0 \tan^2\gamma, \tag{7.66}$$

where $S_0 = \pi z_0^2 \tan^2\gamma$ is the square of $\partial\Omega_0$.

From equality of the expressions (7.65) and (7.66), we obtain

$$\sigma_0 = \frac{2\sqrt{3}}{9} \tan\gamma \left\{ \sigma_y^{(2)} \left(c^{(2)}\right)^{3/2} + \sigma_y^{(1)} \left[1 - \left(c^{(2)}\right)^{3/2} \right] \right\}, \tag{7.67}$$

where $c^{(2)} = \tan^2\gamma_0 / \tan^2\gamma$ is the volume fraction of the filament.

From other hand, σ_0 can be calculated directly by averaging the longitudinal component of stress:

$$\sigma_z^{(a)} = \sigma_r^{(a)} \cos\theta - \sigma_{r\theta}^{(a)} \sin\theta \tag{7.68}$$

on surface $\partial\Omega_0$ as

$$\sigma_0 = \frac{1}{S_0} \iint\limits_{\partial\Omega_0} \sigma_z^{(a)} dS_0. \tag{7.69}$$

Then we have the next boundary condition on $\partial\Omega_0$ based on the expressions (7.67) and (7.69):

$$\frac{1}{S_0} \iint\limits_{\partial\Omega_0} \left(\sigma_r^{(a)} \cos\theta - \sigma_{r\theta}^{(a)} \sin\theta \right) dS_0 = \frac{2\sqrt{3}}{9} \tan\theta \left\{ \sigma_y^{(2)} \left(c^{(2)}\right)^{3/2} + \sigma_y^{(1)} \left[1 - \left(c^{(2)}\right)^{3/2} \right] \right\},$$

$$\text{at } z = r\cos\theta = z_0.x \tag{7.70}$$

Analytical Solution

Let's consider analytical solution of original boundary-value problem (7.54–7.64), (7.70). Kinematically accessible field of rates could be defined from incompressibility equation (7.58) and boundary condition (7.64) as

$$v^{(a)} = -v_0 \frac{z_0^2}{r^2} \frac{p^{(a)}(\theta)}{\cos^3\theta}, \tag{7.71}$$

where the function $p^{(a)}(\theta)$ takes into account a friction effect at the interface matrix and filament. In the case of absence of friction ($m = 0$), we obtain $p^{(a)}(\theta) = 1$. From axial symmetry of the problem, it is also followed that $p^{(a)}(\theta) = 1$ at any m. Based on expression (7.71) the strain rates could be written as

$$\dot{\varepsilon}_r^{(a)} = \frac{\partial v^{(a)}}{\partial r} = 2v_0 \frac{z_0^2}{r^3} \frac{p^{(a)}(\theta)}{\cos^3\theta}; \tag{7.72}$$

$$\dot{\varepsilon}_\theta^{(a)} = \dot{\varepsilon}_\phi^{(a)} = \frac{v^{(a)}}{r} = -v_0 \frac{z_0^2}{r^3} \frac{p^{(a)}(\theta)}{\cos^3\theta}; \tag{7.73}$$

$$\dot{\varepsilon}_{r\theta}^{(a)} = \frac{1}{2r}\frac{\partial v^{(a)}}{\partial \theta} = -\frac{v_0}{2}\frac{z_0^2}{r^3} \frac{\partial_\theta p^{(a)}(\theta) + 3p^{(a)}(\theta)\tan\theta}{\cos^3\theta}, \tag{7.74}$$

where designation ∂_θ defines derivative on θ.

Intensity of the strain rates, $\dot{\varepsilon}_{\text{int}}^{(a)}$, is calculated in the form:

$$\begin{aligned}\dot{\varepsilon}_{\text{int}}^{(a)} &= \frac{2}{3}\sqrt{(\dot{\varepsilon}_r^{(a)} - \dot{\varepsilon}_\theta^{(a)})^2 + 3(\dot{\varepsilon}_{r\theta}^{(a)})^2} \\ &= -2\frac{v^{(a)}}{r}\sqrt{1 + \frac{3}{4}\tan^2\theta + \frac{1}{2}\frac{\partial p^{(a)}(\theta)}{p^{(a)}(\theta)}\tan\theta + \frac{1}{12}\left[\frac{\partial p^{(a)}(\theta)}{p^{(a)}(\theta)}\right]^2}.\end{aligned} \tag{7.75}$$

Intensity of the strains, $\varepsilon_{\text{int}}^{(a)}$, could be determined by integrating $\dot{\varepsilon}_{\text{int}}^{(a)}$ on time along current lines. Taking into account that $dz = v^{(a)}\cos\theta dt$ and $z = r\cos\theta$, we obtain

$$\varepsilon_{\text{int}}^{(a)} = \int_0^t \dot{\varepsilon}_{\text{int}}^{(a)} dt = \int_{z_0}^{z} \frac{\dot{\varepsilon}_{\text{int}}^{(a)}}{v^{(a)}\cos\theta} dz = -2\ln\left(\frac{z}{z_0}\right)\sqrt{1 + \frac{3}{4}\tan^2\theta + \frac{1}{2}\frac{\partial p^{(a)}(\theta)}{p^{(a)}(\theta)}\tan\theta + \frac{1}{12}\left[\frac{\partial p^{(a)}(\theta)}{p^{(a)}(\theta)}\right]^2}. \tag{7.76}$$

The solution (7.76) for $\varepsilon_{\text{int}}^{(a)}$ allows one to calculate the yield strength of bi-material composite after drawing on the base of known empirical relations of toughening for included materials.

The stress components satisfying yield condition (7.56) could be written into parametric form:

$$\sigma_r^{(a)}/\sigma_y^{(a)} = C_1^{(a)}\ln r + f^{(a)}(\theta) + \sqrt{1 - [g^{(a)}(\theta)]^2}; \tag{7.77}$$

$$\sigma_\theta^{(a)}/\sigma_y^{(a)} = \sigma_\phi^{(a)}/\sigma_y^{(a)} = C_1^{(a)}\ln r + f^{(a)}(\theta); \tag{7.78}$$

$$\sigma_{r\theta}^{(a)}/\sigma_y^{(a)} = g^{(a)}(\theta)/\sqrt{3}. \tag{7.79}$$

where $C_1^{(a)}$ are constants of integrating, the function $g^{(a)}(\theta)$ takes into account influence of friction at the interface "matrix-filament". In the case of absence of friction ($m = 0$), $g^{(a)}(\theta) = 0$.

By substituting the expressions (7.77–7.79) into equilibrium equations (7.54), (7.55) and boundary conditions (7.59), (7.61–7.63), we obtain two value-boundary problems:

$$\text{(I)}\ \partial_\theta g^{(a)}(\theta) + 2\sqrt{3}\sqrt{1 - [g^{(a)}(\theta)]^2} + g^{(a)}(\theta)\text{ctg}\theta + \sqrt{3}C_1^{(a)} = 0; \tag{7.80}$$

$$g^{(2)}(0) = 0; g^{(1)}(\gamma_0)/g^{(2)}(\gamma_0) = \lambda; g^{(1)}(\gamma) = m; C_1^{(1)}/C_1^{(2)} = \lambda. \tag{7.81}$$

$$\text{(II)}\ \partial_\theta f^{(a)}(\theta) = -\sqrt{3}g^{(a)}(\theta); \tag{7.82}$$

$$f^{(1)}(\gamma_0)/f^{(2)}(\gamma_0) = \lambda \tag{7.83}$$

Here $\lambda = \sigma_y^{(2)}/\sigma_y^{(1)}$.

Approximate asymptotic solution of I problem will be presented in the next section. The solution of II problem is calculated by using $g^{(a)}(\theta)$ and $C_1^{(a)}$ directly integrating Eq. (7.82) with account of the boundary conditions (7.70) and (7.83).

In order to determine the function $p^{(a)}(\theta)$, we substitute expressions (7.72–7.74) into the flaw equation (7.57). By taking into account the boundary condition (7.60), we have the following value-boundary problem:

$$\partial_\theta p^{(a)}(\theta) + \left[\frac{2\sqrt{3}g^{(a)}(\theta)}{\sqrt{1 - [g^{(a)}(\theta)]^2}} + 3\tan\theta\right]p^{(a)}(\theta) = 0; \tag{7.84}$$

$$p^{(2)}(0) = 1; p^{(1)}(\gamma_0)/p^{(2)}(\gamma_0) = 1. \tag{7.85}$$

The solution of the problem (7.84), (7.85) has the form:

$$p^{(a)}(\theta) = \exp\left\{-\int\left[\frac{2\sqrt{3}g^{(a)}(\theta)}{\sqrt{1 - [g^{(a)}(\theta)]^2}} + 3\tan\theta\right]d\theta + C_2^{(a)}\right\}, \tag{7.86}$$

where

$$C_2^{(2)} = 0; C_2^{(1)} = \exp\left\{\int\left[\frac{2\sqrt{3}g^{(1)}(\theta)}{\sqrt{1 - [g^{(1)}(\theta)]^2}} - \frac{2\sqrt{3}g^{(2)}(\theta)}{\sqrt{1 - [g^{(2)}(\theta)]^2}}\right]d\theta\right\}_{\theta=\gamma_0}. \tag{7.87}$$

One of important parameters is the drawing stress, σ_d, presenting itself an averaged characteristic of the process. It may be simply measured experimentally and used to verify theoretical results.

The stresses $\sigma_d^{(a)}$ at the exit of filament $\partial\Omega_1$ is defined as

$$\sigma_d^{(a)} = \sigma_1^{(a)} + \Delta\sigma_1^{(a)}, \tag{7.88}$$

where $\sigma_1^{(a)}$ is the longitudinal component of stresses at $\partial\Omega_1$ calculated by formula (7.68) at $z = r\cos\theta = z_1$, $\Delta\sigma_1^{(a)}$ is the additive defined by discontinuity in the rate field.

Let's consider an element of cross-section with square dS_1 crossing the boundary $\partial\Omega_1$. The discontinuity in the rate field along tangential direction equals to $|\Delta v_1| = v_1\tan\theta$, where v_1 is the specimen rate after exit from plastic zone. The rate of energy dissipation at the surface dS_1 is found as

$$d\dot{E}_1 = \frac{\sigma_y^{(a)}}{\sqrt{3}}|\Delta v_1|dS_1. \tag{7.89}$$

The rate of changing additive drawing energy required to create the discontinuity in the rate field is found as

$$d\dot{E}_{1d} = \Delta\sigma_1 v_1 dS_1. \tag{7.90}$$

By equating expressions (7.89) and (7.90), we find

$$\Delta\sigma_1 = \frac{\sigma_y^{(a)}}{\sqrt{3}}\tan\theta. \tag{7.91}$$

Finally, the drawing stress, σ_d, is calculated by averaging $\sigma_d^{(a)}$ on surface $\partial\Omega_1$ as

$$\sigma_d = \frac{1}{S_1}\iint\limits_{\partial\Omega_1} \sigma_d^{(a)} dS_1, \tag{7.92}$$

where $dS_1 = z_1^2(\tan\theta/\cos^2\theta)d\theta d\varphi$, $S_1 = \pi z_1^2\tan^2\gamma$ is the square of $\partial\Omega_1$.

Asymptotical Calculation

We shall search asymptotic solution of the value-boundary problem (7.80, 7.81). With this aim to present unknown functions $g^{(a)}(\theta)$ in the form of series on degrees of angle θ:

$$\begin{aligned} g^{(1)}(\theta) &= a_0^{(1)} + a_1^{(1)}(\gamma - \theta) + a_2^{(1)}(\gamma - \theta)^2 + \ldots;\ g^{(2)}(\theta) \\ &= a_0^{(2)} + a_1^{(2)}\theta + a_2^{(2)}\theta^2 + \ldots \end{aligned} \tag{7.93}$$

By substituting the series (7.93) into Eq. (7.80) and equating factors at the same degrees of θ and $\gamma - \theta$ we receive

$$a_0^{(2)} = a_2^{(2)} = 0;\ a_1^{(2)} = -\frac{\sqrt{3}}{2}(2 + C_1^{(2)});$$
$$a_3^{(2)} = \frac{2\sqrt{3}}{3} + \frac{17\sqrt{3}}{24}C_1^{(2)} + \frac{3\sqrt{3}}{16}(C_1^{(2)})^2;\ldots \tag{7.94}$$

$$a_0^{(1)} = m;\ a_1^{(1)} = m\mathrm{ctg}\gamma + 2\sqrt{3(1 - m^2)} + \sqrt{3}C_1^{(2)});$$

$$a_2^{(1)} = \frac{2\sqrt{3}(1-2m^2)\mathrm{ctg}\gamma - 6mC_1^{(1)} - (11m - \sqrt{3}C_1^{(1)}\mathrm{ctg}\gamma - 2m\mathrm{ctg}^2\gamma)\sqrt{1-m^2}}{2\sqrt{1-m^2}};$$

$$\begin{aligned} a_3^{(1)} = &-\{10/\sqrt{3} + \sqrt{3}(C_1^{(1)})^2 - (19/\sqrt{3})m^2 + 3\sqrt{3}m^4 + 2m(2 - m^2)ctg\gamma C_1^{(1)} \\ &- [\sqrt{3} - (10/\sqrt{3})m^2 + 2\sqrt{3}m^4]ctg^2\gamma + [(11 - 5m^2)C_1^{(1)}/\sqrt{3} + m(43 - 31m^2)ctg\gamma/6 \\ &- \sqrt{3}(1 - m^2)C_1^{(1)}ctg^2\gamma/2 + m(1 - m^2)ctg^3\gamma](1 - m^2)^{1/2}\}(1 - m^2)^{-3/2};\ \ldots \end{aligned} \tag{7.95}$$

The integrating constants $C_1^{(a)}$ are calculated from the boundary conditions (7.81).

The region of applicability of the solution (7.93) could be approximately estimated by using Pade approximants (PA) method [84]. In many cases, application of PA allows one to significantly improve convergence of asymptotical series (AS). So, AS ensures well calculation accuracy in the region where it agrees with corresponding diagonal PA. Rearranging series (7.93) in PA of order [2/2], we obtain

$$g^{(1)}(\theta) = \frac{c_0^{(1)} + c_1^{(1)}(\gamma - \theta) + c_2^{(1)}(\gamma - \theta)^2}{d_0^{(1)} + d_1^{(1)}(\gamma - \theta) + d_2^{(1)}(\gamma - \theta)^2};$$
$$g^{(2)}(\theta) = \frac{c_0^{(1)} + c_1^{(1)}\theta + c_2^{(1)}\theta^2}{d_0^{(1)} + d_1^{(1)}\theta + d_2^{(1)}\theta^2}, \tag{7.96}$$

where

$$c_0^{(a)} = a_0^{(a)}(a_2^{(a)})^2 - a_0^{(a)}a_1^{(a)}a_3^{(a)};$$
$$c_1^{(a)} = a_1^{(a)}(a_2^{(a)})^2 - (a_1^{(a)})^2a_3^{(a)} - a_0^{(a)}a_2^{(a)}a_3^{(a)} + a_0^{(a)}a_1^{(a)}a_4^{(a)};$$
$$c_2^{(a)} = a_0^{(a)}(a_3^{(a)})^2 + (a_2^{(a)})^3 + (a_1^{(a)})^2a_4^{(a)} - 2a_1^{(a)}a_2^{(a)}a_3^{(a)} - a_0^{(a)}a_2^{(a)}a_4^{(a)}, \tag{7.97}$$

$$d_0^{(a)} = (a_2^{(a)})^2 - a_1^{(a)}a_3^{(a)};\ d_1^{(a)} = a_1^{(a)}a_4^{(a)} - a_2^{(a)}a_3^{(a)};$$
$$d_2^{(a)} = (a_3^{(a)})^2 - a_2^{(a)}a_4^{(a)}. \tag{7.98}$$

A comparison of AS (7.93) with PA (7.96) for specific example conducted in [154] showed that in common case, accuracy of the asymptotical solution decreased at increasing the drawing angle, γ, friction factor, m, and difference between properties of components ($\lambda \gg 1$ or $\lambda \ll 1$).

7.2.5 Mathematical Modeling of Cold Rolling

The cold rolling[3] conducted at the room temperature is directed too reducing the cross-section of the sample by a squeezing action when it passes between two rotating rolls. If the sample is a thin strip with the thickness/width ratio of at least ten, then the process is essentially a plane strain problem excepting only narrow zones near the sample edges.

Mathematical modeling of rolling includes a set of problems. Deformation process is non-steady when sample enters into the roll gap and become steady after the sample has passed fully through the roll gap. The developed model should provide good description of finite plastic strains, enforce isochoric character of plastic deformation and simulate friction at the sample-die interface. Finite strain theories [68, 696, 1596], based on multiplicative decomposition of deformation gradient, should be incorporated into rolling models, that it has been fulfilled in [379]. Following [186], we obtain below necessary evolution equations.

Neutral Point Position

Based on mechanics of continuum [1078], continuity equation in local form, applied to the considered sample, has the form in spatial coordinates:

$$\frac{d\rho}{dt} = \rho \mathrm{div}\, \mathbf{v}, \tag{7.99}$$

where ρ is the material density, $\mathbf{v}$ is the velocity vector and t is the time. In the case of the material incompressibility (i.e. ρ is a constant), the continuity equation reduces to the following one:

$$\mathrm{div}\, \mathbf{v} = 0. \tag{7.100}$$

In material coordinates, Eq. (7.99) is found as

$$\frac{d}{dt}(\rho J) = 0, \tag{7.101}$$

[3] In the case of heavy reductions in cross-section, the rolling process could be carried out at elevated temperatures, defining the hot rolling.

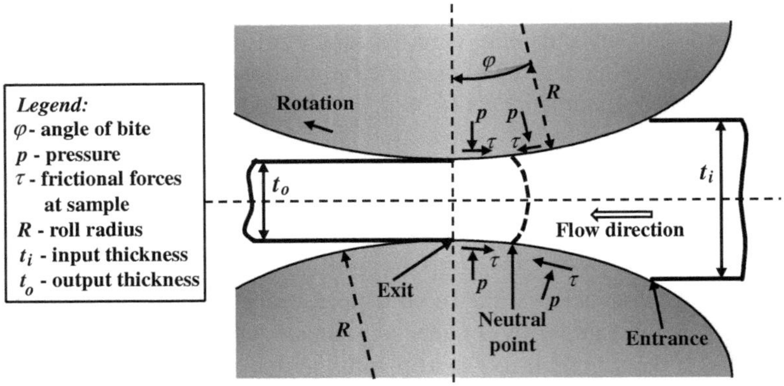

Fig. 7.16 The definition of geometry at rolling [186]

where J is Jacobian of deformation gradient expressed in the form (where V is the volume):

$$J = \frac{dv}{dV}, \tag{7.102}$$

By introducing $\rho = \text{const}$, we obtain from (7.101) $J = 1$.

At the entrance to the roll gap, material is pulled between rolls by the frictional forces (see Fig. 7.16). Obviously, the direction of frictional forces states that velocity of the sample is slower than the roll speed at the interface. Due to we shall neglect volume change caused by elastic strain and consider the process as incompressible (i.e. $J = 1$). This means when material flaws to the exit, velocity should continuously grow and the end material speed will be faster than the roll speed. As a sequence, direction of frictional forces at the exit is opposite to the frictional forces at the entrance. Owing to the material flaws slower at the entrance and faster at the exit, somewhere between these two points the velocities of material and roll are equal (this point is called *neutral point*). In this point, the frictional forces change direction of action. The disposition of the point is unknown in advance defining a difficulty in formulation of boundary conditions.

Multiplicative Finite Strain Elastoplasticity

Finite strain elasto-plastic deformation is described by a multiplicative decomposition of strain gradient into elastic (e) and plastic (p) parts:

$$\mathbf{F} = \mathbf{F}^e\mathbf{F}^p. \tag{7.103}$$

Velocity gradient is:

$$\mathbf{L} = \dot{\mathbf{F}}\mathbf{F}^{-1}. \tag{7.104}$$

It is divided into elastic and plastic velocity gradients:

$$\mathbf{L} = \mathbf{L}^e + \mathbf{L}^p, \tag{7.105}$$

where

$$\mathbf{L}^e = \dot{\mathbf{F}}^e(\mathbf{F}^e)^{-1};\ \mathbf{L}^p = \mathbf{F}^e\dot{\mathbf{F}}^p(\mathbf{F}^p)^{-1}(\mathbf{F}^e)^{-1}. \tag{7.106}$$

As a strain measure several tensors could be used. For example, Hencky of logarithmic strain tensor has the form:

$$\mathbf{E}^e = \ln \mathbf{U}^e, \tag{7.107}$$

where $\mathbf{U}^e$ is the elastic right stretch tensor obtained from polar decomposition theorem:

$$\mathbf{F}^e = \mathbf{R}^e\mathbf{U}^e, \tag{7.108}$$

where $\mathbf{R}^e$ is the elastic rotation tensor. Stress conjugate to Hencky strain tensor is rotated tensor $\bar{\boldsymbol{\tau}}$:

$$\bar{\boldsymbol{\tau}} = (\mathbf{R}^e)^t\boldsymbol{\tau}\mathbf{R}^e. \tag{7.109}$$

Estimation of variables describing the elasto-plastic process ($\mathbf{F}$, $\mathbf{F}^p$, $\boldsymbol{\tau}$, σ) completely determines whole the strain process. The last variable σ is the isotropic hardening to being a scalar function of equivalent plastic deformation, i. e. $\sigma = \sigma(e^p)$. In order to define these state variables, a constitutive equation is required for $\boldsymbol{\tau}$, while for $\mathbf{F}^p$ and σ evolution equations are necessary. The next consideration solves these problems.

Helmholtz free energy per unit mass is defined by [1078]:

$$\psi = u - s\theta, \tag{7.110}$$

where s is the entropy, θ is the temperature, u is the internal energy. Time derivative is

$$\dot{\psi} = \dot{u} - \dot{s}\theta - s\dot{\theta}. \tag{7.111}$$

From first law of thermodynamics, we have

$$\rho\frac{du}{dt} = \boldsymbol{\tau} : \mathbf{D} + \rho r - \nabla\cdot\mathbf{q}, \tag{7.112}$$

where $\boldsymbol{\tau}$ is Cauchy stress tensor, $\mathbf{D} = \text{sym}\ (\mathbf{L})$ is the strain rate tensor, r is the internal heat source and $\mathbf{q}$ is the heat flux. Skew-symmetric part of $\mathbf{L}$ is assumed to be zero into framework of this theory. Local form of Clausius–Duhem inequality is

$$\frac{ds}{dt} \geq \frac{r}{\theta} - \frac{1}{\rho}\nabla\cdot\left(\frac{\mathbf{q}}{\theta}\right). \tag{7.113}$$

This equation is reduced to

$$\rho(-\dot{\theta}s - \dot{\psi}) + \boldsymbol{\tau} : \mathbf{D} - \frac{\mathbf{q}}{\theta} \cdot \text{grad}\theta \geq 0, \tag{7.114}$$

Introducing free energy and entropy per unit volume:

$$W = \rho\psi, \; S = \rho s \tag{7.115}$$

gives

$$-(\dot{\theta}S + \dot{W}) + \boldsymbol{\tau} : \mathbf{D} - \frac{\mathbf{q}}{\theta} \cdot \text{grad}\theta \geq 0. \tag{7.116}$$

In isothermal case ($\dot{\theta} = 0$):

$$-\dot{W} + \boldsymbol{\tau} : \mathbf{D} \geq 0. \tag{7.117}$$

By dividing D into elastic and plastic parts we obtain

$$-\dot{W} + \boldsymbol{\tau} :(\mathbf{D}^e + \mathbf{D}^p) \geq 0. \tag{7.118}$$

From elastic work conjugacy [599]:

$$\boldsymbol{\tau} : (\mathbf{D}^e + \mathbf{D}^p) = \bar{\boldsymbol{\tau}} : (\dot{\mathbf{E}}^e + \overline{\mathbf{D}}^p), \tag{7.119}$$

where

$$\overline{\mathbf{D}}^p = \text{sym}(\overline{\mathbf{L}}^p) = \text{sym}(\dot{\mathbf{F}}^p(\mathbf{F}^p)^{-1}) \tag{7.120}$$

we have

$$-\dot{W} + \bar{\boldsymbol{\tau}} : \dot{E}^e + \bar{\boldsymbol{\tau}} : \overline{\mathbf{D}}^p \geq 0. \tag{7.121}$$

Then, after dividing the free energy into elastic and plastic parts:

$$W(\mathbf{E}^e, e^p) = W^e(\mathbf{E}^e) + W^p(e^p). \tag{7.122}$$

we obtain

$$\left(\bar{\boldsymbol{\tau}} - \frac{\partial W^e}{\partial \mathbf{E}^e}\right) : \dot{\mathbf{E}}^e + \bar{\boldsymbol{\tau}} : \overline{\mathbf{D}}^p - \dot{W}^p \geq 0. \tag{7.123}$$

Due to (7.123) must hold for all admissible processes state, then

$$\bar{\boldsymbol{\tau}} = \frac{\partial W^e}{\partial \mathbf{E}^e}; \; D = \bar{\boldsymbol{\tau}} : \overline{\mathbf{D}}^p - \dot{W}^p \geq 0. \tag{7.124}$$

Since $\overline{\mathbf{D}}^p = \text{sym}(\overline{\mathbf{L}}^p)$ we use (7.120) and inequality (7.124) to define evolution equation for $\mathbf{F}^p$. Actual stress state $\bar{\boldsymbol{\tau}}$ maximizes dissipation function D and is subjected to constraints:

$$\Phi \leq 0;\ \dot{e}^p\Phi = 0;\ \dot{e}^p \geq 0, \tag{7.125}$$

where Φ defines yield surface. It is represented by

$$\Phi = \sqrt{\frac{3}{2}}\|\text{dev}\bar{\tau}\| - \sigma. \tag{7.126}$$

Necessary condition for such maximum of D is

$$\frac{\partial D}{\partial \bar{\tau}} - \dot{e}^p \frac{\partial \Phi}{\partial \bar{\tau}} = 0. \tag{7.127}$$

These conditions are classical Kuhn-Tucker conditions are widely used in the convex mathematical programming literature. Following equations are obtained from (7.124) and (7.126):

$$\frac{\partial D}{\partial \bar{\tau}} = \overline{\mathbf{D}}^p;\ \frac{\partial \Phi}{\partial \bar{\tau}} = \sqrt{\frac{3}{2}}\frac{\text{dev}\bar{\tau}}{\|\text{dev}\bar{\tau}\|}. \tag{7.128}$$

By using Eqs. (7.127) and (7.128) we obtain:

$$\overline{\mathbf{D}}^p = \sqrt{\frac{3}{2}}\dot{e}^p\overline{\mathbf{N}};\ \sqrt{\frac{2}{3}\overline{\mathbf{D}}^p : \overline{\mathbf{D}}^p} = \dot{e}^p. \tag{7.129}$$

Here $\overline{\mathbf{N}}$ is the normal to the yield surface in rotated stress space. By using (7.120) and (7.129) we obtain evolution equation for $\mathbf{F}^p$. Finally, evolution equation for σ is defined in common form:

$$\dot{\sigma} = H \cdot \dot{e}^p, \tag{7.130}$$

where $H = H(e^p)$ is the hardening modulus.

The presented theory has been implemented into finite element model and used for modeling cold rolling of some metallic materials [186].

7.2.6 Finite-Element Modeling of Deformation Processes

Rolling and Pressing of BSCCO/Ag Tapes

The finite element method (FEM) is capable of obtaining qualitative estimations of parameters found by material deformation during BSCCO/Ag tape processing. However, the FEM application to these problems is a complicated task still not widely applied. For instance, the limiting factors for this method are complicated geometry of the composite and great time required to obtain numerical results with necessary accuracy. The application of coarse meshes can decrease the simulation time. However, these meshes could lead to simulation problems and insufficient detailing of results.

In [1077, 1546, 1659], drawing of monocore wire is modeled by two-dimensional axis-symmetric models. The constitutive equations describing the powder flow are stated using the Drucker–Prager model with an elliptical cap criterion. Processing parameters, such as the die angle and degree of reduction in each drawing step, are shown to influence the density of the powder in drawn wire. The distribution of density, being high at the silver/powder interface and lower in the center, is in agreement with test data. FEM modeling of the multi-filament wire drawing has not been found in literature, probably because a full three-dimensional model is necessary.

Numerical simulation of flat rolling also requires a 3D model to describe the deformations in length, width and thickness directions. Simplified 2D simulation can be made, assuming zero deformation in either the width or the length direction. Assuming zero width strain, the pressure distribution along the roll gap is modeled in [413, 1025, 1509]. In this case, the pressure profile forms either a friction hill or a friction valley, depending on the roll diameter and degree of thickness reduction. These parameters also influence the shear strains in the strain zone. When material flow in the cross-section is analyzed in a 2D model, zero strain in the length direction must be assumed [445]. This technique enables the prediction of filament geometry and density, incorporating the influence of wire geometry and friction.

The input data for numerical properties are essential for the precision of numerical results. Reference [610] presents an extensive investigation of the mechanical properties of BSCCO powder. The powder is evaluated applying a combination of fracture tests and triaxial strain tests. As a result, it is concluded that the Drucker–Prager model is not capable of describing the yield surface of BSCCO powder in detail. Reference [113] shows how the yield surface can be determined by a few relatively simple tests combined with a Drucker–Prager conical cap model. It is demonstrated that an FEM simulation based on this approach gives a rather good prediction of the density in the individual filaments for small reductions of tape thickness in a 2D model.

In more details three FEM approaches for simulation of flat rolling are considered below [446]. The next different approaches are applied, namely (i) 2D pressing in a mesh with 50×50 elements, (ii) 3D pressing between non-rotating rolls in a mesh with $17 \times 26 \times 60$ elements (number of elements in width, height and length, respectively) and (iii) full 3D rolling in the same mesh as in the case of the 3D pressing. The superconducting filaments have rectangular shape and lie in a square matrix of pure silver, surrounded by an alloyed silver sheath. Both silver materials are described by Von Mises yield criterion with the flow stress parameters ($\sigma = C\varepsilon^n$) given in Table 7.2. The constitutive plasticity model, describing the powder, is the Drucker–Prager model with a conical cap (see Fig. 7.17). This model is a rough approximation to the real yield surface, but it includes the most important properties of the powder, which are pressure-dependent yield stress, volumetric strain and material hardening. In the $(p, \sqrt{J_2})$ plane, the yield surface consists of two intersecting lines, shown in Fig. 7.17. The Drucker–Prager failure surface is written as

Table 7.2 The model parameters for silver, alloy and BSCCO powder

Parameters		Silver	Silver alloy	BSCCO powder
Young's modulus (MPa)		57.32	57.32	13.79
Poisson's ratio		0.38	0.38	0.2
$\sigma = C\varepsilon^n$	C (MPa)	320	350	–
	n	0.3	0.11	–
η		–	–	1.2
ξ		–	–	0.29
p_t, (MPa)		–	–	12
$p_c = a\{\exp[b(\rho - \rho_0)] - 1\}$	a (MPa)	–	–	27
	b	–	–	1
Relative density (ρ_0)		–	–	0.6

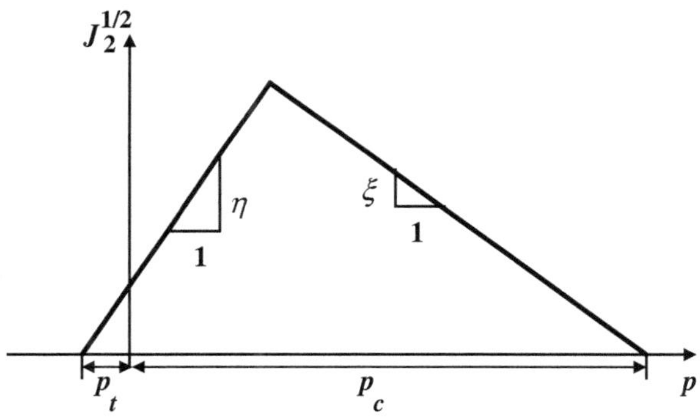

Fig. 7.17 Drucker–Prager model with a conical cap [446]

$$F_s = \sqrt{J_2} - \eta(p + p_t) = 0, \tag{7.131}$$

where J_2 is the second invariant of the stress tensor, η is the slope of the failure line and p_t is the yield stress in pure, hydrostatic tension. The cap yield surface is a line with a negative slope:

$$F_c = \sqrt{J_2} + \xi(p - p_c) = 0, \tag{7.132}$$

where ξ is the slope of the cap line and p_c is the hydrostatic compression yield stress. The model parameters are given in Table 7.2 [610].

All three models calculate higher densities in the center than at the edges and overestimate the relative density of filaments. In Fig. 7.18 the relation between tape thickness and tape width is shown for 2D and 3D FEM simulations of rolling and pressing and compared with test data. As it is followed from Fig. 7.18, the 3D rolling simulation fits very well with the test results. At the same time, the FEM simulations of both models overestimate the widening.

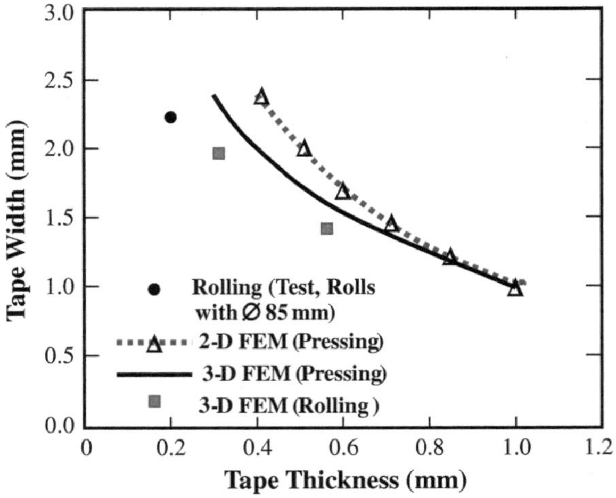

Fig. 7.18 Tape width versus tape thickness for experiments with ∅ 85 mm rolls and 2D and 3D FEM simulations [446]

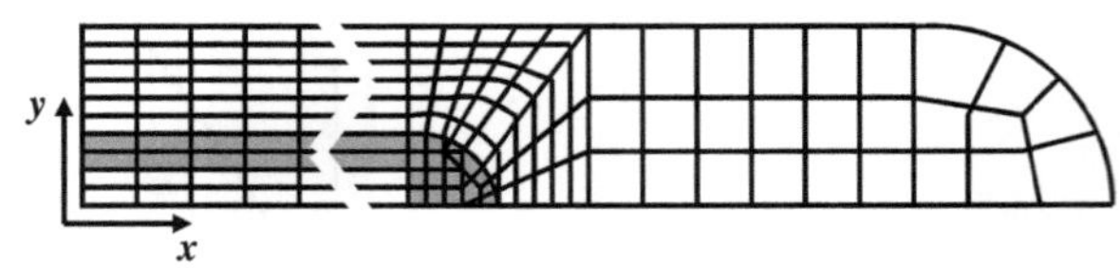

Fig. 7.19 Mesh used for FEM computation. The entire mesh contains 1500 nodes and 250 elements

Thermal Cycling of BSCCO/Ag Tapes

FEM is also applied to investigate thermal cycling of BSCCO/Ag tapes. For example, in [1256], a simulation of thermal stresses is carried out in monocore tape under the following assumptions:

- The BSCCO/Ag tape is stress-free at the sintering temperature (1113 K).
- The mechanical properties of both components are isotropic.
- Silver behaves elastically and plastically while BSCCO is brittle.
- The Baushinger effect is neglected, that is, the magnitude of the yield stress of silver is assumed to be the same in tension and compression.
- The composite is symmetrical about the *x*- and *y*-axes.
- The tape is infinitely long so that plane strain conditions apply.
- The BSCCO core center is assumed to be 65 % dense and the outer core is assumed to be 85 % dense.

Figure 7.19 [1256] shows a section of the mesh used for the thermal cycling analysis. The *x*-axis is parallel to the width of the tape (3.5 mm), the *y*-axis is parallel to the height of the tape (200 μm) and the *z*-axis is parallel to the length, assuming to be infinite. Both the central part of the core tape and external part of the region corresponding to the silver sheath are presented by white color. The gray part defines the external part of the superconductor or interface. The mechanical properties of BSCCO and Ag are shown in Table 7.3, where E is Young's modulus, ν is Poisson's ratio, K_p is the slope of the stress–strain curve

Table 7.3 Mechanical properties of Ag and BSCCO [1256]

Parameters	BSCCO	Ag
α (K^{-1})	13.6×10^{-6}	21.9×10^{-6}
E (GPa), 100 % dense	127.0	71.0
E (GPa), 85 % dense	83.8	–
E (GPa), 65 % dense	54.1	–
σ_y (MPa) at 300 K	–	12.6
K_p (GPa) at 300 K	–	0.57
ν	0.14	0.37
σ_y (MPa) at 77 K	–	13.2
K_p (GPa) at 77 K	–	0.7

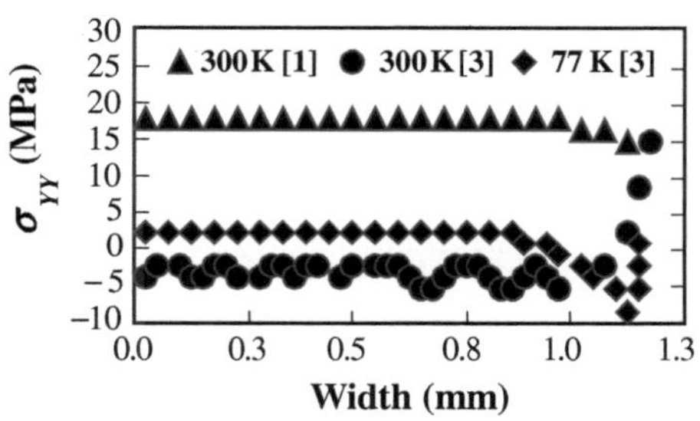

Fig. 7.20 FEM, showing the variation of σ_{yy} with thermal cycling across the width of BSCCO core. Cycles are denoted by the *numbers* in *square brackets* [1256]

after yielding and σ_y is the yield stress. Figure 7.20 shows the variation of σ_{yy} with thermal cycling across the width of the tape. The stress state is fairly uniform in the center of the core except at the edges, where the curvature of the core can act as a stress concentrator.

Chapter 8
Modeling of YBCO Oxide Superconductors

8.1 Modeling of 123 Phase Solidification from Liquid

8.1.1 Heterogeneous Mechanism

Initially, for peritectic reaction (basis of melt processing), *heterogeneous mechanism* has been suggested, that is, a formation of 123 phase in direct contact of melt and properitectic ("green") 211 phase followed by the growth of 123 crystallites in account of the component diffusion through product layer [783]. Shortcoming of this mechanism results in the following causes:

(1) Calculated growth rate of 123 phase, limited by diffusion of slowest component ($D_{Y^{3+}} = 10^{-11}$ cm^2/s), should be approximately 4 μm/h [103], which is some order of magnitude less than test data [103, 1441, 1442]. In this case, the peritectic reaction, which occurs in diffusion mechanism, should be completed after formation of 123 layer around 211 particles.
(2) It is impossible to describe YBCO platelet structure, because diffusion of components should mostly impede in the direction of dominant growth, that is, in *ab*-plane.
(3) Parallel disposition of platelets is not explained, because arbitrary initial orientation of particles of the 211 properitectic phase into liquid should lead to the absence of anisotropy and to arbitrary orientation of 123 crystallites, presented in Fig. 8.1a.

8.1.2 Models Based on Yttrium Diffusion in Liquid

In realization of melt-processing for preparation of YBCO large-granular samples with improved superconducting properties, it is important a priori to obtain a grain growth in the direction of the gradient to maximize the ratio of "external" thermal

I. A. Parinov, *Microstructure and Properties of High-Temperature Superconductors*, 407
DOI: 10.1007/978-3-642-34441-1_8, © Springer-Verlag Berlin Heidelberg 2012

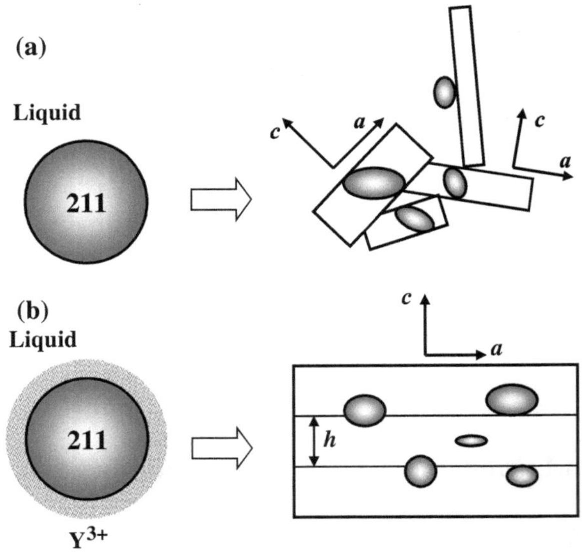

Fig. 8.1 Models of heterogeneous formation of nuclei at 211 particles [783] (**a**) and of solubility of 211 phase into liquid, assisted by homogeneous formation of nuclei (**b**) [292, 764]

gradient (G) to solidification rate (R). From this point of view, it seems better to transfer the sample across a large thermal gradient. However, in this case a large part of the sample stays for a long time at elevated temperature, where Y_2BaCuO_5, $BaCuO_2$ and CuO_2 co-exist. Moreover, an undesirable change of the geometric shape of the sample is possible by liquid flow. Thus, a good compromise between G and R is required, because the G/R ratio governs the stability of the solidification process. The thermodynamic aspects of this problem considered in [292, 764] have led to a model of 211 phase solubility in liquid and *homogeneous formation of nuclei* (see Fig. 8.1b). This model is based on the following assumptions:

(1) Ion concentration of Y^{3+} attains a maximum in liquid near the surface of 211 particle and decreases sharply at the mid-distance of neighbor 211 inclusions. When the distance exceeds this length, the ion concentration of Y^{3+} decreases asymptotically down to a value corresponding to the peritectic melt.
(2) Increasing of yttrium concentration in the liquid enhances a possibility of 123 phase solidification because of metastability of the melt and existence of figurative point in the two-phase region "123 solid phase–melt."
(3) Constitutional undercooling in the system, leading to supersaturation of the melt by yttrium, is found (i) by "external" thermal gradient, (ii) by undercooling in comparison with saturated melt and (iii) by decreasing of solidification temperature for melt, saturated by yttrium relatively to peritectic melt.

The theoretical analysis leads us to introduce a criterion connected to conception of undercooling and to compare the G/R ratio with the relation, $m_L(C_S-C_L)/D_L$ [1205], where $C_S(C_L)$ is the ion concentration of Y^{3+} in the solid (liquid) phase at the interface level; D_L is the diffusion coefficient of the ion of Y^{3+} and m_L is

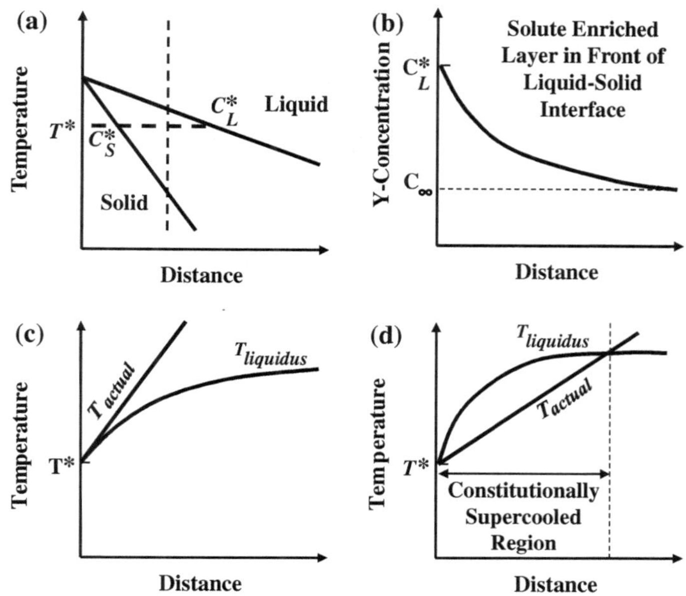

Fig. 8.2 (**a**) Temperature and (**b**) concentration profiles, showing the constitutional underco-oling, (**c**) conditions for a plane front, (**d**) unstable case, where imposed temperatures are lower than the temperature of the liquidus [486]. T^* is the initial interface temperature

the slope of the liquidus curve in the temperature–composition phase diagram. Depending on the various situations presented in Fig. 8.2 two cases are possible:

(1) If *G/R* ratio is high, that is, *G* is high, the slope of the curve giving the temperature at the interface level is high, whatever the position of this interface. The comparative behavior of the temperature at the interface level and the temperature of the liquidus is given in Fig. 8.2c. The equation

$$G/R \geq m_L(C_S - C_L)/D_L \tag{8.1}$$

is continuously satisfied. The interface temperature evolution is stable in time, and the solidification front is flat because the interface temperature follows the temperature of the liquidus.

(2) If the *G/R* ratio is relatively small, inequality (8.1) is not satisfied, causing instability of the interface temperature (see Fig. 8.2d), because it should be continuously re-adjusted to that of the liquidus. Therefore, the solidification front is no flat, but has cellular (dendritic) shape or consists of equiaxed blocky (see Fig. 8.3), because the interface temperature fluctuates between the two curves, leading alternatively to a liquid or a solid phase.

The above results have been used to explain directed solidification of YBCO after peritectic reaction [1441]. The diffusion of yttrium is necessary to re-compose 123 phase in the liquid, that is, for a non-classical peritectic reaction, which is

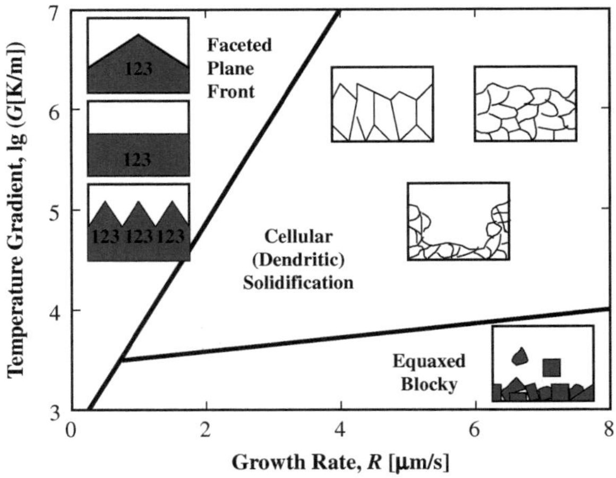

Fig. 8.3 Effects of temperature gradient and growth rate on the morphology of the solidification interface [292]

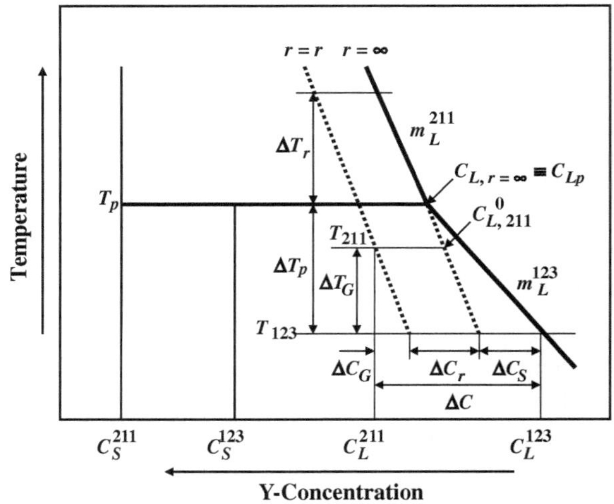

Fig. 8.4 Schematic phase diagram, showing the undercooling and the concentration differences as the driving force for the diffusion of yttrium during the peritectic re-composition of 123 phase [1230]

governed by diffusion through melt and additionally demands to attain triple point between primary, secondary and liquid phases [788]. The driving forces inside the diffusion zone, leading to a migration of yttrium, correspond to concentration gradients. Two phenomena, presented in Fig. 8.4, have been established for these concentration gradients [293, 1209]: (i) the change of the chemical potential at the "211 phase-liquid" interface caused by the curvature of the 211 particles and (i) constitutional undercooling (ΔT_S), which corresponds to the temperature difference between the actual temperature of the solidification interface and the temperature of the "211 phase-liquid" interface.

The first phenomenon explains a change of 211 particles during melt transformation in 123 crystal. The change of chemical potential, caused by the curvature of the 211 particles, relates to the undercooling, ΔT_r, derived from the well-known Gibbs–Thomson relationship:

$$\Delta T_r = 2\Gamma_{211}/r, \tag{8.2}$$

where Γ_{211} is Gibbs–Thomson coefficient ($=\sigma/\Delta S$), σ is the interface energy between the 211 particle with radius, r, and the liquid, ΔS is the volumetric entropy of melt. Then, the composition difference of yttrium, ΔC_1, which is considered to be a driving force for diffusion, is expressed using the phase diagram (see Fig. 8.4), as

$$\Delta C_1 = \Delta C_r = C^{211}_{L(r)} - C^{211}_{L(r=\infty)} = \frac{2\Gamma_{211}}{r m^{211}_L}, \tag{8.3}$$

where $C^{211}_{L(r)}$ and $C^{211}_{L(r=\infty)}$ are the yttrium concentrations of liquid at the 211 particle and at the 123 interface, respectively; m^{211}_L is the liquidus slope of the 211 phase.

The constitutional undercooling is stated using the second composition change ΔC_2, given by extending towards low temperatures the "211 phase-liquid" liquidus line [765]:

$$\Delta C_2 = \Delta C_S = \left(\frac{1}{m^{123}_L} - \frac{1}{m^{211}_L}\right)\Delta T_S. \tag{8.4}$$

Introduction of a third parameter, ΔC_3 taking into account the imposed gradient, completes this model [765]. This term is directly given by the distance, z, between the solidification interface and the considered 211 particle:

$$\Delta C_3 = \Delta C_G = \frac{Gz}{m^{123}_L}, \tag{8.5}$$

where m^{123}_L is the liquidus slope of the 123 phase. Thus, total difference of the yttrium concentration, governing the solidification process of the 123 phase is equal to

$$\Delta C = \Delta C_1 + \Delta C_2 + \Delta C_3. \tag{8.6}$$

The next development of the model is connected with the assumption of the 211 particles to be spheres and consideration of the concentration difference in yttrium between the 211 phase and the solidification front as the driving force for the diffusion of yttrium in the liquid phase (see Fig. 8.5) [765].

Assume that the distance between 211 particles is equal to 2δ, and the 211 phase remains in final structure. The concentration differences of yttrium near solidification front of 123 phase due to solution of 211 particles are shown in Fig. 8.5b. Moreover, we assume that the 211 particles are in interfacial equilibrium with their surrounding liquid and are large enough that the effect of radius of

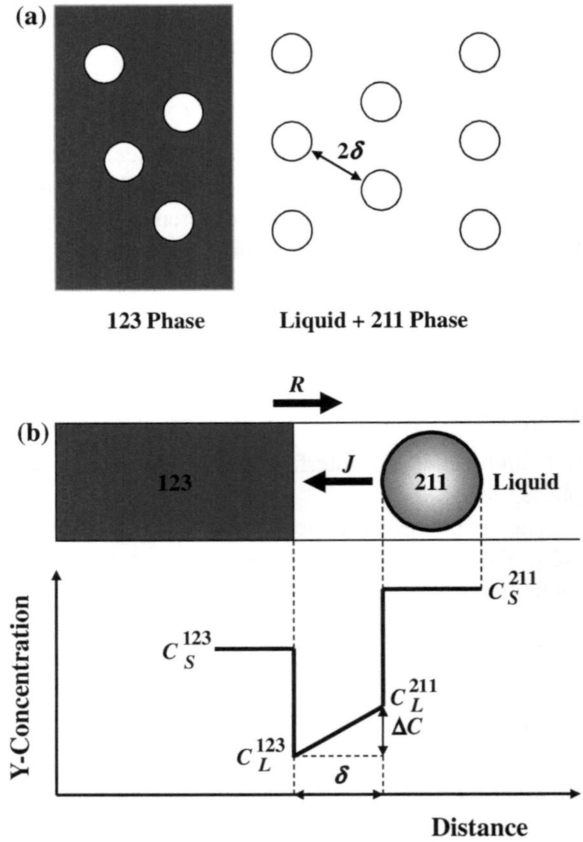

Fig. 8.5 (**a**) Model of peritectic solidification and (**b**) yttrium concentration profile near solidification front of 123 phase due to the dissolution of 211 particles. J is the flux of yttrium ions, which is necessary for growth of the 123 phase

curvature on melting point is negligible.[1] The phase diagram in the peritectic region, presented in Fig. 8.4, shows that the temperature of the growing 123 front is undercooled by an amount ΔT_p below the peritectic temperature, T_p, where

$$\Delta T_p = \Delta T_G + \Delta T_S + \Delta T_C, \tag{8.7}$$

where ΔT_G is the depression of the integrated temperature, resulting from the temperature gradient, G; ΔT_S is the maximum constitutional supercooling ahead of the interface (i.e., temperature difference between the equilibrium liquidus and the actual temperature at $x = \delta$) and ΔT_C is the temperature depression, resulting from the deviation in yttrium concentration at the 211 particle interface from that of the peritectic liquid composition, C_{Lp}. From Fig. 8.5b

$$\Delta T_G = G\delta. \tag{8.8}$$

[1] If the particles are sufficiently small, this radius of curvature effect could be a significant, even a major, defining driving force of mass transport, leading to 123 phase solidification [766].

The yttrium concentration in equilibrium with the 123 interface (at $x = 0$, $T = T_{123}$) is C_L^{123}, and that at $x = \delta$ and $T = T_{123} + \Delta T_G$ in equilibrium with the 211 particles is $C_{L,211}^0$. The dependence of the yttrium distribution for $x > \delta$ is given by the equilibrium 211 liquidus. Equating yttrium solvent rejected from the growing 123 front to that diffusing into the liquid gives

$$-R(C_S^{123} - C_L^{123}) = -D_L\left(\frac{C_{L,211}^0 - C_L^{123}}{\delta}\right), \tag{8.9}$$

where R is the growth rate of the solidification front. A relation between interface undercooling (ΔT_p) and growth rate, R, can be derived, assuming $C_S^{123} - C_L^{123} \approx C_S^{123} - C_{Lp}$. Moreover, linear liquidus lines, shown in Fig. 8.4, result in $C_L^{123} = C_{Lp} - \Delta T_p/m_L^{123}$, and $C_{L,211}^0 = C_{Lp} - (\Delta T_p - \Delta T_G)/m_L^{211}$. Substitution of these expressions in (8.9) and making use of (8.8) gives

$$\Delta T_p = \left[\frac{R\delta}{D_L}\left(C_S^{123} - C_{Lp}\right) - \frac{G\delta}{m_L^{211}}\right]\frac{m_L^{123}m_L^{211}}{m_L^{211} - m_L^{123}}. \tag{8.10}$$

Thus, ΔT_p is linearly proportional to R, when G is sufficiently small. Similarly, the next relation can be written between ΔT_S and R:

$$\Delta T_S = m_L^{123}\frac{R\delta}{D_L}\left(C_S^{123} - C_{Lp}\right) - G\delta. \tag{8.11}$$

The quantity ΔT_S is the constitutional supercooling at $x = \delta$, and it is the maximum constitutional supercooling in the semisolid region. In solidification of this type, there must always be finite constitutional supercooling in front of the growing interface in order to create the compositional driving force for diffusion from the particle surface to the crystal interface. Assuming an existence of maximum undercooling [292] and noting that the distance between the solidification front and considered 211 particle, $z = \delta$, we obtain from (8.11) a relation defining the maximum growth rate, R_{max}, compatible with a plane solidification front:

$$R_{\max} = \frac{D_L}{z\left(C_S^{123} - C_{Lp}\right)}\left[\frac{(\Delta T_S)_{\max} + Gz}{m_L^{123}}\right]. \tag{8.12}$$

Thus, the highest solidification rates compatible with a steady-state growth are obtained theoretically at the following conditions: (i) with larger undercooling, (ii) with higher temperature gradient and (iii) with smaller size of 211 particles after peritectic decomposition (z is directly dependent on the particle size). Equation (8.12) shows that the maximum solidification rate is inversely proportional to the distance z between the 211 particles, resulting from the peritectic decomposition of 123 phase. This suggests a decrease in the 211 particle size and the addition of very fine 211 powder to the starting 123 precursor, because an increase of the volume ratio of the 211 phase in the melt for a given particle size results in a reduction of the interparticle spacing z. In fact, such an addition of 211 to 123 offers multiple advantages:

(1) It allows the geometric shape of the sample to be kept at high temperature owing to the formation of a solid skeleton.
(2) Diffusion of yttrium favors the supply of this species at solidification.
(3) It compensates the loss of yttrium in the recombination process, where entrapment of 211 particles takes place.
(4) Addition of properitectic 211 dispersion refines the grain size of the 211 precipitates, coming from the peritectic decomposition of the 123 phase (properitectic 211 particles act as nucleation centers).
(5) It improves the mechanical properties of the textured 123 sample.
(6) The 211 inclusions, in excess trapped in the textured 123 material, act as pinning centers, increasing the critical current.

Compare the terms $(\Delta T_S)_{max}$ and Gz, taking into account that in a classical furnace $z \approx 1\ \mu\text{m}$ and $G < 100$ K/cm [356]. Then, it is obvious that $Gz \ll (\Delta T_S)_{max}$ and the solidification rate appears practically insensitive to the temperature gradient and is proportional to the undercooling, ΔT_S. It should be noted that the case of YBCO system, where the slopes of the liquidus curves are quite different for 211 and 123 phases, leading to relatively large driving forces, appears favorable for sample texturing.

The directed observations of 123 phase solidification using an IR camera have found a rate of solidification close to 10^{-7} m/s and estimated the diffusion coefficient of yttrium in the liquid by the value 6×10^{-11} m^2/s [525]. From a practical point of view this solidification rate appears quite low. From (8.12), it is governed by two parameters: ΔT_S and z. During a slow cooling rate, r, in a thermal gradient, G, because of the translation, x, of the "liquid–solid" interface in this gradient the undercooling is given versus time as [356].

$$\Delta T_S = \Delta T_{S0} + rt - Gx \tag{8.13}$$

This equation shows that solidification at a constant temperature ($r = 0$) is possible but is limited by the thermal gradient. In this case, only a large initial undercooling, ΔT_{S0}, can lead to relatively large superconducting crystals (monodomains).

8.1.3 Models Based on Interface Phenomena

The models [293, 764, 1209] considered in the previous section seem in good agreement with test results, but they do not take into account the interface kinetics processes, assuming the crystallization of a pure 123 phase even though some 211 inclusions are always entrapped in the textured 123 material. Moreover, addition of properitectic 211 dispersion to 123 phase can provoke yttrium supersaturation at

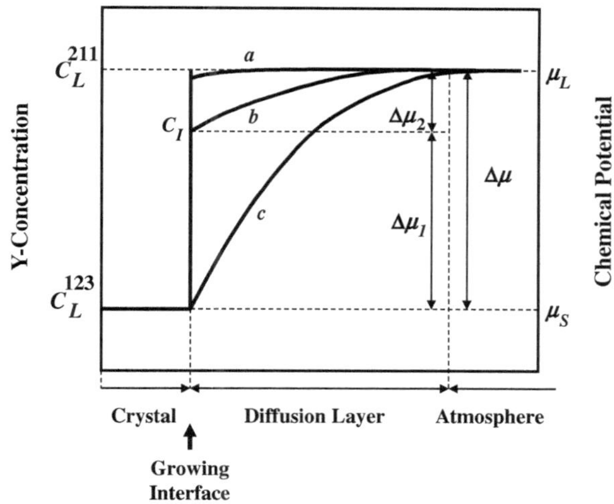

Fig. 8.6 Yttrium concentration profiles in front of the growing interface [1230] and differences in chemical potential in the diffusion layer: (**a**) kinetics governed by interface phenomena, (**b**) kinetics governed by both diffusion and interface phenomena and (**c**) kinetics governed by diffusion

the solidification interface. Then, the diffusion rate of yttrium is rapid compared with the propagation of the solidification front, so the diffusion of yttrium is no longer the limiting factor for the growth of 123 phase. Therefore, it is followed to introduce in consideration interface kinetics phenomena. The supersaturation can be presented as

$$\sigma = \frac{C_I - C_L^{123}}{C_L^{123}}, \tag{8.14}$$

where C_I is the yttrium concentration at the "123 phase-liquid" interface and C_L^{123} is the yttrium concentration at equilibrium. Three cases can be distinguished (see Fig. 8.6): (i) the growth rate of 123 phase is governed by the yttrium diffusion (curve *c*); (ii) the growth rate of 123 phase is under mixed control conditions (curve *b*) and (iii) the growth rate of 123 phase is governed by the interface reaction (curve *a*).

A model taking into account the interface phenomena [1230] introduces two rates of solidification (in the *ab*-plane and in the *c*-axis direction): $R_{ab} = k_a\sigma^2$, and $R_c = k_c\sigma$, where k_a and k_c are the kinetic coefficients for the *a*- (or *b*-) and *c*-directions, respectively. The effect of the kinetic coefficient on the growth rate as a function of the undercooling can be taken into account, using either square power law dependence or linear one on the supersaturation [1230]. As followed in Fig. 8.7, a linear dependence *vs* supersaturation leads to a quasi-linear dependence of the growth rate versus supercooling. On the contrary, in the case of a square power law dependence versus supersaturation, both behaviors are possible, depending on the *k* value: for $k < 10^{-4}$ cm/s the growth rate would show a square power law dependence versus supersaturation, and for $k > 10^{-3}$ cm/s the growth rate would demonstrate a linear dependence versus undercooling, irrespective of the square law dependence of the growth rate versus supersaturation.

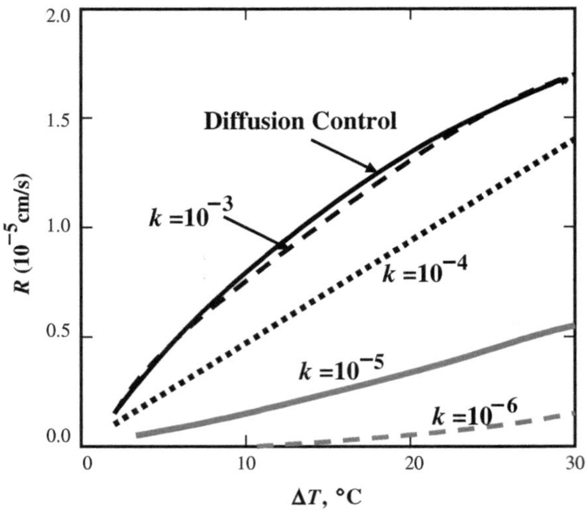

Fig. 8.7 Effect of kinetic coefficient, k, on growth rate, R, as a function of the undercooling in the case of a linear dependence on the supersaturation [1230]

In the case of pure 123 phase, a good agreement with experiment [1230] is observed for R_{ab} and R_c versus the undercooling. Hence, one can conclude that growth of 123 phase is generally controlled simultaneously by yttrium diffusion and interface kinetics processes. If a 211 properitectic phase is added to 123 phase before the melt process, for a given undercooling the larger amount of 211 phase can supply more yttrium diffusion to the growing front, corresponding to the growth rate of solidification process. At the same time, the change from a quadratic dependence of R_{ab} versus supersaturation to a linear dependence (and vice versa for R_c), observed in [1230], is indeed unexpected if one assumes that the yttrium flux reaching the interface is proportional to the 211 phase concentration. Obviously, this means that the growth mechanisms in the considered directions are different for both compositions, in spite of identical growth conditions. Thus, one can modify the growth rate and its anisotropy, changing solidification process and superconductor composition. The particle motion near an advancing solidification front can be described in terms of an interfacial energy relationship between a particle, solid and liquid in the framework of the pushing/trapping mechanism of the particle by the solidification front (see Fig. 8.8) [1722]:

$$\sigma_{PS} = \sigma_{PL} + \sigma_{SL}, \tag{8.15}$$

where σ_{PS}, σ_{PL} and σ_{SL} are particle/solid, particle/liquid and solid/liquid interfacial free energies, respectively. According to this criterion, a particle present at a growing solid phase is trapped when $\sigma_{PS} < \sigma_{PL} + \sigma_{SL}$. In contrast, when $\sigma_{PS} > \sigma_{PL} + \sigma_{SL}$, a particle is pushed out from a growing solid by the fast diffusion of the liquid.

This interfacial energy criterion has been applied to melt-processed YBCO samples in order to explain trapping of 211 particles within 123 crystals [764, 853, 855, 861]. However, this is not enough because it assumes the interaction of the

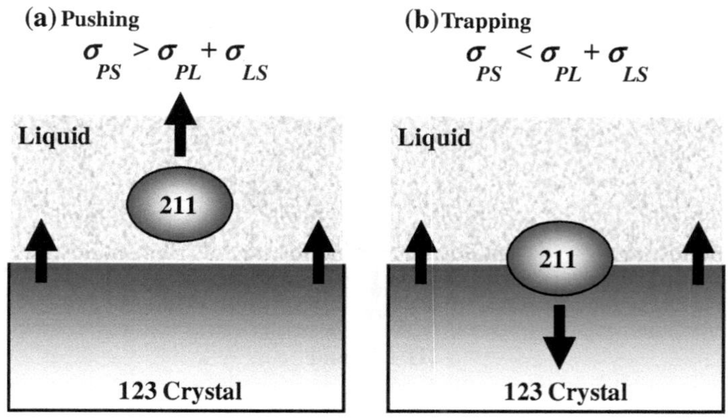

Fig. 8.8 Schematic of interfacial energy criteria on particle pushing and trapping

front with only one particle pushed out or trapped by the interface. Moreover, the above criterion should be modified for anisotropically growing crystals of melt-processed YBCO samples. The modified interfacial energy relationship is given as [855].

$$\sigma_{PS(hkl)} = \sigma_{PL} + \sigma_{S(hkl)L}, \tag{8.16}$$

where $\sigma_{PS(hkl)}$, σ_{PL} and $\sigma_{S(hkl)L}$ are particle/(*hkl*) interface, particle/liquid and (*hkl*) interface/liquid interfacial free energies, respectively. Then, in the isotropically growing system or directionally growing interface, 211 particles will be pushed out or trapped within a 123 domain with a random mode (see Fig. 8.9). In this case, for an anisotropically growing 123 crystal, trapping criteria for each growing plane are different.

The crystal structure of 123 domains at the peritectic temperature has a tetragonal symmetry. Therefore, there are three main growth interfaces of (100), (010) and (001) planes (see Fig. 8.10). In the tetragonal crystal structure, the atomic arrangement of a (100) plane coincides with that of the (010) plane. But the atomic arrangement of a (001) plane is different from that of the (100)/(010) plane. The boundary condition of particle trapping is inferred from the interfacial energy difference among growing (*hkl*) planes. The criteria for 211 particle trapping in a melt-processed YBCO system are given as [851]

$$\Delta\sigma_{(hkl)} = \sigma_{PS(hkl)} - \left(\sigma_{PL} + \sigma_{S(hkl)L}\right); \tag{8.17}$$

$$\Delta\sigma_{(100)} = \Delta\sigma_{(010)} \neq \Delta\sigma_{(001)}; \tag{8.18}$$

$$\Delta\sigma_{(101)} = \Delta\sigma_{(011)} \neq \Delta\sigma_{(110)} \neq \Delta\sigma_{(111)}. \tag{8.19}$$

From the interfacial energy model and experimental observations of 211 patterns [773, 853, 855, 1737], it is followed that the crystallographic planes of X-like 211 tracks are the six diagonal planes of 123 domain, which divide the cubic space

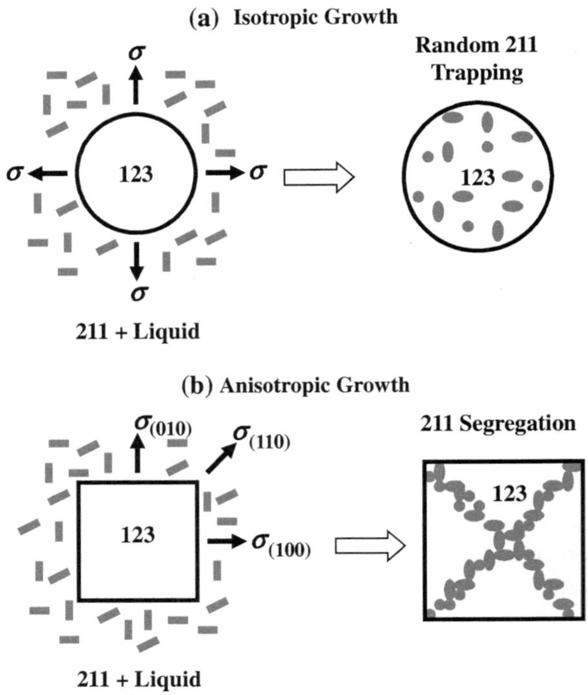

Fig. 8.9 Schematic of particle trapping modes in isotropically (**a**) and anisotropically (**b**) growing systems

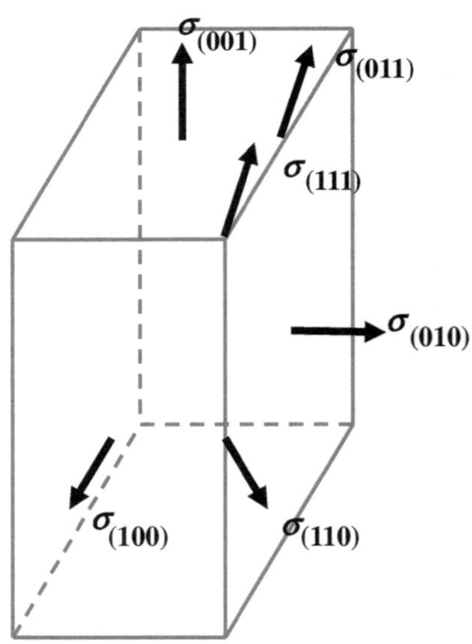

Fig. 8.10 Boundary planes for 211 particle trapping, based on an interfacial energy model

into three pairs of pyramids (see Fig. 8.11). These diagonal planes may vary with the aspect ratio of grown 123 domains. For example, there may be [110] planes for a cubic domain or [103] planes for an orthorhombic domain [855] (see Fig. 8.12).

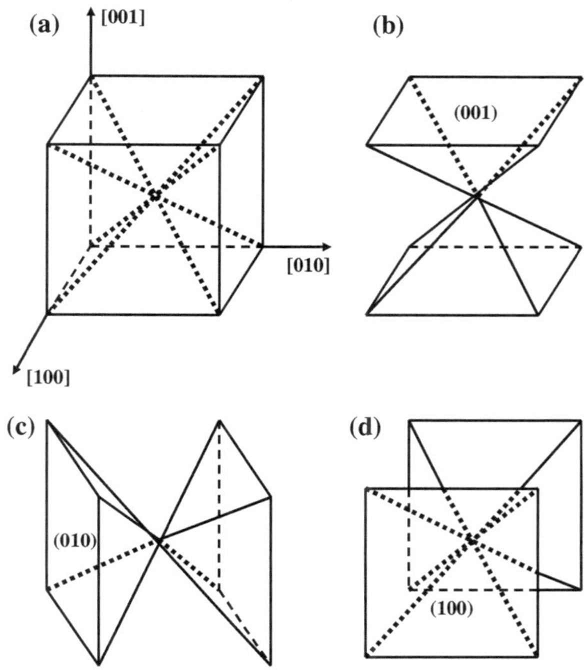

Fig. 8.11 (**a**) Three-dimensional demonstration of the planes, where 211 particles are trapped in a stoichiometric YBCO system, (**b**) the 211-free spaces and (**c,d**) the 211-filled spaces in a 211 phase excess system

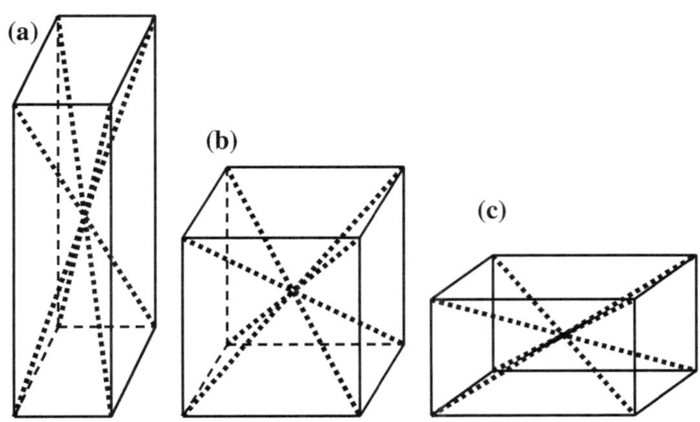

Fig. 8.12 Variation of diagonal planes as a function of grain anisotropy

In the case of planar segregation of 211 phase, a pair of pyramids having (001) growth fronts corresponds to 211-free spaces, while the other two pairs of pyramids having (100) and (010) growth fronts are filled with 211 particles.

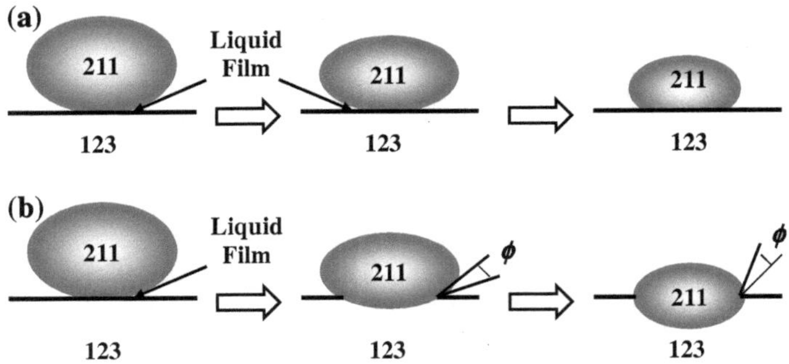

Fig. 8.13 Schematic of two different cases for 211 particle trapping within growing 123 interface [773]. Dihedral angle, ϕ, between 211 and 123 phases is (**a**) $\phi = 0°$ and (**b**) $\phi > 0°$

The crystallographic alignment of 211 particles may also be explained in terms of an interfacial energy relationship between the (*uvw*) plane of a 211 particle, (*hkl*) plane of interface and liquid [851]. The corresponding criterion is given as

$$\sigma_{P(uvw)S(hkl)} = \sigma_{P(uvw)L} + \sigma_{S(hkl)L}, \tag{8.20}$$

where $\sigma_{P(uvw)}$ is an interfacial energy of the (*uvw*) plane of a 211 particle. In this case, these particles have a polygonal shape with faceted surfaces with interfacial anisotropy.

An alternative way to express the interfacial energy criteria is the liquid wetting angle (ϕ) at advancing 123 interfaces. The trapping criterion of the 211 particle can be expressed with the liquid wetting angle and has been proposed for a melt-infiltrated YBCO system [773]. As schematically shown in Fig. 8.13, two different situations can exist for particle trapping. In the case of $\phi = 0°$, a 211 particle should be pushed toward the liquid, because the liquid film is present between the 211 particle and the growth front. In contrast, when $\phi > 0°$ and the 211 particle is not dissolved completely in the liquid, the particle is expected to be easily trapped within the 123 domain.

8.1.4 Models of Platelets-Like Growth of 123 Phase

The model of platelets-like growth of 123 phase, which is more faster in the *ab*-plane compared with the *c*-axis direction in the limit of single-crystalline material and taking into account interaction with 211 particle, is presented in [14]. The 211 particle may contribute to the gap-formation process, when the growth front bi-passes this particle, and the 123 material does not envelop the 211 particle completely by growing at nearly identical rates on either side of the particle (see Fig. 8.14d). In this case, the resultant microstructure consists of platelets of identical orientation separated by gaps, which are filled in with the rejected liquid.

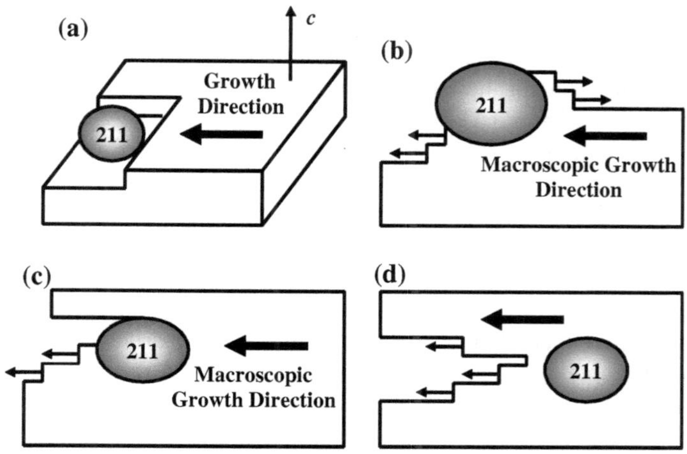

Fig. 8.14 Schematic of possible effect of 211 particles on the growth of melt-processed YBCO crystal: (**a**) 123 platelet abuts a growing platelet; (**b**) heterogeneous nucleation occurs at the "platelet–211 particle–liquid" junction; (**c**) once sufficient growth along the *c*-direction occurs such that the 211 particle can be bypassed, rapid lateral growth will occur; (**d**) this process results in a gap, which may not heal completely [14]

The generalization of this model [561] describes the formation of linear and plane gaps between 123 platelets and rejection of liquid at domain boundaries in the case of 211 particle trapped by 123 matrix. As is well known, there is a potential possibility of formation not only the $YBa_2Cu_3O_{7-x}$ (123) phase, but the $BaCuO_2$ (011) phase from liquid, which does not demonstrate superconducting properties and has solidification temperature about 1015 °C, that is, near to start formation of the 123 phase [60]. Thus, solidification of either 123 phase or 011 phase is possible [544]. The situation, connected with faster solidification of barium cuprate, may be explained on the basis of the phase diagram and taking into account the above-described mechanism of the 123 phase formation [14, 561]. The relatively small (compared with the process of solidification of the 011 phase) rate of establishment of equilibrium between liquid (L) and solid 123 phase can lead to thermodynamic equilibrium in each of these components. The liquid will demonstrate independence because of low concentration of yttrium that is caused by low rate of change between the solid and liquid phases. In this case, a homogeneous formation of 123 nuclei is impossible, and the figurative point, describing melt, displaces into $L + BaCuO_2 + CuO$ region, resulting in the solidification of barium cuprate.

There are two causes for breaking of usual equilibrium between liquid and solid 123 phase: (i) a growth of the 211 particles (due to their different sizes and well-known Ostwald ripening), reducing interphase interface with melt, decreasing yttrium flux into liquid, leading to envelopment of these particles by solidification front with rejection of the yttrium-depleted melt at the intergranular boundaries and finally initiating a de-lamination of the system; (ii) a decreasing of the thermal stability of 123 nuclei in the liquid, leading to absence of dominating 123 solidification even at sufficient yttrium concentration [545]. These causes change the stoichiometry of liquid in the case of trapping of large 211 particles [561], and even solidification of pure 011 phase [545].

The development of the model [764] for the case of low *G/R* ratio is carried out in [1503]. The microstructure close to the quenched solid–liquid interface exhibits bridges of 123 material between the solidifying 123 interface and 211 particles. In order to describe their morphologies, a combination of both phenomena, namely a peritectic reaction being mediated by the liquid and a peritectic transformation of the 211 particles, being linked to the solidification front via bridges of 123 phase, is necessary. Note that all the above models neglect peritectic transformation [789]. The principal difference of the model [934] from the models [293, 764, 1209] is the account of influence on the process of the 123 phase formation of Lifschitz–Zlyozov boundary effects [491, 1020, 1582, 1719], and action of the capillary attraction forces between moving front of the 123 phase solidification and 211 particles into liquid.

The entire *local* process of the engulfment of 211 particles into the solidifying interface can be explained, considering four steps: (i) liquid-phase diffusion-controlled growth, following the temperature gradient, if the 211 particle is far away from the phase boundary (see Fig. 8.15a); (ii) bridge formation, when the 123 interface faces an increased Y^{3+} concentration gradient, when being approached by a 211 particle (see Fig. 8.15b); (iii) peritectic surface reaction during the engulfment process (see Fig. 8.15c); and (iv) peritectic transformation (negligible effect compared to previous steps) (see Fig. 8.15d).

Due to the peritectic character of the 123 phase creation, this phase needs a Y-concentration that is not provided by the melt, being in equilibrium with the 211 phase. Therefore, as in classical nucleation theory, a depletion zone arises and the growth of the 123 phase is driven by a concentration gradient in the depletion zone, δ, close to the 123 interface. At the same time, the dissolving 211 particles maintain a medium yttrium concentration in liquid, c_m, corresponding to the Ostwald ripening theory.

Bridge formation starts when the depletion zone ahead of the 123 phase boundary and dissolution region of the 211 particle, begin to overlap (see Fig. 8.15b). The increased concentration gradient leads to an accelerated growth of the 123 phase toward the 211 particles, resulting in a bridge.

The growing bridge, reaching the 211 interface, defines the start of the peritectic reaction (in its original sense) [686], which subsequently covers the surface of the 211 particle with solid 123 material (see Fig. 8.15c). Once the 211 surface is covered, any further formation of 123 phase is governed by peritectic transformation, that is, strongly limited by diffusion in the solid and negligible (Fig. 8.15d).

As example of the above model, we explain the 1:1 correlation between 211 particle size and thickness of the 123 platelets, observed in [782]. A possible explanation is based on two assumptions: (i) the plate-like growth of the 123 phase

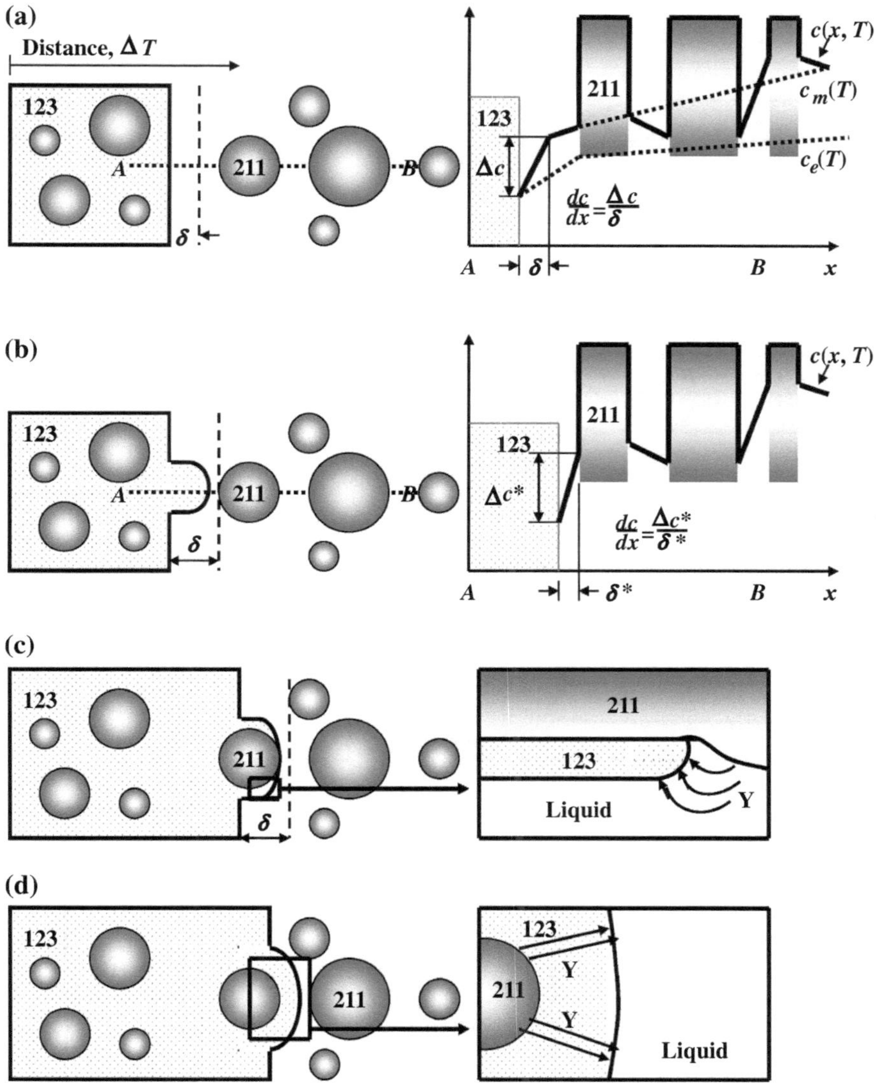

Fig. 8.15 Model of *local* influence of 211 particles on the growth morphology of 123 phase. The parameter $c(x, T)$ corresponds to the *local* yttrium concentration, c_e represents the equilibrium solubility, and c_m the mean concentration, corresponding to the Ostwald ripening theory; dc/dx denotes the concentration gradient at the 123 interface. (**a**) Liquid diffusion-controlled growth, (**b**) bridge formation, (**c**) peritectic reaction and (**d**) peritectic transformation [1503]

is caused by the strong anisotropic growth rates [14], $v_{ab} \gg v_c$, and (ii) the engulfment of the 211 particles is provided by the growing 123 matrix.

The anisotropy of the growth rates leads to a preferred growth of the *ab*-planes parallel to the temperature gradient. At the same time, the low growth rate along

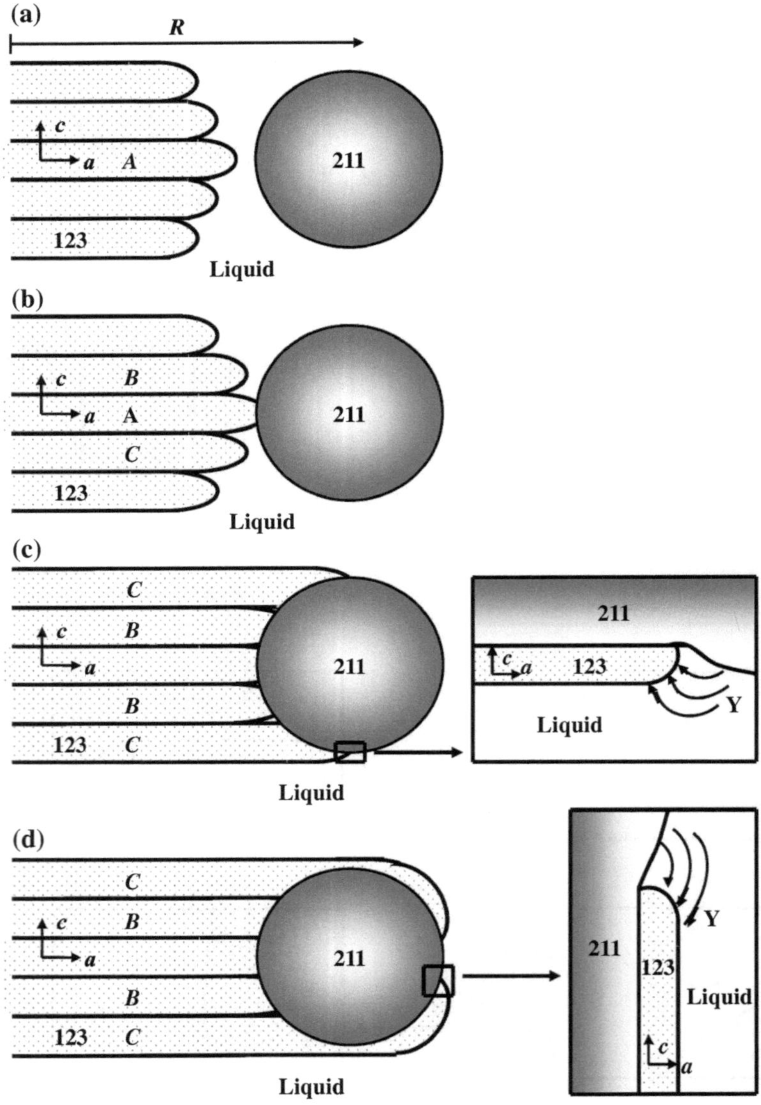

Fig. 8.16 Competing growth of 123 platelets near a 211 particle, leading to quasi-single crystalline material via a zipper-like mechanism. (**a**) The platelet closest to the 211 particle forms a bridge, (**b**) the *c*-axis growth of platelet *A* is in competition with the fast *ab*-plane growth of the adjacent platelet *B*, (**c, d**) explain enlarged platelet thickness of the outer platelets after passing the particle [1503]

the *c*-axis direction results in a morphological instability, leading to residual melt enclosed in planar defects between the *ab*-platelets. The platelet *A* closest to the 211 particle (see Fig. 8.16a) is the first to be influenced by the spherical diffusion

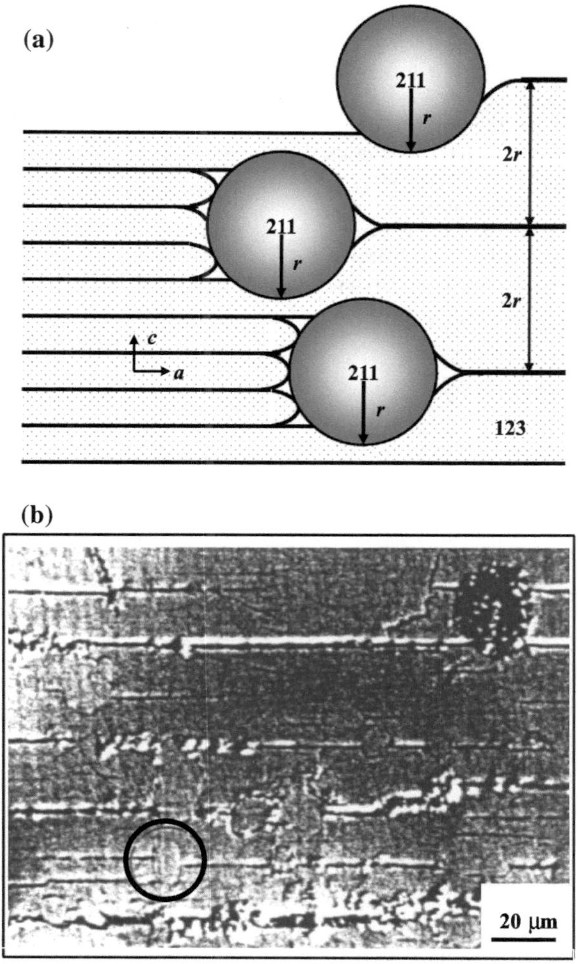

Fig. 8.17 (**a**) A zipper mechanism and (**b**) experimental evidence for this mechanism [1503]

region of the 211 particle and therefore will form a bridge as explained above. Next growth in all (*a, b* and *c*) directions of this platelet *A* is governed by the peritectic reaction, occurring along the 211 particle surface. In particular, the *c*-axis growth of platelet *A* is in competition with the fast *ab*-plane growth of the adjacent platelet *B* (see Fig. 8.16b). While the growth of platelet *B* is additionally accelerated, when approaching the 211 particle, the *c*-axis growth of platelet *A* is more and more decelerated because supply of yttrium is more and more hindered by the growing peritectic reaction layer. Considering the outer platelets *C*, note there is only competitive growth with one platelet (e.g., platelet *B*), giving rise to enhanced growth as well as being parallel to the *ab*-planes and the *c*-axis. This leads to an enlarged platelet thickness of the outer platelets after passing the particle (see Fig. 8.16c and d). This process automatically results after some iterations in a platelet thickness, corresponding to the mean particle diameter,

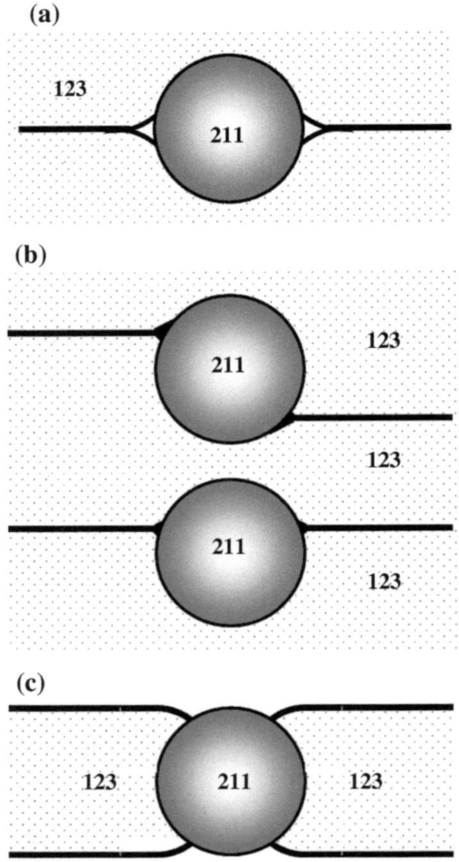

Fig. 8.18 123 pseudo-grains (platelets), passing a 211 particle of approximately the same dimensions. Predominant growth, when passing the particle is in the paper plane in the case (**a**) and in a plane perpendicular to the paper plane, but parallel to the *ab*-planes, in the cases (**b**) and (**c**) [1503]

which is supported by experimental observation presented in Fig. 8.17b. Once the thickness is reached, further growth of the 123 phase will be determined either by two platelets, each passing one side of the 211 particle (see Fig. 8.18a) or one platelet passing the whole particle, respectively (see Fig. 8.18b).

Note that even a hypothetic planar interface would change to a cellular morphology, when interacting with 211 particles also yielding platelet dimensions according to the 211 particle diameter (see Fig. 8.19). The zipper-like mechanism, acting in this process, results in an oriented growth of multiple connected platelets to a quasi-single crystal. Thus, the considered mechanism of solidification explains both a very high rate of the 123 phase creation and an existence of sufficiently sharp boundary between 211 inclusions and 123 matrix [1209]. The models [14, 1503] also explain a coincidence of 211 particle sizes with thickness of 123 platelets, observed in [782].

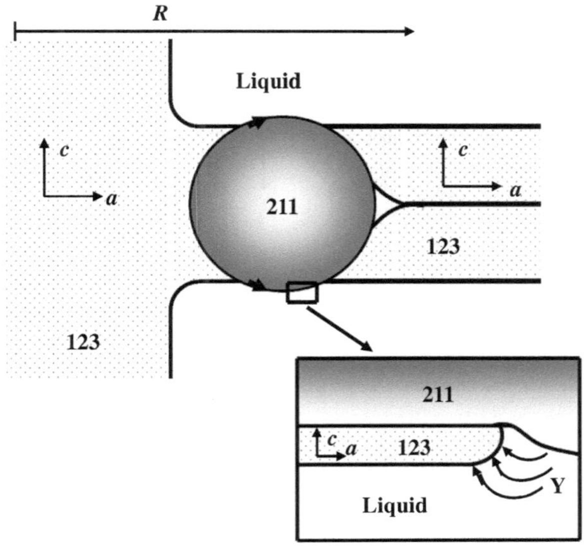

Fig. 8.19 Change in the growth morphology of a planar 123 interface, evoked by a 211 particle [1503]

8.1.5 Modeling of Solidification Kinetics

Only thermodynamic and chemical representations cannot totally describe growth processes in the Y(*RE*)BCO compounds. It is also necessary to include in consideration a kinetic process of the 123 phase solidification. As is followed from tests [1736], during the growth processes, the residual phases are pushed and distributed along the growing steps and sometimes are trapped in the 123 matrix, leading to the formation of microcracks. Their shape correlates with the rapid lateral growth that occurs in the *ab*-plane. This can lead to the formation of multi-grain domains as schematically drawn in Fig. 8.20.

Computer simulation of 2D growth kinetics of 123 front near 211 particles in the *ac*- and *ab*-plane is carried out in [1736] on the basis of Eden kinetics model (see Appendix E) for stochastic cell-by-cell growth of compact clusters. On a square lattice, we take a box of size L with periodic boundary conditions in the abscissa direction. The growing crystal is represented by the set of occupied sites on the grid (see Fig. 8.21). At each step of the growth, the set of empty sites in contact with the growing surface defines the perimeter of the solidification front (the circles show these sites). A site of the perimeter is randomly occupied, following the probability rule described below. This defines a new perimeter configuration, and the same process is repeated. In such a model, an occupied site can represent a unit cell or a cluster of 123 cells.

It is known that an anisotropic growth rule simulates precipitation processes [1122], as well as the growth of cell colonies [93]. In order to simulate an anisotropic growth front in the *ab*- and *ac*-planes, we consider two variants of the Eden model [1736]:

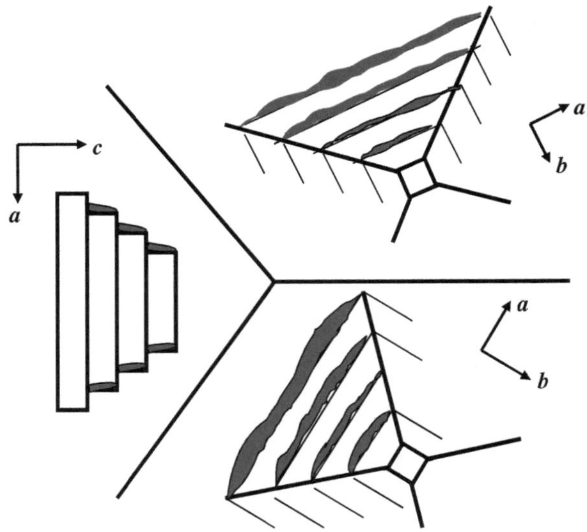

Fig. 8.20 A pyramid-like picture, sketching the formation of a polycrystalline region in the solidification process. Crystallographic orientations are emphasized and residual phase (*grey regions*) along growing steps [1736]

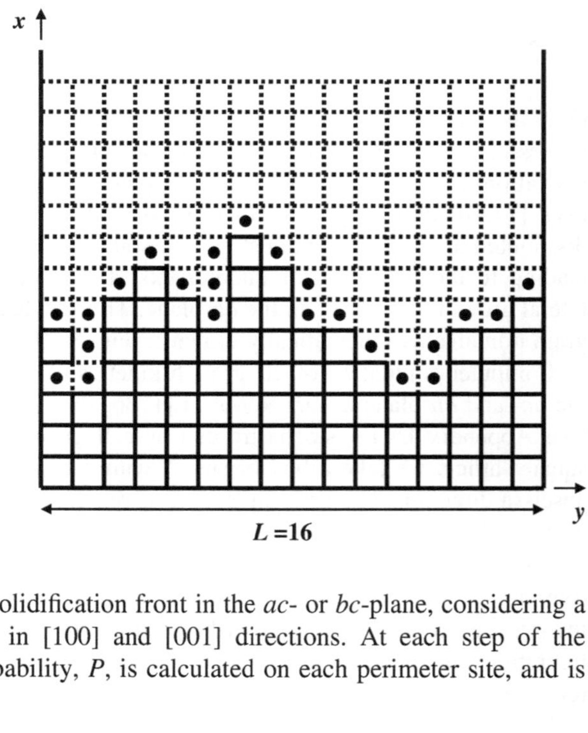

Fig. 8.21 Crystal front simulated using the isotropic Eden model. The *black circles* illustrate the perimeter sites

(1) *Model I* simulates the solidification front in the *ac*- or *bc*-plane, considering a square lattice oriented in [100] and [001] directions. At each step of the growth, a growing probability, *P*, is calculated on each perimeter site, and is given by

$$P \sim \exp(p_a N_a + p_c N_c), \tag{8.21}$$

where N_a and N_c are the number of occupied nearest neighbors (*nn*) in the *a* and *c* directions, respectively; p_a and p_c are the anisotropic growth parameters. Then, the set of these probabilities (defined on the perimeter) is renormalized in the interval [0, 1]. A random number generator chooses the growing site as in a Monte-Carlo simulation. This site is then occupied, defining a new perimeter with the next repetition of the whole process.

Note that the exponential law simulates a curvature effect (or a so-called Gibbs–Thomson effect) [1804]. When $p_a > p_c$, the growth probabilities are more important in empty sites linked to the crystal, following the *a*-axis direction. It results in a faster growth in the [100] direction than in the [001] direction, simulating an anisotropic solidification process in the *ac*-plane. When $p_a = p_c = 0$, the growth process reduces to a simple Eden's model. When $p_a < p_c$, the growth is trivial: the front remains flat and parallel to the substrate (i.e., to the [001] direction).

(2) *Model II* simulates the solidification front in the *ab*-plane. In order to simulate the [110] as the fast growth direction, the growth probability, *P*, calculated on each site of a square lattice is given by

$$P \sim \exp(p_{nn}N_{nn} + p_{dnn}N_{dnn}), \tag{8.22}$$

where N_{nn} and N_{dnn} are the number of occupied nearest neighbor sites (*nn*) in the [100] or [010] directions and diagonal nearest neighbor sites (*dnn*) in the [110] directions, respectively; p_{nn} and p_{dnn} are the growth parameters. The interaction with the diagonal nearest neighbors (*dnn*) is introduced in this model in order to take into account the diagonal fast growth directions in the *ab*-planes. When $p_{nn} = p_{dnn} = 0$, the model also reduces to the simple Eden's model.

In the both models, the presence of 211 particles can be simulated, avoiding the growth process to be achieved in circle-like regions of the square lattice. Thus, the 211 particles play a passive role, in contrast to, for example, the models [293, 764, 1209], in which these particles provide with yttrium solidification front of 123 phase.

The numerical results [1736], obtained on the basis of above models, estimate anisotropic effects of grain growth in the *ab*-plane ($g_{110}/g_{100} \sim 10$) and in the *ac*-plane ($g_{100}/g_{001} \sim 50$), where g_{hkl} is the growth probability in the crystallographic *hkl* directions. Replacing particles by spins [66], a magnetic field effect can be simulated and the magnetic texture growth processes could be further analyzed [348].

8.1.6 Multi-phase Field Method

In order to minimize the processing time as well as improve texture of YBCO melt-processed bulks, total information about solidification isotherm distribution in time and in space is necessary. This also defines qualitative estimations of a grain growth. The consideration of two-level scaling (in this case the macroscopic study of thermal fields and microscopic grain growth processes) decreases calculation time significantly. The macroscopic simulation of thermal fields in the YBCO samples, sintered into a Bridgman furnace in existence of the thermal insulation layer between the furnace and sample, has been carried out in [1533]. In order to simulate YBCO microstructure formation, three different approaches could be used, namely (i) Monte-Carlo technique [1300, 1324, 1327]; (ii) cellular automata models [512]; and (iii) phase field method [216, 1140]. In this section, we consider in more detail the latter method. The phase field conception has been applied to multi-phase system [1605], and also to the microscopic simulation of the 123 platelets growth at the existence of the 211 particles [1533].

In the multi-phase field model, describing evolution of interphase boundary, each phase is identified with individual phase field, and phase transformations between any neighbor pairs are considered depending on their own characteristics. The phase field method [265, 1792], based on the Ginzburg–Landau theory of phase transformations [725, 1006, 1502, 1569], was applied to the study of structural phase transformations of the "solid–solid" type in [264]. The theoretical basis was the functional of the local density of free energy, depending on an order parameter of the system and its partial derivatives on the spatial coordinates. The order parameter can be a scalar function, for example, concentration of solid component in the investigation of the phase transformations of the "solid–liquid" type, which changes from 0 (in liquid) to 1 (in solid) [1605], was limited by the parameters, p_i, describing only local phase state of the system. In this case, each phase field is identified either with melt or single 123 grain or 211 particle. Their time evolution is described by the system of non-linear parabolic partial differential equations [1605]:

$$\dot{p}_i = \sum_{k(k \neq i)}^{n} \frac{1}{\tau_{ik}} \left\{ \varepsilon_{ik}^2 \left(p_k \nabla^2 p_i - p_i \nabla^2 p_k \right) - \frac{p_i p_k}{2a_{ik}} \left[p_k - p_i - 2m_{ik}(\Delta T_{ik}) \right] \right\}, \quad (8.23)$$

where the point, as usual, denotes the time derivative; the parameters τ_{ik}, ε_{ik}, a_{ik} and m_{ik} can be defined through the test values (μ_{ik} is the mobility, σ_{ik} is the surface energy and λ_{ik} is the interface thickness) as [1792]

$$\tau_{ik} = \frac{L_{ik} \lambda_{ik}}{T_{ik} \mu_{ik}}; \quad (8.24)$$

$$\varepsilon_{ik}^2 = \lambda_{ik} \sigma_{ik}; \quad (8.25)$$

$$a_{ik} = \frac{\lambda_{ik}}{72\sigma_{ik}}; \quad (8.26)$$

$$m_{ik} = \frac{6a_{ik}L_{ik}(T_{ik} - T)}{T_{ik}}. \quad (8.27)$$

They are calculated through a set of parameters, describing the phase transformation ($i \to k$), namely (i) the equilibrium temperature of the phase transition, T_{ik} (e.g., the melting temperature, T_m, in solidification processes); (ii) the heat release during phase transition, L_{ik} (e.g., the melting heat in solidification problems); (iii) the driving force of the phase transition, m_{ik}, defined by deviation from equilibrium; and (iv) the difference, ΔT_{ik}, between the temperature of local cooling of the phase interface and the temperature, corresponding to a condition for the local equilibrium of i and k phases. These equations are coupled with a diffusion equation, determining the local concentration of yttrium at any position and time. It may by shown that in simulation of the superconducting structure solidification, the solution of constitutive equation for i phase is required to find only near interface boundary with neighbor phase. This set of differential equations is solved by a finite difference technique, using rectangular grid [1614].

Evolution of single phase, growing by isotropic way (e.g., liquid droplets in gas media), leads during interaction of its structure elements to the formation of 120°-angles in triple points. In the case of multi-phase systems, the triple junctions between different phases can be triple points of the "solid-liquid-gas" type. Then, considerable difference between surface energies for the systems "melt-solid", "solid–gas" and "gas–melt" leads to formation of triple points with angles different from 120°.

Different crystallographic directions cause anisotropic behavior on account of crystallographic structure. The investigation of phase transition from isotropic melt to anisotropic solid leads to corresponding change of equations for phase fields. This is attained by either introduction of kinetic factor or diffusion coefficient for phase field depending on the orientation of phase boundary to the crystallographic directions of neighbor grains or phases. The multi-phase conception may also be used to consider different grains with various spatial orientations, belonging to the same phase. With this aim in [1614], all orientations are divided into 10 orientation classes (the Pott's model), and each class is identified with own order parameter. Two- and three-dimensional computing simulation of grain growth with different crystallographic orientations has shown a dominant grain growth along the direction of the solidification front propagation.

The couple consideration of the phase field equations for thermal and solubility fields allows the process of microstructure formation to be modeled in actual peritectic systems (e.g., YBCO) [986]. The computer simulation results show coupling of the peritectic phase growth with a solubility of the properitectic phase. Obviously, the properitectic particles dissolve in liquid phase, according to their diameters (or surface curvature), increasing local concentration of yttrium in surrounding melt. As has been shown above, a microstructure of forming superconductor does not have not flat front, but creates bridges between solidification front and properitectic particles [1503]. When the 211 particles are sufficiently great, the time for their total dissolution before interaction with 123 front is insufficient. As a result, they are trapped by the 123 matrix and can serve as pinning centers (see Fig. 8.22) [1533]. Based on the phase field conception [1614], the computer simulation of a single particle in a population of 211 particles ahead of a growing 123 front reveals a "virtual pushing" of the particle, displaying in its initial coarsening before gradual and final dissolution (see Fig. 8.23) [1505]. The computer results show that a dissolution/re-precipitation mechanism at least contributes to the displacement of the 211 particles ahead of a growing 123 front.

Another example of computer simulation [1504, 1533] relates to the geometrically caused grain growth. In this case, the grain selection is found by the conditions of their anisotropic growth. This example may be useful for processing optimization of superconducting tapes. Moreover, the effects of nuclei grains sizes can be investigated in this way. Computer simulation of 211 particles, defining superconducting properties of YBCO and forming X-like precipitation, has been carried out in [1504]. This yttrium enrichment of the 123 grain diagonals leads to conclusion observed in experiment that the precipitation of 211 particles is controlled by the 123 phase evolution. Because yttrium concentration is greater at the edges and in the angles of growing 123 crystallite, the driving force of the 211 phase dissolution will be smaller in these sites. Then, the dispersed 211 particles dissolve more slowly and hence are easily trapped along diagonal directions. The neglected small effects of volume diffusion in solid state and corresponding minimum peritectic transformation cause a stability of these particles in the final microstructure [1504]. The simulation supports the possibility of effective variation by parameters especially in relation to geometric composition of superconductor. The geometrically caused grain growth does not increase the solidification rate of single 123 grains, but optimizes the processing rate of conductor in a whole, considering directed solidification of great number of grains simultaneously with

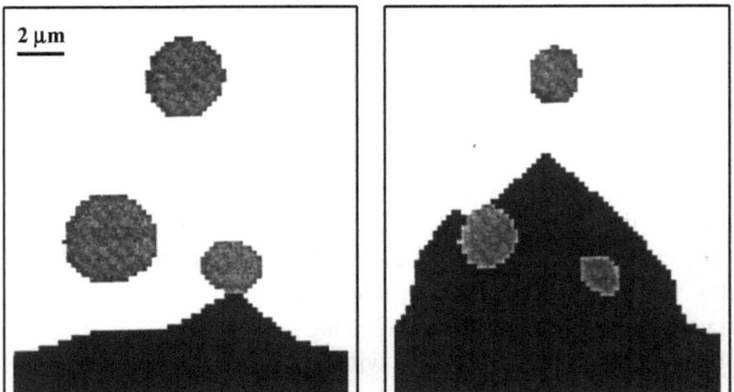

Fig. 8.22 The microscopic simulation results, defining engulfment of 211 particles and bridge formation [1533]

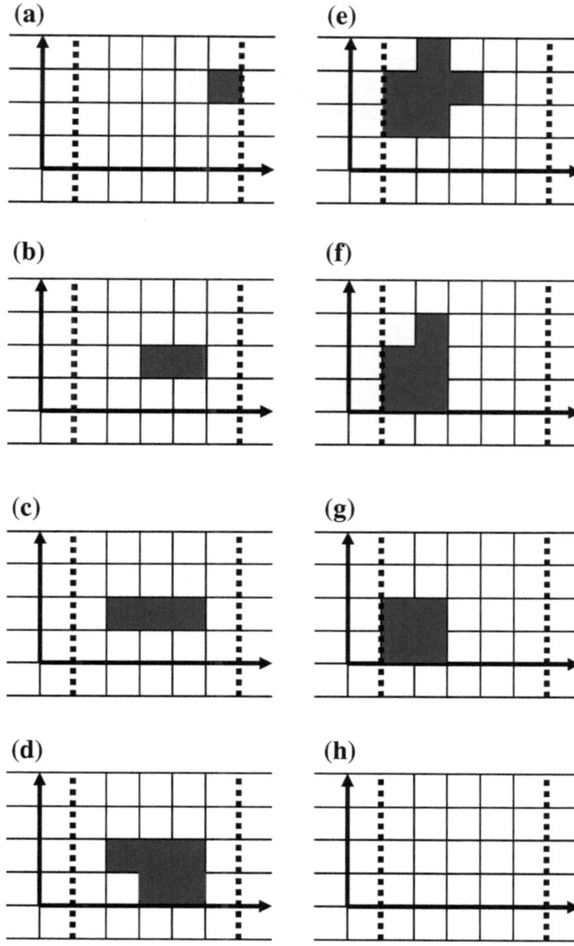

Fig. 8.23 211 particle ahead of a peritectic solidification front (the front grows from *right* to *left*). The motion of the particle is shown relative to the grid in combination with initial coarsening and then final dissolution [1505]

grain nucleation in some sites. The selection of 123 grains for this process occurs, using anisotropic growth conditions. Of particle interest is the grain selection distance, required for the well-oriented grains to overgrow the misaligned ones (see Fig. 8.24) [1533]. In this case, the growth of grains, having maximum rates and crystallographic orientations parallel to the main direction of the solidification front propagation, accelerates. In contrast, the growth of grains, perpendicular to the solidification front, slows down. That selection can be used in order to increase processing rate and critical current density in the melt-processed YBCO. Thermodynamic parameters, used in calculations, are presented in Table 8.1 [1504]. These parameters are either well-known data for YBCO family or estimations obtained for related materials.

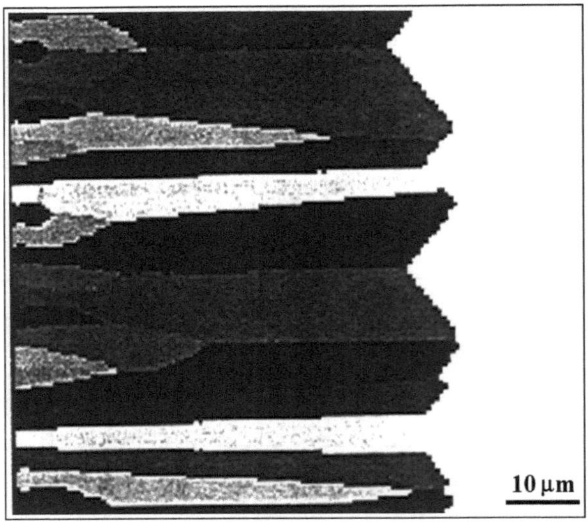

Fig. 8.24 Effect of grain selection investigated by computer simulation method and using phase field approach. Different 123 grains are visualized by different *grey levels* [1533]

Table 8.1 Thermodynamic parameters used in modeling

Parameter	Value
Latent heat	1000 J/cm^3
Diffusion coefficient	10^{-6} cm^2/s
Peritectic temperature	1288 K
Cooling temperature	37 K
Yttrium concentration in melt near 211 particle	0.26 mol%
Yttrium concentration in melt near 123 interface	0.14 mol%
Surface energy	10^{-6} J/cm^2
Linear kinetic factor	0.186 cm/s K
123 interface thickness	1.018×10^{-4} cm

Thus, the phase field method allows some thermodynamic phases in global non-equilibrium, taking into account local couple interactions of neighbor phases near the point of thermodynamic equilibrium, to be considered. The superposition of these couple interactions causes characteristics of triple point (i.e., multi-phase equilibrium). Then, the solution of equation for multiphase field defines the phase transition kinetics, which co-relates with limited diffusion during dissolution or with surface energy of interfaces. The equilibrium conditions for phase interface can be selected from existing thermodynamic databases. Numerous phase diagrams for metallic alloys and ceramics define broad potential applications of this method for prediction of HTSC microstructure evolution.

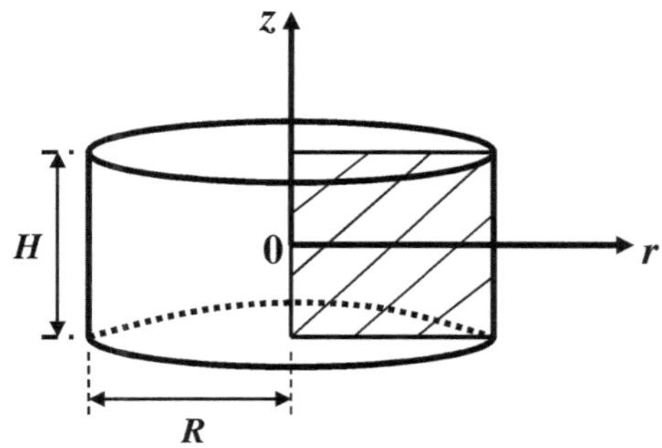

Fig. 8.25 Axisymmetric superconducting cylinder considered in this analysis

8.2 Stress–Strain State of HTSC in Applied Magnetic Fields

HTSC can trap great magnetic fields (more 10 T) at sufficiently high shielding currents. In this case, there are significant stresses and strains, induced by the Lorentz force between the shielding current and the magnetic field, which is capable to form and propagate cracks. One-dimensional stress distributions in the totally magnetized superconductor have been considered [787, 1422]. Two-dimensional (i.e., axi-symmetric 3D) problem using numerical methods has been investigated in [1649, 1711].

Here, we consider 2D problem for axisymmetric cylinder, presented in Fig. 8.25. In the cylindrical coordinate system, the following equations are obtained for radial and axial displacements, u and v from the force balance condition for an infinite-small element [1712, 1713]:

$$F_r = \frac{-E}{(1+v)(1-2v)}\left[(1-v)\frac{\partial^2 u}{\partial r^2} + \frac{(1-2v)}{2}\frac{\partial^2 u}{\partial z^2} + \frac{(1-v)}{r}\frac{\partial u}{\partial r} - (1-v)\frac{u}{r^2} + \frac{1}{2}\frac{\partial^2 w}{\partial r \partial z}\right]; \tag{8.28}$$

$$F_z = \frac{-E}{(1+v)(1-2v)}\left[(1-v)\frac{\partial^2 w}{\partial z^2} + \frac{(1-2v)}{2}\frac{\partial^2 w}{\partial r^2} + \frac{(1-2v)}{2r}\frac{\partial w}{\partial r} + \frac{1}{2}\frac{\partial^2 u}{\partial r \partial z} + \frac{1}{2r}\frac{\partial u}{\partial z}\right], \tag{8.29}$$

where F_r and F_z are radial and axial Lorentz forces, v and E are Poisson's ratio and Young's modulus, respectively. Non-zero components of the strain tensor are defined as

$$\varepsilon_r = \frac{\partial u}{\partial r};\ \varepsilon_\theta = \frac{u}{r};\ \varepsilon_z = \frac{\partial w}{\partial z};\ \gamma_{\mathrm{rz}} = \frac{\partial u}{\partial z} + \frac{\partial w}{\partial r}, \tag{8.30}$$

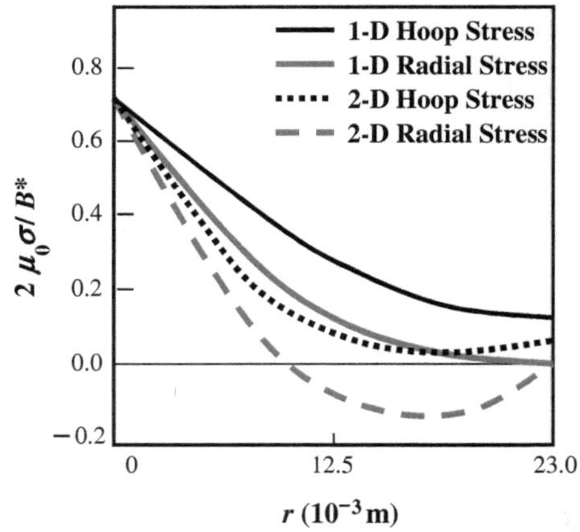

Fig. 8.26 Radial (σ_r) and hoop (σ_θ) stresses versus radial component (at $z = 0$) in 1D and 2D models [1712]

where ε_r, ε_θ, ε_z and γ_{rz} are radial, hoop, axial and shear strains, respectively.

Corresponding components of the stress tensor are defined as

$$\sigma_r = \frac{E}{(1+v)(1-2v)}\left[(1-v)\frac{\partial u}{\partial r} + v\left(\frac{u}{r} + \frac{\partial w}{\partial z}\right)\right]; \tag{8.31}$$

$$\sigma_\theta = \frac{E}{(1+v)(1-2v)}\left[v\frac{\partial u}{\partial r} + (1-v)\frac{u}{r} + v\frac{\partial w}{\partial z}\right]; \tag{8.32}$$

$$\sigma_z = \frac{E}{(1+v)(1-2v)}\left[v\frac{\partial u}{\partial r} + v\frac{u}{r} + (1-v)\frac{\partial w}{\partial z}\right]; \tag{8.33}$$

$$\sigma_{rz} = \frac{E}{2(1+v)}\left(\frac{\partial u}{\partial z} + \frac{\partial w}{\partial r}\right). \tag{8.34}$$

Under a condition of $\partial/\partial z = 0$, (8.28)–(8.34) reduce to 1D problem considered in [1422]. Macroscopic electromagnetic phenomena in the HTSC are described by the Maxwell equations as

$$\nabla \times \boldsymbol{E} = -\partial \boldsymbol{B}/\partial t; \quad \nabla \times \boldsymbol{B} = \mu_0 \boldsymbol{J}; \quad \boldsymbol{J} = \boldsymbol{J}_{SC} + \boldsymbol{J}_{ex}, \tag{8.35}$$

where μ_0, $\boldsymbol{E}$ and $\boldsymbol{B}$ are the magnetic permeability in air, the electric and the magnetic fields, respectively. The magnetic field is caused by the shielding current $\boldsymbol{J}_{SC}$ and the external current $\boldsymbol{J}_{ex}$. A type-II superconductor in a quasistatic field is well described, using the standard critical state model [1624]. Constitutive relationships between the shielding current density $\boldsymbol{J}_{SC}$ and the electric field $\boldsymbol{E}$ are obtained from force balance on a vortex as

$$\boldsymbol{J}_{SC} = J_c(|\boldsymbol{B}|)\boldsymbol{E}/|\boldsymbol{E}|, \qquad \text{if } |\boldsymbol{E}| \neq 0; \tag{8.36}$$

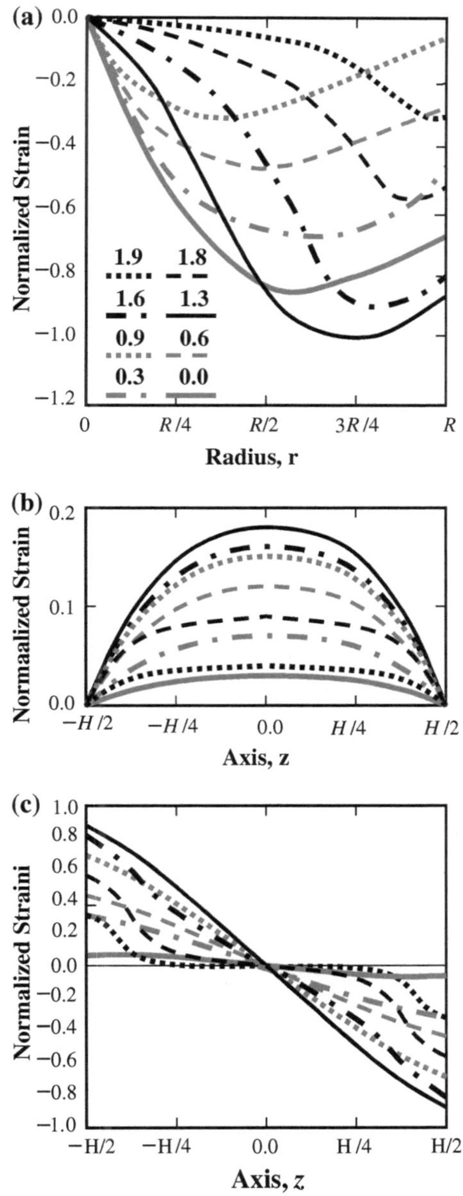

Fig. 8.27 The normalized strains, when the external field is reduced from 2.0 to 0.0T: (**a**) shear strains (γ_{rz}/γ_{max}) on the upper surface of a bulk HTSC, (**b**) and (**c**) hoop strain ($\varepsilon_\theta/\gamma_{max}$) and shear strain ($\gamma_{rz}/\gamma_{max}$) on the side surface of a bulk HTSC [1713]

$$\partial \boldsymbol{J}_{SC/\partial t} = 0, \qquad \text{if } |\boldsymbol{E}| = 0. \tag{8.37}$$

When the electric field $\boldsymbol{E}$ is induced in a local region by change of the magnetic field from (8.35), shielding currents are induced from (8.36). If there is no electric field by the shielding effect, the situation of currents is not changed from (8.37).

While the critical current density J_c has a strong dependence on the magnetic field, the Bean model [1201][2] is applied to the present analysis to clarify the basic properties of the stress–strain state. Self-consistent solutions which satisfy the non-linear equations (8.36) and (8.37) can be obtained, using a numerical iterative technique [1624, 1649, 1711] and the following boundary conditions: $\sigma_r = 0$ on the side surface of the cylinder (at $r = R$), and $\sigma_z = 0$ on the upper and lower surfaces of the cylinder (at $z = \pm H/2$). After the Lorentz force calculation from the shielding current distribution at each time step [1649], the finite difference method is applied to solve (8.28) and (8.29). In the iterative calculations, using successive over-relaxation method, the displacements u and w are found in by using turns the boundary conditions until they are converged to definite value. Then, the strain distributions are calculated from (8.30), and the stresses depending on the obtained displacements of u and w are found from (8.31) to (8.34).

Numerical results are obtained for cylinder with the geometrical parameters: $R = 23.0$ mm and $H = 15.0$ mm, Young's modulus and Poisson's ratio: $E = 95.9$ GPa and $\nu = 0.14$, respectively. In the Bean model, a standard value of the critical current density is $J_c = 1.0 \times 10^8$ A/m^2 [1712, 1713]. Full magnetization with field cooling is obtained when shielding currents are induced in the whole volume of the HTSC (this occurs when the external field is reduced from 2.0 to 0.0 T).

Maximum trapped field B^* in the sample center for 1D and 2D models are 2.9 and 1.7 T, respectively [1712]). Two-dimensional distributions of the radial (σ_r) and hoop (σ_θ) stresses in the sample center (at $z = 0$) are compared with 1D solutions in Fig. 8.26. The distributions are normalized with corresponding value of B^*. Obviously, largest stresses are obtained in the center of the bulk. Figure 8.27 presents normalized strains on the side, upper and lower surfaces of the cylinder (all strains are normalized with the maximum shear strain, γ_{max}, when the external field is 1.3 T). The distributions change as the external field is reduced from 2.0 to 0.0 T. The large shear strain (γ_{rz}) is obtained by the large Lorentz force between distributed shielding currents and the large external field in the first half of the magnetization. After that, the shear strain distribution reduces as the external field is decreased in the magnetization process.

[2] According to the Bean model [110], for cylindrical body: $\partial B_z/\partial r = \text{const}$, yielding a *linear variation* of B_z with the r-coordinate. In the similar Kim model [868, 869], it is assumed that $B_z \partial B_z/\partial r = \text{const}$, yielding a *parabolic variation* of the function $B_z(r)$.

Chapter 9
Modeling Conductive and Elastic Properties of Superconductive Composites

Definition of composite properties (in particular, for superconductors) in dependence on the microstructure and phase features is one of the main problems in mechanics of composite materials. Simplest models operate with properties of single components (or phases) and their concentrations. More complex computation methods take into account different micromechanical effects (e.g. interactions of adjacent components, non-ideal contacts between them, etc.). Due to sharp difference of conductivity of normal and superconducting components of superconductive composites, and also geometrical features of various phases, the asymptotic approaches to solve the problems become very effective and attractive. These studies of conductive and mechanical properties of superconductive composites are fulfilled by taking into account the influence of inclusion coverage (for fibers and grains) on the composite conductivity, effects of non-ideal contacts of inclusions and matrix, structure irregularity and cluster formation. In these cases, for joining the solutions are used Pade two-point approximations and method of asymptotically-equivalent functions.

The investigation methods of conductive and mechanical properties of composites include, in particular statistical mechanics of multi-particle systems, analysis of correlation functions, percolation theory, mathematical modeling, using the variation bounds and interaction between different properties. Modern research methods and corresponding results for conductivity of superconductive composites will be present in first part of this chapter. The corresponding results for elastic properties of the composites will be discussed in second part. The couples between conductive and elastic properties will be investigated in third part of the chapter. Based on the asymptotic approaches the numerical results for conductive and mechanical properties of superconductive composites will be obtained in fourth part of the chapter by taking into account influence of inclusion coverage (for fibers and grains) on the composite conductivity, effects of non-ideal contacts between inclusions and matrix, structure irregularity and cluster formation. The modeling approach, presented in [471], will find itself application in some simplified form in the next chapters in modeling fracture processes and corresponding strength properties, in particular based on definition of effective properties of the high-temperature superconductors.

I. A. Parinov, *Microstructure and Properties of High-Temperature Superconductors*,
DOI: 10.1007/978-3-642-34441-1_9, © Springer-Verlag Berlin Heidelberg 2012

9.1 Effective Conductivity of Superconductive Composites

The studies of percolation features of composites allow microstructure modeling to design superconductive systems with optimal transport properties. It is known, that the critical exponents for conductive properties in general case differ from their lattice values due a singular distribution of 'necks' coupling single cells near percolation threshold [466]. Let phase 1 is conductive medium and phase 2 corresponds to void space with infinity-small conductive. Then there is the percolation threshold (critical volume fraction) $c_{2p} = 1 - c_{1p}$ that, when $c_2 > c_{2p}$ the system ceases to support a conductivity that leads to discontinuity of the phase 1. In this case the critical exponent t for effective conductivity σ_0 can be defined as

$$\sigma_0 \sim (c_1 - c_{1p})^t \text{ at } c_1 \to c_{1p} + 0; \tag{9.1}$$

9.1.1 Two-Phase Composites with Phase-Interchange Properties

It should be noted that in difference from a lattice percolation, the conductivity threshold and the threshold of standard *Bernoulli* connectivity generally do not coincide with estimation of the percolation threshold in continuum. Consider the case of simultaneous percolation (bi-continuity) both phases in two-phase composite. In particular, this example is the case of two-dimensional (2D) two-phase model of the chessboard type (black squares corresponds to a superconductive phase and white squares define a normal-conductive phase) belonging to the material class with phase-reverse symmetry. Two-phase continuous system of dimension d is bi-continuous one at $c_{2p} < c_2 < 1 - c_{2p}$ for $d \geq 2$ under condition that the percolation threshold $c_{2p} < {}^1\!/\!_2$ [1697]. Obviously, that the bi-continuity in 2D case may be achieved with more difficulties compared with the dimensions $d \geq 3$. However the introduction of criterion of the long-range connectivity allows decreasing c_{2p} lower than ½ and introducing the bi-continuity. Because, conductivity could be occurred through common angle points of adjacent superconductive cells coupled only by the diagonal connection then the percolation threshold in this case is found as $c_{2p} = 1 - p_c^s \approx 0.4073$ (where p_c^s is the standard percolation threshold in square lattice by taking into account only nearest neighbors). At the same time, the *Bernoulli* connectivity threshold for superconductive phase is equal to $c_{2p} = p_c^s \approx 0.5927$. So, this system will be bi-continuous at $p_c^s < c_2 < 1 - p_c^s$. The result firstly has been obtained in [1552], where noted importance of conductivity through angle points.

Relations coupling effective properties of two-phase heterogeneous material with effective properties the same microstructure but with phase-interchange components are named by the *phase-interchange relations*. When these strict relations are applied to wide class of heterogeneous materials then they can ensure useful tests for analytical and numerical estimations of effective properties. In [842] it has been treated heterogeneous material in 2D statement consisting of arbitrary lattice of the same parallel cylinders symmetrical both as in x-direction as in y-direction. It has been proved that the effective conductivity in the x-direction σ_0^x is coupled with the effective conductivity of the phase-interchange composite in the y-direction σ_0^y as

$$\sigma_0^x(\sigma_1, \sigma_2)\sigma_0^y(\sigma_2, \sigma_1) = \sigma_1\sigma_2. \tag{9.2}$$

In this equation the first argument in σ_0^x (or σ_0^y) defines conductivity of matrix phase (or inclusion) and the second argument signs conductivity of inclusion (or matrix phase). In [415] it has been shown that the effective conductivity of 2D two-phase composites with phase distributions, which were statistically equivalent one to other, defined by mean geometrical value $(\sigma_1\sigma_2)^{1/2}$, where $\sigma_i (i = 1, 2)$ was the conductivity of i-th phase. Then these results have been generalized [1129] and, in particular it has been shown that Eq. (9.2) applied to any 2D two-phase composite (ordered or not) until x and y to be principal axes of effective conductivity tensor $\boldsymbol{\sigma}_0$. In this case the geometrical symmetry is not required. If composite is macro-isotropic then

$$\sigma_0(\sigma_1, \sigma_2)\sigma_0(\sigma_2, \sigma_1) = \sigma_1\sigma_2. \tag{9.3}$$

Further if two various composites have phase distributions which to be statistically equivalent one to other then $\sigma_0(\sigma_1, \sigma_2) = \sigma_0(\sigma_2, \sigma_1) = \sigma_0$ and Eq. (9.3) leads to that the effective conductivity is given via mean geometrical value of the phase conductivities [415], that is, $\sigma_0 = (\sigma_1\sigma_2)^{1/2}$.

In the case of 2D polycrystal consisting of identical monocrystals, and effective conductivity tensor $\boldsymbol{\sigma}_0$ with the property of rotational symmetry (to being isotropic tensor), the effective conductivity $\sigma_0 = (\sigma_a\sigma_b)^{1/2}$, where σ_a, σ_b are the principal values (or eigenvalues) of the conductivity tensor. Moreover, the effective conductivity does not depend on microstructure details. In 3D case that equality is absent and instead arise upper and lower bounds. Simplest bounds for macroisotropic polycrystals are the mean-arithmetic bound and mean-harmonic one:

$$\left[\frac{1}{3}\left(\frac{1}{\sigma_a} + \frac{1}{\sigma_b} + \frac{1}{\sigma_c}\right)\right]^{-1} \le \sigma_0 \le \frac{1}{3}(\sigma_a + \sigma_b + \sigma_c). \tag{9.4}$$

Except very special cases of isotropic 2D two-phase media, there are no phase-interchange equalities for effective elastic moduli. In general case, as it has been shown [531], the phase-interchange relations for isotropic two-phase media in 2D and 3D cases have forms of inequalities.

9.1.2 Conductivity of Composites with Periodic Arrays of Superconductive Inclusions

Lord Rayleigh [1049] was first who considered effective conductivity the periodic lattice of spheres based on simple cubic grid. In the next years this problems has been studied by numerous investigators which used the Rayleigh method. Specific results have been achieved in the case of arbitrary volume fraction of spheres for simple cubic (sc) structures [1120], and also for body-centered cubic (bcc) and face-centered cubic (fcc) lattices [1117]. Asymptotic formula in the Rayleigh form for these three cubic systems is found (for dimensionless effective conductivity $\lambda_0 = \sigma_0/\sigma_1$) as

$$\lambda_0 = 1 + \frac{3\beta c_2}{1 - \beta c_2 - a_1 \beta^2 c_2^{10/3}/(1 + 2\beta/7)}, \tag{9.5}$$

where for sc-lattice $a_1 = 1.6772$, for bcc-lattice $a_1 = 0.073886$, for fcc-lattice $a_1 = 0.0060503$, and β at any dimension d is determined in the form:

$$\beta = \frac{\sigma_2 - \sigma_1}{\sigma_2 + (d-1)\sigma_1}. \tag{9.6}$$

Asymptotic solutions of higher orders have been obtained for sc-structure [1120] and two other types of cubic lattices [1117]. Then corresponding asymptotic formulae have been determined for 2D case of disc-shaped inclusions [1352]. For discs disposed in nodes of square lattice (square array) we have

$$\lambda_0 = 1 + \frac{2\beta c_2}{1 - \beta c_2 - 0.305827\, \beta^2 c_2^4}, \tag{9.7}$$

and for discs disposed in nodes of triangle lattice (hexagonal array) we have

$$\lambda_0 = 1 + \frac{2\beta c_2}{1 - \beta c_2 - 0.075422\, \beta^2 c_2^6}. \tag{9.8}$$

In [841] it has been stated that for volume fraction near to maximum values of the inclusion package, the interaction between spheres concentrates in the regions neighboring to contact point. The local analysis of the region between two almost touching spheres together with the known disposition of nearest neighbors allows one to define principal contribution of the inclusion interactions for all system. For simple cubic array of identical superconductive spheres into matrix with conductivity σ_1 near percolation threshold of the particle phase $c_{2p} = \pi/6$, the dimensionless effective conductivity λ_0 is found as

$$\lambda_0 = -\frac{\pi}{2} \ln\left(c_{2p} - c_2\right). \tag{9.9}$$

In [841] it has been also obtained corresponding asymptotic solution for the dimensionless effective conductivity λ_0 of square array of identical superconductive circular cylinders into matrix with conductivity σ_1 near percolation threshold of the particle phase $c_{2p} = \pi/4$ as

$$\lambda_0 = \frac{\pi^{3/2}}{2}\left(c_{2p} - c_2\right)^{-1/2}. \tag{9.10}$$

9.1.3 *Effective Medium Approximation for Definition of Effective Conductivity of Superconductive Composite*

Effective medium approximations apply to estimate effective properties into broad range changing the volume fraction and also phase properties. However, these approximations can take into account only very simple microstructure information, for example an inclusion concentration and shape. Therefore, while the effective medium approximations can provide qualitative trends in behavior of effective properties for dispersed media they cannot make quantitative forecasts for general cases and hence it is important to understand the bounds of applicability these models in relation to specific microstructures.

The approximations of Maxwell type are one of typical approximations for effective medium. In the case of spherical inclusions in 3D approximation the approximate scheme has been assumed [1106]. At arbitrary spatial dimension d and spherical non-interactive inclusions with conductivity σ_2 disposing into matrix with conductivity σ_1 the next relation called *Maxwell approximation* is carried out for effective conductivity σ_0 [1697]:

$$\frac{\sigma_0 - \sigma_1}{\sigma_0 + (d-1)\sigma_1} = c_2\left[\frac{\sigma_2 - \sigma_1}{\sigma_2 + (d-1)\sigma_1}\right]. \tag{9.11}$$

Equation (9.11) coincides with well-known optimal *Hashin–Shtrikman bounds* that coincides with the proper lower bound if $\sigma_2 \geq \sigma_1$, and with the proper upper bound if $\sigma_2 \leq \sigma_1$. In the case of superconductive spheres in relation to matrix ($\sigma_2/\sigma_1 = \infty$), Eq. (9.11) takes the form:

$$\lambda_0 = \frac{\sigma_0}{\sigma_1} = \frac{1 + (d-1)c_2}{1 - c_2}. \tag{9.12}$$

Thus, the composite becomes superconductive at trivial percolation threshold $\left(c_{2p} = 1\right)$.

Then, firstly, *self-consistent approximation* for the case of spherical inclusions has been assumed in [191] and then developed in [973]. In this case an effect of whole material outside inclusion is caused by homogeneous medium (matrix), whose effective conductivity is unknown value and should be calculated. With this aim it is required that the perturbations of homogeneous field introduced by inclusions are equal to zero in average. In the case of macroisotropic composite consisting of M various types of the spherical inclusions with volume fractions c_i and conductivities σ_i ($i = 1,\ldots, M$), this leads to the following equation:

$$\sum_{i=1}^{M} c_i \frac{\sigma_i - \sigma_0}{\sigma_i + (d-1)\sigma_0} = 0. \tag{9.13}$$

This equation defines self-consistent approximation and as it is evident to be square-root equation for effective conductivity σ_0. For two-phase case ($M = 2$) its solution takes the form:

$$\sigma_0 = \frac{\alpha + \sqrt{\alpha^2 + 4(d-1)\sigma_1\sigma_2}}{2(d-1)}, \tag{9.14}$$

where $\alpha = \sigma_1(dc_1 - 1) + \sigma_2(dc_2 - 1)$ when both $\sigma_i > 0$. The restrictions of applicability of the self-consistent approach to actual media are connected with the next causes: (i) one does not include information on spatial distribution of inclusions or possible correlations between properties of adjacent inclusions, that creates difficulties in its application in the case of composites without phase-inverse symmetry, (ii) this approximation assumes existence of effective medium only outside the spherical inclusion considered, that it is evident impossible in the package case of the same spheres because between them always there are voids, (iii) violation of the self-consistent approximation occurs at its application to dispersed structures with strong-contrast phase conductivities. In the case of infinite contrast it is predicted no trivial but error percolation threshold. For example, in the case of two phases (9.14) gives the simple relation in the limit case of superconductive inclusions (phase 2):

$$\lambda_0 = \frac{\sigma_0}{\sigma_1} = \frac{1}{1 - dc_2} \text{ at } \sigma_2/\sigma_1 = \infty. \tag{9.15}$$

Thus, the phase 2 percolates (that is the composite becomes superconductive) at the critical volume fraction $c_{2p} = 1/d$ independently of microstructure details. However, it is evident that the percolation threshold c_{2p} should change into wide limits from one composite to other being very sensitive to microstructure.

The *differential approximations* based on effective medium are other popular approximations, whose scheme firstly has been introduced in [191]. In difference from self-consistent approach the differential approximations do not consider each phase symmetrically. However, the both approximate approaches use the same idea of incremented homogenization for two-phase composite in which indices 1 and 2 define matrix phase and inclusions, respectively. By assuming that the effective conductivity $\sigma_0(c_2)$ is known at one value of c_2, consider $\sigma_0(c_2)$ as intrinsic conductivity of composite and assume that $\sigma_0(c_2 + \Delta c_2)$ presents an effective conductivity of the composite after that the small fraction $\Delta c_2/(1 - c_2)$ of the composite with the intrinsic properties has been substituted by inclusions of phase 2. In mean, the fraction $\Delta c_2/(1-c_2)$ of the composite with the intrinsic properties should be replaced by the material phase 2 with aim to change whole fraction of phase 2 up to value of $c_2+\Delta c_2$. Then in the case of d-dimensional spherical inclusions for effective conductivity σ_0 we obtain

$$\sigma_0(c_2+\Delta c_2)-\sigma_0(c_2)=\sigma_0 c_2\left[\frac{\sigma_2-\sigma_0(c_2)}{\sigma_2+(d-1)\sigma_0(c_2)}\right]\frac{\Delta c_2}{1-c_2}d. \tag{9.16}$$

In the limit of $\Delta c_2 \to 0$ this relation transforms into differential equation:

$$(1-c_2)\frac{d\sigma_0}{dc_2}=\sigma_0\left[\frac{\sigma_2-\sigma_0}{\sigma_2+(d-1)\sigma_0}\right]d \tag{9.17}$$

with initial condition $\sigma_0\ (c_2=0)=\sigma_1$. The analytical solution of Eq. (9.17) is determined as

$$\left(\frac{\sigma_2-\sigma_0}{\sigma_2-\sigma_1}\right)\left(\frac{\sigma_1}{\sigma_0}\right)^{1/d}=1-c_2. \tag{9.18}$$

The differential approximation defines preservation of continuity of the original matrix material in final composite [1839].

In the case of superconductive phase 2 in relation to phase 1 Eq. (9.18) takes the form:

$$\lambda_0=\frac{\sigma_0}{\sigma_1}=\frac{1}{(1-c_2)^d}\ \text{at}\ \sigma_2/\sigma_1=\infty. \tag{9.19}$$

Figure 9.1 compares Maxwell, self-consistent and differential approximations in the case of existing superconductive phase into two-phase composite. In this case the numerical results are determined by formulae (9.12, 9.15) and (9.19),

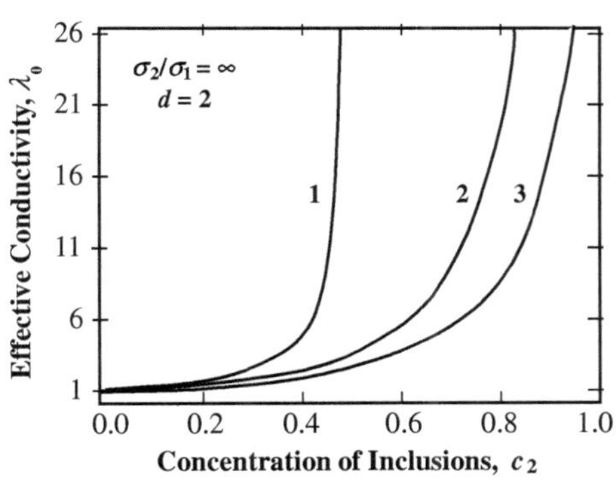

Fig. 9.1 Comparison of three models of effective medium for effective conductivity (*1* self-consistent approximation, *2* differential approximation, *3* Maxwell approximation) in dependence on concentration of superconductive inclusions for 2D two-phase composite. The results correspond to formulae (9.12, 9.15) and (9.19)

respectively. The behavior these three approximations in the case coincides qualitatively with the behavior of corresponding approximations for shear modulus in existence of absolutely rigid phase. Obviously, as the Maxwell approximation as the differential one badly fit for estimation of composites with inclusions consisting of great conglomerates.

9.1.4 Cluster Expansions

Maxwell formula [1106] for effective conductivity of dispersed spheres assumes absence their mutual interactions and to be sufficient for first-order presentation of solution on the sphere concentration c_2. These estimations may be modernized by taking into account couple, triplet, etc. mutual interactions. The problem is solved by using the *method of cluster expansions*. While in general case it is impossible to exactly calculate effective properties of arbitrary dispersed inclusions taking into account all orders of expansion on powers of c_2, the expansions for small density (non-interacting) inclusions present a good basis for verification of approximate theories and test data.

For a dispersion of the same mutual-interacting spheres in first approximation of the cluster expansions method, the effective conductivity σ_0 for any dimension d has the next form [1697]:

$$\sigma_0 = \sigma_1 + d\sigma_1\beta_{21}c_2 + O\left(c_2^2\right), \qquad (9.20)$$

where $\beta_{21} = \frac{\sigma_2 - \sigma_1}{\sigma_2 + (d-1)\sigma_1}$. In the case of $d = 3$ this result coincides with Maxwell approximation (15) for first-order expansion on powers of c_2. For superconductive spheres in relation to matrix ($\sigma_2/\sigma_1 = \infty$), Eq. (9.20) becomes as

$$\lambda_0 = \frac{\sigma_0}{\sigma_1} = 1 + dc_2 + O\left(c_2^2\right). \qquad (9.21)$$

By taking into account mutual-interacting inclusions [774], using the method [101], in which the integral describing mutual interaction of two solids was made absolutely convergent, it was estimated effective conductivity of dispersed spheres taking into account second-order expansion on powers of c_2. A broadening this renormalization technique in order to round conditionally convergent integrals and calculate terms of higher order has been carried out in [775], where estimated the values of σ_0 for all terms of the expansion.

In the case non-overlapping spheres the approach [774] allows one to obtain for superconductive inclusions (in the case of $d = 3$) the following estimation:

$$\lambda_0 = 1 + 3c_2 + 4.51c_2^2 + O\left(c_2^3\right). \qquad (9.22)$$

By using other function of radial distribution relative to 'clearly divided' dispersed inclusions the next estimation has been also obtained [1697]:

$$\lambda_0 = 1 + 3c_2 + 3c_2^2 + O(c_2^3). \tag{9.23}$$

It is not surprising that when spheres dispose sufficiently near one to other the effective conductivity is greater in comparison with the case of the clearly divided spheres. The approximate estimation for non-overlapping superconductive discs ($d = 2$), similar to (9.22) has been obtained in [1354]:

$$\lambda_0 = 1 + 2c_2 + 2.74c_2^2 + O(c_2^3). \tag{9.24}$$

In 2D estimation factor 2.74 is replaced by 2.

The case of overlapping spheres has been studied in dependence on σ_2/σ_1 [1694]. In the limit case of totally overlapping spheres for superconductive inclusions ($\sigma_2/\sigma_1 = \infty$) has been obtained the estimation:

$$\lambda_0 = 1 + 3c_2 + 7.56c_2^2 + O(c_2^3). \tag{9.25}$$

The comparative results for non-overlapping and totally overlapping spheres (Eqs. (9.22) and (9.25)) are present in Fig. 9.2. Thus, in the case of account of the second-order expansion on sphere concentration, the clustering effect causes increasing the effective conductivity for superconductive inclusions at fixed value of c_2. Similar cluster expansions could be also applied to estimation of effective elastic moduli.

9.1.5 Strong Contrast Expansions

In [189], it has been proposed a procedure obtaining an expansion for perturbation caused by phase contrast with application of rational function for effective conductivity σ_0 in 3D case of two-phase isotropic medium. This procedure used the expansion of phase conductivities on powers of rational functions. This result has

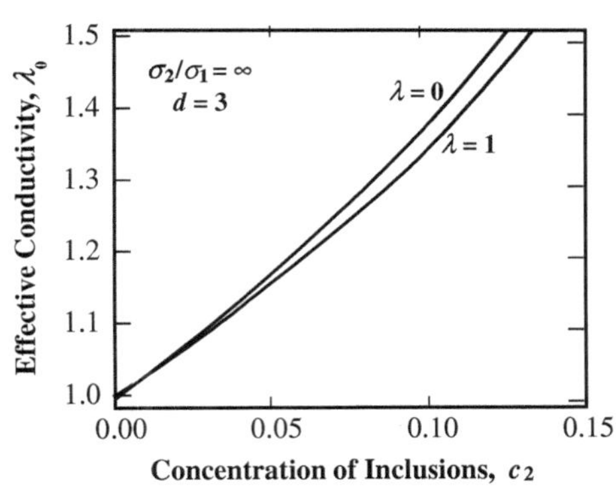

Fig. 9.2 Comparison effective conductivity for systems with non-overlapping ($\lambda = 1$) and totally overlapping ($\lambda = 0$) superconductive spheres for 3D two-phase composite. The results correspond to formulae (9.22, 9.25)

been expanded to isotropic medium of arbitrary dimension d [1693]. By using third term of the expansion the effective conductivity defining exact estimation in the case when the inclusions do not form cluster structure is found in the form [1693]:

$$\lambda_0 = \frac{1 + (d-1)c_2\beta_{21} - (d-1)c_1\zeta_2\beta_{21}^2}{1 - c_2\beta_{21} - (d-1)c_1\zeta_2\beta_{21}^2}, \tag{9.26}$$

where three-point parameter ζ_2 is determined for various structures in dependence on the inclusion concentration c_2 (see Table 9.1 [501]).

Numerical results for three-point approximation (9.26) in the case of different arrays of superconductive inclusions ($\sigma_2/\sigma_1 = \infty$) are present in Fig. 9.3. It is evident that formula (9.26) ensures excellent coincidence of the results up to the maximal published data on the inclusion concentration (up to 95 % of volume fraction). So, in this case, it is stated sensitivity of σ_0 to microstructure features in 3D case. The conclusion is not proper for Maxwell approximation (9.11) which is equivalent to Hashin–Shtrikman lower bound being no sensitive to specific lattice (see Fig. 9.3).

In the case of self-consistent approach for high phase contrast similarly we obtain three-point approximation [1697]:

$$c_2\frac{\sigma_0 + (d-1)\sigma_1}{\sigma_0 - \sigma_1} + c_1\frac{\sigma_0 + (d-1)\sigma_2}{\sigma_0 - \sigma_2} = 2 - d + (d-1)(c_2 - \zeta_2)\beta_{21}, \tag{9.27}$$

Table 9.1 Three-point parameter ζ_2 in dependence on inclusion concentration c_2 for sc, bcc and fcc arrays of spheres and also for square and hexagonal arrays of ordered infinite-long circular cylinders

	Three-point parameter ζ_2				
c_2	sc	bcc	fcc	Square array	Hexagonal array
0.10	0.0003	0.0000	0.0000	3.398×10^{-5}	8.380×10^{-8}
0.20	0.0050	0.0007	0.0004	6.117×10^{-4}	6.034×10^{-6}
0.30	0.0220	0.0031	0.0021	3.540×10^{-3}	7.855×10^{-5}
0.40	0.0678	0.0107	0.0078	1.306×10^{-2}	5.149×10^{-4}
0.45	0.1104	0.0184	0.0136		
0.50	0.1738	0.0307	0.0232	3.833×10^{-2}	2.357×10^{-3}
0.60		0.0796	0.0619	9.965×10^{-2}	8.798×10^{-3}
0.66		0.1381	0.1095		
0.70			0.1596	0.2473	2.958×10^{-2}
0.71			0.1756		
0.76				0.4314	6.057×10^{-2}
0.78				0.5229	7.722×10^{-2}
0.80					9.888×10^{-2}
0.89					0.3409
0.90					0.4010
0.905					0.4364

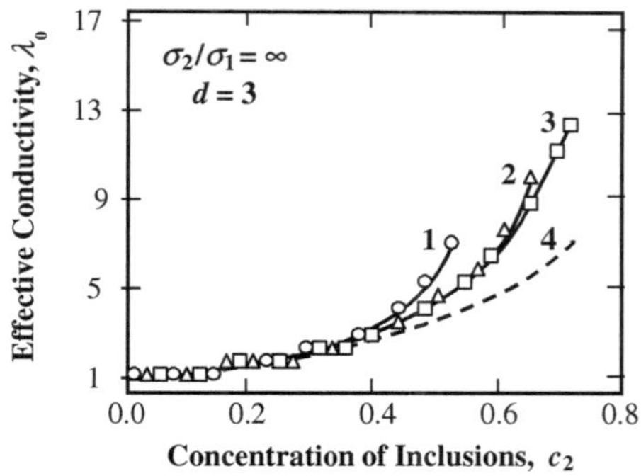

Fig. 9.3 Effective conductivity of composites in dependence on concentration of superconductive spherical inclusions: the *solid curves* correspond to three-point approximations (9.26) for *1* sc-lattice, *2* bcc-lattice, *3* fcc-lattice; *circles*, *triangles* and *squares* determine corresponding numerical results [1117, 1120]. Maxwell approximation (9.12) equivalent to Hashin–Shtrikman lower bound is shown by dashed line 4

and for superconductive inclusions ($\sigma_2/\sigma_1 = \infty$), Eq. (9.27) gives the dependence:

$$\lambda_0 = \frac{1 + (d-1)(c_2 - \zeta_2)}{c_1 - (d-1)\zeta_2}, \tag{9.28}$$

at condition $\zeta_2 \leq [d/(d-1)]c_1$. This condition is satisfied in 2D case for any values of ζ_2. Relation (9.28) states that phase 2 percolates (and the composite becomes superconductive) at $c_1 = (d-1)\zeta_2$. This percolation threshold in difference on the predicted threshold from (9.13) depends on microstructure due to three-point parameter ζ_2. For materials with symmetry and spherical cells Eq. (9.28) predicts the same percolation threshold that formula (9.13), that is $c_2 = 1/d$.

9.1.6 Strong Bound Estimations

Theoretical methods differing from variation approaches could be also applied to define the bounds of effective conductivity tensor $\boldsymbol{\sigma}_0$. The analytical method for $\boldsymbol{\sigma}_0$ [124] has stated analytical properties based on these effective characteristics in dependence on the properties of single phases. The *Pade-approximations method* [1154] is closely connected with the analytical method because the conductivity tensor $\boldsymbol{\sigma}_0$ presents Stieltjes function. Other theoretical approach named the *translation method* (or *compensated compactness method*) has been developed to state the bounds for tensor $\boldsymbol{\sigma}_0$ [1059]. Broad set of experimental and theoretical methods and also computational approaches (in particular, by using finite-element software ANSYS and ACELAN) for definition conductive and other structure-sensitive properties of superconductive and piezoelectric materials and composites has been presented in the author's monographs [8, 471, 1307, 1312, 1362–1364].

For macroisotropic 3D two-phase composites have been obtained the probable bounds defining effective conductivity and including information on volume fraction of the components [649]. One-point (trivial), two-point, three-point and four-point (which use two-point, three-point and four-point probable function, respectively) d-dimensional Hashin–Shtrikman bounds for σ_0 at $\sigma_2 \geq \sigma_1$ for these composites have been determined [136] and present below.
One-point bounds:

$$\left\langle \sigma^{-1} \right\rangle^{-1} \leq \sigma_0 \leq \langle \sigma \rangle. \tag{9.29}$$

Two-point bounds:

$$\sigma_L^{(2)} \leq \sigma_0 \leq \sigma_U^{(2)}, \tag{9.30}$$

where

$$\sigma_L^{(2)} = \langle \sigma \rangle - \frac{c_1 c_2 (\sigma_2 - \sigma_1)^2}{\langle \hat{\sigma} \rangle + (d-1)\sigma_1}; \ \sigma_U^{(2)} = \langle \sigma \rangle - \frac{c_1 c_2 (\sigma_2 - \sigma_1)^2}{\langle \hat{\sigma} \rangle + (d-1)\sigma_2}; \tag{9.31}$$

$\langle \widehat{(\ldots)} \rangle = (\ldots)_1 c_2 + (\ldots)_2 c_1$, and designation $\langle (\ldots) \rangle$ defines mean value of corresponding parameter.
Three-point bounds:

$$\sigma_L^{(3)} \leq \sigma_0 \leq \sigma_U^{(3)}, \tag{9.32}$$

where

$$\sigma_L^{(3)} = \langle \sigma \rangle - \frac{c_1 c_2 (\sigma_2 - \sigma_1)^2}{\langle \hat{\sigma} \rangle + (d-1)\langle \sigma^{-1} \rangle_\zeta^{-1}}; \ \sigma_U^{(3)} = \langle \sigma \rangle - \frac{c_1 c_2 (\sigma_2 - \sigma_1)^2}{\langle \hat{\sigma} \rangle + (d-1)\langle \sigma \rangle_\zeta}; \tag{9.33}$$

$$\langle (\ldots) \rangle_\zeta = (\ldots)_1 \zeta_1 + (\ldots)_2 \zeta_2; \zeta_1 + \zeta_2 = 1. \tag{9.34}$$

Four-point bounds:

$$\sigma_L^{(4)} \leq \sigma_0 \leq \sigma_U^{(4)}, \tag{9.35}$$

where

$$\sigma_L^{(4)} = \sigma_1 \left\{ \frac{1 + [(d-1)c_2 - \gamma_2/\zeta_2]\beta_{21} + (1-d)[c_1\zeta_2 + c_2\gamma_2/\zeta_2]\beta_{21}^2}{1 - [c_2 + \gamma_2/\zeta_2]\beta_{21} + [c_1(1-d)\zeta_2 + c_2\gamma_2/\zeta_2]\beta_{21}^2} \right\}; \tag{9.36}$$

$$\sigma_U^{(4)} = \sigma_2 \left\{ \frac{1 + [(d-1)c_1 - \gamma_1/\zeta_1]\beta_{12} + (1-d)[c_2\zeta_1 + c_1\gamma_1/\zeta_1]\beta_{12}^2}{1 - [c_1 + \gamma_1/\zeta_1]\beta_{12} + [c_2(1-d)\zeta_1 + c_1\gamma_1/\zeta_1]\beta_{12}^2} \right\}; \tag{9.37}$$

$$\gamma_1 - \gamma_2 = (d-2)(\zeta_2 - \zeta_1); \; \beta_{ij} = \frac{\sigma_i - \sigma_j}{\sigma_i + (d-1)\sigma_j}; \; i \neq j. \tag{9.38}$$

Three-point bounds (9.32) depend not only on the phase concentrations c_i, but on the three-point microstructure parameters ζ_i being multi-dimensional integrals consisting of three-point probable functions [1693, 1696]. Because $\zeta_i \in [0, 1]$ then bounds (9.32) improve Hashin–Shtrikman two-point bounds. At $\zeta_2 = 0$ bounds (9.32) coincide and equal to two-point lower bound (9.31). At $\zeta_2 = 1$ bounds (9.32) coincide and equal to two-point upper bound (9.31). At the same time four-point bounds (9.35) depend on c_i, ζ_I and four-point microstructure parameters η_i, stated by four-point probable functions [1150]. Note, that in 2D case the four-point parameters transform in zero and hence the four-point bounds will be only defined by parameters c_i and ζ_I. The improved lower three-point bound is determined for the case $d = 3$ in the form [1693]:

$$\frac{\sigma_L^{(3)}}{\sigma_1} = \frac{1 + (1 + 2c_2)\beta_{21} - 2(c_1\zeta_2 - c_2)\beta_{21}^2}{1 + c_1\beta_{21} - (2c_1\zeta_2 + c_2)\beta_{21}^2}. \tag{9.39}$$

Usual upper and lower bounds (for example in the case of contrast or cluster phases) generally divergent one with other in the limit of infinite phase contrast. In the case of superconductive phase 2 relatively phase 1 ($\sigma_2/\sigma_1 = \infty$) upper bounds of finite order aspire to infinity because in the consideration take into account realizations in which phase 2 percolates while in reality one is not percolating. This is sequel of that the limited microstructure information introducing the bounds is to be insufficient for statement of the phase connectivity. Therefore, it is useful to obtain bounds no diverging into utmost case of infinite phase contrast.

The concept of "secure" sphere for creation of trial fields have been proposed in [843] to obtain limits for effective toughness of suspensions. The same idea has been proposed to define bounds of effective conductivity [1699]. In this case, it is treated statistically isotropic system of the same spheres with radius R and conductivity σ_2 into matrix with conductivity σ_1. After circulation i-th sphere by greater "secure" sphere, the area of concentric shell between the actual sphere and the "secure" sphere contains phase 1 only. Then by considering trial fields into "secure" spheres and into matrix region without "secure" spheres, using principles of minimum of whole energy and additional energy for this composite structure there are stated lower and upper bounds for σ_0. For any dimension d in the case of superconductive inclusions of concentration c_2 the upper bound is found in the form [1697]:

$$\lambda_0 \leq 1 + 2dc_2 \int_0^\infty \frac{x^d}{x^d - 1} RH_P(x)dx, \tag{9.40}$$

where $2x = r/R$ is the dimensionless distance measured from center of each sphere, $H_P(x)dx$ is the probability that center of arbitrary particle into coordinate

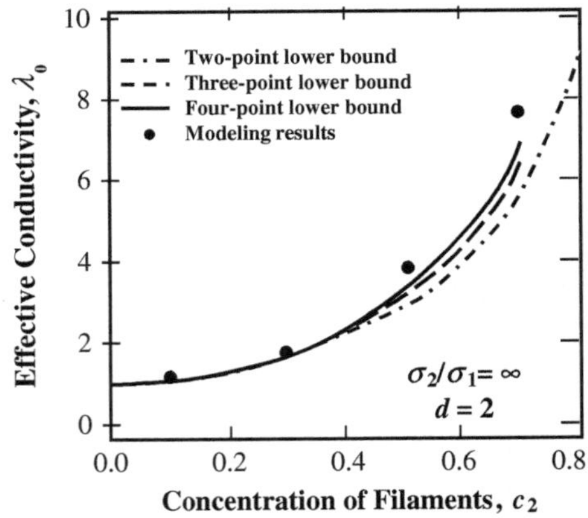

Fig. 9.4 Lower bounds for effective transverse conductivity versus concentration of ordered the same superconductive cylinders. The results for point approximations are determined by formulae (9.30–9.38), and modeling data from [268, 865]

system connected with center of nearest particle disposes into range between x and $x + dx$. Lower bound in whole is defined by usual three-point lower bound. Note, that the factor before $H_P(x)$ in (9.40) contains pole at $x = 1$ (when the spheres touch). So, in difference of usual upper bounds, the upper bound (9.40) in the case of "secure" spheres remains finite for superconductive inclusions under condition of sufficiently rapid convergence of $H_P(x)$ to zero at $x \to 1$ in order to make the integral to be convergent. Upper bound (9.40) will give non-trivial result in the case absence of touching the spheres one to other which dispose into regular lattice with minimal distance between them equal to $2bR$.[1] In this case we obtain

$$2R\, H_P(x) = \delta(x - b). \tag{9.41}$$

By using the results for H_P in the case of various arrays of inclusions, are determined non-trivial bounds for "secure" spheres under condition their touching [1699]. The bounds of "secure" spheres types for σ_0 have been stated in [196] by using the *method of complex variables*. However, these bounds were obtained in terms of minimal distance between all adjacent inclusions (but no via H_P).

9.1.7 Improved Estimations of Range Bounds for Effective Conductivity of Superconductive Composites

More exact estimations of microstructure parameters allow improving three- and four-point bounds for effective conductivity σ_0. In particular, the results obtained for effective transverse conductivity for equilibrium arrays of ordered the same superconductive cylinders ($\sigma_2/\sigma_1 = \infty$) in 2D case and for superconductive spheres in 3D case are present in Figs. 9.4 and 9.5, respectively. Obviously, that in

[1] Dimensionless parameter b is found below from formula (9.43).

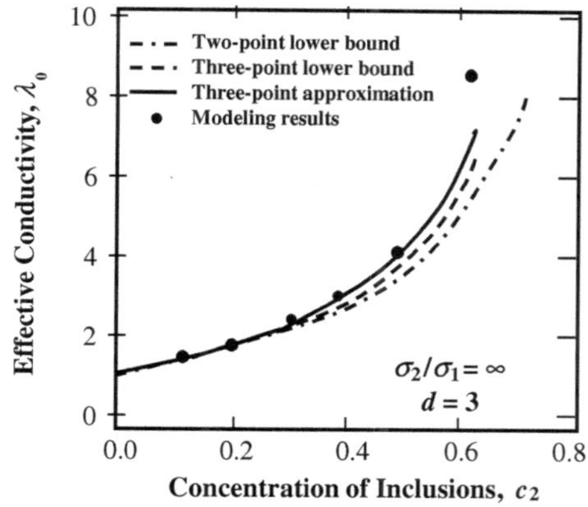

Fig. 9.5 Bounds for effective conductivity in dependence on concentration of superconductive inclusions for arbitrary equilibrium arrays of spheres. Results for two- and three-point lower bounds are found by formulae (9.30, 9.31, 9.26), respectively; improved three-point approximation (9.39) obtained in [1146] and modeling data in [866]

the case of superconductive inclusions all upper bounds aspire to infinity. Superconductive discs cannot form clusters for the treated range of volume fraction that is the conductive phase to be discontinuous. Really, for arbitrary equilibrium models of rigid spheres at any spatial dimension the contacts between particles occur only at maximal density of arbitrarily jammed particles. Figure 9.5 compare two- and three-point bounds for 3D model of superconductive spheres obtained in [866, 1146] and also three-point approximation (9.26). Best lower bound provides a good estimation of σ_0, and three-point approximation to be absolutely exact.

The prediction of effective conductivity for systems with overlapping spheres is more interest because of non-trivial cluster formation. Figure 9.6 compares four-

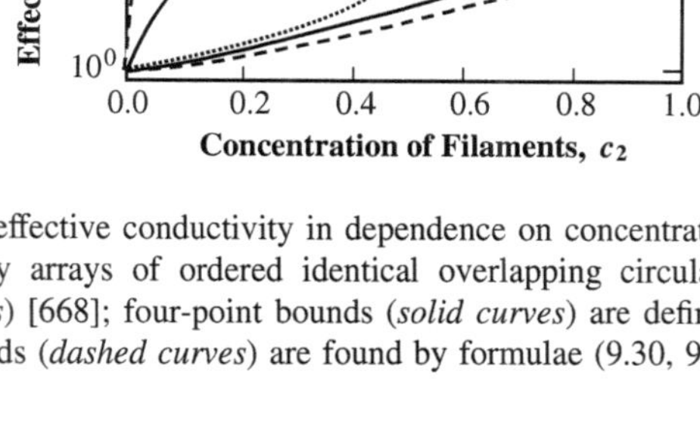

Fig. 9.6 Bounds for effective conductivity in dependence on concentration of superconductive filaments for arbitrary arrays of ordered identical overlapping circular discs compare with modeling results (*dots*) [668]; four-point bounds (*solid curves*) are defined by formulae (9.35–9.38); two-point bounds (*dashed curves*) are found by formulae (9.30, 9.31)

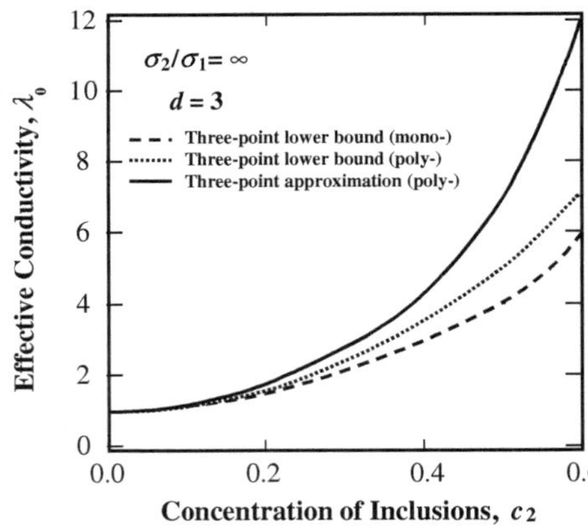

Fig. 9.7 Three-point lower bounds (9.39) of effective conductivity for equilibrium systems of superconductive mono- and poly-dispersed spheres compared with exact three-point approximation (9.26) for poly-dispersed spheres

point bounds for the same overlapping discs (the percolation threshold $c_{2p} \approx 0.68$) [1698], in the case of $\sigma_2/\sigma_1 = 1000$ with results of modeling [668]. For range $0 \le c_2 \le 0.4$ the lower bound provides a good estimation by assuming that proper cluster size Λ_2 of phase 2 is relatively small. The increasing c_2 from 0.4 up to 0.9 initiates crossover from lower bound to upper one corresponding to increasing Λ_2, which becomes macroscopically great at achievement of percolation threshold. The upper bound provides a good estimation for test data at $0.9 \le c_2 \le 1$ assuming that proper cluster size Λ_1 of phase 1 is relatively small. Figure 9.7 compares for mono-dispersed and poly-dispersed superconductive spheres three-point lower bound obtained in 3D case at $\sigma_2 \gg \sigma_1$ [1152], which improves lower bound (9.33). In the poly-dispersed case is used three-point approximation (9.26). The effect of poly-dispersity leads to increasing σ_0.

As it is shown by formula (9.40) for upper bound of "secure" spheres at any concentration of superconductive inclusions c_2 the bound could remain finite in difference from usual bounds. These bounds have been estimated [1699] for equilibrium sphere arrays in which each sphere with conductivity σ_2 and diameter D was covered by thin layer of matrix material with conductivity σ_1 so that diameter of each composite sphere equals to D_0. That coverage is required for preventing a mutual touching of internal spheres. It is interesting to study effects of diameter $D_0 > D$ and volume fraction $\tilde{c}_2 = (D_0/D)^3 c_2$, where c_2 is actual concentration of spheres. By estimating the function of probability density for nearest neighbors $H_P(x)$ at dimensionless distance $(D_0/D)x$ for volume fraction of inclusion $\tilde{c}_2$, we show the bounds of "secure" spheres for superconductive spheres in Fig. 9.8 for several values of dimensionless thickness of the coverage $\in = (D_0 - D)/D$. It is evident, that the bounds become sharper due to increasing $\in$. For cubic lattices of spheres in the case of dimension d by using estimation (9.41) for function $H_P(x)$ and upper bound of "secure" spheres (9.40) we obtain

$$\lambda_0 = \frac{\sigma_0}{\sigma_1} \le 1 + d\frac{b^d}{(b^d - 1)} c_2 \text{ at } \sigma_2/\sigma_1 = \infty. \tag{9.42}$$

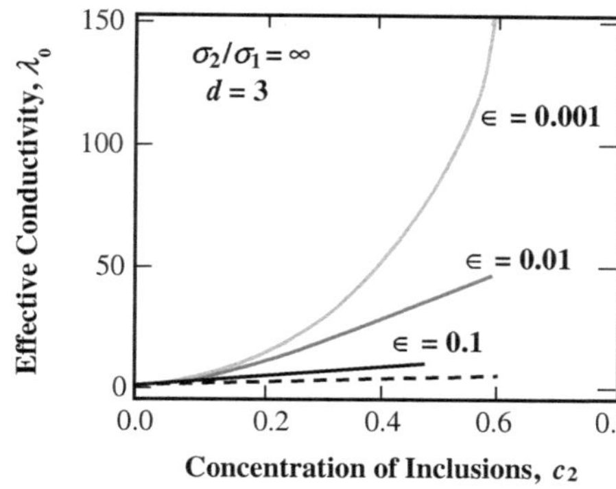

Fig. 9.8 Upper bounds of effective conductivity (9.40) in model of 'secure' spheres for arbitrary equilibrium systems of superconductive inclusions for various values of dimensionless coverage thickness ∈ [1699] (*solid lines*); *dash line* shows Hashin–Shtrikman lower bound

Here b is minimal distance between particles which depends on volume fraction as

$$b^d = c(d)c_2^{-1}, \tag{9.43}$$

where $c(d)$ is the constant stated by dimension d and lattice type. For example, we obtain for d-dimensional simple cubic lattice

$$c(d) = \frac{\pi^{d/.2}}{2^d\Gamma(1 + d/2)}. \tag{9.44}$$

The bound (9.42) is exact while of first order on c_2, and relatively sharp at moderately big volume values of superconductive inclusions as for $d = 2$ as for $d = 3$. Moreover, one is exact on all orders of c_2 as for $d = 1$ as for $d = \infty$.

9.2 Effective Elasticity of Superconductive Composites

Similar to the conductivity considered in Sect. 9.1, it is known, that the critical exponents for elastic properties in general case also differ from their lattice values due a singular distribution of 'necks' coupling single cells near percolation threshold [466]. Let phase 1 is elastic medium and phase 2 corresponds to void space with infinity-small elastic properties. Then there is the percolation threshold $c_{2p} = 1 - c_{1p}$ that, when $c_2 > c_{2p}$ the system ceases to support a mechanical load that leads to discontinuity of the phase 1. In this case the critical exponent f for effective Young modulus E_0 can be defined as

$$E_0 \sim (c_1 - c_{1p})^f \text{ at } c_1 \to c_{1p} + 0. \tag{9.45}$$

9.2.1 *Effective Medium Approximations for Definition of Effective Elastic Moduli*

Maxwell approximation, self-consistent approach and differential approximations based on effective medium are also used to study elastic moduli. For example, for $M-1$ different spheres ($M \geq 2$) with volume fractions c_i, bulk moduli K_i and shear moduli G_i, where $i = 2,\ldots, M$ (value of the index $i = 1$ corresponds to matrix parameters), the Maxwell approximation for dimension d leads to the relations [1697]:

$$\frac{K_0 - K_1}{K_0 + 2G_1(d-1)/d} = \sum_{i=1}^{M} c_i \frac{K_i - K_1}{K_i + 2G_1(d-1)/d}; \tag{9.46}$$

$$\frac{G_0 - G_1}{G_0 + H_1} = \sum_{i=1}^{M} c_i \frac{G_i - G_1}{G_i + H_1}; \; H_1 = \frac{G_1[dK_1/2 + (d+1)(d-2)G_1/d]}{K_1 + 2G_1}. \tag{9.47}$$

When all inclusions are stiffer (softer) compared to matrix phase, Eqs. (9.46) and (9.47) coincide with lower (upper) Hashin–Shtrikman–Walpole bounds [1697]. At $d = 3$ these formulae are equivalent to Mori–Tanaka relations [1190] based on other approach [1791]. Comparison of relations for effective bulk and shear moduli obtained into framework of Maxwell approximations in the case of macroanisotropic composite, consisting of $M-1$ various types one-directional isotropic ellipsoidal inclusions with the same shape, and test data has shown that these estimations could not be used at volume fraction of the inclusions greater than 20–30 % [127].

Self-consistent approximations for effective elastic moduli are totally similar to the relations obtained for effective conductivity. Firstly, the approximations have been obtained [199] by using other approach than above one discussed for the effective conductivity. Additional approach for self-consistent approximations has been realized in [126]. In the case of macroisotropic composites consisting of M various types of spherical inclusions with volume fractions c_i, bulk moduli K_i and shear moduli G_i, ($i = 1,\ldots, M$), an effect of whole material without inclusion of ith type creates homogeneous structure (matrix) with effective bulk modulus K_0, to being unknown. Then by using the condition of equality in mean to zero the perturbations of homogeneous field, the effective bulk modulus K_0 is selected in according with the condition of self-consistency thus the perturbations initiated by each type of the inclusions in average become zero. This leads to the next relation which is similar to Eq. (9.13) for effective conductivity:

$$\sum_{i=1}^{M} c_i \frac{K_i - K_0}{K_i + 2G_0(d-1)/d} = 0. \tag{9.48}$$

This self-consistent estimation of the effective bulk modulus K_0 depends on unknown effective shear modulus G_0. By following to the same procedure that in the case of the bulk modulus, the self-consistent approximation for the effective shear modulus is found as

$$\sum_{i=1}^{M} c_i \frac{G_i - G_0}{G_i + H_0} = 0, \tag{9.49}$$

where

$$H_0 = \frac{G_0[dK_0/2 + (d+1)(d-2)G_0/d]}{K_0 + 2G_0}. \tag{9.50}$$

Bruggeman ideas [191] have been used to obtain solutions for effective elastic moduli K_0 and G_0 [160, 1118] similar to above differential approximations for conductivity. Analogously to the case of conductivity for effective elastic moduli of two-phase composite (indices 1 and 2 correspond to matrix and inclusion) for dimension d we obtain the differential equation similar to (9.17):

$$(1 - c_2)\frac{dK_0}{dc_2} = \left[K_0 + \frac{2(d-1)}{d}G_0\right]\frac{K_2 - K_0}{K_2 + 2G_0(d-1)/d} \tag{9.51}$$

with initial condition $K_0\ (c_2 = 0) = K_1$. The corresponding equation for effective shear modulus G_0 is determined as

$$(1 - c_2)\frac{dG_0}{dc_2} = (G_0 + H_0)\frac{G_2 - G_0}{G_2 + H_0} \tag{9.52}$$

with initial condition $G_0\ (c_2 = 0) = G_1$, where H_0 is found by relation (9.50). Differential approximations (9.51, 9.52) are realized because they present the special case of generalized differential scheme assumed in [1248] and hence dispose between Hashin–Shtrikman bounds.

9.2.2 Strong Bound Estimations for Elastic Properties

Theoretical methods differing from variation approaches could be also applied to define the bounds of stiffness tensor $\boldsymbol{C}_0$. The analytical method for stiffness $\boldsymbol{C}_0$ [812] has stated analytical properties based on these effective characteristics in dependence on the properties of single phases. The translation method has been also developed to state the bounds for tensor $\boldsymbol{C}_0$ [1060]. Broad set of experimental and theoretical methods and also computational approaches (in particular, by using finite-element software ANSYS and ACELAN) for definition elastic properties, of superconductive and piezoelectric materials and composites has been presented in the author's monographs [8, 471, 1307, 1312, 1362–1364].

The strong d-dimensional Hashin–Shtrikman bounds for effective elastic moduli for two-phase isotropic composites are similar to corresponding bounds for effective conductivity and defined [1697] as present below.

One-point bounds:

$$\langle K^{-1}\rangle^{-1} \le K_0 \le \langle K\rangle;\ \langle G^{-1}\rangle^{-1} \le G_0 \le \langle G\rangle. \tag{9.53}$$

Two-point bounds:

$$K_L^{(2)} \le K_0 \le K_U^{(2)}, \tag{9.54}$$

where

$$K_L^{(2)} = \langle K\rangle - \frac{c_1c_2(K_2-K_1)^2}{\langle\hat{K}\rangle + 2[(d-1)/d]G_1};\ K_U^{(2)} = \langle K\rangle - \frac{c_1c_2(K_2-K_1)^2}{\langle\hat{K}\rangle + 2[(d-1)/d]G_2}; \tag{9.55}$$

$$G_L^{(2)} \le G_0 \le G_U^{(2)}, \tag{9.56}$$

where

$$G_L^{(2)} = \langle G\rangle - \frac{c_1c_2(G_2-G_1)^2}{\langle\hat{G}\rangle + H_1};\ G_U^{(2)} = \langle G\rangle - \frac{c_1c_2(G_2-G_1)^2}{\langle\hat{G}\rangle + H_2}; \tag{9.57}$$

$$H_i = G_i\left[\frac{dK_i/2 + (d+1)(d-2)G_i/d}{K_i + 2G_i}\right]. \tag{9.58}$$

Three-point bounds:

$$K_L^{(3)} \le K_0 \le K_U^{(3)}, \tag{9.59}$$

where

$$K_L^{(3)} = \langle K\rangle - \frac{c_1c_2(K_2-K_1)^2}{\langle\hat{K}\rangle + 2[(d-1)/d]\langle G^{-1}\rangle_\zeta^{-1}}; \tag{9.60}$$

$$K_U^{(3)} = \langle K\rangle - \frac{c_1c_2(K_2-K_1)^2}{\langle\hat{K}\rangle + 2[(d-1)/d]\langle G\rangle_\zeta}; \tag{9.61}$$

$$G_L^{(3)} \le G_0 \le G_U^{(3)}, \tag{9.62}$$

where

$$G_L^{(3)} = \langle G\rangle - \frac{c_1c_2(G_2-G_1)^2}{\langle\hat{G}\rangle + \Xi};\ G_U^{(3)} = \langle G\rangle - \frac{c_1c_2(G_2-G_1)^2}{\langle\hat{G}\rangle + \Theta}. \tag{9.63}$$

In the case of three-point bounds for effective transverse shear modulus for fiber composites in 2D case ($d = 2$), the parameters Ξ and Θ satisfy to relations [1697]:

$$\Xi = \left(2\langle 1/k\rangle_\zeta + \langle 1/G\rangle_\eta\right)^{-1}; \ \Theta = \frac{2\langle k\rangle_\zeta \langle G\rangle^2 + \langle k\rangle^2 \langle G\rangle_\eta}{\langle k+2G\rangle^2}. \tag{9.64}$$

Here $k \equiv K$ for $d = 2$ and

$$\langle(\ldots)\rangle_\eta = (\ldots)_1\eta_1 + (\ldots)_2\eta_2; \ \eta_1 + \eta_2 = 1. \tag{9.65}$$

In 3D case ($d = 3$), the parameters Ξ and Θ satisfy to relations [1156]:

$$\Xi = \frac{\langle 128/K + 99/G\rangle_\zeta + 45\langle 1/G\rangle_\eta}{30\langle 1/G\rangle_\zeta \langle 6/K - 1/G\rangle_\zeta + 6\langle 1/G\rangle_\eta \langle 2/K + 21/G\rangle_\zeta}; \tag{9.66}$$

$$\Theta = \frac{3\langle G\rangle_\eta \langle 6K + 7G\rangle_\zeta - 5\langle G\rangle_\zeta^2}{6\langle 2k - G\rangle_\zeta + 30\langle G\rangle_\eta}. \tag{9.67}$$

While the three-point bounds for the bulk modulus depend on the component concentrations c_i and three-point microstructure parameters ζ_i [1693, 1696], the three-point bounds for the shear modulus depend additionally on other three-point microstructure parameters η_i, which at $d = 2$ and $d = 3$ are defined by multi-dimensional integrals including three-point probability functions obtained in [1151] and [1150], respectively.

Four-point bounds for elastic moduli of 3D two-phase composites have been constructed [1156]. In this case the bounds for bulk modulus include three various microstructure parameters, and the bounds for shear modulus includes eight parameters.

9.3 Couples Between Conductivity and Elastic Properties

Statement of these couples is especially useful in the case of simpler test determination of one property compared to other. Because effective properties of arbitrary media reflect definite morphological information on the medium then it could be expected that can pick out useful information on effective property given by using exact definition (experimental or theoretical) other effective property even when their corresponding constitutive equations are not coupled one with other. The couple relations provide estimation methods for possible range of values which could have other effective properties (in this case an achieved region is stated into multi-dimensional space of the properties). So, they have important applications for design of multi-functional composites.

It may be shown that the effective conductivity and effective elastic moduli are generally coupled one with other even while the first property is calculated from Laplace equation and the second ones are determined by vector equations of equilibrium. It has been shown for arbitrary d-dimensional isotropic two-phase medium [1152] that if phase bulk moduli K_i are equal phase conductivities σ_i then effective bulk modulus K_0 is limited from above by effective conductivity σ_0. This result has been generalized on more common microstructure case with arbitrary topology demonstrating phases with non-negative Poisson's ratio [1695]. At condition $K_2/K_1 \leq \sigma_2/\sigma_1$ for dimensionless effective bulk elastic modulus and effective conductivity has been stated the following dependence:

$$K_0/K_1 \leq \sigma_0/\sigma_1. \tag{9.68}$$

Moreover, it has been stated upper bound for effective shear modulus G_0 (into terms of σ_0 and effective Poisson's ratio ν_0) and weaker bound determined only by effective conductivity σ_0:

$$\frac{G_0}{K_1} \leq \frac{\sigma_0}{\sigma_1}\frac{d[1-\nu_0(d-1)]}{2(1+\nu_0)}; \quad \frac{G_0}{K_1} \leq \frac{\sigma_0}{\sigma_1}\frac{d}{2}. \tag{9.69}$$

Above-mentioned bounds have interesting applications. For compressible phase 1 ($K_1 < \infty$) with finite conductivity ($\sigma_1 < \infty$) and non-compressible superconductive phase 2 ($K_1 = \infty, \sigma_1 = \infty$), it has been stated the next conclusion [1152]: inequality (9.68) assumes that the composite could not be non-compressible and possess finite conductivity. However, this does not interdict that the composite will be compressible and superconductive. The corresponding imaginary structure is present in Fig. 9.9, where thin black superconducting couples of non-compressible phase connect big black blocks allowing the composite to be superconductive. Due to the black phase has finite shear modulus, the thin links will bend under influence of hydrostatic loading creating the compressible composite.

In order to obtain sharp bounds of effective properties for composites is used translational method [1060]. The method is based on obtaining corresponding bounds for a comparative medium, whose local properties differ from original medium by tensor of constant translation being quasi-convex. This method has been applied to define sharp bounds of couples of the effective properties (σ_0, k_0) and (σ_0, G_0) in 2D case of two-phase isotropic composites with different structures in dependence on component concentrations c_i [530]. These bounds cover definite regions into planes σ_0–k_0 and σ_0–G_0 realized by specific microstructures and so

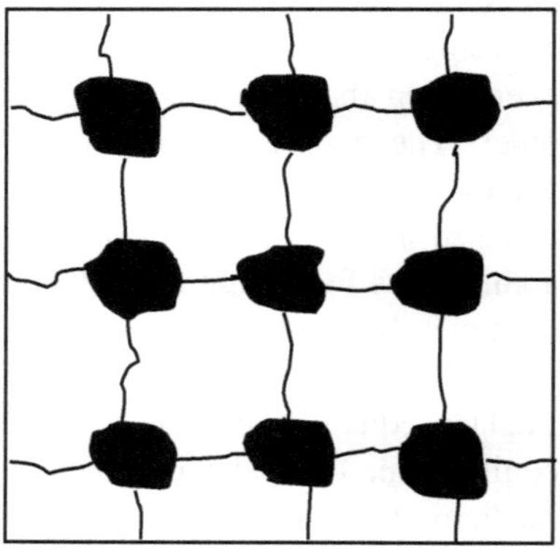

Fig. 9.9 Scheme of compressible superconductive composite consisting of non-compressible superconductive phase (*black color*) and compressible normal phase (*white color*)

being optimal. In special example of absolutely rigid and superconductive phase 2 ($k_2/k_1 = \infty$, $G_2/G_1 = \infty$, $\sigma_2/\sigma_1 = \infty$) the bounds are determined by the next inequalities [1697]:

$$\sigma_0 \geq \sigma_{1*}^{\infty};\ k_{1*}^{\infty} \leq k_0 \leq k_{1*}^{\infty} + \max\left[\frac{k_1 + G_1}{2\sigma_1}; \frac{2k_2G_2}{(k_2 + G_2)\sigma_2}\right]\left(\sigma_0 - \sigma_{1*}^{\infty}\right); \tag{9.70}$$

$$\sigma_0 \geq \sigma_{1*}^{\infty};\ G_{1*}^{\infty} \leq G_0 \leq G_{4*}^{\infty} + \max\left[\frac{k_1 + 2G_1}{4\sigma_1}; \frac{k_2G_2}{(k_2 + G_2)\sigma_2}\right]\left(\sigma_0 - \sigma_{1*}^{\infty}\right), \tag{9.71}$$

where

$$\sigma_{1*}^{\infty} = \frac{1 + c_2}{c_1}\sigma_1;\ k_{1*}^{\infty} = \frac{k_1 + G_1c_2}{c_1}; \tag{9.72}$$

$$G_{1*}^{\infty} = \frac{k_1G_1(1 + c_2) + 2G_1^2}{(k_1 + 2G_1)c_1};\ G_{4*}^{\infty} = \frac{k_1c_2 + 2G_1}{2c_1}. \tag{9.73}$$

In difference from Hashin–Shtrikman upper bounds for k_0 and G_0 upper bounds (9.70) and (9.71) do not divergent to infinity if σ_0 remains finite in this case of infinite contrast. Note that lower bounds for elastic moduli are independent on conductivity and coincide with corresponding Hashin–Shtrikman lower bounds. Moreover, upper bounds for k_0 and G_0 can depend on ratio of infinite moduli because very small quantity (that is the volume fraction of order $1/k_2$ and $1/\sigma_2$) of very rigid conductive phase can lead to finite effective properties.

In the case of superconductive absolutely rigid cylinders (phase 2) located in cells of triangle lattice (hexagonal array) disposed into matrix (phase 1) and satisfying to conditions: $k_2/k_1 = \infty$, $G_1/k_1 = G_2/k_2 = 0.4$ and $\sigma_2/\sigma_1 = \infty$, the exact results for effective conductivity [1352] together with relations (9.70) and (9.71) introduce bounds for effective moduli. In the case of additional condition that the phase 1 defines inclination of upper bounds in (9.70, 9.71), that is $(k_1 + G_1)/(2\sigma_1) \geq 2k_1G_2/[(k_2 + G_2)\sigma_2]$, Fig. 9.10 shows that the cross-property bounds for

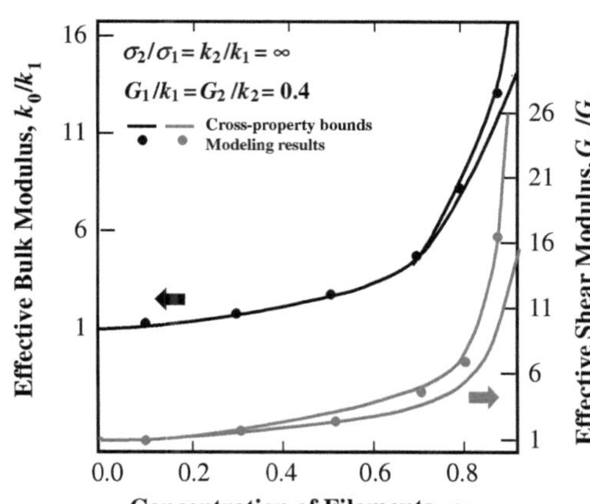

Fig. 9.10 Comparison of cross-property bounds for effective bulk modulus (9.70) and shear modulus (9.71) for hexagonal arrays of superconductive cylinders with modeling results [424]

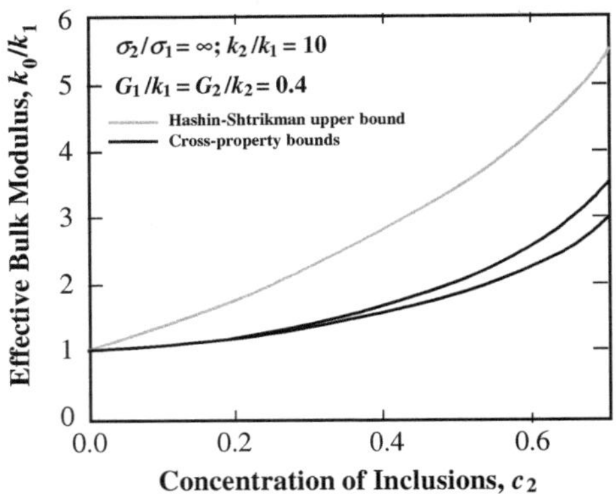

Fig. 9.11 Cross-property bounds for effective bulk modulus for arbitrary arrays of superconductive non-overlapping cylindrical inclusions based on data for conductivity [865]

effective elastic moduli very well coincide with corresponding model data[2] [424]. At the same time, in this case the standard variation upper bounds of the effective properties (for example, Hashin–Shtrikman upper bounds) divergent to infinity because they do not introduce information that the absolutely rigid phase is in fact to be discontinuous. On the contrary, the cross-property upper bounds use information that the phase with infinite contrast of properties is discontinuous due to using conductive properties.

In the case of arbitrary array of superconductive cylinders ($\sigma_2/\sigma_1 = \infty$) for $k_2/k_1 = 10$, $G_1/k_1 = G_2/k_2 = 0.4$ the effective volume moduli are present in Fig. 9.11. It is evident that the cross-property upper bound provides significant improvement in comparison with Hashin–Shtrikman upper bound for k_0.

In order to obtain the cross-property upper bounds of effective elastic and conductive properties of two-phase isotropic composites in 3D case the translational method has been used [454]. In the case of absolutely rigid and superconductive phase 2 ($K_2/K_1 = \infty$, $G_2/G_1 = \infty$, $\sigma_2/\sigma_1 = \infty$) these bounds are determined by the following inequalities [1697]:

$$\sigma_0 \geq \sigma_{1*}^{\infty};\ K_{1*}^{\infty} \leq K_0 \leq K_{1*}^{\infty} + \max\left[\frac{3K_1 + 4G_1}{9\sigma_1};\ \frac{6K_2G_2}{(3K_2 + 4G_2)\sigma_2};\ \frac{2G_2}{3\sigma_2}\right](\sigma_0 - \sigma_{1*}^{\infty}); \tag{9.74}$$

where

$$\sigma_{1*}^{\infty} = \frac{1 + 2c_2}{c_1}\sigma_1;\ K_{1*}^{\infty} = \frac{3K_1 + 4G_1c_2}{3c_1}. \tag{9.75}$$

In difference from Hashin–Shtrikman upper bounds stated for K_0 [128], upper bounds (9.74) do not divergent to infinity if σ_0 remains finite in this case of infinite contrast. Note that the lower bound for K_0 does not depend on conductivity and coincides with corresponding Hashin–Shtrikman lower bound.

---

[2] Note, that only upper bounds for elastic moduli contain information on conductivity.

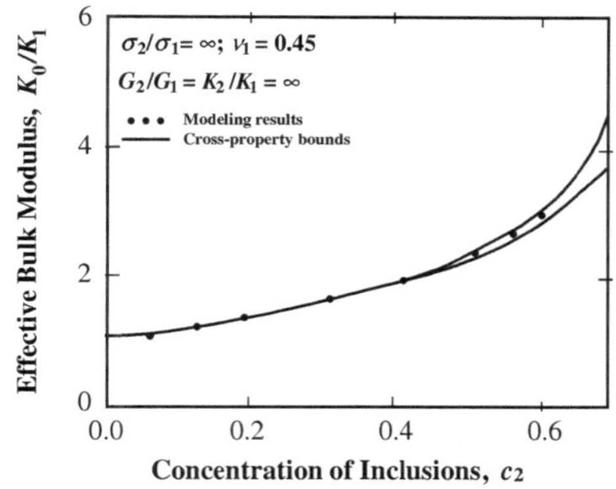

Fig. 9.12 Comparison of cross-property bounds (9.74) for effective bulk modulus (fcc-array of superconductive spheres) with modeling results [1240]

In the case of cubic array of superconductive spheres exact results [1120] for effective conductivity of that array together with inequalities (9.74) could be applied to obtain elastic moduli bounds. For example, in the case of fcc-array of superconductive absolutely rigid spherical inclusions (phase 2) disposed into matrix (phase 1) and satisfying to conditions $K_2/K_1 = \infty$, $G_2/G_1 = 0.4$ and $\nu_1 = 0.45$, the bounds are determined by inequalities (9.74). Moreover, by assuming that the phase 1 defines inclination of the upper bound in (9.74), we obtain dependence of effective bulk modulus on inclusion concentration (see Fig. 9.12). In the figure are also present comparative results of Ref. [1240]. It should be noted, that only upper bound contains information on conductivity of the composite. The results show that at $c_2 \leq 0.5$ the cross-property upper bounds predict the bulk modulus of composite in practice exactly. In the case of higher concentration of inclusions coincidence with existing results is very well. Typical upper bounds (e.g. Hashin–Shtrikman upper bound) are divergent to infinity because they cannot take into account that the absolutely rigid phase is discontinuous in fact. On the contrary, the cross-property upper bound use important topological information on discontinuity of the phase with infinite contrast of properties by taking into account an information on conductivity of the composite.

In the case of equilibrium array of non-overlapping superconductive spheres in [1120] have been obtained results in dependence on volume fraction of inclusions [866] (see Fig. 9.13). It is evident, that the results for the upper bound improve significantly corresponding results for Hashin–Shtrikman upper bound.

9.4 Asymptotical Methods for Definition of Properties of Superconductive Composites

As it is known, high-temperature superconductive (HTSC) mono- and multi-filamentary tapes, wires, coated conductors and cables are filamentary composites in which superconductive brittle filaments are rounded by metallic (or from alloy)

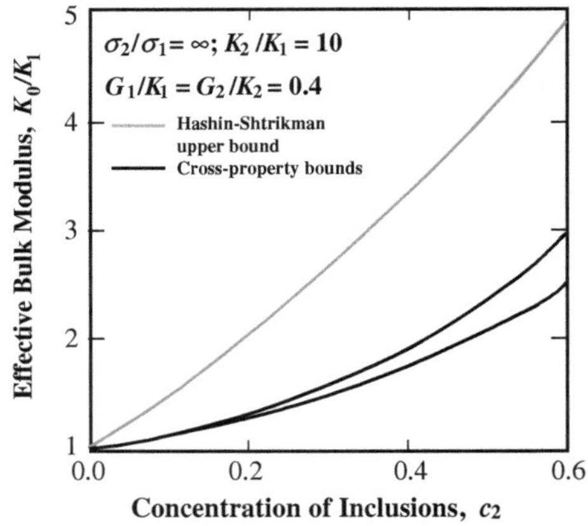

Fig. 9.13 Cross-property bounds for effective bulk modulus for arbitrary arrays of superconductive non-overlapping spherical inclusions based on data for conductivity [866]

matrix. Due to infinite conductivity of superconductive component, in order to define effective properties it is convenient to use asymptotical approaches. In this case the introduced small parameters are defined not only by physical features of composite but proper relative sizes of filaments and matrix. Introduction, so-called 'slow' and 'fast' coordinates, defining solutions in macro-scale and micro-scale, respectively, allows one to effectively divide original boundary-value problem and take into account local oscillations that are present in final solution as corrections determined by order of small structural parameter. By introducing the periodicity cell characterizing the composite structure we obtain the required ingredient finalizing a formal statement of the problem. Similar asymptotical approach will be also developed below for granular superconductive composites.

9.4.1 Conductivity of Filamentary Composites

Consider filamentary composites consisting of matrix Ω_1 and filaments Ω_2. It is obvious, that effective conductivity factor $\sigma_0 = \sigma_{33}$ in longitudinal direction x_3 could be determined by using mixture rule as

$$\sigma_{33} = (1 - c_2)\sigma_1 + c_2\sigma_2, \tag{9.76}$$

where c_2 is the volume fraction of filaments, σ_i are the conductivity factors of the matrix ($i = 1$) and filaments ($i = 2$).

Then we treat the problem of calculation of the effective conductivity $\sigma_0 = \sigma_{11} = \sigma_{22}$ into transverse area x_1x_2. In the stationary case the constitutive equation stated on the base of the conductivity equations and conservation equations for components $i = 1, 2$ has the form:

$$\sigma_i \nabla^2_{xx} u_i = -f_i, \tag{9.77}$$

where $\nabla_{xx}^2 = \sum_k \frac{\partial^2}{\partial x_k^2}$ is the Laplace operator, σ_i are the conductivity factors of matrix ($i = 1$) and filaments ($i = 2$), u_i and f_i are the corresponding potentials and densities of volume sources, $k = 1, 2$. At the interface $\partial\Omega$ of the components there are considered the conditions of ideal contact (equality of potentials, and equality of fluxes), namely:

$$\{u_1 = u_2\}_{\partial\Omega}, \left\{\sigma_1 \frac{\partial u_1}{\partial \boldsymbol{n}} = \sigma_2 \frac{\partial u_2}{\partial \boldsymbol{n}}\right\}_{\partial\Omega}, \tag{9.78}$$

where $\partial/\partial \boldsymbol{n}$ are the derivatives on normal to $\partial\Omega$. Select in composite two spatial scales: (i) microscale determined by the distance l between centers of adjacent filaments and (ii) macroscale characterized by the specimen size L. Then, the small parameter $\varepsilon = l/L$ will define the order of the composite heterogeneity.

The solution of the boundary-value problem (9.77, 9.78) can be present as the expansion:

$$u_i = u_0(\boldsymbol{x}) + \varepsilon u_i^{(1)}(\boldsymbol{x},\boldsymbol{y}) + \varepsilon^2 u_i^{(2)}(\boldsymbol{x},\boldsymbol{y}) + \cdots, \tag{9.79}$$

where $\boldsymbol{x} = \sum_k \boldsymbol{x_k e_k}, \boldsymbol{y} = \sum_k \boldsymbol{y_k e_k}$, $\boldsymbol{e}_k$ are the basis vectors of Cartesian coordinate system, $x_k = x_k$ and $y_k = \varepsilon^{-1} x_k$ are the 'slow' and 'fast' coordinates. The first term u_0 presents averaged part of the solution, changing in macroscale of whole specimen and no depending on 'fast' coordinates ($\partial u_0/\partial y_k$). The following terms $u_i^{(j)} (j = 1, 2, 3, \ldots)$ introduce the corrections order of ε^j and describe local oscillations in microscale. Due to the medium periodicity $u_i^{(j)}$ satisfy also to the relative periodicity condition.

By taking into account the medium periodicity and using the perturbation method of boundary shape [614], it could be obtained first approximation for solution of the problem in polar coordinates (r, θ), introduced into periodicity cell ($r^2 = y_1^2 + y_2^2$; $\tan\theta = y_2/y_1$) [154] as

$$u_i^{(1)} = \left(C_{1i} r + C_{2i} r^{-1}\right) \frac{\partial u_0}{\partial \boldsymbol{n}}; \ \frac{\partial}{\partial \boldsymbol{n}} = \frac{\partial}{\partial x_1}\cos\theta + \frac{\partial}{\partial x_2}\sin\theta, \tag{9.80}$$

where

$$C_{11} = \frac{(\lambda_2 - 1)\chi}{D}; \ C_{21} = -\frac{C_{11}A^2}{\chi}; \ C_{12} = -\frac{(\lambda_2 - 1)(1 - \chi)}{D}; \ C_{22} = 0; \tag{9.81}$$

$$D = \lambda_2 + 1 - \chi(\lambda_2 - 1); \ \chi = \left[\frac{A}{R(\theta)}\right]^2 = \frac{c_2}{c_{2\max}}\left[\frac{R_0}{R(\theta)}\right]^2; \ R(\theta) = \frac{R_0}{\cos\delta}. \tag{9.82}$$

Here A is the filament radius, $R_0 = L/2$ is the radius of inscribed circle, $\lambda_2 = \sigma_2/\sigma_1$ is the dimensionless conductivity of filaments, $c_{2\max}$ is the geometrically-maximal volume fraction of filaments, $c_2 = \pi A^2/S_0$, S_0 is square of the periodicity cell in 'fast' coordinates, $\delta = \theta - \theta_0 < 1$. In the case of square lattice

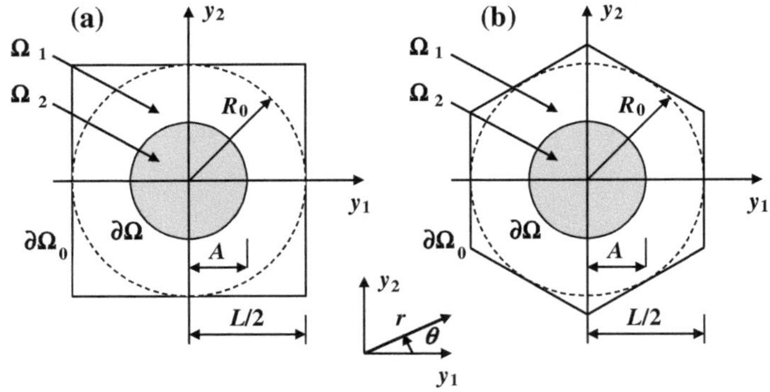

Fig. 9.14 Periodicity cell in square (**a**) and hexagonal (**b**) lattices

(see Fig. 9.14a) we obtain: $-\pi/4 \leq \delta \leq \pi/4$, $\theta_0 = \pi n/2$ $(n = 0,\ 1,\ 2,\ldots)$, $c_{2\max} = \pi/4 = 0.7853$, $S_0 = L^2$ and for hexagonal lattice (see Fig. 9.14b) it is carried out: $-\pi/6 \leq \delta \leq \pi/6$, $\theta_0 = \pi n/3$, $c_{2\max} = \pi\sqrt{3}/6 = 0.9068$, $S_0 = \sqrt{3}L^2/2$.

By assuming the same asymptotical order of conductivities for the components $\sigma_2/\sigma_1 = O(\varepsilon^0)$, taking into account recurrent sequence of boundary-value problems on the periodicity cell, including microscopic equations of conductivity and microscopic equations of ideal contacts, and applying the averaging operator on area of the periodicity cell, $S_0^{-1}\iint_{\Omega_0}(\cdot)dS$, where $dS = dy_1dy_2$, we obtain the averaged equation [154] as

$$\sigma_1 \iint\limits_{\Omega_1} \left[\nabla^2_{xx}u_0 + \nabla^2_{xy}u_1^{(1)}\right] dS + \sigma_2 \iint\limits_{\Omega_2} \left[\nabla^2_{xx}u_0 + \nabla^2_{xy}u_2^{(1)}\right] dS = -\iint\limits_{\Omega_1} f_1 dS - \iint\limits_{\Omega_2} f_2 dS. \tag{9.83}$$

By substituting into (9.83) the relation for $u_i^{(1)}$ defined from (9.80) and (9.82) we obtain macroscopic conductivity equation with order of ε^0 in the form:

$$\sigma_0 \nabla^2_{xx} u_0 = -f_0, \tag{9.84}$$

where $f_0 = (1 - c_2)f_1 + c_2 f_2$ is the averaged density of volume sources. The effective conductivity σ_0 could be calculated after computation of integrals in Eq. (9.83). By using numerical integration for dimensionless effective conductivity $\lambda_0 = \sigma_0/\sigma_1$ in the case of superconductive filaments ($\sigma_2 = \infty$), we obtain the dependences on concentration of superconductive components c_2 for square and hexagonal lattices (see Fig. 9.15), which are compared with relative theoretical dependences calculated by using Rayleigh method [1352]. Figure 9.15 presents also the test results [1352] of conductivity in hexagonal lattice of superconductive cylinders. In the case of conductive cylinders with very high conductivity

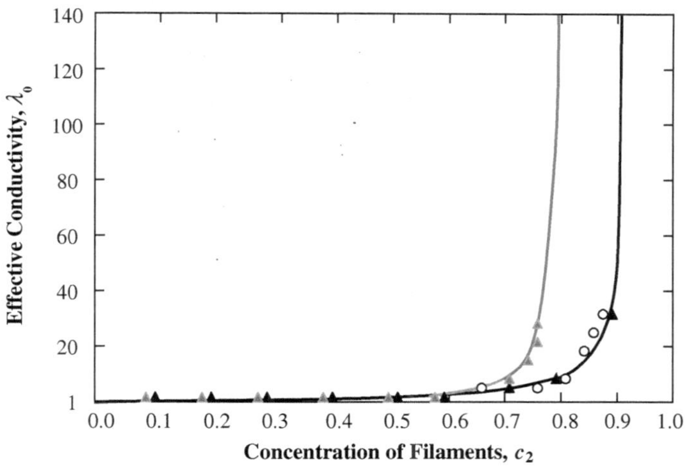

Fig. 9.15 Effective conductivity of composites in dependence on concentration of superconductive filaments: *grey* and *black curves* correspond to solution (9.80) for square and hexagonal lattices; *triangle* show theoretical data [1352] and *circles* present test data [1352]

$(\sigma_2 \to \infty)$ aspiring to contact $(c_2 \longrightarrow c_{2\max})$, it has been also found the next asymptotical formula [1352]:

$$\lambda_0 = M\pi \left\{ 2\frac{\ln \lambda_2}{\lambda_2} + \sqrt{2\left(\frac{c_{2\max}}{c_2}\right)^{1/2} - 1} \right\}^{-1}, \tag{9.85}$$

where for square lattice $M = 1$ and for hexagonal lattice $M = \sqrt{3}$.

Effect of Filament Coverage on Composite Conductivity

In the case when filaments and matrix are divided ones with other by coverage layer Ω_3, the periodicity cell takes the form presented in Fig. 9.16. The original conductivity equation coincides with Eq. (9.77), in which $i = 1, 2, 3$. At the 'matrix-coverage' interfaces $\partial\Omega_1$ and 'inclusion-coverage' interfaces $\partial\Omega_2$ the conditions of ideal contact are selected to be similar conditions (9.78), namely:

$$\{u_1 = u_3\}_{\partial\Omega_1}, \left\{\sigma_1 \frac{\partial u_1}{\partial \boldsymbol{n}} = \sigma_3 \frac{\partial u_3}{\partial \boldsymbol{n}}\right\}_{\partial\Omega_1}, \{u_2 = u_3\}_{\partial\Omega_2}, \left\{\sigma_2 \frac{\partial u_2}{\partial \boldsymbol{n}} = \sigma_3 \frac{\partial u_3}{\partial \boldsymbol{n}}\right\}_{\partial\Omega_2}. \tag{9.86}$$

By using the averaging method to boundary-value problem (9.77, 9.86) the approximate solution of the problem for the periodicity cell may be written in form (9.80), where factors C_{ki} are found as

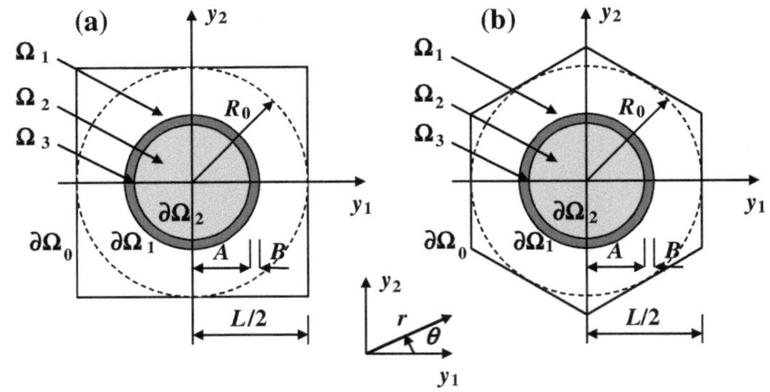

Fig. 9.16 Periodicity cell in square (**a**) and hexagonal (**b**) lattices

$$
\begin{aligned}
C_{11} &= -\chi(c_2+c_3)\left[-\lambda_2 c_3 - \lambda_3(2c_2+c_3) + \lambda_2\lambda_3(2c_2+c_3) + \lambda_3^3 c_3\right]/D;\\
C_{12} &= \{\lambda_2[(\chi+1)c_2+c_3]c_3 + \lambda_3\left[3(\chi-1)c_2c_3 - 2(1-\chi)c_2^2 + \chi c_3^2\right]\\
&\quad + \lambda_2\lambda_3\left[(1-3\chi)c_2c_3 + 2(1-\chi)c_2^2 - \chi c_3^2\right] + \lambda_3^2[(1-\chi)c_2 - \chi c_3]c_3\}/D;\\
C_{13} &= \{\lambda_2\left[(\chi-1)c_2c_3 + 2c_2^2 + \chi c_3^2\right] + \lambda_3\left[(3\chi-1)c_2c_3 + 2\chi c_2^2 + \chi c_3^2\right]\\
&\quad + \lambda_2\lambda_3\left[(1-3\chi)c_2c_3 + 2(1-\chi)c_2^2 - \chi c_3^2\right] + \lambda_3^2[(\chi+1)c_2 - \chi c_3]c_3\}/D;\\
C_{21} &= -C_{11}A^2/\chi;\ C_{22} = 0;\ C_{23} = 2A^2 c_2(c_2+c_3)(\lambda_2-\lambda_3)/D;
\end{aligned}
\tag{9.87}
$$

$$
\begin{aligned}
D &= -\lambda_2[(1+\chi)c_2+\chi c_3]c_3 - \lambda_3\left[(1+3\chi)c_2c_3 + 2(1+\chi)c_2^2 + \chi c_3^2\right]\\
&\quad + \lambda_2\lambda_3\left[(3\chi-1)c_2c_3 + 2(\chi-1)c_2^2 + \chi c_3^2\right] + \lambda_3^2[(\chi-1)c_2 + \chi c_3]c_3,
\end{aligned}
\tag{9.88}
$$

where $\lambda_3 = \sigma_3/\sigma_1$ is the dimensionless conductivity of the coverage layer conductivity; $c_3 = c_2 h(2+h)$ is the coverage volume fraction, $h = B/A$ is the dimensionless thickness of the coverage.

The effective conductivity σ_0 is obtained after calculation of integrals in averaged equation [154]:

$$
\begin{aligned}
&\sigma_1 \iint\limits_{\Omega_1} \left[\nabla_{xx}^2 u_0 + \nabla_{xy}^2 u_1^{(1)}\right] dS + \sigma_2 \iint\limits_{\Omega_2} \left[\nabla_{xx}^2 u_0 + \nabla_{xy}^2 u_2^{(1)}\right] dS + \sigma_3 \iint\limits_{\Omega_3} \left[\nabla_{xx}^2 u_0 + \nabla_{xy}^2 u_3^{(1)}\right] dS\\
&= -\iint\limits_{\Omega_1} f_1 dS - \iint\limits_{\Omega_2} f_2 dS - \iint\limits_{\Omega_3} f_3 dS.
\end{aligned}
\tag{9.89}
$$

Then we find local increments $U_3^{(1)}$ and $U_3^{(2)}$ to the averaged electrical potential u_0 at both sides $\partial\Omega_1$ and $\partial\Omega_2$ of the coverage:

$$U_3^{(1)} = [u_3 - u_0]_{\partial\Omega_1} = \varepsilon u_3^{(1)}\Big|_{r=A+B} + O(\varepsilon^2)$$
$$= a\left[C_{13}(1+h) + \frac{C_{23}}{A^2(1+h)}\right]\frac{\partial u_0}{\partial \boldsymbol{n}} + O(\varepsilon); \tag{9.90}$$

$$U_3^{(2)} = [u_3 - u_0]_{\partial\Omega_2} = \varepsilon u_3^{(1)}\Big|_{r=A} + O(\varepsilon^2) = a\left(C_{13} + \frac{C_{23}}{A^2}\right)\frac{\partial u_0}{\partial \boldsymbol{n}} + O(\varepsilon), \tag{9.91}$$

where $a = \varepsilon A$ is the inclusion radius in 'slow' coordinates.

In composites with well-conductive filaments ($\lambda_2 \gg 1$) the most flux initiates at interfaces of the components. The fluxes $q_3^{(1)}$ and $q_3^{(2)}$ at the interfaces $\partial\Omega_1$ and $\partial\Omega_2$ determined along normal to the boundary are calculated as

$$q_3^{(1)} = -\sigma_3\left[\frac{\partial u_0}{\partial \boldsymbol{n}} + \frac{\partial u_3^{(1)}}{\partial r}\right]_{r=A+B} + O(\varepsilon) = q_0\frac{\lambda_3}{\lambda_0}\left[1 + C_{13} - \frac{C_{23}}{A^2(1+h)^2}\right] + O(\varepsilon); \tag{9.92}$$

$$q_3^{(2)} = -\sigma_3\left[\frac{\partial u_0}{\partial \boldsymbol{n}} + \frac{\partial u_3^{(1)}}{\partial r}\right]_{r=A} + O(\varepsilon) = q_0\frac{\lambda_3}{\lambda_0}\left[1 + C_{13} - \frac{C_{23}}{A^2}\right] + O(\varepsilon), \tag{9.93}$$

where $q_0 = -\sigma_0\partial u_0/\partial \boldsymbol{n}$ is the averaged flux via boundary $\partial\Omega$ in macroscale.

In the case of thin coverage ($h \to 0$) at $\lambda_3 < \lambda_2$ the contact conditions of matrix with inclusions satisfy to the model of soft (badly conductive) boundary defining the flux continuity ($q_3^{(1)} = q_3^{(2)}$) and the jump of electrical potential ($U_3^{(1)} > U_3^{(2)}$). On the contrary case, at $\lambda_3 > \lambda_2$ it is realized the model of rigid (well-conductive) boundary stating the flux discontinuity ($q_3^{(1)} > q_3^{(2)}$) and equality of the electrical potentials ($U_3^{(1)} = U_3^{(2)}$). In this case take place the next limit transitions:

$$\text{at } \lambda_3 \to 0 \text{ we have } U_3^{(1)} \to U_3|_{\lambda_2=0};\ U_3^{(2)} \to U_3|_{\lambda_2=\infty};\ q_3^{(1)} = q_3^{(2)} \to q_3|_{\lambda_2=0} = 0; \tag{9.94}$$

$$\text{at } \lambda_3 \to \infty \text{ we have } U_3^{(1)} = U_3^{(2)} \to U_3|_{\lambda_2=\infty};\ q_3^{(1)} \to q_3|_{\lambda_2=\infty};\ q_3^{(2)} \to q_3|_{\lambda_2=0} = 0. \tag{9.95}$$

The results for dimensionless effective conductivity λ_0, obtained by using three-phase model [647] and compared with above asymptotical model lead to under-estimated value of λ_0 at $\lambda_3 \to \infty$, because the three-phase model does not take into account spatial geometry of the filament package.

Effect of Non-Ideal Contact of Filaments and Matrix

Effect of the non-ideal contact between filaments and matrix could be modeled by assuming that the volume fraction of coverage $c_3 \to 0$ and its dimensionless conductivity $\lambda_3 \to 0$. In this case in dependence on the ratio λ_3/c_3 may be estimated different adhesion order for the composite components. With this aim assume that $h \to 0$, $c_3 = c_2h(2 + h) \to 0$ and introduce dimensionless couple parameter a:

$$\lambda_3 = (1 - a)c_3/a, \text{ at } 0 \leq a \leq 1. \tag{9.96}$$

The value of $a = 0$ corresponds to ideal contact, and $a = 1$ defines the case of total absence any contact between the components. Moreover, flux q_3 via the 'matrix-filament' interface $\partial\Omega$ is proportional to the jump of electrical potential $\Delta u_3 = [u_1 - u_2]_{\partial\Omega}$:

$$q_3 = -\left(\sigma_{ef}/b\right)\Delta u_3, \tag{9.97}$$

where $b = \varepsilon B$ is the coverage thickness in 'slow' coordinates, σ_{ef} characterizes the interface conductivity.

At the non-ideal contact ($a = 0$) $\Delta u_3 = 0$, and in total absence any contact ($a = 1$) $q_3 = 0$. The last case corresponds to the composite with absolutely non-conductive filaments (or voids). As a result, it has been estimated influence of the couple parameter a in the case of superconductive inclusions ($\lambda_2 = \infty$) on effective parameter λ_0 and flux q_3 (at $\theta = \theta_0$) at the 'matrix-filament' interface for the filament concentration $c_2 = 0.6$. The plots of these dependences are present in Fig. 9.17.

Filamentary Composites of Irregular Structure

At accidental perturbation of periodical disposition of filaments into composite, we present each filament center into circle with diameter d. By this the circles

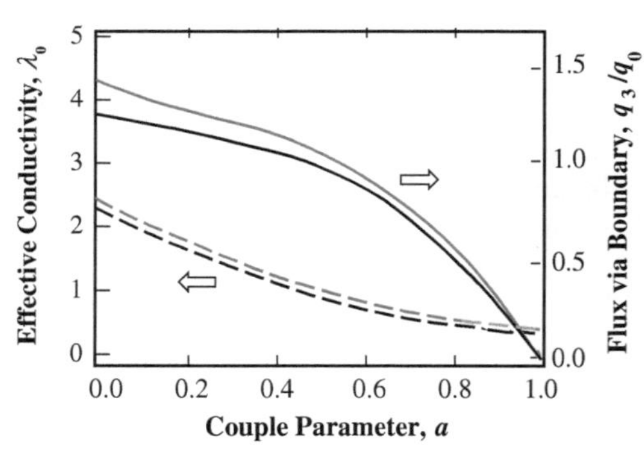

Fig. 9.17 Influence of couple parameter on effective conductivity and flux via 'matrix-filament' interface: *grey* and *black curves* correspond to square and hexagonal lattice, respectively

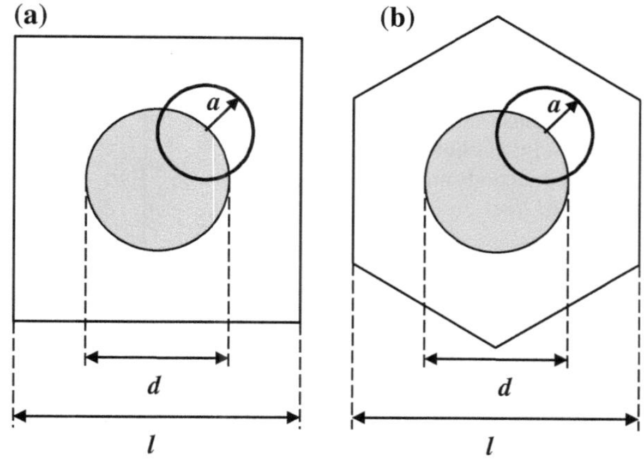

Fig. 9.18 Periodicity cell for filamentary composites of irregular structure in square (**a**) and hexagonal (**b**) lattices

compose square lattice (Fig. 9.18a) or hexagonal one (Fig. 9.18b) with period l. Parameter $\beta = d/l$, $0 \le \beta \le \beta_{\max}$ defines order of the structure irregularity. In the case when adjacent inclusions touch one to other the maximal value is calculated as $\beta_{\max} = 1 - (c_2/c_{2\max})^{1/2}$. It is assumed absence of the inclusion overlapping and no formation of cluster structure. If the filament conductivity is greater than the matrix conductivity ($\lambda_2 > 1$),[3] then there are introduced the corresponding bounds for dimensionless effective conductivity λ_0:

$$\lambda_L \le \lambda_0 \le \lambda_U. \tag{9.98}$$

In the regular case (at $\beta = 0$) the effective conductivity has been defined in dependence on the filament concentration c_2 in the form [934]: $\lambda_0|_{\beta=0} = \Lambda_0(c_2)$. The strong lower bound λ_L is determined by the solution for corresponding regular structure in the form:

$$\lambda_L = \Lambda_0(c_2). \tag{9.99}$$

The upper estimation λ_U may be defined by substitution of accidently disposed filaments with radius a by periodic lattice of filaments with radius $a + d/2$ (so-called *method of safe spheres* [1455]):

$$\lambda_U = \Lambda_0\left[\left(\sqrt{c_2} + \beta\sqrt{c_{2\max}}\right)^2\right]. \tag{9.100}$$

Other known variation Hashin–Shtrikman estimations [646] give the next relations for extreme estimations:

$$\lambda_L = 1 + \frac{c_2}{(\lambda_2 - 1)^{-1} + (1 - c_2)/2}; \ \lambda_U = \lambda_2 + \frac{1 - c_2}{(1 - \lambda_2)^{-1} + c_2/(2\lambda_2)}. \tag{9.101}$$

---

[3] The solution for contrary case ($\lambda_2 < 1$) is obtained by using the formula $\lambda_0|_{\lambda_2=\lambda} = \frac{1}{\lambda_0|_{\lambda_2=1/\lambda}}$ [842].

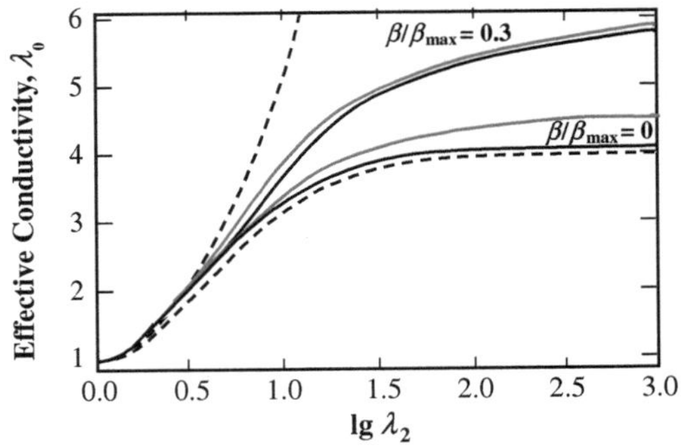

Fig. 9.19 Effective conductivity of irregular composites: *grey* and *black curves* correspond to square and hexagonal lattice, respectively; Hashin–Shtrikman bounds are shown by *dashed lines*

The numerical results presented in (Fig. 9.19) for $c_2 = 0.6$ at very high conductivity of filaments ($\lambda_2 \to \infty$), significantly improve corresponding Hashin–Shtrikman estimations described by dashed lines.

Cluster Conductivity of Filamentary Composites

At accidental disposition of filaments, an increasing their concentration c_2 at proper critical value (percolation threshold) $c_2 = c_{2p}$ leads to formation of infinite cluster with fractal structure from contacting links of the inclusions. According to model [1568] its conducting skeleton may be presented by a lattice with period (correlation radius) R_c:

$$R_c = \frac{1}{\left|c_2 - c_{2p}\right|^{v}} \text{ at } c_2 \to c_{2p}, \tag{9.102}$$

where l is the distance between the centers of contacting inclusions, v is the index of the correlation radius. Value of R_c defines maximal size of finite clusters at $c_2 < c_{2p}$. Effective characteristics σ_0 the composite near the percolation threshold ($c_2 \to c_{2p}$) are defined by asymptotical dependences of type $\sigma_0 \sim \left|c_2 - c_{2p}\right|^t$, where t is the critical exponent of corresponding critical properties. The critical exponents depend on the spatial dimension. In 2D case we have $c_{2p} = 0.5$ and $v = 1.33$.

Below we obtain qualitative estimations for composite with superconductive filaments whose fraction aspires to the percolation threshold from below ($\sigma_2 = \infty$, $c_2 \to c_{2p} - 0$) [154].

Consider the region with proper size $R_c + l$, into of which to be cluster with maximal length R_c. Electrical resistivity of the region is equal

$$\rho_0 = \rho_1 \frac{l}{R_c + l}, \tag{9.103}$$

where $\rho_1 = 1/\sigma_1$ is the electrical resistivity of matrix and the effective conductivity is approximately determined as

$$\sigma_0 = \frac{1}{\rho_0} \approx \frac{\sigma_1}{\left(c_{2p} - c_2\right)^{\nu}}. \tag{9.104}$$

This formula allows one to state the singularity order no pretending on strict account of numerical coefficients.

At small concentration of filaments ($c_2 \to 0$) the dimensionless effective conductivity in 2D case is defined by Garnet formula [522] as

$$\lambda_0 = \frac{\sigma_0}{\sigma_1} = 1 - 2c_2 \left[\frac{(\sigma_1 + \sigma_2)}{(\sigma_1 - \sigma_2)} + c_2\right]^{-1}. \tag{9.105}$$

For superconductive inclusions ($\sigma_2 = \infty$) we obtain

$$\lambda_0 = \frac{\sigma_0}{\sigma_1} = \frac{1 + c_2}{1 - c_2}. \tag{9.106}$$

By joining relations (9.104) and (9.105), we define approximate analytical solution corresponding to any value from range $0 \le c_2 \le c_{2p}$. With this aim it is used the method of asymptotically equivalent functions [47], allowing one to present the effective conductivity as

$$\lambda_0 = \frac{\sigma_0}{\sigma_1} = \frac{a_0 + a_1\xi + \xi^2(1 + \xi)^{\nu}}{1 + b_1\xi + \xi^2 c_{2p}^{\nu}}, \tag{9.107}$$

where $\xi = c_2/(c_{2p} - c_2)$. Asymptotic of relation (9.107) coincides with (9.104) at $c_2 \to c_{2p}$. Unknown factors a_0, a_1 and b_1 are calculated from the condition of coincidence of function expansion (9.107) into series on powers of c_2 at $c_2 \to 0$ and function expansion (9.106) into series on powers up to order $O\left(c_2^2\right)$, inclusive. As a result, we obtain $a_0 = 1, a_1 = 1 + c_{2p} + \left(1 - c_{2p}^{\nu}\right)/\left(2c_{2p}\right), b_1 = 1 - c_{2p} + \left(1 - c_{2p}^{\nu}\right)/\left(2c_{2p}\right)$.

The numerical results for composite with superconductive filaments are present in Fig. 9.20. Here, the solid line corresponds to formula (9.107), the dashed line 1 is determined by formula (9.106), and the dashed line 2 defined by formula (9.104).

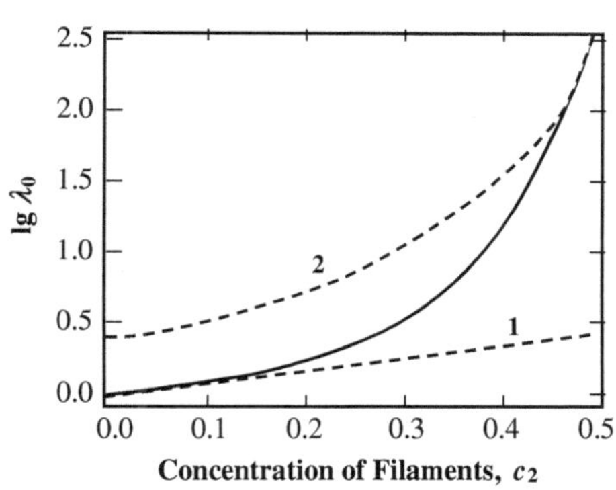

Fig. 9.20 Effective conductivity of composite with superconductive filaments

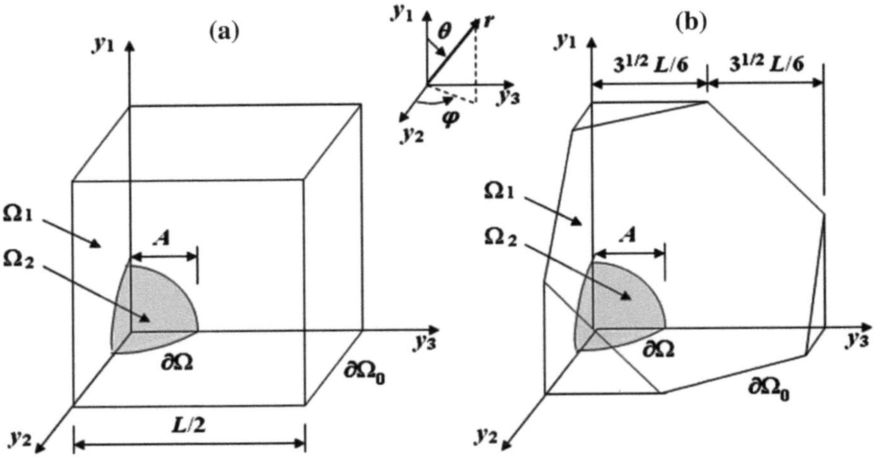

Fig. 9.21 Periodicity cell of granular composites in sc-lattice (**a**) and bcc-lattice (**b**)

9.4.2 Conductivity of Granular Composites

Consider granular composites consisting of matrix Ω_1 and spherical inclusions Ω_2. The periodicity cells (their 1/8 parts) for simple cubic (sc) and body-centered cubic (bcc) lattices are shown in Fig. 9.21. In order to define effective conductivity $\sigma_0 = \sigma_{11} = \sigma_{22} = \sigma_{33}$ consider constitutive equation (9.77), where $i = 1, 2, 3$ and conditions of ideal contact (9.78). The solution of boundary-value problem (9.77, 9.78) is present again into form of expansion (9.79).

By taking into account the medium periodicity, using the method of perturbation of the boundary shape [614], we obtain first approximation for solution of the problem into spherical coordinates (r, θ, φ), introduced into periodicity cell $\left(r^2 = y_1^2 + y_2^2 + y_3^2;\ \tan\varphi = y_3/y_2;\ \cos\theta = y_1/r\right)$, in the form [154]:

$$u_i^{(1)} = \left(C_{1i}r + C_{2i}r^{-2}\right)\frac{\partial u_0}{\partial \boldsymbol{n}};\ \frac{\partial}{\partial \boldsymbol{n}} = \frac{\partial}{\partial x_1}\cos\theta + \frac{\partial}{\partial x_2}\sin\theta\ \cos\varphi + \frac{\partial}{\partial x_3}\sin\theta\ \sin\varphi, \tag{9.108}$$

where

$$C_{11} = \frac{(\lambda_2 - 1)\chi}{D};\ C_{21} = -\frac{C_{11}A^3}{\chi};\ C_{12} = -\frac{(\lambda_2 - 1)(1 - \chi)}{D};\ C_{22} = 0; \tag{9.109}$$

$$D = \lambda_2 + 2 - \chi(\lambda_2 - 1);\ \chi = \left[\frac{A}{R(\theta, \varphi)}\right]^3 = \frac{c_2}{c_{2\max}}\left[\frac{R_0}{R(\theta, \varphi)}\right]^3. \tag{9.110}$$

Table 9.2 Function $R(\theta, \varphi)$

φ	θ	$R(\theta, \varphi)$
sc-lattice		
From 0 up to $\pi/4$	From 0 up to $\operatorname{atan}\left(\frac{1}{\cos\varphi}\right)$	$\frac{R_0}{\cos\theta}$
	From $\operatorname{atan}\left(\frac{1}{\cos\varphi}\right)$ up to $\pi/2$	$\frac{R_0}{\sin\theta\cos\varphi}$
bcc-lattice		
From 0 up to atan(1/2)	From 0 up to $\operatorname{atan}\left[\frac{1}{2(\sin\varphi+\cos\varphi)}\right]$	$\frac{2R_0}{\sqrt{3}\cos\theta}$
	From $\operatorname{atan}\left[\frac{1}{2(\sin\varphi+\cos\varphi)}\right]$ up to $\operatorname{atan}\left[\frac{2}{\cos\varphi-2\sin\varphi}\right]$	$\frac{\sqrt{3}R_0}{\cos\theta+\sin\theta(\sin\varphi+\cos\varphi)}$
	From $\operatorname{atan}\left[\frac{2}{\cos\varphi-2\sin\varphi}\right]$ up to $\pi/2$	$\frac{2R_0}{\sqrt{3}\sin\theta\cos\varphi}$
From atan(1/2) up to $\pi/4$	From 0 up to $\operatorname{atan}\left[\frac{1}{2(\sin\varphi+\cos\varphi)}\right]$	$\frac{2R_0}{\sqrt{3}\cos\theta}$
	From $\operatorname{atan}\left[\frac{1}{2(\sin\varphi+\cos\varphi)}\right]$ up to $\pi/2$	$\frac{\sqrt{3}R_0}{\cos\theta+\sin\theta(\sin\varphi+\cos\varphi)}$

Here A is the grain radius, $R_0 = L/2$ is the radius of inscribed sphere, $\lambda_2 = \sigma_2/\sigma_1$ is the dimensionless grain conductivity, $c_{2\max}$ is the geometrically maximal volume fraction of grains, $c_2 = 4\pi A^3/(3V_0)$, V_0 is the volume of the periodicity cell into 'fast' coordinates. For 1/16-part of the cell ($0 \le \theta \le \pi/2$, $0 \le \varphi \le \pi/4$) presented in Fig. 9.21, relations for $R(\theta, \varphi)$ are shown in Table 9.2. At other values of angles θ and φ the function $R(\theta, \varphi)$ continues periodically. For sc-lattice we obtain (see Fig. 9.21a) $c_{2\max} = \pi/6 = 0.5235$, $V_0 = L^3$ and for bcc-lattice (see Fig. 9.21b) $c_{2\max} = \pi\sqrt{3}/8 = 0.6801$, $V_0 = 4\sqrt{3}L^3/9$.

By applying the averaging operator on region of the periodicity cell $V_0^{-1}\iiint_{\Omega_0}(\cdot)dV$, where $dV = dy_1dy_2dy_3$, we obtain the averaged equation [154]:

$$\sigma_1 \iiint_{\Omega_1} \left[\nabla_{xx}^2 u_0 + \nabla_{xy}^2 u_1^{(1)}\right] dV + \sigma_2 \iiint_{\Omega_2} \left[\nabla_{xx}^2 u_0 + \nabla_{xy}^2 u_2^{(1)}\right] dV$$
$$= -\iiint_{\Omega_1} f_1 dV - \iiint_{\Omega_2} f_2 dV. \tag{9.111}$$

By substituting into (9.111) the relation for $u_i^{(1)}$ from (9.108) to (9.110), we obtain macroscopic equation of conductivity for estimation of the effective conductivity σ_0.

The numerical results for dimensionless effective conductivity $\lambda_0 = \sigma_0/\sigma_1$ in the case of superconductive inclusions are shown in Fig. 9.22. Moreover, there are present theoretical data obtained by using Rayleigh method [1117, 1120] and test

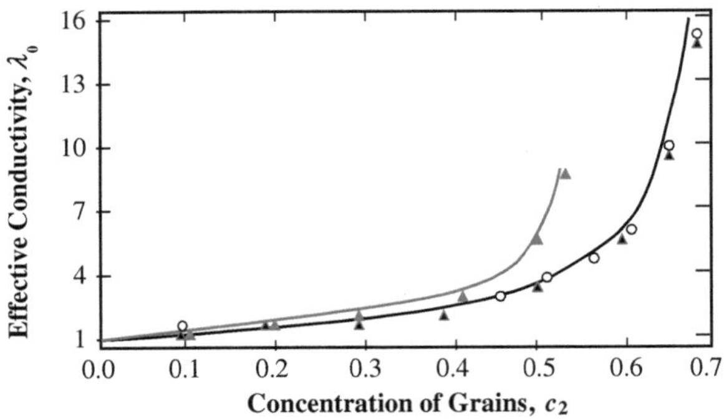

Fig. 9.22 Effective conductivity of composites in dependence on concentration of superconductive grains: *grey* and *black curves* correspond to solution (9.108) for sc-lattice and bcc-lattices; *triangles* show theoretical data for sc-lattice [1120] and bcc-lattice [1117], and *circles* present test data for bcc-lattice [1117]

results measuring electric conductivity of bcc-lattice consisting of superconductive spheres [1117].

Asymptotic of λ_0 in the case of strong interaction of adjacent superconductive grains is present in Fig. 9.23. The solution (solid curves) good coincides with asymptotical formulae [102] (dashed curves) obtained for two very well conductive spheres aspiring to contact:

$$\lambda_0 = M_1 \ \ln \zeta - M_2 \text{ at } \lambda_2 = \infty, c_2 \to c_{2\max}; \tag{9.112}$$

$$\lambda_0 = M_1 \ \ln \lambda_2 - M_2 \text{ at } \lambda_2 \to \infty, c_2 = c_{2\max}, \tag{9.113}$$

where $\zeta = 1/\left[1 - (c_2/c_{2\max})^{1/3}\right]$, $M_1 = \pi/2$, $M_2 = 0.7$ for sc-lattice and $M_1 = \pi\sqrt{3}/2$, $M_2 = 2.4$ for bcc-lattice.

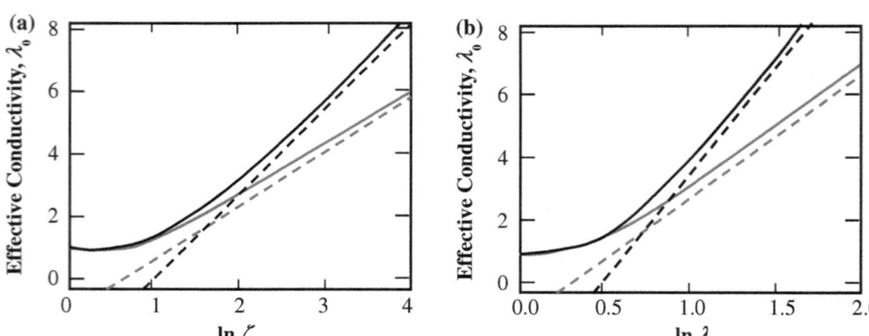

Fig. 9.23 Effective conductivity of composites in the cases: (**a**) $\lambda_2 = \infty$, $c_2 \to c_{2\max}$; (**b**) $\lambda_2 \to \infty$, $c_2 = c_{2\max}$. *Solid grey* and *black lines* correspond to solutions defined for sc-lattice and bcc-lattices, *dashed straight lines* present corresponding asymptotical results [102]

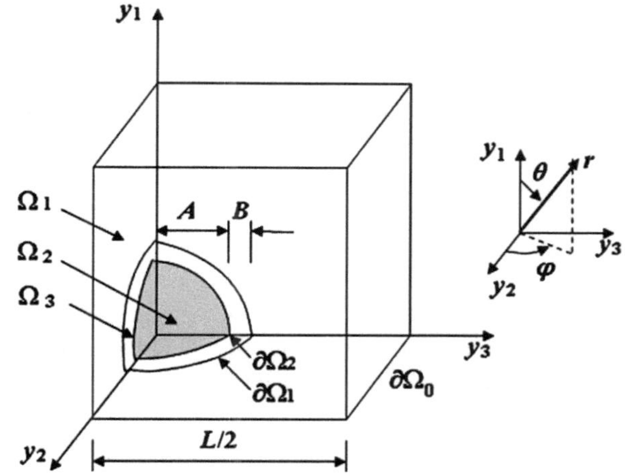

Fig. 9.24 Periodicity cell of granular composites in sc-lattice

Coverage Grain Effect on Composite Conductivity

In the case when simple cubic lattice of spherical grains and matrix are separated ones with other by coverage layer Ω_3, the periodicity cell (its 1/8 part) takes form shown in Fig. 9.24. Original equation of conductivity coincides with Eq. (9.77), in which $i = 1, 2, 3$. At 'matrix-coverage' interface $\partial\Omega_1$ and 'inclusion-coverage' interface $\partial\Omega_2$ conditions of ideal contact (9.86) are selected.

By using the averaging method for boundary-value problem (9.77, 9.86), approximate solution of the problem for periodicity cell may be written in form (9.80), where factors C_{ki} are determined as

$$\begin{aligned}
C_{11} &= -\chi(c_2+c_3)\left[-\lambda_2 c_3 - \lambda_3(3c_2+2c_3) + \lambda_2\lambda_3(3c_2+c_3) + 2\lambda_3^2 c_3\right]/D \\
C_{12} &= \{\lambda_2[(\chi+2)c_2+\chi c_3]c_3 + \lambda_3\left[5(\chi-1)c_2c_3 + 3(\chi+1)c_2^2 + 2\chi c_3^2\right] \\
&\quad +\lambda_2\lambda_3\left[3(1-\chi)c_2^2 - \chi c_3^2 + (1-4\chi)c_2c_3\right] + \lambda_3^2[2(1-\chi)c_2 - 2\chi c_3]c_3\}/D; \\
C_{13} &= \{\lambda_2\left[\chi c_2c_3 + 3c_2^2 + \chi c_3^2\right] + \lambda_3\left[(5\chi-2)c_2c_3 + 3\chi c_2^2 + 2\chi c_3^2\right] \\
&\quad +\lambda_2\lambda_3\left[(1-4\chi)c_2c_3 + 3(1-\chi)c_2^2 - \chi c_3^2\right] + \lambda_3^2[2(1-\chi)c_2 - 2\chi c_3]c_3\}/D; \\
C_{21} &= -C_{11}A^3/\chi;\; C_{22} = 0;\; C_{23} = 3A^3c_2(c_2+c_3)(\lambda_2-\lambda_3)/D;
\end{aligned} \tag{9.114}$$

$$\begin{aligned}
D &= -\lambda_2[(2+\chi)c_2+\chi c_3]c_3 - \lambda_3\left[(4+5\chi)c_2c_3 + 3(2+\chi)c_2^2 + 2\chi c_3^2\right] \\
&\quad + \lambda_2\lambda_3\left[(4\chi-1)c_2c_3 + 3(\chi-1)c_2^2 + \chi c_3^2\right] + \lambda_3^2[2(\chi-1)c_2 + 2\chi c_3]c_3,
\end{aligned} \tag{9.115}$$

where $\lambda_3 = \sigma_3/\sigma_1$ is the dimensionless conductivity of the coverage layer, $c_3 = c_2h(3 + 3\,h + h^2)$ is the coverage volume fraction, $h = B/A$ is the dimensionless thickness of the coverage.

The effective conductivity σ_0 is calculated computing integrals in the averaged equation:

$$\sigma_1 \iiint\limits_{\Omega_1} \left[\nabla^2_{xx} u_0 + \nabla^2_{xy} u_1^{(1)}\right] dV + \sigma_2 \iiint\limits_{\Omega_2} \left[\nabla^2_{xx} u_0 + \nabla^2_{xy} u_2^{(1)}\right] dV + \sigma_3 \iiint\limits_{\Omega_3} \left[\nabla^2_{xx} u_0 + \nabla^2_{xy} u_3^{(1)}\right] dV$$
$$= - \iiint\limits_{\Omega_1} f_1 dV - \iiint\limits_{\Omega_2} f_2 dV - \iiint\limits_{\Omega_3} f_3 dV. \tag{9.116}$$

The local electrical potentials $U_3^{(1)}$ and $U_3^{(2)}$, and also the fluxes $q_3^{(1)}$ and $q_3^{(2)}$ at interfaces $\partial\Omega_1$ and $\partial\Omega_2$ are determined again by using formulae (9.90–9.92), in which A^2 is substituted on A^3.

For thin coverage ($h \to 0$) at $\lambda_3 < \lambda_2$ the conditions of matrix-inclusion contact satisfy to the model of badly conducting boundary $\left(q_3^{(1)} = q_3^{(2)}, U_3^{(2)} > U_3^{(2)}\right)$. On the contrary, at $\lambda_3 > \lambda_2$ is realized the model of well conducting boundary $\left(q_3^{(1)} > q_3^{(2)}, U_3^{(2)} = U_3^{(2)}\right)$. Again, the results for dimensionless effective conductivity λ_0, based on three-phase model [647], which does not take into account spatial package of grains in comparison with above asymptotical model, lead to underestimated value of λ_0 at $\lambda_3 \to \infty$.

Effect of Non-Ideal Contact of Grains and Matrix

Assume that the volume fraction of coverage $c_3 \to 0$ and its dimensionless conductivity $\lambda_3 \to 0$. Introduce dimensionless couple parameter a (9.96) and suppose that $h \to 0$, $c_3 = c_2 h(3 + 3\,h + h^2) \to 0$. Again, the value of $a = 0$ corresponds to ideal contact, and $a = 1$ defines total absence any contact between components. In this case the flux q_3 via 'matrix-grain' interface $\partial\Omega$ is proportional to the jump of electrical potential Δu_3 (9.97). At ideal contact ($a = 0$) $\Delta u_3 = 0$, and at total absence any contact between components ($a = 1$) $q_3 = 0$. The last case corresponds to composite with absolutely non-conducting grains (or voids). As a result, it may be estimated influence of the couple parameter a in the case of superconductive inclusions ($\lambda_2 = \infty$) on dimensionless effective conductivity λ_0 and flux q_3 (at $\theta = \theta_0$, $\varphi = \varphi_0$). The plots these dependences are shown in Fig. 9.25.

Granular Composites of Irregular Structure

At accidental perturbation of periodical disposition of grains into composite, we present any grain centre into sphere of diameter d (in this case these spheres compose sc-lattice or bcc-lattice with period l). Parameter $\beta = d/l$, $0 \le \beta \le \beta_{\max}$ defines order of the structure irregularity. In the case when adjacent inclusions

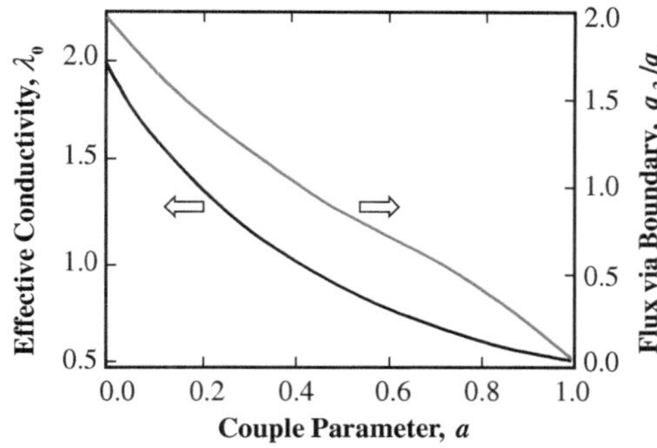

Fig. 9.25 Effect of couple parameter on effective conductivity and flux via 'matrix-grain' interface

touch one to other the maximal value is calculated as $\beta_{\max} = 1 - (c_2/c_{2\max})^{1/2}$. It is assumed absence of the inclusion overlapping and no formation of cluster structure. If the grain conductivity is greater than the matrix conductivity ($\lambda_2 > 1$), then there are introduced the corresponding bounds for dimensionless effective conductivity λ_0 (9.98).

In the regular case (at $\beta = 0$) the effective conductivity has been defined in dependence on the grain concentration c_2 in the form [934]: $\lambda_0|_{\beta=0} = \Lambda_0(c_2)$. The strong lower bound λ_L is determined by the solution for corresponding regular structure in the form:

$$\lambda_L = \Lambda_0(c_2). \tag{9.117}$$

The upper bound λ_U may be defined by substitution of accidently disposed grains with radius a by periodic lattice of grains with radius $a + d/2$ by using again the method of safe spheres [1455]:

$$\lambda_U = \Lambda_0\left[\left(c_2^{1/3} + \beta c_{2\max}^{1/3}\right)^3\right]. \tag{9.118}$$

Other known variation Hashin–Shtrikman bounds [649] give the next relations for extreme estimations:

$$\lambda_L = 1 + \frac{c_2}{(\lambda_2 - 1)^{-1} + (1 - c_2)/3}; \quad \lambda_U = \lambda_2 + \frac{1 - c_2}{(1 - \lambda_2)^{-1} + c_2/(3\lambda_2)}. \tag{9.119}$$

More precise estimations (9.117) and (9.118), shown by solid curves in Fig. 9.26 for $c_2 = 0.4$ at high conductivity of grains are significantly better than corresponding Hashin–Shtrikman bounds presented by dashed lines.

Cluster Conductivity of Granular Composites

At accidental disposition of grains and proper their concentration c_2 they can form clusters. The correlation radius is determined by relation (9.102), in which for 3D

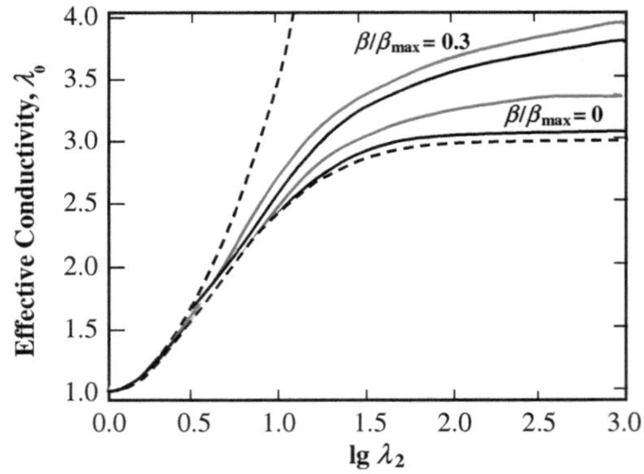

Fig. 9.26 Effective conductivity of irregular composites: *grey* and *black curves* correspond to sc-lattice and bcc-lattices, *dashed lines* present Hashin–Shtrikman bounds

case $c_{2p} = 0.16$ and $v = 0.85$ [344]. For composite with superconductive inclusions, whose fraction aspires to percolation threshold from below ($\sigma_2 = \infty$, $c_2 \to c_{2p} - 0$) effective conductivity σ_0 in approximate form may again be calculated by using formula (9.104). At small concentration of grains ($c_2 \to 0$) the dimensionless effective conductivity λ_0 is also defined by using Garnet formula [522] as

$$\lambda_0 = \frac{\sigma_0}{\sigma_1} = 1 - 3c_2 \left[\frac{(2\sigma_1 + \sigma_2)}{(\sigma_1 - \sigma_2)} + c_2 \right]^{-1}. \tag{9.120}$$

For superconductive inclusions ($\sigma_2 = \infty$) we obtain

$$\lambda_0 = \frac{\sigma_0}{\sigma_1} = \frac{1 + 2c_2}{1 - c_2}. \tag{9.121}$$

By joining relations (9.104) and (9.121), the approximate analytical solution coincides formally with relation (9.107), in which the factors are now found as $a_0 = 1, a_1 = 1 + 2c_{2p} + \left(1 - c_{2p}^{v}\right)/(3c_{2p}), b_1 = 1 - c_{2p} + \left(1 - c_{2p}^{v}\right)/(3c_{2p})$.

The numerical results for composite with superconductive grains, whose concentration no greater than the percolation threshold, are shown in Fig. 9.27. Here, the solid curve corresponds to formula (9.107), dashed curve 1 is stated by formula (9.121), and dashed curve 2 is defined by formula (9.104).

9.4.3 *Effective Elastic Properties of Composites*

During procedure defining longitudinal shear modulus G_L of filamentary composite, it is considered deformation of pure shear into planes x_1x_3 and x_2x_3 under action of tangential stresses σ_{13} and σ_{23} at zero other stresses. Then, the equilibrium equations into displacements u, v, w corresponding to coordinates x_1, x_2, x_3 for ith component of the composite have the form:

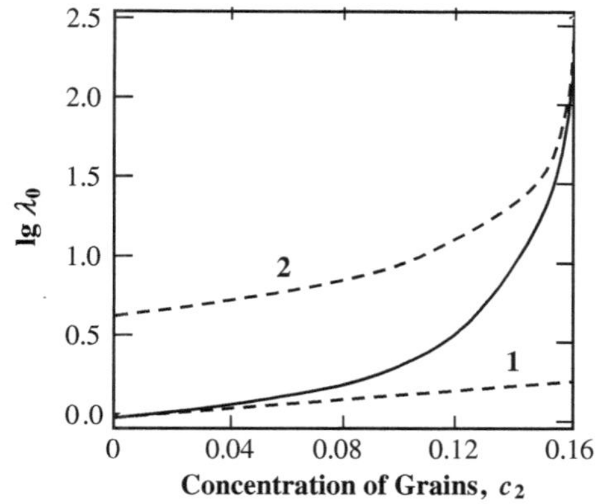

Fig. 9.27 Effective conductivity of composite with superconductive grains

$$G_i\left(\frac{\partial^2 w_i}{\partial x_1^2}+\frac{\partial^2 w_i}{\partial x_2^2}\right)=-f_i, i=1,2. \tag{9.122}$$

where f_i is the component (longitudinal) of bulk force along direction of x_3.

At the interface $\partial\Omega$ there are considered the conditions of ideal contact of the components of the composite consisting in equality their longitudinal displacements:

$$\{w_1 = w_2\}_{\partial\Omega} \tag{9.123}$$

and tangential stresses along normal $\boldsymbol{n}$ to $\partial\Omega$:

$$\left\{G_1\frac{\partial w_1}{\partial \boldsymbol{n}}=G_2\frac{\partial w_2}{\partial \boldsymbol{n}}\right\}_{\partial\Omega}. \tag{9.124}$$

The boundary-value problem (9.122–9.124) is mathematically coincides with the conductivity problem (9.77, 9.78). Therefore, all results obtained in Sect. 9.4.1 for effective conductivity σ_0 directly carry over on effective shear modulus G_L at strain along longitudinal direction.

The effective bulk modulus K_T and effective shear modulus G_T at deformation in transverse direction may be obtained by using poly-dispersed model [648] and three-phase model [284], respectively. They do not take into account geometry of composite and lead to underestimated values for effective moduli in the case of high stiffness and high concentration of inclusions. Similar solutions for effective elastic moduli of granular composites could be obtained by using the same models [284], no taking into account the type of spatial package of grains and assuming that the spherical inclusions are uniformly distributed into matrix and do not form clusters.

Chapter 10
Computer Simulation of HTSC Microstructure and Toughening Mechanisms

10.1 YBCO Ceramic Sintering and Fracture

It is quite obvious that the spectrum of structure-sensitive properties of superconducting ceramics is caused directly by essential inhomogeneous structure, consisting of superconducting grains, secondary phases, pores and microdefects, as a rule disposed on intergranular boundaries. The microstructure formation and fracture occur during sintering, causing internal (residual) stresses and during the material loading by different thermo-mechanical and electromagnetic fields. Based on computer simulation, a joint study of sintering, cooling and fracture of the structure-heterogeneous material [825, 827, 1296] allows a prediction and an optimization of superconductor properties depending on the parameters, including composition, heat rate, initial porosity of material. In Chap. 6 fundamentals of HTSC computer monitoring have been discussed. In this section, an example of $YBa_2Cu_3O_{7-x}$ gradient sintering is considered [120], and proper toughening mechanisms are investigated. The present analysis could be used, in particular, to study heterogeneous mechanism of the YBCO structure formation and to model the cracking processes in *ab*-plane.

10.1.1 Sintering Model of Superconducting Ceramic

During gradient sintering up to microstructure formation, there are the following transformations in the material (Fig. 10.1) [120]:

(1) the primary crystallization, which consists of formation and growth of facetted crystallites of superconducting phase into powder media of the sample ($T \sim$ 800–900 °C);
(2) the formation of non-superconducting phase around crystallites due to active local thermo-diffusion ($T \sim$ 900–920 °C);
(3) the sample shrinkage and formation of isolated (closed) porosity owing to a melting of the non-superconducting phase and its following pushing from

I. A. Parinov, *Microstructure and Properties of High-Temperature Superconductors*,
DOI: 10.1007/978-3-642-34441-1_10, © Springer-Verlag Berlin Heidelberg 2012

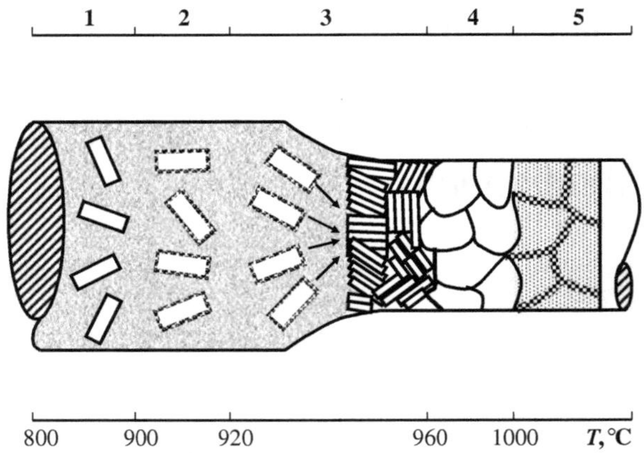

Fig. 10.1 The physical model of microstructure formation of YBCO ceramic during sintering in thermal gradient. The next process zones are shown of primary crystallization (1); non-superconducting phase formation (2); shrinkage (3); abnormal grain growth (4) and structure decomposition (5)

aggregates of superconducting crystallites into pores and at the sample surface ($T \sim 920$–940 °C);

(4) the secondary re-crystallization, accompanied by breaking of intergranular buffer layers and by creation of big grains with irregular shape ($T \sim 960$–980 °C);

(5) the structure decomposition ($T > 1000$ °C).

The proposed computer model generalizes the research results of gradient sintering of the ferroelectric [825, 827, 1296] and YBCO [1319] ceramics. It is assumed that temperature in the furnace changes linearly with coordinate. The modeling consists of successive consideration of following stages: (i) the propagation of a heat front in a powder compact with definition of temperature distribution in the sample; (ii) the press-powder crystallization into region of the sintering temperature that is determined by heat front, in conditions of the actual gradient sintering of YBCO [120]; (iii) the sample shrinkage with formation of a closed porosity; (iv) the secondary re-crystallization, which is an abnormal grain growth under conditions of actual inhibition due to existence of a mixed secondary phase. In this consideration, the action regions of the above processes are found by temperature distributions, corresponding to present physical model. The modeling procedure of the forming HTSC microstructure during the sample heating has been written in detail in Sect. 6.4, and the initial parameters, used in computer experiments, are presented in Table 10.1. The finite-difference method is used to solve the first main problem for quasi-linear equation of thermal conduction (see Appendix C.1) for zero initial and corresponding boundary conditions. In this case, the region, considered in *ab*-plane, is presented by a 2D lattice with 1000 square cells of characteristic size, δ. Every cell is either a grain nucleus of 123 phase or void. The cells with the same numbers form corresponding grain or void (see Fig. 10.2).

The secondary re-crystallization is modeled on the basis of the Wagner–Zlyosov–Hillert's model of abnormal grain growth [1]. In the YBCO ceramic, it is caused by admixture phases at intergranular boundaries. As followed from [120], the material of intergranular buffer layers is $BaCuO_2$-CuO system with inclusions of admixtures. In any time t, the abnormal grain growth occurs under condition [1]:

Table 10.1 Initial parameters used in computer simulation

Parameter	Value	Unit of measurement
Particle properties		
Mean diameter (d)	1.5×10^{-6}	m
Thermal conductivity (λ_1)	18	W/(m K)
Blackness order (ε)	1	–
Density, (ρ)	3200	kg/m^3
Heat capacity (c_V)	0.8	kJ/(kg K)
Air properties		
Atmospheric pressure (H)	1010	GPa
Thermal conductivity (λ_g)	0.0241	W/(m K)
Adiabatic index (γ)	1.4	–
Prandtl's criterion (Pr)	0.7	–
Accommodation factor (a)	0.97	–
Length of molecular free run (Λ_c)	0.65×10^{-7}	m

$$\left|1/R_c - 1/R_j\right| > I_R/2, \tag{10.1}$$

where

$$R_c = \sum_{j=1}^{n_c} f_j R_j^2 / \sum_{j=1}^{n_c} f_j R_j;\ f_j = n_j/N_T;\ \sum_{j=1}^{n_c} f_j = 1;\ \sum_{j=1}^{n_c} n_j = N_T; \tag{10.2}$$

n_c is the size-class number of the width ΔR, containing n_j grains with radius R_j in j-class of the discrete space (R, t); $I_R = 6f_V/(\pi r)$ is the inhibition; f_V is the volumetric fraction of secondary phase particles; r is the mean radius of particles. The inhibition, I_R, depending on the particle radius and on the fraction of secondary phase, governs abnormal grain growth that allows one to define the microstructure and strength parameters in dependence on the secondary phase characteristics.

Assume that mass-transfer from one grain to another occurs according to the mechanism of normal crystal growth on non-singular surfaces (i.e., the boundaries with sufficiently high concentration of steps and demonstrating maximum free surface energy) [1642]. The breaks covering these facets, causes a possibility of new particles addition practically in any site of the surface. Number of breaks does not constraint the re-crystallization rate, and the facets grow self-perpendicular.

Numerical algorithm for modeling of the abnormal grain growth includes the following procedures:

(1) the definition of all nearest neighbors for each grain;
(2) the statement of the nearest neighbors pair (i, j) with $\max\limits_{1 \le i,j \le n_c} \left|1/R_i - 1/R_j\right|$;
(3) the growth of larger from grains (i, j) at the expense of other;
(4) the verification of the conditions: $\left|1/R_c - 1/R_j\right| \le I_R/2; j = 1, 2, \ldots, n_c$ and $1/R_{min} - 1/R_{max} \le I_R$;

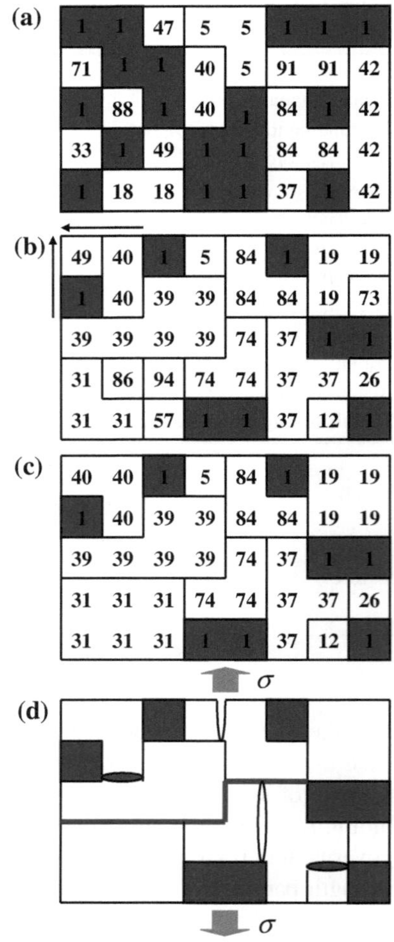

Fig. 10.2 Scheme of 2D fragment of YBCO ceramic structure (*ab*-plane) in PC: (**a**) crystallization (porosity is denoted by 1 and shown by *gray color*); (**b**) shrinkage (*arrows* show its directions); (**c**) abnormal grain growth and (**d**) macrocrack propagation (which is depicted by *gray line*) into model structure (where *white color* shows microcracks formed during the sample cooling, microcracks into process zone have *gray color*). *Numbers* denote a sequence of nucleation of present grains during sintering

(5) the end of grain growth in the case of fulfillment, at least of one of the conditions (4) or corresponding change of the parameters: n_j, n_c, N_T, f_j, R_j, R_c with the following repetition of the steps (1)–(5).

An example of model microstructure fragment before and after abnormal grain growth is presented in Fig. 10.2b, c.

10.1.2 Ceramic Cracking During Cooling

The residual stresses form around 211 inclusions into large-grain 123 matrix during cooling due to thermal mismatch in their behavior. Big internal stresses can be created at the tetragonal-orthorhombic phase transition in the large-grain matrix

Table 10.2 Some properties of 123/211 composite

Parameter	$YBa_2Cu_3O_7$ (*c*-axis)	$YBa_2Cu_3O_7$ (*ab*-plane)	Y_2BaCuO_5	References
E (GPa)	143	182	213	[1463]
ν	0.255	0.255	0.25[a]	[565]
α	3.2×10^{-5}[b]	0.86×10^{-5}[b]	1.24×10^{-5}	[998, 1227]
K_{Ic} ($MPa{\cdot}m^{1/2}$)	0.8	0.32	–	[401]

[a] The typical value for oxides
[b] This includes thermal expansion and oxygen increase

(the size of single 123 crystallites can attain some millimeter [703]) that leads to fracture of 123 matrix and to formation of defects. In this case, a density of damage and of defect structures is the function of annealing time at transition from tetragonal to orthorhombic phase.

Thus, the peculiarities of the YBCO ceramic microcracking are connected with the tetragonal-orthorhombic phase transition, causing a spontaneous deformation due to the phase mismatch, and also the thermal expansion anisotropy (TEA) of grains. Due to macroscopic heterogeneity of 211 particles in 123/211 composite, the thermal expansion factors of 123 (α_{123}) and 211 (α_{211}) phases differ significantly (see Table 10.2). As a result, the 123/211 composite has different average values of thermal expansion factors in various parts of the sample:

$$\alpha_{123/211} = \alpha_{123} V_{123} + \alpha_{211} V_{211}. \tag{10.3}$$

These heterogeneities lead to complicated picture of residual stresses after sample cooling from sintering temperature. Microcracks form in 123/211 polycrystalline structure under the influence of internal stresses between 123 grains during sintering due to anisotropy in their thermal expansion [562]. The 123 grain has a larger compression along the c-axis direction compared to the a- and b-axes during cooling (owing to both thermal expansion and oxygen content increase). This creates residual stresses, leading to tension of the sample along the c-axis.

In contrast to ferroelectric ceramics, where deformation phase mismatch is the main cause of microcracking [1296], we consider both effects in the proposed model of HTSC ceramics. The tetragonal-orthorhombic phase transition (occurred for $YBa_2Cu_3O_{7-x}$ at $T \sim$ 600–700 °C and accompanied by oxygen diffusion [879, 1254]) leads to alignment of twinning platelet grains in *ab*-plane, which is perpendicular to c-axis. However, twinning is practically absent, leading to alignment along the c-axis. The shrinkage along this direction is not compensated by twinning, causing microcracking parallel to (001) [56]. Moreover, a change of elastic stiffness components C_{22}, C_{23}, C_{55} at phase transformation leads to spontaneous deformation of crystalline lattice [573].

Due to this, may be estimated effect of TEA by cooling and spontaneous deformation, caused by the phase transition on the formation of intergranular microcracks. The critical length of cracked boundary, l_{sc}, can be estimated as [498]

$$l_c^s = \beta_0 [K_c^b (1 + \nu)/(E\varepsilon)]^2. \tag{10.4}$$

Here, K_c^b is the fracture toughness of grain boundary; $\beta_0 \approx 3.5$ is the test constant; E is Young's modulus; ν is Poisson's ratio; ε is the deformation of intergranular boundary due to TEA or deformation phase mismatch. In the case of the TEA effect, $\varepsilon = \Delta\alpha\Delta T$, where $\Delta\alpha$ is the thermal expansion factors difference, defined by extreme values of α; ΔT is the temperature difference at cooling. In the case of deformation phase mismatch, ε is the spontaneous deformation caused by change of the elastic stiffness components.

Selecting $\Delta\alpha = 5 \times 10^{-6}\ \mathrm{K}^{-1}$ [498] and $\Delta T = 625$ K (in order to exclude the tetragonal-orthorhombic phase transition) [390], we have for the case of TEA $\varepsilon \sim 3.1 \times 10^{-3}$. Comparing with deformation caused by the phase transition ($\varepsilon \sim 10^{-4}$ [574]), it may be concluded that the main factor defining a micro-cracking of superconducting ceramic is the TEA. Therefore, we consider below only this case.

The criterion of microcrack formation at a boundary with length, l, is selected, taking into account a grain misorientation [499]:

$$l/l_c^s \geq 2/[1 + \cos(2\Theta_1 - 2\Theta_2)], \tag{10.5}$$

where Θ_i ($i = 1, 2$) is the angle between the axis of maximum compression in i-grain and the grain boundary plane. In order to estimate these angles, we use the Monte-Carlo procedure [336]. Consider that the microcracks propagate along the grain boundary after their nucleation at triple junctions. The microcracks are arrested at the neighboring junctions because the adjacent boundary facets usually subjected to internal compression [499]. The computer algorithm of intergranular microcracking during cooling includes the following steps:

(1) the definition of all triple points in model structure;
(2) the definition of sizes of all intergranular boundaries, crossing the triple points;
(3) the re-consideration of the boundaries and their replacement by microcracks in the case of fulfillment of the microcracking criterion (10.5) for corresponding boundary.

10.1.3 Formation of Microcracks around 211 Particles in 123 Matrix

Thermal Stresses on 123/211 Interface

During cooling from the sintering temperature, inside 211 particles and around them in 123 matrix, the thermal stresses form as a result of the thermal expansion factors difference between 123 and 211 phases. Consider the spherical coordinate system $\{r, \Theta, \Phi\}$ with origin coinciding with the 211 particle center. Then, depending on hydrostatic stress P_0 applied to the particle of radius, R_{211}, the radial $\sigma_{123R}(r)$ and tangential stresses $\sigma_{123\Theta}(r) = \sigma_{123\Phi}(r)$ in the matrix (for $R_{211} < r < \infty$) can be defined as [1535]

Table 10.3 Stresses obtained on 123/211 interface

Stress	σ_{123R} (GPa)	$\sigma_{123\Theta}$ (GPa)	$\sigma_{123\Phi}$ (GPa)
c-axis	−2.0	1.0	–
ab-plane	0.43	−0.215	−0.215

$$-2\sigma_{123\Theta}(r) = -2\sigma_{123\Phi}(r) = \sigma_{123R}(r) = P_0(R_{211}/r)^3; \quad (10.6)$$

$$P_0 = \frac{(\alpha_{211} - \alpha_{123})\Delta T}{(1 + \nu_{123})/2E_{123} + 2(1 - 2\nu_{211})/E_{211}}. \quad (10.7)$$

For convenience, it is proposed that $\alpha_{211} > \alpha_{123}$; $\Delta T = T_s - T_0$ is the temperature difference at cooling (where T_s and T_0 are the sintering temperature and room temperature, respectively). The stresses on 123/211 interface are independent of the particle size R_{211}. They decrease according to the law of $(R_{211}/r)^3$ and attain 12.5 % from maximum value for $r = R_{211}$. Selecting necessary values from Table 10.2 and taking into account the magnitude of $T_s = 925$ °C, we obtain stresses on the 123/211 interface, presented in Table 10.3.

The obtained tangential stress in the *c*-direction (acting on equator of the 211 particle) is sufficiently great and capable to form *ab*-microcracks around the 211 particles. However, because the 211 particles are under compression, *ab*-microcracks do not propagate through them, and the 123/211 interface does not fail unless the size difference between two neighboring particles is too large. Because the radial stress in *ab*-plane is tensile, but both tangential stresses are compressive, then taking into account higher fracture toughness in the planes, which are parallel to the *c*-axis, it may be expected that microcracks, introduced by the 211 particles will not be parallel to the *c*-direction. General picture of change of the value and sign for stresses σ_{123R} and $\sigma_{123\Theta}$ in the 123 matrix on the 123/211 interface is presented schematically in Fig. 10.3. The complicated picture of the stress state in the 123/211 composite leads to alternative opinions about acting stresses and 123 lattice distortions near 211 particles. In particular, some authors neglect stress state

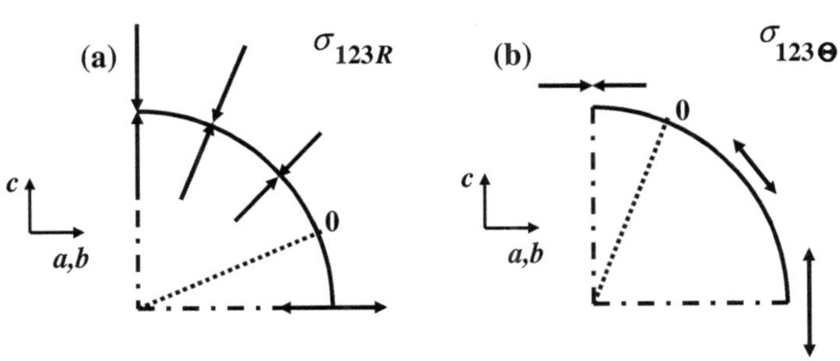

Fig. 10.3 Schematic representations of: (**a**) radial and (**b**) tangential stresses in 123 matrix at 123/211 interface. Sizes and directions of *arrows* demonstrate values and also action regions of tensile and compressing stresses

Fig. 10.4 Microcrack pattern around 211 particles. The microcracks in *ab*-plane avoid 211 particle poles and deflect in radial direction [364]

around the 211 particles [1216]. Other authors define perturbations of the 123 matrix far from the 211 particles or higher density of crystalline defects near some 123/211 interfaces [70]. The considerable indirect experimental supporting of thermal stresses, existing around 211 particles, is provided with the observed processes of de-twinning or predomination of one twin variant along [100] directions in the case of movement of the twin boundaries due to thermal stresses, caused by 211 (422) particles [367, 368]. The stresses required for the motion of twin boundaries and estimated by the value of 50 MPa [367] are well in agreement with obtained magnitudes in *ab*-plane. The presented analysis of stress state is also supported by observed distribution of microcracks around the 211 particles (see Fig. 10.4). The high radial compressive stress on poles of the 211 particle inhibits microcrack growth into *ab*-plane near the poles. Therefore, *ab*-microcracks avoid the 211 particle poles and return to propagation along radial direction in existence of tangential tensile stresses, according to Figs. 10.3 and 10.4.

Microcracking Criterion in ab-Plane

Formation of the *ab*-microcracks in the case when the particle demonstrates a smaller thermal expansion compared to matrix demands existence of a microcrack nucleus on the 123/211 interface and supply by energy, which is necessary for the defect growth. This is provided by elastic energies of the particle and surrounding matrix. In isotropic case, total stored energy per unit volume can be defined as [337]

$$U_T = P_0^2 \pi R_{211}^3 \left[\frac{1 + \nu_{123}}{E_{123}} + \frac{2(1 - 2\nu_{211})}{E_{211}} \right]. \qquad (10.8)$$

In this case, only part of the stored energy can be used to form *ab*-microcracks. Using the 123 constants for *c*-direction, it may be assumed that the energy approximately equals third part of the total energy, U_{Tc}. Then, the necessary, but insufficient condition of microcrack formation is that the energy of creation of new surface, $U_s = \gamma_s A$ (where γ_s is the effective surface energy of matrix and A is the square of the newly formed surface) must not exceed $U_{Tc}/3$, that is

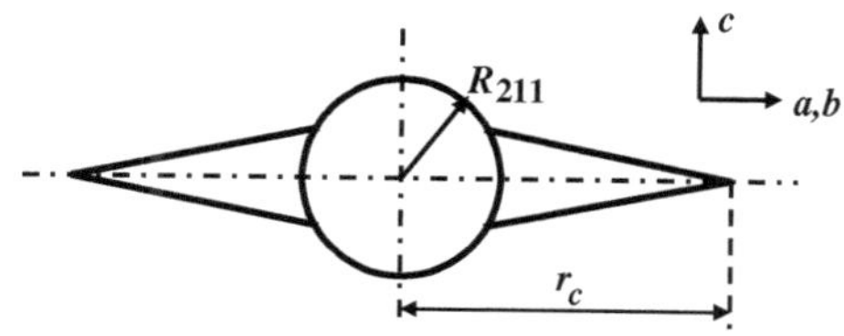

Fig. 10.5 Schematic representation of *ab*-microcrack formation due to release of elastic energy introduced by 211 particle

$$U_{Tc}/3 \geq U_s. \tag{10.9}$$

Taking into account $\gamma_s^{ab} = (K_{Ic}^{ab})^2/E^c$ (the upper indexes here and below point the *a*, *b* and *c* directions) and selecting the corresponding data from Table 10.2, we obtain $\gamma_s^{ab} = 7.16 \times 10^{-7}$ MPa·m. The square *A* is found from Fig. 10.5 as $A = 2\pi(r_c^2 - R_{211}^2)$. Considering the usual distance between 211 particle, which is equal to $4R_{211}$ (for 20 mol% of the 211 particles) [364], it is necessary to equate the value of r_c to $2R_{211}$ in order to create the *ab*-microcrack, connecting two adjacent 211 particles. Then we have

$$U_S = 2\pi[(2R_{211})^2 - R_{211}^2]\gamma_s = 6\pi R_{211}^2\gamma_s. \tag{10.10}$$

The existing energy is proportional to R_{211}^3; at the same time the energy absorbed during cracking is proportional to R_{211}^2. Combining (10.8)–(10.10), we define the critical size of the particle, R_{c211}, as

$$R_{c211} \geq \frac{18\gamma_s^{ab}}{(P_0^c)^2\left[(1+\nu_{123})/E_{123}^c + 2(1-2\nu_{211})/E_{211}\right]}. \tag{10.11}$$

Substituting the obtained values of P_0^c and γ_s^{ab} and also elastic constants from Table 10.2 for the 123/211 composite, we define $R_{c211} = 0.24$ μm. Another estimation of the critical particle size for microcracking in *ab*-plane (giving the value < 1 μm [367]) agrees well with the value (R_{c211}) obtained in the present analysis. The strong proof that the *ab*-plane defects are not growth-related defects is that they were not observed in the tetragonal 123 phase [370, 856] as well as in the tetragonal Nd-123 system with small 422 particles (see Fig. 10.6) [369].

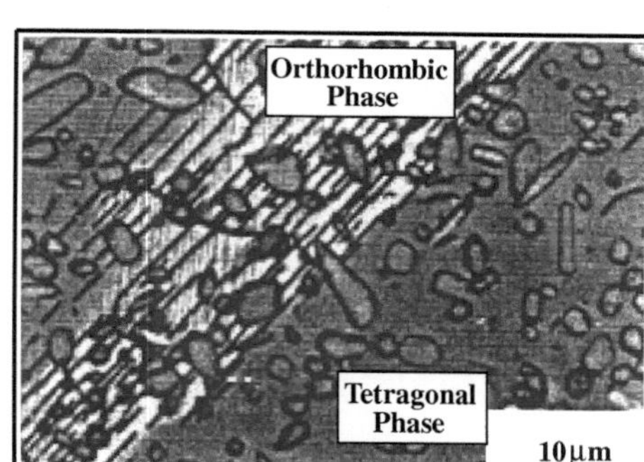

Fig. 10.6 Dense *ab*-microcracks in the orthorhombic Nd-123 phase and no *ab*-microcracking in the tetragonal region. The case of small 422 particles [369]

10.1.4 Fracture Features at External Loading

The sufficiently high thermo-mechanical and electromagnetic loading can lead to fracture of material. In order to estimate critical fracture parameters, we use a model of crack growth in anisotropic body [944]. Consider a polycrystalline aggregate of grains with hexagonal shape, containing an annular flaw (Fig. 10.7). The probabilistic condition of misorientation of the adjacent grains leads to initiation of the greatest compression in the central grain, while the surrounding grains will be subject to tension. It is assumed that the central grain is forced into a cavity of diameter, $D = 2R$, surrounded by the annular flaw with length of S. In this case, we shall take into account effects of elastic stress concentration from external tension, σ, and thermal strains on the grain boundaries, $\varepsilon = \Delta\alpha\Delta T$. Then, the summary stress intensity factor (SIF) due to concentration of the thermal stresses and applied external loading is found by

$$K_{tot} = K_t^r + K_t^\theta + K_a. \tag{10.12}$$

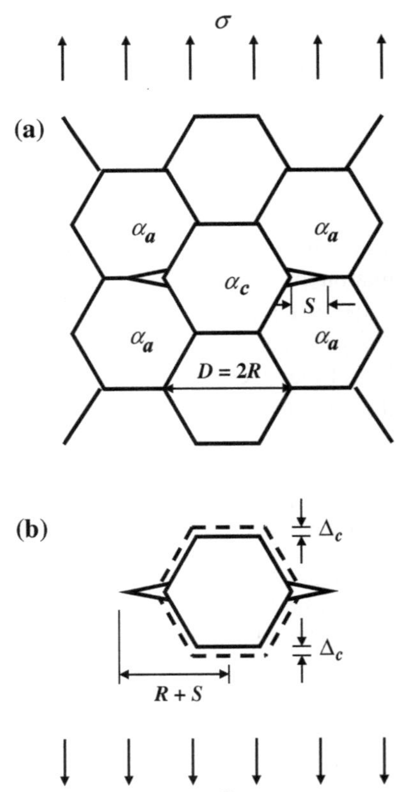

Fig. 10.7 Polycrystalline aggregate, containing annular flaw: (**a**) hexagonal grain under stress; (**b**) crack opening due to TEA (α_a and α_c are the thermal expansion factors along a- and c-axes, respectively)

The first term is determined by the radial component of the thermal stress concentration from the external side of spherical boundary [944, 1319] as

$$K_t^r = \frac{2\sigma_t R^{1/2}}{\pi^{1/2}}(1+S/R)^{1/2}\left\{1-\left[1-\frac{1}{(1+S/R)^2}\right]^{1/2}\right\}, \tag{10.13}$$

and the second term is found by the tangential component [944, 1319] as

$$K_t^\theta = \frac{2\sigma_t R^{1/2}}{\pi^{1/2}}\left[1-\frac{1}{(1+S/R)^2}\right]^{1/2}\frac{1}{(1+S/R)^{3/2}}. \tag{10.14}$$

Here, $\sigma_t = \frac{2E_{0m}\Delta\alpha\Delta T}{3(1-\nu_{0m})}$ is the thermal stress; E_{0m}, ν_{0m} are Young's modulus and Poisson's ratio for porous cracked material, respectively. Chapter 9 has presented common theories for calculation of effective elastic properties of the superconductive composites. From methodical viewpoint here and further in the book we shall be restricted by more simplified models. In order to define elastic constants of porous superconductor, we use the modified cubic model [694]

$$E_{0m} = E_m(1-P_V^{2/3});\ \nu_{0m} = \nu_m(1-P_V^{2/3}), \tag{10.15}$$

where P_V is the volumetric fraction of closed porosity; E_m, ν_m are the elastic constants of cracked no-porous ceramic. The elastic constants E_m, ν_m can be calculated from the following relationships [1443]:

$$\frac{E_m}{E} = 1\bigg/\left[1+\frac{16(1-\nu^2)(10-3\nu)}{45(2-\nu)}\beta_m\right]; \tag{10.16}$$

$$\frac{\nu_m}{\nu} = \frac{1+[(16/45)(1-\nu^2)/(2-\nu)]\beta_m}{1+[(16/45)(1-\nu^2)(10-3\nu)/(2-\nu)]\beta_m}, \tag{10.17}$$

taking into account arbitrary-oriented microcracks. Here E, ν are the elastic moduli of defectless material; $\beta_m = N_m a_m^3/V_m$ is the microcrack density in the considered volume, V_m; N_m and a_m are the number and characteristic size of microcracks, respectively.

The SIF, caused by applied stress, K_a, is defined as [944, 1319]

$$K_a = \frac{2\sigma R^{1/2}}{\pi^{1/2}}(1+S/R)^{1/2}\Phi_e, \tag{10.18}$$

where

$$\Phi_e = \left[1-\frac{1}{(1+S/R)^2}\right]^{1/2}\left\{1+\frac{(4-5\nu_{0m})}{2(7-5\nu_{0m})}\frac{1}{(1+S/R)^2}+\frac{9}{2(7-5\nu_{0m})}\left[1+\frac{(1+S/R)^2-1}{3}\right]\frac{1}{(1+S/R)^4}\right\}. \tag{10.19}$$

We define strength of anisotropic body from the critical conditions of a crack initiation: $K_{tot} = K_c^b$ (K_c^b is the fracture toughness of grain boundary) and $\sigma = \sigma_f$ as

$$\sigma_f = \frac{K_c^b}{\Phi_e}\left[\frac{\pi}{2D(1+S/R)}\right]^{1/2} - \sigma_t\frac{\Phi_t}{\Phi_e}, \tag{10.20}$$

where σ_f is the critical stress;

$$\Phi_t = 1 - \left[1 - \frac{1}{(1+S/R)^2}\right]^{1/2}\left[1 - \frac{1}{2(1+S/R)^2}\right]. \tag{10.21}$$

Finally, we write a condition for critical displacement at crack opening (see Fig. 10.7b): $\Delta_t = \Delta_c$ (where Δ_t is the linear thermal expansion of the central grain) as [944]

$$D\Delta\alpha\Delta T = \frac{2(1-v_{0m}^2)K_c^b}{E_{0m}}\left(\frac{R+S}{\pi}\right)^{1/2}. \tag{10.22}$$

Then, the critical grain size, D_c, defining further crack growth, can be obtained, suggesting $D = D_c$ and defining a positive root of quadratic Eq. (10.22) as [1319]

$$D_c = b + (b^2 + 4Sb)^{1/2}; \quad b = \frac{(1-v_{0m}^2)^2(K_c^b)^2}{\pi E_{0m}^2(\Delta\alpha\Delta T)^2}. \tag{10.23}$$

Then, we select average grain of model structure as grain size, R, and define value of S (the length of nucleus microcrack) through mean microcrack length, a_m, formed during the ceramic cooling. Here and below, an absence of the microcrack interaction is assumed.

At microcrack development in brittle materials, there are processes caused by the material structure properties, which lead to the material toughening. The effects of microcracking, crack branching and bridging and so on render significant, but no simple effect on change of strength properties and fracture resistance. The considerable correction of effects of different toughening mechanisms may occur, taking into account their joint action. A domination of one from the mechanisms is determined by microcrack distribution, and their density causes the toughening value.

10.1.5 Crack Tip Shielding Introduced its Geometry

Crack front interaction with secondary-phase particles or grains into matrix causes its deflections from a straight growth direction leading to tortuosity of the crack. In common case, the tortuosity introduces a local mixed-mode I + II + III at the crack front in any case of applied remote loading. In order to describe the crack stability under the mixed-mode loading various criteria into framework of linear-elastic fracture mechanics have been proposed [46]. The rough crack faces behind the tortuous crack front in ceramics do not experience any significant sliding contact. Due to the crack-wake friction is responsible for somewhat higher fracture toughness obtained at the remote sliding modes II and III compared with the value for the mode I [1071], then we can assume that $K_{Ic} \approx K_{IIc} \approx K_{IIIc}$ along tortuous crack front of remote mode I loading. The simplest stability criterion in energetic form

$$G_{eff} = G_I + G_{II} + G_{III} \tag{10.24}$$

can be applied, where G_{eff} is the effective crack driving energy. The equivalent relation in force approach through SIF has the form:

$$K_{eff} = \sqrt{K_I^2 + K_{II}^2 + \frac{1}{1-\nu} K_{III}^2}. \tag{10.25}$$

In the case of a long straight crack with single kinked tip, we have

$$K_{eff} = \cos^2(\varphi/2) K_I, \tag{10.26}$$

where φ is the kink angle. Obviously, $K_{eff} < K_I$ for $\varphi \neq 0$. This inequality holds for any spatially complex crack front. The geometrically induced shielding (GIS) belongs to so-called extrinsic shielding mechanisms [1439]. Generally, the fracture resistance can be divided into two categories of intrinsic and extrinsic toughening. The first mechanism represents the inherent matrix resistance caused by the atomic bond strength or macroscopic strength and plasticity. The modifications of chemical composition and/or thermal treatment are capable to improve the intrinsic fracture toughness. At the same time, the processes of kinking, branching or bridging of the crack front introduced by microstructure features are typical extrinsic mechanisms. They reduce driving crack force and increase intrinsic crack resistance. So, the common fracture toughness can be expressed as a sum of the intrinsic and extrinsic values of toughness:

$$K_{Ic} = K_{Ici} + \sum K_{Ice}. \tag{10.27}$$

The standard procedure for calculation of K_{Ic} usually use the sample with straight planar crack no taking into account extrinsic shielding effects caused by micro-heterogeneities. The general expressions for GIS in the cases of brittle and quasi-brittle fracture could be obtained on the base of Faber-Evans approach [458]. Following [1377], for brittle fracture we have

$$K_{Ici} = \left(\frac{\bar{g}_{eff,r}}{R_A}\right)^{1/2} K_{Ic}, \tag{10.28}$$

where K_{Ic} and K_{Ici} are the measured and intrinsic values of fracture toughness, $\bar{g}_{eff,r}^{1/2} = \bar{k}_{eff,r}$ is the mean effective k-factor for the tortuous crack front normalized

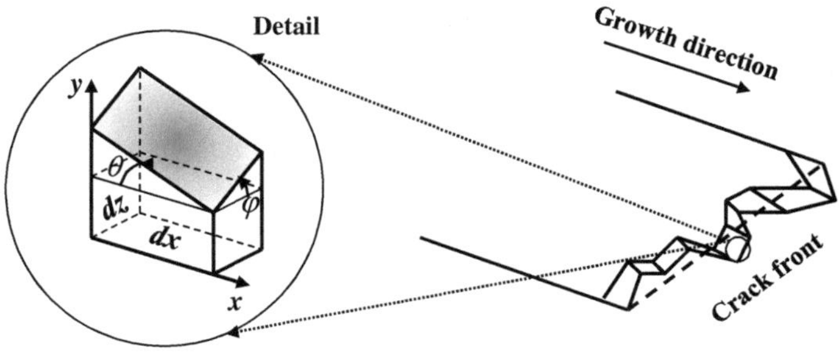

Fig. 10.8 Scheme of the tortuous crack front and its segment [1378]

to the remote K_I ($k_{eff,r} = k_{eff}/K_I$) and R_A is the roughness region of the fracture surface.

In order to explain the expression (10.28), we consider a cracked sample with thickness B and intrinsic resistance G_{Ici} under remote mode I loading (see Fig. 10.8). In the case of no GIS the measured fracture toughness G_{Ic} (or K_{Ic}) is equal to its intrinsic value, i. e. $G_{Ic} = G_{Ici}$ (or $K_{Ic} = K_{Ici}$).

When the crack front is microscopically tortuous, a variable local mixed mode I + II + III characterized by g_{eff} (or k_{eff}) is defined at each point along the crack front. At the external loading under increasing remote loading G_I, the proportionality $g_{eff} \sim G_I$ (or $k_{eff} \sim K_I$) must be valid. So, the ratio $g_{eff,r} = g_{eff}/G_I$ can be introduced as independent on G_I by dependent on the crack front tortuosity. By designing through G_{uI} the crack driving force at the moment of an unstable elementary propagation dx of the crack front and taking into account that the value is equal to the conventionally measured fracture toughness G_{Ic}, we obtain the elementary energy release rate caused by the creation of a new crack surface square $dxdz$ is equal to $G_{uI}dxdz$. At the same time, local elementary energy release rate at the crack front tortuosity is

$$g_{eff}dxdz = g_{eff,r}G_{uI}dxdz. \tag{10.29}$$

Then, the total energy is necessary to create a new surface region Bdx along the crack front can be written as

$$dW = G_{uI}dx \int_0^B g_{eff,r}dz. \tag{10.30}$$

Figure 10.8 shows that the actual new elementary surface region $dS = R_A Bdx > Bdx$, since

$$R_A = \frac{1}{B} \int_0^B \frac{dz}{\cos\varphi(z)\cos\theta(z)}. \tag{10.31}$$

In Eq. (10.31) R_A is the roughness of the fracture surface and $dxdz/(\cos\varphi\cos\theta)$ is the square of hatched rectangle in Fig. 10.8. Due to G_{uI} is the intrinsic resistance to crack growth, the total fracture energy is defined as

$$dW = G_{uI} dS = G_{Ici} R_A B dx. \tag{10.32}$$

From Eqs. (10.30) and (10.32) one obtains

$$G_{uI} \equiv G_{Ic} = \frac{R_A}{\bar{g}_{eff,r}} G_{Ici}, \tag{10.33}$$

where

$$\bar{g}_{eff,r} = \frac{1}{B} \int_0^B g_{eff,r} dz. \tag{10.34}$$

Usually, $G_{Ic} \geq G_{Ici}$, because $\bar{g}_{eff,r} \leq 1$ and $R_A \geq 1$. Due to $G_{Ic}/G_{Ici} = (K_{Ic}/K_{Ici})^2$, Eq. (10.34) reduces to Eq. (10.28).

Values of $\bar{g}_{eff,r}$ and R_A must be estimated numerically or analytically. One of these models is the pyramidal model of tortuous crack front [1377], which states that GIS effect could be effectively estimated by realizing the next three steps:

(i) building a realistic model of the crack front on the base of 3D determination of fracture surface roughness;
(ii) calculation of local normalized SIFs k_{1r}, k_{2r} and k_{3r} along the crack front;
(iii) definition of the effective SIF $k_{eff,r}$.

The first step could be realized due to 3D reconstruction of fracture morphology. The second problem may be solved, for instance by using boundary element method [754]. The third step could be attained by using analytical estimations based on the pyramidal model of crack front [1378, 1379]. This model is based on a pyramid-like approximation of the tortuous crack front, whose each element is defined by the twist angle Φ and the highest tilt angle Θ_m towards the macroscopic crack plane (see Fig. 10.9). The profile roughness R_L (measured along the crack front) and the periodicity λ_{pl} (λ_{pp}) measured parallel (perpendicular) to the crack front are associated with the angles Φ and Θ_m with help of the following expressions:

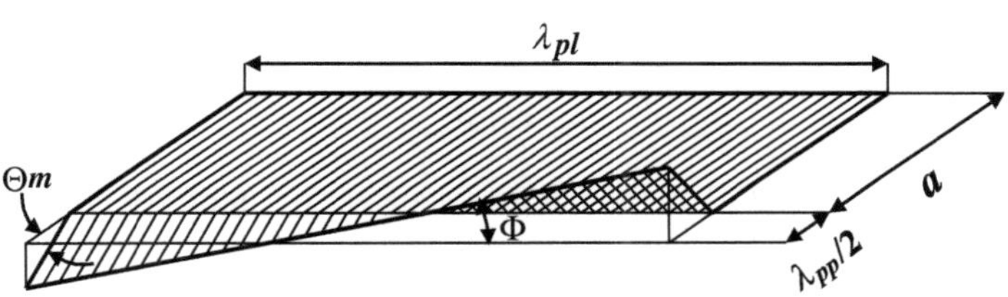

Fig. 10.9 A periodic element of the pyramidal model of tortuous crack front

$$\lambda_{pp} \tan \Theta_m = \lambda_{pl} \tan \Phi; \quad R_L = \cos^{-1} \Phi. \tag{10.35}$$

The characteristic periodicities λ_{pl} and λ_{pp} could be defined either by the Fourier analysis of roughness morphology measured at corresponding sites of the fracture surface or identified by using relative microstructure periodicity (for instance, average grain size). Then the effective SIF $k_{eff,r}$ (normalized to the remote value of K_I) at each point of pyramidal front can be calculated by using Eq. (10.25) with the next approximate analytical expressions for local SIFs:

$$\begin{aligned} k_{1r} &= \cos(\Theta/2)[2v \sin^2 \Phi + \cos^2(\Theta/2) \cos^2 \Phi]; \\ k_{2r} &= \sin(\Theta/2) \cos^2(\Theta/2); \\ k_{3r} &= \cos(\Theta/2) \sin \Phi \cos \Phi [2v - \cos^2(\Theta/2)]. \end{aligned} \tag{10.36}$$

The results based on expressions (10.36) are sufficiently accurate at the $\lambda_{pp} \ll 2a$, where a is the pre-crack size. Then, the effective factor $\bar{k}_{eff,r}$, averaged on the periodic crack front geometry, combined of the same pyramidal elements, is calculated as [1377]

$$\bar{k}_{eff,r} = \frac{\pi - 2}{2\Theta_m(2R_L + \pi - 4)} \int\limits_{-\Theta_m}^{\Theta_m} \left(k_{1r}^2 + k_{2r}^2 + \frac{k_{3r}^2}{1 - v} \right) d\Theta. \tag{10.37}$$

10.1.6 Microcracking Process Zone Near Macrocrack

In the vicinity of growing macrocrack, a microcracking process zone initiates, which leads to the macrocrack shielding and change of the ceramic fracture toughness [454, 498, 928]. Both phenomena are caused by microcrack distribution. The compliance of the process zone increases the fracture toughness; however, the microcracks, directly neighboring the macrocrack tip, decrease fracture resistance and toughness of material [454, 498]. The critical number of microcracks per volume unit, N_c^*, defining initiation of coalescence in the vicinity of the crack, depends on proper microcrack size, a_m [1763]:

$$N_c^* = 9/(64a_m^3). \tag{10.38}$$

Define the width of the process zone, $2h_m$, as [207]

$$h_m = a_m I_m^2, \tag{10.39}$$

where $I_m^2 = \beta_m E_m/(E - E_m)$ is the parameter of elastic interaction of the microcracks; $\beta_m = N_m a_m^3/V_m$ is the microcracking density in the considered sample volume, V_m; N_m is the number of microcracks [1443].

The finite-element analysis [928] shows that the stress state of macrocrack to a greater degree promotes growth of microcracks, which are parallel to the macrocrack propagation, and to a lesser degree promotes microcracks of any orientation ahead in the direction of the macrocrack propagation. The stress state of macrocrack also impedes microcracks, which are perpendicular to the macrocrack and disposed at the side from direction of its growth. Moreover,the process zone size in the crack plane is approximately two times smaller than in perpendicular direction. These features are taken into account in computer simulation of microcracking into process zone and in definition of its extent. The simulation of macrocrack growth along intergranular boundaries (see also a preliminary discussion in Sect. 6.7) is carried out on the basis of the graph theory [823, 824], using the Viterbi's algorithm [147]. During transition from one sequence to other, the process zone width from (10.39) and all triple points in the zone inside are estimated. Then, we model microcracks on proper boundaries with length of l, using the critical facet size: $l_c = 0.4 l_c^S$ [498]. The procedure of the microcrack modeling repeats the algorithm used in Sect. 10.1.2 at the stage of spontaneous cracking. In this case, the following are taken into account: the microcracks do not cross the macrocrack and the condition, defining an absence of the microcrack coalescence: $N_m / V_m < N_c^*$.

Obviously, the microcracking density, β_m, increases together with the grain size that guaranties increasing of the process zone size, h_m.[1] As has been shown by calculation of I_m^2 in dependence on the Young's modulus ratio E_m/E [209] at increasing density of microcracking, the function I_m^2 attains a maximum value, corresponding to the critical density of microcracks, $\beta_m = \beta_m^c$. This value divides the zones of the material toughening and the crack amplification.

The fracture toughness change due to the alternative trends, caused by the spontaneous cracking and by the process zone of microcracking (Fig. 10.10), can be estimated, using Cherepanov–Rice's J-integral [1430]. We obtain in the case of stationary macrocrack [498]

$$\frac{K_{Ic}^{\infty}}{K_{Ic}^{0}} = (1-\chi_m)\left\{\frac{1-[(16/45)(1-\nu^2)(10-3\nu)/(2-\nu)]\chi_\infty}{1-[(16/45)(1-\nu^2)(10-3\nu)/(2-\nu)]\chi_m}\right\}^{1/2}, \qquad (10.40)$$

where K_{Ic}^{∞}, K_{Ic}^{0} are the ceramic fracture toughness at existence and absence of microcracks, respectively; χ_m, χ_∞ are the fraction of cracked boundaries in the process zone and in the sintered ceramic. The first multiplier in (10.40) defines a change of local fracture toughness, K_{Ic}^{l}/K_{Ic}^{0}, in the crack tip. In calculations for χ_m and χ_∞, the following formulas are used:

---

[1] However, in contrast to infinite growth of the process zone, stated in [928], even for grain sizes, which are smaller than the critical one, in reality the process zone must be finite. This contradiction is the consequence of the selected model in [928] for array of hexagonal grains, in which all grain boundaries have the same length, l, and on each of them the triple point (microcrack nucleus) exists. Therefore, for $l \to l_c^S$ the condition $h_m \to \infty$ is reached. Our model is free from the above shortcoming.

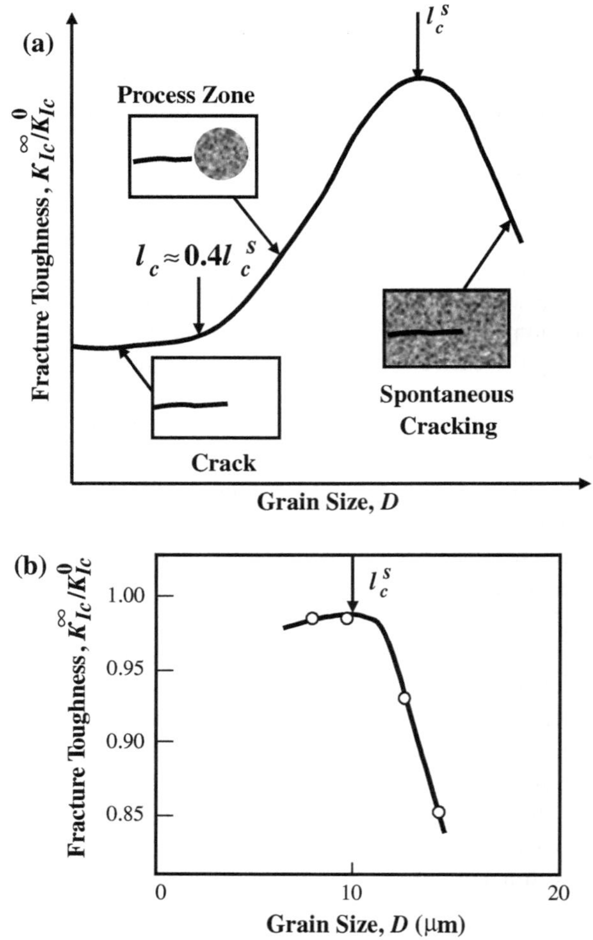

Fig. 10.10 (**a**) Fracture toughness of ceramic *versus* grain size [498] and (**b**) the simulation results obtained for $PbTiO_3$ ceramic

$$\chi_m = \frac{N_m S}{2h_m L N_g};\ \chi_\infty = \frac{N_\infty}{N_g}, \tag{10.41}$$

where N_∞, N_m is the microcrack number after spontaneous cracking and the boundary number in the sample area, S, respectively; L is the length of the macrocrack path.

Block-scheme of one realization of the computational process, describing secondary recrystallization (Sect. 10.1.1), cooling microcracking (Sect. 10.1.2), and estimation of fracture toughness change taking into account microcracking into process zone (Sect. 10.1.6) is presented in Fig. D.4. A simulation veracity of computational algorithm of the secondary recrystallization has been found by direct verification on test example. The errors of computer experiments at doubling

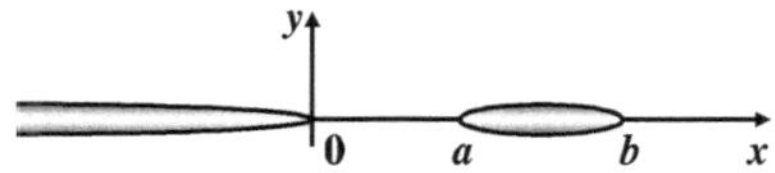

Fig. 10.11 Scheme of interaction of half-infinite crack with collinear microcrack

of number of the considered cells to be into range of 2 % in all variants calculation of the averaged parameters.

The macrocrack shielding caused by microcracking in process zone is estimated, using a model of crack growth under monotonous increasing loading. By assuming that microcracks are the array of isotropic oriented disk-like cracks, it may be obtained [994] that

$$\frac{K_I^0}{K_I^\infty} = \left[\frac{(1-\nu^2)E_m}{(1-\nu_m^2)E}\right]^{1/2}. \tag{10.42}$$

In this case, Young's modulus, E_m, and Poisson's ratio, ν_m, for cracked material may be estimated, using a model based on averaging of strains on micro-volumes, containing misoriented microcracks [1443] from (10.16) and (10.17).

The coalescence effect leading to decreasing of the crack growth resistance is estimated, using a coalescence model of half-infinite crack (having tip in the point $x = 0$) with collinear microcrack (having tips in the points: $x = a$, and $x = b$, $a < b$) that is shown in Fig. 10.11. The collinear microcrack is equivalent to an array of microcracks, distributed in the layer with thickness of $h = N^{-1/3}$, which is measured along normal direction to the fracture plane (N is the mean number of microcracks per volumetric unit). In this case, $a/b = 1 - A$, where $A = N^{2/3}\langle c^2\rangle$ is the fraction of microcracking square in the fracture plane, and $\langle c^2\rangle$ is the mean cracking square, projected on the fracture plane. Then, for the condition of quasi-static crack growth, we obtain [1448]:

$$K_c/K_c^0 = [aE_m/(bE)]^{1/2}C, \tag{10.43}$$

where K_c is the SIF, defining the crack coalescence; K_c^0 is the intrinsic fracture toughness of ceramic without microcracks; $C = K(k)/E(k)$; $K(k)$, $E(k)$ are the elliptic integral of 1 and 2 type, respectively; $k = (1 - a/b)^{1/2}$.

Numerical results, obtained for the ceramics without voids, possessing the perovskite structure (which HTSC also have): $BaTiO_3$ ($E = 120$ GPa, $\nu = 0.25$) and $PbTiO_3$ ($E = 80$ GPa, $\nu = 0.3$), depending on the inhibition parameter of abnormal grain growth (see Table 10.4), lead to the following conclusions.

The increasing of mean grain radius, $\overline{R}$, occurs with the decreasing of the inhibition parameter that enhances the fraction of spontaneous-cracked facets and also (that is not obvious) the diminution of microcracking fraction in the process zone. The values of h_m/l_c^S are shown in Table 10.4 for initial stage of the macrocrack growth when the process zone size is stated by structure parameters of

Table 10.4 Strength properties of $BaTiO_3$ and $PbTiO_3$ ferroelectric ceramics [1328]

Ceramic	$I_R\overline{R}_0$	$\overline{R}/\overline{R}_0$	χ_∞	χ_m	h_m/l_c^S	K_{Ic}^l/K_{Ic}^0	K_{Ic}^∞/K_{Ic}^0	K_I^0/K_I^∞
$BaTiO_3$	1.0	1.00	0.121	0.106	1.53	0.894	0.878	0.910
	0.8	1.10	0.161	0.062	1.57	0.938	0.835	0.944
	0.6	1.18	0.183	0.059	1.65	0.941	0.810	0.946
	0.4	1.33	0.187	0.052	1.70	0.948	0.806	0.952
$PbTiO_3$	1.0	1.00	0.121	0. 099	1.56	0.901	0.880	0.912
	0.8	1.10	0.161	0.056	1.59	0.944	0.844	0.947
	0.6	1.18	0.183	0.056	1.67	0.944	0.821	0.947
	0.4	1.33	0.187	0.049	1.73	0.951	0.817	0.951

sintered ceramic. Taking into account the critical grain size for Al_2O_3 ceramic ($D_c^S = 400$ μm [928]), which has strength properties related with the $BaTiO_3$ ferroelectric ceramic, we obtain the process zone size, $h_m = 306$–340 μm that coincides with analogous test value ($h_m \approx 300$ μm), fixed for aluminum-oxide ceramic [208].

The numerical results for different grain sizes of $PbTiO_3$ ferroelectric ceramic (see Fig. 10.10b) [1296] support qualitative trends of characteristic behavior of the fracture toughness depending on microcracking, pointed in Fig. 10.10a [454]. The spontaneous cracking during cooling is the factor determining the fracture toughness change, K_{Ic}^∞/K_{Ic}^0, for grain sizes which are greater than the critical size. Numerical results for ferroelectric ceramics show that in account of the frontal zone of microcracking, the value of the fracture toughness is equal to 0.80–0.88, which is near to 0.9, obtained for Al_2O_3 [454]. The local toughening, K_{Ic}^l/K_{Ic}^0, to the greatest degree defines the fracture process at minimum spontaneous cracking. Finally, the trends in the macrocrack shielding by the process zone coincide with the known results [994]. The decreasing of the stress intensity at the macrocrack tip due to the microcracking increasing enhances the macrocrack shielding. As the numerical results show, the analogous trends are proper to HTSC ceramics.

10.1.7 Crack Branching

The $YBa_2Cu_3O_{7-x}$ ceramic samples heated to the temperature 700–930 °C have demonstrated strong crack branching at intergranular boundaries [1458]. The crack branching due to microcracks in process zone, surrounding the crack tip, enhances the fracture energy. Non-regularity of crack front, causing an increasing of ceramic toughness, often occurs in the case of very high rate of the crack tip. A crack branching in Al_2O_3 ceramic has been observed due to TEA of the grains [1808] that is also proper to superconducting ceramics. The growing crack can absorb microcracks and, on the other hand, the microcracking causes crack branching.

In the modeling of macrocrack, propagating into superconducting ceramic structure along intergranular boundaries, the models based on the graph theory can

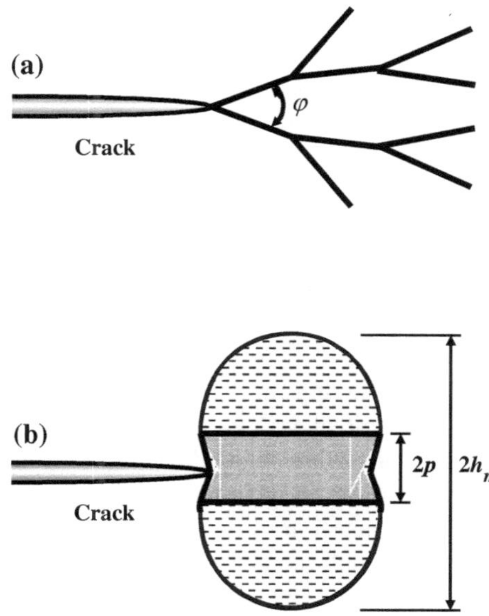

Fig. 10.12 (**a**) Model of crack branching and (**b**) macrocrack tip, showing the zones of crack branching ($2p$) and microcracking near crack tip ($2h_m$)

be used, which have been realized in investigation of PZT ferroelectric ceramics [822, 824, 827]. As has been noted above, a process zone of microcracking forms in crack vicinity due to TEA of grains, creating the macrocrack shielding (the process zone size, $2h_m$, is found from (10.39)). The simulation procedure of microcracks into the process zone repeats the algorithm presented in the previous section. By assuming an existence of a branching crack zone with width, $2p$, in the crack tip (see Fig. 10.12b), apparent fracture energy can be found as [514]

$$\frac{\gamma_{gb}}{\gamma_0} = \left[1 - \frac{\pi\beta_m}{2(3 - \pi\beta_m)}\right]^{-1} \left\{ [2\cos(\varphi/2)]^{\{\log 2/\log[2\cos(\varphi/2)]-1\}} - \frac{p\beta_m}{a_m(1 - \beta_m\pi/2)^{1/2}} \right\}, \tag{10.44}$$

where γ_{gb}, γ_0 are the fracture energy of grain boundary with and without the macrocrack branching; φ is the branching angle. Note that (10.44) includes a theoretically possible limit case of the microcracking absence ($\beta_m = 0$) taking place, in practice, at creation of fine-grain structure due to corresponding selection of the ceramic processing technique:

$$\frac{\gamma_{gb}}{\gamma_0} = [2\cos(\varphi/2)]^{\{\log 2/\log[2\cos(\varphi/2)]-1\}}. \tag{10.45}$$

Moreover, the fracture toughness change may be estimated in the case of crack branching as

$$K_c/K_c^0 = (\gamma_{gb}/\gamma_0)^{1/2}, \tag{10.46}$$

where K_c is the critical SIF, caused by the crack branching; K_c^0 is the fracture toughness of the ceramic without crack branching. Note that β_m is the summary density of microcracks formed during cooling and in the process zone, $2h_m$. In addition, in general, the zones of the crack branching and microcracking near the crack tip do not coincide with each other ($h_m \neq p$). Obviously, the actual toughening effects demand fulfillment of the condition $h_m > p$.

10.1.8 Crack Bridging

Other toughening mechanism for ceramics with TEA of grains is connected with enhancement of fracture resistance to growing crack due to the formation and fracture of bridges behind the crack front (crack bridging) (see Fig. 10.13). The TEA creates residual compressive stresses, σ_R, keeping the grains that restraint the crack opening in the sites of their disposition. The toughness change as a function of crack length, c (*T-curve*) due to this toughening mechanism may be estimated on the base of the following assumptions [122]:

(1) the intergranular fracture is only considered;
(2) the grain structure is unimodal and uniform with equal probability of distribution of the compressive and tensile internal stresses (compressed grains form bridges, and tensile grains define a matrix);
(3) the solution is constructed in approximation of a "weak shielding", that is, the effect of compressive stresses on crack is taken into account in the SIF balance, but is neglected in the relationship for displacements of crack surfaces at its opening;
(4) the conception of a "geometric similarity" is used, that is, YBCO ceramic structure changes only in scale without significant alteration of the grain geometry, but parameters, characterizing the geometrical features of the microstructure, are invariant.

Then, T-curve can be constructed as a function of growing crack length, c, as [122]

$$
\begin{aligned}
&T(c)=T_0-\psi\sigma_R c^{1/2};\quad c\le d\\
&T(c)=T_0-\psi\sigma_R c^{1/2}\left[1-\frac{2p_D}{3\sigma_R}\left(\frac{c}{c_*}\right)^{1/4}\left(1-\frac{d^2}{c^2}\right)^{3/4}\right];\quad d\le c\le c_0;\\
&T(c)=T_0-\psi\sigma_R c^{1/2}\left\{1-\left(1-\frac{d^2}{c^2}\right)^{1/2}+\left(\frac{c_0}{c}\right)^{1/2}\left(1-\frac{d^2}{c_0^2}\right)^{1/2}\left[1-\frac{2p_D}{3\sigma_R}\left(\frac{c_0}{c_*}\right)^{1/4}\left(1-\frac{d^2}{c_0^2}\right)^{1/4}\right]\right\}\\
&\qquad+\psi p_M c^{1/2}\left[\left(1-\frac{d^2}{c^2}\right)^{1/2}-\left(\frac{c_0}{c}\right)^{1/2}\left(1-\frac{d^2}{c_0^2}\right)^{1/2}\right]\left\{1-\frac{1}{2}\left[\left(\frac{c}{c_*}\right)^{1/2}\left(1-\frac{d^2}{c^2}\right)^{1/2}+\left(\frac{c_0}{c_*}\right)^{1/2}\left(1-\frac{d^2}{c_0^2}\right)^{1/2}\right]\right\};c_0\le c\le c_*\\
&T(c)=T_0+0.5\psi p_M c_*^{1/2};\quad c\ge c_*,
\end{aligned}
\tag{10.47}
$$

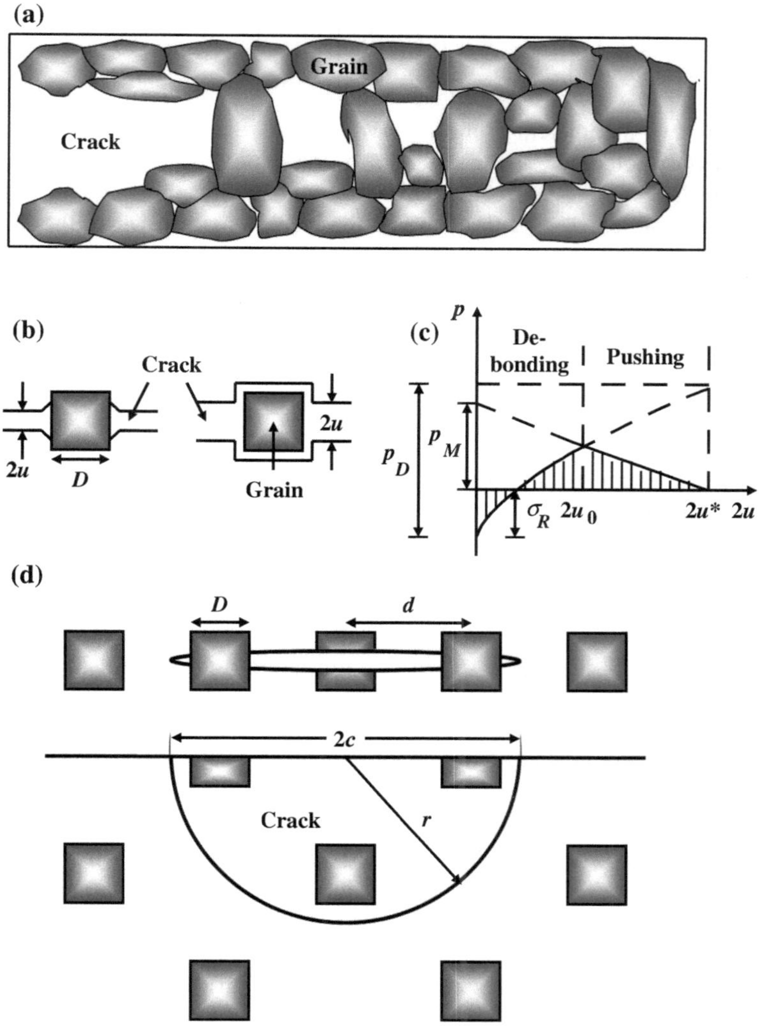

Fig. 10.13 Formation and fracture of bridges behind growing crack front: (**a**) schematic structure fragment; (**b**) two interaction stages of advancing crack with grain-bridge (de-bonding with matrix and pushing of grain by the crack surfaces); (**c**) stress distribution, $p(u)$, inhibiting the crack opening and acting at interfaces; (**d**) penny-shaped crack into region of the crack bridging (sight with side and upper)

where $p_D = 2(\in_L \mu \sigma_R E_{0m})^{1/2}$ are the shear stresses, forming at the grain de-bonding; c_0 is the crack length, corresponding to a cross of the regions of de-bonding and pushing from matrix of the grains, restraining the crack; $\psi = 1.24$ is a geometrical parameter, corresponding to a penny-shaped surface crack; $\in_L$ is the deformation at the fracture of the links; μ is the factor of the sliding friction. The value of c_0 is calculated from the equation:

$$c_0^2 - c_* \left(\frac{p_D}{2p_M}\right)^4 \left\{ \left[1 + \frac{4p_M(p_M + \sigma_R)}{p_D^2}\right]^{1/2} - 1 \right\}^4 c_0 - d^2 = 0. \qquad (10.48)$$

There are two parameters of the *T-curve* that are required for maximization to attain necessary tolerance of the ceramic to crack growth. These are T_∞/T_0 and c_*/d, where $T_0 = (2\gamma_0 E_{0m})^{1/2}$ is the intrinsic toughness of ceramic, but a toughness at the steady-state crack growth (T_∞) and the crack length in which the far most links from the crack tip begin fracture (c_*) are calculated as

$$T_\infty = T_0 + 0.5\psi p_M c_*^{1/2}; \qquad (10.49)$$

$$c_* = \left(\frac{\in_L E_{0m} d}{2\psi T_0}\right)^2, \qquad (10.50)$$

where d is the mean spacing between bridging grains, selected as equal to characteristic grain size of the ceramic structure; $p_M = 2 \in_L \mu\sigma_R$ are the stresses caused by the sliding friction; $\gamma_0 = \gamma_s - \gamma_{gb}(1 - f_m)/2$ is the specific surface energy; γ_s is the surface energy in transgranular fracture; γ_{gb} is the fracture energy of grain boundary; f_m is the summary fraction of the cracked facets during the ceramic cooling and in the process zone.

Source of the *T*-curve is a successive formation and fracture of bridges in material that act as inhibiting links behind the crack tip. Secondary phases at grain boundaries (in our case, the $BaCuO_2$-CuO system with admixture inclusions [120]) play a special role in quantitative interpretation of the *T*-curve. On one hand, they lead to brittlement of grain boundaries and to decreasing their toughness, but on the other hand they inhibit grain growth, leading to the increasing of ceramic strength. In addition, for related ceramics, there are phenomena which point to a possible stabilization of the crack growth [307]: (i) significant increase of bore applied load after the elastic limit is exceeded, followed by a load drop to non-zero stress (see Fig. 10.14b); (ii) the erratic crack advance is caused by local heterogeneities in the fracture resistance (see Fig. 10.15); (iii) the discontinuous crack traces are regions of unruptured or frictionally locked material which are the restraining ligaments and which center on large grains (see Fig. 10.16); and (iv) the fracture of the grains-bridges occurs in transgranular mechanism at the primary intergranular failure. These phenomena witness on the existence of the processes of the bridges formation behind the fracture front that states an effective toughening mechanism. Due to this, we use results of [307] below and identify three regions of behavior of the penny-shaped crack with the radial coordinate c (see Fig. 10.13d), namely (i) the crack motion does not restraint, and $T = T_0$ for short cracks ($c < d$), where d is the average distance between adjacent grains-bridges; (ii) the zone of ligaments is active in the region $d \leq c \leq c^*$, and $T > T_0$ for intermediate cracks; (iii) the rupture of the couples occurs at sites far from the crack tip, leaving a zone of width $c^* - d$ for long cracks ($c > c^*$). The latter zone expands outward together with the crack. In this case, the toughness reaches a

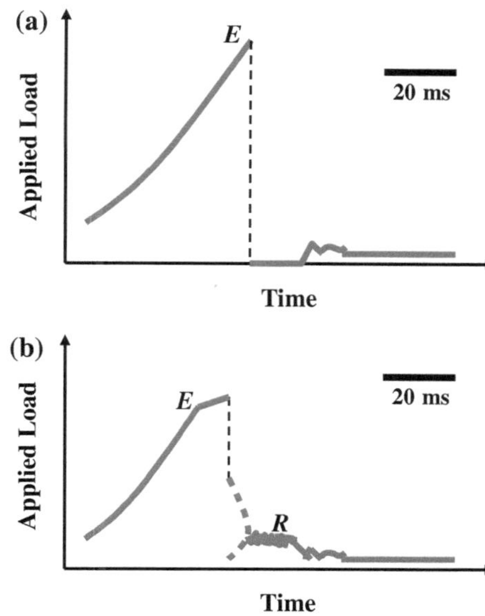

Fig. 10.14 Applied loading as function of time: (**a**) typical response for brittle material, showing spontaneous fall of the loading to zero after exceeding of the elastic threshold (*E*), and (**b**) response for material at stable crack motion, showing significant increasing bore loading out of elastic limits with following fall of the applied stress down to some non-zero level (*R*) [307]

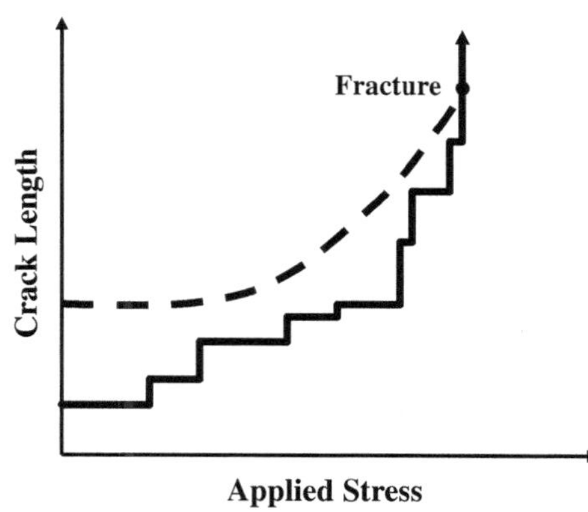

Fig. 10.15 Crack growth *versus* applied stress (*continuous line*) for Al_2O_3 sample, demonstrating erratic crack advance, caused by local heterogeneities in fracture resistance. *Dashed curve* corresponds to material with uniform fracture resistance (this line is reduced to the same fracture stress) [307]

maximum value for steady-state crack growth. This toughness corresponds to the crack length, $c*$, which is calculated as [1300]

$$c* = \frac{\beta}{2}\left\{1 + \left[1 + 4\left(\frac{d}{\beta}\right)^2\right]^{1/2}\right\}, \tag{10.51}$$

where $\beta = \left(\frac{E_{0m}u*}{\psi T_0}\right)^2$; $2u*$ is the opening of the crack with length $c*$.

Then, we consider the constitutive equations (the stress-extension dependencies) for inhibiting ligaments of three types [1326], namely: (i) $\sigma(u) = -\sigma^*(u/u^*)$ for elastic ligaments, where σ^* is the value of the peak restraining stress exerted by

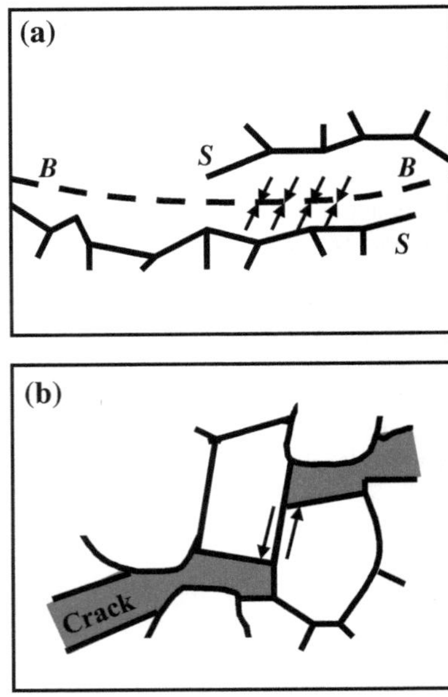

Fig. 10.16 (**a**) Ruptures of crack path, localized at big grains in Al_2O_3 ceramic, which are regions of non-fractured material, forming links that limit the crack growth; *S* the crack surfaces and *BB* the continuous crack into bulk, depicted by *dashed curve*; the stresses, which close the crack and caused by bridges at the crack surfaces, are shown by *arrows*; (**b**) grains connected one to other by friction forces; the stresses, closing the crack and caused by the crack surface friction, are also depicted by *arrows* [307]

the ligament; (ii) $\sigma(u) = -\sigma^*(1 - u/u^*)^m$ for restraints caused by the compressive grains-bridges due to TEA ($m = 1$), and (iii) frictional ligaments ($m = 2$). Then, the corresponding microstructure contributions to toughness values are estimated as $T_\mu = 0$ at $c < d$, and $T_\mu = T_\infty - T_0$ at $c > c^*$ (for all three types of the ligaments). In the intermediate zone ($d \leq c \leq c^*$), the corresponding contributions are estimated for elastic ligaments as

$$T_\mu = (T_\infty - T_0)\left[\frac{c^*(c^2 - d^2)}{c(c^{*2} - d^2)}\right]^{1/2}, \tag{10.52}$$

for ligaments due to TEA ($m = 1$) and frictional ligaments ($m = 2$) as

$$T_\mu = (T_\infty - T_0)\left\langle 1 - \left\{1 - \left[\frac{c^*(c^2 - d^2)}{c(c^{*2} - d^2)}\right]^{1/2}\right\}^{m+1}\right\rangle. \tag{10.53}$$

Following [308], we use superposition method in the framework of the force approach and put a local internal stress, caused by microstructure effects of fracture resistance on uniform driving force of crack. Then, total SIF is $K = K_a + K_r$, where $K_a = \psi\sigma_a c^{1/2}$ is the applied SIF from indentation; σ_a is the homogeneous external stress. The local SIF is defined as $K_r = \chi P/c^{3/2}$, where $\chi = 0.004(E_{0m}/H)^{1/2}$ is the constant, depending on the contact geometry and elastic–plastic properties; H is the material hardness; P is the contact loading. We assume a power dependence of toughness on the crack length [308], $T = T_0(c/d)^\tau$, where $0 < \tau < 0.5$ (the toughening is absent for $\tau = 0$, and the catastrophic fracture does not reach for $\tau = 0.5$). Then, from equilibrium condition $K = T$ and the transition condition from stable crack growth to unstable one, $dK/dc = dT/dc$, we find the critical crack length, c_{cat}, and the applied loading, σ_{cat}, at which the fracture becomes unstable or catastrophic [308, 1072]:

$$c_{cat} = \left[\frac{4\chi P d^{\tau}}{T_0(1-2\tau)}\right]^{2/(2\tau+3)}; \tag{10.54}$$

$$\sigma_{cat} = \left[\frac{T_0(2\tau+3)}{4\psi d^{\tau}}\right] c_{cat}^{(2\tau-1)/2}. \tag{10.55}$$

Then, the inert strength, $\sigma_m(P)$, may be estimated in the kinetic effects absence, considering a condition of unlimited fracture of the ceramic under loading of P due to indentation by using Vickers pyramid [122]. Then, the applied tensile stress, $\sigma_a(c)$, as function of the crack length, c, is calculated, using the toughness magnitude, T, as

$$\sigma_a(c) = \frac{1}{\psi c^{1/2}}\left[T(c) - \frac{\chi P}{c^{3/2}}\right]. \tag{10.56}$$

Generally (for non-cubic, ferroelectric and superconducting ceramics), the function of $\sigma_a(c)$ shows two peaks, which are divided by the value, $c = d$ (see Fig. 10.17b) [1303, 1316]. The first maximum (for $c < d$) is associated with the residual contact field that dominates. The second maximum (for $c > d$) is associated with the bridging that dominates. The appropriate barrier heights depend on the external load P. When crack overcomes the first barrier (for $c < d$), the crack becomes unstable. However, the crack grows spontaneously up to catastrophic failure if the second barrier (for $c > d$) turns out below the first barrier. Otherwise, the macroscopic fracture is possible only if the load to be sufficient that the crack overcomes the second barrier. Hence, the ceramic strength, σ_m, is found by the greater maximum. Note also that the transition point to the horizontal section of the curve $T(c)$ in Fig. 10.17a corresponds to the crack length, $c = c^*$ [1319].

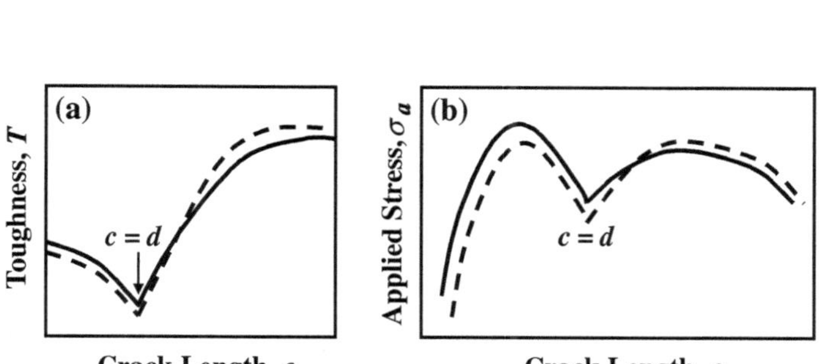

Fig. 10.17 Qualitative trends in the toughness (**a**) and in the applied loading (**b**) as functions of crack length. *Solid* and *dashed curves* demonstrate the dependences before and after grain growth, respectively

10.1.9 Some Numerical Results

In order to obtain numerical results, the known parameters of $YBa_2Cu_3O_{7-x}$ and some data for related ceramics have been used. A necessary number of the computer realization has been established on the basis of stereological approach. The corresponding procedure for definition of unbiased estimation of any stereological characteristic [271] has been presented in Sect. 6.6. We consider the mean grain radius as this stereological characteristic in the computer simulations. Young's modulus is defined as $E = 3\,K(1 - 2\nu)$, using the volume modulus, $K = \gamma c/(\beta V)$, where $\gamma = 1.35$ is Grüneisen's constant; $c/V = 2.72$ MPa/K is the characteristic volume heat; $\beta = 5.86 \times 10^{-5}$ K^{-1} is the mean volume thermal expansion factor; $\nu = 0.21$; $K_c^b = 1$ MPa·m$^{1/2}$ [562]. Selecting $\delta = l_{sc}$, we estimate an effect of the initial sample porosity before sintering, C_p^0, and the inhibition parameter of grain growth, I_R, on different strength characteristics. The obtained numerical results are shown in Figs. 10.18,10.19,10.20 and in Table 10.5 [1319].

The calculations show a growth of the mean grain size with the decreasing of C_p^0; at the same time, the insignificant alteration of closed porosity is observed in the considered cases. This causes a formation of longer microcracks at smaller values of C_p^0. However, the effect of grain growth stagnation and existence of a smaller number of triple points (nuclei of microcracks) state the insignificant change of the microcracking density $\beta_m^{(1)}$ (after cooling) with attainment of its minimum value for $C_p^0 = 30$ %.

Joint effects of microcracking and grain size on strength properties are found by the ratio *S/R*. This parameter owing to above results prevails compared to dependence on the closed porosity of the sample. This explains, at first view, the paradoxical result of the fracture stress, σ_f, increasing with the initial porosity, C_p^0 (see Fig. 10.18). The last result follows from the concept that the fracture stress increases with decreasing of the grain size. In addition, the same trend retains in the dependence of σ_f on *S/R* (at fixed value of the grain size) [944].

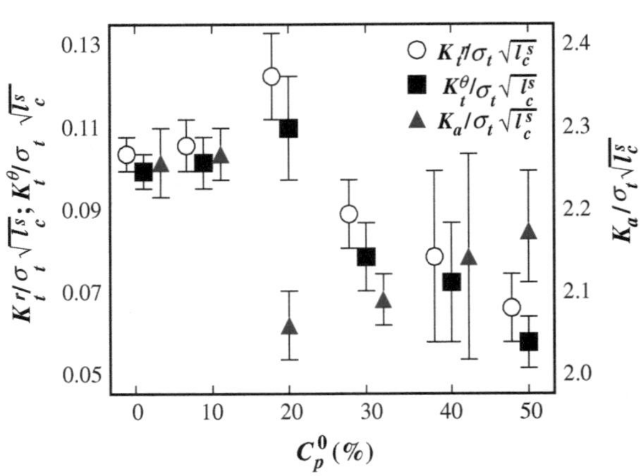

Fig. 10.18 SIF *versus* initial porosity of precursor sample, C_p^0. The deviation of results corresponding to interval change of magnitude of the abnormal grain-growth inhibition parameter ($I_R l_c^s = 0.4 - 1.0$) is shown

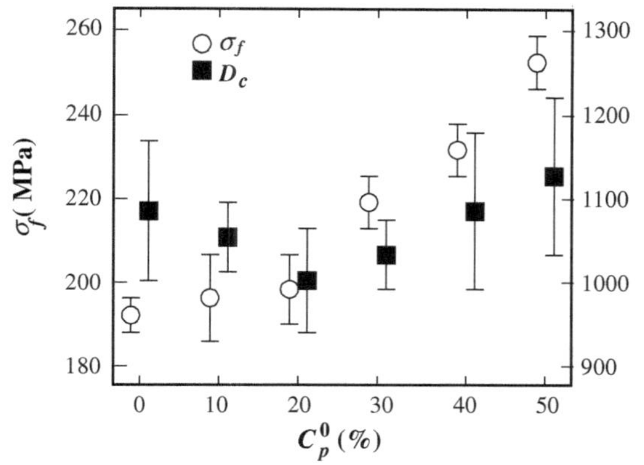

Fig. 10.19 Fracture stress, σ_f, and critical grain size, D_c, causing the further crack growth *versus* initial porosity of precursor sample, C_p^0 ($I_R l_c^s = 0.4 - 1.0$)

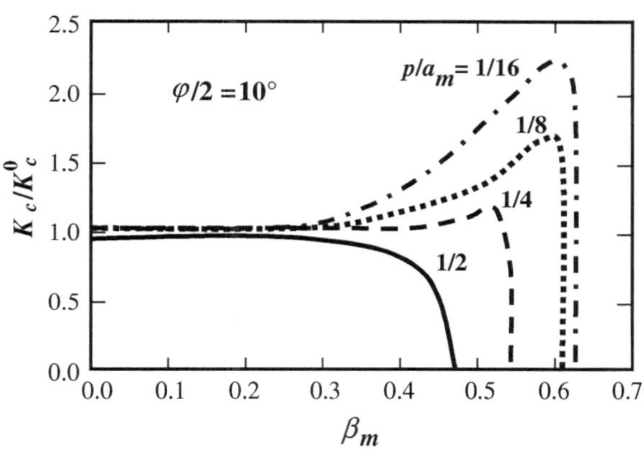

Fig. 10.20 Fracture toughness, K_c/K_c^0, *versus* microcracking density, β_m, for different sizes of process zone for crack branching (p/a_m) at $I_R D_0 = 2.0$

The obtained values of the fracture stress are near the test data observed by the electronic microscopy method for $YBa_2Cu_3O_{7-x}$ ceramic ($\sigma_f = 300$ MPa) [917]. The SIF dependencies (K_t^r,K_t^θ, K_a) and the critical grain size for further crack growth (D_c) as function of C_p^0, presented in Figs. 10.18 and 10.19, reach extremes in intermediate points of the interval values of C_p^0. This coincides with the observed trends for non-cubic [944] and ferroelectric [1368] ceramics. It should be noted that the SIF, corresponding to the tangential component of thermal stresses (K_t^θ) is smaller than the SIF caused by the radial component (K_t^r). Due to K_t^θ cannot be independently used to define a critical condition of cracking. Similarly, the spontaneous microcracking during cooling to be significant phenomenon of I mode of fracture in the ceramics with non-cubic symmetry, which demonstrate brittle fracture [995].

The numerical results (Fig. 10.20 and Table 10.5) show that fracture toughness increases together with the crack branching angle and at interaction of macrocrack with microcracks in the short process zone of the crack branching. However, in the case of the process zone increasing, the dominating influence of the crack

Table 10.5 Fracture toughness change at crack branching ($I_R D_0 = 2.0$)

Properties	$C_p^0 = 0$ %	$C_p^0 = 10$ %	$C_p^0 = 20$ %	$C_p^0 = 30$ %	$C_p^0 = 40$ %	$C_p^0 = 50$ %
$\beta_m^{(1)}$	0.17	0.17	0.13	0.12	0.15	0.17
K_c/K_c^0, $p/a_m = 1/8$	1.04	1.04	1.03	1.03	1.03	1.04
K_c/K_c^0, $p/a_m = 1/3$	0.99	0.99	1.00	1.00	1.00	0.99
K_c/K_c^0, $p/a_m = 1/2$	0.96	0.96	0.97	0.97	0.97	0.96
$\beta_m^{(2)}$	0.28	0.31	0.32	0.33	0.39	0.44
K_c/K_c^0, $p/a_m = 1/8$	1.08	1.09	1.10	1.10	1.14	1.19
K_c/K_c^0, $p/a_m = 1/3$	0.98	0.98	0.98	0.97	0.96	0.92
K_c/K_c^0, $p/a_m = 1/2$	0.90	0.88	0.87	0.86	0.77	0.62

$\beta_m^{(1)}$ is the microcracking density after cooling; $\beta_m^{(2)}$ is the summary density of microcracks, formed during cooling and in the process zone near macrocrack.

branching is replaced by the microcracking effect, decreasing the fracture toughness. The transition from toughening to the crack amplification is defined by the process zone size, $p/a_m \approx 1/3$ (for the branching angle, $\varphi/2 = 10°$). The weak dependence of K_c/K_c^0 on C_p^0 at fixed value of p/a_m is caused by the small microcracking density, $\beta_m = 0.12$–0.17 (after the sample cooling). An increase of $\beta_m = 0.28$–0.44 (on account of damage into the microcracking process zone) leads to formation of the fracture toughness dependence on the initial sample porosity. The analysis of the dependence of K_c/K_c^0 on β_m (see Fig. 10.20) shows that an effect of the crack amplification ($K_c/K_c^0 < 1$) takes place for all sizes of the crack branching zone and exists in the case of the microcrack coalescence, at $\beta_m \approx 0.5$. The last value is usually selected as the critical one for initiation of the crack coalescence in ceramics [498].

In order to estimate the effect of the crack bridging, we use the next data [122, 307, 1296]: $\epsilon_L = 0.1$; $\mu = 1.8$; $\gamma_s \approx 2\gamma_{gb} = 6$ J/m^2; $\sigma_R = 100$ MPa; $H = 19.1$ GPa. We take into account in the present case only microcracks formed during cooling. The microcracking, caused by macrocrack propagation, changes quantitative results but qualitative ones, presented in Fig. 10.17, are retained. In modeling of abnormal grain growth, the characteristic values of $D_0/l_c^s = 2.20$ and 2.84 (the case of no porous ceramic) have been estimated preliminarily for two values of the grain growth inhibition parameter, $I_R D_0 = 2.0$ and 1.2 (D_0 is the grain size before beginning of the grain growth). As is followed from Fig. 10.17a, the effect of the grain size on the change of T-curve is not simple. This is explained by an additional microcracking effect of intergranular boundaries. In the field $c < d$, an enhancement of the value d will lead to corresponding diminishing of left part of the $T(c)$ curve and at some value of parameter d, the $T(c)$ curve crosses the c-axis before fulfillment of the condition: $c = d$. Then, primarily existing microcracks and residual stresses, formed during the ceramic processing, will lead to unstable macrocrack advancing without dependence on applied load. In this case, the spontaneous microcracking prevails over the toughening effects, caused by the crack bridging, and states the ceramic strength. The critical value of the parameter d is estimated from the condition: $T = 0$, at $c = d$. Hence, for the sizes of grains-bridges

$$d \geq (T_0/\psi\sigma_R)^2, \quad (10.57)$$

a positive effect of crack bridging is exceeded by the effects of "deleterious" microcracking.

Other condition for definition of the critical value of the parameter d can be stated from (10.4). Assuming that the sizes of grains and facets per grain are subjected to normal distribution, we obtain [336]: $D_c^s \approx 2l_c^s$. Then, the mean separation between bridging grains, d, coinciding with the characteristic grain size can be estimated as

$$d < 2\beta_0[K_c^b(1+\nu)/(E\varepsilon)]^2. \quad (10.58)$$

The selection of the condition for d as well as for fixing of the grains-bridges into ceramic microstructure should be carried out on the base of experimental data. Then note that the de-bonding region ($d \leq c \leq c_0$) is very short ($c_0/d \leq 1.01$) and is not observed practically in Fig. 10.17. Thus, pushing of grain by the surfaces of the growing crack mainly contributes to the increasing of fracture resistance on account of existence of the residual thermal stresses. This conclusion is supported by theoretical analysis and numerical results of [122, 1747].

The parameters characterizing a capability of the YBCO ceramics to fracture resistance in the considered cases ($D_0/l_c^s = 2.20$ and 2.84) are, respectively, equal to $T_\infty/T_0 = 2.01$ and 2.17 (from (10.49)); $c^*/d = 15.6$ and 17.1 (from (10.50)). They define a significant effectiveness of present process as the toughening mechanism of superconducting ceramics. Generally, these parameters increase with grain size growth (see Fig. 10.17) for the same initial porosity, C_p^0, that corresponds to experimental observations [1747].

By using the above computational approach, the microstructure parameters required to estimate trans- and intergranular fracture can be calculated [1300, 1326]. The numerical data, characterizing the YBCO microstructure at different initial porosity, C_p^0, are given in Table 10.6. The computational procedure includes the next stages. First, we define the grain-boundary intercept length, λ, and ratio of the largest grain size in the material to the average one, η. Then, the grains-bridges are selected at the lattice directly (in every case there are 10 % of grains from those, that are similar to the experimental data for related compositions [1747]). These are the greatest grains and we select the value of d so that it is equal to the average size of the grains-bridges. Next, the transgranular fracture of ligaments and failure of the intergranular boundaries between bridges are simulated (see Fig. 10.21b), using the graph theory for definition of the shortest crack route at the lattice of the intergranular boundaries [147]. After that, the proportion of transgranular failure, A_T, is found. The sections of the transgranular failure are determined using the characteristic size of the bridge, but the sections of the intergranular fracture are defined using the direct crack path along grain boundaries. Further, we choose the parameter u^* for the above ligaments, namely: $u^*/\lambda \sim 0.01$ for elastic and frictional restraints [307] and $u^*/\lambda \sim \eta^2\varepsilon_T$ for compressive traction [1633], where the strain for YBCO due to the TEA is

Table 10.6 Some model parameters for YBCO ceramic

Properties	$C_p^0 = 0\ \%$	$C_p^0 = 20\ \%$	$C_p^0 = 40\ \%$	$C_p^0 = 60\ \%$
η	1.498	1.636	1.758	1.842
A_T	0.312	0.293	0.274	0.442
d/δ	3.423	3.270	3.028	2.646
u^*/λ	0.0070	0.0083	0.0096	0.0105
c^*/d	10.56	11.05	11.92	13.62
	5.32	7.66	10.98	15.00

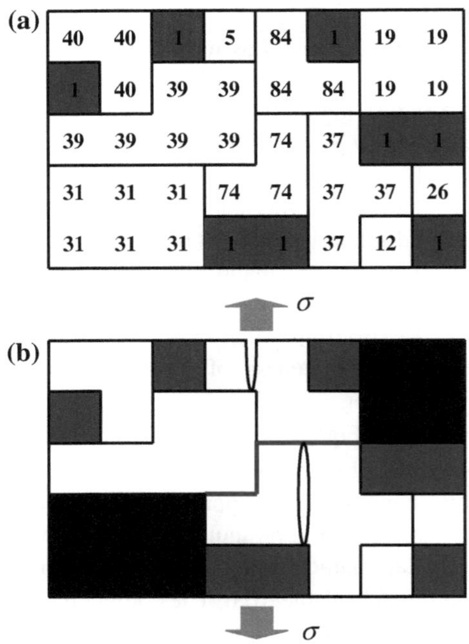

Fig. 10.21 (**a**) Representation of YBCO ceramic structure fragment in PC (voids are denoted by 1 and depicted by *gray color*), and (**b**) macrocrack propagation (*gray line*) into model structure between two grains-bridges depicted by *black color*

$\varepsilon_T = 3.1 \times 10^{-3}$ (see Sect. 10.1.2). Finally, we determine the values of c^* and $T_\mu/(T_\infty - T_0)$ from (10.51)–(10.53).

The ratio of the largest grain to the average one (η) rises with initial porosity, showing a growth of the heterogeneity of grain sizes. This is due to the outstripping decrease of the average grain size compared with the highest one. Further, there is growth of the grain size and grain-bridge with the initial porosity decrease. Moreover, the numerical data give dependence, $d = k\lambda$ $(1 < k < 2)$, as being similar to alumina ceramics [307]. The intergranular fracture is decreased with the initial porosity growth, because high sinuosity of a crack trajectory corresponds to the coarse-grained structures. However, the fraction of the transgranular fracture (A_T) does not change monotonously due to the relative equalizing of the transgranular failures in the various cases of C_p^0 and owing to a rise of the structural parameter (η) with the initial porosity. Thus, the most fine-grained structure with C_p^0= 60 %, but possessing the greatest value of η, has the increased toughening effects, which correlate with increased area fractions of transgranular failure [307]. On the other hand, this result supports a known thesis about necessary to characterize a polycrystalline structure by a distribution of grain sizes, but no their maximum (or minimum) value.

The parameters which characterize the ability of superconducting ceramics to resist failure are ratios of $T_\mu/(T_\infty - T_0)$ and c^*/d. The numerical results for the latter (obtained from (10.51)) are shown in Table 10.6 where the upper line corresponds to the elastic and frictional ligaments, but lower line relates to the compressive traction due to TEA. The observed growth of c^*/d together with the initial porosity for all ligaments is caused by appropriate rise of the structure heterogeneity parameter, η, and coincides with experimental data for different ceramics [122, 307, 1747]. In Fig. 10.22, there are microstructure contributions of $T_\mu/(T_\infty - T_0)$ due to the crack bridging mechanism as function of the crack length c for the above types of ligaments, and $C_p^0 = 0$ %. Obviously, the main contribution to the toughening is found by the compressive traction due to TEA, and the least one corresponds to the elastic links. These trends are supported by the experiments [122, 1747] as well.

The critical parameters c_{cat} and σ_{cat} (see (10.54) and (10.55)) are shown in Fig. 10.23 depending on the load P and on different toughening powers τ for characteristic grain size, $D_0/\delta = 2.20$. The results show an increasing of the critical crack length, c_{cat}, and a decreasing of the critical applied stress, σ_{cat}, with growth of the contact load, P, without dependence on features of abnormal grain growth [1319]. However, the changes of these parameters from the toughening power have different characters.

Thus, the inhibition parameter of grain growth (I_R) and the mean spacing between bridging grains (d) are parameters which should be taken into account in the calculation and optimization of microstructure and strength parameters of HTSC ceramics. The first parameter is stated by the history of the ceramic processing and by its composition (i.e., thermal treatment, admixture additions in

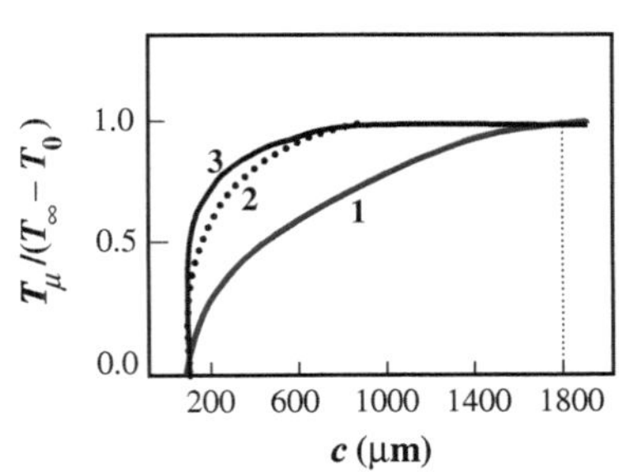

Fig. 10.22 Microstructure contributions of $T_\mu/(T_\infty - T_0)$ to the HTSC ceramic from elastic (1), frictional (2) and compressive (3) ligaments *versus* crack length, c

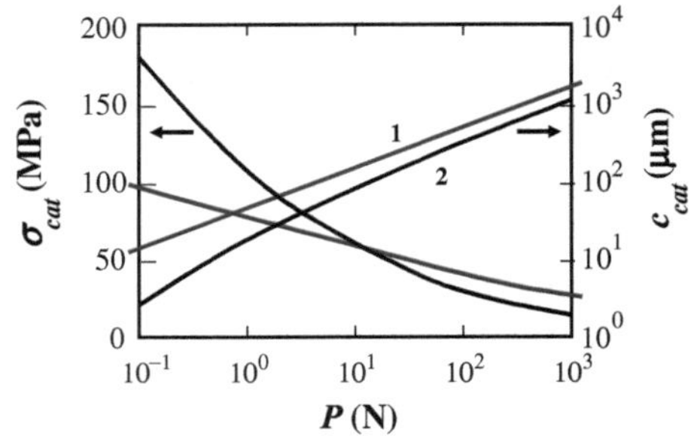

Fig. 10.23 Critical parameters σ_{cat} and c_{cat} *versus* contact load P for two toughening powers: $\tau = 0.25$ (1) and $\tau = 0.10$ (2)

press-powders). The second one is the key parameter of the bridging mechanism, which states a transition from the toughening effects to the amplification of crack. At the same time, the account of these parameters does not reject but, on the contrary, proposes to use in calculations other characteristics, in particular, the toughness of intergranular boundaries (T_0), residual stresses (σ_R) and coefficient of the sliding friction at pushing of grains by surfaces of growing crack (μ). Hence, a design of YBCO ceramic microstructure, which is optimum from the view of material strength, is connected with introducing of the grains-bridges on the path of the possible macrocrack propagation at corresponding suppression of "deleterious" microcracking in this zone. This assumes a requirement to form superconducting grains with maximum permissible sizes, which do not exceed the critical value of the spontaneous cracking, and with distribution, demonstrating a maximum possible parameter of the structure heterogeneity, η. The "deleterious" microcracking is formed owing to the grain growth that can be regulated by admixture phases, pulled out during sintering at intergranular boundaries. As a result, it is possible to predict properties of the sintered ceramics even at the earlier stages of the ceramic fabrication in dependence on, for example, the sintering technique, thermal treatment and parameters of secondary phases.

10.2 Crack Shielding Processes in Ferroelastics and Ferroelectrics

It is well known that YBCO ceramics possess ferroelastic properties owing to domain (platelet) structure. Therefore, a sufficiently great attention has been spared to description of microstructure and residual stresses of YBCO polycrystals in the framework of the numerical method for description of twinning processes [56, 462, 1353] and also to computer simulation of twins [1760].

Single crystallites in polycrystalline ferroelastics (YBCO) and ferroelectrics ($BaTiO_3$, PZT, etc.) at structure phase transition are subjected to twinning, which is a result of the elastic energy minimization. The twin formation leads to significant decreasing of quasi-uniform stresses in grain volume at simultaneous formation of considerable heterogeneous stresses localized near intergranular and

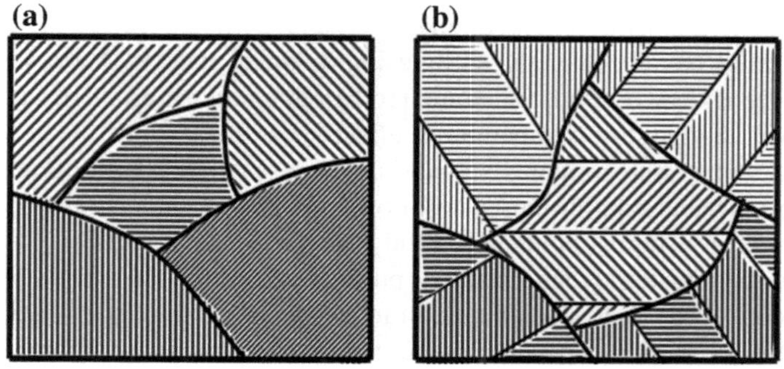

Fig. 10.24 (**a**) Simple platelet twinning in YBCO and (**b**) strip structure of twins in $BaTiO_3$

interphase boundaries [56, 451, 1353]. A twinning as a sequence of phase transformation, causing a small structure change of unit cell, can be described as mechanical twinning [56, 826] or as martensitic transformation [359, 452]. The transition accompanied by diffusion, in some cases, also leads to the twinning.[2] The mechanical twinning in ferroelastics can be compared with formation of ferroelectric domains. In the latter case, an energy decrease for electric field is accompanied by the energy diminution for domain walls that leads to minimization of total energy of the system. In the ferroelastics, this summary energy, including elastic energy and energy of domain walls, also becomes minimal. The domain walls displacement can be accompanied by change of the crystallite shape and serves the sources of residual stresses. From the view of influence on fracture toughness, a change of domain structure in vicinity of crack under mechanical loading presents a special interest. This process affects the crack growth. A relationship of domain structures (see Fig. 10.24) and twinning features [56, 1353] prompts a possibility of comparative analysis of the ferroelectrics and ferroelastics behavior. An influence of the twinning processes on fracture toughness of these samples and appropriate effects on fracture resistance are estimated in Sects. 10.2.2–10.2.4.

10.2.1 Ferroelastics and Ferroelectrics with Effect of Negative Stiffness of Components

It should be assumed that materials near proper phase transitions will demonstrate negative stiffness in microscale due some causes. Superconductive ferroelastics [1472] and ferroelectrics [214] near structural phase transitions possess stiffness components which achieve minimal values or aspire to zero at critical temperature.

[2] In particular, the tetragonal–orthorhombic phase transition in YBCO superconductor demands oxygen diffusion into crystalline lattice [473].

Lower critical temperature the domain or band structure forms. These phase transformations have been analyzed by using modified Landau theory [460, 461], in which free energy depends on strain, temperature and material constants. It has been shown that at low temperatures effective stiffness may be negative. At decreasing temperature initiates spontaneous deformation which corresponds to shear from unstable equilibrium state to one of stable points of minimal energy, and the material acquires domain or band structure.

Poly-domain material demonstrates positive stiffness. The formation of the band structure has been studied into framework of elastic model of the solid medium [897], in which dependence of stress on strain has no monotonous section. Domain walls require energy for their formation, therefore as single domains may be particles of sufficiently small sizes (from micrometer up to centimeter scale in dependence on material) [1472]. It should be assumed that the single domains will demonstrate negative stiffness. Experimental results leading to conclusion about existence of negative stiffness have been published [967, 968]. Foam materials under strong compression demonstrate macroscopic positive stiffness. By this they create band structures under strong compression and this band structure corresponds to stability of individual cells [1445]. Transformation to band structure is similar to bulging edges of the foam cell which also leads to non-linear dependence $\sigma - \varepsilon$. High-temperature superconductive (HTSC) foams [1421] could find sufficiently wide application in modern electro-energetic and demand detailed studies their conductive and mechanical properties.

As it is known, for elastic two-phase composite consisting of isotropic phases the stiffness of Voigt-Reuss composite presents strong upper and lower bounds for Young modulus at given phase concentration. Hashin–Shtrikman equations [651] state strong bounds for elastic stiffness of macroisotropic composites. These bounds are narrower than Voigt-Reuss bounds.

The structure corresponding to Voigt composite (upper bound of Voigt-Reuss composite) has phases disposed along direction of applied loading, so each phase is subjected to the same strain. The Voigt bound (mixture rule) is defined in 1D model as

$$E_0 = E_1 V_1 + E_2 V_2, \tag{10.59}$$

where E_0, E_1, E_2 are Young moduli (stiffness) of composite and the both phases, V_1 and V_2 determine volume fractions of phases ($V_1 + V_2 = 1$). It is evident, that for Voigt structure defined by Eq. (10.59), the phase with negative stiffness simply decreases a whole the composite stiffness.

The structure corresponding to Reuss composite (lower bound of Voigt-Reuss composite) has phases disposed in perpendicular direction to applied loading, so each phase is subjected to the same stress. The Reuss bound is defined in 1D model as

$$\frac{1}{E_0} = \frac{V_1}{E_1} + \frac{V_2}{E_2} \text{ or } S_0 = S_1 V_1 + S_2 V_2 \tag{10.60}$$

where S_0, S_1, S_2 are the compliance of the composite and the both phases. If stiffness one of the phase this composite is negative (that is the phase has also negative compliance), then at addition the negative compliance of the phase to positive compliance with similar value of other phase, it may be achieved near-to-zero compliance of the whole composite (or aspiring to infinity the stiffness of the composite). Thus, the great values of stiffness should initiate in the composite structure in which the phases with positive and negative stiffness are almost balanced owing to that the interface between components moves much faster than applied loading providing large strain energy in each phase.

The Reuss model including one element of each phase is stable only in conditions of restraint of the phase deformations. Moreover, great negative stiffness is possible in the restrained system however a region for which stiffness of Reuss composite aspires to positive infinity to be unstable [967].

Following [1315], we investigate into framework of elasticity theory behavior of composite with required negative stiffness. With this aim, we consider spherically symmetric problem presenting a spherical matrix of isotropic material containing a spherical inclusion other phase at spherically symmetric boundary conditions. Into linear-elastic statement of the problem [683] effective volume modulus K_0 for the composite sphere is determined as

$$K_0 = \frac{a_1^3 K_1 \alpha_1 - 4G_1\beta_1/3}{a_1^3\alpha_1 + \beta_1}, \tag{10.61}$$

where a_i are the radiuses, K_i and G_i are the volume and shear moduli, α_i and β_i are the unknown constants calculated from boundary conditions, $i = 1, 2$ (here indices 1 and 2 define matrix and inclusion, respectively).

Consider two types of the problems.

(1) *Homogeneous radial stress applied to the boundary of spherical matrix* $\sigma_{rr}(a_1) = \sigma$. Other conditions: (i) radial displacement in center of spherical inclusion $u_r(0) = 0$, and (ii) continuity of σ_{rr} and u_r on matrix-inclusion interface $r = a_2$. We define from boundary conditions the unknown constants (in particular, α_1 and β_1), and then taking into account dependence $2G = 3\,K(1 - 2v)/(1 + v)$, where v is Poisson's ratio, obtain K_0 as

$$\begin{aligned} &\alpha_1 = a_1^3\sigma(3K_2 + 4G_1)/\gamma\ ;\ \beta_1 = -3a_1^3a_2^3\sigma(K_2 - K_1)/\gamma; \\ &\gamma = 12a_2^3G_1(K_2 - K_1) + 3a_1^3K_1(3K_2 + 4G_1) \end{aligned}, \tag{10.62}$$

$$K_0 = \frac{K_2(1 + 2V_2) + 2K_1(1 - V_2)(1 - 2v_1)/(1 + v_1)}{(1 - V_2)K_2/K_1 + 2(1 - 2v_1)/(1 + v_1) + V_2}. \tag{10.63}$$

From here, the bulk modulus of the composite becomes infinite at negative bulk modulus of inclusion determined as

$$K_2 = -\frac{K_1}{(1 - V_2)}\left[\frac{2(1 - 2\nu_1)}{(1 + \nu_1)} + V_2\right]. \tag{10.64}$$

When $\gamma \to 0$, the stress–strain state and displacement field become infinite. Due to Eq. (10.62) it takes place at the next relation between moduli:

$$K_2 = -\frac{2(1 - V_2)(1 - 2\nu_1)}{[2V_2(1 - 2\nu_1) + 1 + \nu_1]}K_1. \tag{10.65}$$

(2) *Homogeneous radial displacement applied to the boundary of spherical matrix* $u_r(a_1) = u$. Other conditions coincide with conditions (i) and (ii) of first problem. In this case the bulk modulus of the composite is also calculated from (10.63), becomes infinite at condition (10.64), and the unknown constants have the forms:

$$\begin{aligned} \alpha_1 &= a_1^2 u(3K_2 + 4G_1)/\delta\ ;\ \beta_1 = -3a_1^2 a_2^3 u(K_2 - K_1)/\delta\ ; \\ \delta &= -3a_2^3(K_2 - K_1) + a_1^3 K_1(3K_2 + 4G_1). \end{aligned} \tag{10.66}$$

The relation for moduli is obtained from condition $\delta = 0$ as

$$K_2 = -\frac{K_1}{(1 - V_2)}\left[\frac{2(1 - 2\nu_1)}{(1 + \nu_1)} + V_2\right]. \tag{10.67}$$

One coincides with relation (10.64), defining the condition of infinity for bulk modulus of the composite. So, in intrinsic more stable the boundary-value problem into displacements, the extremely great values of effective bulk modulus may be achieved without condition that obtained elastic fields become singular. Moreover, relation (10.63) shows that into conditions of given displacements, it may be obtained maximally great negative bulk modulus of the composite at corresponding selection of negative values of the inclusion bulk modulus. In order to achieve maximal positive values of the composite bulk modulus (this first of all is expedient from practical view), it may be shown [967], that inclusion bulk modulus K_2 should be some more negative compared with bulk modulus K_0 of the composite from Eq. (10.63). However, the determined linear-elastic equation states that all elastic fields become infinite in the composite sphere already at the value of bulk modulus achieving the value of the composite bulk modulus K_0.

By considering the same problem into framework of formulating finite strains with using constitutive equations presenting generalization to 3D statement of problem [1258] of the constitutive equations for 'harmonic material' (firstly considered in [790]), it may be proved [967] that in difference from the linear-elastic case, the fields calculated into framework of the finite strain conditions will always remain finite in the whole composite. This result exists even in the case when values of modulus K_2 achieve critical negative value which defines switch of effective volume modulus of the composite from very great negative values to very big positive ones.

This proposes that in actual elastic material in which could be expected existence some non-linearity in stress–strain state, the value of K_2 may be decreased below critical negative value predicted by linear-elastic analysis without forming singular behavior of the composite. In this case the composite will possess great positive bulk modulus at infinity-small deformations. Similar result occurs for the problem with given boundary stresses.

For 3D composites subjected to Hashin–Shtrikman formulation [651] and satisfactory for known microstructures for which the exact analytical solutions are fulfilled into framework of elastic theory, we obtain

$$G_0 = G_1 + \frac{V_2}{1/(G_2 - G_1) + 6(K_1 + 2G_1)V_1/[5(3K_1 + 4G_1)G_1]}; \tag{10.68}$$

$$K_0 = K_1 + \frac{V_2(K_2 - K_1)(3K_1 + 4G_1)}{(3K_1 + 4G_1) + 3V_1(K_2 - K_1)}. \tag{10.69}$$

Exact solution (10.68) for shear modulus is achieved into framework of hierarchical layer morphology [494]. At the same time, formula (10.69) for bulk modulus is exact solution for hierarchical microstructure of covered spheres with various sizes for which ratio of radii of external layer and coverage has proper value for all inclusions [645]. The both exact solutions remain satisfactory also for negative stiffness of components. As it is known, in the case of rigid inclusions a small their concentration increases weakly the composite stiffness. However, if the inclusions have proper negative stiffness then the stiffness of that composite rises arbitrary aspiring to infinity.

Then, it may be shown that the variation principle providing estimations for all elastic moduli of arbitrary composites remains satisfactory in the case when one of phases of the composite has negative stiffness. The estimations obtained on the base of the variation principle also confirm that the composite stiffness may be made maximally great at corresponding selection of moduli of the phase with negative stiffness. In the variation approach developed in [650], it has been introduced homogeneous 'comparison body' with moduli no depending on coordinates. The stated variation principle satisfied even in the case when phases of the composite had negative values of stiffness until tensors of elastic moduli of actual composite and comparative one had usual symmetry of indices and in condition that the tensor of comparative moduli was selected so that for this solid existed some Green's function. In [1802], it has been stated tensor of effective moduli for arbitrary composite with isotropic distribution of phases. The tensor presents variation estimation which is satisfactory for composites possessing phase with negative stiffness. In the case of two-phase composite consisting of isotropic matrix containing arbitrary distribution of isotropic inclusions of arbitrary forms, the variation estimations for bulk and shear moduli of the composite are determined in the forms [967]:

$$K_0 = \frac{4G(V_1K_1 + V_2K_2) + 3K_1K_2}{4G_1 + 3(V_1K_2 + V_2K_1)}; \tag{10.70}$$

$$G_0 = \frac{G(9K + 8G)(V_1G_1 + V_2G_2) + 6(K + 2G)G_1G_2}{G(9K + 8G) + 6(K + 2G)(V_1G_2 + V_2G_1)}, \tag{10.71}$$

where the parameters without lower indices present corresponding comparative moduli. Note, that for composites with positive-definite phases the standard Hashin–Shtrikman lower (upper) bound is defined from relations (10.70) and (10.71) by selecting the comparative moduli to be equal minimal (maximal) values of the component moduli.

By considering a composite consisting of positive-definite matrix containing inclusions with negative stiffness the obvious selection for the comparative moduli is to select the matrix moduli. In this case Eq. (10.70) and (10.71) show that the variation estimations of bulk and shear moduli can become arbitrary great at proper (negative value) selection of inclusion moduli achieving infinite values at

$$K_2 = -\frac{4G_1 + 3V_2K_1}{3V_1}; \; G_2 = -G_1\frac{3(3 + 2V_2)K_1 + 4(2 + 3V_2)G_1}{6(K_1 + 2G_1)V_1}, \tag{10.72}$$

Note, that these results suggest that, if search for material with maximal bulk modulus then inclusions should be subjected to the condition of strong ellipticity. In this case the bulk modulus of inclusions should has value near to that which is defined by Eq. (10.70), and shear modulus should be selected positive and sufficiently great in order to carry out the condition of strong ellipticity ($K_2 > -4G_2/3$, that will be demonstrated below).

As it easily saw the above-obtained formulae (10.63), (10.69), (10.70) are identical taking into account that the comparative modulus in the last formula is equal to corresponding matrix modulus. This conclusion fulfills also for formulae (10.68) and (10.71), again if the comparative moduli in the last formula are equal to corresponding matrix moduli. Thus, these formulae describe the same behavior of effective moduli in different situations. Therefore, the conclusion on design of the composites with high stiffness possessing phase with proper negative stiffness is applied to all considered cases with in fact quantitatively the same selection of negative moduli obtained in various formulae.

The negative stiffness leads to violation of usual dependence between force and displacement in deformed solid because in the last case the direction of force applied to the solid coincides with the strain direction by assuming that the restoration force will restore neutral state of the deformed solid. The negative stiffness defines unstable equilibrium and hence a positive supply of energy at equilibrium. The negative stiffness differs from negative Poisson's ratio.[3] For isotropic materials Poisson's ratio changing into range from -1 up to 0.5

[3] For example, the foams which expand in transverse cross-section when they are subjected to tension relates to the materials with negative Poisson's ratio. Negative Poisson's ratio is demonstrated by also several mono-crystals and layered structures [965]. Hierarchical layered structures [1701] can demonstrate values of Poisson's ratio achieving -1 [1153]. The foam materials with negative Poisson's ratio possessing non-convex cellular structure have been already designed and processed [963]. Moreover, there have been developed and prepared similar

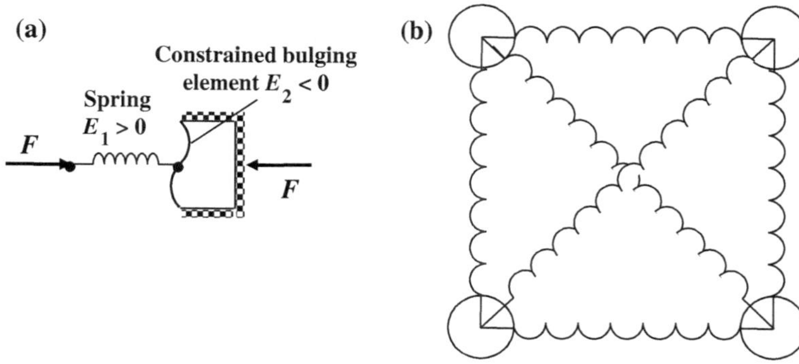

Fig. 10.25 (**a**) Element with negative stiffness (effective modulus E_2) presented by constrained bulging element consecutively disposed with spring possessing positive stiffness (effective modulus E_1) in Reuss model; (**b**) cell consisting of rigid-rotatable nodes with pre-strained link springs. At sufficient pre-deformation shear modulus of the cell is negative

associates with stability of the material, at the same time, bulk materials with negative stiffness are unstable.

The examples of negative stiffness are characterized by proper structures [109]. In particular, the sequence of elements constrained by S-shaped configuration losing stable under loading relates to that structure (see Fig. 10.25a) [967]. A proper longitudinal loading that structure should leads to snapping this element. The condition of negative stiffness is unstable, but all system could be stabilized at a longitudinal constraint, for example due to its unification in rigid block. The negative stiffness forms also in single-cell model foam materials. For example, flexible tetrakaidecahedral models demonstrate non-monotonous dependence of strain on applied force under compression [1445]. In this case, the cells bulge inwardly during high compressive deformation causing geometrical non-linearity. The negative stiffness forms also in cellular structure shown in Fig. 10.25b [967], consisting of rigidly rotatable nodes connected by pre-deformed springs. At proper pre-deforming effective shear modulus of that cell becomes negative. That 2D lattice structure can demonstrate negative Poisson's ratio [964] and even negative shear modulus in condition of sufficient pre-straining.

Instability of materials including components with negative stiffness may be understood following way. If influence by using small force on block of materials possessing positive stiffness then the block will resist to strain due to initiation of opposite force. At the same time, in the case of negative stiffness the block carries the force in direction of the influence forming divergent unstable state.

(Footnote 3 continued)
2D structures [985, 1812]. The value of Poisson's ratio which is equal to -1 could be achieved in chiral cellular structure [1398].

Consider infinite-small homogeneous isotropic linear-elastic deformations from non-constrained state. The strain energy is positive-definite if and only if shear modulus and Poisson's ratio satisfy to the conditions [1679]:

$$G > 0; \; -1 < \nu < 0.5. \tag{10.73}$$

The materials subjected to conditions (10.73) lead to unique solutions for mixed boundary-value problems in which are found any proper combinations of surface stresses and displacements. It has been shown [682] that the uniqueness is the necessary condition for incremented stability, and that materials lead to stable solutions for broad class of mixed boundary-value problems. Based on this, it may be assumed a possibility to process materials with negative Poisson's ratio which should be stable in absence of external constraints.

For boundary-value problems under given displacements the uniqueness of solution [311] defines incremented stability [1048] at strong ellipticity of elastic moduli:

$$G > 0; \; -\infty < \nu < 0.5 \text{ or } 1 < \nu < \infty \tag{10.74}$$

These conditions are lesser constraining compared with conditions (10.73). By using Lame elastic constant λ conditions of strong ellipticity (10.74) are determined as

$$G < 0; \quad \lambda + 2G > 0. \tag{10.75}$$

Then for bulk modulus $K = -\frac{2G(1+\nu)}{3(1-2\nu)} = \lambda + 2G/3$ the condition of strong ellipticity takes the form: $-4G/3 < K < \infty$. These inequalities assume that the condition of strong ellipticity allows volume modulus to be negative. In violation of strong ellipticity a material can demonstrate instability connected with formation of bands of the heterogeneous deformations [897].

Achievement of extremely large shear modulus of composite requires negative shear modulus of inclusions which does not violate the condition of strong ellipticity in the inclusions. However, violation of the strong ellipticity does not guarantee a stability loss of inclusions. The experiments show than in superconductive ferroelastics [1472] and ferroelectrics [214] expenditure of energy due to formation band (or domain) structure suppress the process forming that structure when the material particles to be sufficiently small. Thus, the instability criterion based only on theory of elasticity may predict instability in regimes its actual absence. Hence, negative stiffness is not excluded at all by physical laws. The solids with negative stiffness are unstable if they have free surfaces. They can be stabilized if they are constrained rigidly or by matrix of elastic composite. At the same time solid medium with negative volume modulus will stable under deformation constraints. Negative shear modulus of solid medium can form domain (band) instability associated with loss of strong ellipticity but that instability does not form always when violates strong ellipticity.

The composites with inclusions of negative stiffness may be used by investigating properties of single domains of the superconductive ferroelastics and ferroelectrics. In this case no need that the inclusions were temperature-sensitive superconductive ferroelastics because pre-stressed or pre-strained elements also may be used for this aim.

Thus, sharp (anomalous) changes of composite properties could be predicted in the case when inclusions have negative modulus close in value to the matrix modulus. Similar anomalies, in particular have place in hierarchical structures of Hashin–Shtrikman type and in various composites of 'matrix-inclusion' type [1697]. The anomalies assume singular behavior of composites with linear-elastic components and also containing components in which a non-linearity is determined by geometric shape of the inclusions. If matrix demonstrates non-linearity in constitutive equations then anomaly of the composite modulus becomes finite. If components are linear-viscoelastic [966] then the anomaly in behavior of the composite modulus also to be finite and accompanied by high peak in viscoelastic damping $\tan\delta$.

The present analysis into framework theory of elasticity does not allow one to divide effects of various instability types in the composites possessing phase with negative stiffness. For example, in relation to the band instability it is known from theory of elasticity that in violation of strong ellipticity a formation of the band structure is predicted based on classic theory of elasticity. However, as it has been pointed above the tests show that this band instability suppresses in particles of sufficiently small sizes due to energetic conditions of domain formation. Thus, there is necessary additional physical basis outside frameworks theory of elasticity (for example, connected with experimental measurement of energy expenditure) to theoretically predict possible regimes of that instability.

Extreme elastic and viscoelstic behaviors of composites (elastic moduli and viscoelastic damping) possessing the component with negative stiffness have been demonstrated for composites with dispersed ferroelastic inclusions in quasi-static regime at frequencies in one order of magnitude lower than natural frequency of oscillations [968]. In the case of small concentration of inclusions there were observed extreme values of mechanical damping $\tan\delta$ and stiffness of composite (1 % by volume of vanadium dioxide inclusions into tin matrix) in temperatures near of phase transition of the ferroelastic inclusions far higher than corresponding values for each component [968]. However, theory predicts in this case for particles with infinite stiffness, the composite would be only 2.1 % stiffer than the matrix. If the particle stiffness aspires to zero then the composite would soften by 1.9 % and have no change in $\tan\delta$.

10.2.2 Fracture Features in Domain Structure of Ferroelectric

In order to consider fracture features in the domain structure of ferroelectric, we study a possibility of sub-critical crack in PZT monocrystal, growing parallel to internal 180°-domain or phase boundary in the structure of bi-material [1299, 1316]. Three mechanisms of the crack braking at 90°-domain boundaries are studied in [821, 826], namely (i) due to change of the crack orientation, (ii) due to the crack interaction with non-coherent boundary and (iii) due to the crack propagation through the boundary with absorbed impurities. The crystallite with layer domain structure or co-existing phases will be present as a composite system, which consists of homogeneous layer with thickness h and half-infinite substrate (see Fig. 10.26). Let their elastic properties (E, ν) be the same and the homogeneous tensile stresses, acting in the layer, $\sigma_0 = E\varepsilon$, are found by microstrain ε, caused by the thermal and phase properties of the ferroelectric material. The stress analysis shows that the most possible crack initiation occurs in substrate and it extends parallel to the internal interface of the composite. We assume a stable (sub-critical) character of the crack growth. When the crack is localized in substrate at the depth λh below the interface, then the layer above the crack is subjected to a bend and some elastic strain energy retains after crack propagation. In order to estimate this energy, we take into account that the contributions of the strain energy due to the compressive force P and bending moment M are additive [1676]. Then, using the theory of composite beam and based on the approach of [395], we obtain the asymptotic strain energy release rate, G, for steady-state crack with length c as

$$G = \frac{\partial}{\partial c}\left(\frac{P^2 c}{2EA} + \frac{M^2 c}{2EIh^3}\right), \tag{10.76}$$

where the dimensionless moment of inertia, I, of the beam per unit layer thickness and the effective cross-section, A, are

$$I = (\lambda + 1)^3/12; \quad A = h(1 + \lambda); \quad \lambda = z/h. \tag{10.77}$$

The load P and the bending moment M (per unit thickness) caused by uniform tensile stress σ_0 are defined as

$$P = \sigma_0 h; \quad M = \sigma_0 h^2 \lambda/2. \tag{10.78}$$

Then, we have from (10.76)–(10.78):

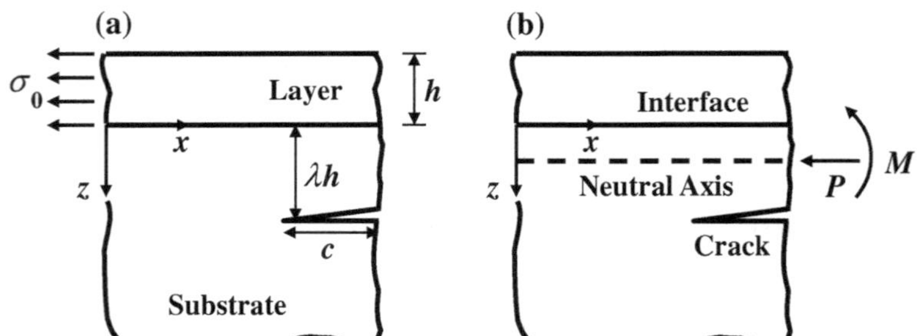

Fig. 10.26 Used composite structure: (**a**) tension σ_0 is caused by microstrain ε in the homogeneous layer; (**b**) equivalent system with applied force P and bending moment M

$$G = \frac{\sigma_0^2 h(4\lambda^2 + 2\lambda + 1)}{2E(\lambda + 1)^2}. \tag{10.79}$$

Further, the dimensional analysis gives relations for modes I and II stress intensity factors, K_{I} and K_{II}, assuming that each of them has own contributions caused by the force P and the moment M [395]:

$$K_I = C_1 P h^{-1/2} f(\lambda) + C_2 M h^{-3/2} g(\lambda); \tag{10.80}$$

$$K_{II} = C_3 P h^{-1/2} f(\lambda) + C_4 M h^{-3/2} g(\lambda), \tag{10.81}$$

where C_i are unknown constants and $f(\lambda)$, $g(\lambda)$ are unknown functions. An assumption about simplicity of the functions $f(\lambda)$ and $g(\lambda)$ is supported by the computational results of [395]. Comparing (10.76) and (10.79) with the relationship: $GE = K_I^2 + K_{II}^2$, connecting the strain energy release rate with the SIFs, we obtain:

$$(C_2^2 + C_4^2)g^2 = 1/2I;\ (C_1^2 + C_3^2)f^2 = h/2A. \tag{10.82}$$

Assume that the factors C_i contribute only to the constant terms of (10.82), while the functions $f(\lambda)$ and $g(\lambda)$ contribute to the variables; then (10.82) are reduced to

$$(C_2^2 + C_4^2) = 1/2;\ (C_1^2 + C_3^2) = 1/2;\ g = I^{-1/2}; f = (h/A)^{1/2}. \tag{10.83}$$

In order to define finally the constants C_i, it is necessary to suggest a strict solubility of the problem for the case of either $P = 0$ or $M = 0$. The validity of the last condition in the present formulation of the problem and also the construction and computational solution of the corresponding integral equation for SIF at half-infinite crack propagation near free surface have been represented in [1676]. The unknown constants have the values: $C_1 = 0.434$ and $C_3 = 0.558$. Besides, there is an additional condition for unknown constants: $C_1C_2 + C_3C_4 = 0$. Then, we obtain from (10.83), $C_2 = C_3 = 0.558$; $C_4 = -C_1 = -0.434$; and the relationships for K_{I} and K_{II} are obtained from (10.80) and (10.81), substituting the corresponding parameters, as

$$\frac{K_I}{\sigma_0 h^{1/2}} = \frac{1}{(1+\lambda)^{1/2}}\left(0.434 + 0.966\frac{\lambda}{\lambda + 1}\right); \tag{10.84}$$

$$\frac{K_{II}}{\sigma_0 h^{1/2}} = \frac{1}{(1+\lambda)^{1/2}}\left(0.558 - 0.752\frac{\lambda}{\lambda + 1}\right). \tag{10.85}$$

As is followed from test data for different bi-material systems and the loading schemes [395, 1676], the crack path, corresponding to the simple Mode I stress intensity demonstrates a surprising stability, and fracture resistance values, obtained under conditions of the Mode I, have a fine reproducibility compared with the results for mixed mode. Furthermore, as the condition of steady-state crack into ferroelectric domain structure, we select the following one:

$$K_{II} = 0. \tag{10.86}$$

Hence, from (10.85), the depth λ^*, at which the crack has a steady-state trajectory, namely: $\lambda^* = z^*/h = 2.876$ is found. The existence of the asymptotic limit permits to calculate the critical layer thickness, h^*, below which the complete fracture is inhibited. This value is obtained by equating the asymptotic value of $K_{\rm I}$ to the corresponding value of the fracture toughness, $K_{\rm Ic}$. In the case of the crack growth into substrate along the steady-state trajectory, $K_{\rm II} = 0$, (10.85) provides the prediction:

$$h^* = 0.755(1 + \lambda^*)(K_{Ic}/\sigma_0)^2. \tag{10.87}$$

The above assumption about stable crack growth in ferroelectric crystallites demands additional discussions and grounds. First, in [571], it has been shown that a stress concentration near crack tip in the region of the morphotropic boundary at existence of the tetragonal (*T*) and rhombohedral (*R*) phases can initiate phase transitions in the ceramic grains. It has been proved that for short cracks ($c < c^*$), at any great stresses, an initiation of phase transformations near crack tip is more advantageously compared to its catastrophic growth. For long cracks ($c > c^*$), the calculation gives a higher critical load compared to Griffith's formula. Secondly, in [204], the mechanism of local phase transition near heterogeneities has been considered, in particular at crack tip. It is shown that an interaction of order parameter with strains leads to the fact that the stresses, concentrating at heterogeneities, can transform the media in other phases [204]. Thus, an occurring of local transformational plasticity in the vicinity of the crack tip is possible, dragging its growth. Thirdly, a non-linear decreasing of the mechanical strength has been observed at a tension with an increasing of a poling field in the poling of the ferroelectric on the base of $PbTiO_3$ [578]. This non-linear behavior is explained by non-stationary processes of microcracks nucleation and growth due to their interaction with re-constructing domain structure and defects. In particular, it is assumed that the vacancies, released from connections after disappearance of domain walls during poling, can form "clouds" near tips of the cracks with critical size and restrain their growth. Finally, the cracks in $BaTiO_3$ monocrystals which have been observed near indenter imprint at the (001) facet and grown along the cleavage planes of {100}, {110} and {111} can restrain up to total stop at strong fixed 90°-domain boundaries which are similar to the twin boundaries [821]. Hence, it is seemed to be correct a conclusion about principle possibility of stable, sub-critical crack growth into domain structure of ferroelectric.

As has been shown by experiments [950], the strains inside crystallites in paraelectric phase change from 5×10^{-5} to 2×10^{-4} in the morphotropic transition regions in dependence on the sintering temperature of PZT ceramics. Similar strains in grains of the *T*-phase are independent of temperature within test error. They exceed the microstrains in the paraelectric phase, attaining the values $\sim(6 \pm 3) \times 10^{-4}$. The microstrains in the crystallites of *T*- and *P*-phases coincide in the order of magnitude. Therefore, in computer simulations, we consider the microstrain range: $\varepsilon = 5 \times 10^{-5} - 9 \times 10^{-4}$, for definition of the uniform tensile stress σ_0, leading to the crack propagation.

Select the values of Young's modulus, E = 63–100 GPa [1368], and also the fracture toughness of PZT monocrystals, K_{Ic}. In relation to the latter, note that the poling processes lead to the fracture toughness anisotropy as of whole ceramic, as of single crystallites. In the ferroelectric ceramics, the value of K_{Ic} attains maximum value along the poling direction and minimum value in the perpendicular direction [1318, 1368]. However, anisotropy of the tetragonal ferroelectric monocrystals shows another character, namely: the fracture toughness in a plane perpendicular to the direction of spontaneous poling exceeds significantly the toughness values in other directions. Fracture in this plane is not advantageous energetically, because it leads to a greater surface density of charges with different signs at the opposite surfaces of the crack [821]. This fracture character is experimentally supported by prevailing growth of microcracks through grain bulk in the poling direction [578]. These features of anisotropic behavior can be explained: in the polarized ceramic, a significant part of domains remains oriented in arbitrary direction, not contributing to the anisotropy of K_{Ic}. Moreover, for fine-grain ceramics, an intergranular fracture is proper, which excludes the crack interaction with domain structure [1368]. Due to complex behavior of the fracture toughness, we use in calculations the experimental values: K_{Ic} = 0.4–1.4 MPa m$^{1/2}$ [1368].

Finally, we define the layer thickness as $h = nh_d$, where $h_d \approx 0.2$ μm is the equilibrium domain width for grain size, D = 10 μm, based on the simulation of the stress sources at the boundaries of polydomain grains, using continuously distributed dislocations [1353]; n is the domain number in crystallite under uniform tension, σ_0. Obviously, in order to the coordinate z^* for steady-state crack path to be into grain, it is necessary to select h_{max} = 3.4 μm, then $n_{max} = 17$.

Further, (10.84) and (10.87) give $K_I = 0.585\sigma_0 h^{1/2}$, and $h^* = 2.926(K_{Ic}/\sigma_0)^2$, where K_I is the SIF for steady-state crack growth at the depth of $\lambda^* h$, and h^* is the limit layer thickness, defining a restraint fracture. The selection of the parameters, corresponding to the maximum of K_I and minimum of h^*, gives $\varepsilon = 9 \times 10^{-4}$; K_{Ic} = 0.4 MPa m$^{1/2}$; h = 3.4 μm; E = 100 GPa. As a result, we obtain K_I = 0.097 MPa m$^{1/2}$, h^* = 57.8 μm. Obviously, $K_I \ll K_{Ic}$ and $h^* \gg D$ for the considered PZT parameters. Hence, the stable (i.e., subcritical) crack growth is impossible parallel to or along the interfaces in the considered crystallites. The nucleated crack at this boundary will propagate catastrophically until it meets a fixed 90°-domain boundary. Thus, there may be stable crack growth in ferroelectric crystallites only in the cases of crack braking by the 90°-domain boundaries [821, 826].

10.2.3 Domain Structure and Fracture of Ferroelectric Ceramic

It is well known that the fracture toughness anisotropy of ferroelectric ceramic (FC) is the sequence of structure changes during poling. There are three main causes of formation of the fracture toughness anisotropy [1368]: (i) the toughness anisotropy of single crystallites (due to the crack interaction with the boundaries of twins), the greater part of which is oriented along the poling direction; (ii) the anisotropy of internal (residual) stresses into micro-volume of piezoelement, leading at the crack propagation to higher values of K_{Ic} in poling plane and to lower values in perpendicular plane; (iii) the plastic deformation of twinning near crack tip. The investigations of (i) and (ii) cases have been carried out using computer simulations in [826, 1318]. Here, we consider the twinning effects on the strength properties, taking into account the ceramic ferrohardness and initial porosity on example of PZT ceramics, sintered under thermal gradient at different initial press-powder porosities. The FC structure is described by lattice with square cell size, δ. After ceramic fabrication (i.e., sintering, shrinkage, cooling and poling process) a microstructure contains the grains, voids, microcracks at the intergranular boundaries and domain structure. We estimate a microcrcaking by poling due to domain re-orientations which differ from the 180° ones (otherwise the residual stresses do not occur). In this case, the spontaneous strains, ε, for different ferrohardness degrees have been defined by experiments [472] as $\varepsilon_{I} = 14 \times 10^{-3}$, $\varepsilon_{II} = 11 \times 10^{-3}$ and $\varepsilon_{III} = 8 \times 10^{-3}$ (where indexes I, II and III are concerned with compositions, possessing by ferrohardness, mean ferrohardness and ferrosoftness, respectively). Then, the critical length of the cracked boundary, l_c^s, and the facet length subject to cracking, l, for each composition are found by (10.4) and (10.5).

After localization of the microcracks at the intergranular boundaries, we model stress sources at the polydomain grain boundaries by the continuously distributed effective dislocations and obtain an equilibrium domain width as [1353]

$$d = \left[\frac{\pi^3 D(1-v)\sigma_d}{4.2GS_0^2(2-v)}\right]^{1/2}, \tag{10.88}$$

where D is the mean grain size independent of the ferrohardness degree, σ_d is the energy of 90° domain boundary, G is the shear modulus, S_0 is the spontaneous shear strain and v is Poisson's ratio. Note that the critical grain size, $D*$, lower of which there is no domain twinning, may be found under condition of $D = d$ (a structure of mono-domain grains) as

$$D* = \frac{\pi^3(1-v)\sigma_d}{4.2GS_0^2(2-v)}. \tag{10.89}$$

Further, it is known that FC poling causes a concentration decrease in domain boundaries in the poled sample as compared with the non-poled samples. The structure transition of the twinning type under stress near crack tip can provoke a reverse process of the domain boundaries density rise, but a twinning process zone, forming around propagating crack, will impede the crack. Obviously, the process zone size depends on the composition ferrohardness and the ceramic piezoelectric properties. Moreover, this size is also defined by the crack propagation direction. It should be noted that the maximum influence is rendered on the crack, growing along residual poling direction. In contrast, along normal one this structure change is restrained by the stress state near crack tip, that is, the crack propagation is realized without an additional energy loss, and the twinning process zone around the crack does not form.

Around initiated macrocracks favorably oriented to poling direction, there are twinning process zones with thickness h_d. Then, a toughening due to the twinning may be estimated in the crack shielding terms. The shielding, corresponding to the crack growth nucleation is represented by the solution for stationary crack, but the crack nucleation toughness, K_c, has form [359]:

$$K_c = (E_m/H)^{1/2} K_c^0, \qquad (10.90)$$

where $K_c^0 = (2\gamma_0 E_m)^{1/2}$ is the critical SIF, required for the crack growth and reflecting a fracture resistance of twin planes, γ_0 is the fracture energy, H is Vickers hardness, E_m is Young's modulus for cracked ceramic, calculated by (10.16), where β_m is the microcracking density, formed by poling. By neglecting the closed porosity effects, the steady-state crack toughening due to twinning is estimated as [359]

$$\frac{K_{SS}}{K_c^0} = \left\{1 + \left(\frac{E_m}{H} - 1\right)\left[\alpha^2 \ln\left(\frac{h_d}{d}\right) - \frac{\sigma_0^2 h_d}{\left(K_c^0\right)^2}\right]\right\}^{1/2}, \qquad (10.91)$$

where K_{SS} is the steady-state crack toughness, $\alpha^2 = 3/8\pi$, σ_0 is the yield stress or the threshold stress, defining the twinning nucleation near crack tip.

The process zone width, h_d, for different ferrohardness degrees is found, using a calculation of the parameter, $0 < \xi < 1$, reflecting a stability degree of piezoelectric composition to de-poling (i.e., the ferrohardness degree) [928]. These results are based on the energy balance of the modeled process and on application of the finite element method to calculate the zone of feasible microstructure transformations at the crack tip. For PZT ceramic compositions I, II and III, respectively, we obtain the values [928]: $4h_d/D = 0.2$ (for $\xi = 0.3$), 0.5 (for $\xi = 0.5$) and 1.2 (for $\xi = 0.7$).

On account of a necessary number of computer realizations, defined on the basis of stereological method [271], in order to obtain numerical results, we use known data for PZT ceramics and related materials: $E = 70$ GPa, $v = 0.25$ [1368], $H = 3$ GPa [1367], $G = 20$ GPa, $S_0 = 0.01$, $\sigma_d = 3 \times 10^{-3}$ J/m^2 [1353].

The cell size, δ, is selected to be equal to the critical size, l_c^s, for composition with mean ferrohardness. Then, selecting $\gamma_0 = \gamma_s - \gamma_{gb}\,(1 - f_m)/2$, where γ_s is the surface energy for bulk body ($\gamma_s \approx 2\gamma_{gb} = 6$ J/m^2 [1296]) and f_m is the fraction of facets cracked in the poling process, we calculate the values of D, β_m and f_m using computer simulation. Finally, the threshold stress, $\sigma_0 = 20$ MPa, is used which has been estimated in the experiments on impact loading of $BaTiO_3$ ceramic [935]. The numerical results show that the observed decreasing of the grain size, D, with increasing of the initial porosity, C_p^0, leads to a decreasing of the width of equilibrium domain, d, and the process zone size, h_d. The latter also decreases at rise of

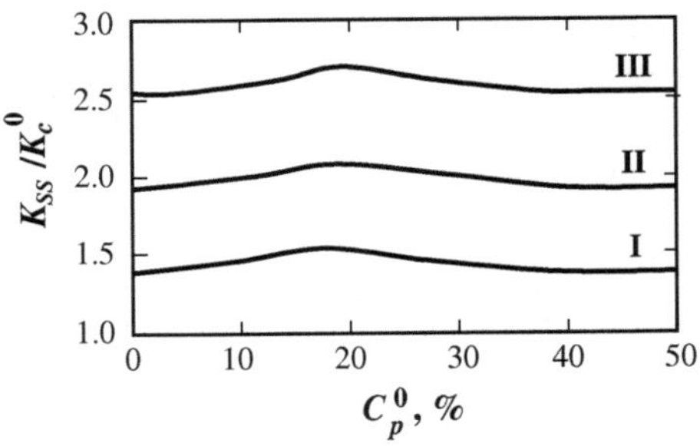

Fig. 10.27 Fracture toughness (K_{SS}/K_c^0) at steady-state crack *versus* initial porosity (C_p^0) for different ferrohardness of FC

the FC ferrohardness. In the toughening due to twinning, the value of K_c/K_c^0, defining start of the crack growth, changes in the limits: 4.06 ± 0.15, 4.30 ± 0.15 and 4.74 ± 0.08, for I, II and III compositions, respectively, at the initial porosity $C_p^0 = 0 - 60$ %. Similar dependencies of toughening, K_{SS}/K_c^0, on the ferrohardness and the initial porosity in the case of the steady-state crack, are shown in Fig. 10.27 [1297]. The results for the case of mean ferrohardness are close to the ratio of the fracture toughness in existence of twinning to its absence in γ-TiAl [359]. In this case, the known growth of the twinning process zone and fracture toughness with increasing of FC ferrosoftness [1368] is supported. FC toughening, caused by the twinning process, is a more effective toughening mechanism compared to microcracking near crack tip [827], to crack branching and crack bridging [1298], and also at phase transformations due to coexistence of rhombohedral and tetragonal ferroelectric phases in the region of morphotropic transitions. The obtained maximum toughening in the range of change of the initial porosity (at $C_p^0 = 20$ %) could be explained by mutual superposition of complex microstructure effects and is qualitatively supported by similar results for the toughening mechanism of microcracking near crack tip [1368].

10.2.4 Influence of Domain Switching Near the Crack on Fracture Toughness

In order to study the influence of the ferroelectric domain switching on changes in crack shielding and fracture resistance, we consider the fracture of ceramic taking into account the poling, re-poling and wake energy process effects by using Kreher-Pompe's model [936, 937] based on the energy balance method combined with a suitably simplified domain switching model. For a steady-state crack propagating together with a crack tip process zone of domain switching (see Fig. 10.28), the energy balance is presented as [936]

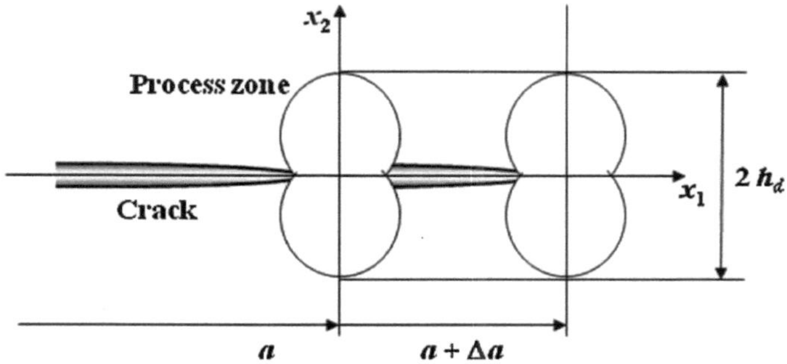

Fig. 10.28 Steady-state crack propagation with crack tip process zone of domain switching. Electrical field is applied perpendicular to plane x_1x_2 and tension mechanical stress is directed along axis x_2

$$\Delta W = \Delta W_\gamma + \Delta W_d. \tag{10.92}$$

Here $\Delta W = G\Delta a$ is the macroscopic energy supply defined by applied electrical (E) and mechanical (σ) fields, and the change of internal energy outside the crack tip zone, $G = G\ (\sigma, E, a)$ is the release rate of the strain energy, Δa is the crack size difference. The applied electric field E is constructed to be parallel to the crack front, i.e., the poling direction is also parallel to this front and the release rate of the strain energy is dependent only on the applied stress σ perpendicular to the crack plane.

The criterion of crack growth is $G = G_c$, where $G_c = (1 - \nu^2)K_{Ic}^2/E_0$ is the critical release rate of the strain energy, K_{Ic} is the fracture toughness, E_0 is the Young's module measured under constant electric field. Then, $\Delta W_\gamma = 2\gamma_b\Delta a$ is the energy which is available locally at the crack tip and which can be consumed at the creation of new fracture surfaces; $\Delta W_d = 2h_d\Delta a(w_d + w_w)$ is the net energy which is dissipated or stored in the nonlinear process zone around the crack tip, where w_d is the energy dissipated per unit volume when the material element is transformed from the state outside the process zone into a state in the wake far behind the crack tip; w_w is the energy stored per volume in the wake region due to a possible residual stress state existing behind the crack tip. Thus, Eq. (10.92) of energy balance reduces to the following relationship:

$$G_c = 2\gamma_b + 2h_d(w_d + w_w), \tag{10.93}$$

which is defined as the sum of the extrinsic toughness and contribution (positive or negative) caused by the process zone. Similar relationships are also obtained for various crack problems, for example in Refs. [201, 936, 937]. The introduction of a simplified description of hysteretic FC behavior based on internal variables that measure the relative frequencies of a set of six representative domain orientations allows one to analyze the relevant processes of each material volume element near the crack tip, namely (i) poling due to electric field, (ii) domain switching under mechanical and electric loads, and (iii) partial re-switching into internal stress state in the wake region [938]. Then, Eq. (10.93) may be presented as

$$G_c = 2\gamma_b + 2h_d w_d^0 f\left(\frac{E}{\overline{E}_c}, \Delta E, \eta_m, \eta_{me}\right), \tag{10.94}$$

here $\overline{E}_c = (E_{c^+} + E_{c^-})/2$, $\Delta E = (E_{c^+} - E_{c^-})/2\overline{E}_c$ $(0 \le \Delta E \le 1)$, $w_d^0 = 1.5\varepsilon^{s0}\bar{\sigma}_c$, and the relative energy densities are

$$\eta_m = \frac{3E_0\varepsilon^{s0}}{2(1-\nu^2)\bar{\sigma}_c}, \quad \eta_{me} = \frac{3\varepsilon^{s0}\bar{\sigma}_c}{2P^{s0}\overline{E}_c}, \tag{10.95}$$

where $\bar{\sigma}_c = (\sigma_{c^+} + \sigma_{c^-})/2$, ε^{s0}, and P^{s0}denotes the maximal values of remnant strain and polarization, respectively. The parameters $\sigma_{c^+}(\sigma_{c^-})$ and $E_{c^+}(E_{c^-})$ are shown in Fig. 10.29, which also shows the simplified nonlinear constitutive material laws for homogeneous uniaxial loading of ferroelectric and ferroelastic materials. The curves begin from a non-poled isotropic initial state in a linear regime. The ferroelastic (ferroelectric) transition occurs by achieving a certain coercive stress or electric field. This process continues until saturation is attained at σ_{c^+} or E_{c^+}. Unloading from the intermediate state results in a certain remnant strain and polarization values lower the maximal values depicted in Fig. 10.29. As shown in [938], the function f could be explicitly given and for any material the condition 2/3 $\le f \le$ 2 fulfilled.

In order to obtain numerical results, we use some typical results for PZT and $BaTiO_3$ ferroelectric ceramics [907, 1297], namely $E_0 = 70$ GPa, $\nu = 0.35$, $\bar{\sigma}_c =$ 40 MPa, $\varepsilon^{s0} = 1 \times 10^{-3}$, $\overline{E}_c = 1$ kV/mm, $P^{s0} = 0.05$ C/m^2, $\Delta E = 0.5$. Estimation of fracture toughness requires knowledge of the size of the process zone h_d. For the PZT ceramic compositions I, II and III, respectively, we again as in Sect. 10.2.3 use the next values [929]: $4h_d/D = 0.2$ (for $\xi = 0.3$), 0.5 (for $\xi = 0.5$), and 1.2 (for $\xi = 0.7$), where D is the mean grain size independent of the ferrohardness degree.

A necessary number of computer realizations are established on the basis of the stereological method [271]. Then, the square cell size Δ is selected equal to the

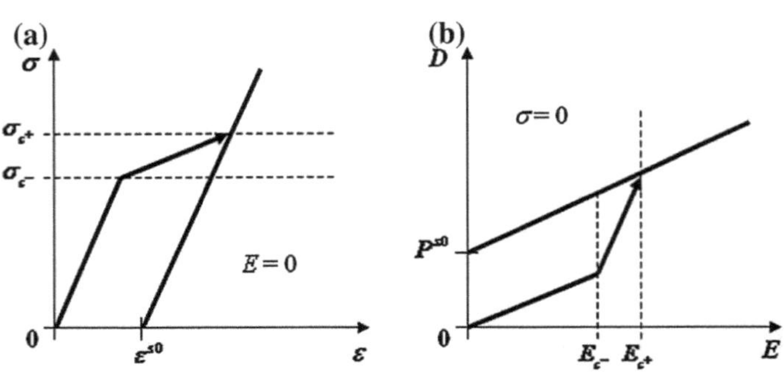

Fig. 10.29 Schematic (**a**) non-linear uniaxial mechanical (stress σ *versus* strain ε) and (**b**) electrical (dielectric displacement D *vs* electric field E) behavior of ceramic

Table 10.7 Computer simulation results

Properties		Initial porosity C_p^0 (%)					
		0	10	20	30	40	50
D/l_c^s		2.04	1.92	1.84	1.71	1.69	1.61
D/D^0		1.00	0.94	0.90	0.84	0.83	0.79
h_d, μm	Ferrohard ceramic ($\xi = 0.3$)	9.54	8.98	8.60	7.99	7.90	7.53
	Mean-ferrohard ceramic ($\xi = 0.5$)	23.84	22.44	21.51	19.99	19.75	18.82
	Ferrosoft ceramic ($\xi = 0.7$)	57.23	53.86	51.62	47.97	47.41	45.17

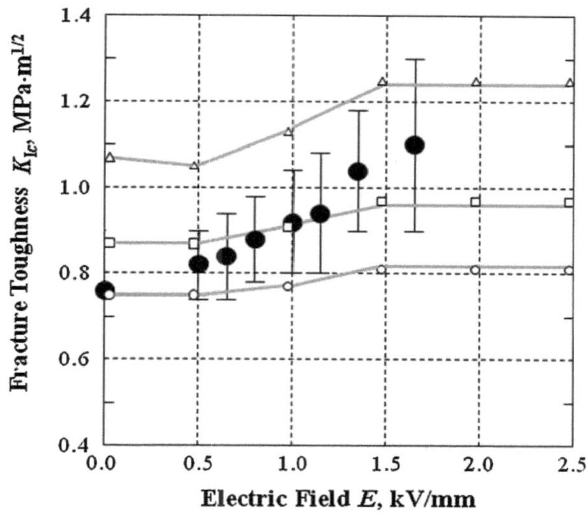

Fig. 10.30 Fracture toughness K_{ic} of ferroelectric ceramics dependent on applied electric field E. *Black dots* represent the experimental results for barium titanate ceramics [907], the open shapes denote results of computer simulations for PZT ceramics at the $C_p^0 = 0$; triangles represent ferrosoft composition, rectangles—composition with mean ferrohardness, circles—ferrohard composition

critical size l_c^s for composition with mean ferrohardness. For quantitative comparison, we need the grain boundary energy γ_b of FC without any dissipative contribution connected with domain re-orientation. The value of $\gamma_b = 3$ J/m^2 could be used with this aim [1297]. The obtained numerical results are present in Table 10.7 and Fig. 10.30 [1314].

Thus, it follows from Table 10.7 that the size of the process zone h_d qualitatively coincides with the corresponding results obtained for barium titanate ceramics possessing near properties, namely in the theoretical model of the ferroelectric domain switching [938] $h_d = 25$–35 μm and in experimental observations [1134], where a characteristic maximal half-width of the process zone of about 40 μm is obtained. Figure 10.30 also demonstrates qualitative coincidence with the experimental results [907], again nearest FC with mean ferrohardness. Figure 10.30 also allows an estimate of the degree of toughening due to ferroelectric domain switching, in comparison with the obtained results with the intrinsic fracture toughness K_{Ic}^0 (for examples for barium titanate ceramics $K_{Ic}^0 = 0.48$ MPa m$^{1/2}$ [907]). The numerical results for PZT compositions with various ferrohardness qualitatively also coincide with the results of other modeling [1297].

10.2.5 Thermodynamics of Martensitic Transformation in HTSC

Now, investigate the twinning processes in YBCO superconductor [1320]. As has been noted above, due to the grain TEA, the cracking of HTSC ceramics occurs during cooling that deteriorates mechanical and strength properties. At the same time, a twinning process zone, forming around an advancing crack, as may be suggested similar to ferroelectric ceramics, should lead to the crack shielding and to the fracture toughness increasing. In this case, a relaxation of internal stresses of second type, nucleated in the HTSC, occurs in two ways [473]: (i) due to the crack nucleation, which is proper for temperatures, $T \leq 500$ °C, and (ii) owing to the martensitic process of re-construction of the grain domain structure, that is most intensive at $T \geq 500$ °C.

During oxygen annealing of HTSC, two characteristic sites are observed at the curve of oxygen parameter (x) in dependence on time (t). At the initial "fast" stage of the process, where an accumulation of oxygen by cuprate follows to diffusive regime, the dependence "x–t" is subjected to parabolic law. At the next "slow" stage, the kinetics of reaction are found by relaxation of elastic stresses, and oxidation follows to $\partial x/\partial t \approx$ const regime, that is, to linear law [473, 474]. As has been shown in [473], $YBa_2Cu_3O_{7-x}$ compound possesses the "shape memory effect" that speaks about martensitic mechanism of the stress relaxation at the slow stage of the material oxidation. Therefore, it may be assumed that deletion of the internal stresses of second type initiated near growing macrocrack can occur, according to the martensitic mechanism in account of the energetically advantageous re-construction of domain structure.

We shall assume that the stresses, introduced in the twinning process, to be shear. The normal stresses, defining a boundary microcracking, give a secondary contribution in the energy of plastic deformation and do not take into account here.

Estimate the critical number of twins, η_c, in spherical grain with radius R, corresponding to the martensitic transformation. With this aim, consider alterations in the thermodynamic potential, which accompany formation of the twinned martensite. They include the increment of the mechanical potential, $\Delta\Phi_m$, consisting of change of the strain energy and interaction energy, the increment of the surface energy, $\Delta\Phi_S$, and also the increment of the chemical potential, ΔF_c. The last is independent of geometrical parameters of twins, and for spherical particle with radius R, the increment of the chemical free energy per crystallite is found by [1394]

$$\Delta F_c = -(4/3)\pi R^3 \Delta F_0, \tag{10.96}$$

here

$$\Delta F_0 = P_V \delta_{ij} \Delta e_{ij}^T = P_V \Delta V, \tag{10.97}$$

where P_V is the stress, leading to the transformation; Δe_{ij}^T is the difference of components of the strain tensor; ΔV is the volumetric deformation at the transformation; and Kronecker's delta-function is found as

$$\delta_{ij} = \begin{cases} 1, & i = j; \\ 0, & i \neq j. \end{cases} \tag{10.98}$$

The surface work is defined by changes occurring at the martensite interface and during formation of the twin boundaries. The total increment of the surface energy, $\Delta\Phi_S$, for spherical grain is calculated as [451]

$$\Delta\Phi_S = 4\pi R^2 \Delta\Gamma_i + (2/3)\pi R^2 (\eta - 1)(\eta + 1)\Gamma_t/\eta, \tag{10.99}$$

where $\Delta\Gamma_i$ is the difference of the surface energy; Γ_t is the energy of twin boundary; $\eta = 2R/d$ is the number of twins per grain; d is the twin extent (see Fig. 10.31).

The difference of the mechanical energy has two terms, namely (i) due to the twinning, $\Delta\Phi_T^m$, and (ii) owing to macroscopic change of the grain shape, $\Delta\Phi_m^m$. The component caused by the twinning is calculated using Eshelby's method [449] for study of the tangential stress state, developing in the martensite plates, consisting of twins. Then, for Poisson's ratio ($v = 0.2$), we have [451]

$$\Delta\Phi_T^m = \frac{1.6\pi R^3 G\gamma_T^2}{3\eta}, \tag{10.100}$$

where G is the shear modulus, γ_T is the shear deformation at the twinning. This term is determined by the tangential stresses, an action of which is restrained by the zone close to the interface "parent phase–twinning product". Moreover, due to neglectfulness by the normal stresses, it may be assumed that (10.100) presents lower boundary for possible values of $\Delta\Phi_T^m$.

Macroscopic component includes the strain energy and the energy of interaction with applied stress p_{ij}^A. We shall be limited by only dilatation component of the applied stress, p^A (neglecting deviator component) and while the fraction of transformed particle contains a set of planes, formed by shear, there are altogether alternative planes subjected to shear of opposite sign. Therefore, appreciable macroscopic shear strains apparently are difficultly expected even in the case of finite number of twins. Then, $\Delta\Phi_m^m$ is found as [455]

$$\Delta\Phi_m^m = (4/3)\pi R^3 \Delta V (0.14 E \Delta V - p^A), \tag{10.101}$$

where $E = 2G(1 + v)$ is Young's modulus.

Taking into account (10.96), (10.99)–(10.101), we obtain a change of the total potential due to the transformation as

$$\frac{\Delta\Phi}{(4/3)\pi R^3} = -\Delta F_0 + \frac{6\Delta\Gamma_i}{\eta d} + \frac{(\eta^2 - 1)}{\eta^2 d}\Gamma_t + 0.14 E \Delta V^2 - p^A \Delta V + \frac{0.4}{\eta} G\gamma_T^2. \tag{10.102}$$

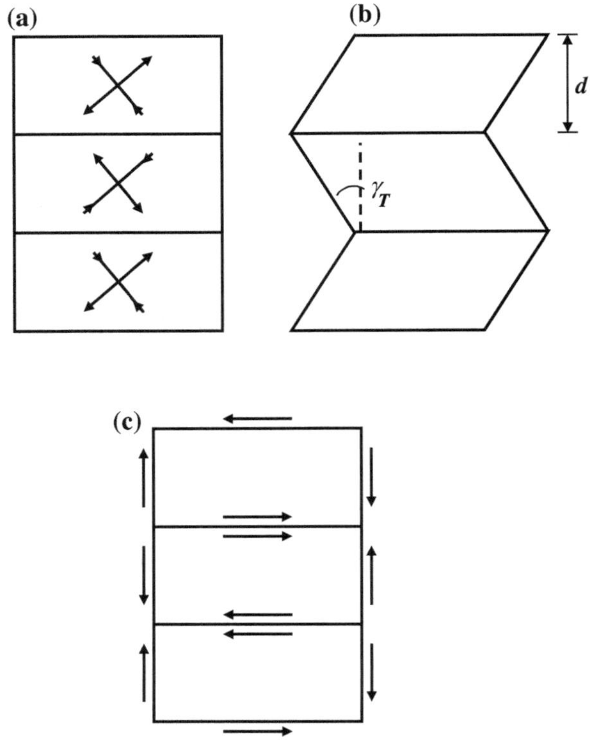

Fig. 10.31 (**a**) No restraint shape of particle before start of twinning; the directions of tension and compression along main axes, characterizing transformation are shown by *arrows*; (**b**) the same particle after twinning (γ_T is the shear angle, d is the twin extent); (**c**) restoration of initial shape, using uniform tangential stresses ($\tau = G\gamma_T/2$), shown by *arrows*

Thus, the transformation will be defined by four main parameters, namely (i) the chemical free energy (ΔF_0), (ii) the grain radius (R), (iii) the extent of twins (d), and (iv) the applied hydrostatic component of stress (p^A).

In order to estimate a critical value of $\eta_c = 2R_c/d$, corresponding to the transformation (where R_c is the critical grain radius), we equate the change of the total potential, $\Delta\Phi$, to zero. In this condition, there is minimum value of the critical number domains per grain, η_c, at which the transformation will start. Then, we have from (10.102):

$$-\Delta F_0 + \frac{6\Delta\Gamma_i}{\eta_c d} + \frac{(\eta_c^2 - 1)}{\eta_c^2 d}\Gamma_t + 0.14E\Delta V^2 - p^A\Delta V + \frac{0.4}{\eta_c}G\gamma_T^2 = 0. \quad (10.103)$$

Next, estimate the twin extent, d, using a modeling of the stress source at the boundaries of polydomain grains by the continuously distributed dislocations as [1353]

$$d = \left(\frac{\pi^3\sigma_d D}{12.6GS_0^2}\right)^{1/2}, \quad (10.104)$$

where S_0 is the spontaneous shear strain; σ_d is the energy of 90°-domain wall. Note that the square root dependence $d \sim D^{1/2}$ has also been predicted for partially stabilized ZrO_2 at $D \gg d$ [451]. Moreover, the critical grain size, D^*, lesser than which the domain twinning does not occur, is estimated from the condition, $D = d$. Hence

$$D^* = \frac{\pi^3 \sigma_d}{12.6 G S_0^2}. \tag{10.105}$$

This value corresponds to the grain, which has no sufficient elastic energy to render significant influence on the agreement of domain wall, possessing the energy σ_d.

Substituting (10.104) into (10.103) and assuming $p^A = 0$ (i.e., an absence of applied stresses), we obtain finally the equation for calculation of the critical value of η_c.

10.2.6 About Toughening of Superconducting Ceramics

Today, a special value has consideration of stable (sub-critical) crack growth at study of fracture processes in the framework of microstructure fracture mechanics. Differing from unstable (catastrophic) fracture, satisfying to the classic Griffith–Irwin's theory, it can take into account internal mechanisms of fracture resistance, which are intrinsic for the material structure, and defines essentially true strength properties. In the framework of the force approach by using the SIF for given fracture mode (I, II, III or their combination), the constitutive relationships, defining equilibrium states of crack growth at the pointed stages of its propagation, have the form [308]:

$$dK/dc > dK_c/dc \textit{ for unstable fracture}; \tag{10.106}$$

$$dK/dc < dK_c/dc \textit{ for stable fracture}, \tag{10.107}$$

where c is the length of the growing crack; K_c is the critical value of SIF, defining the crack start (i.e., the ceramic fracture toughness). If there are internal mechanisms, contributing in the change of the crack growth, then effective driving force, acting in the crack tip (K), can be presented as a sum of applied external load (K_a) and internal residual stresses (K_i):

$$K = K_a + \sum_i K_i. \tag{10.108}$$

The terms of K_i in (8.73) can interpret the shielding and anti-shielding (leading to the crack amplification) influences on transmission of external stresses to the crack tip depending on their signs (i.e., negative or positive, respectively). In particular, the microcracking near macrocrack, crack branching and crack bridging, studied in Sects. 10.1.6–10.1.8, lead to the shielding effects. At the same time, the microcrack–macrocrack coalescence causes a decreasing of fracture resistance (the anti-shielding effect).

Now, consider directly the toughening of HTSC ceramic, caused by the processes of twinning near growing crack. In this case, the hydrostatic component of stress, required to introduce the transformation, is obtained from the condition $\Delta\Phi \leq 0$ and can be written using (10.102) as

$$p^A \Delta V \geq -\Delta F_0 + \frac{6\Delta\Gamma_i}{\eta d} + \frac{(\eta^2 - 1)}{\eta^2 d}\Gamma_t + 0.14E\Delta V^2 + \frac{0.4}{\eta}G\gamma_T^2. \quad (10.109)$$

The hydrostatic stress near the crack tip is subjected to the non-equality [451]:

$$p^A \leq \frac{2K(1+v)}{3\sqrt{2\pi r}}\cos(\Theta/2), \quad (10.110)$$

where $\{r, \Theta\}$ are the polar coordinates connected with the crack tip. Then, using non-equalities (10.109) and (10.110), we obtain the distance of r_c^* from the crack tip at which single grain will be subjected to the transformation. We have for $v = 0.2$:

$$2\pi r_c^* = \left\{\frac{0.8K\Delta V d\eta^2\eta_c^2\cos(\Theta/2)}{[2(3\Delta\Gamma_i + 0.2G\gamma_T^2 d)\eta\eta_c - \Gamma_t(\eta_c + \eta)](\eta_c - \eta)}\right\}^2. \quad (10.111)$$

Then, the size of the transformation zone, r_c, can be calculated, using (10.111) at $K = K_c$ and maximum value of the reached toughness (for $\Theta \approx \pi/3$) [452, 1087]. The latter suggests that each particle is subjected to sufficient shear stress, which introduces the martensitic transformation under condition of thermodynamic restriction (10.103) (at $p^A = 0$), superimposed on the particle.

Note that the value of the angle $\Theta \approx \pi/3$ divides the zones of the toughening and the crack amplification near the crack tip. The transformed particles at the crack front increase K_{tip} (the stress concentration at the crack tip), while the particles disposing into the region, $|\Theta| > \pi/3$, lead to the opposite effect.

Finally, consider the toughening, increasing fracture resistance of HTSC ceramics due to twinning in the process zone with width $2r_c$, surrounding the advancing crack (see Fig. 10.32). Taking into account the superconductor behavior, caused by the hysteresis curve "σ–ε", we can estimate the toughening for the steady-state crack, caused by the twinning as [201, 359]

$$K_{SS} = \sqrt{\left(K_c^0\right)^2 + E\Delta J_{SS}}, \quad (10.112)$$

where $\Delta J_{SS} = 2\int_0^{r_c} U(y)dy$ is the increment of fracture resistance; U is the square under the curve "σ–ε", calculated at each value of y-coordinate on the basis of the governing law for the material subjected to the twinning.

Equation (10.112) is reduced to the following form for one-axis curve "σ–ε" [359]:

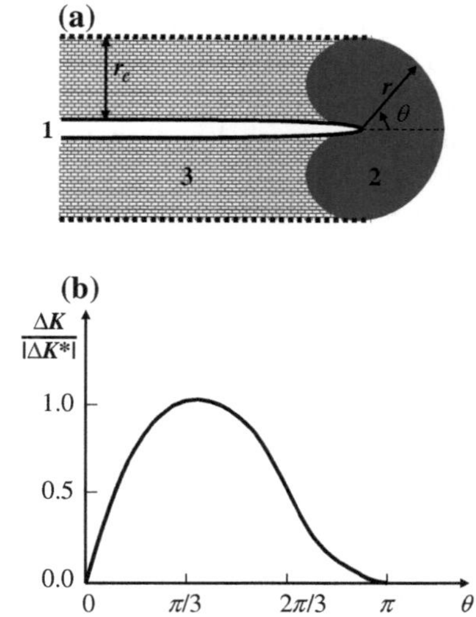

Fig. 10.32 Effect of transformation zone, forming near crack (1), on ceramic toughening: (**a**) frontal (2) and wake (3) zones of toughening (the pointed parameters are discussed in text); (**b**) fracture toughness change, connected with disposition of transformed particle; ΔK is the SIF difference due to total development of the transformation zone, that is, at great values of $\Delta c/r_c$ [452]

$$\frac{K_{SS}}{K_c^0} = \left\{1 + \frac{(E-H)}{H}\left[\frac{3}{8\pi}\ln\left(\frac{r_c}{d}\right) - \left(\frac{\sigma_0}{K_c^0}\right)^2 r_c\right]\right\}^{1/2}, \quad (10.113)$$

where K_{SS} is the fracture toughness of the superconductor for the steady-state crack; $K_c^0 = (2\gamma_0 E)^{1/2}$ is the ceramic toughness without the twinning process; γ_0 is the fracture energy; H is the hardness factor of the ceramic at indentation; σ_0 is the threshold stress, stating a transition to non-elastic material behavior due to nucleation of the twinning near the crack tip.

In order to obtain numerical results, we use the known parameters for $YBa_2Cu_3O_{7-x}$, namely: $G = 41$ GPa, $\gamma_T \sim S_0 = 0.018$, $\Gamma_t \sim \sigma_d = 0.01$ J/m^2 [451]. In more detail, we discuss a selection of the values for the volumetric deformation, ΔV, and the threshold stress, σ_0 ($\sim P_V$). As has been shown in [572], the spontaneous (non-elastic) deformation initiates at temperature $T_S = 920$ K, resulting in the structural phase transition from tetragonal phase (*P4/mmm*) to orthorhombic (*Pmmm*). This deformation is defined by the components C_{22}, C_{23} and C_{55} of the elastic stiffness matrix. Lower the temperature of T_S (into orthorhombic phase), the formed domain structure of HTSC can re-switch from one spontaneous deformed oriented state to the other by external shear load, σ. The shear modulus attains minimum value for $\sigma = \sigma_c$ (where σ_c is the coercive stress), because the number of twins is maximum in this case, and the sample is most compliant to mechanical loading. The number of twins decreases at $\sigma > \sigma_c$, leading to greater stiffness of the sample and to increasing of G. Due to this, we select as the threshold value the next one: $\sigma_0 = \sigma_c$. The presented temperature dependencies of σ_c [572] in the vicinity of the phase transition give the value of $\sigma_c \approx 3$ MPa. This is much less than the corresponding value of the threshold stress, causing the twinning (20 MPa) that is calculated in [935] for $BaTiO_3$ samples, tested on impact load at different rates of the loading, and also than the value for partially stabilized ZrO_2 (4 GPa) [1394]. Finally, the spontaneous deformation jump at the phase transition, we suggest, is equal to $\Delta V = 10^{-4}$, according to experimental results of [572]. This is also much less than the corresponding values for $BaTiO_3$ (54×10^{-4}) [1368] and for ZrO_2 (57×10^{-3}) [1394].

Substitute the necessary values in (10.103) (at $p^A = 0$), calculating the preliminary values of the twin extent, d, from (10.104) for the observed grain sizes of YBCO, $D = 10$–100 μm. The solution (10.103) for the above range of the grain sizes leads to the value of $\eta_c \to 0$. Then, from (10.111) and (10.112) an absence of real toughening for HTSC ceramics, caused by the twinning processes, is found. This is explained by very small magnitudes of ΔV and σ_0 ($\sim P_V$), defining a spontaneous strain in the YBCO, compared to the corresponding values for partially stabilized ZrO_2 [1394] and ferroelectric ceramic $BaTiO_3$ [934, 1368], where the twinning processes play the most important rule in material toughening. Thus, the toughening (or crack amplification) at the fracture of $YBa_2Cu_3O_{7-x}$ superconducting ceramic is caused by microcracking due to deformation or/and thermal anisotropy, and by the processes, connected with microcracking (in particular crack branching and crack bridging), but no transformations of the martensitic type.

10.3 Toughening Mechanisms for Large-Grain YBCO

10.3.1 Model Representations

Two main computer simulation approaches, namely Monte-Carlo simulations [1319, 1326] and phase field method [1504, 1533, 1614], are usually used to study microstructure transformations, which accompany the YBCO fabrication. The phase field method has been discussed in Sect. 8.1.6. Here, we consider a modified Monte-Carlo scheme. The simulation of the YBCO precursor microstructure will be based on the re-crystallization model for ferroelectric ceramics [825, 1328] and on the study of thermal conductivity in mixes and structure-heterogeneous solids [408]. Formation of the final YBCO microstructures by using seeds at existence of dispersed 211 phase will be found by the Monte-Carlo techniques [36, 1610, 1607].

A precursor microstructure after YBCO pellet modeling is represented by discrete grid with 2000 square cells of characteristic size δ. Each grid cell is assigned by a number (between 1 and Q), corresponding to the orientation of the grain in which one is embedded. All grains have different orientations, defined during the pellet microstructure simulation, in which lesser numbers correspond to earlier initiated crystallites. This liquidates a neighborhood of the grain of the same orientation. Then, the cells which have neighbors with unlike orientation lie at the grain boundary; in the other case, they are placed into grains. The grain boundary energy is specified, defining an interaction between nearest neighboring lattice sites as [36, 1610, 1607]

$$E = -J\sum_{nn}(\delta_{S_iS_j} - 1), \tag{10.114}$$

where S_i is one of the Q orientations on site i ($1 \leq S_i \leq Q$), δ_{ij} is the Kronecker's delta. The sum is taken over all nearest neighbor sites (nn), J is a positive constant that sets the energy scale of the simulation. Note that the orientation dependence of the energy of a straight boundary segment exhibits a ratio of the maximum to minimum boundary energy of $2/\sqrt{3}$ for triangular lattice and $\sqrt{2}$ for square one. Both ratios are close to unity, that is, the grain boundary energy is nearly isotropic and only weakly grid dependent [36].

In simulation of boundary motion kinetics by the Monte-Carlo techniques, a grid cell is selected at random, and a new trial orientation is randomly changed to one of the other grain orientations. Then, the energy alteration caused by change in orientation (ΔE) is evaluated. The transition probability (P) is given by [36]

$$P = \begin{cases} \exp(-\Delta E/k_BT), & \Delta E > 0; \\ 1, & \Delta E \leq 0, \end{cases} \tag{10.115}$$

where k_B is Boltzmann constant, T is the temperature. A re-orientation of a site at a grain boundary corresponds to boundary migration. A boundary segment moves with rate, related to the local chemical potential difference (ΔE_i) as [36]

$$v_i = C[1 - \exp(-\Delta E_i/k_BT)], \tag{10.116}$$

where factor C is found by the boundary mobility and symmetry of the grid. The simulations for the cases of $T \approx 0$ and $T \approx T_m$, where T_m is the melting point, have shown similar results [1610]; therefore, we use $T \approx 0$, below.

The ($N - N_p$) attempts of re-orientations, where N is the number of lattice cells and N_p is the total particle number, are arbitrarily used as unit of time and defined as one Monte-Carlo step (MCS) per site. In order to incorporate particles of 211 phase into the model corresponding cells are assigned by orientation which are distinct from all grain orientations. It is suggested that the particle concentration and sites initially arbitrarily selected are fixed during simulation and are independent from any attempts of cell re-orientation. The insertion of the crystalline seed into microstructure is done by replacing the grains at the center of the microstructure with one large square grain with characteristic size considerably greater than the mean grain radius of the YBCO precursor pellet.

In the computer simulations, we consider three cases of YBCO microstructure evolution after primary re-crystallization (see Fig. 10.33):

(1) particle dispersion of 211 phase into matrix of 123 phase;
(2) insertion of large grain (seed) into matrix of 123 phase;
(3) insertion of large grain (seed) into matrix of 123 phase with dispersed 211phase.

As a condition of the computation stop in the first and third cases, an existence of one particle at any intergranular boundary at least (that corresponds to complete

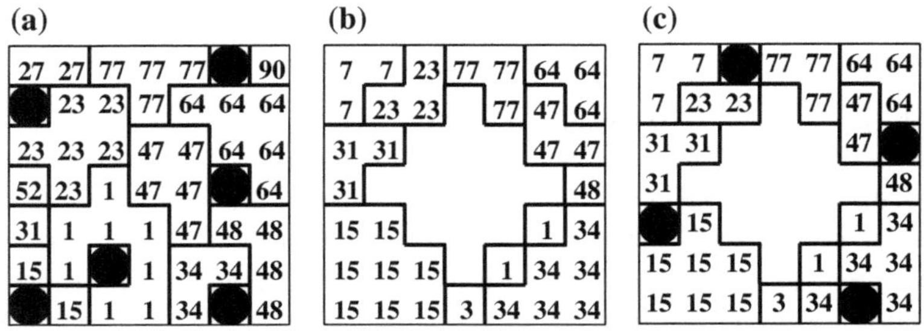

Fig. 10.33 Three models of YBCO microstructure evolution, used in computer simulations: (**a**) dispersion of 211 phase (*black*) in 123 matrix, (**b**) large-grain seed (*white*) in 123 grain matrix and **c** 211 dispersion and grain-seed in 123 matrix

pinning of microstructure with dispersed particles [1610, 1607]) is selected. In the second case, the computations are finished when some grain radius attains the seed grain size.

Block-scheme of one realization of the computational process describing three cases of evolution of YBCO microstructure after primary recrystallization is presented in Fig. D.5. A simulation veracity of computational algorithm of the secondary recrystallization has been found by direct verification on test examples. In order to obtain statistically representative results, we use approach [271]. The errors of computer experiments at doubling of number of the considered cells to be into range of 2 % in all variants calculation of the averaged parameters. Then, by using comparatively small grain aggregates together with stereological method accelerates computations due to the operation with smaller arrays of variables and obtains necessary statistics.

10.3.2 Effect of 211 Particles on YBCO Fracture

As has been shown in the previous chapters, a special effect on the 123 matrix fracture may be exerted by dispersed inclusions of the 211 phase. The platelet structure of 123 superconducting phase covering the 211 single inclusions during solidification from melting via a zipper-like mechanism promotes the formation at the 123/211 phase interface of the defects-faults and increased dislocation concentration. Therefore, the existence of secondary phase inclusions causes the alteration of the strength properties indirectly and directly. In the first case, these inclusions influence formation of certain microstructures during material fabrication with corresponding fracture resistance. In the second case, these inclusions define the acting toughening mechanisms, fracture toughness and strain energy release rate. Therefore, we consider the possible mechanisms of toughening in the case of the composite structure $YBa_2Cu_3O_{7-x}/Y_2BaCuO_5$.

Crack Deflection

As the 123/211 interface strength is less than YBCO matrix strength, a crack deflection around dispersed particles (see Fig. 10.34) in the form of crack tilt and twist is possible. Then the driving force of the crack, which orients randomly to the main tension vector, is described by the local stress intensity factors k_1, k_2 and k_3, corresponding to the opening, sliding and tearing fracture modes. The crack driving force, governed by the strain energy release rate, G, is found as [458]

$$EG = k_1^2(1 - v^2) + k_2^2(1 - v^2) + k_3^2(1 + v), \tag{10.117}$$

where E and v are the elastic modules.

In the case of applied SIF of mode I, K_I, the local SIFs of k_i^t (i = 1, 2, 3) for tilted crack are given as [458]

$$k_1^t = F_{11}(\Theta)K_I; \quad k_2^t = F_{21}(\Theta)K_I, \tag{10.118}$$

where

$$F_{11}(\Theta) = \cos^3(\Theta/2); \quad F_{21}(\Theta) = \sin(\Theta/2)\cos^2(\Theta/2). \tag{10.119}$$

For twisted crack the local SIFs of $k_i^T (i$ = 1, 2, 3) are found by [458]

$$k_1^T = F_{11}(\Phi)k_1^t + F_{12}(\Phi)k_2^t; \quad k_3^T = F_{31}(\Phi)k_1^t + F_{32}(\Phi)k_2^t, \tag{10.120}$$

where

$$\begin{aligned} F_{11}(\Phi) &= \cos^4(\Theta/2)[2v\sin^2\Phi + \cos^2(\Theta/2)\cos^2\Phi]; \\ F_{12}(\Phi) &= \sin^2(\Theta/2)\cos^2(\Theta/2)[2v\sin^2\Phi + 3\cos^2(\Theta/2)\cos^2\Phi]; \\ F_{31}(\Phi) &= \cos^4(\Theta/2)\sin\Phi\cos\Phi[\cos^2(\Theta/2) - 2v]; \\ F_{32}(\Phi) &= \sin^2(\Theta/2)\cos^2(\Theta/2)\sin\Phi\cos\Phi[3\cos^2(\Theta/2) - 2v]. \end{aligned} \tag{10.121}$$

The computation of toughening due to crack deflection is the statistical problem, which is connected with averaging of the driving force in all possible tilt and twist angles. The effect of particle volume fraction, V_f, which causes an initial crack path alteration, is equal to [458]

$$G_c/G_c^m = 1 + 0.87V_f, \tag{10.122}$$

where G_c and G_c^m are the fracture toughness of composite and matrix, respectively. The inclusions with higher characteristic ratio have stronger effects, that is, spherical particles create a lesser toughening compared with disk-shaped and rod-shaped ones. Toughening effect attains saturation at $V_f \approx 0.2$. An increased dimension of loading generally causes the increase of crack driving force due to the complexity of stress state and decrease of toughening effects.

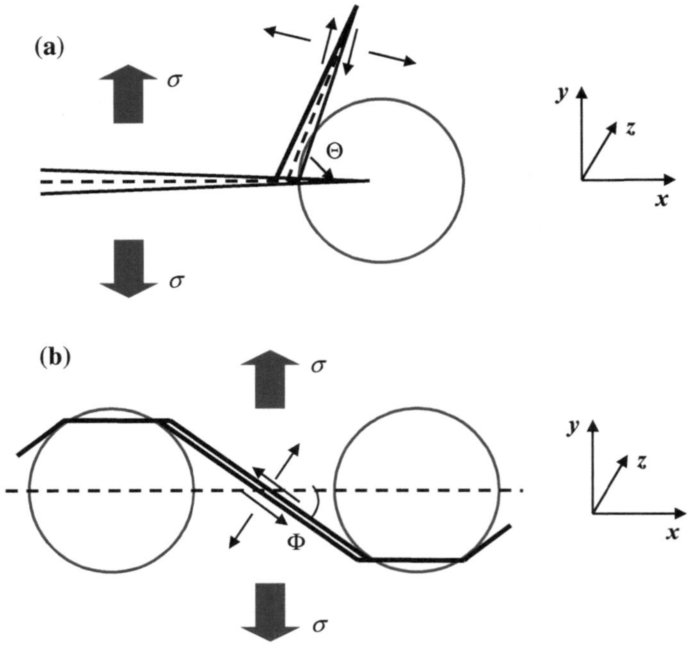

Fig. 10.34 (**a**) Tilt and (**b**) twist of crack at its interaction with inclusion

Crack Pinning by Particles

In the case of small-scale crack bridging in which the bridging zone size (L) is small compared with the crack length, sample sizes and distance from crack tip to the sample boundaries, the crack pinning by elastic particles with size $2a$ (see Fig. 10.35) is possible. Then, the toughening factor, Λ, may be found as [200]

$$\Lambda \equiv \frac{K_c/K_{cm}}{\sqrt{\omega(1-c)}} = \left\{1 + \frac{\pi S^2 ac(1-c^{1/2})}{2K_{cm}^2}\right\}^{1/2}; \quad \omega = \frac{E(1-\nu_m^2)}{E_m(1-\nu^2)}; \quad c = (a/b)^2, \tag{10.123}$$

where K_c and K_{cm} are the critical SIFs for crack growth in composite and in matrix, respectively; S is the particle strength; b is the radius of penny-shaped crack with central pinning particle; E and ν are effective elastic moduli of composite (E_m and ν_m are the elastic matrix constants). Thus, the toughening is increased with inclusion size and strength.

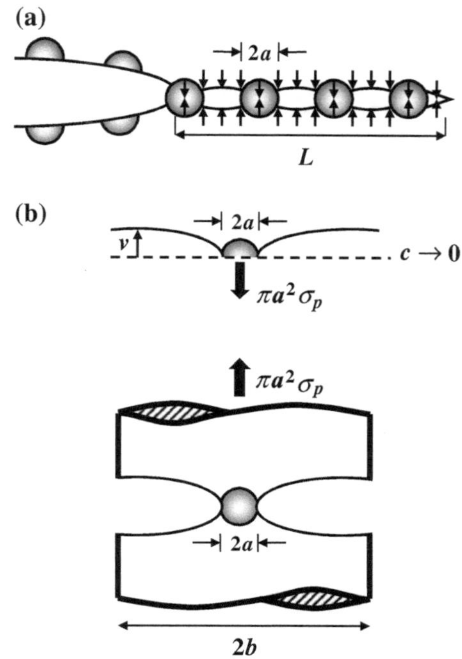

Fig. 10.35 (**a**) Crack pinning by particles and (**b**) crack opening; v is the crack opening rate; σ_p is the stress, loading the particle

Toughening due to Periodically Distributed Inclusions

The internal stress state caused by periodically distributed inclusions is almost sinusoidal at the mean plains between inclusion layers, where a finite crack is placed along the x-axis. Dislocation techniques have been applied to inclusions, modeled by centers of shear, compression, anti-plane and extension (dilatation) (see Fig. 10.36 [1010]). These are the following results:

(1) For shear centers:

$$\Delta K_{II} = 1.6\frac{\mu V_f e_{xy}^T \lambda^{1/2}}{(1-v)}; \quad a \ll \lambda;\ b_x = he_{xy}^T. \tag{10.124}$$

(2) For horizontal compression centers:

$$\Delta K_{I} = 1.6\frac{\mu V_f e_{xx}^T \lambda^{1/2}}{(1-v)}; \quad a \ll \lambda;\ b_x = he_{xx}^T. \tag{10.125}$$

(3) For vertical compression centers:

$$\Delta K_{I} = 2.2\frac{\mu V_f e_{xx}^T \lambda^{1/2}}{(1-v)}; \quad a \ll \lambda;\ b_x = he_{xx}^T. \tag{10.126}$$

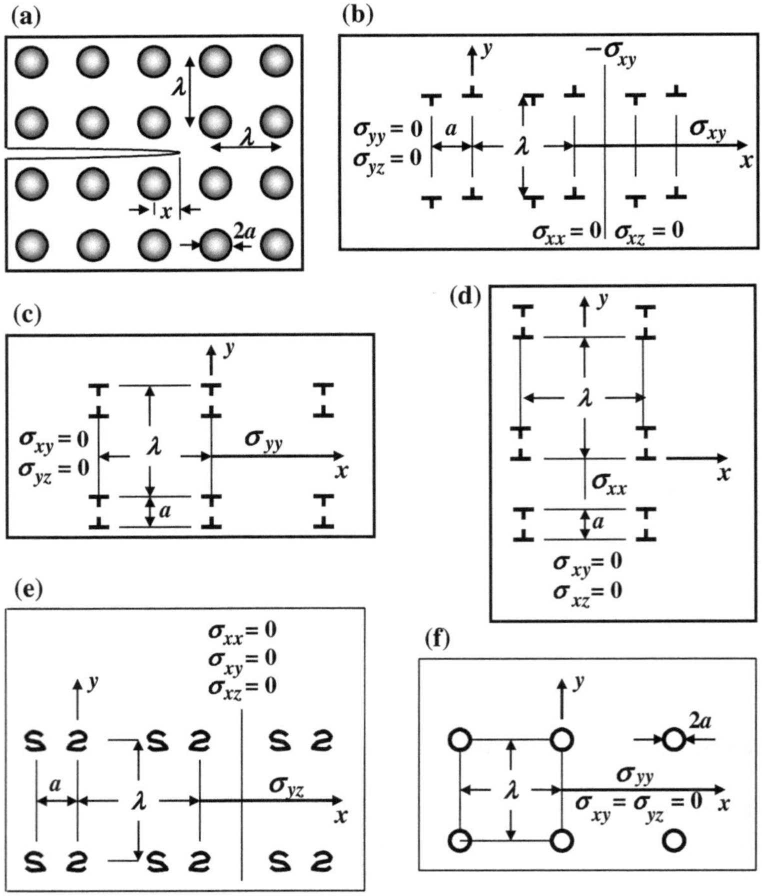

Fig. 10.36 (**a**) Crack toughening by periodical array of inclusions; dislocation structures (with Burgers vector, ***b***), modeling the inclusions: (**b**) shear centers; (**c**) horizontal compression centers; (**d**) vertical compression centers; (**e**) anti-plane shear centers and (**f**) extension (dilatation) centers

(4) For anti-plane shear centers:

$$\Delta K_{III} = 0.4\mu V_f e_{yz}^T \lambda^{1/2}; \quad a \ll \lambda;\ b_x = he_{yz}^T. \tag{10.127}$$

(5) For extension (dilatation) centers:

$$\Delta K_I = 0.8 \frac{EV_f e^T \lambda^{1/2}}{(1 - 2\nu)}; \quad a \ll \lambda;\ e^T = \Delta V/V. \tag{10.128}$$

Thus, the SIF difference of corresponding mode for periodically distributed inclusions is proportional to the elastic modulus, volume fraction of the inclusions, deformation mismatches ($e^T_{\alpha\beta}$) and square root of the inclusion spacing (λ). However, it is independent of the crack length, that is, toughening due to the particle dispersion is more effective at the crack initiation compared with its growth.

10.3.3 Some Numerical Results

The numerical results presented in Table 10.8 [1327] have been found for some microstructure properties in all possible variants that are found by the 211 particle concentration ($f = 0.1$ and 0.2) and the seed size ($S_S \approx 2S_0^m$ and $3S_0^m$, where S_S is the seed area and S_0^m is the maximum grain square after primary re-crystallization before beginning of the 123 grain growth). The final microstructures with the dispersed 211 particles demonstrate dependencies similar to $\overline{S} \approx 4a/f$, where $\overline{S}$ is the mean pinned grain area, a is the particle square. This corresponds to the topological ideas developed in the computer simulations of the grain pinning by dispersed particles [1610]. The behavior of outstripping grain growth around seed coincides well with the computational results of large grain growth in the microstructure of normal grain growth [1607] and the growth morphology of YBCO specimens with seed, solidified at unfavorable temperatures [1038]. Moreover, the present computer simulations may be characteristic to the pushing process of the 211 particles due to 123 front advancing [1038]. The numerical results show a correlation between the grain area, $\overline{S}$, and superconducting field per grain, $\overline{S}_{SR}$.

10.4 Fatigue Fracture

Fatigue fracture is very important phenomenon from practical viewpoint. At the level of micromechanics the fatigue fracture could be understood as a sequence of some stages, namely: nucleation of cracks, stable growth of short cracks, stable growth of long cracks and unstable failure. In this section we consider the most important for HTSC oxides and composites fracture stages. First part is devoted to mechanisms of nucleation and propagation of short cracks under remote Mode I loading. The growth of short cracks is governed by shear stresses in crystallographic slip systems inclined under approximately 45° to the maximal principle stress. They grow in a local mixed mode I + II and their growth mechanisms are completely differ from the corresponding mechanisms for long cracks.

After an initial stage incorporating crystallographic and transient growths of short cracks the fatigue cracks incline to direction perpendicular to the maximal principal stress, i.e. nearly Mode I loading. We consider a micromechanical description of all important accompanying phenomena. When the shear components of the applied stresses are dominant, both short and long cracks can propagate macroscopically under shear loading Modes II or III. In this stage, the processes and mechanisms of crack closure have an important value.

Table 10.8 Numerical results

Parameters	$f = 0.0$ $S_S \approx 2S_0^m$	$f = 0.0$ $S_S \approx 3S_0^m$	$f = 0.1$ $S_S = 0$	$f = 0.1$ $S_S \approx 2S_0^m$	$f = 0.1$ $S_S \approx 3S_0^m$	$f = 0.2$ $S_S = 0$	$f = 0.2$ $S_S \approx 2S_0^m$	$f = 0.2$ $S_S \approx 3S_0^m$
$\bar{S}/\delta^2$	13.89	23.81	45.45	50.03	62.50	25.01	29.41	31.25
$\bar{S}_{SR}/\delta^2$	13.89	23.81	40.91	45.08	56.25	20.05	23.53	25.00

10.4.1 Nucleation and Growth Mechanisms of Short Cracks

The fatigue cracks predominantly originate at the surface [1376], i.e. at the boundaries with extremally high different properties of separated media. During the first loading cycles progressive changes in dislocation and damage structures start to proceed. The damage processes begins at the sites of cyclic strain localization, so-called persistent slip bands (PSBs) and results in the formation of sharp surface slip patterns, so-called persistent slip markings (PSMs). The slip localization in surface thin bands can be caused by several mechanisms. In particular, in f.c.c. and b.c.c. polycrystals subjected to cyclic loading the PSBs are often formed inside the surface grains. The PSBs consist of hard walls (high density of edge dislocation dipoles) and soft channels (low density of screw dislocations which glide and cross-slip) creating well-knownn ladder-like structure [1059].

The amount of plastic slip inside the slip bands and PSBs could be modeled by considering an elongated bulk inclusion embedded into matrix presenting the whole polycrystal [1340]. By using Eshelby solution for bulk inclusions [448] one can show that the primary slip in vicinity of the free surface is greater than in the bulk, where the strain is constrained by the grain boundary surrounded by a hard elastic material [1493]. The shear strain within the surface slip bounds can be found as [1377]

$$\gamma_p = r(1-\nu)\frac{L}{h}\left(1+\frac{h}{L}\right)^2\frac{(\Sigma/2-\tau_{crs})}{G}, \tag{10.129}$$

where L is the length of the slip band (usually equals to the grain size), $r = 1.9$, ν is the Poisson's ratio, G is the shear modulus, Σ is the applied stress and τ_{crs} is the critical resolved shear stress in the channel [1493].

The high local plasticity in slip band and channel PSBs creates a surface microroughness in the form extrusions and intrusions [1380]. The movement of screw dislocations with jogs generates vacancies in the channels. The counterbalancing flux of atoms into the channels forms extrusions at the free surface, the volume of which is far greater than that of intrusions. On the other hand, the volumes near the surface and close to the outer boundary of the channels becomes depleted by atoms and enriched by vacancies. This forms thin intrusions next to extrusions [1381]. High stress concentrations around extrusions and at the tips of intrusions cause a formation of short surface cracks which begin to grow inside the bulk along the slip bands of PSBs.

There are some models explaining possible mechanisms of initial growth stages of short cracks [492, 1376, 1380]. Because the crack nuclei are strictly aligned with the slip planes of PSBs their next growth occurs along these planes. This means that the cracks propagate under complex of the shear and tensile normal stresses (the mixed mode I + II). Always existing in HTSC the oxide components are fractured in a result of deformations in the adjacent polycrystal leading to re-oxidation and brittlement of the sample. The process of repeated fracture and re-oxidation is a central moment of slip-oxidation models widely used for description of propagation of short and long cracks [492, 1376].

The simplest model of the short crack growing along a single slip plane is present in Fig. 10.37. In the tensile half-cycle, the edge dislocations are emitted (or absorbed at) the crack tip generating a new failure surface on one of the crack face. The size of new surface is defined by number of dislocations, the Burgers vector and the crack front length. By assuming simultaneous surface oxidization (brittlement), the dislocations returning during the unloading cycle cannot remove the new fracture surface on the other crack face, the size of which is also equal to the number of returning dislocations, the Burgers vector and the crack front length. Under constant loading amplitude the number of returning dislocations to the crack tip is approximately coincides with the number dislocations generated during loading. Hence, the crack extension per cycle is equal to the cyclic crack tip opening displacement.

Then the crack growth rate da/dN could be assumed approximately proportional to the frequency of the oxide component fractures, which is again proportional to the plastic strain γ_p in the slip band. Due to the crack growth rate can be assumed to be proportional to $(L/h)^m$, where $m \in (0.3, 0.8)$ [492]. This also means that the growth rate of short cracks is greater in the long PSBs embedded in large surface grains.

The crack growth of, so-called microstructurally short cracks (MSCs) could be retarded by various microstructural barriers of different strength. The maximal length of MSCs is determined by the distance b_s of the strongest barrier such as grain or phase boundary. The growth rate of MSCs is often presented as [1148]

$$da/dN = A\Delta\varepsilon_p^l(b_s - a)^k, \tag{10.130}$$

where $\Delta\varepsilon_p$ is the applied plastic strain range, and A, l and k ($k \leq 1$) are material parameters. The expression defines a proportionality between the applied and the local plastic strains and arrest of MSCs at the barriers ($da/dN = 0$ at $a = b_s$). Note that MSCs can grow at significantly lower applied stress compared to long cracks, i. e. at far lower the fatigue limit. It is mainly connected with absence of crack closure effects. At the further propagation of MSCs, the applied stress must be raised to reinitiate the crack into adjacent grains. When the crack overcomes the grain boundaries, it begins to be spatially tortuous, i.e. crack faces become rougher and the growing friction forces decelerate the crack growth under the shear mode II. After this stage the crack inclines to the plane perpendicular to the applied stress direction and is subjected to mode I component. The cracks occurring in the transient stage are called physically short cracks (PSCs). As a rule, MSCs become to be PSCs after crossing one or two grain boundaries. At the same time PSC transforms to a long crack after passing more than ten grain boundaries.

10.4.2 Crack Closure Mechanisms

The fatigue crack growth behavior near the threshold of the applied SIF is caused by the chemical composition and microstructure features. As it has been noted in Sect. 10.1.5 various crack shielding mechanisms could be divided into intrinsic and extrinsic ones [1439]. A common scheme of extrinsic mechanisms is depicted in Fig. 10.38, where applied $\Delta K = K_{max} - K_{min}$ ($K_{min} < 0$) includes two main components: the closure range ΔK_{cl} and the effective one ΔK_{eff}. The first range consists of three parts: the oxide-induced closure (OIC), the roughness-induced closure (RIC) and the plasticity-induced closure (PIC). At the same time, ΔK_{eff} includes the geometrically-induced shielding ΔK_{br} (crack kinking and branching) and the intrinsic range ΔK_{in}. While all the crack closure mechanisms act in the wake zone of crack, the crack kinking and branching operate at the crack tip. The crack remains to be closed not only in the compressive part of the loading cycle, but to some extent also at the tensile step of the loading.

The total crack closure effect is stated by a sum of the PIC and RIC components defining main influences on long cracks. The total contact shielding ratio is determined as [1377]

$$\frac{\delta_{cl}}{\delta_{\max}} = C\eta\sqrt{R_S^2 - 1} + \frac{3\eta(R_S - 1)}{2\sqrt{6} + 6(R_S - 1)} + 2C, \qquad (10.131)$$

where δ_{cl} is the summary closure distance for the PIC and RIC components, δ_{max} is the maximal crack opening displacement, C is a constant, the parameter η

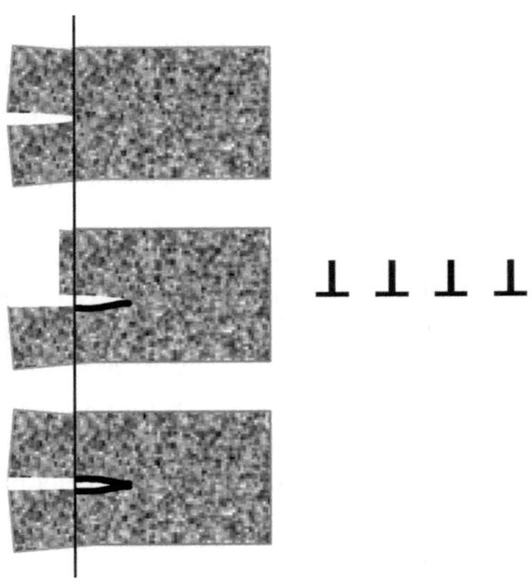

Fig. 10.37 The single-slip model of short crack growth

determines the weight of the RIC component similarly to that of the geometrical shielding term in the brittle fracture case (see Sect. 10.1.5), R_S is the surface roughness ($R_S = 1$ at the zero closure distance for the RIC component in the case of microscopically straight cracks, i.e. microstructurally short fatigue cracks in polycrystals or shear cracks in single crystals).

The effective crack driving force ΔK_{eff} can be calculated by using expression (10.131) and relations $K_{eff} = K_{max} - K_{cl}$; $K_{max} = \Delta K/(1 - R)$ in the form

$$\Delta K_{eff} = \left(1 - C\eta\sqrt{R_S^2 - 1} - \frac{3\eta(R_S - 1)}{2\sqrt{6} + 6(R_S - 1)} - 2C\right)\frac{\Delta K}{1 - R}. \qquad (10.132)$$

The last relation can be directly compared with known experimental data.

10.5 Small Cyclic Fatigue of YBCO Ceramics

It is known that the fracture toughness of YBCO is significantly smaller than that of typical ceramics, for example, Al_2O_3, partially stabilized ZrO_2 or SiC [1724]. Therefore, it is very important to consider fracture processes in YBCO, in particular, under cyclic loading. In this section the small cyclic fatigue fracture model for YBCO is stated, based on the microstructure dissimilitude effect [251], using joint considerations of the superconducting ceramic manufacture and fracture for the YBCO samples [1324]. Special attention is devoted to the correct definition of specific fracture energy, taking into account porosity, microcracking and cooling features.

10.5.1 Model Representations

Remind that typical techniques for melt-processed YBCO ceramics consist of the precursor powder preparation, formation of the so-called "green" sample, sintering and cooling of the material. The sample cooling and re-crystallization initiate a formation of defects and microcracks due to the deformation mismatches of the 211 and 123 phases, thermal expansion anisotropy of grains and tetragonal–orthorhombic phase transition. Subsequent electromagnetic and thermo-mechanical loading of the ceramic in devices contributes to the damage development (or their initiation) and to corresponding deterioration of the conductive properties of YBCO.

By modeling of the processes which accompany the ceramic fabrication and fracture, the general research scheme, presented in the previous sections, is realized:

(1) The precursor is represented by 2D grid with 1000 square cells of the characteristic size, δ. Each cell is either a grain nucleus of the 123 phase or a void. Initial disposition of the voids is found by Monte-Carlo procedure, described in Sect. 6.4.

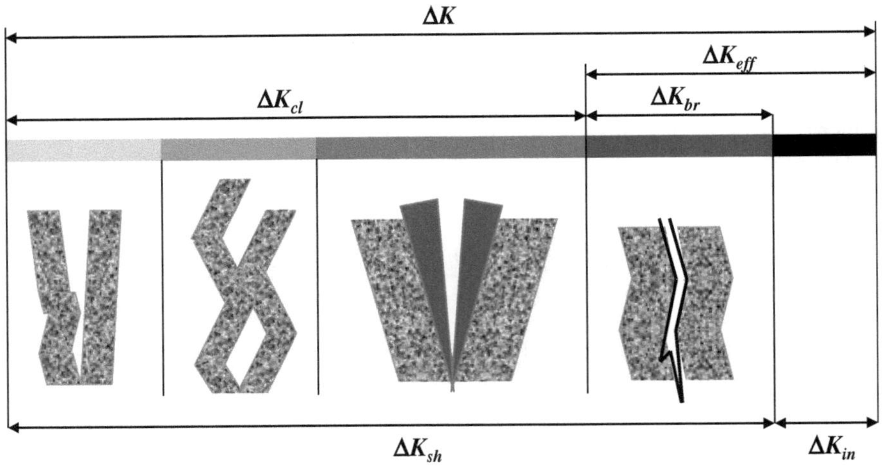

Fig. 10.38 A common scheme of extrinsic mechanisms taking place during the fatigue crack propagation. The designations are explained into text [1377]

(2) A modeling of the heat front propagation during sintering is carried out. It is assumed that a temperature in gradient furnace changes only along one coordinate with constant rate *v*.
(3) By description of the ceramic structure re-crystallization, the Monte-Carlo procedure is also used. Here, we neglect the material shrinkage and grain growth, assuming that during formation of closed porosity there are no considerable material densification (see Fig. 10.39).
(4) The microcracks nucleate at triple junctions during sample cooling (in the same conditions for all considered structures) from the melting temperature down to the room temperature. The cooling process consists of a slow decrease (with rate of 1 °C/h) of temperature from 1100 °C down to 970 °C and then quenching, for example, in liquid nitrogen bath [703].

10.5.2 Microstructure Dissimilitude Effect

As has been noted, the short fatigue cracks (i.e., cracks that intersect some grains) frequently grow much faster than long ones under other equal conditions, in particular due to the higher stress concentration factors and the decreased closing of the cracks. Microstructure dissimilitude is caused by impossibility to average the strength and fracture properties on the many grains and the existence of the dependence on the crack size. As has been demonstrated, the microstructure dissimilitude can lead to anomalous rapid growth of the small cracks, altering: (i) the intrinsic fracture toughness and cyclic crack growth resistance in the process zone and also (ii) the local crack driving force (ΔK) [250].

Consider a short crack which violates microstructure similitude and uses the modified Barenblatt–Dugdale's (BD) crack tip model that is based on the linear elastic fracture mechanics. The classical BD-crack with process zones (presenting extended cohesive parts at the crack tips) may be replaced by the superposition of two loading configurations (see Fig. 10.40), reflecting a local cleavage and general fracture. Then, the fracture toughness increment of large cracks (ΔK_{Ic}) can be divided into two terms for cyclic loading:

$$\Delta K_{Ic} = \Delta K_c + \Delta K_{cd}, \tag{10.133}$$

where ΔK_c is the fracture toughness increment of small cracks, and ΔK_{cd} represents the local increment of the fracture toughness, induced as a result of microstructure dissimilitude and grain misorientation. Then, ΔK_{cd} for an elastic crack, subjected to a partial tensile load increment on the crack surfaces ($\Delta\sigma_{cd}^m$), is given as [1639]

$$\Delta K_{cd} = \frac{2\sqrt{2}}{\pi}\Delta\sigma_{cd}^m\sqrt{\pi b}, \tag{10.134}$$

where $\Delta\sigma_{cd}^m$ is taken to be $2\sigma_{cd}^m$, so that

$$\Delta\sigma_{cd}^m = 2\sigma_c(1 - \sigma_c^m/\sigma_c). \tag{10.135}$$

In (10.134) and (10.135), b is the length of the region over which the crack tip cleavage occurs (the process zone size), σ_c is the macroscopic fracture strength of the specimen, σ_c^m is the local cleavage strength in a grain, containing the crack tip. Substituting (10.134) and (10.135) in (10.133), we obtain

$$\Delta K_{Ic} = \Delta K_c\left[1 + \frac{4\sqrt{2}}{\pi}(\sigma_c/\Delta\sigma_\infty)\sqrt{b/a}(1 - \sigma_c^m/\sigma_c)\right], \tag{10.136}$$

where a is the half-length of the crack without process zone, and $\Delta\sigma_\infty$ is the increment of the applied remote stress. Similarly, it can be estimated that

$$K_{Ic} = K_c\left[1 + \frac{2\sqrt{2}}{\pi}\frac{\sigma_c}{\sigma_\infty}\sqrt{b/a}(1 - \sigma_c^m/\sigma_c)\right]. \tag{10.137}$$

Assuming that for small and large cracks there are relationships: $K_c = \left[\frac{\gamma_F^{(1)}}{1-\nu_1^2}E_1\right]^{1/2}$, and $K_{Ic} = [\frac{\gamma_F^{(0)}}{1-\nu_0^2}E_0]^{1/2}$, respectively, we obtain from (10.136) and (10.137)

$$\Delta K_{Ic} = \Delta K_c[1 + 2(S - 1)\sigma_\infty/\Delta\sigma_\infty], \tag{10.138}$$

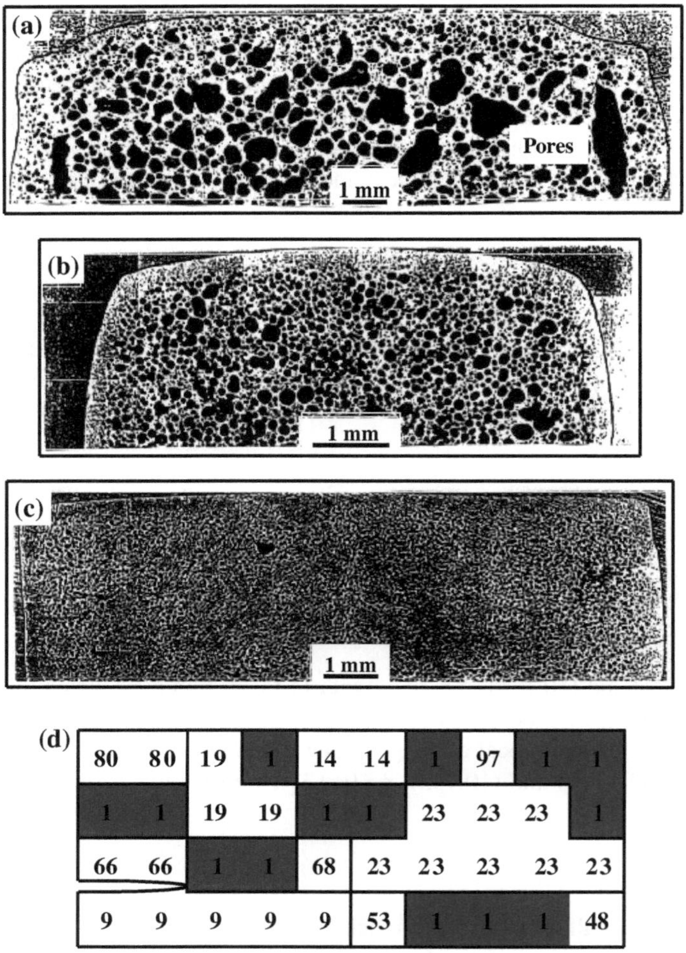

Fig. 10.39 Micrographs of polished cross-sections of YBCO samples, prepared at the heat rates: (**a**) 30 °C/h (**b**) 20 °C/h and (**c**) 10 °C/h [1040]; (**d**) fragment of model microstructure; porosity is shown *by gray color*, microcrack is present at intergranular boundary

where

$$S = \left[\frac{\gamma_F^{(0)} E_0 (1 - \nu_1^2)}{\gamma_F^{(1)} E_1 (1 - \nu_0^2)} \right]^{1/2}. \tag{10.139}$$

Effective elastic moduli E_1 and ν_1 for solid with pores and microcracks can be estimated, for example, using the dependencies of strain characteristics for elastic bodies on the concentrations and sizes of spherical pores and disc-like cracks in the self-consistent differential method as [416]

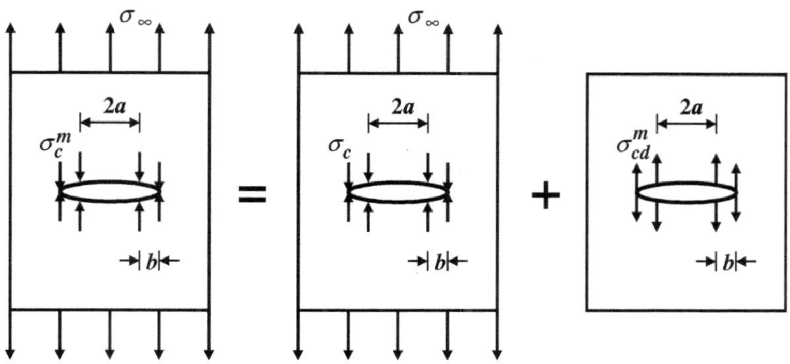

Fig. 10.40 Barenblatt–Dugdale's crack, presenting short crack, which disturbs microstructure similitude. One may be studied in the framework of linear elastic fracture mechanics, introducing additional stresses at the crack surfaces ($\sigma_{cd}^{m} = \sigma_c - \sigma_c^{m}$) that are caused by microstructure dissimilitude

$$E_1 = E_p \exp(-\alpha_1 \eta_c);\ \nu_1 = \nu_p \exp(-\alpha_2 \eta_c); \tag{10.140}$$

$$E_p = E_0 (1 - \eta_p)^{\beta_1};\ \nu_p = 0.2 + (\nu_0 - 0.2)(1 - \eta_p)^{\beta_2}, \tag{10.141}$$

where $\eta_c = (a_c/r_c)^3$, $\eta_p = 1 - \rho/\rho_0$ are the damage parameters, determined through the average size of the disc-like cracks (a_c) and the average distance between them (r_c); ρ and ρ_0 are the material and crystalline cell density, respectively; α_i, β_i ($i = 1, 2$) are given constants; E_0, ν_0 are the elastic moduli of the material without defects; E_p, ν_p are the elastic moduli of sample with voids.

10.5.3 Fracture Energy and Microstructure Features

Obviously, the correct definition of fracture energy, taking into account a microstructure of superconductor, has significance for understanding of the dynamical crack problem, generally, and the cyclic fractures, particularly. We present the energetic balance condition in the usual form:

$$J = 2\gamma, \tag{10.142}$$

where

$$J = \int_S \left[(U + T) n_1 - \sigma_{jk} n_k \frac{\partial u_j}{\partial x_1} \right] dS;\ 2U = \sigma_{jk} \varepsilon_{jk};\ 2T = \dot{u}_j \dot{u}_j, \tag{10.143}$$

where σ_{jk}, ε_{jk}, u_j, n_k are the components of the stress and strain tensors, vector of the displacement and external normal to surface of S, surrounding a crack tip, respectively; γ is Griffith's parameter. A dot above a symbol implies the material time derivative. There is summation on the repeated indexes from 1 to 3 in (10.143). In the static case, the condition (10.142) reduces to the algebraic relation between physical parameters. The energetic balance condition in the form (10.142) is also applied to the dynamical case. However, the assumption of the *J*-integral as critical parameter in the criterion of crack propagation [186] reflects only a static nature of the condition (10.142). Therefore, in the dynamical case of the rectilinear flaw with size, $2L(t)$, it is necessary to take into account that the *J*-integral depends on the time t and is caused by the selection of the $L(t)$. The *J*-integral is equal to constant value of 2γ only when a law of the flaw growth for the $L(t)$ satisfies (10.142) in any time. After calculation of the *J*-integral, this condition reduces to non-linear differential equation, which defines the crack growth during time. Thus, a statement of the dynamical crack problem must include the condition (10.142), which identifies the flaw growth with crack advancing during fracture of solid. Then specific fracture energy, γ_F, to being a limit ratio of the work on separation of atom couples at the fracture surface F to its square at decreasing of this surface to zero, demonstrates a local character. The value of γ_F should be significantly distinct from the Griffith's parameter γ, which is a result of the γ_F averaging on the fracture surface. Therefore, the physical proper statement demands to replace γ through γ_F in condition (10.142). The energetic balance condition as dynamical condition for brittle fracture can be represented as [814]

$$J = \frac{\sigma_\infty^2 \pi(1-v^2)\sin^2\Theta_0}{E} L(t)\left\{F_1^2\left[\frac{\dot{L}(t)}{c_2}\right]\sin^2\Theta_0 + F_2^2\left[\frac{\dot{L}(t)}{c_2}\right]\cos^2\Theta_0\right\} = 2\gamma_F, \tag{10.144}$$

where E and v are the elastic moduli; σ_∞ is the remote tension; Θ_0 is the angle between the load and crack direction at the time t; c_2 is the rate of transverse waves; F_1 and F_2 are the monotonically decreasing functions, determined in the non-stationary problem. It may be shown [814] that the crack growth is irregularly accelerated and gives the relationship:

$$L(t) = L(0) + \frac{a_c t^2}{2} - \frac{L(0)}{3}\left[\frac{a_c t^2}{2L(0)}\right]^2 + O\left\{\left[\frac{a_c t^2}{2L(0)}\right]^3\right\}, \tag{10.145}$$

where the acceleration a_c is found as

$$a_c = \frac{\pi(1-v)}{8\rho\gamma_F}\frac{\sigma_\infty^2\sin^2\Theta_0}{\alpha_1(v)\sin^2\Theta_0 + \alpha_2(v)\cos^2\Theta_0}; \; L(0) = \frac{2E\gamma}{\sigma_\infty^2\pi(1-v^2)}. \tag{10.146}$$

The parameters $\alpha_1(v)$ and $\alpha_2(v)$ are also found, solving a non-stationary problem. Note that (10.143) defines all features of sub-critical crack behavior, in particular, the crack tilt, twist and branching. The value of γ_F is calculated in the case of fatigue fracture, using both material microstructure and cyclic loading. Estimate this parameter taking into account the structural elements of YBCO (i.e., voids and microcracks), which are small to compare with macroscopic scale of stress–strain state. Because the microcracks and voids, as a rule, are localized at the grain boundaries which are regions of a lesser dense atomic packing compared with grain phase, the curve of interatomic interactions can be approximated by the sinusoid [48]:

$$\sigma(\varepsilon) = \begin{cases} \sigma_a \sin 2\pi\varepsilon, & \varepsilon \in [0; 0.5]; \\ 0, & \varepsilon \in [0.5; \infty], \end{cases} \tag{10.147}$$

where $\varepsilon = r/r_0 - 1$, r_0 is the mean equilibrium distance between atoms, $\sigma_a = E_0/2\pi$ is the theoretical strength of atomic link. Then, we obtain for transgranular fracture:

$$\gamma_F^{(0)} = \frac{1}{2}\int_{r_0}^{\infty} \sigma[\varepsilon(r)]dr = \frac{1}{2}\int_{r_0}^{3r_0/2} \sigma_a \sin 2\pi(r/r_0 - 1)dr = \frac{E_0 r_0}{4\pi^2}. \tag{10.148}$$

In order to define the mean interatomic spacing at intergranular boundary, r_1, we take into account a proportional dependence between adjacent atoms and the absolute temperature T. Then, $r/r_0^* - 1 = \alpha T$, where α is the thermal expansion factor, r_0^* is the equilibrium spacing between the atoms at the $T = 0$ K. Further, in order to define the temperature field in the ceramic sample during cooling from the sintering temperature (T_1) down to the room temperature (T_0), with no destroying of generality, we study the next 1D initial boundary problem:

$$\frac{\partial T}{\partial t} = b_t^2 \frac{\partial^2 T}{\partial x^2}, \; b_t^2 = \frac{k}{\rho c}; \tag{10.149}$$

$$\frac{\partial T}{\partial x} = \pm h(T - T_0), \quad \text{for } x = \pm l; \tag{10.150}$$

$$T = T_1, \quad \text{for} \quad t = 0, \tag{10.151}$$

where c is the specific heat, k is the thermal conductivity factor, h is the thermal expansion factor, 2 l is the sample length. The solution of this problem is determined as

$$T(x,t) = T_0 + \sum_{m=1}^{\infty} A_m \exp(-\lambda_m^2 b_t^2 t) \cos \lambda_m x, \tag{10.152}$$

where $A_m = 2\sin\lambda_m l/(\lambda_m l + \sin 2\lambda_m l)$, and eigenvalues, λ_m, have the following approximate values: $\lambda_m = m\pi(1/l - h/\pi^2)$. Retaining only the first term of the quickly convergent series (10.151), we have finally:

$$\begin{aligned} \gamma_F^{(1)}/\gamma_F^{(0)} &= E_1 r_1/E_0 r_0 \\ &= \exp(-\alpha_1 \eta_c)(1 - \eta_p)^{\beta_1}[1 + \alpha(T_1 - T_0) 2 \sin\lambda_1 l/(\lambda_1 l + \sin 2\lambda_1 l)]. \end{aligned} \tag{10.153}$$

When $\lambda_1 l$ is small, (10.153) reduces to

$$\gamma_F^{(1)}/\gamma_F^{(0)} = \exp(-\alpha_1\eta_c)(1-\eta_p)^{\beta_1}[1+2\alpha(T_1-T_0)/3]. \qquad (10.154)$$

10.5.4 Some Numerical Results

The computer simulation results have been obtained for known parameters of YBCO [416, 703]: $T_1 = 970$ °C, $T_0 = 20$ °C, $\alpha = 2 \times 10^{-5}$ K^{-1}, $\alpha_1 = 1.65$, $\alpha_2 = 1.47$, $\beta_1 = 1.96$, $\beta_2 = 0.23$, $\nu_0 = 0.22$, $E_0 = 64$ GPa. Additionally, the values of η_c and η_p are stated by the microstructure modeling, using the formulae: $\eta_c = Na_c^3$ and $\eta_p = C_p$, where N is the crack number per volume unit, C_p is the closed porosity. The numerical results are shown in Fig. 10.41 for different values of the initial porosity, C_p^0, and the heat rate of the sample, v [1324]. The increased rate at the same initial porosity gives a more fine-grain structure, smaller microcracking of grain boundaries and greater value of $\gamma_F^{(1)}/\gamma_F^{(0)}$. Generally, the decreasing heat rate leads to finer sizes of voids and to their homogeneity in sintered samples that improve the YBCO conducting properties (see Fig. 10.39). However, excessive rate decreasing may increase the liquid phase loss. This would be undesirable to a highest degree from viewpoint of stoichiometry consideration [1040]. The increasing initial porosity (at the same heat rate) defines the finer-grain structure with corresponding decreasing of a spontaneous microcracking of intergranular boundaries during cooling. At the same time, this leads to the smaller values of $\gamma_F^{(1)}/\gamma_F^{(0)}$ and proper increasing of $\Delta K_{Ic}/\Delta K_c$. It is obvious the closure of the small cracks under unloading compared with the large crack case is insignificant. Then, the maximum toughening, based on the steady-state crack conditions, can be achieved after a finite crack extension [1609]. This can lead to an overestimation of the toughening in consideration of the cyclic fatigue and, therefore, it should be calculated directly, taking into account the microstructure dissimilitude effects.

Block-scheme of one realization of the computational process describing the cases of microstructure dissimilitude is presented in Fig. D.6. The errors of computer experiments at doubling of number of the considered cells to be into range of 2 % in all variants calculation of the averaged parameters.

10.6 Residual Thermal Stresses in YBCO/Ag Composite

As has been shown in previous chapters, the silver inclusions are widely used to improve both superconducting and mechanical properties of HTSC. Consider effects of residual stresses in this section which form around inclusions in YBCO/Ag composite after cooling from the peritectic temperature down to room temperature and are caused by the great difference in thermal expansion factors of the metal particles and the ceramic matrix. Plastic deformations in ductile silver inclusions should be considered in an estimation of residual stresses around silver particles. The finite element method (FEM) has been applied for evaluating such difficult problem [1066]. The using of ANSYS general-purpose software (or other FEM software, for example ACELAN [8]) helps to estimate effects of the distribution of the residual stresses on features of macrocrack growth in this composite. For the matrix and inclusion properties given in Table 10.9 and also for the thermal difference ΔT between the peritectic temperature of YBCO/Ag (970 °C) and room temperature (20 °C), two silver round inclusions with a diameter $D = 10$ μm and the distance between them, $l = 5$ μm, are considered on the isotropic *ab*-plane of the YBCO matrix (see Fig. 10.42a).

Figure 10.42b shows the distribution of the residual stresses calculated by the FEM and indicates that expansive (compressive) stresses create after cooling of the composite due to the TEA of the matrix and inclusions. Figure 10.43 explains the mechanism of the crack deflection between two inclusions. Tensile and compressive stresses remain around the silver particles in the *r*-axis and θ-axis of the polar coordinate system, respectively. First, the tensile stress along the *r*-axis assists the crack growth. However, when it reaches a gap between the silver particles, the angle between the vectors of the residual stresses and the crack growth is changed. This means that the tensile (compressive) residual stress promotes (suppresses) the crack growth along the *y*-axis (*x*-axis). Therefore, the direction of the crack growth is altered by the residual stresses around the silver particles, according to the energy minimum condition. This also means that I mode of fracture is changed to the mixing mode (both I and II modes) by this deflection. Such crack deflections ensure the possibility of toughening of the brittle YBCO matrix and increasing its fracture toughness due to the increasing of the deflected crack path.

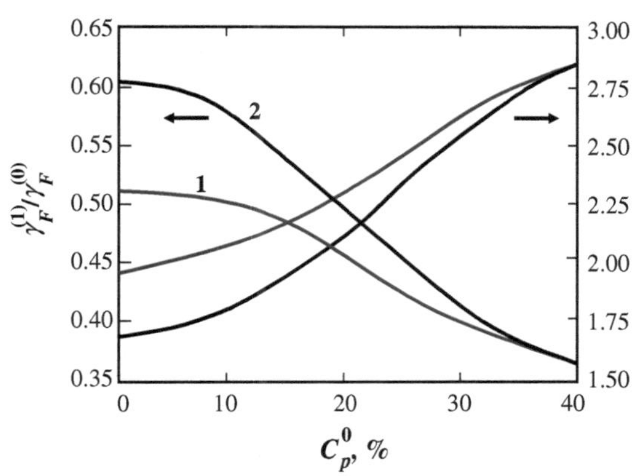

Fig. 10.41 Relative values of specific fracture energies, $\gamma_F^{(1)}/\gamma_F^{(0)}$, and fracture toughness increments, $\Delta K_{Ic}/\Delta K_c$, *versus* ceramic porosity, C_p^0, at $2\sigma_\infty = \Delta\sigma_\infty$. For two heat rates, $v = 10$ °C/h (1) and $v = 20$ °C/h (2)

10.7 Toughening of Bi-2223 Bulk, Fabricated by Hot-Pressing Method

Models for processing and fracture of Bi-2223 ceramic which is obtained by hot-pressing are discussed in this section. Computer simulation is applied to phenomena occurring during sintering, cooling and following fracture due to the macrocrack growth. The effects of Ag particles dispersed into the Bi-2223 matrix on some strength properties are studied [933].

10.7.1 Microstructure Formation by Processing

The sintering and cooling models for oxide superconductor have been discussed in detail in the previous sections. Therefore, we discuss in some detail the model of abnormal grain growth, caused by the sintering conditions (i.e., pressure and temperature [1642]) and by existence of secondary phases (e.g., $CaCuO_2$ and CuO), defining the Bi-2223 microstructure formation [1140]. For different ceramics [1296, 1319], the Wagner–Zlyosov–Hillert's model [1] can be used. However, this model does not take into account the effects of texture on grain growth, which in this case is controlled by the inhibition parameter, depending on volume fraction and size of the secondary phase particles. At the same time, experiments have shown that the primary re-crystallization, as a rule, initiates a texture [2]. Therefore, in modeling of secondary re-crystallization, we consider the dependence of intergranular boundary energy and its mobility on grain misorientation, using a corresponding texture component for each grain. For simplicity, we limit the texture components to two (*A* and *B*), which are quite sufficient to describe considerable number of test data. In particular, we could model the growth processes for considered anisotropic Bi-2223 superconductor along the *c*-direction and in the *ab*-plane. Mass transfer between crystallites is based on the following assumptions [423]:

(1) The migration rate of a boundary between two grains i and j is given by

$$v_{ij} = m_{ij}p_{ij} = 2\gamma_{ij}(1/R_i - 1/R_j), \tag{10.155}$$

where p_{ij} is the driving force, m_{ij} and γ_{ij} are the mobility and boundary energy, $(1/R_i - 1/R_j)$ is the average curvature for this grain boundary and $M_{ij} = 2m_{ij}\gamma_{ij}$ is the grain growth diffusivity.

(2) All grains of the same size and orientation experience the same growth rate (homogeneity condition).

(3) The grains, surrounding a given grain, are distributed randomly with respect to size and orientation.

Then, the grain growth rate of the size class i with a texture component *A* (or *B*) without the growth stagnation is found by [423]

Table 10.9 Material properties used in FEM computations

Material	Thermal expansion factor ($\times 10^{-6}$ K^{-1})	Young's modulus (GPa)	Yield strength (MPa)	Maximum elongation
YBCO (*a*-axis)	13 [836]	182 [565]	–	–
Ag	21 [1135]	80 [619]	50 [619]	0.5 [619]

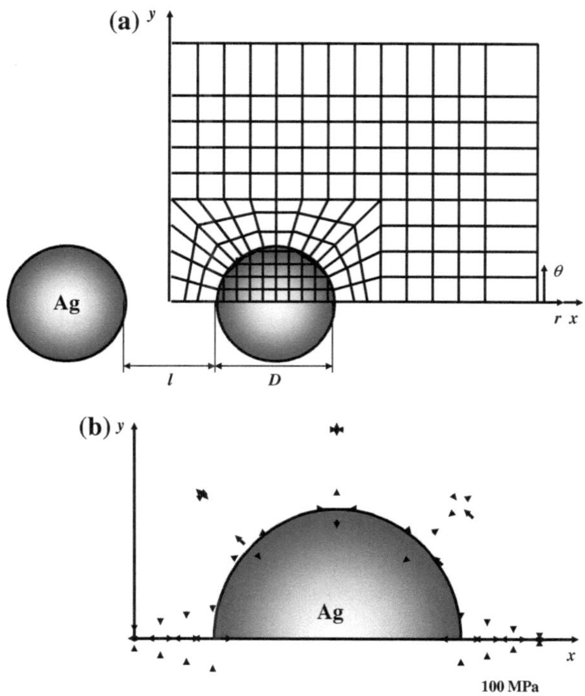

Fig. 10.42 (**a**) Mesh pattern for FEM calculation around silver particles dispersed in YBCO matrix and (**b**) maximum and minimum principal stress distributions around the silver particles in the 123 matrix [1066]

$$dR_i^{A(B)}/dt = M_*^{A(B)}[1/R_*^{A(B)} - 1/R_i^{A(B)}], \tag{10.156}$$

where $M_*^{A(B)}$ and $R_*^{A(B)}$ are the integrated diffusivity and integrated critical radius, respectively, controlling the grain growth of the component *A* (or *B*). Note that the value of $R_*^{A(B)}$ is the intermediate grain size of the component *A* (or *B*) between increasing and decreasing grains, and $M_*^{A(B)}$ determines the rate of these processes [423]:

$$M_*^A = \frac{F^A\langle R_A^2\rangle M^{AA} + F^B\langle R_B^2\rangle M^{AB}}{F^A\langle R_A^2\rangle + F^B\langle R_B^2\rangle}; \quad M_*^B = \frac{F^A\langle R_A^2\rangle M^{BA} + F^B\langle R_B^2\rangle M^{BB}}{F^A\langle R_A^2\rangle + F^B\langle R_B^2\rangle}; \tag{10.157}$$

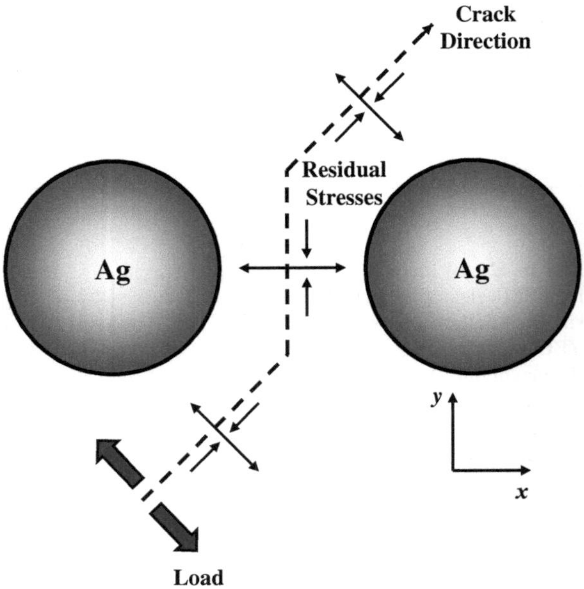

Fig. 10.43 Schematic illustration of the relationship between residual stress and crack propagation [1066]

$$R_*^A = \frac{F^A \langle R_A^2 \rangle M^{AA} + F^B \langle R_B^2 \rangle M^{AB}}{F^A M^{AA} \langle R_A \rangle + F^B M^{AB} \langle R_B \rangle}; \quad R_*^B = \frac{F^A \langle R_A^2 \rangle M^{BA} + F^B \langle R_B^2 \rangle M^{BB}}{F^A M^{BA} \langle R_A \rangle + F^B M^{BB} \langle R_B \rangle}, \tag{10.158}$$

where $M^{HK}(H,K = A \text{ or } B)$ is the grain-growth diffusivity of a boundary between grains of the orientation classes H and K. The grain fraction of the size class i with the orientation A (or B) is defined as

$$F_i^{A(B)} = n_i^{A(B)}/N_G; \; \sum_{i=1}^{N_S} (n_i^A + n_i^B) = N_G; \; \sum_{i=1}^{N_S} (F_i^A + F_i^B) = 1; \; F^{A(B)} = \sum_{i=1}^{N_S} F_i^{A(B)}, \tag{10.159}$$

where N_S and N_G are the total number of size classes and grains, respectively; $n_i^{A(B)}$ is the number of grains per unit volume of the size class i with orientation A (or B); $\langle R \rangle$ and $\langle R_{A(B)} \rangle$ are the mean radius of grains in the whole system and that with the orientation component A (or B), respectively:

$$\langle R \rangle = \sum_{i=1}^{N_S} F_i R_i; \quad \langle R_{A(B)} \rangle = \sum_{i=1}^{N_S} F_i^{A(B)} R_i^{A(B)} / F^{A(B)}; \tag{10.160}$$

$$\langle R^2 \rangle = \langle R \rangle^2 + \sigma^2; \quad \langle R_{A(B)}^2 \rangle = [\langle R_{A(B)} \rangle]^2 + \left[\sigma^{A(B)}\right]^2, \tag{10.161}$$

where σ and $\sigma^{A(B)}$ denote the standard deviations. The size class for a given grain is found by the cell number, containing in this grain. The condition for abnormal grain growth in the grain size class i with the orientation A (or B) in the space (R, t), taking into account (10.156), is found as [423]

$$\left|1/R_*^{A(B)} - 1/R_i^{A(B)}\right| > I_R/2, \tag{10.162}$$

where $I_R = 6f_v/(\pi r)$ is the value of the grain-growth stagnation, f_v and r are the volume fraction and mean radius of the secondary phase particles. It is assumed that the stagnation parameter is independent of the grain orientation and is calculated as described above. This parameter, as well as the critical radii of the components A and B, governs the abnormal grain growth. This circumstance enables us to define the HTSC ceramic properties and their dependencies on the secondary phase characteristics.

The size parameters, which are necessary for calculations, are found in the simulation of the primary re-crystallization. The orientations A and B are distributed between grains, using the Monte-Carlo procedure, and they do not change during growth. Moreover, it is assumed that mass transfer between the grains of different orientations is absent. As an example, we consider the case of the next diffusion parameters for the orientation classes [2]: $M^{AB} = M^{BA} = 2M^{AA} = 2M^{BB}$. This sufficiently arbitrary selection is explained by the absence of reliable test data for the Bi-2223 ceramic. The mass transport between grains is simulated in accordance with the grain growth mechanism at the non-singular surfaces [1642].

The computational algorithm for abnormal grain growth consists of the following steps:

(1) The distribution of the orientations H (where $H = A$ or B) between the grains that are formed after primary re-crystallization.
(2) The definition of all neighbors for every grain of both orientation classes.
(3) The determination of adjacent grain couple in each orientation class (i^H, j^H) with $\max\limits_{1 \le i^H, j^H \le N_S} \left|1/R_i^H - 1/R_j^H\right|$.
(4) The growth of the larger grains from the (i^H, j^H) at the expense of smaller ones.
(5) The checking of the conditions: $\left|1/R_*^{A(B)} - 1/R_i^{A(B)}\right| \le I_R/2$, where $i^H = 1 \ldots N_S$.
(6) End of the grain growth in the corresponding component H, if the conditions (5) have been satisfied; else a change of the corresponding parameters in (10.157)–(10.161) and fulfillment of the steps: (2)–(6) again.

The simulation of intergranular cracking due to cooling is carried out as before, using (10.4) and (10.5) and the procedure described in detail in Sect. 10.1.2.

10.7.2 Bi-2223 Toughening by Silver Dispersion

As has been noted above, an addition of Ag ductile phase dispersion to the Bi-2223 ceramic causes a considerable increasing in the fracture resistance of the superconductor compared to that of untoughened matrix. It is known, that the main mechanism responsible for enhanced toughness of brittle composites with ductile particles appears to be the crack bridging by the ductile phase. Here, we limit ourselves to the most important case for HTSC, when an increasing of the ceramic toughness is independent of the particle size and ductile strength. This corresponds to the state, when the ductile flow has occurred in a considerable zone near the crack tip. Then, the toughness increasing due to the ductile particles (stationary crack case) can be estimated as [1571]

$$\sqrt{1-\nu_{ef}^2}\frac{K_c}{K_c^0} = \sqrt{3}\left[1+\frac{10(1-\nu_{ef}^2)f_p}{(7-5\nu_{ef})(1-f_p)}\right]^{1/2}, \tag{10.163}$$

where K_c and K_c^0 are the fracture toughness with and without toughening, respectively; f_p is the ductile particulate concentration, defined by the fraction of this phase, intercepting the macrocrack path; ν_{ef} is effective Poisson's ratio, defined by the relative concentration of the Bi-2223 and Ag phases, and also by the intergranular microcracks, formed during the composite processing. Under conditions of the modified cubic model, we obtain [694]:

$$\nu_{ef} = (1-f_m^{2/3})\nu_c + f_m^{2/3}\frac{[\nu_m f_m^{1/3}/E_m + \nu_c(1-f_m^{1/3})/E_c]}{[f_m^{1/3}/E_m + (1-f_m^{1/3})/E_c]}, \tag{10.164}$$

where the indexes m and c correspond to the metal inclusions and the ceramic matrix. For the cracked matrix with a microcrack density, β_{cr}, Poisson's ratio, ν_c, and Young's modulus, E_c, are expressed as [1443]

$$\frac{\nu_c}{\nu_0} = \frac{1+[(16/45)(1-\nu_0^2)/(2-\nu_0)]\beta_{cr}}{1+[(16/45)(1-\nu_0^2)(10-3\nu_0)/(2-\nu_0)]\beta_{cr}}; \tag{10.165}$$

$$\frac{E_c}{E_0} = 1/\left[1+\frac{16(1-\nu_0^2)(10-3\nu_0)}{45(2-\nu_0)}\beta_{cr}\right], \tag{10.166}$$

where ν_0 and E_0 are the intrinsic elastic modules.

Since the metallic inclusions have a greater thermal expansion than the ceramic matrix [909], the residual tension (σ_R) occurs in the metal, but the ceramic is compressed. This internal stress state affects the toughening, because the compressive stresses in matrix must be exceeded within the bridging zone before beginning of crack opening. The additive increment of the toughness is estimated as [1571]

$$\Delta G_R \approx \alpha f_p \sigma_R u^*, \tag{10.167}$$

where α ($\sim$0.25) is a factor that depends on the precise nature of the function $\sigma = \sigma(u)$; u^* is the total crack opening when the ductile material fails (Fig. 10.44a). For cylindrical metal particles in a non-hardening material, the axial residual stress, σ_R^z, can be obtained as [715, 1571]

$$\frac{\sigma_R^z}{E_m \Delta\alpha \Delta T^*} = \frac{3}{3(1-2v_m) + 2(1+v_c)(E_m/E_c)}, \tag{10.168}$$

where $\Delta\alpha$ is the thermal expansion factor difference of the phases; ΔT^* is the cooling thermal range, in which the rapid creep provides relaxation.

It has been suggested in computer simulations that the silver inclusions are localized at the triple junctions, where there are usually microdefect sites, healed by the Ag. The necessary parameters for Ag particles are given elsewhere [1359]. The optimum Ag volume concentration in the Bi-2223 bulk is assumed to be $f_m = 0.2$ [909]. Finally, we modeled the intergranular macrocrack path (see Fig. 10.44b), using Viterbi's algorithm for graphs [1296, 1328], taking into account the grain structure and processing microcracks.

Block-scheme of one realization of the computational process describing the microstructure formation and fracture with estimation of fracture toughness change is presented in Fig. D.7. The errors of computer experiments at doubling of number of the considered cells to be into range of 2 % in all variants calculation of the averaged parameters. Statistically reliable results during computer simulations are obtained again by application of the stereological method [271].

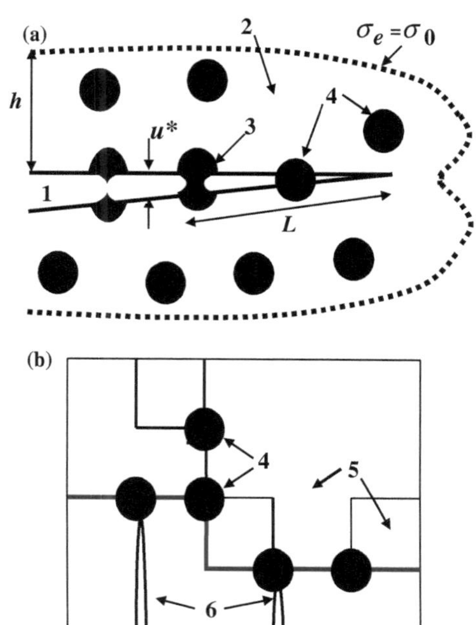

Fig. 10.44 (**a**) Scheme of crack *(1)* bridging by intercepted ductile silver particles *(4)*; the process *(2)* and stretch *(3)* zones of plastically distorted particles are shown (h and L are the corresponding zone sizes; u^* is the residual crack opening at link failure and σ_0 is the yield strength) [1571]; (**b**) a model fragment of Bi-2223/Ag composite with the Bi-2223 grains *(5)* and cooling microcracks *(6)*; macrocrack is denoted by *gray line*

Table 10.10 Numerical results [933]

D/δ	β_{cr}	l_{cr}/b	f_p	K_c/K_c^0
2.40	0.15	1.51	0.12	1.96
2.84	0.22	1.62	0.10	1.92

In order to define more accurate estimations of the effects of dispersed Ag particles on the strength properties, we assumed that other toughening mechanisms, considered above, are absent. We obtain numerical results for values of $I_R D_0 = 2.0$ and 1.2 (first and second line in Table 10.10, respectively), where D_0 is the grain size before the grain growth. In the macrocrack simulation which propagates along the intergranular boundaries of the Bi-2223 matrix, the value of f_p is found, taking into account the macrocrack length, the number of triple junctions (or Ag particles) in the crack path and the relationship between the defect size ($2a$) and critical boundary size (l_c^s,) namely: $a = 0.1 l_c^s$, which is the same as that for a ceramic with TEA [450]. As it follows from (10.167), the value of f_p is proportional to the toughness difference, ΔG_R. At the same time, a decreasing of the parameter I_R leads to increasing of the matrix grain size and corresponding enhancement of the microcracking density (β_{cr}). Moreover, the number of triple points also decreases with the increasing of the relative macrocrack size l_{cr}/b, where l_{cr} is the macrocrack size, and b is the specimen width due to the longer path, required around the larger grains. The latter causes a decreasing of f_p and corresponding decreasing ΔG_R. Finally, the smaller values of β_{cr} increase the elastic moduli and together with the increased concentration of f_p lead to increasing K_c/K_c^0 at decreasing of the grain size.

Chapter 11
Mechanical Destructions of HTSC Josephson Junctions and Composites

As it has been shown in Sect. 1.3.8, Josephson effects are connected with a behavior of weak links of superconductors. According to the classification of high-temperature superconducting Josephson junctions (HTSC JJs), presented in Sect. 2.5, special interest from view of strength and fracture toughness is excited by the junctions with intrinsic barriers or interfaces, formed by the intergranular boundaries with different crystallographic orientations, and also HTSC JJs with extrinsic interfaces in the fabrication of which, the artificial normal metallic or insulating barriers are used. The strength problems of the JJs, having the extrinsic interfaces (see Fig. 11.1), cause their small effective superconducting area as compared with the geometrical square and can lead to the large parameter spreads. The microstructure destruction can be found by some causes, namely by the deformation (lattice) mismatches and thermal expansion anisotropy [26, 895], by exceeding the critical film thickness [537, 1492], by the misorientation effects [537], by the rough or damaged interfaces [1492], etc.

The use of buffer layers (e.g., CeO_2, MgO, YSZ, ZrO_2, $SrTiO_3$ [1563]) has enabled the increase of superconducting and transport properties of HTSC composite structures due to diminution of TEA and lattice mismatches between HTSC film and substrate and also owing to decrease of a chemical reactivity of the substrate. Nevertheless, the problems of critical mechanical behavior for these laminated structures under conditions of existence of residual stresses and external loads remained in the center of attention of HTSC JJ researchers and engineers.

This chapter presents a set of models for estimation of strength properties for HTSC composites of S–I–S and S–N–S (where S is the superconductor, I is the insulator, N is the normal metal) types, based on the consideration of interfaces, taking into account TEA, geometric and material parameters, external loads and residual stresses. Moreover, features of the mechanical damage of HTSC composites and proper fracture resistance mechanisms are considered [1302, 1304].

I. A. Parinov, *Microstructure and Properties of High-Temperature Superconductors*,
DOI: 10.1007/978-3-642-34441-1_11,

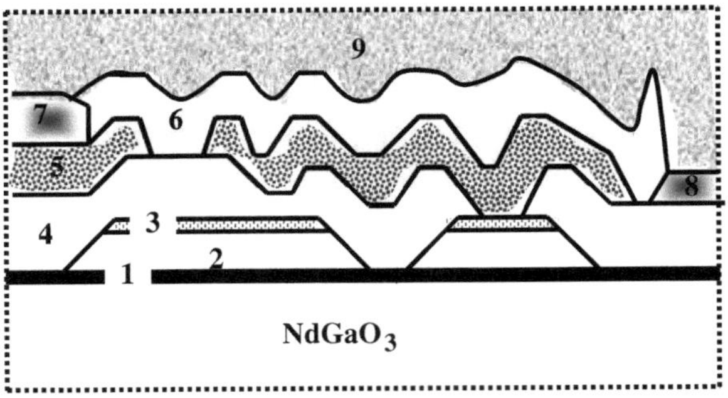

Fig. 11.1 Cross-section of HTSC JJ, consisting of nine layers, on $NdGaO_3$ substrate: *1* buffer layer (30 nm), *2* YBCO base layer (150 nm), *3* transitional superconducting layer (150 nm), *4* $SrTiO_3$ epitaxial dielectric (150 nm), *5* YBCO top layer (150–200 nm), *6* non-epitaxial silicon nitride dielectric (250 nm), *7* silver contact layer (600 nm), *8* molybdenum resistor (90 nm) and *9* silver wiring layer (600 nm) [1219]

11.1 Interface Fracture

The interface roughness in HTSC bimaterial and a crack growth near and at the interface introduce inevitably a mixed loading mode. In this case, there are some morphological features of fracture [453, 779, 1561], shown in Fig. 11.2, namely (i) an interface fracture, (ii) a crack growth into more brittle component, (iii) an alternative fracture along interface or between the interface and adjacent material, and (iv) a crack deflection from one interface to other. The experiments on various materials have shown that the crack path is found by both the ratio of interface fracture energy to fracture energy of more brittle component (Γ_i/Γ_2) and the phase angle of loading, $\Psi = \mathrm{arctg}(K_{\mathrm{II}}/K_{\mathrm{I}})$, where K_{I} and K_{II} is the SIF of I and II mode, respectively. On the other hand, this angle can be coupled with the ratio of displacements—the shear to the opening (i.e., to the ratio of *v/u*) due to the insertion of Dunders parameters (α and β) as [410]

$$\Psi = \arctan(v/u) - \in_0 \ln r - \arctan(2 \in_0), \tag{11.1}$$

where

$$\in_0 = \frac{1}{2\pi} \ln\left(\frac{1-\beta}{1+\beta}\right); \tag{11.2}$$

$$\alpha = \frac{\mu_1(1-v_2) - \mu_2(1-v_1)}{\mu_1(1-v_2) + \mu_2(1-v_1)}; \ \beta = \frac{\mu_1(1-2v_2) - \mu_2(1-2v_1)}{2[\mu_1(1-v_2) + \mu_2(1-v_1)]}; \tag{11.3}$$

μ_k and v_k (k = 1, 2) are the shear modulus and Poisson's ratio, respectively, for k-component and r is the distance measured from the crack tip at the interface. The parameter of α together with the dependence of Γ_i/Γ_2 on Ψ enables to separate the

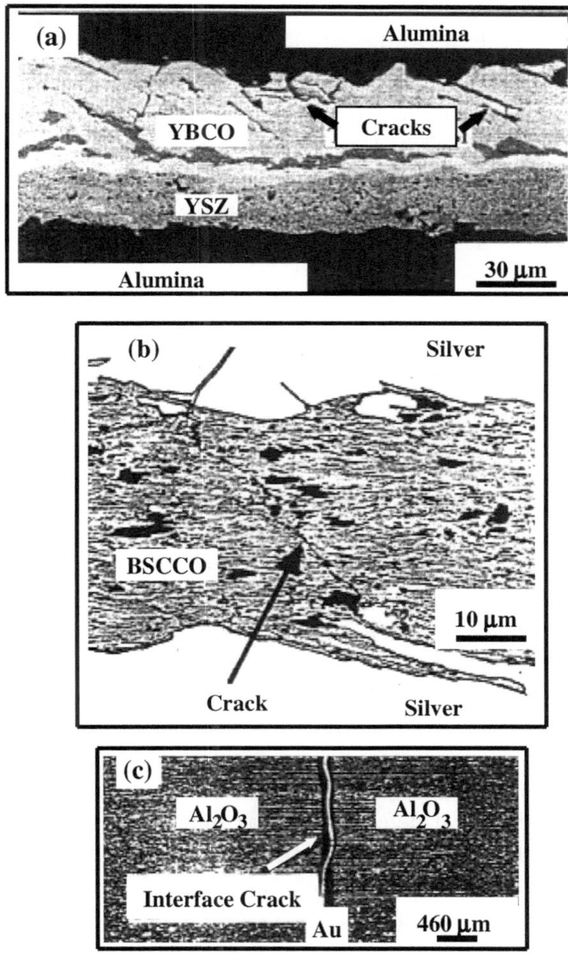

Fig. 11.2 Characteristic morphology of possible fractures at and near interfaces into HTSC composite structures: (**a**) interface fracture and an alternative fracture between the interface and adjacent material (SEM cross-sectional micrograph of YBCO thick film, processed on YSZ barrier layer on alumina [1561]), (**b**) crack growth into more brittle component (back-scattered electron image of transversal cross-section of BSCCO/Ag tape [779]) and (**c**) crack in Al_2O_3 bonded with Au, showing a crack alternating between interfaces [453]

areas of the interface crack and the brittle substrate fracture in the case of an initial crack at the interface.

Then, the fracture behavior and the interface fracture energy are very sensible to a sign of the phase angle at the great difference of the fracture energies for both components ($\Gamma_1 \gg \Gamma_2$). In the case of the positive value of Ψ, there are both regimes, namely the interface cracking and a crack deflection into more brittle component, depending on the parameter α. The second case (when $\Psi < 0$) is more interesting. Therein, the greater value (i.e., Γ_1) is compared with the interface fracture energy Γ_i. As the condition $\Gamma_1 \gg \Gamma_i$ prohibits crack propagation away from the interface, then there are two cases. For the low material strength, a plastic bluntness of the crack at the interface occurs and the failure features are caused by

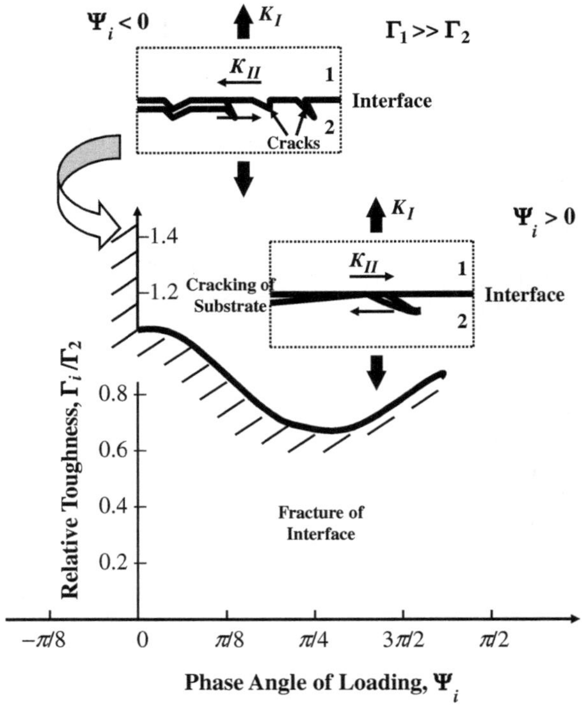

Fig. 11.3 Fracture at interface into bi-material for $\Gamma_1 \gg \Gamma_2$ (result in the case of $\alpha = \beta = 0$). When phase angle, $\Psi < 0$, the interface fracture demonstrates near-boundary segments of material free from cracks [453]

the toughening mechanisms, including an initiation of voids at the interface.[1] In the contrary case, the stress state of the interface crack interacts with microcracks and structural defects which as rule exist in the brittle material and provoke a growth of microcracks along the direction to the interface. This causes a saw-tooth fracture, with chips of the brittle material attached to the interface (see Fig. 11.3).

Further, because the complete smoothness of the interfaces is impossible (e.g., see Fig. 11.4a), an estimation of the JJs fracture resistance depending on the interface roughness is the actual problem. The crack surfaces, growing along the interface, contact each other either at the roughness or at the facets. In this case, it is possible to obtain different values of the interface fracture resistance, which grow with the phase angle of loading, Ψ. These effects have been observed and estimated for different brittle materials [456, 776]. In particular, a comparative analysis of microstructure properties and fracture parameters, caused by the mixed loading mode, has been fulfilled for the fine-grain ($PbTiO_3$) and coarse-grain ($BaTiO_3$) ferroelectric ceramics [1321]. The decrease of the strain energy release rate (or crack shielding) $\Delta G = G - G^t$ (where G and G^t are the values of the strain energy release rate connected with applied load and at the crack tip,

[1] At the same time, note that those intergranular voids formed, for example, by thermo-mechanical treatment during a multi-stage processing of monocore Bi-2223 tapes, may become the main cause of critical current diminution in the case of prolonged final annealing [1301].

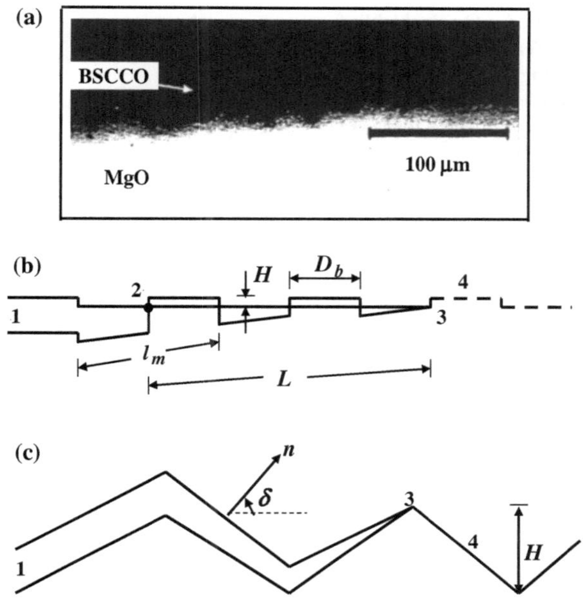

Fig. 11.4 **a** Proper interface between melted Bi-2212 thick film and polycrystalline MgO substrate (optical micrograph of cross-section [1236]), and two models (**b**) and (**c**) used in the analysis of crack at rough interface [456, 776]. The numbers denote crack (*1*), last contact point (*2*), crack tip (*3*) and interface (*4*)

respectively) can be estimated using two models: (i) the contact zone of the crack surfaces taken into account, but without account of Coulomb's friction [456], and (ii) the inclination angle of the faceted interface, δ, is taken into account [776]. We have for the first model (see Fig. 11.4b) [456]:

$$\Delta G/G = [1 - \lambda^{-2}(\alpha)]\tan^2\Psi/(1 + \tan^2\Psi), \tag{11.4}$$

where $\alpha = (L/l_m)/\ln[1/\sin(\pi D_b/2l_m)]$. The values of the function $\lambda(\alpha)$ for various values of α have been tabulated in [200]. The length of the contact zone, L, is found by $L = (\pi/32)[EH/(1-\nu^2)K_I]^2$. Here l_m is the spacing between the facet centers, D_b is the facet length and H is the height of the interface step.

The numerical approximation is obtained, taking into account the typical geometry of undulating interfaces, assuming that $D_b/l_m \sim 1/2$; $H/l_m \sim ½$, equating K_I to the fracture toughness, K_{Ic} (which is approximately equal to 1 MPa · $m^{1/2}$ for a broad set of ceramics), and introducing a material parameter $\chi = E^2H/(1-\nu^2)K_{Ic}^2$, which defines the length of the contact zone, L. Note that the parameter χ in total causes a fracture behavior in this microstructure consideration. Therefore, one should be measured with a high accuracy in order to obtain acceptable results for the HTSC JJs. For example, the value of $\chi \approx 100$ has been found for the glass—polymer interface [456].

We obtain for the second model (Fig. 11.4c) [776]:

$$\Delta G/G = (\cos\delta\sin\Psi - \sin\delta\cos\Psi)^2, \quad \Psi > \delta \quad H \gg a, \tag{11.5}$$

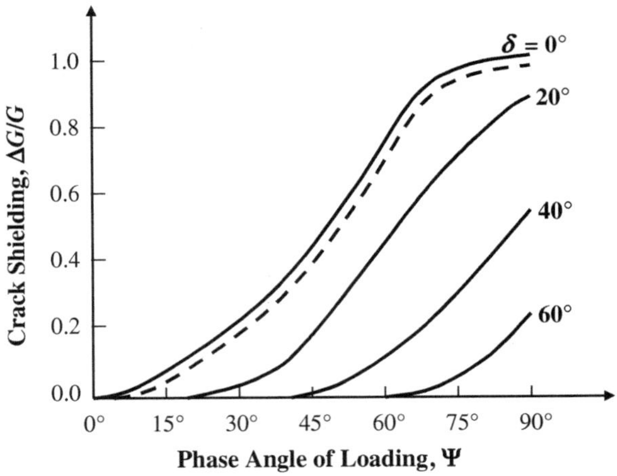

Fig. 11.5 Crack shielding ($\Delta G/G$) versus phase angle of loading (Ψ). The *dashed curve* corresponds to the model of contact zone ($\chi \approx 100$), and the *solid curves* are obtained due to the model for different inclination (δ) of facet sides

where a is the crack size. Then, the condition of facet contact in the second model [776] coincides with the condition of a single crack deflection in the first model [456]. In the case of existence of the facet contacts, the second model as compared with the first one takes into account, additionally, an interaction between various crack deflections. The proper dependencies of the crack shielding ($\Delta G/G$), caused by the interface roughness, on the phase angle of loading (Ψ) and on the different inclinations of the facet sides (δ) are shown in Fig. 11.5. The obtained trends of the crack shielding growth with the phase angle of loading agree well with experimental observations of the interface crack in different ceramics, in which the steady-state crack trajectory is caused by the condition, $K_{II} = 0$ [453, 456, 776, 1321].

11.2 Thin Films on Substrates

During processing of JJ, the residual stresses can initiate in thin films adherent to substrates due to the deformation mismatches or/and thermal expansion anisotropy. The residual stresses in YBCO film structures, caused by different causes (e.g., lattice mismatches, thermal expansion, non-stoichiometry) can reach, in whole, the value of $\sim$100 MPa [560]. A relaxation of these stresses can also lead to the formation of defect structures, in particular the dislocation mismatches form gradually with increase of the film thickness and accumulation of elastic stresses. Due to the high anisotropy of microstructure properties, thin films are subjected to the microcracking and twinning in the specific directions. For example, generally, the cracks form into (001) plane, and spacing between them increases with the film thickness. First, crack occurs in the film at attainment of the critical thickness (H_c), depending on thermo-elastic stress of the lattice disagreement (σ) and fracture toughness (K_c) as $H_c \sim (K_c/\sigma)^2$ [1028]. The distinct feature of film from bulk crystals is a cellular domain structure inside single grains with the cell boundaries parallel to the (100) and (010) planes. The cells differed by mutual replacement of directions of the a- and b-axes and coupled coherently similarly to twins. Together with external electric and magnetic fields, the internal stresses lead to different damage, developing in the film-substrate system, in particular to the film de-cohesion from substrate. The de-cohesion mechanism depends on the residual stress sign (i.e., tension or compression) and existence of the stress gradient. In the film tension, de-laminations could initiate at the sample edges and propagate into brittle substrate parallel to interface [395, 718]. Moreover, a transverse film cracking which nucleates at the free surface is possible [717, 718], with subsequent its de-lamination and buckling [153]. Under conditions of compression, the film de-lamination and buckling are possible, too [153]. The investigations of mechanical and electromagnetic properties of HTSC films are complicated due to (i) structure defects, (ii) high anisotropy of elastic moduli, and (iii) existence of structure phase transformations [1028].

In order to estimate strength parameters and study fracture features, the composite beam theory is applied, assuming an existence of the initial microdefect crack at the sample edge. Consider the model sample, presented in Fig. 11.6a, which is equivalent to the case of uniform distribution of the thermal stress, $\sigma_0 = E_f \Delta\alpha \Delta T$, where E_f and ν_f are elastic moduli for the film, $\Delta\alpha$ is the difference of the thermal expansion factors between the film and substrate and, ΔT is the temperature difference, covered at cooling. Note that the thermal strain, $\varepsilon = \Delta\alpha\Delta T$, must be replaced in the case of the deformation mismatches by the strain, depending on the crystallographic properties of the system. Then, the corresponding SIFs can be found by [395]

$$K_I = 0.434 P h^{-1/2} (\Sigma + \lambda)^{-1/2} + 0.558 M (I h^3)^{-1/2}; \tag{11.6}$$

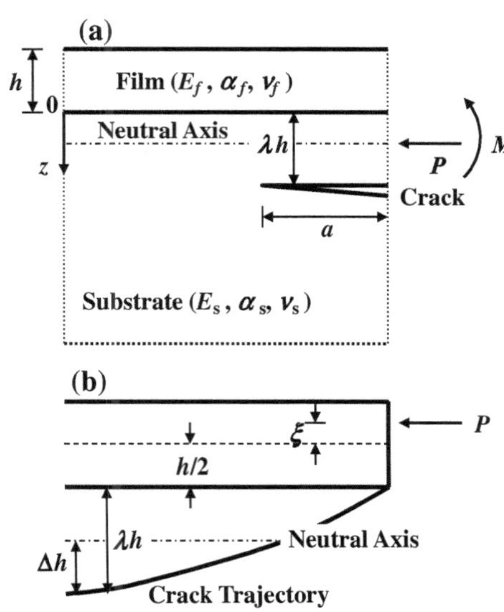

Fig. 11.6 Models of steady-state crack into substrate, taking into account, (**a**) thermal expansion difference between film and substrate, and (**b**) existence of stress gradient in the film

$$K_{II} = 0.558Ph^{-1/2}(\Sigma + \lambda)^{-1/2} - 0.434M(Ih^3)^{-1/2}, \tag{11.7}$$

where h is the film thickness, $\Sigma = E_f/E_s$ (in the plane stress case), and $\Sigma = E_f(1 - v_s^2)/E_s(1 - v_f^2)$ (in the plane strain case); E_s and v_s are the elastic moduli for the substrate; $I = \left\{\Sigma[3(\Delta - \lambda)^2 - 3(\Delta - \lambda) + 1] + 3\Delta\lambda(\Delta - \lambda) + \lambda^3\right\}/3$ is the dimensionless moment of inertia; $\Delta = (\lambda^2 + 2\Sigma\lambda + \Sigma)/2(\lambda + \Sigma)$; $\lambda = z/h$; $P = \sigma_0 h$ is the load; and $M = \sigma_0 h^2\lambda(\lambda + 1)/2(\Sigma + \lambda)$ is the bending moment (per unit thickness).

The steady-state crack path into brittle substrate parallel to the interface is found by the condition $K_{\rm II} = 0$. Then, equating the SIF of I mode to the fracture toughness of the substrate ($K_{\rm I} = K_{{\rm I}s}$), we obtain the critical layer thickness (h_s) as

$$h_s = 0.755(\Sigma + \lambda_s)(K_{Is}/\sigma_0)^2, \tag{11.8}$$

where λ_s is the relative depth, governing the steady-state crack path, for which $K_{\rm II} = 0$. At the smaller value of the layer thickness, complete fracture is inhibited,

Consider also the case of the stress gradient, existing on the film thickness (see Fig. 11.6b). In this case, σ_0 is the mean stress in the film and $M = P[(\lambda + 0.5 - \Delta)h + \xi]$, where ξ is the distance from the film center to the force action line in (11.7). Similar to the previous model the parameters h_s and λ_s are found depending on the value of ξ, that is, on the given stress gradient.

Then the simple equations can be obtained for toughness parameters in the case of small-scale yielding of the substrate under condition of steady-state transversal cracking of the film. In this case, the strain energy release rate (G_{ss}) and corresponding SIF (K_{ss}) are found as [717]

$$G_{ss}E_f/\sigma^2 h = \pi F(\Sigma);\ K_{ss}/\sigma\sqrt{h} = \sqrt{\pi F(\Sigma)}, \tag{11.9}$$

where σ is the film tension (residual or/and applied); the function $F(\Sigma)$ is given in Table 11.1 [717] for various values of $\Sigma = \Sigma_f/\Sigma_s$. Then, the typical de-laminations, occurring from the notch orthogonal to the tension direction, are classified in Fig. 11.7a, and b. In the case (a), there is a de-lamination of open type, which grows together with the longitudinal cracking of the film. In the case (b), the longitudinal cracks are absent. The de-lamination remains to be closed and "pocket-like", and the buckling takes place due to Poisson's effect under sufficiently-high applied stress. The form of this de-lamination can be approximated by ellipse with semi-axes a and b.

Generally, the film de-lamination from the substrate and its buckling in compression as well as under tension can be studied in the framework of the stability theory in fracture mechanics with consideration of different stages for crack-

Table 11.1 Function $F(\Sigma)$ in dependence on ratio of elastic moduli, Σ

Σ	4	3	2	1	1/2	1/3
$F(\Sigma)$	0.79	0.75	0.70	0.62	0.57	0.54

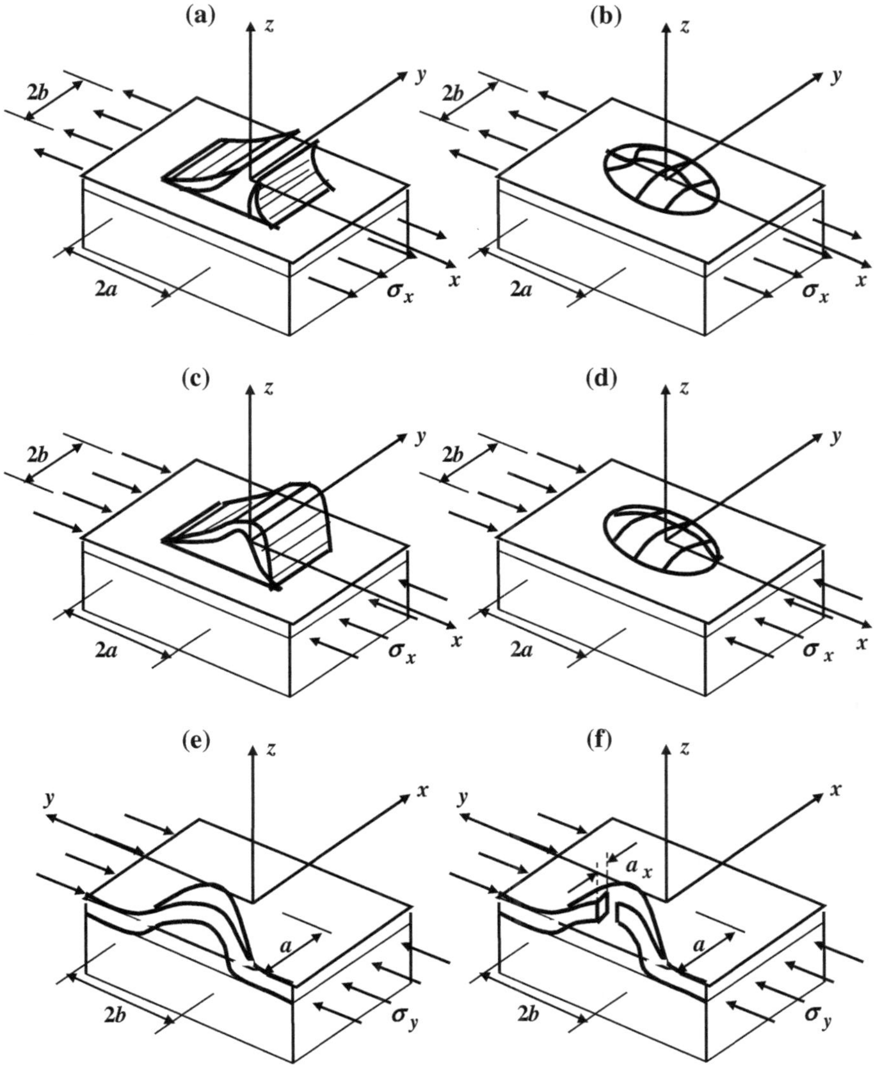

Fig. 11.7 Typical de-laminations in film-substrate structure (see explanation in the text) [153]

shaped defects [153]. The typical de-laminations under compression are shown in Fig. 11.7c–f. By sufficiently-high compression, the de-lamination opens and is accompanied by a buckling, which is similar to the cylindrical beam-like bending (case (c)). Case (d) represents a close de-lamination in the form of ellipse with semi-axes a, and b. The so-called edge de-laminations are presented by cases (e) and (f) and can be approximated by half-ellipse. In case (f), the secondary crack can be observed, which grows across the de-lamination as a result of the de-lamination bending.

In the case of the beam approximation, the critical buckling strain, $\in_c$, can be estimated for acting compressive stresses in the film and substrate (σ^f_{xx} and σ^s_{xx}, respectively) as [1560]

$$\in_c = \frac{\Omega_c}{12}\left[3 + \frac{(1-2v_s)}{(1-v_s)}\Omega_c + \frac{2}{3}\Omega_c^2 + \frac{(3-4v_s)}{12(1-v_s)^2}\Omega_c^3\right]\left(1+\frac{\Omega_c^2}{6}\right)^{-1}, \quad (11.10)$$

where

$$\Omega_c^3 = \frac{12(1-v_s)^2}{\Sigma(3-4v_s)}; \qquad \Sigma = \frac{\sigma^f_{xx}}{\sigma^s_{xx}} = \frac{E_f(1-v_s^2)}{E_s(1-v_f^2)}. \quad (11.11)$$

11.3 Step-Edge Junctions

The strength problems, related to above ones arise, considering an inclined interface crack in the step-edge JJs and in the S–N–S edge junctions with ground planes. It is known that the tapered edges of the base electrode with edge angles, being lesser than 45°, are important to avoid a grain boundary formation in the counter-electrode [727]. On the other hand, the shallower edge angles have revealed problems connected with a void formation on the edge angles, being lesser than 15° [728]. So, in order to optimize the strength and damage behavior of these HTS JJs, it is useful to consider a crack problem, when the crack surfaces contact at the kinks. In this case, the SIFs of I and II modes at the crack tip (i.e., K_I^t and K_{II}^t) differ from corresponding applied SIFs by the values, which depend on the kink angle (β), the kink amplitude and Coulomb friction factor (μ) (see Fig. 11.8a). This geometry and a replacement of the inclined frictional force, F, acting at the crack surfaces, by the homogeneously distributed tractions, applied to the segment, $2 \in d$ (see Fig. 11.8b), allow us to estimate a contact effect through normal and shear forces. In turn, it permits to determine the contributions of this effect on the SIFs at the crack tip and at the corresponding displacements of the crack surfaces.

Further, it is possible to consider the conditions of sliding and locking of the crack surfaces via terms of the phase angle of loading (Ψ), the friction angle ($\Phi = \arctan \mu$) and the kink angle (β). An analysis, which has been carried out for $K_{\mathrm{I}} > 0$ [456], allowed one to distinguish the main types of the crack behavior at the crack asperity (see Fig. 11.9). Then, based on the conditions of a contact and a frictional locking, the crack shielding (ΔG) may be estimated in the case of the sliding contact due to the facet contact as [456]

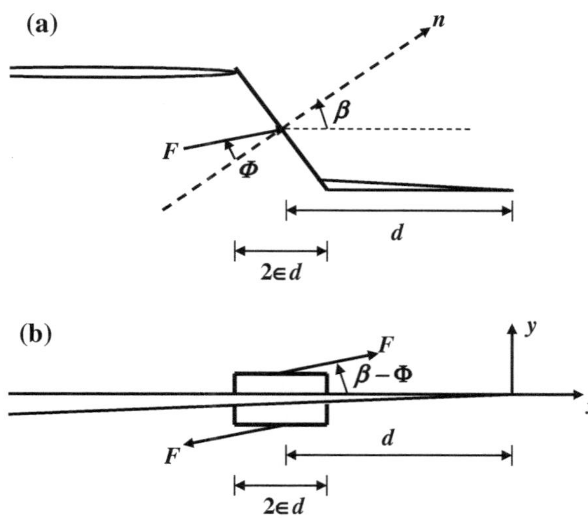

Fig. 11.8 Model of interface crack [456] in step-edge HTSC JJ used in order to analyze effects of the crack shielding

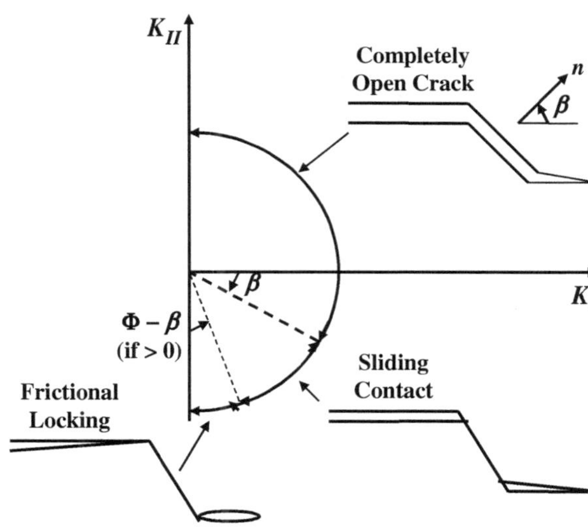

Fig. 11.9 Chart of possible crack behavior types, connected with crack asperity, taking into account loading modes [456]

$$\frac{\Delta G}{G} = 2h(\in)\frac{(\sin\beta + \cos\beta\tan\Psi)[\sin(\beta - \Phi) + \cos(\beta - \Phi)\tan\Psi]}{\cos\Phi(1 + \tan^2\Psi)} - \frac{h^2(\in)(\sin\beta + \cos\beta\tan\Psi)^2}{\cos^2\Phi(1 + \tan^2\Psi)}, \qquad (11.12)$$

where $0 < \beta < \pi/2$ (for $\pi/2 < \beta < \pi$, Φ must be replaced by $(-\Phi)$), and due to the crack locking as

$$\Delta G/G = 1 - [1 - h(\in)]^2, \qquad (11.13)$$

where $h(\in) = f(\in)/g(\in)$, but the functions $f(\in)$ and $g(\in)$ are found by

$$f(\in) = (\sqrt{1+\in} - \sqrt{1-\in})/\in; \tag{11.14}$$

$$g(\in) = \frac{1}{2\in}\left[\sqrt{1+\in} - \sqrt{1-\in} + \left(1+\frac{\in}{2}\right)\ln\left(\frac{\sqrt{1+\in}+1}{\sqrt{1+\in}-1}\right)\right.$$
$$\left. - \left(1-\frac{\in}{2}\right)\ln\left(\frac{1+\sqrt{1-\in}}{1-\sqrt{1-\in}}\right)\right]. \tag{11.15}$$

The comparison of $h(\in)$ with the function, defined from exact solution of the problem on a microcrack ahead of a macrocrack [1447], shows that the proposed model slightly overestimates a crack shielding, caused by the facet contact. The dependencies of the crack shielding ($\Delta G/G$) on the various parameters for above two mechanisms are presented in Figs. 11.10 and 11.11. The crack behavior due to the facet contact is more interesting. In the both examples of this case, namely without and with a friction (i.e., for $\Phi = 0°$ and 45°, respectively), the crack shielding ($\Delta G/G$) shows non-monotonous dependencies on the phase angle of loading (Ψ). Initially, the value of $\Delta G/G$ grows and then one diminishes with increase of Ψ. A friction presence displaces the maxima of the curves in the direction of the increase of Ψ, restoring usual tendency of a growth of the crack shielding with the phase angle of loading. At the same time, a displacement of the maxima in the direction of the decrease of Ψ occurs, increasing the kink angle β. The latter corresponds to diminution of the edge angle (α) of the step-edge junction, since $\alpha = 90° - \beta$.

Then a non-zero friction may introduce the contrary mechanism to the crack shielding, namely the crack amplification ($\Delta G < 0$). This is impossible in the case

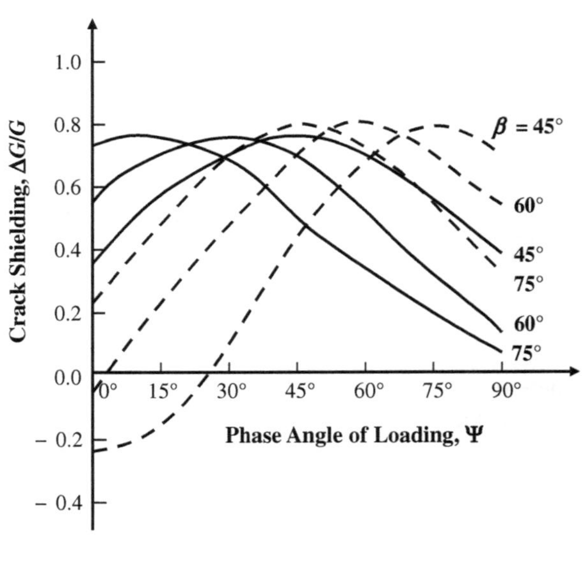

Fig. 11.10 Crack shielding ($\Delta G/G$) as functions of phase angle of loading (Ψ), kink angle (β) and friction angle (Φ). *Solid curves* correspond to the case of friction absence ($\Phi = 0°$), and *dashed curves* find the case of $\Phi = 45°$

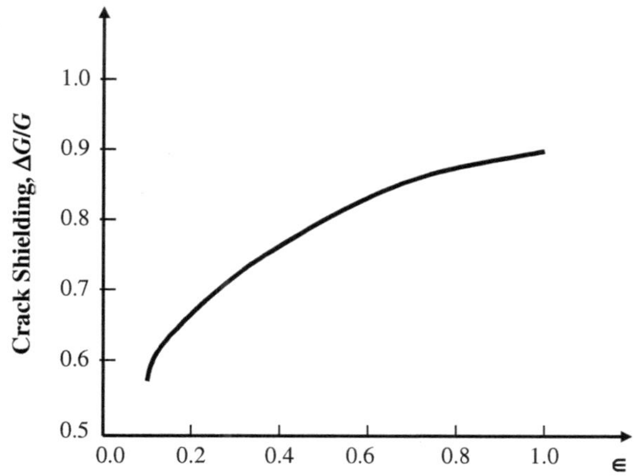

Fig. 11.11 Crack shielding ($\Delta G/G$) versus parameter $\in$ in sliding contact (effect of the crack locking)

of $90° < \beta < 180°$, when $\Delta G/G$ grows monotonously with Ψ (into interval of its alteration from 0° up to 90°) [456]. Thus, the crack, which climbs on a step, is safer as compared with the crack that goes down from the step. Therefore, the maximum values of the crack shielding for the present type of HTSC JJs may be found, thanks to an optimum and simultaneous selection of the parameters β, Φ and Ψ in accordance with the trends, depicted in Fig. 11.10. In the case of the crack locking, a monotonous growth of the crack shielding takes place with the parameter $\in$ (see Fig. 11.11), which finds the step size at given values of β and d. The greater edge size corresponds to the greater microcrack ahead of a macrocrack in the case of the inclined interface, at $\beta = \text{const}$ and at the invariable crack shielding. Then, even the small structural defects (with a high probability initiated at the interfaces during HTSC JJ processing, for example, due to the thermal anisotropy or/and crystallographic mismatches) may be sufficient in order to achieve an actual crack shielding, at the small edge size.

11.4 Transverse Fracture

During HTSC JJ processing and a device work under high electromagnetic fields, a nucleation and a propagation of microcrack-like defects along the direction perpendicular to the layers of the HTSC composite is possible (e.g., see Fig. 11.12). On the whole, fracture picture and fracture resistance are not only caused by the material properties and by the geometrical parameters of thick (matrix) layers and thin (buffer) layers, but even rather by the failure features at the interfaces. Decohesion (or de-bonding) processes at the material interfaces may be caused by the residual stresses, formed during the HTSC JJ processing. In this case, the internal stresses may be increased by the supplementary stresses due to the thermal change and/or Lorentz forces.

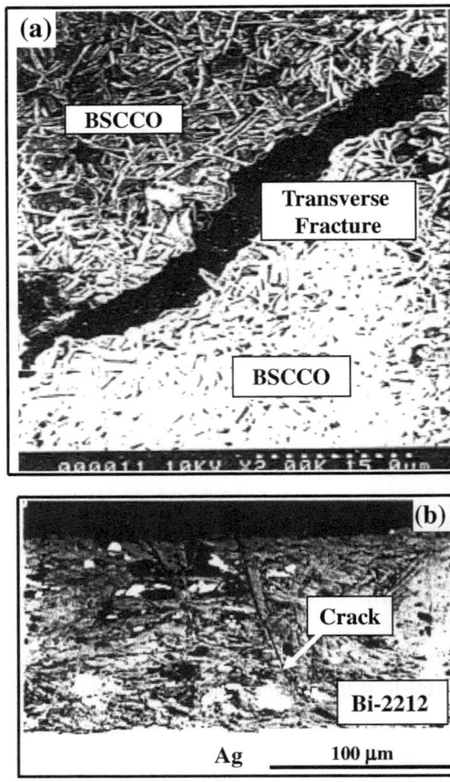

Fig. 11.12 (**a**) SEM micrograph of transverse fracture of BSCCO/Ag rod [1255] and (**b**) SEM image of cross-section of Bi-2212/Ag film with transverse crack [470]

The mechanical damage, nucleated by de-cohesion processes, causes an immediate diminution of the functional properties of the junction, decreasing steadily with time that limits reliability and longevity of the device. At the same time, a partial de-cohesion can play a positive role in the case of a transversal fracture of the HTSC layer composite, increasing its fracture resistance. Hence, it is necessary to estimate different strength parameters and fracture toughness, which should be obtained in the framework of fracture mechanics, applied to the brittle matrix–fiber composites, when a matrix crack is bridged by the one-axially aligned reinforcing fibers [457, 1086, 1675]. The use of buffer layers in the HTSC JJs is connected, in particular with attempts to significantly decrease the thermal and lattice mismatches. Therefore, an absence of residual stresses will be assumed below. In this case, the damage processes are caused by external tensile stresses (σ_∞) remotely applied and parallel to the fibers (or analogously modeled buffer layers).

First, note some general features connected with the above fracture mechanism. The main idea of the toughening of the brittle matrix (including, the case of the HTSC composites) by the brittle fibers is linked with the processes of de-bonding and sliding at the interfaces (see Fig. 11.13). A de-bonding at the matrix–fiber

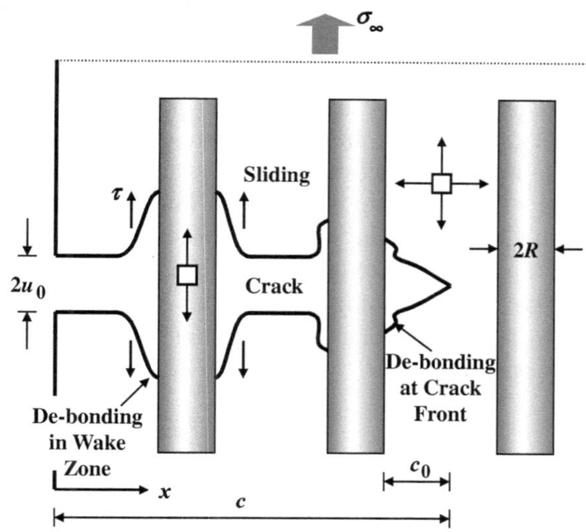

Fig. 11.13 Schematic representation of transverse matrix crack in layered HTSC composite

interface should be more favorable compared with a fiber fracture at the matrix crack front in the case of the crack, inhibited by the fibers. The de-bonding is more probable than the fiber fracture, if the interface fracture energy is sufficiently small as compared with the fiber fracture energy. Then, the HTSC composite, which to be under the de-bonding conditions, demonstrates pushing effect of the broken fibers by the crack surfaces. At the same time, an alternative fracture mechanism causes a growth of the matrix crack through the fibers without a de-bonding. So, a non-catastrophic fracture mode could be obtained in the composites with weak interfaces and high-strength fibers (or buffer layers). For this sub-critical fracture mode, the numerous matrix damages prior to a failure of the fibers are proper. Then, the complete strength of the composite is caused by the fracture of the fibers and by the following processes of the broken fiber pushing. The catastrophic mode is found by the fiber fracture in the wake of the main matrix crack during its propagation. In this case, the complete strength is limited by the single dominant crack and is found by a fracture resistance curve (i.e., dependence of toughness on crack size) [457]. Then, a sliding of the broken fibers or thin buffer layers inevitably introduces a fiber pushing by the crack surfaces. This process demands the study of fiber strength statistics, which are usually found by Weibull's distribution with the shape and scaling parameters m and S_0, respectively [1675]. A decrease of m corresponds to the more broad distribution of the fiber strength, that is, to the fracture of greater number of the fibers far from the matrix crack front. This defines an increase of the fiber pushing zone size. High median fiber strength is another useful property with a view of the fracture resistance growth. It is found by the large values of S_0, and causes the sub-critical fracture mode.

For different applications of HTSC JJs, it is more interesting to consider a short crack, that is, the case, when the entire crack contributes in a stress concentration, and a stress required for the crack growth is sensitive to its size. Moreover, it is important to estimate critical parameters at the transition to the steady-state cracking (i.e., to long crack), when the stresses at the crack tip grow with an applied tension but are independent on the whole crack size. The effects of buffer layers on fracture of the HTSC composite can be estimated, introducing as the stresses, which close the crack surfaces, as the stresses in the buffer layers, playing role of the bridges between the crack surfaces [1086]. The corresponding decrease of SIF at the crack tip is calculated from these surface attractions, using standard Green's function. The crack growth criterion is found, equating the SIF at the crack tip to intrinsic toughness of the matrix without reinforcements (K_c^M). Then an analysis of the pushing buffer layers from the matrix, based on the results of [1086], enables to find a relationship between the closure pressure (p) and the crack opening (u) as

$$p = \left[u\tau V_b^2\, E_b(1+\eta)/R\right]^{1/2}, \tag{11.16}$$

where $\eta = E_b V_b / E_m V_m$, $2R$ is the buffer layer thickness, τ is the sliding frictional stress at the interface, E_m and E_b correspond to Young's moduli of the matrix and buffer layer, and $V_m = 1 - V_b$ is the volume fraction of the matrix.

An approximate analytical solution for the short crack can be obtained, assuming that the crack profile at small crack sizes (c) does not differ greatly from that of the crack, subjected to uniform pressure. Then the crack opening is found as [1086]

$$u(x) = 2(1 - v_c^2)K^L c^{1/2}(1 - x^2/c^2)^{1/2}/E_c\pi^{1/2}, \tag{11.17}$$

with the limiting displacement ($2u_0$) and the equilibrium stress (σ_∞), defined in dependence on the crack size, respectively, as

$$u_0 = \sigma_\infty^2 R/\tau V_b^2 E_b(1+\eta); \tag{11.18}$$

$$\sigma_\infty/\sigma_m = (1/3)(c/c_m)^{-1/2} + (2/3)(c/c_m)^{1/4},\ \mathrm{c} \le \mathrm{c}_0, \tag{11.19}$$

where $E_c = E_m V_m + E_b V_b$, $v_c = v_m V_m + v_b V_b$, $K^L = K^M E_c/E_m$, K^L and K^M are the SIFs of the composite and of the matrix, respectively. The crack size ($c = c_0$), being transitional to the steady-state crack, and the corresponding equilibrium stress ($\sigma_\infty = \sigma_0$), which is independent on the crack size, can be obtained as

$$c_m = 2\left[K_c^M E_m V_m^2(1+\eta)R/I^2\tau V_b^2 E_b(1-v_c^2)\right]^{2/3}; \tag{11.20}$$

$$\sigma_m = (6.7/\Omega)\left[I^2\left(1-v_c^2\right)\left(K_c^M\right)^2 \tau E_b V_b^2 V_m (1+\eta)^2/E_m R\right]^{1/3}; \tag{11.21}$$

for straight crack:

$$\sigma_0/\sigma_m = 1.02;\ c_0/c_m = 1.88;\ \Omega = \pi^{1/2};\ I = 1.20 \tag{11.22}$$

for penny-shaped crack:

$$\sigma_0 = \sigma_m;\ c_0 = c_m;\ \Omega = 2/\pi^{1/2};\ I = 2/3. \tag{11.23}$$

Then, the steady-state toughness increment is found as [1449]

$$\Delta G_c = 2\int_0^{u_0} p(u)du, \tag{11.24}$$

and we have, taking into account (11.16),

$$\Delta G_c = (4/3)\alpha_0 u_0^{3/2}, \tag{11.25}$$

where $\alpha_0 = \left[\tau V_b^2 E_b(1+\eta)/R\right]^{1/2}$. Limiting by the case of the penny-shaped crack and substituting $\sigma_\infty = \sigma_m$ into (11.18), we have finally:

$$\Delta G_c = 4\sigma_m^3/3\alpha_0^2. \tag{11.26}$$

On the other hand, inserting the buffer layer strength (S) into (11.16) and (11.24), another equation for ΔG_c may be found [457]. Application of the specific dependence of ΔG_c (on the parameter u_0 or S) is very important in order to experimentally determine the value of ΔG_c, namely to select the type of strength test, the shape and sizes of experimental sample. Then note that (11.20), (11.21) and (11.26) include considerable number of material parameters; therefore, it is necessary to solve the problem of multi-parametric optimization for their optimum selection with account of the specific loading conditions and possible damages.

11.5 HTSC Systems of S–N–S Type

HTSC JJs of S–N–S type include laminated compositions of the “ceramic–metal–ceramic” kind. First, consider a pair of the edge cracks growing symmetrically and parallel to the metallic buffer layer dividing two half-infinite brittle half-planes

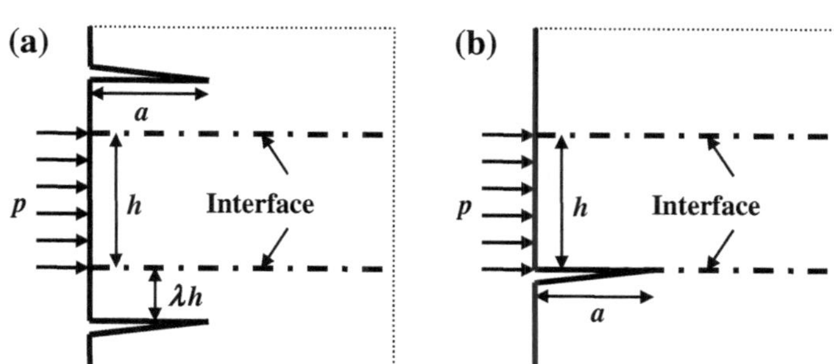

Fig. 11.14 Model representations of damage in composites of the “ceramic–metal–ceramic” type: (**a**) pair of steady-state cracks into brittle substrates, and (**b**) interface crack

(see Fig. 11.14a). Then, the SIFs determining the crack path may be estimated using uniform pressure in the metal, $p = E_m(\Delta\alpha_m - \Delta\alpha_c/2)\Delta T/(1 - \nu_m)$, where $\Delta\alpha_c = \alpha_{c_1} - \alpha_{c_2}$, $\Delta\alpha_m = \alpha_m - \alpha_{c_1}$. The following are analytical representations for the case of the steady-state cracks and like ceramics [228]:

$$K_I/p\sqrt{h} = -0.26(1 + 2\lambda)^{-1/2};\ K_{II}/p\sqrt{h} = 0.43(1 + 2\lambda)^{-1/2}, \tag{11.27}$$

where λh is the distance from the cracks to the corresponding interface and h is the thickness of the metal layer. At the greater thermal expansion of the metal, that is, for $\alpha_m > \alpha_c$ (e.g., Ag), $K_I < 0$, and edge cracks cannot grow along interface, and only after coalescence near edge are able to propagate into ceramic away from the interface. At smaller thermal expansion of the metal, that is, for $\alpha_m < \alpha_c$ (e.g., Mo), $K_I > 0$ and a fracture along interface is possible, when ductile layers are subjected to large residual stresses and has a requisite thickness [718]. Therefore, the metals, which possess small thermal expansion, are undesirable for bond integrity in HTSC JJs.

In the case of single interface crack (see Fig. 11.14b), it may be shown for ceramic layers with like properties that the SIFs are maximal near the sample edge. The equations for long crack ($a \geq 3\ h$) are [228]

$$K_I/p\sqrt{h} = \Phi_0(h/a)^{1/2};\ K_{II}/p\sqrt{h} = \omega(h/a)^{1/2}, \tag{11.28}$$

where $\Phi_0 \approx \pm\ 0.47$ and $\omega \approx 0.73$, with negative sign of Φ_0 for $\alpha_m > \alpha_c$. The values of the SIFs for short crack ($a/h \ll 1$) are given as

$$K_I/p\sqrt{h} = -0.99(a/h)^{1/2};\ K_{II}/p\sqrt{h} = 0.63(a/h)^{1/2}. \tag{11.29}$$

The comparison of the solutions for single crack and pair of cracks defines three different regions, namely (i) the SIFs for both cases are equal at $a \ll h$, (ii) the SIFs for single crack have larger values at $a < h$ due to the shielding of the crack pair, and (iii) the SIFs for the crack pair exceed those for the single crack at $a > h$ because of full release of the strain energy in the metallic layer by the crack pair.

In the case of unlike ceramics, the additive SIF, caused by the difference in their properties, should be superimposed on the solution for like ceramics. Obviously, this value is the SIF for body subject to the equal but opposite surface forces on either side of the bond plane. The crack growth in the case of small-scale yielding is found by the summary value of the SIFs from residual and external stresses. We obtain $K_I > 0$ for the edge cracks at $\alpha_m < \alpha_c$, and the residual stress state will define the interface strength, S, corresponding to maximum stress for the crack growth as [228]

$$S = E_m G_c/[2\sqrt{\pi}(\Phi_0 + \sqrt{\Phi_0^2 + \omega^2})ph(1 - \nu_m^2)], \tag{11.30}$$

where G_c is the fracture resistance of the interface governed by the phase angle of loading, $\Psi \approx \pi/6$. Then, we obtain, $K_I < 0$ at $\alpha_m > \alpha_c$ and the edge crack growth is initially inhibited, but fracture nucleates from center. The interface strength in this case can be given as [228]

$$S \approx 0.5\sqrt{\pi E_m G_c / a_0 (1 - \nu_m^2)} - p/10, \tag{11.31}$$

where G_c is the fracture resistance governed by $\Psi = 0$ and a_0 is the radius of the interfacial crack, which controls the fracture.

11.6 Toughening Mechanisms

The toughening mechanisms in the S–N–S layered systems are connected with the interface geometry, inducing the crack deflection and the out-of-plane microcracking that constructs the links—bridges, which increase toughness and fracture resistance at the stage of sub-critical crack growth along the ceramic–metal interfaces. Microcracks–voids, formed during processing (e.g., by using photolithographic techniques, combined with evaporation and diffusion bonding) and localized at the ceramic–metal interfaces, under loading promote the bridging bulges and stretching of the metal film, which can inhibit the crack growth. In this case, a cracking both along plain side of the patterned region and along the patterned side of the metal film is possible.

As has been shown by the experiments conducted for glass–copper interfaces, the toughness increase is nearly comparable for the samples, where the crack is driven along the plain interface, wherein the crack deflection is absent, and where the crack grows at the patterned interface, wherein the toughening results from both the crack deflection and bridging [1261]. The *R*-curve behavior (i.e., fracture resistance versus crack extension) states in the second case that the crack bridging is the dominating mechanism for the crack tip shielding (compared with the crack deflection) in the interface toughening. The corresponding toughening mechanisms, connected with the crack bridging and deflection, are shown in Fig. 11.15. The estimation of the toughening value in the steady-state crack, induced by stretching of the metal layer, which remains to be joined with the brittle substrate at one side (see Fig. 11.15a), can be derived, assuming each metallic ligament stretches in the shape of arc with height H, width d and period λ. Then, the toughness increment (ΔG_c^{br}) is found through thickness h, yielding strength σ_y and plastic strain ϵ_p for metallic layer [1261]:

$$\Delta G_c^{br} = \sigma_y h \epsilon_p = \sigma_y h (8d/3\lambda)(H/d)^2. \tag{11.32}$$

The toughening in the case of the fracture along the patterned interfaces due to the crack deflection can be estimated on the base of the model for 2D crack with repeated kink segments (see Fig. 11.15b) as [1261]

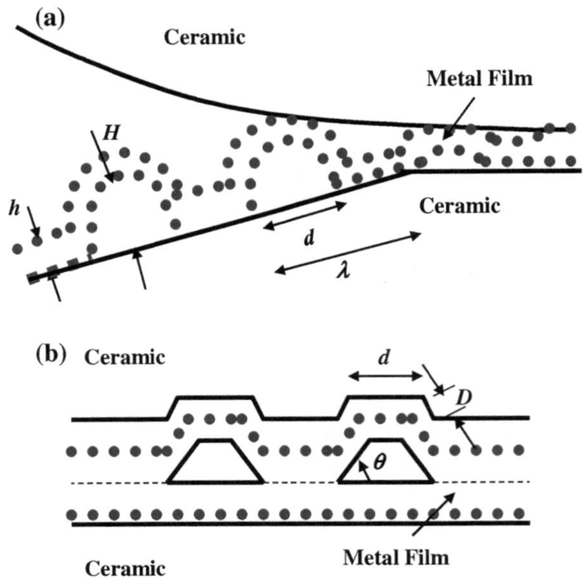

Fig. 11.15 Two models of interfacial toughening due to (**a**) crack bridging under plastic stretching of metal film, and (**b**) crack deflection

Table 11.2 Material parameters used in calculations

Property	YBCO	$SrTiO_3$	Al_2O_3	Ag	Mo
α ($\times$ 10^{-6}, K^{-1})	16.0	9.4	8.3	18.9	5.3
E (GPa)	64	300	380	75	315
ν	0.22	0.23	0.25	0.37	0.31
σ_y (MPa)	–	–	–	25	570

$$\Delta G_c^{def} = G_0\left[\left(\frac{D+d}{D\cos^2(\theta/2)+d}\right)^2 - 1\right], \tag{11.33}$$

where G_0 is the intrinsic toughness, taking into account the inclined and flat interfaces; D and d are the deflected and undeflected crack segment lengths, respectively, and θ is the angle of the crack deflection. Finally, it should be noted that a modest effect of the toughening (compared to the macroscopic crack deflection) occurs from much finer scale surface roughness, causing both the interfacial crack tilts and twists, which inhibit the crack extension and induce greater the plastic stretching of the bridging metal ligaments as the main crack extends [1261].

The known material parameters (see Table 11.2) enable to estimate some numerical results (see Table 11.3) for composite structures, which are proper for

Table 11.3 Numerical results

Composition	h_s/h	$K_{ss}/\sigma h^{1/2}$	ϵ_c	S, MPa	$\Delta G_c^{br}/G_0$	$\Delta G_c^{def}/G_0$
YBCO/$SrTiO_3$	4.9	–	1.41	–	–	–
YBCO/Al_2O_3	0.8	–	1.60	–	–	–
YBCO/Ag	–	1.38	–	–	–	–
YBCO/Mo	–	1.28	–	–	–	–
YBCO/Ag/$SrTiO_3$	–	–	–	585	0.02	0.02
YBCO/Mo/$SrTiO_3$	–	–	–	380	0.51	0.02

HTSC JJs. We use in the computations the following relationships for geometrical parameters: $d/\lambda = 2/3$, $H/d = 0.1$, $D/d = 0.15$ and values of $\theta = 35°$, $G_c = G_0 = 2$ J/m^2 [1261] $K_{Is} = 1$ MPa · m$^{1/2}$, $a = a_0 = 3\ h = 300$ nm, $\Delta T = 800$ K.

11.7 Charts of Material Properties and Fracture

The numerical results may also be obtained for different HTSC composites of the S–I–S type using (11.1)–(11.3), (11.20), (11.21) and (11.26). However, even initially, there are great spreads of the values for various mechanical and strength properties, which caused the actual structure of these materials.[2] In particular, the elastic moduli for HTSC ceramics can be on 1–2 orders of magnitude smaller than the moduli of the same superconducting crystals. For example, in the case of BSCCO, the fracture toughness (K_c) changes in the range 0.5–3.0 MPa · m$^{1/2}$, but Young's modulus lies in the limit 54.1–230 GPa [1255]. Therefore, the computations based on (11.20, 11.21) and (11.26) lead to significant differences for the limit values of K_c^M and E_m at other fixed parameters (see Table 11.4). These results are obtained in the simplest case, considering two matrix layers (BSCCO) and one buffer layer (MgO) with the same thickness (i.e., $V_m/V_b = 2$), selecting other parameters as $v_m = 0.2$ [1055], $v_b = 0.36$, $E_b = 290$ GPa [87]. The corresponding shear modules are found as $\mu_k = E_k/(v_k + 1)$, where $k = 1, 2$. This example is also interesting from the view of the HTSC processing due to the very small chemical reaction between MgO and BSCCO melt [1236].

The numerical results lead to qualitative trends in change of some normalized parameters in dependence on the fracture toughness (K_c) and Young's modulus (E_m). For example, for the penny-shaped crack into superconducting matrix (see Table 11.4), the crack size, corresponding to its transition to the steady-state crack ($c_0 = c_m$), grows together with the fracture toughness or/and Young's modulus. In this case, the equilibrium stress ($\sigma_0 = \sigma_m$) enhances with K_c but diminishes with increase of E_m. The fracture toughness increment for the steady-state crack (ΔG_c)

[2] The same statement relates also to materials used in HTSC JJs of the S–N–S type, some properties of which are presented in Tables 11.2 and 11.3.

Table 11.4 Some numerical results for penny-shaped crack

Properties	$c_m(\tau/R)^{2/3}$	$\sigma_m(R/\tau)^{1/3}$	$\Delta G_c(R/u_0^3\tau)^{1/2}$
$K_c = 0.5$ MPa $\cdot$ m$^{1/2}$; $E_m = 54.1$ GPa	3.516	6.408	14.507
$K_c = 0.5$ MPa $\cdot$ m$^{1/2}$; $E_m = 230$ GPa	5.362	2.298	9.655
$K_c = 3.0$ MPa $\cdot$ m$^{1/2}$; $E_m = 54.1$ GPa	11.608	21.158	14.507
$K_c = 3.0$ MPa $\cdot$ m$^{1/2}$; $E_m = 230$ GPa	17.704	7.588	9.655

decreases with increase of E_m. The normalizing multiples in the considered dependencies include tangential stress at the interface, taking into account the sliding friction (τ) and the limiting displacement at the crack opening ($2u_0$). These parameters state the features of transverse fracture and characterize the material interfaces (in particular, the de-bonding processes), and therefore, preliminary experimental definition of these is necessary.

Similar spreads of data exist for YBCO and other ceramics and metals used in HTSC JJs processing. The main causes, which find these broad ranges of material properties, are the following ones: porosity, microcracks, damages, domain and crystallographic structures, etc. The effects of porosity and microcracking on elastic moduli of YBCO and BSCCO ceramics may be taken into account, in particular, using the self-consistent differential method [933, 1319]. In any case, mechanical properties every time should be selected with account of specific microstructure, loading and fracture features of materials. So, it is necessary to optimize elastic moduli of both material components in the estimation of Dunders parameter α, defined in (9.3).[3] A number of parameters are necessary to be given in the estimation of critical properties of the HTSC-layered composites using (11.20, 11.21) and (11.26). In order to solve the above selection problem, the material property charts [61] may be useful. Figure 11.16 presents that chart, in which one property is plotted against another on logarithmic scales. This displays properties in very clear form and ensures to represent additional fundamental relationships in each chart (in particular, the critical release rate of strain energy, $G_{Ic} = K_{Ic}^2/E$, is shown additionally in Fig. 11.16). These charts help to select range for any property and to state criteria for the estimation of threshold material behavior that are used in HTSC JJs processing, taking into account specific loading conditions, thermal and electromagnetic treatments and so on.

When the crack grows through polycrystalline or composite structure, a significant change of the surrounding material characteristics occurs with increase of the crack size and local stresses at the crack front. Numerous toughening mechanisms are initiated, namely microcracking and twinning near crack, crack branching and bridging, phase transformations and domain re-orientations,

[3] The parameter α, in particular, is very important in the estimation of the fracture character, because its increase diminishes the fracture of interface cracking compared with the substrate failure in the plot of dependence of the relative toughness (Γ_i/Γ_s) on the phase angle of loading (Ψ) [453].

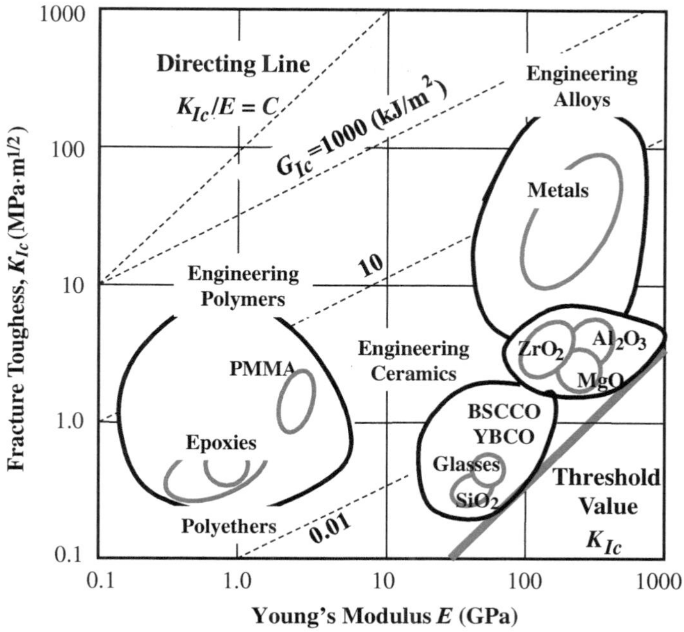

Fig. 11.16 Material property chart representing K_{Ic} versus E on logarithmic scales. The guide lines K_{Ic}/E = const help in fracture design of materials. Additional property, $G_{Ic} = K_{Ic}^2/E$, is shown

interaction of crack with microcracks and mesostructure elements, etc. Due to these, it is complicated to obtain the quantitative estimations of crack growth rate and hence to define a time before fracture or sample durability for given stress (σ) and temperature (T). Thus, it is very difficult to distinguish between action of various mechanisms of fracture resistance and also material failure features. The fracture charts, based on experimental data and plotted for many materials, solve this problem in some way and help to select material properties, taking into account the fracture type. The boundaries in the fracture charts define equal contributions of adjacent regions, which correspond to action of specific fracture mechanism. They displace at alteration of microstructure characteristics of the given material. The fracture charts for b. c. c. metals (e.g., Mo) and f. c. c. metals (e.g., Ag) differ significantly (see Fig. 11.17a and b). F. c. c. metals demonstrate only ductile fracture after initial significant plastic deformation. At the same time, b. c. c. metals show brittle fracture at low temperatures.

In the case of silver [62], the material fracture occurs in creep mechanism at low tensile stresses ($\sigma/E < 10^{-3}$). Nucleation of voids and boundary sliding take place in intergranular fracture. Transgranular fracture is accompanied by the void nucleation at inclusions and the void growth together with the creep strain. The strain localizes in weaken cross-section of the sample that leads to growth of local stresses and rupture. High stresses cause ductile fracture also with the void

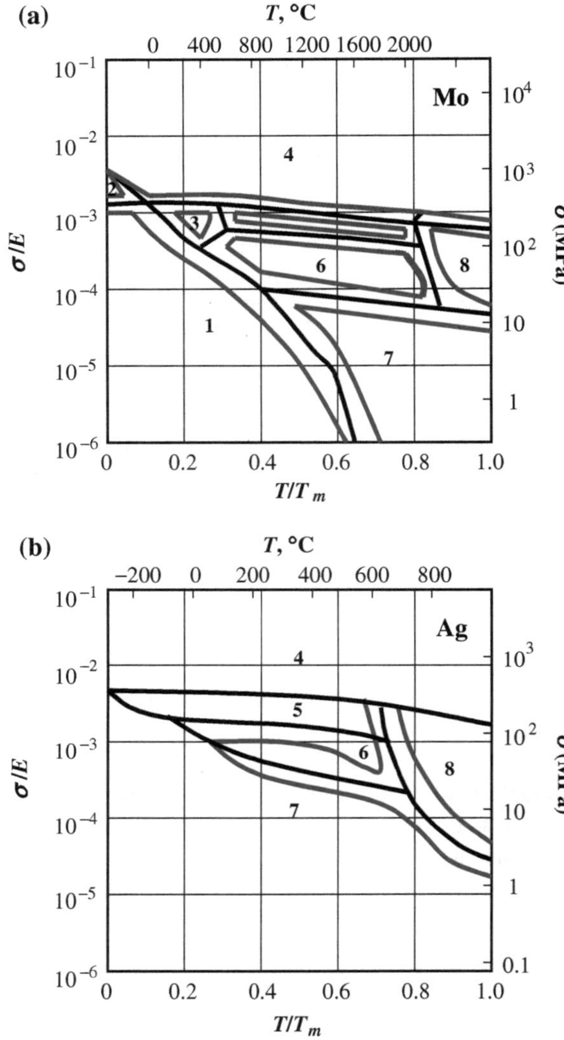

Fig. 11.17 Fracture charts for Mo (**a**) and Ag (**b**). Numbers denote the next fracture types: *1* cleavage I, *2* cleavage II, *3* cleavage III, *4* dynamic fracture, *5* ductile fracture, *6* transgranular and *7* intergranular creep fracture, 8 rupture. *Gray curves* show limit changes of interfaces

nucleation at inclusions. At increasing of the stresses, fracture occurs in the mechanism, which is proper for polycrystals, when nucleation and growth of cracks and voids at the grain boundaries are facilitated. Finally, the fracture occurs in dynamical regime with propagation of stress waves in the sample bulk.

In the case of molybdenum [511], three regions of brittle fracture by cleavage are added in the fracture chart, namely (I) for the nuclei of cracks, preliminary existing in crystal; (II) for the nuclei of cracks, forming at coalescence of dislocations in initial stage of plastic strain (microplasticity), and (III) when the plastic strain attains a significant level (~ 10 %), but the fracture takes place owing to the cleavage. Similar fracture charts for high-temperature superconductors can significantly help in design and processing of HTSC JJs with optimum characteristics. Finally, we note Refs. [781, 1813] as successful examples of development of the system analysis methods including creation of the material property charts used for estimation of critical behavior of the HTSC materials and composites.

11.8 Concluding Remarks

The results for different toughening mechanisms, actually acting in HTSC microstructures, and also the features of the fracture resistance alterations due to the growth of crack-like defects have been presented in Chaps. 10 and 11. They have confirmed an existence of the numerous and non-simple effects. At the same time, today in the applied superconductivity, the simplified approaches to estimation of the strength and fracture toughness are widespread. For example, the following relation is widely applied [366, 1002, 1464, etc.]:

$$\sigma_f = YK_c/c^{1/2}, \tag{11.34}$$

where σ_f is the fracture stress, Y is a geometric constant, K_c is the fracture toughness and c is the flaw size.

However, it is well known (see, e.g., [294, 452, 687, etc.]) that (11.34) has been stated in the framework of the classical Griffith-Irwin's approach [576, 760] for *homogeneous linear-elastic solid under homogeneous loading*. This equation could be applied *only* under conditions of real dominating crack with size considerably greater than microstructure scale and in the absence of internal (residual) stresses. Obviously, an account of these macrocracks is no interesting for HTSC, because the superconducting and transport properties very rapidly degrade even in an insignificant microcracking. Moreover, as has been shown above, the HTSC microstructure is very heterogeneous with heterogeneity scale, being compared with characteristic microcrack size. Finally, the HTSC compositions undergo nonhomogeneous loading during working and possess considerable residual stresses, initiated by processing. Note also that another fracture mechanisms act under small fracture stresses. Hence, the use of (11.34) could lead to incorrect conclusions, for example, "K_{Ic} can be improved [in *RE*BCO] by the refinement of the *RE*-211 particle size" (see [1464], p. 111), or "K_c is independent of flaws in the material (cracks, pores…) and, therefore represents a true property of the Y-123/Y-211 compound" (see [1002], p. 2074). Violation of the dependence between σ_f and K_c, given by (11.34), is evident, for example, from Fig. 10.17. Obviously, (11.34) could be applied only in the absence of acting toughening and crack amplification mechanisms and under conditions of other restrictions described above.

The coming certification of HTSC systems will demand correct definition of the strength properties. The combination of the above considered toughening and crack amplification mechanisms could exert on fracture resistance not only summary effect but rather multiplication effect. Therefore, for every actual toughening (or crack amplification) mechanism, ratio K_c/K_c^0 or G_c/G_c^0 should be defined, where K_c (G_c) is the critical SIF (critical strain energy release rate) due to actual toughening (or crack amplification) mechanism and $K_c^0(G_c^0)$ is the corresponding intrinsic parameter without toughening (or crack amplification).

Then the total fracture resistance (in the force and energy approach, respectively) should be found as

$$K_c^{tot}/K_c^0 = \prod_{i=1}^{n} K_c^{(i)}/K_c^0; \tag{11.35a}$$

$$G_c^{tot}/G_c^0 = \prod_{i=1}^{n} G_c^{(i)}/G_c^0, \tag{11.35b}$$

where n is the common number of the toughening and crack amplification mechanisms.

The research of effectiveness of the HTSC toughening mechanisms should be accompanied by the study of the HTSC conductivity because it is often these effects which are contrary to each other (see, e.g., [933, 1327]). The modeling of the current-carrying properties of HTSC system is the main aim of the final chapter of the monograph.

Chapter 12
Modeling of Electromagnetic and Superconducting Properties of HTSC

12.1 Modeling of Intercrystalline Dislocations

There are two basic approaches for description of intercrystalline structures [77], namely (i) the structural unit approach, focuses on the atomic arrangement at the intergranular boundary and (ii) the intergranular boundary dislocation (IBD) approach, based on the periodic strain field that is observed at many intergranular boundaries. Both models use a *coincidence site lattice* (CSL) description of the intergranular boundary geometry. While these models are equivalent [88], each one has its own advantages in description of the intercrystalline structure. Nevertheless, today studies of intergranular boundaries in HTSCs focus primarily on the IBD description.

The structure of intergranular boundary may be presented, consisting of two different types of parallel conducting channels, disposed along the boundary and defined by dislocation structure (see Fig. 12.1). One of them possesses approximate structure and properties of superconducting crystal, associated with the regions between dislocations. Other channel is normal one or demonstrating a weak superconductivity, and it is compared with dislocation cores or their elastic deformations [663]. When the misorientation angle, θ, increases then the dislocation density increases, the interdislocation spacing diminishes which leads to gradual inhibition of the superconducting channels and to forcing of the weak link behavior. Therefore, an understanding of IBD nature is very important for statement of transition from the strong link behavior to the weak one.

The IBD models are based on the concept that the grain boundary free energy is particularly low for a certain set of special misorientation relationships, θ [*hkl*]. In this case, the boundaries with θ [*hkl*] values other than special ones relax to a configuration in which sites of a low-energy structure are preserved by a localized rotation of the crystals. The macroscopic value of θ [*hkl*] and its deviation from a low-energy misorientation are produced by the introduction of a regular array of dislocations that rotates one crystal relative to the other on a macroscopic scale. On the microscopic scale, these dislocations separate the sites of boundary with

I. A. Parinov, *Microstructure and Properties of High-Temperature Superconductors*, DOI: 10.1007/978-3-642-34441-1_12, © Springer-Verlag Berlin Heidelberg 2012

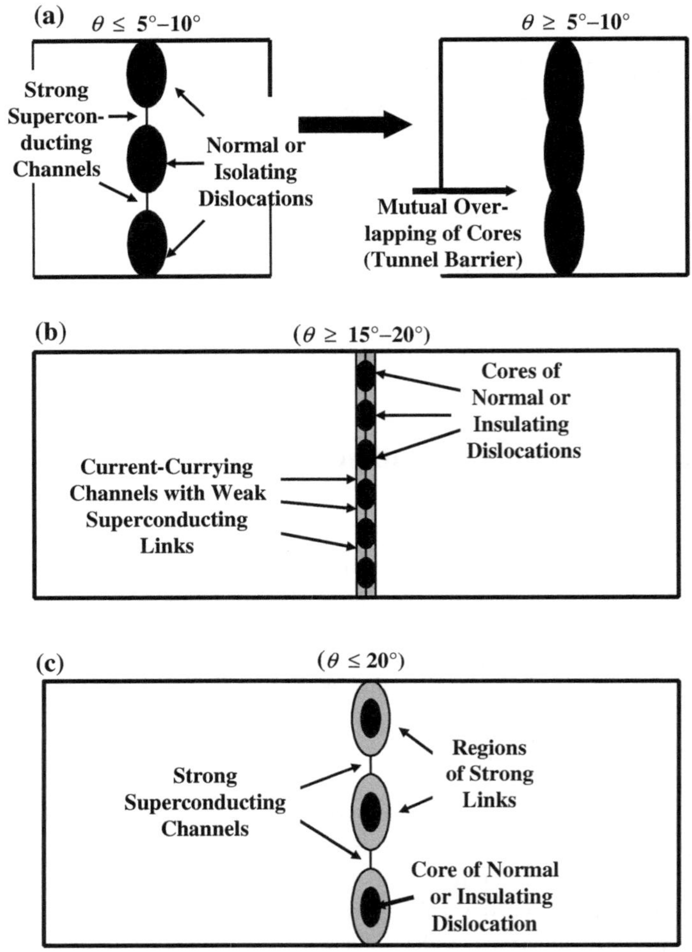

Fig. 12.1 Three models explaining intergranular values of J_c in thin YBCO films with [001] direction *versus* intergranular boundary structure: (**a**) model of dislocation cores before and after their partial recovering (Dimos model) [279, 374], (**b**) model of Dayem's bridge [1480] and, (**c**) model of channel with strong links [477]

low-energy structure. Thus, most intergranular boundaries are presented as being divided into regions of "good fit," having low-energy atomic structure, separated by a net of "bad fit" of IBDs. As the difference between the actual misorientation relationship and the low-energy one increases, the spacing of the dislocations decreases and the fraction of bad fitting region increases. A CSL construction is used to identify the low-energy misorientations and describe the structure periodicity of the boundary. The IBD model suggests that low energy is associated with relatively short-period CSLs or, equivalently, low-coincidence index,

Σ (where Σ is the fraction of lattice sites of one crystal that is coincident with the other [693]).

CSLs can form in orthorhombic and tetragonal crystals only when the squares of the lattice parameters all form rational ratios (i.e., a^2:b^2:c^2 are rational numbers) [872]. This condition is rarely met in real crystal structures. The approach selected to apply the coincidence concept to these systems is to impose a small strain on the lattice in the boundary vicinity. This strain forces the lattice parameters to meet the rational ratios condition, and the lattice can form CSLs. The CSLs, resulting under strained conditions often are called *constrained CSLs* (CCSLs). Usually, the additional strain is attributed to additional sets of dislocations that are localized in the boundary and become part of the IBD structure. Thus, the IBD structure has two components one to produce the correct axial ratios and other to accommodate deviations from low-energy misorientations. Because the strain-producing and deviation-accommodating IBDs can have opposite signs, the dislocation density may not increase monotonously with the deviation from the exact CCSL misorientation relationship. As a result, the minimum spacing may not exist at the CCSL misorientation and widely spaced IBD structures may accommodate rather large deviations from lattice coincidence [872]. This leads to a boundary division into seemingly different patches.

There are at least two reasons to explore the (C)CSL/IBD model. First, it is necessary to identify the fundamental origin of the weak-link problem in order to discern whether one is intrinsic to the boundary structure or extrinsic. The second cause is connected with identification of certain types of intergranular boundaries other than low-angle boundaries that have specific properties [77].

Because the structures of boundaries that lie parallel to low-index planes of the CCSL have the shortest wavelength periodicity possible for that misorientation relationship, their energies are expected to be minimum and faceting onto those planes might be expected. In this case, the modeling of intercrystalline structure is simplified in strongly faceting structures, because macroscopic boundaries with a variety of boundary planes can be considered as mixtures of some boundary facets, each of which has a fixed structure. The CSL and CCSL conceptions can be used to optimize intergranular boundary behavior. At the same time, they are not the governing structures due to the following reasons [77]:

(1) Because the unit-cell dimensions are allowed to vary to achieve coincidence, relatively low Σ misorientations (where the coincidence index Σ presents itself the fraction of lattice sites of one crystal that are coincident with the other [1222]) densely populate θ [*hkl*] space. As a result, for many observed boundary misorientation relationships there are several nearby CCSLs that may be used to the boundary description.
(2) The types of relaxations that occur in these boundaries often assume statement of compromise between the magnitude of the local strain, required to achieve coincidence, the achievement of low Σ values and the minimization of strain energy due to dislocations.

(3) The choice of an appropriate CCSL is complicated by the absence of a suitable estimation for possible deviations from the exact coincidence misorientation that reasonably can be referenced back to a particular CCSL. This criterion must state the compromise between strains, low Σ, the mechanism by which the strain is accommodated and the possible non-monotonic dependence of the dislocation spacing on $\Delta\theta$.

The elastic strains, produced by the IBD arrays in individual facets, can be estimated using the equations derived for isolated pure-edge dislocations in an isotropic medium [1782] and assuming reasonability of this approximation for elastically isotropic material in the *ab*-plane [1707]. Then, the net strain field produced by a finite array of dislocations can be calculated using the superposition method. The numerical results obtained for a boundary containing seven equally spaced edge dislocations show that the material volume within about 20 nm of the facet junction can deform ~0.8 % or more. In comparison, the orthorhombic to tetragonal phase transformation is accompanied by an analogous strain ~1 % [1707].

Conventional model [693] describes a symmetric low-angle grain boundary (GB) by a chain of edge dislocations with the Burgers vector perpendicular to the GB plane. However, the structure of GBs in HTSC can be more complicated due to existence of partial dislocations and facets at GBs, long-range strain fields and compositional variations near GBs [77, 190]. The IBD structures influence the change of J_c due to an alteration of local composition and hole concentration, additional electron scattering and significant strains near the dislocation cores. The models [417, 1177, 1480] have been proposed, in which the current channels in GBs are described as arrays of parallel point contacts, which demonstrate weak-link behavior if their width becomes smaller than the superconducting coherence length, ξ [1021]. The theoretical description of the rapid decreasing $J_c(\theta)$ with increasing θ is usually ascribed by the strain-induced compositional suppression of superconducting order parameter near dislocation cores or in the layer of some thickness near GBs [77, 417, 1177, 1480]. Due to insufficiency of these phenomenological models, another model is presented in [608], which describes $J_c(\theta)$, taking into account the GB dislocation structure. It is assumed that the strains and excess ion charge of the GB dislocation structure can locally induce a dielectric phase near dislocation cores and cause progressive overall suppression of the superconducting order parameter with θ in a narrow layer of the order of the screening length near GBs. The model provides an intrinsic mechanism for the rapid decreasing of $J_c(\theta)$ with θ, solving the Ginzburg–Landau equation [343, 1006], which describes well the practically important temperature range for HTSC ($T > 77$ K).

Dislocation model [1125] computes a stress state at intergranular boundaries of three types, namely symmetrical and anti-symmetrical tilt ones with periodical system of edge dislocations, and also symmetrical tilt boundaries with arbitrary system of the edge dislocations. The angle dependencies of intercrystalline critical currents are calculated in these computations. Moreover, it is shown that adequate description of superconducting features of the boundaries can be obtained only in the framework of the arbitrary distributed dislocations.

The models of local increasing of the critical temperature (T_c), compared to corresponding values of the bulk critical temperature (T_{c0}) due to structure defects, causing the long-range strains, are presented in [607]. It is shown that the strain-induced T_c variation on defects is markedly enhanced in HTSCs due to small ξ, high T_c and strong anisotropic pressure dependence of the bulk critical temperature [24, 169, 286, 521, 589, 797, 1126, 1127, 1241, 1253, 1499, 1645, 1780, 1789, 1797]. The latter reflects the characteristic bell-like dependence of T_c on the holes concentration [1790], which changes near defects due to local lattice distortions and electro-neutrality condition. These distortions $\varepsilon(x, y)$ can be quite strong, causing local plastic deformations or structure transformations around dislocation arrays, where the local holes concentration can vary from the critical value, c_s at which $T_c = 0$ up to the optimum concentration, c_m which corresponds to the maximum possible critical temperature, $T_{cm} = \max T_c(\varepsilon)$ in a *deformed* sample. This can give rise to localized superconducting regions coupled by the proximity effect above the bulk critical temperature, T_{c0}. Note that the localized deformations around specific structure defects can be much stronger than those accessible in experiment by applying uniaxial stress which is limited by the overall mechanical strength of the sample. In this case, even the hydrostatic pressure tests show a substantial increase of T_c under pressure from several degrees for optimally doped Bi-2212 single crystals up to ~20 K in Hg-1223 [286, 1253]. The effect markedly increases in under-doped HTSCs [24].

The increasing of T_c is estimated in [607] for edge dislocations, low-angle grain boundaries and metastable linear dislocation arrays, taking into account anisotropic strain dependence of T_c in the *ab*-plane and proximity effect, defining the superconducting state at intergranular boundaries. Moreover, the compositional changes are estimated due to the strain states, caused by defects, and effect of the T_c variations on magnetic flux pinning and magnetic granularity.

Being dependent on the sample strain, the T_c changes can be directly affected by applied stress state, which alters the dislocation distribution, GBs, microcracks and so on. In turn, this makes the effect of T_c increasing dependent on the particular deformation pre-history, which can give rise to localized metastable superconducting states on defects above the bulk T_c. For example, maximum ΔT_c is defined by dislocation walls of finite length which can exist in individual crystallites. However, these macrodefects are metastable and can disappear after annealing or redistribution of dislocations in the remnant strain fields.

The strains around defects can also significantly suppress T_c, especially near dislocation walls [607] or high-angle boundaries. Depending on the direction of current, this can manifest itself either as additional pinning or weak links because of the anisotropic flux pinning typical for planar crystalline defects [601]. For the perpendicular current, $\boldsymbol{J}$, the defect behaves as a weak link which locally diverts $\boldsymbol{J}(\boldsymbol{r})$ as it has been observed in magneto-optical experiments [1337]. For parallel current, the dislocation wall enhances flux pinning by producing a deep potential well for vortices [607]. Such a well can trap many vortices, if the length of the dislocation wall, $2L > (\phi_0/H)^{1/2}$, where ϕ_0 is the quantum of magnetic flux and H is the applied field. Moreover, the local non-stoichiometry near GBs or dislocation arrays can significantly increase the strain-induced T_c variations at GBs as compared to the material inside grains.

12.2 Current-Limiting Mechanisms and Grain Boundary Pinning

A dual role of the grain boundaries, which causes increasing of the magnetic flux pinning and, at the same time, effectively block or divert macroscopic supercurrent flow is discussed in [599]. As it has been shown [374], the high-angle GBs, microcracks, stacking faults and other planar defects can lead to significant decreasing of the superconducting order parameter, and thus to diminution of connectivity of the superconducting grains. As a result, HTSCs often demonstrate electromagnetic granularity, which causes significant decrease of current-carrying properties of the superconductor [261, 582, 981, 1337, 1391]. Moreover, GBs in HTSC are strong current-limiting defects [977] because of the exponential decreasing of the critical current: $J_b(\theta) = J_0\exp(-\theta/\theta_0)$ across tilt GBs with the misorientation angle, θ, between adjacent crystallites, where $\theta_0 = 4$–$5°$ [374].

Low-angle GBs are inhomogeneous on different scales. On a nanoscale, they are formed by edge dislocations spaced by the distance, $d = b/2\ \sin(\theta/2)$, where b is the Burgers vector (see Fig. 12.2). Besides GBs, in HTSC there are macroscopic facet structures with periods, $D_p \sim 100$–1000 Å, which are much longer than the superconducting coherence length, ξ, along with long-range strains and local non-stoichiometry on the scale of about D_p both across and along GBs [219].

Due to the proximity of HTSC to the antiferromagnetic metal–insulator transition, they demonstrate the strong strain dependence, $T_c = T_{c0} - C_a\varepsilon_a - C_b\varepsilon_b$ (where C_a and C_b define the T_c change under one-axis compression along the a- and b-axis, respectively, for deformations ε_a and ε_b), which increases into dielectric regions of the order of characteristic size b near dislocation cores. These strain-induced dielectric core regions reduce the current-carrying cross-section of

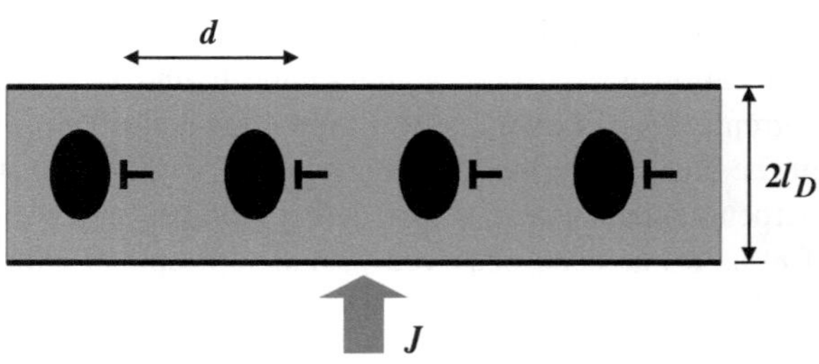

Fig. 12.2 Channel structure of a low-angle GB formed by dielectric dislocation core regions (*black*) and charge space layer (*gray*)

GBs, dividing it into a set of current channels. For the anisotropic in-plane dependence $T_c(\varepsilon)$ in YBCO (where $C_a = -C_b$ due to the effect of the CuO_2 chains), these channels have form, very different from channels for the nearly isotropic dependence $T_c(\varepsilon)$ in BSCCO crystals (for which $C_a = C_b = 300$ K) [608]. Moreover, the long-range deformations, caused by faceting can lead to macroscopic modulations of T_c and J_c along grain boundary on the length of order D_p [219].

Another current-limiting mechanism is due to that the dislocation structure leads to an overall lattice expansion near GB in the layer of thickness of order d, where the ion density N_0 is reduced by $\delta N \sim \zeta < \varepsilon^2 > N_0$, where $\zeta \sim 1$–2 is Grüneisen's number [599]. The screening of the excess GB ion charge by carriers (holes) causes the shift of the electric potential on GB, $\Phi_0(\theta) \propto \theta$, proportional to the GB dislocation density and thus the misorientation angle, θ. This leads to the chemical potential variation at the $\delta\mu(\theta) \propto \theta$, that causes T_c for $\theta \sim 15$–$20°$. In turn, the hole depletion and zone bending near a GB define strong superconductivity suppression in the space-charge layer with thickness of the double Thomas–Fermi screening length, $2l_D \approx \xi$ [608, 678, 679].

As a result, current transport through GBs is determined by the two factors: (i) strain-induced dielectric regions, which impede current flow and (ii) suppression of the superconducting order parameter in the current channels due to charging effects. When the angle θ increases, the chemical potential $\mu(\theta)$ on GB changes toward the under-doped insulating state, leading to progressive suppression of J_b with θ. Calculations of $J_b(\theta)$ are carried out in [608] on the base of the Ginzburg–Landau equations. They show that above "transistor" model of GB describes well the observed exponential decreasing of $J_b(\theta)$ in HTSC.

The current transport in polycrystals in a magnetic field of B is determined by vortex dynamics and pinning, which are significantly affected by the channel structure of low-angle GBs. The increase in the misorientation angle, θ, directly alters the structure of GB vortices. For very small θ, vortices on a GB are Abrikosov's (A) vortices, localized on the dielectric dislocation core regions. At higher θ, the width of the GB current channels becomes of order of the coherence length ξ, so the normal vortex core disappears turning into a Josephson core, which is a 2π-phase kink of length, $l \approx \xi J_d/J_b$, along GB, where J_d is the decoupling current density [599]. An intermediate Abrikosov vortex with Josephson core (AJ-vortex) [597] exists in the region of θ, where the phase core size, l, is smaller than the London penetration depth, λ. This condition ($l < \lambda$) holds in a rather wide region, $J_d/k < J_b(\theta) < J_d$, where $k = \lambda/\xi \sim 100$ is Ginzburg–Landau parameter. For higher θ, $J_b(\theta)$ decreases below of J_d/k, and the AJ vortex turns into the Josephson vortex of length $\lambda_J = (c\phi_0/16\pi^2\lambda J_b)^{1/2}$ [599]. This continuous A → AJ → J vortex transition with increase in θ directly affects vortex pinning on GB.

Vortices on grain boundaries are pinned due to both interaction with heterogeneities along GBs (core pinning) and magnetic interaction with GBs and bulk A-vortices. The pinning force, f, is highly anisotropic with respect to the current direction. If $\boldsymbol{J} \parallel$ GB-direction, bulk A-vortices, spaced by $L < l$ from a GB are strongly pinned due to their magnetic interaction with the GB. This causes the inverse dependence, $J_c \sim \xi J_d/D$, on the grain size, D, at low field, B, provided that GBs do not block current flow, that is, $J_c < J_b$ [601]. This behavior of $J_c(D)$ alters at higher values of B, either because the global value of $J_c(B)$ is now limited by $J_b(B)$ or pinning becomes collective if the Larkin pinning correlation length, $L_c(B) > D$.

When a current crosses GBs, the global value of J_c in polycrystals is found by a de-pinning of vortices, displacing along the GBs network. The core pinning of such vortices is caused by their interaction with modulations of the local current density, $J_b(\boldsymbol{x})$, which are found by current channels or facet structure. For $J_b(\boldsymbol{x}) = J_{b0} + J_a \cos(q\boldsymbol{x})$, the AJ-vortices on a GB become de-pinned, if the component of J perpendicular to the GB exceeds $J_b^{\parallel}$, which is found as [1337]

$$J_b^{\parallel} = 2\pi J_a q l \exp(-ql) \tag{12.1}$$

The critical current density, $J_b^{\parallel}(T, B)$, is maximum for the optimum period of the pinning potential, $L_{opt} = 2\pi l(T, B)$. For fixed q, the optimum pinning occurs in the region of T and B, where $ql(T,B) \approx 1$. Note, that the pinning of GB vortices parallel to the grain boundary dislocations has been observed in [362]. However, because of the large size of the AJ phase core ($l \approx \xi J_d/J_b \gg \xi$), the pinning of AJ-vortices is much weaker than that of bulk A-fluxons. As a result, GBs become channels of preferential motion for AJ- or J-vortices.

AJ-vortices, moving along the GB network experience the magnetic potential, $U(\boldsymbol{x}) = -\phi_0 H(\boldsymbol{x})/4\pi$, produced by strongly pinned bulk A-vortices. Here, the magnetic field along a GB varies as $H(\boldsymbol{x}) \approx B + H_a \cos(2\pi \boldsymbol{x}/a)$, where $a = (\phi_0/H)^{1/2}$ in the inter-vortex spacing and the amplitude, $H_a \approx \phi_0/(2\pi\lambda)^2$ is of the order of the lower critical field, H_{c1}. Thus, the magnetic pinning force, $f^{\parallel} \sim \phi_0 H_a \sin(2\pi x/a)/a \propto H^{1/2}$, increases with H, leading to a non-monotonous field dependence of the global current density, $J(H)$ (so-called "fish-tail" effect) [601] and matching peaks in the dependence, $J_c(H)$ [129]. These matching peaks have also been observed in YBCO bi-crystals at fields $H_n = \phi_0 n^2/D^2$, for which the inter-vortex spacing is commensurated with the GB facet structure [219].

12.3 Vortex Structures and Current Lines in HTSC with Defects

The observed influences of defects on flux pattern lead to a classification of the defects into two groups [900]:

(1) The "extended" defects, formed by a weakly superconducting material (typical size between some micrometers up to millimeters), which force the currents to flow along them, thus leading to a facilitation of flux penetration along

the defects (these defects form channels for easy flux penetration into the sample).

(2) The small obstacles of non-superconducting material (typical size of micrometers or even less), where the currents can flow around them. There is no enhancement of the flux penetration, but a parabolic discontinuity line of the current is formed.

In Figs. 12.3 and 12.4, schematic current patterns (the critical currents are shown by the arrows) for both defect classes are presented. The characteristic feature of the first type (see Fig. 12.3) is an extended defect line, where flux penetrates the sample preferentially. This leads to the formation of a fingerlike flux pattern. Flux density gradients always are pointed away from the defect line and at the end of the flux front the currents are forced to a U-turn. At the beginning of the defect line, the currents form a broad turn towards the defect line as the currents always flow along the sample edges [1523] and also along the defect lines.

Figure 12.4 shows the current flow around a small obstacle of the second type, here symbolized as a round object. When the currents come close to this defect, they are forced to a slight turn and now follow the circumference of the obstacle. All the points, where the current lines turn around, lie on a parabola, which can be seen in the flux distributions. The defect forms a "bottleneck" for the flux penetration that may trigger flux jumps [1523]. If the small obstacle has a rectangular shape, the beginning of the parabola will be linear (forming an angle of 45° [1523]), which in some distance from the defect again becomes as in Fig. 12.4.

The sufficiently large density of the defects localizes an area of interaction of two adjacent defects (see Fig. 12.5). At small fields, the current flow is similar to Fig. 12.3. During flux penetration in an increasing external magnetic field the flux fronts meet each other in some moment. This leads to complete change of the current pattern, because the currents can now flow along both defect lines and form new smaller current loops. In this case, the length scale of the currents changes and the sample is divided magnetically into smaller pieces. If the external magnetic field is raised further, the parts of the sample will behave completely independent

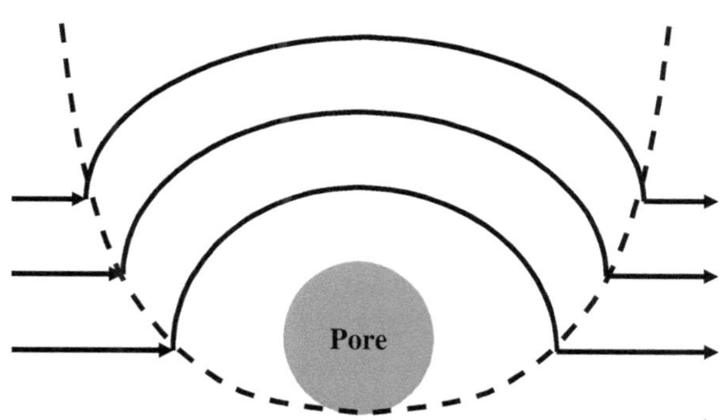

Fig. 12.3 A defect of linear type, symbolized by the *dashed curve* (current flow around cylindrical pore), leads to increasing flux penetration along it [227]. Therefore, the current flow is found by the flux density gradients. In contrast to the defect, shown in Fig. 12.4, there is no bottleneck between the normal flux front and the flux penetration along the defect line

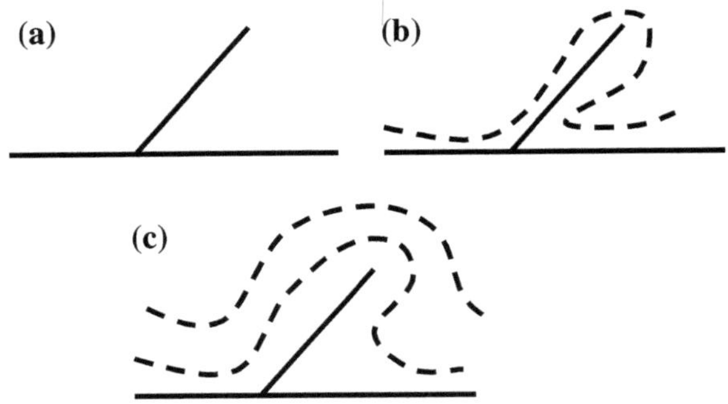

Fig. 12.4 A schematic drawing of the current distribution in the vicinity of a cylindrical cavity near the superconductor edge, located at the bottom of the figure. The current density spatially constant is shown by equidistant current lines (*dashed curves*). When the magnetic flux, penetrating from the sample edge, reaches the cavity, the current lines form sharp bends in order to flow around the hole on arcs of concentric circles. These bending points lie on a parabola, but the penetrating flux to the inside of the parabola always passes the defect, forming a "bottleneck" for the vortices [900]

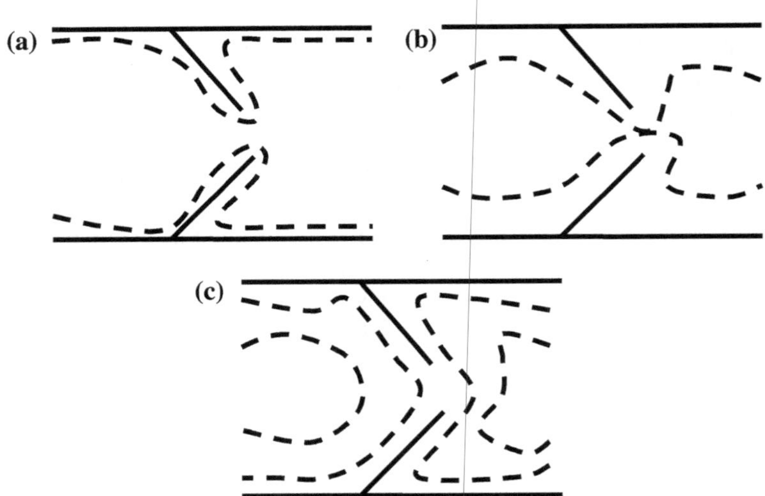

Fig. 12.5 At high defect density, two adjacent defects can be considered in sample (**a**). The flux penetration and current flow are identical to those of Fig. 12.3 as long as the external magnetic field is small (**b**). With rising field, the vortices penetrating the sample from both sides may merge together (**c**). This causes the current flow along both defect lines, leading to formation of a current loop [900]

from each other. This leads to magnetically induced *granularity* of the sample, with the defect lines acting as weak links at higher magnetic fields.

Figure 12.6 presents schematically a crystal with many defects. At moderate fields, where Meissner's phase still resides in the sample, the current flow is the same as that shown in Fig. 12.3, and many "flux fingers" are formed along the

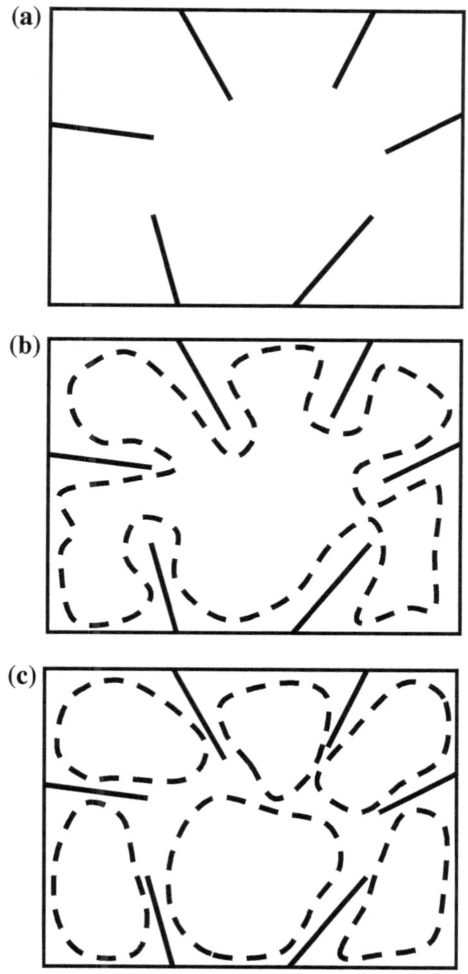

Fig. 12.6 A schematic representation of many defect lines in a rectangular sample (**a**). In moderate external fields (**b**), the flux penetrates easily along the defect lines and only a small flux penetration along the sample edges occurred. The critical currents (shown by *dashed curves*) flow around the defects. If the external magnetic field is sufficiently large (**c**), the flux penetration along the defect lines merges together and current loops are formed, both for the critical currents and for the eventually remaining shielding currents [900]

defect lines. With sufficiently large external magnetic field, many current sub-loops are formed. For this sample, the critical state model, assuming a homogeneous flux distribution is not used. One will severely underestimate the critical current density of such sample, if the critical current is determined from its magnetic moment (with the tacit assumption of a single current loop, shielding the whole sample). It should be noted that, once a field or current distribution is disturbed, the pattern stays like this even up to high external magnetic fields.

The observed various flux distributions permit to classify the defects in HTSC through defect influence on the flux patterns [900]. In all cases, intrinsic (crystallographic) defects should be distinguished from extrinsic ones, which are caused by external influences (i.e., by cracks, scratches, cutting and etching):

(1) To the extended defects of the first type belong the intrinsic defects, such as twin boundaries [383, 384, 1500, 1716, 1717, 1757] and the tilt boundaries found in Bi-2212 single crystals, caused by intergrowths of Bi-2223 phase [903]. Moreover, extrinsic defects, such as cracks in the sample surface, which are typical defects for most Bi-2212 single crystals and of melt-processed YBCO sample [659, 1211], and also the defects caused by inhomogeneous film growth on substrates, containing scratches, belong to this type. However, the growth steps, observed in many single crystals, do not influence the current patterns [903]. Grain boundaries also belong to the extended defects, but they act as channels for easy flux penetration even at very small external magnetic fields. This separates a sample magnetically into smaller pieces.
(2) Above-described small obstacles of the second type in HTSC are caused by small flux droplets, observed in thin films and single crystals, by islands of the a-axis-oriented growth in the c-axis-oriented films, by small particles of foreign phases and by extrinsic defects, created owing to irregular cut or etching of the sample.

The presented classification characterizes all defects found in HTSCs. However, this approach is only valid for nearly homogeneous, monocrystalline samples, where a homogeneous flux front will appear in the ideal case. If an internal granularity is presented in the sample, the flux distribution is disturbed by the immediate flux penetration along the grain boundaries. Therefore, the analysis can only be fulfilled for individual grains. Then, the flux pinning in superconductors with high critical current density is also found by a dense network of planar crystalline defects. In this case, there are two essential features of HTSCs, namely (i) the most effective pinning can be caused by the dense network of planar defects parallel to the flux lines, however (ii) unlike the case of arbitrary distributed point pins, the network of planar defects can block or divert the macroscopic current flow, if the tunneling superconducting current density (j_c) through these defects is smaller than J_c, determined by flux pinning. For $J_c > j_c$, the pinning structure can lead to the magnetic granularity, which manifests itself in a drop of the transport J_c due to the appearance of closed current loops within macroscopic crystalline grains, where densities of circulating magnetization currents become larger than J_c [324, 900].

Note also the importance of geometry of the planar defect network. For example, in twin domains with various orientations of the crystallographic axes, the planar defects can divert the local current flow, limiting the macroscopic value of J_c owing to reduction of the current-carrying cross-section area [488, 962, 1011]. On the other hand, in a random network of planar defects with the density of the pinning centers, exceeding the percolation threshold, the supercurrent must cross several defects, which would increase J_c owing to flux pinning. This arbitrary defect network can have a complicated topological structure, in which only a small fraction of the defects belongs to the path of percolation [1610]. Generally, both regimes, in which the planar defects act as pinning centers or block (or divert) the current flow co-exist. Their relative contributions depend on both j_c and the geometry of the pinning network. The theoretical study of flux pinning due to a network of planar high-j_c crystalline defects, which are modeled by the corresponding Josephson contacts is carried out in [601]. A solution of equations of the non-local Josephson electrodynamics [597, 606] for a vortex parallel to the planar defect leads to the calculation of the magnetic field distribution and the transverse pinning force between the vortex and defect. The longitudinal pinning force of vortices along the defect is determined by both their magnetic interaction with pinned transgranular fluxons and local inhomogeneities of the defect. It is shown that the longitudinal component is much smaller than the transverse one, leading to the preferential flux motion along the paths of percolation formed by planar defects.

In highly anisotropic-layered superconductors or artificial superlattices, the interlayer superconducting coupling can be so weak that the coherence length in the direction perpendicular to the layers (ξ_c) becomes smaller than the interlayer spacing (s). In this case, the discreteness of the superconductor at the atomic level directly defines the structure of the vortex core [202, 1758]. There are several models, taking into account the layered structure of superconductor, for example Lawrence–Doniach's model depicting a stack of thin superconducting layers connected by Josephson interaction [992], or the models of S–N–S superlattices consisting of alternating superconducting (S) and normal (N) layers [296, 932, 1758]. An important feature of these models is that, due to weak Josephson interlayer coupling, the maximum supercurrent density (j_c) between the layers is much smaller than the interlayer de-pairing current density (j_d). As a result, the maximum current density that can be locally generated by a vortex parallel to the layers is limited by j_c. Due to the short coherence length ($\xi_c < s$) in layered HTSCs planar defects parallel to the ab-planes can significantly reduce interlayer coupling, limiting the current flow along the c-axis and rising the magnetic field penetration along the defects. In this case, the methods of non-local Josephson electrodynamics can be used to study both static and moving vortex (including non-linear area of Josephson core) at planar defect in layered HTSC and for the description of their structure [600].

The effect of macroscopic defects on localized magnetic flux (fluxons) and critical force of de-pinning is found by the defect size and type, and also by its disposition at grain boundary (or at Josephson junction). The case of the defect width of order of the Josephson penetration depth is considered in [1611]. Using the sine–Gordon model for defective Josephson junction, critical current as a function of magnetic field in the case of asymmetrically disposed defect is calculated. Moreover, the investigation of the interaction between fluxons and defects defines coercive forces of pinning or de-pinning a fluxon from a defect. Consideration, using an analysis of the different mode stability in dependence on the defect disposition at zero magnetic field, finds the bounds of these modes due to two factors, namely (i) the instability at the junction boundaries away from the defect and (ii) the instability due to the fluxon trapping or de-trapping by the defect. When the defect localizes near one of the junction edges, both criteria contribute (in definition) to the instability. In general case, there is coupling between defects and junction edges (surface defects), especially in the case of the numerous defects.

12.4 Non-Linear Current in Superconductors with Obstacles

Following [602], consider features of non-linear current in HTSC systems with existence of obstacles. Macroscopic electrodynamics of type II superconductors in the mixed state is found by the pinning and thermally activated creep of vortex structures, increasing a weakly dissipative critical current, irreversible magnetization and slow current relaxation (flux creep). These phenomena manifest themselves on length scales much longer than the Larkin's pinning correlation length (L_c), on which the critical state is formed [146]. On this macroscopic scale ($L \gg L_c$), the material property, defining the behavior of superconductor in electromagnetic fields, is the next non-linear local relation between the electric field, $\boldsymbol{E}$, and current density, $\boldsymbol{J}$:

$$\boldsymbol{J} = (\boldsymbol{E}/E)J(E), \tag{12.2}$$

where $\boldsymbol{E}(\boldsymbol{r}, t)$ and $\boldsymbol{J}(\boldsymbol{r}, t)$ correspond to the macroscopic electric field and current density, averaged over all relevant intrinsic scales of pinned vortex structure. Consider an isotropic E–J dependence stated by (12.2), that, in particular models the nearly two-dimensional current flow in the *ab*-plane of layered HTSCs. Equation (12.2) combined with the Maxwell equations

$$\partial_t \boldsymbol{B} = -\nabla \times \boldsymbol{E}; \quad \nabla \times \boldsymbol{H} = J(\boldsymbol{E}), \tag{12.3}$$

permit to calculate the evolution of heterogeneous distributions $\boldsymbol{E}(\boldsymbol{r}, t)$ and $\boldsymbol{B}(\boldsymbol{r}, t)$ and, thus, to describe macroscopic magnetic, transport and relaxation phenomena in superconductors.

A type-II superconductor in the mixed state demonstrates a highly nonlinear $E(J)$ dependence lower a critical current density, J_c, which separates regimes of magnetic flux flow (at $J > J_c$) and its creep (at $J < J_c$) (see Fig. 12.7). The critical behavior of $E(J)$ at $J < J_c$ is determined by thermally activated vortex creep as

$$E(J) = E_c \exp\left[-\frac{U(J,T,B)}{T}\right], \tag{12.4}$$

where $U(J, T, B)$ is the activation barrier, depending on the current density, J, temperature, T, and magnetic induction, B. Here, E_c is the conditional criterion for electric field, which defines corresponding value of J_c through the relation, $U(J_c, T, B) = 0$. For example, the vortex glass/collective creep models [146] determines a dependence, $U = U_c[(J_c/J)^{\mu} - 1]$, for small values of $J \ll J_c$, which is observed in transport and magnetization measurements in HTSCs [1837]. Similar, though less singular logarithmic dependence $U = U_c \ln(J_c/J)$ corresponds to power-law

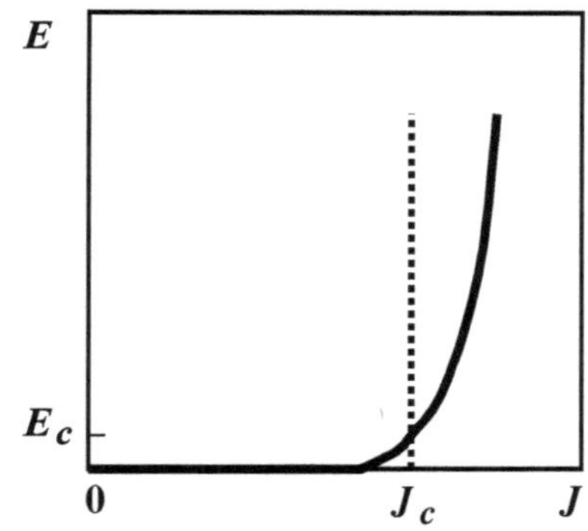

Fig. 12.7 $E(J)$ characteristic of a type-II superconductor

E–J dependencies, $E = E_c(J/J_c)^n$ with $n(T, B) = U_c/T \sim 3$–30, being a well approximation, especially for layered BSCCO [1244, 1735, 1885].

When the distribution $\boldsymbol{J}(\boldsymbol{r})$ changes on spatial scales greater than L_c, a superconductor can be considered as a highly non-linear, heterogeneous conductor with a local characteristic of $\boldsymbol{E}(\boldsymbol{J}, \boldsymbol{r})$. For example, (12.3) describes non-linear transport current in superconductors with macroscopic obstacles or current percolation in HTSC polycrystals with grain size $\gg L_c$ (see Fig. 12.8). This case is important to understand the current-limiting mechanisms of HTSCs, which in addition to grain boundaries, often contain other macrodefects, for example, secondary phases, microcracks and areas of local non-stoichiometry on scales of order 10–1000 μm [977, 978]. These obstacles, blocking current flow, cause macroscopically heterogeneous distributions of transport and magnetization currents [220, 1337, 1382, 1392, 1524–1526].

In turn, even relatively weak heterogeneity of local current density, $\boldsymbol{J}(\boldsymbol{r})$, can lead to exponentially strong variation of the electric field, $\boldsymbol{E}(\boldsymbol{r}) \propto \boldsymbol{J}\exp[-U(J)/T]$, radically changing global properties of HTSC, $J(E, B, T)$, observed in experiments. Thus, the behavior of $J(E, B, T)$ in superconductor can differ from local characteristics, described by (12.2) and defined by thermally activated vortex dynamics and pinning on mesoscopic scales (at $L < L_c$). This difference increases due to the non-linearity of $E(J)$, makes the effective current-carrying cross-section dependent on T, B and E [566, 595, 596, 604, 609, 652, 689, 997, 1602, 1616].

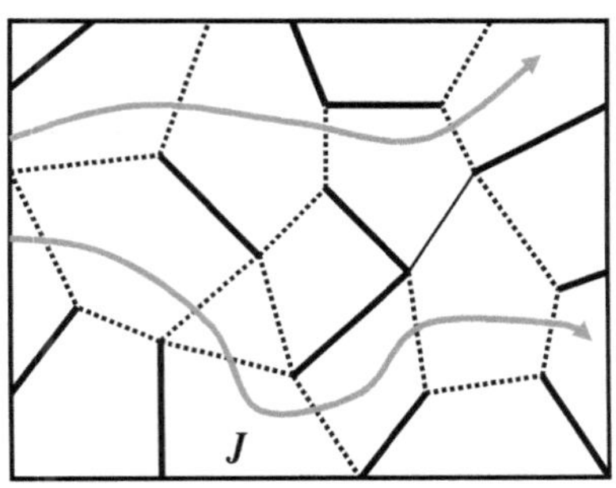

Fig. 12.8 Percolation of current flow in HTSC polycrystals. High-angle grain boundaries, blocking the current flow, are shown by *solid lines*, but low-angle grain boundaries, "transparent" for current flow, are denoted by *dotted lines*

As examples of two-dimensional current flows for various cases, which could be considered as elements of a more general percolation network, presented in Fig. 12.8, we note proper cases, shown in Fig. 12.9 [602]. These geometries are standard in experimental studies of resistive states in superconductors, thin-film superconducting electronic circuits and HTSC conductors for power applications. In all cases presented in Figs. 12.8 and 12.9, the electric field and current density are heterogeneous on macroscales and demonstrate singularities near the sharp edges and corners. Calculation of the global *E–J* characteristic demands the solution of highly non-linear (12.2) and (12.3). For large values of n, even weak spatial changes of $\boldsymbol{J}(\boldsymbol{r})$ around planar defects in Fig. 12.8 lead to power variation of electric field, $E(r) = E_c[J(r)/J_c]^n$. Together with singularities of $\boldsymbol{J}(\boldsymbol{r})$ and $\boldsymbol{E}(\boldsymbol{r})$ at the edges, this significantly complicates numerical analysis of (12.2) and (12.3).

An approximate method for the definition of current distribution in superconductor is given by Bean's critical state model [111, 227], which replaces the real curve $E(J)$ in (12.2) by a step-wise dependence $\boldsymbol{J} = \boldsymbol{E}J_c/E$ for $E > 0$ and $E = 0$ for $J < J_c$ (see Fig. 12.7). This model can be considered as a limit of the power-law dependence, $E = E_c(J/J_c)^n$ for $n \to \infty$ [173–178]. It enables analytical solutions for some simple cases, presented in Fig. 12.10. These solutions demonstrate characteristic "discontinuity" lines (*d*-lines), along which the current abruptly alters direction. Another feature of this model is that the defects with finite size can disturb current in infinite range, as it take places for current flow around a cylindrical void, in which parabolic *d*-lines extend to infinity [227, 1527] (see Fig. 12.10b). Both the zero thickness of the *d*-lines and the infinite spatial scale of current perturbations are followed from the zero resistivity for the idealized *E–J* characteristic in the whole range, $0 < J \le J_c$. In this case (12.3), describing steady state, become an ill-defined mathematical problem, which reduces to the only condition of current continuity, div $\boldsymbol{J} = 0$, when $J \le J_c$. These conditions can be satisfied by many various current distributions for a given sample geometry,

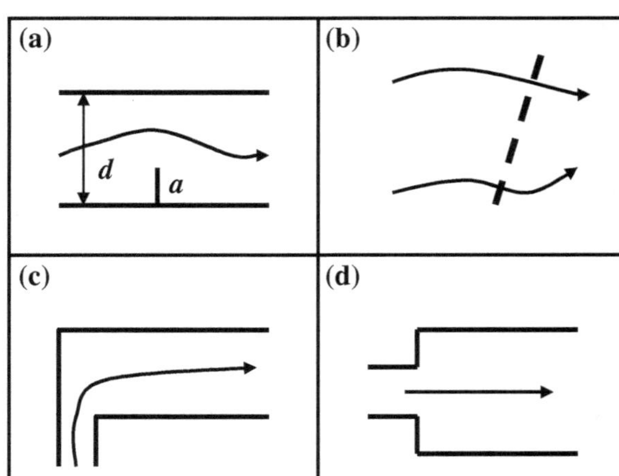

Fig. 12.9 Characteristic cases of two-dimensional non-linear current flow: (**a**) superconducting film with strong current-limiting planar defect (microcrack at high-angle grain boundary), (**b**) current flow through faceted grain boundary with alternating segments of various J_c values, (**c**) current injector (or magnetic flux transformer) and (**d**) microbridge

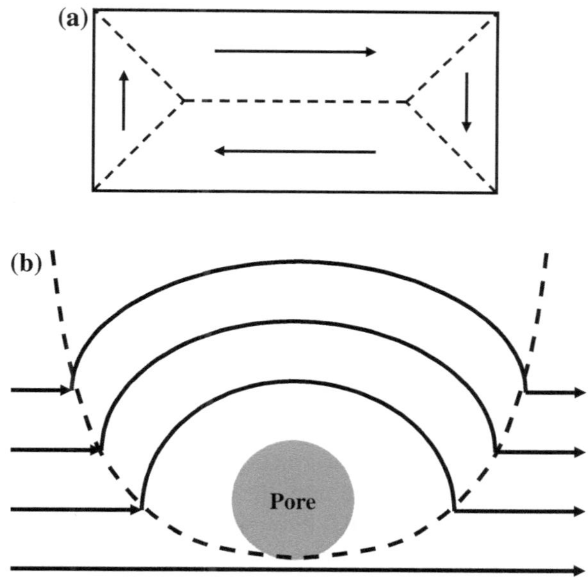

Fig. 12.10 Magnetization current flow, predicted by Bean model: (**a**) long slab in parallel magnetic field and (**b**) current flow around cylindrical void [227]. *Dashed lines* show parabolic *d*-lines, on which $\boldsymbol{J}$ sharply changes direction. Current streamlines between the *d*-lines are circular and centered in the void

because the selection of Bean's solutions for $\boldsymbol{J}(\boldsymbol{r})$ is found by corresponding initial conditions.

The inherent hysteresis of the critical state makes the Bean model inappropriate for calculations of steady-state transport current flow around planar obstacles, presented in Figs. 12.8 and 12.9, for which the current-carrying cross-section varies along a superconductor. In this case, the current path inevitably breaks into the regions of critical state, $J = J_c$ and in sub-critical areas of $0 < J < J_c$. At the same time, the specific distribution of these regions depends on initial conditions, and thus cannot be calculated by solving the steady-state equations, div $\boldsymbol{J} = 0$, and $J \leq J_c$. This unphysical situation is caused because Bean's model ignores the necessary condition, $\nabla \times \boldsymbol{E} = 0$, which can only be satisfied taking into account the highly nonlinear *E–J* characteristic stated by (12.2). The non-zero electric field, although being small due to the power-law dependence at $J < J_c$, plays an important role, because now the solution of (12.2) can be calculated into framework of a well-defined mathematical problem, for which each non-zero value of $\boldsymbol{J}$ corresponds to a certain value of $\boldsymbol{E}$. The account of E eliminates the infinite extent of current perturbations around local heterogeneities (see Fig. 12.10b), the zero thickness of the *d*-lines and multiple solutions of the Bean model, fixing a unique steady-state transport current distribution for given boundary conditions. However, the account of the non-linear dependence, $E(J)$, depends on the solution of the next non-linear equation for scalar potential, φ [602]:

$$\operatorname{div}\left[\frac{\nabla\varphi}{|\nabla\varphi|}J(|\nabla\varphi|)\right] = 0, \tag{12.5}$$

which is obtained, substituting $E = -\nabla\varphi$ into (12.2) and then into equation, div $\boldsymbol{J} = 0$.

An analytical distribution of two-dimensional steady-state transport current flow in superconductors, taking into account their nonlinear $E(J)$ characteristics, can be found using a *hodograph transformation* [598, 602, 605], initially used to describe compressible gas flow [254, 972, 1149, 1532]. In [598], the hodograph method was applied to describe current flow in anisotropic superconductors. Using this method, analytical steady-state solutions of (12.2) and (12.3) for current flow past a planar defect were obtained for the power-law E–J characteristics [605]. The essence of this method is the following [602]. Instead of considering highly non-linear equation for $\varphi(\boldsymbol{r})$ in the coordinate space, the variables are altered and $\varphi(\boldsymbol{r})$ is expressed as a function of the electric field, $\boldsymbol{E}$:

$$\varphi(\boldsymbol{r}) \to \varphi(\boldsymbol{E}). \tag{12.6}$$

The hodograph transformation (12.6) reduces the non-linear (12.5) for $\varphi(\boldsymbol{r})$ to a *linear* one for $\varphi(\boldsymbol{E})$. In turn, the equation for $\varphi(\boldsymbol{E})$ can be further reduced to other well-stated linear partial differential equations, such as Thomas–Fermi or London equations, whose known solutions can be used to obtain exact solutions for two-dimensional current flows in superconductors with defects [602, 605].

The hodograph method enables to resolve the ambiguities of the Bean's model and to study analytically the features of two-dimensional non-linear transport current flow. In particular, it gives a unique steady-state distribution, $\boldsymbol{J}(\boldsymbol{r})$ for a given sample geometry and shows that current flow breaks into domains with different orientations of $\boldsymbol{J}$, separated by current domain walls reminiscent of the *d*-lines of the Bean's model. However, there are important differences between the current domain walls and the *d*-lines. First of all, the domain walls have an internal structure and varying width, which depends both on the value of n and on the geometry of current flow. For example, for the void in Fig. 12.10b, the width of the domain walls increases with the distance from the void. This broadening of the domain walls provides the decay of current perturbations, caused by the defect size a on a *finite* distance, $L_\perp \sim an$, much longer than a, if $n \gg 1$. Moreover, the current domain walls described by exact hodograph solutions remain different from the phenomenological *d*-lines even in the critical state limit, at $n \to \infty$. It may be shown that current flow near domain walls satisfies the equation, $\nabla \times \boldsymbol{E} = 0$.

As the results of [602] show, a planar obstacle of size a greatly increases electric field disturbances over distance $L_\perp \sim an$ perpendicular to current flow or $L_\parallel \sim an^{1/2}$ along the current flow, causing long-range interaction between defects and strong effect of the sample geometry. Indeed, even a sparse network of planar defects can significantly reduce the effective current-carrying cross-section if the average spacing between the defects is smaller than $L \sim an$. For typical values of $n \sim 30$, this non-linear blockage of transport current by defects can occur at rather small concentration of obstacles, which is only a few percent of the geometrical cross-section of a superconductor. This effect can have important applications for current percolation in superconductor. For example, a microcrack in a film with thickness, d (see Fig. 12.9a), can cause a strong local peak of $E(x)$ in YBCO coated conductor, even for rather small defects, $a > d/n \ll 1$. In turn, such local peaks of voltage and dissipation change the global current–voltage characteristic of the whole conductor, similar to the effect of macroscopic random heterogeneities [609].

Measuring the scales $L_{\perp}$ and $L_{\parallel}$ of magnetic-flux disturbance, one can estimate both the value of $n \approx (L_{\perp}/L_{\parallel})^2$, and the defect size, $a \sim L_{\perp}/n$. The excessive dissipation caused by defects is another important feature of the strong non-linearity of $E(J)$. As a rule, HTSCs contain many macroscopic current-limiting defects (cracks and high-angle grain boundaries), contributing considerably to the overall *ac*-losses and energy dissipation. For example, for a sparse array of planar defects of size approximately a, spaced by $l \gg a$, the excessive dissipation per unit volume can be estimated as [602], $n^{3/2}(a/l)^2 J_0 E_0$, which attains the order of the background bulk dissipation (corresponding to uniform sample without defects), $J_0 E_0$, if the geometrical parameter, $l < an^{3/4}$. Then, the overall dissipation will be stated by defects, if they occupy rather small fraction $\sim n^{-3/4}$ (of the sample geometrical cross-section), which is few percents for superconductors with typical values of $n \sim$ 20–30. The excessive dissipation due to defects can also trigger local thermal instabilities in HTSCs [606]. These instabilities develop, mostly in the regions of increased heat generation (for $E > E_0$), which dominate the overall dissipation. Other types of orientation current instabilities can be caused by the high anisotropy of HTSCs, when current is forced to flow along the c-axis by microcracks [220, 1382] or a "brick-wall" structure of in-plane tilt grain boundaries [203].

Equation (12.3) for steady-state current flow become

$$\nabla \times \boldsymbol{E} = 0; \quad \nabla \times \boldsymbol{H} = \boldsymbol{J}. \tag{12.7}$$

In the case of the transformer or elbow geometry (see Fig. 12.11), the hodograph transformation method enables to compare the behavior of current

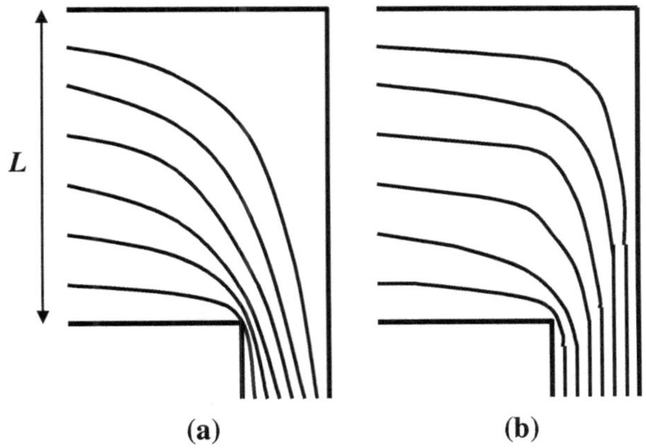

Fig. 12.11 Current streamlines for a transformer: (**a**) normal metal ($n = 1$) and (**b**) superconductor ($n = 30$) [495]

streamlines for normal metal ($n = 1$) and superconductor ($n = 30$), leading to the important differences [495]. The metal streamlines become increasingly compressed near the corner ($J \propto 1/x^{1/2}$ and $E \propto 1/x^{1/2}$, where x is the distance from the corner). The current density for a superconductor is only weakly divergent at the corner with $J \propto 1/x^{1/n}$ ($n \gg 1$), while the electric field singularity is strongly increased ($E \propto 1/x$). There are several similarities with behavior, predicted by the Bean's critical state model, namely (i) in the limit of $n \to \infty$, the current density displays no singularities, even for flow around a sharp corner, and (ii) the current flow breaks up into well-defined domains. At the same time, in contrast to Bean's model, the equilibrium current density does not remain uniform in a sample with obstructions. Figure 12.11b presents three different flow regions: two asymptotic regions, where current flow is straight and uniform, and an intermediate region, where the streamlines point in radial outward, as if from a point-source, located at the top right corner of the current lead. In this intermediate region, the current density is not uniform, but the horizontal width of the point-source region is approximately $2L/\pi$.

Figure 12.12 shows the current streamlines for the two current blocking constrictions of different sizes. In Fig. 12.12a, 90 %-crack shows a point-source flow domain, similar to the transformer geometry. Again, the current disturbance has a horizontal width of approximately $2L/\pi$. In Fig. 12.12b, 10 %-crack exhibits the long-range nature of current disturbances near obstruction. Slightly below the defect, J is enhanced above its unperturbed value by an average factor of about $(1 + 1/n^{1/2}) \approx 1.18$, for $n = 30$. For a defect of size a, streamlines typically return to their unperturbed, uniform state beyond a characteristic length scale of $an^{1/2}$. At the same time, significant perturbations of J extend across the entire width of the sample in Fig. 12.12b. On the opposite side from the defect, the current density is still enhanced by 9 % over its unperturbed value.

The current density for finite n has a weak singularity at the end of a defect. Thus, a region of flux-flow will be nucleated, leading to detrimental consequences for the macroscopic values of J_c. Figure 12.13 presents the flux-flow boundary and current streamlines near a planar defect. For a defect with size a, which is much longer than the flux-flow region, the boundary is perfectly circular, with a radius of

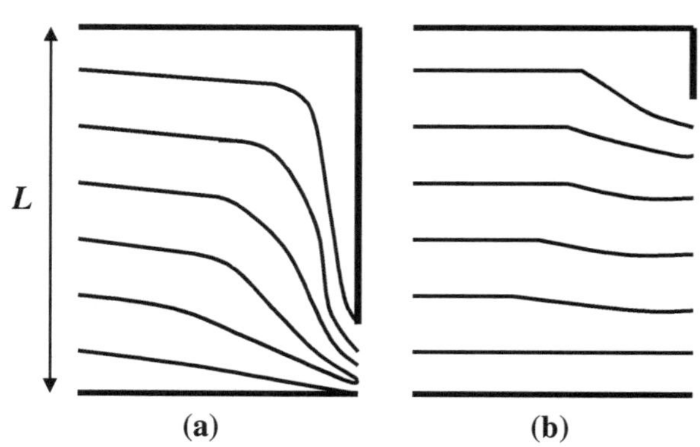

Fig. 12.12 Current streamlines for two constriction geometries ($n = 30$) [495]

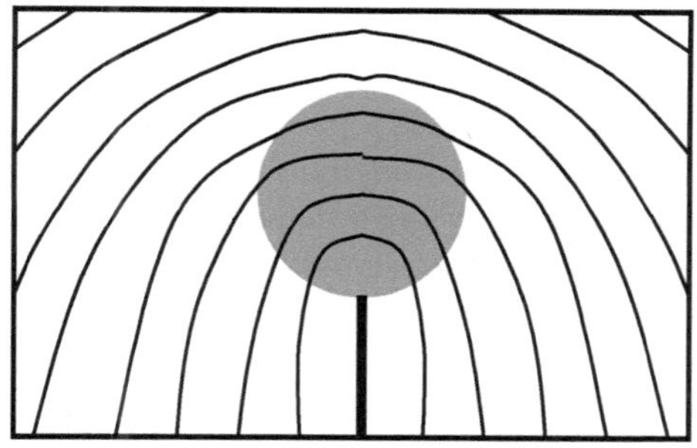

Fig. 12.13 Current streamlines around a crack. *Gray region* represents flux flow [495]

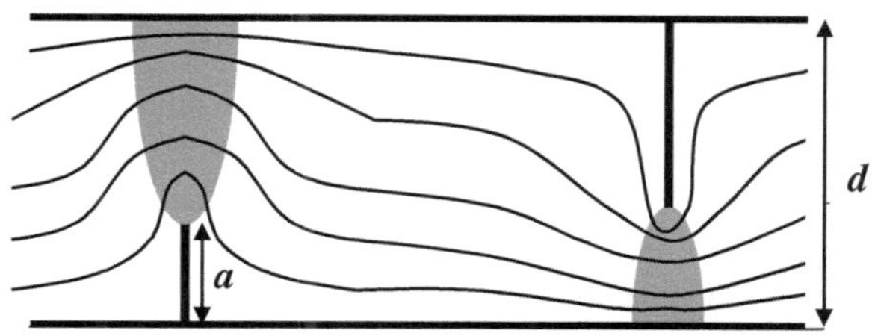

Fig. 12.14 Current flow around planar defects [603]

order $(E/E_0)an^{1/2}$, where E_0 is the electric field far from the defect. In the limit of $n \to \infty$, the boundary just touches the end of the defect.

In the case of current flow around planar defects (high-angle grain boundaries, microcracks, etc.) in a film (see Fig. 12.14) for $n \gg 1$, the distribution $\boldsymbol{J}(x, y)$ breaks into orientation current flow regions of characteristic width, $w = 2\sqrt{ad/\pi}$, for $a \ll d$ [603]. These regions are separated by the regions of nearly uniform current flow (presented by white color). A narrow region of highly enhanced electric field (gray color) extends from the defect end. In this region, the distribution $E(x)$ on the film side opposite to the defect is given as [603]

$$E(x) \approx E_0 + (E_m - E_0)\exp(-x^2/l^2), \tag{12.8}$$

$$l = 2(d - a)/\sqrt{\pi n}, \tag{12.9}$$

where $E_m = [d/(d-a)]^n E_0/\sqrt{e}$ and E_0 is the uniform electric field in the white regions of Fig. 12.14.

The narrow peaks of $E\ (x)$ with width, $l \sim d/n^{1/2} \ll d$, and $E_m \gg E_0$ near planar defects give rise to the local excess voltage, $\Delta V = (E_m - E_0)l\pi^{1/2}$, as

$$\Delta V = \frac{2E_0 d}{\sqrt{ne}}\left(\frac{d}{d-a}\right)^{n-1}. \tag{12.10}$$

The factor $[d/(d-a)]^{n-1}$, for $n \gg 1$ significantly enhances ΔV, even at $a \ll d$. As a result, a sparse array of defects, spaced by L defines the overall voltage on the sample, when $\Delta V > LE_0$, that is, the defect size of a exceeds a_c:

$$a > a_c = d\left[1 - \left(2d/\sqrt{enL}\right)^{1/(n-1)}\right]. \tag{12.11}$$

Equation (12.11) gives $0.1d < a_c < 0.2d$ at $10d < L < 100d$, for $n = 30$. Therefore, the strong non-linearity of $E(J)$ significantly enhances the blockage of current flow by small defects ($a_c \ll d$), even if they occupy only a small fraction of the geometrical cross-section of a superconductor.

Finally, using a non-linear effective medium theory [609], the global averaged characteristics $\overline{J}(E, B, T)$ of superconductors with weak macroscopic random heterogeneities can be calculated. In this case, $\overline{J}(E, B, T)$ is related to $J(E, B, T)$ of a macroscopically uniform superconductor as [603]

$$\overline{J} = \left(1 - \eta^{2/3}/2p\right)J, \tag{12.12}$$

where heterogeneities are quantified by a small dimensionless parameter η. Due to the strongly non-linear behavior of $E = E_c\exp[-U(J, T, B)/T]$, the parameter η depends on E, T and B, so even weak macroscopic random disorder can qualitatively alter the behavior of the global characteristic $\overline{J}(E, B, T)$as compared to $J(E,B, T)$. For example, weak heterogeneities ($\eta \ll 1$) can lead to non-monotonous dependence of the critical current density J_c on the magnetic field B (to the "fish-tail" effect), while strongly affecting the curvature of the plot $\ln E - J$ [603].

12.5 Double Critical-State Model for Type-II Superconductors

The double critical-state model for type-II superconductivity has been initially proposed by Clem and Perez-Gonzalez [297, 1349–1351] and formulated as a three-dimensional model, characterizing non-linear relation between electric field, electric current, perpendicular and parallel components of electric field to magnetic flux in bulk type-II superconductors. In this model, it is postulated that electric field, $\boldsymbol{E}$, should be presented in the form: $\boldsymbol{E} = \rho_\perp \boldsymbol{J}_\perp + \rho_{\|}\boldsymbol{J}_{\|}$, where $\boldsymbol{J}_\perp$ and $\boldsymbol{J}_{\|}$ are the perpendicular and parallel components of electric current density, $\boldsymbol{J}$, to the magnetic flux density, $\boldsymbol{B}$, respectively. The resistivity $\rho_\perp$ and $\rho_{\|}$ satisfy for corresponding critical values of $J_{c\perp}$ and $J_{c\|}$ to the following conditions: (i) if $|\boldsymbol{J}_\perp| < J_{c\perp}$, then $\rho_\perp = 0$; (ii) if $|\boldsymbol{J}_{\|}| < J_{c\|}$, then $\rho_{\|} = 0$.

In [297, 1349–1351], it has been considered a case, when rotating parallel magnetic field applied to infinite superconductive plate, that is, the model could be formulated as one-dimensional model in space. Then Badia and López [79–81] included the Clem–Perez–Gonzalez model in their generalized theoretical statement of the problem in order to study the problems of critical state demonstrating a variational structure of the model. Into framework of their variational formulation, they analyzed the double critical-state model for one-dimensional geometry of finite plate.

Kashima [831] modified the original model of the double critical-state by adding term $\rho_0 \boldsymbol{J}$ with resistivity ρ_0, which disappeared, if value of $|\boldsymbol{J}|$ was lesser then specific critical value of J_{c0}: $\boldsymbol{E} = \rho_{\perp}\boldsymbol{J}_{\perp} + \rho_{\parallel}\boldsymbol{J}_{\parallel} + \rho_0\boldsymbol{J}$. The additive term $\rho_0\boldsymbol{J}$ makes energy density to be coercive defining the obtained constitutive relation relatively to the current density, $\boldsymbol{J}$, in the way, when there is a solution of the problem, formulated in the form of a variational inequality. The convex optimization problem, obtained in the form totally discrete formulation, admits existence of its unique minimizer. By selecting enough great critical current density J_{c0} and resistivity ρ_0, the term $\rho_0\boldsymbol{J}$ is considered as arising of great resistivity after jump-like transition from superconductive state to normal one. However, the conducted mathematical modeling has shown that the resistivity ρ_0 disappeared always in this case and the term $\rho_0\boldsymbol{J}$ did not define the model nature.

The numerical analysis of formulation of the macroscopic models for critical state in the form of variational inequalities for type-II superconductors has been initiated by Prigozhin [1402, 1403], which proposed sub-differential formulation of the Bean critical state model [111], proved its well-posedness formulation and carried out intensive investigations of two-dimensional case. Then, it has been estimated in two-dimensional case an error of finite-element approximation of the Bean model, formulated in the form of variational inequalities [431, 432]. Barrett and Prigozhin [96] obtained dual formulation of the Bean model in the terms of electric field as the conjugate variable of magnetic field and proved convergence of the finite-element approximation of their dual formulation. Elliott and Kashima [430] fulfilled three-dimensional finite-element analysis of the critical-state models formulated in the form of variational inequalities which defined a magnetic field and current density around bulk type-II superconductor.

The sub-differential formulation of the Bean critical state model introduces a restriction consisting in that the electric field must be always parallel to electric current. This condition is carried out for some geometric configurations for which the component $\boldsymbol{J}_{\parallel} = 0$, for instance axially symmetric superconductors or superconductive thin films under perpendicular applied field. However, a redistribution of the pinned magnetic flux caused by Lorentz force could lead to electric field which is no parallel to current in the common three-dimensional case. In order to study the macroscopic behavior of electro-magnetic field around a bulk type-II superconductor in general three-dimensional configuration, it is necessary to use a mathematical model taking into account a whole vector character of current density $\boldsymbol{J}$, and develop a numerical method for discretization of this model.

In the class of non-linear Ohm's laws, for example Bean model; modified Bean model, proposed by Bossavit [159] and model of power law of Rhyner [1428], electric field $\boldsymbol{E}$ is considered as the sub-differential operator of convex energy density on $\boldsymbol{J}$. For these models, the corresponding formulation of variational inequalities in the terms of unknown magnetic field is the gradient system determined by the sub-differential operator of convex energy functional and existence of unique solution for evolutionary variational inequality. This is defined directly by application of the theorem on unique solvability of non-linear evolutionary system proved by Brezis [182]. However in the Kashima formulation of the model [831], it is necessary to decompose current density into parallel and perpendicular components relatively to magnetic flux. In this case the energy functional depends not only on electric field but magnetic flux and is not convex in relation to unknown magnetic flux. Therefore, the existence theorem of Brezis could be not applied to the considered case and analysis of existence of the solution should be different. Kashima [831] showed a solvability of this problem by applying the Schauder theorem for fixed point together with the unique existence theorem for non-linear evolutionary system defined by sub-differentials which are dependent on time [590, 845, 1849]. He proved a theorem on definition of the sub-differential into class of energy functional including energy defining the main formulation of the model. By this, Faraday's law and non-linear Ohm's law could be stated in the superconductor in the sense of almost everywhere. At the numerical realization of the obtained equations, the spatial discretization has been fulfilled by using boundary finite elements of lowest order at the tetrahedral grid. At the discretization on time, it has been applied on-step scheme and general discrete formulation has been reduced to solution of the minimization problem. In order to introduce into the model restrictions which are free of rotations imposed on magnetic field outside superconductor, it has been introduced scalar magnetic potential and proposed hybrid "magnetic field—scalar potential" formulation, which was equivalent to original model.

12.6 Current Percolation and Pinning of Magnetic Flux in HTSC

Experimental studies of GBs show that their critical currents (I_c) could vary over orders of magnitude [375]. Then, it may be expected that supercurrent flow in polycrystalline samples at least near the critical state, would be percolated in nature that is supported by magneto-optic investigations of Bi-2223/Ag tapes [1338]. As a result, the behavior of current flow in the case of a distribution of boundaries with a sufficiently widespread of I_c may be understood. It is possible to apply the percolation theory to two directions of superconductor's studies:

(1) Transport in polycrystalline superconductors is characterized by links (or breaking of those) between adjacent grains that can be described by percolation process [298, 358, 1459, 1610]. This approach is applied to sintered HTSC bulks [196, 933, 1165, 1300, 1324, 1327], to Bi- 2223/Ag tapes [1280, 1400, 1602] and to two-dimensional coated conductors [652, 705, 898, 1456, 1457, 1603].
(2) Flow of Abrikosov's vortices through a superconductor can be also regarded as percolation process, which is governed physically by local trapping (or untrapping) of vortices [1822,1886].

The square [652, 898, 933, 1280, 1300, 1324, 1327], hexagonal [1456, 1457, 1603, 1862] and cubic [1204, 1400] meshes are used for different numerical models. On the theoretical level, percolation methods have dealt mainly with linear circuit elements and focus on the critical behavior (i.e., on power laws) near the percolation threshold [1887]. At the same time, the case of a mesh of resistive elements with arbitrary non-linearity, where individual resistive elements have been removed at random has been considered [1616].

It may be shown using the renormalization-group theory that *I*–*V* characteristic follows a power-law behavior near the percolation threshold. However, these renormalization-group calculations do not tell us anything about the value of the critical current or the shape of *I*–*V* curve away from the critical region. The metal–insulator problem (in a certain sense similar to the superconductor–metal problem) has been studied away from the percolation threshold [341]. The case of a linear medium containing a small admixture of non-linear elements has also been studied. This model is applied to the study of composites, formed by non-linear impurities, embedded in a linear host [1000, 1007, 1828]. Theoretical predictions of *I*–*V* characteristics are made using the Clausius–Mossotti approximation [1007] or the effective-medium solution [1828]. Both cases are limited to the case of weak non-linearity or a low density of non-linear elements. References [996, 997] are more close to understanding of the entire *I*–*V* characteristic for a strongly non-linear system. In these works, a Ginzburg–Landau model of conduction in polycrystalline superconductors and a simpler model of non-linear resistor network is considered. The latter provides a reasonable and relatively simple representation of the physical properties of a superconductor with weak couples.

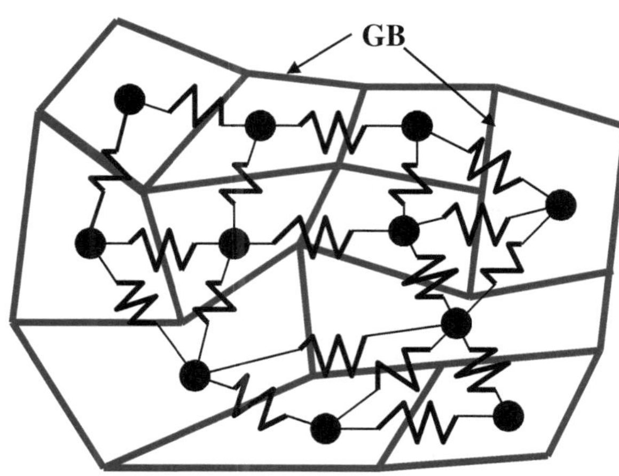

Fig. 12.15 The reduction of a polycrystalline superconducting network to an equivalent resistor network. The irregularly shaped regions are superconducting grains taken as equipotentials. All potential drops take place across GBs, shown as nonlinear resistors [652]

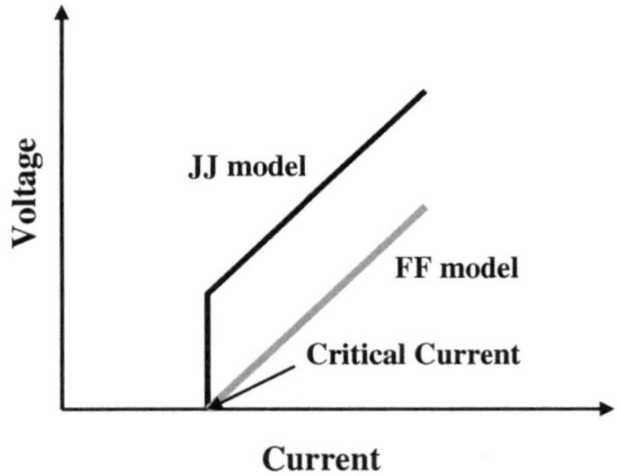

Fig. 12.16 *I–V* characteristics for Josephson junction (JJ) and flux-flow (FF) models. The first is considered to be appropriate for low fields and the second for high fields [652]

12.6.1 Model of Non-Linear Resistor Network

It is assumed that the current flows through a random array of microcrystallites, separated by grain boundaries that act as weak links. Following [652], we construct on the first step an equivalent resistor network. For simplicity, it is assumed that each grain is perfectly conducting (i.e., they are equipotential areas). This allows one to separate, at least temporarily, the two current-limiting factors (flux pinning and weak links). Therefore, the obtained results will be limited to cases in which the total magnetic field (applied plus self) is small enough that there is no flux, penetrating into the grains. Flux only penetrates the GBs.

Consider the grains as nodes of above network, but replace each boundary by a non-linear resistor (see Fig. 12.15). In terms of graph theory, the graph of boundaries is replaced by its dual graph, with relation being one-to-one [58]. We shall assume that the geometrical randomness is less important than the randomness in the electric resistance of the resistors.

Investigate the current flow through a granular superconductor, focusing on two models of the GBs, applicable to the cases of Josephson junctions in low and in high external fields, respectively.[1] Each boundary is found by its *I–V* characteristic. V is some non-linear function of I for every resistor, but each resistor may be different. Generally, the resistors are described by a probability distribution, $P[V(I)]$. Consider, for better understanding of the results, binary distributions consisting of a mixture of weak and strong couples, which are identical to each other in each group. The corresponding concentrations of the links are denoted by p and $q = 1 - p$.

Divide the present model into two kinds corresponding to (i) weak-field Josephson junction (JJ) and (b) strong-field flux-flow (FF) boundaries.

[1] Why actual boundaries should behave in this way is explained in [1021].

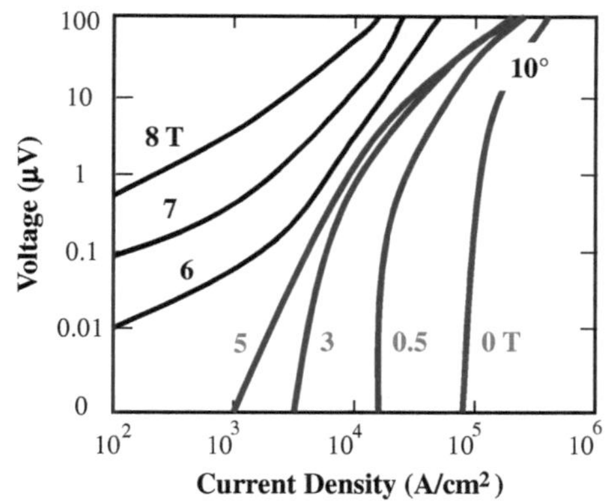

Fig. 12.17 Proper behavior of test *I–V* characteristics for 10°-GB in YBCO at different applied fields. At low fields, *I–V* characteristic is similar with the JJ model. As the applied field increases, there is a crossover at 5.5 T to a behavior that is similar to the FF model [652]

Binary JJ model. For the JJ case, the weak (strong) links with corresponding concentration, $p(1-p)$, have a critical current, $I_{c1}(I_{c2})$. If $I < I_{ci}$ in such resistor, then $V = 0$; if $I > I_{ci}$, then $V = IR_i$, where $R_i = \text{const}$ $(i = 1, 2)$ and $I_{c1} < I_{c2}$. Note that V is a *discontinuous* function of I for both types of links (see Fig. 12.16).

Binary FF model. For the FF case, the weak (strong) links with corresponding concentration, $p(1-p)$, have a critical current, I_{c1} (I_{c2}). If $I < I_{ci}$ in such resistor, then $V = 0$; if $I > I_{ci}$, then $V = (I - I_{ci})R_i$, where $R_i = \text{const}$ $(i = 1, 2)$ and $I_{c1} < I_{c2}$. Here, V is a *continuous* function of I for both types of links (see Fig. 12.16).

Both models differ at microscopic scale. The JJ model represents a situation of low current, and zero applied field. The Josephson junction is assumed to be superconducting for small currents and normal at high currents, when the voltage drop is just ordinary Ohmic loss. There is a discontinuous change between the two regimes. The FF model is intended to simulate weak links because flux pinning in them is weak. In this case, I_c of the link represents the de-pinning current. The losses take place for $I > I_c$ at the movement of flux along the boundary. Experimental measurements on individual GBs show a crossover between the two sorts of behavior. Figure 12.17 shows *I–V* characteristics for a 10°-GB. At low applied field, the boundary behavior is similar with the JJ model, while at high field it is closer to the FF model. The considered non-linear network can be an approximation to an actual JJ array in the over-damped limit. The JJ model can represent a set of over-damped junctions with no pinning, in which the McCumber parameter, $\beta_J \equiv \hbar R^2/(2eI_{c0}C)^{1/2} < 1$, where R and C are the resistance and capacitance of the junctions. Then, output I^{tot} represents the *dc*-component of the actual output value of real over-damped junctions. In turn, the FF model is the case, where pinning of Josephson vortices dominates the transport. The result of numerical realization of two models is V^{tot} (I^{tot}) dependence for the system as a whole. Thus, we feed a fixed current at one end and collect it at the other and measure the voltage drop.

Let there be a distribution of non-linear resistors on an infinite mesh, where each resistor has some critical current and occupation probability.[2] This implies that the critical current can be found as

$$I_c^{tot} = \min_S \sum_{l \in S} I_c^{(l)}, \tag{12.13}$$

where the minimum is taken over *all* surfaces that separate the electrodes [1429]; l are the wires that pierce S and $I_c^{(l)}$ is the critical current of wire l. Because I_c^{tot} depends only on $I_c^{(l)}$, (12.13) states the dependence of total critical current of the mesh only upon the critical currents of individual elements and not upon their dissipative behavior at $I > I_c$.

Consider two adjacent infinite surfaces S and S^*, which separate the electrodes. The maximum supercurrent that can be carried through S, considered entirely by itself, is:

$$I_c^{(\infty)} = \sum_j p_j I_{cj}. \tag{12.14}$$

However, it is not clear that the current can penetrate from S to S^*. Indeed, we expect the critical current of the infinite mesh to be less, than $I_c^{(\infty)}$, defining an upper bound for I_c^{tot}, independent of model and system dimensionality.

In order to obtain a better approximation for I_c^{tot}, consider a binary model, in which there are only two types of resistors (I_{c1} and I_{c2}) with occupation probabilities, p and $q = 1 - p$, respectively.[3] Equation (12.14) for this model gives

$$I_c^{tot} = pI_{c1} + qI_{c2}. \tag{12.15}$$

Consider $M \times N$ square mesh and suppose there is a current, $I > NI_{c1}$, transported horizontally across the mesh. Divide the resistors into two groups. Those resistors, which dispose perpendicular to the average current flow direction, are called by "row resistors". These rows are connected by what we will call "column resistors", that is, the resistors dispose parallel to the current direction (see Fig. 12.18). The current flow can be considered as a superposition of two current distributions: NI_{c1}, running directly through the mesh (along the column resistors) and a percolating current, $I - NI_{c1}$. Considering the percolating current, we can "subtract off" an I_{c1} resistor from each column resistor (a modified dilute resistor network is shown in Fig. 12.18). The columns of this lattice have holes with probability, p, and $I_{c3} = I_{c2} - I_{c1}$ critical current resistors with probability, $q = 1 - p$. This modified mesh has a critical current, I_{cM}. Then, the total critical current is $I_c^{tot} = I_{c1} + I_{cM}$.

[2] We take a mesh constant of unity to avoid distinguishing between currents and current densities.

[3] Because, it is considered only the critical current, the exact I–V characteristic of these resistors is unimportant.

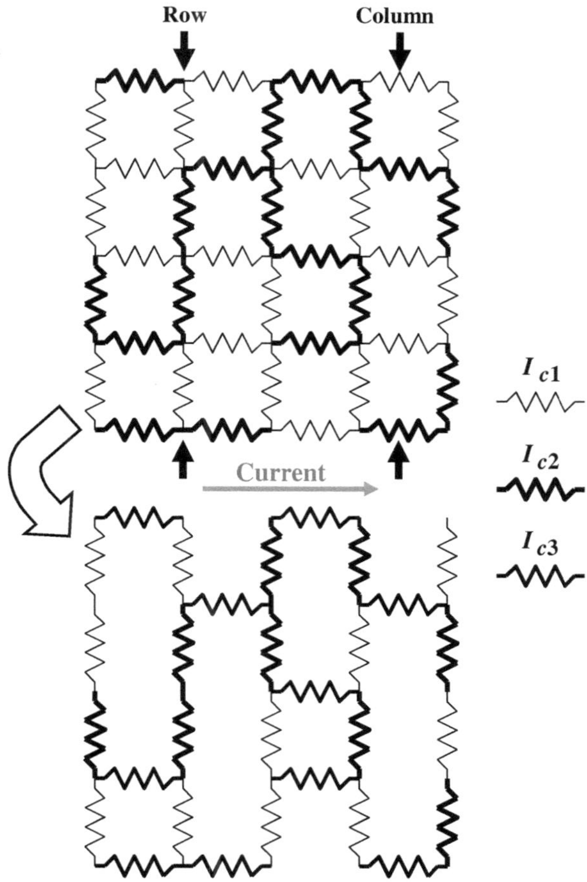

Fig. 12.18 Scheme of two-dimensional network for current flowing to the *right*. Resistors are divided into those *parallel* (column) and *perpendicular* (row) to the current. Then, the actual mesh is reduced to a dilute network, subtracting out a uniform current to the *right* [652]

The value of I_{cM} is found by two factors. First, the amount of current that can be transported across any given column of the mesh, and second, whether or not that current can be re-distributed to the I_{c3} resistors in the next column. There are two various regimes of behavior distinguished by whether I_{c3} is less or greater than I_{c1}.

When $I_{c3} < I_{c1}$ (or $I_{c2} < 2I_{c1}$), it is easy enough to re-distribute the current between columns. Any current flowing through a column resistor can be shunted sideways along a row and re-distributed to the next column as long as another column current does not get in the way. We can approximate the probability that a given column current can be re-distributed along a row without being interfered with in the following fashion. It is equal to the probability that there is an I_{c3} resistor directly across the given column current, plus the sum of probabilities that there is a path at a site s steps along the row, and no paths either coming in or out of the row before that point. By only summing over steps to the right (or left) of the given column current, we avoid the problem of interfering with the paths of other incoming column currents. This re-distribution probability can be written as [652]

$$p_r = q\sum_{s=0}^{\infty}(1-q)^{2s} = \frac{q}{2q-q^2}. \tag{12.16}$$

Taking into account a volume fraction q of I_{c3} row currents, we write down the averaged critical current density as

$$I_c^{tot} = I_{c1} + (I_{c2} - I_{c1})qp_r = I_{c1} + (I_{c2} - I_{c1}) \times \frac{q^2}{2q-q^2}. \tag{12.17}$$

The obtained approximate formula (12.17), in particular, neglects the possibility that the percolation current goes backward. However, we expect this path to make a smaller contribution than the ones calculated, because it corresponds to relatively rare configurations of resistors. In particular, approximation (12.17) should be well for the infinite mesh, if $I_{c3} < I_{c1}$ [652].

When $I_{c3} > I_{c1}$, the presented calculation will fail, because it is no longer easy to shunt the current sideways along a row of resistors and hence the redistribution probability will be different. Similar to above, a different critical current is calculated as [652]

$$I_c^{tot} = I_{c1} + q\left[I_{c1}(p_r - p_r^*) + (I_{c2} - I_{c1})p_r^*\right], \tag{12.18}$$

where

$$p_r^* = q\sum_{s=0}^{\infty}(p^2q)^s = \frac{q}{1-p^2q}. \tag{12.19}$$

Then, assume that the distribution of resistive elements is known for a nonlinear resistor network. A voltage drop (which will be calculated below for the entire mesh) occurs if a current is imposed across the network. The correct distribution of currents in the mesh can be found by solving Kirchoff's equations, which uniquely define the current distribution in the network. In a network of linear resistors, this reduces a set of linear equations to solution, for which there are many standard methods. We are not aware of any general method for nonlinear resistors. The method, developed in [689, 996, 997], can be applied only for the JJ case. The approach for the study of any current densities, which is used below, satisfies current conservation and transport the imposed current across the mesh.

We begin the solution of this problem, choosing a distribution that is thought to be close to the actual one. This choice must conserve current at each mesh node and also transport the imposed current across the lattice. One possibility would be a distribution where the current flows uniformly through the mesh. Obviously, this initial approximation needs to be modified. We do this by superimposing circulation currents on top of the initial distribution as shown in Fig. 12.19. While these circulation currents contribute nothing to the net transport of current they change the current path. Current conservation is obviously satisfied for all values of the

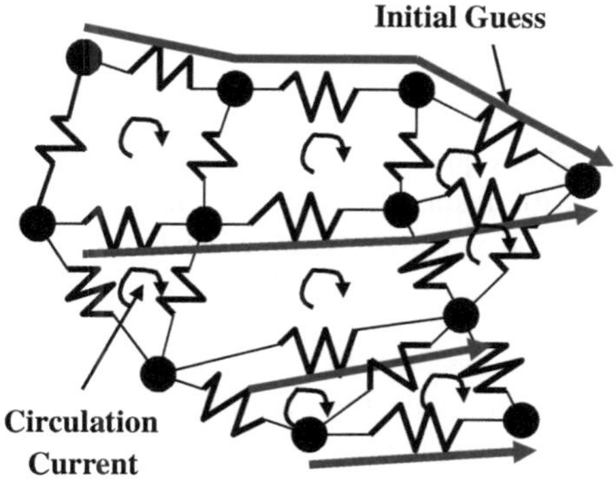

Fig. 12.19 Initial guess for the current distribution and imposition of circulation currents [652]

circulation currents, which we will consider as free variables. Thus, the search method can study all possible currents.

The requirement to calculate the current distribution, satisfying the voltage Kirchoff's law, can be expressed in terms of a minimization principle. If, for a given resistor, we define the quantity

$$g_j = \int_0^{I_j} V_j(i)di, \tag{12.20}$$

where $V_j(i)$ is the voltage-current characteristic of the j-resistor (assuming to be known) and I_j is the current flowing across the j-resistor, then the solution of Kirchoff's laws is equivalent to minimizing

$$G = \sum_j g_j, \tag{12.21}$$

where the sum is over all resistors in the lattice. We prove this by varying G with respect to the currents in such a way that current conservation is preserved. With this aim, find a variation of G with respect to a circulation current:

$$\frac{\delta G}{\delta C_k} = \sum_j V(I_j)\frac{\delta I_j}{\delta C_k} = 0. \tag{12.22}$$

Because the circulation current, C_k, flows around a specific loop, variations in C_k only affect the currents along that loop. An additional effect to be unit line in C_k:

$$\frac{\delta I_j}{\delta C_k} = \begin{cases} 1, & \text{for } I_j \text{ on } k\text{-loop}; \\ 0, & \text{otherwise}, \end{cases} \tag{12.23}$$

where the sign is positive in the direction of C_k. Then,

$$\frac{\delta G}{\delta C_k} = \sum_{j \in \text{loop}} V_j = 0. \tag{12.24}$$

This equation is valid for any loop in the mesh, that is, the minimization of G with respect to the circulation currents is equivalent to the solution of the Kirchoff's voltage law. Note that, in the case of linear resistors, G reduces to representation $G = \frac{1}{2}\sum_j R_j I_j^2$. Thus, for the linear case or any network, where all the voltages have the same power-law dependence upon current, the minimization of G is equivalent to minimization of the power loss of the mesh (a familiar result). An equivalent minimization principle for an applied voltage can be formulated, minimizing the sum:

$$\sum_j h_j = \int_0^{V_j} I_j(v)dv, \tag{12.25}$$

where the sum is over the resistors and the free variables are the voltages at each node [1616].

In order to understand what the effect finite size has on the critical current, consider a binary model on a square lattice. Equation (12.14) states the bounds for an infinite mesh as

$$I_{c1} \le \frac{I_c^{tot}}{N} \le pI_{c1} + qI_{c2}. \tag{12.26}$$

At the same time, this upper bound decreases for a finite mesh. We have for $M \times N$ lattice of nodes a resistor mesh with $M = N - 1$ columns and N resistors in each column (see Fig. 12.18). Investigate possible fluctuations. The occupation probabilities may be p and $q = 1 - p$. However, obviously, there are great fluctuations from average values in a smaller mesh. Calculate the expectation value for the maximum number of I_{c1} resistors in any of the M columns. This value defines the limiting factor for the critical current. In order to find this expectation value, it is necessary to know the probability $P\ (a)$ that the maximum number of I_{c1} resistors in any of the M columns is equal to a (out of N).

First, define $f(a)$ to be the probability that a specific column has a resistors of the I_{c1}-type in it:

$$f(a) = \frac{N!}{a!(N-a)!} p^a (1-p)^{N-a}. \tag{12.27}$$

Then, define $P_M(x,\ a)$ as the probability that x columns (out of M) have a resistors of the I_{c1}-type in them:

$$P_M(x,a) = \frac{M!}{x!(M-x)!} f(a)^x [1 - f(a)]^{M-x}, \tag{12.28}$$

where $0 \leq x \leq M$. Finally, the probability $P\,(a)$ is given as

$$P(a) = \begin{cases} \sum\limits_{x=1}^{M} \left[P_M(x,a) \prod\limits_{i=a+1}^{N} P_{M-x}(0,i) \right], & a \neq N; \\ \sum\limits_{x=1}^{M} P_M(x,a), & a = N. \end{cases} \tag{12.29}$$

The obtained equation states the probability that x columns in M have a resistors of R_1 multiplied by the probability that there are no columns (in the $M - x$ remaining) that have more than a resistors of R_1 in them, summed over x. Then, we use $P(a)$ to calculate the expectation value of a:

$$\overline{a} = \sum_{a=0}^{N} aP(a) / \sum_{a=0}^{N} P(a). \tag{12.30}$$

Finally, we get the critical current for the mesh as

$$I_c^{tot} = \frac{\overline{a} I_{c1} + (N - \overline{a}) I_{c2}}{N}. \tag{12.31}$$

The present analysis is only relevant for small lattices, because for very large meshes a will be very close to p. Thus, the critical current is limited by the finite size of the mesh and in that the critical current for the infinite case would be larger.

The most important result of the computational realizations is that the overall shape of the I–V characteristics for individual elements is preserved when the elements are combined into a common mesh. Figure 12.20 compares 10 × 10 mesh composed of JJ resistors with one of FF ($p = 0.5$, $I_{c1}/I_{c2} = 0.75$ and $R_1/R_2 = 0.75$ in both cases) [652]. Obviously, the resulting overall characteristics are very different. This in itself is not surprising because the individual components have various properties. However, the fact that the overall characteristics should resemble that of the individual elements so closely is surprising. Statistical averaging, even with strong non-linearity, does not wash out the underlying input.

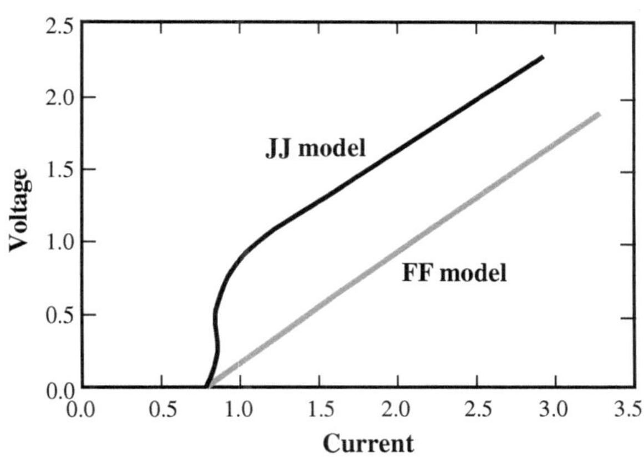

Fig. 12.20 Comparison of I–V characteristics for binary JJ and FF models on 10 × 10 percolated mesh. Both models possess the same distribution of microscopic critical currents and normal state resistances ($p = 0.5$, $I_{c1}/I_{c2} = 0.75$ and $R_1/R_2 = 0.75$). The *shape* of the individual circuit elements I–V characteristics is preserved, when they are combined into a mesh

12.6.2 Simulation of Current Percolation and Magnetic Flux in YBCO Coated Conductors

Despite great progress in texturing of the HTSC layer, grain boundaries remain as the current-limiting factor. It has been shown in tests that dissipation processes have been accompanied by viscous vortex flow through low-angle GBs [697]. Magneto-optical studies of coated conductors (CCs) state the mechanism acting in these materials. Based on this, the model [463] uses the simple, piece-wise linear *I–V* curve of low-angle GBs to calculate the global percolated current transport in CC. Note also that most of the CC simulations are reduced to finding the onset critical currents (I_{c0}) using calculations of "limiting current path" [898, 1456, 1457, 1603].

A method, which enables to calculate current and flux distribution as well as *I–V* curves of two-dimensional grain lattices (square or hexagonal) with arbitrary morphology and GB critical current distribution, is presented in [1603]. Two-dimensional consideration of the current flow in a CC is a good approximation, because the thickness of the HTSC layer is much smaller than the other length scales, and the high degree of sample texture ensures that the current flows essentially within the *ab*-planes. The CC is described in this model as a resistor lattice, namely each GB is represented as a pair of resistors, one of them with zero resistance, but finite current capacity (determined by the GB critical current), and the second one possesses resistance proportional to the flux flow resistivity.[4] The GB critical current is found from the GB misalignment angle using an exponential dependence with a plateau at low angles [1862]:

$$I_c(\alpha) = \begin{cases} TL_{gb}J_{c0}\exp[-(\alpha-\alpha_c)/\alpha_0], & \text{for } \alpha \geq \alpha_c; \\ TL_{gb}J_{c0}, & \text{for } \alpha \leq \alpha_c, \end{cases} \quad (12.32)$$

where L_{gb} is the length of the GB; α_c and α_0 are the critical angles.

Analytically, this problem is reduced to a solution of a set of *linear* equations and inequalities for the currents. This set includes the current-limiting equations for the superconducting resistors, the Kirchoff's equations for each grain and each current loop and also equation for the total current in the considered system. This system of equations is under-determined in all non-trivial cases, that is, there are fewer equations than variables. It is solved, simultaneously minimizing the total dissipation in the lattice for a given current. This type of mathematical problem is well known in economic analysis and can be solved by standard methods of linear optimization [640, 1220].

The considered method enables to analyze both experimentally measured grain morphologies and model structures and sets of hexagons or squares. The former can be potentially used to predict the performance of a CC from the morphology of

[4] A similar approach is usually used in lattice analysis, approximating non-linear transport properties [777, 1358].

its substrate. At the same time, simple model structures are very valuable for fundamental studies of CC properties [1456, 1457]. As a result of experimental analysis of the grain distributions [1862], there is a possibility of the branching of flux channels and also three scaling regimes of *I–V* characteristics are identified, namely (i) the onset superconducting current, I_c, is dominated by finite- size scaling; (ii) the intermediate currents are stated by percolation (or power-law) scaling; and (iii) the high currents are described by the linear, Ohmic scaling.

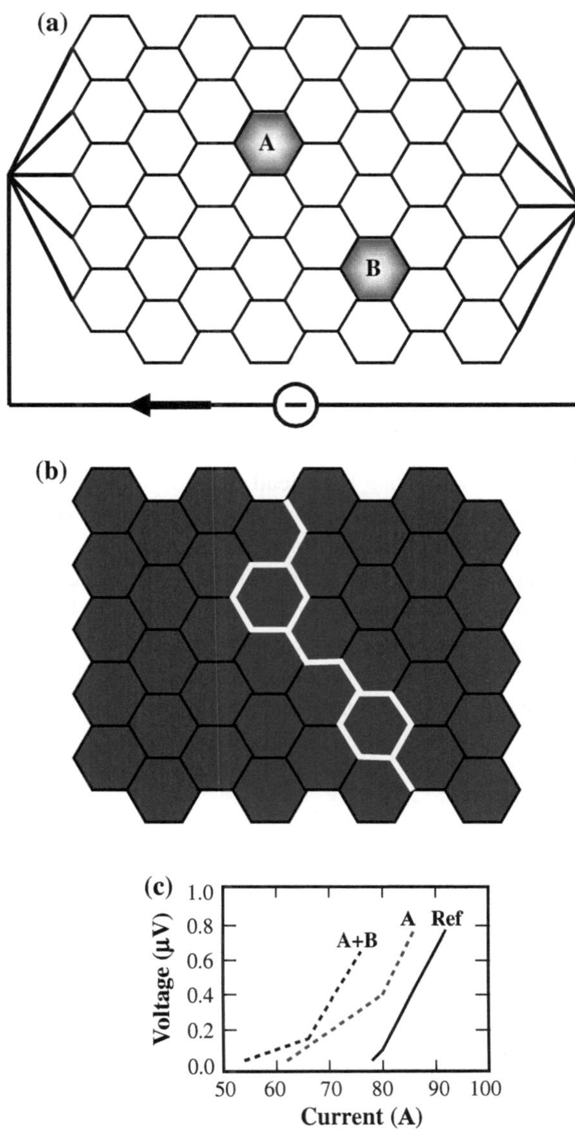

Fig. 12.21 (**a**) Hexagonal model lattice with GB angles between 2 and 3°, except for the grains *A* and *B*, having 45° misorientation; (**b**) flux distribution in the same lattice; and (**c**) *I*–*V* characteristic for the same lattice, labeled "*A* + *B*." Moreover, the curves are shown with only one highly misaligned grain (*A*) and in the case of no misaligned grains (Ref) [1862]

Considerable interest presents a consideration of small model systems with arbitrary distribution of the individual GB misorientation angles or GB critical currents. Figure 12.21 shows an example, where the general grain alignment is determined to be very high (all GB angles <3°), with the existence of two grains (*A* and *B*) with misalignment angles of 45°. This configuration leads to a current distribution with the currents flowing around the misaligned grains.

When the current exceeds the critical current, the first flux line enters the conductor along the misaligned grains (see Fig. 12.21b). The path of the first flux line is also the limiting cross-section for the current flow. In an "undisturbed" lattice of the same topology, the limiting cross-section would be nine GB segments across the conductor. In the case, presented here, the limiting cross-section is reduced to six segments, if one disregards the GB segments of the highly misoriented grains. Accordingly, one expects that the critical current (i.e., the onset current, defined here for non-zero voltage) is decreased by a third.

The *I–V* curves in Fig. 12.21 show that indeed the critical current diminishes from 78 to 54 A, that is, by 32 %. One can also see that the *I–V* characteristics show pronounced kinking, similar to that found for low-angle GB [697]. This kink indicates the onset of a new flux channel through the conductor or alternatively the branching of an existing channel. The effect of the misaligned grains on the value of I_c is quite drastic: 5 % of highly misaligned grains (two grains from 40) lead to 32 % reduction of the critical current. The results of analogous simulation for the lattice with 25 × 50 hexagonal grains are presented in Fig. 12.22 [1862]. In this case, the grain orientation angles have been assigned in a statistical manner, using two Gaussian distributions, centered about 0 and 45°, with their full-width half-maximum set at 5°. They lead to two practically separated percolation systems. The insert in Fig. 12.22 shows that in this large system the critical current strongly decreases even at a small percentage of misaligned grains.

Thus, the *I–V* characteristics in large lattice obey a parameter-independent scaling law typical for percolation processes. In small aggregates, one can find piece-wise linear *I–V* curves, kinked at a point where new flux channels enter the conductor or existing channels branch out. In this case, highly misoriented grains have a strong detrimental effect on the critical currents.

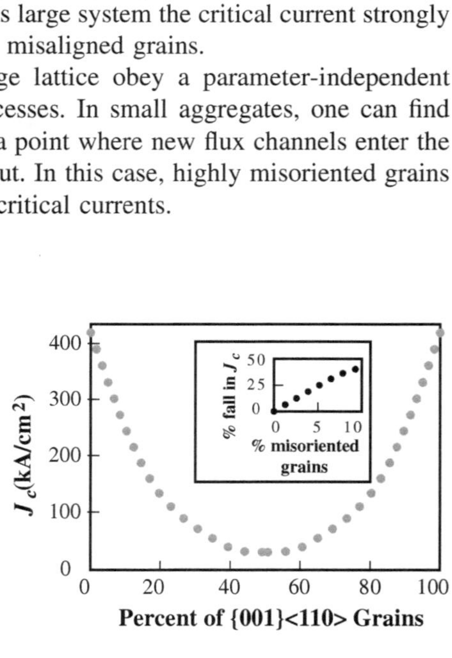

Fig. 12.22 Critical current on lattice of 25 × 50 hexagonal grains in dependence of the percentage of grains with about 45°-misalignment

12.6.3 Modeling of Electromagnetic Properties of BSCCO/Ag Tapes

Based on the actual microstructure of BSCCO/Ag tape, in [1338] the superconducting core is modeled as a stack of relatively well-aligned plate-like grains whose *ab*-planes are approximately parallel to the magnetic field, taking into account the limiting of the intergrain current flow by *c*-axis transport through large-area twist boundaries [203], or by a lattice of low-angle "railway switch" connections [280, 669], or by some other factors. The principal consequence of the planar geometry is the current forcing in the direction perpendicular to magnetic field $\boldsymbol{H}$ (see Fig. 10.23). In this case, the magnetization currents tend to flow in a plane perpendicular to $\boldsymbol{H}_a$ (field parallel to the one of the major axis of the slab and plate-like grains), so that the distribution $\boldsymbol{J}(\boldsymbol{r})$ is effectively two-dimensional, and $\boldsymbol{H}(\boldsymbol{r})$ has only one component $H_z(x, y)$.

Consider the plate-like grain structure, presented in Fig. 12.23, first neglecting the non-superconducting second-phase precipitates and Ag/BSCCO interface irregularities. For the field orientation, shown in Fig. 12.23, the magnetization currents flow macroscopically along the tape, but they must also cross the *ab*-planes at some point in order to permit the current to loop at the end of the tape and to create flow around local barriers. In any case, the intergrain current mostly flows in the plane perpendicular to $\boldsymbol{H}$, either through the large-area GBs normal to the *c*-axis or through small-area low-angle GBs, which separate slightly misoriented *ab*-planes (Fig. 12.23). In this case, the two components $J_x(x, y)$ and $J_y(x, y)$ are related to $H_z(x, y)$ as

$$J_x(x,y) = \frac{\partial H_z}{\partial y};\ J_y(x,y) = -\frac{\partial H_z}{\partial x}. \tag{12.33}$$

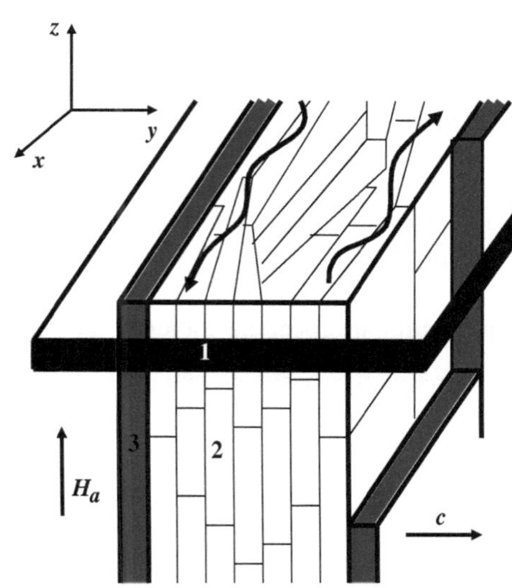

Fig. 12.23 Geometry of sample and magneto-optical indicator film (*1*). Sketched also is the model grain structure of BSCCO core of the "brick wall" type (*2*), covered by silver sheath (*3*). The current streamlines are shown by *thick arrows* for this planar geometry [1338]

The local J_c (x, y) values are found as $J_c = (J_x^2 + J_y^2)^{1/2}$, which yields

$$J_c(x,y) = \sqrt{\left(\frac{\partial H_z}{\partial x}\right)^2 + \left(\frac{\partial H_z}{\partial y}\right)^2}. \qquad (12.34)$$

Therefore, for the specific geometry of magneto-optical (MO) measurements, the current distribution is effectively two-dimensional, permitting a reconstruction of $\boldsymbol{J}(x, y)$ from a real measurement of the $H_z(x, y)$ distribution.

The characteristic spatial scales on which the above two-dimensional scheme can be applied need some discussion. Obviously, second-phase precipitates, intergrowths, cracks, weakly coupled grains and irregularities of the Ag/BSCCO interface can all lead to significant deviations of the local $\boldsymbol{J}(\boldsymbol{r})$ from the average current direction (the vector $\boldsymbol{J}(\boldsymbol{r})$), generally having both perpendicular ($J_\perp$) and parallel ($J||$) components with respect to the local direction of $\boldsymbol{H}(\boldsymbol{r})$. As a result, the field $\boldsymbol{H}(\boldsymbol{r})$ can acquire a tangential component, becoming dependent on z-coordinate, a feature, which is not taken into account by the two-dimensional model. The characteristic size of L of these heterogeneities is estimated by the secondary phase and grain sizes. So, these three-dimensional localized field disturbances strongly decay over a distance larger than L. However, if it is necessary to study the current distribution over macroscopic scale (for $l > L$), the MO image can be averaged over the length of L. After such averaging, any information about current loops smaller than L and random tangential fluctuations of $\boldsymbol{H}(\boldsymbol{r})$ will be lost. At the same time, the macroscopic heterogeneities of $\boldsymbol{J}(\boldsymbol{r})$ over spatial scales larger than L can be extracted from the MO images, using the two-dimensional Eqs. (12.33) and (12.34).

Then, it may be supposed that the transport critical current density (J_c) in the Bi-2223/Ag tapes is limited by weak links at GBs. As it has been noted, supercurrent usually demonstrates two-dimensional behavior. At a grain boundary, weak link breaks from superconducting to non-superconducting state when current density and/or applied magnetic field are sufficiently high. An electric field is generated locally when all weak links break at a whole cross-section of the column perpendicular to the current flow axis. This behavior can be treated as percolation phenomenon based on the stochastic treatment.

For simplicity, consider a single grain as an element in the system in which grains are linked by GBs. Each element has coordination number Z defined by adjacent grains. As it is shown in Fig. 12.24, the system consists of M elements in each column and N elements in each row (here $Z = 4$) [1279]. The stochastic computational procedure includes the following stages:

(1) Among the basic elements there select V % in superconducting state and $(100 - V)$ % in non-superconducting state. All elements are connected by weak links.
(2) When the current applied to the system increases, the breaking of P % weak links occurs (the broken links are selected randomly in the system) and they

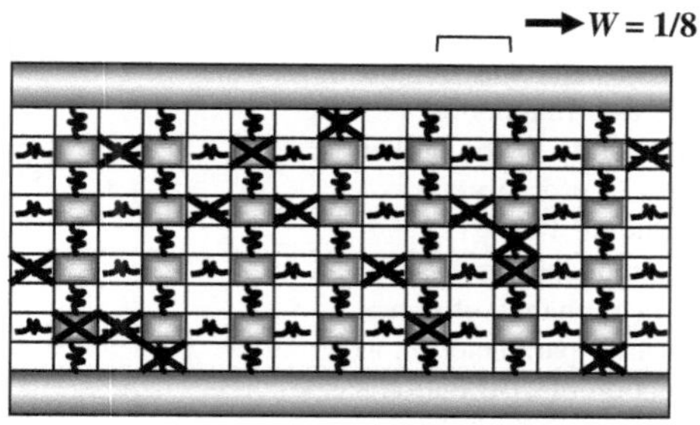

Fig. 12.24 System consisting of M elements in the column and N ones in the row. Each element is connected by weak links with coordination number $Z = 4$

transit in normal state when the current density, crossing weak links, exceeds a critical value, J_{cw}.

(3) Supercurrents cross preferentially the superconducting links. When a sufficiently great number of weak links break, all links, included in a column, are in non-superconducting state. When the current crosses this normal column, an electrical field is generated. The failure of P % from total weak links leads to transition in normal state of W % columns from total number.

(4) In order to consider the effects of grain size, connectivity among grains, etc., a parameter C is introduced, defining the minimum size of elements for keeping of superconducting state.

(5) Assuming Weibull's distribution for estimation of strength of the weak links, the probability P, by which the weak links break, is expressed as a function of current density, crossing the respective weak link:

$$P = k(J - J_{cw}^{\min})^m, \tag{12.35}$$

where m is the distribution shape factor. Hence, we have:

$$J = J_{cw}^{\min} + (P/k)^{1/m}. \tag{12.36}$$

For simplicity, it is assumed that $J_{cw}^{\min} = 0$.

(6) The fraction W of broken columns can be expressed through Weibull's function $f(J_{cw})$ as

$$W(J) = \int_0^J f(J_{cw})dJ_{cw}. \tag{12.37}$$

On the other hand, the derivative of electric field is given as

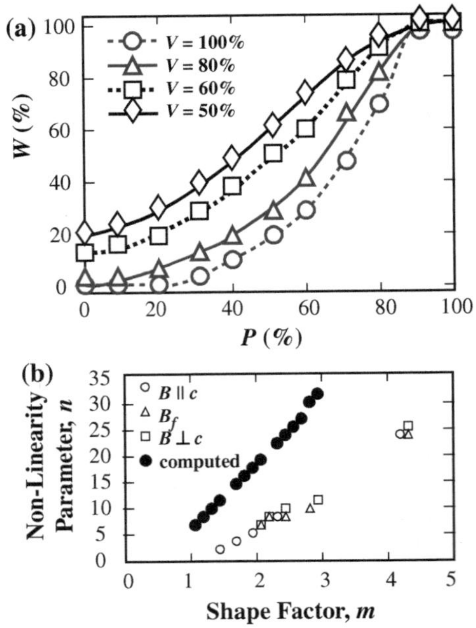

Fig. 12.25 (**a**) Change of degree of broken columns (*W*) as a function of probability of the broken links (*P*) for system, including *V* % superconducting elements, and (**b**) change of non-linearity parameter, *n*, as a function of distribution shape factor, *m* [1279]. The computational results (*black circles*) are compared with experimental data (*white figures*) [1417]

$$\frac{dE}{dJ} = R_f \int_0^J f(J_{cw}) dJ_{cw}, \tag{12.38}$$

where R_f is the resistance for the uniform flux flow. By integrating $W(J)$ with respect to J, the electric field is expressed as

$$E = R_f \int_0^J W(J) dJ. \tag{12.39}$$

Hence, a correlation of the parameters (P, W) with the macroscopic quantities (J, E) is followed.

Figure 12.25a shows W in dependence on P for $M = 400$, $N = 1000$, $C = 2$ and variation of V from 50 to 100 %. In the case of $V = 100$ %, the fraction W of broken columns starts to increase for $P > 20$ %, that means a generation of an electric field. For $V = 60$ %, W has already finite value even for $P = 0$. After calculation of E–J characteristic, so-called non-linearity parameter, n, is found. As it is shown in Fig. 12.25b, the dependence of n on m, calculated in the considered model, has a similarity with the experimental results [1417].

Other approach, based on the statistical distribution of critical current density and used for study of non-linear E–J characteristic, is developed in [1158]. According to this model, the de-pinning probability function (Q) can be described as a function of current density (J) by the next expression:

$$Q(J) = [(J - J_{cm})/J_0]^m, \tag{12.40}$$

where J_{cm} is the percolation threshold identical to the minimum value of J_c; J_0 is the scale parameter, representing half-width of the statistic J_c distribution; and m is the distribution shape parameter. Then, the *E–J* characteristic can be obtained integrating *Q*(*J*) as

$$E(J) = \rho_{FF} \int_0^J [(J_c - J_{cm})/J_0]^m dJ_c, \tag{12.41}$$

where ρ_{FF} is the resistivity for the uniform flux flow [1822]. In particular, this model has permitted to estimate influence of bending strain on transport *E–J* characteristics for Bi-2223 Ag/AgMg-alloy-sheathed tape [877]. Moreover, this approach permits to separate the components, caused by the grain connectivity and flux pinning, originated from the bending strain.

12.6.4 Aging at Mechanical Loading

In design of energetic HTSC cables, an estimation of electric properties of the current-carrying elements presents the significant interest at mechanical aging of the sample, that is, at long cyclic and static loads. A probabilistic approach is one of the main methods for the solution of this problem [1182, 1183]. The *V–I* characteristic can be described, using the Weibull's distribution, and the time behavior of the relevant parameters can be considered during mechanical aging of the cable material, for example, under tensile stress and bending vibration loads. In order to model the sample behavior, a probability function, $g(J_c)$, is introduced, which describes the probability that a cross-section of a superconductor has a given value of the critical current density, J_c [138]. It has been shown that $g(J_c)$ could be analytically derived as the second derivative of the *E–J* characteristic via flux-flow resistivity. This relationship holds, however, for a pure superconductor of type II, that is, neglecting effects of any non-superconducting matrix. Determination of second derivative of *E–J* characteristic can be found numerically [1776]. A best-fitting distribution must be chosen in order to obtain parameters, which are able to describe the function $g(J_c)$, as well as its integral, that corresponds to the cumulative probability, $G(J_c)$. The probability function considered below is the two-parametric Weibull's distribution, which fits both skewed and non-skewed probability density characteristics. The expressions of the two-parametric density and cumulative-probability Weibull's function are found by [844]

$$F(x) = 1 - \exp[-(x/\alpha)^\beta]; \tag{12.42}$$

$$f(x) = (\beta/\alpha)(x/\alpha)^{\beta-1}\exp[-(x/\alpha)^\beta], \tag{12.43}$$

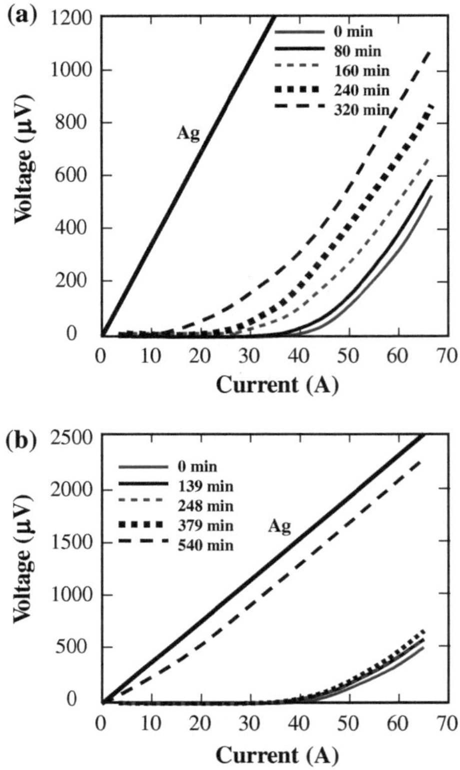

Fig. 12.26 *V–I* characteristics at various aging time for cases of vibration aging (**a**) and tensile-stress aging (**b**) [1183]

where α and β are scale and shape parameters, whereas the random variable x corresponds to the critical current density J_c.

The Weibull's parameters can be estimated in several ways [213]. In particular, they can be calculated on the base of test data, obtained by the prime derivative of the *E–J* characteristic, that is, $G(J_c)$, numerically performed and plotted in the coordinate system $\lg\{-\ln[1-F(x)]\}$ in dependence on $\lg(x)$, which linearizes (12.42), transforming it into a straight line of slope β.

Actually, this probabilistic approach should be only applied to data relevant to the superconductor, while the tested samples have a silver shell, whose contribution is not easily separable from that of the superconductor body. Therefore, the Weibull's function can be used only to the experimentally obtained *V–I* characteristics with the aim to show how it is correlated with mechanical stress aging.

Examples of *V–I* characteristics, measured in aged and non-aged specimens are shown in Fig. 12.26 for vibrated and tensile-stressed samples. The silver shell *V–I* characteristic is also presented in Fig. 12.26 for the sake of comparison. As it

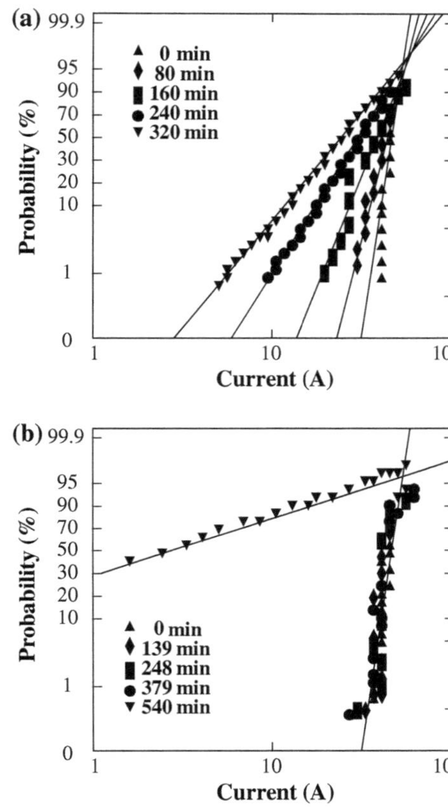

Fig. 12.27 Weibull's plots derived from normalized prime derivative at different aging times for cases of vibration aging (**a**) and tensile-stress aging (**b**)

is followed from this, the mechanical aging process can significantly affect *V*–*I* characteristics. Moreover, the *V*–*I* curves tend to the silver resistance for values of currents monotonously decreasing as aging time increases.

Tensile stress and vibration make aging specimens in a different way. While the former creates local damage cracks only, the latter leads to overall degradation, when silver shell is no more able to endure the applied stress. This approach permits to obtain the Weibull's functions for cases of cyclic and tensile loads (see Fig. 12.27 [1183]) with the normalization factor, defined by the silver characteristics. Moreover, the scale and shape parameters (α, β) of the Weibull's distribution can also be estimated and plotted as a function of aging time (see Fig. 12.28 [1183]). Figures 12.26, 12.27, 12.28 permit to estimate effects of the aging level and type on electrical properties of superconductor. So, the critical current (I_c) at moderately high and low cumulative probability significantly decreases as aging time increases, sharply in the case of tensile stress, progressively for vibration. Similarly, the value of β significantly diminishes with time, reaching the lowest values for the most aged samples.

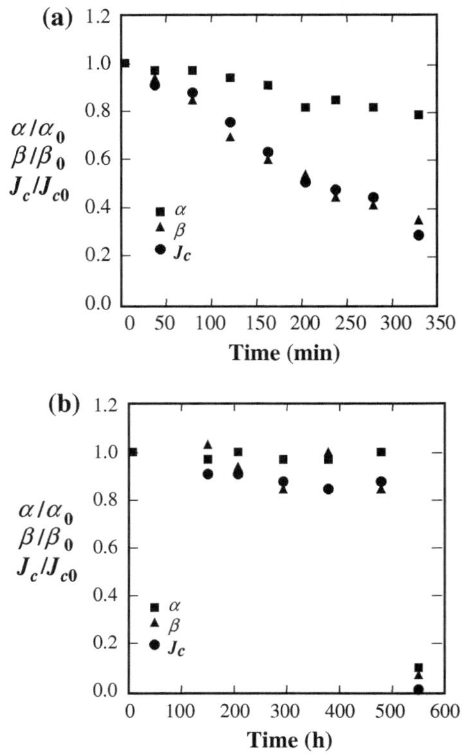

Fig. 12.28 Relative values of scale (α) and shape (β) parameters of the Weibull's distribution and also of critical current density (J_c) *vs* time (t), for one sample in the case of aging due to vibration (**a**) and tensile stress (**b**)

12.6.5 Effective Electrical Conductivity of Superconducting Oxide Systems

Finally, based on the percolation theory, we consider current carrying in the model HTSC systems of YBCO and BSCCO, investigated in Chap. 10, and also study corresponding effective characteristics. Comparison of the results of fracture resistance alteration due to action of different toughening mechanisms, with qualitative features of electrical conductivity, stated in this section permits to estimate an effect (positive or negative) of microstructure features on properties of the considered superconductors.

12.6.5.1 YBCO Ceramic

It is known that electrical conductivity (the inverse value of resistivity) in non-ordered media is proportional to the self-diffusion factor, and hence, the average quadratic deviation of the liquid particles in absence of an external force [559]. Following [1300], consider the YBCO model structure as a percolating cluster with cells, occupied by grains, and by free cells, corresponding to voids.

Obviously, the percolating (or conducting) properties decrease due to the existence of intergranular microcracks and porosity. However, it is also clear that all model structures considered below, stated in Chap. 10, possess a joining, percolating cluster, because the following inequality always fulfills:

$$C_p + f_b \ll p_c, \tag{12.44}$$

where $C_p = N_p/N$ is the closed porosity of the ceramics (N_p is the cell number, occupied by voids and N is the total cell number); $f_b = l_g/l_l$ is the ratio of the cracked facets to the total number of boundaries between cells of the joining cluster (obviously, $f_b < f_m = l_g/l_i$, where l_i is the total length of the intergranular boundaries, because $l_i < l_l$); and $p_c = 0.5927$ is the percolation threshold for a square lattice.[5]

In order to estimate effective electrical conductivity of the model structures (considered in Sects. 10.1.1 and 10.1.2), we modify a known algorithm, called *ant into maze*, applied for diffusion description in irregular media [559]. Take into account together with crystallite phase and voids, an existence of grain boundary microcracks, and also GBs, possessing a smaller conductivity compared to intracrystalline space. Consider a chance movement on only occupied cells (on crystallite phase) of the percolating cluster. At any time, the probable numbers, $p_k \in [0, 1]$ (where $k = 1,\ldots,4$) are generated in all cells, which are adjacent to the main cell. In the case, when the cell under consideration is separated from the main cell by a GB then its probable number is decreased by 0.1 (to design the predominant cluster growth within a grain). If an intergranular boundary has been replaced by a microcrack or the adjacent cell is a void, then the corresponding probable number is found to be zero. The cluster growth results from the occupation of a cell with largest possible number, $p_k \geq p_c$. The cluster growth is impossible when the initial cell is surrounded by voids, microcracks or when all of $p_k < p_c$. Next, all the process is repeated. At each step, including a marking time, the value of t increases by 1. In the time, $t = 0$, some chance cell of the joining cluster begins to move. In time t, the square of distance between initial and final sites is calculated. Then, the modeling is repeated several times, and the mean quadratic displacement, R, associated with the conductivity of the modeled structure is calculated. An example of the cluster growth is shown in Fig. 12.29.

Statistically reliable results are obtained by applying the stereological approach. We obtain, again a necessary number of realizations for the statistical process to find non-skewed estimation of the considered stereological characteristic as [271]

$$n = (200/y)(\sigma_x/\overline{x}), \tag{12.45}$$

[5] Similar modeling can be fulfilled at other regular lattices having the following values of the percolation threshold, p_c, and the shape of unit cell [1697]: 0.6970 (hexagon), 0.5 (triangle), 0.4299 (diamond), 0.3116 (cube), 0.2464 (b.c.c.), and 0.1980 (f.c.c.).

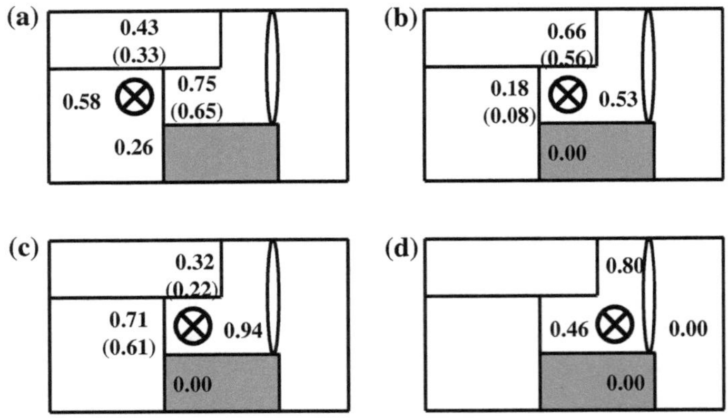

Fig. 12.29 An example of percolating cluster growing in YBCO ceramic. *Numbers* denote probabilities of cell occupation (in *brackets* there are considered probabilities, taking into account a priority, which are compared with percolation threshold, $p_c = 0.5927$). A *cross* corresponds to the cluster cell at previous step. Microcracks, formed during cooling, are shown at GBs; porosity is denoted by *gray color*

where y is the accuracy level (in this case 10 %); σ_x is the variance; $\bar{x}$ is the mean value of the stereological characteristic. At the study of the electric conductivity process, first a growth of a joining cluster on the type of occupying percolation [559] in two-dimensional lattice (25 × 40 cells), forming a path from left to right of its boundary, is considered. As $\bar{x}$, we select the lattice width (=40), and the value of σ_x is calculated by corresponding exceeding of the cluster cell number over lattice width.

The percolation properties of the model microstructures, estimated for $t = 100$, are presented in Table 12.1. The normalized mean quadratic displacement, R/R_0 (where R_0 is the corresponding value for non-porous material), which is proportional to electrical conductivity decreases with growth of initial porosity as well as the square (cell number) of the joining cluster, S/S_0 (S_0 is the cluster area for nonporous material). These data correlate with increasing GBs, l_i/l_l, at the enhancement of the initial porosity, C_p^0, and decreasing of mean grain size, obtained in [1319]. Thus, introducing even small priority for growth of the joining cluster within grain compared to its propagation in adjacent one (in this case, it corresponds to enhancement of probability by 0.1) correlates with experimental data and leads to qualitative confirmation of known results by the results of computer simulation. The presented results point again an important role of GBs

Table 12.1 Some model parameters for YBCO ceramics of various structures

Properties	C_p^0= 0 %	C_p^0= 20 %	C_p^0 = 40 %	C_p^0= 60 %
l_i/l_l	0.503	0.523	0.557	0.630
S/S_0	1.000	0.956	0.872	0.644
R/R_0	1.000	0.866	0.846	0.785

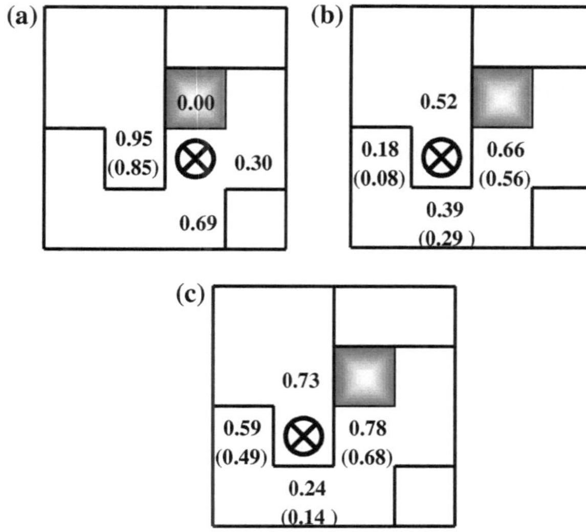

Fig. 12.30 An example of percolating cluster growing in large-grain YBCO sample. *Numbers* denote probabilities of cell occupation (in *brackets* there are considered probabilities, taking into account a priority). A *cross* corresponds to the cluster cell at previous step. *Gray color* shows 211 particle

and need to obtain coarse-granular YBCO structures. More accuracy estimations can be obtained in the framework of this model under condition of quantitative description of the difference of conductive properties in grain and through GB. Obviously, they will depend on features of domain structure of single crystallites, GBs continuity, secondary phases and so on.

12.6.5.2 Large-Grain Melt-Processed YBCO

Taking into account the model structures of the melt-textured large-grain YBCO samples, investigated in Sect. 10.3, it is obvious that their transport properties are found by the existence of non-superconducting (211) and superconducting (123) phases, and also by GBs. Then, in the electrical conductivity model, presented in the previous section, only one position is added, namely in the case of the adjacent cell occupation by the 211 particle, the probable number is replaced by zero. An example of the cluster growth is shown in Fig. 12.30, but the obtained numerical results are presented in Table 12.2 [1327].

A comparison of the obtained effective characteristics with the model results of Sect. 10.3 (see Table 10.8) shows a correlation between the mean pinned grain

Table 12.2 Numerical results

Property	$f = 0.0$ $S_S \approx 2S_0^m$	$f = 0.0$ $S_S \approx 3S_0^m$	$f = 0.1$ $S_S = 0$	$f = 0.1$ $S_S \approx 2S_0^m$	$f = 0.1$ $S_S \approx 3S_0^m$	$f = 0.2$ $S_S = 0$	$f = 0.2$ $S_S \approx 2S_0^m$	$f = 0.2$ $S_S \approx 3S_0^m$
l_i/l_l	0.325	0.235	0.133	0.116	0.111	0.136	0.099	0.083
R/R_0	0.803	1.000	0.649	0.816	0.865	0.452	0.473	0.544
S/S_0	0.979	1.000	0.929	0.957	0.979	0.893	0.900	0.921

area, $\overline{S}$, superconducting field per grain, $\overline{S}_{SR}$ and ratio of the total intergranular boundary length to the total number of GBs between cells of joining cluster, l_i/l_l, on one hand, and change of the seed area, on the other hand.

The percolation properties have been calculated for $t = 100$. The behavior of the mean quadratic displacement, R/R_0, and the joining cluster area, S/S_0 (where R_0 and S_0 are the corresponding values for the case, $f = 0.0$ and $S_S \approx 3S_0^m$), as well as the change in the value of $\overline{S}_{SR}$ coincide with experimental data [545, 1038]. We confirm that YBCO conducting properties decrease with increase in the fraction of the 211 normal particles and with decrease in the seed. At the same time, the considered toughening mechanisms intensify (that in turn, influences indirectly an increasing of current-carrying properties) with an increase of the 211 normal particle size and fraction into 123 superconducting matrix. Finally, note that for completeness of investigation of the percolation properties in this case, it is necessary to take into account the GB microcracking during cooling and misorientaion of grain boundaries, which are caused by indirect effects of dispersed phase and also contribute in final superconducting properties.

12.6.5.3 Effect of Microstructure Dissimilitude

The transport properties of the YBCO model structures, considered in Sect. 10.5, are caused by superconducting granular phase, voids, microcracks and GBs. Again, based on the algorithm *ant into maze* [559], effective characteristics can be calculated, which are presented in Table 12.3 [1324]. An example of the cluster growth is shown in Fig. 12.31.

The numerical results show a correlation between the grain area and the ratio of total intergranular boundary length to total number of boundaries between cells of the joining cluster, l_i/l_l. The percolation properties have been calculated for $t = 100$. The behavior of the mean quadratic displacement, R/R_0, and the joining cluster square, S/S_0 (where R_0 and S_0 are the corresponding values for the case $v = 10$ °C/h), as well as a change of the value of l_i/l_l coincide with test data [1040]. Thus, it is confirmed that YBCO superconducting properties diminish with increased heating rate. This corresponds also to the trends in the changes of the specific fracture energy ($\gamma_F^{(1)}/\gamma_F^{(0)}$) and the microstructure dissimilitude ($\Delta K_{Ic}/\Delta K_c$), following from Fig. 10.41.

Table 12.3 Numerical results

v, °C/h	l_i/l_l	R/R_0	S/S_0
10	0.51	1.00	1.00
20	0.50	0.91	0.87
30	0.39	0.81	0.74

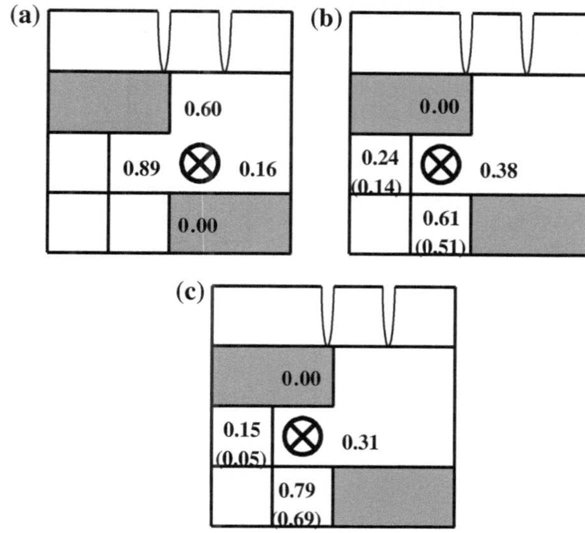

Fig. 12.31 An example of propagation of percolating cluster in the case of microstructure dissimilitude. *Numbers* denote probabilities of cell occupation (in *brackets* there are considered probabilities). A *cross* corresponds to the cluster cell at previous step. *Gray color* shows voids; microcracks are presented at GBs

12.6.5.4 BSCCO Bulks

Considering the Bi-2223 model structure as percolating cluster with cells, occupied by grains, note that the percolation properties are found in this case by GBs, microcracks and texture, describing a connectivity of adjacent grains. The modification of the *ant into maze* algorithm for this example is the following. When the considered cell is separated from the main one by an intercrystalline boundary and the grains have the same orientation, then its probable number is decreased by 0.1 (for designation of the predominant cluster growth within the grain). If these grains have different orientations, then an additional generated probable number is multiplied by the corresponding value of the considered cell, so taking into account the grain connectivity. An example of the cluster growth is shown in Fig. 12.32 and the obtained results are presented in Table 12.4 [933].

The percolation properties have been calculated for $t = 100$. The computed relative increasing of the mean quadratic displacement, R/R_0, and the joining cluster area, S/S_0 (where R_0 and S_0 are corresponding values for the case $D/\delta = 2.40$), together with a decreasing of total GBs length (l_i/l_i^0) in the case of the larger grain growth coincide with test data (e.g., for large-grain melt-processed HTSCs, considered in detail in Sect. 4.4). A comparison of the obtained results with estimations of the strength properties, presented in Sect. 10.7, shows that the denser lattice of the intergranular boundaries and, especially, the misorientation of adjacent grains affect the degradation of transport properties in a greater degree in the case of fine-grain ceramic compared to some increased microcracking of larger-grain structures. Thus, the greater toughening of the Bi-2223 bulks by ductile Ag particles (a fine-grain case) and the increased conductivity (a large-grain case)

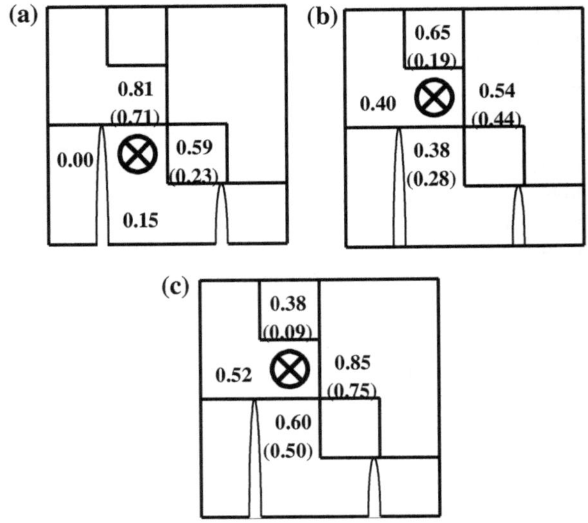

Fig. 12.32 An example of percolating cluster growing in BSCCO bulk. *Numbers* denote probabilities of cell occupation (in *brackets* there are considered probabilities). A *cross* corresponds to the cluster cell at previous step. Microcracks are presented at GBs

Table 12.4 Numerical results

D/δ	l_i^0/l_i	R/R_0	S/S_0
2.40	1.00	1.00	1.00
2.84	0.86	1.85	1.52

demand optimization of both contributions in the estimation of the HTSC properties.

Block-scheme of computational process describing growth of the percolation cluster with estimation of effective characteristic electric conductivity of above four model structures is presented in Fig. D.8. A simulation veracity of the computational algorithm has been found by direct verification on test examples.

Appendix A
Classification of Superconductors

Some main properties (including superconducting ones) of different chemical elements are shown in color inset. Here, a classification of known superconducting compounds is carried out. One includes 14 types of compounds, namely organic superconductors, carbon superconductors, *A*-15 compounds, magnetic superconductors, heavy fermions, oxides without copper, pyrochlore oxides, rutheno-cuprates, high-temperature superconductors, chalcogens, superconductive pnictides and chalcogenides of iron, rare-earth borocarbides, silicon superconductors, MgB_2 and related compounds. The last section is devoted to room-temperature superconductivity.

Organic Superconductors

An idea of realization of organic superconductors arose with postulate of Little [1029] on that what superconductivity organized at interaction of organic column with side molecular chains could take place in organic macro-molecules. On the base of more early predictions of London [1044], Little suggested that the critical temperature of superconductive transition in organic materials should be near the room temperature and even discussed proper biological systems.

Tetrathiafulvalene–Tetracyanoquinonedimethane (TTF–TCNQ) was the first organic superconductor studied in 1973 by Coleman et al. [301]. One-dimensional polymer chains in TTF–TCNQ are built of alternately arranged cyclic molecules of TTF and TCNQ. In solid form with aligned chains this organic material demonstrates metallic conductivity along the direction parallel to the chains at room temperature. TTF–TCNQ is an example of a charge transfer compound (salt) where TTF serves as the cation and TCNQ the anion, its $T_c < 2$ K.

Superconductivity in quasi-one-dimensional (Q1D) organic conductors $(TMTSF)_2PF_6$ has been discovered by Jerome et al. [778]. The molecule of tetramethyltetreselenfulvalne (TMTSF) was synthesized by Bechgaard et al. [114], and used as cation in the charge transfer crystals $(TMTSF)_2X$, where the role of

DOI: 10.1007/978-3-642-34441-1, © Springer-Verlag Berlin Heidelberg 2012

Table A.1 The properties of several organo-metallic compounds with BT = BEDT-TTF [BEDT = bis(ethylenedithio), TTF = tetrathiafulvalene]

No.	Formula	Phase	Pressure (kbar)	T_c (K)	Remarks
1	$(BT)_2ReO_4$	–	4	2.0	
2	$(BT)_2I_3$	α	0	3.3	I_2 doped
3	$(BT)_2I_3$	β_L	0	1.5	
4	$(BT)_2I_3$	β_L	0	2.0	Annealed
5	$(BT)_2I_3$	β_H	0.5	8.1	
6	$(BT)_2I_3$	α_t	0	8.1	
7	$(BT)_2IBr_2$	β	0	2.2–3.0	
8	$(BT)_2AuI_2$	β	0	3.4–5.0	
9	$(BT)_2I_{3-2.5}$	γ	0	2.5	
10	$(BT)_2I_3$	θ	0	3.6	
11	$(BT)_2I_3$	k	0	3.6	
12	$(BT)_4Hg_{2.89}Br_8$	–	0	4.0	
13	$(BT)_4Hg_{2.89}Cl_8$	–	12	1.8	
14	$(BT)_3Cl_2(H_2O)_2$	–	16	2.0	
15	$(BT)_2Cu(NCS)_2$	k	0	10.4	
16	$(BT)_2Cu[N(CN)_2]Br$	k	0	11.6	

anions play small groups of atoms $\chi = PF_6$, ClO_4, SCN and similar to them. In 1983, Parkin et al. discovered superconductivity in quasi-two-dimensional (Q2D) organic conductors (BEDT-TTF)$_4$(ReO_4)$_2$.

The organometallic compounds have double bonds and can provide electrons to form Cooper pairs. There are vibration modes so that a phonon-induced mechanism is feasible. Moreover, the electrons may become attractive due to strong correlations, in which a phonon mechanism is not required.

The organic superconductors relates to type-II superconductors because their London penetration depth of magnetic flux is long due to small density of conductivity electrons, but superconductive coherence length, $\xi = 0.18\hbar v_F/(k_B T)$, is short due to comparatively small Fermi velocity of electrons, v_F, (here $\hbar$ is the Planck constant, k_B is the Boltzmann constant, and T is absolute temperature).

Majority of organic superconductors demonstrate co-existence of superconductivity and antiferromagnetism [509]. In 2007, Brooks [187] argumented proximity of antiferromagnetic and superconductinig states in 1/2 filled systems, and also proximity of charge order phase and superconducting state in 1/4 filled systems. Careful experimental measurements in Q2D-systems collide with solution of issues of interaction and symmetry: on one hand, it is observed strong pairing (in the main phonon) with s-wave (there are possible anisotropic s-wave effects), but on the other hand, there are d-wave antiferromagnetic intermediate interactions.

The critical temperatures, T_c, for BT = BEDT-TTF (BEDT = bis (ethylenedithio), TTF = tetrathiafulvalene) compounds [1577] are given in Table A.1. The properties of these superconductors depend on pressure and on annealing procedure. The critical temperatures of most of these compounds are <10 K. Most high critical temperature $T_c = 14.2$ K for organic superconductors has been obtained in 2003 by Taniguchi in β-(BEDT-TTF)$_2$ICl$_2$ [1661]. However it has been obtained under conditions of hydrostatic pressure of 7 GPa, which is necessary to stabilize superconductive state forming from parent Mott dielectric phase with antiferromagnetic ground state with *Neél temperature* of $T_N = 22$ K. In usual conditions highest temperature is demonstrated by κ-(BEDT-TTF)$_2$Cu([N(CN)$_2$]Br [873]. Some higher critical temperature T_c (>13 K based on data of electric resistance) has been found in related material κ-(BEDT-TTF)$_2$Cu([N(CN)$_2$]Cl at the pressure in several hundreds of bars.

The mechanical loading leads to change of Fermi surface. The stress parallel to conducting planes of quasi-two-dimensional organic superconductor k-(BEDT-TTF)$_2$Cu(SCN)$_2$, causes an increase of T_c and H_{c2}, but the perpendicular stress leads to their decrease. As it is known, superconductivity can be destroyed by a sufficiently high magnetic field. However, this is not always true. The organic superconductor k-(BEDT-TTF)$_2$Cu(SCN)$_2$ demonstrates Fulde–Ferrell–Larkin–Ovchinnikov (FFLO) effect [508, 983], namely superconductivity could be observed only in strong magnetic field. FFLO superconductivity initiates when an electron with spin "up" interacts with a hole with spin "down." In this case, "Cooper pair" of a new type forms, the necessary condition being an external magnetic field directed strictly along BEDT-TTF layers [1578].

In other Q2D organic conductor λ-(BETS)$_2$FeCl$_4$, where BETS = bis-(ethylenedithio)tetraselenafulvalene, the superconducting phase is induced by a magnetic field exceeding 18 T [1723]. Crystalline λ-(BETS)$_2$FeCl$_4$ consists of layers of highly conducting BETS sandwiched between insulating layers of iron chloride. The field is applied parallel to the conducting layers. At the same time this compound, at zero field, is an antiferromagnetic insulator lower 8.5 K. By increasing the applied magnetic field value higher 41 T, this compound becomes metallic at 0.8 K. The dependence $T_c(B)$ has a bell-like shape with a maximum $T_c \approx 4.2$ K near 33 T. It is interesting to note that a magnetic field of only 0.1 T applied perpendicular to conducting BETS-planes destroys the superconducting state. Thus, in this organic conductor, there is a strong superconductivity along the c-axis and weak in the ab-plane (in cuprates, it is the other way round).

New class of organic superconductors based on doped picene $C_{22}H_{14}$ has been discovered in 2010 [1167]. Plane molecule of picene consists of five connected one with other benzol coils and presents small fragment of graphene bordered by hydrogen atoms. As ground in [1167], it has been taken molecular crystal, in which these molecules formed regular periodic structure as atoms in metal. Interaction between molecules of $C_{22}H_{14}$ in the crystal is very weak, that makes possible its intercalation by atoms of other elements. Itself this crystal is dielectric, but after intercalation of enough quantity of alkali-metal atoms, it becomes to be conductor. As it has been discovered in [1167], $K_xC_{22}H_{14}$ with $x \approx 3$ is the type-II superconductor with $T_c = 18$ K and $H_{c1} = 380$ Oe. Obviously, the main cause of this phenomenon is great density of electronic states at the Fermi level caused by weak overlapping of molecular orbitals and hence by narrowness of allowed energy band. Additionally, has been shown that molecular crystals $Rb_xC_{22}H_{14}$ are also to be superconductive.

Carbon Superconductors

Fullerenes present allotropic molecular forms of carbon, in which atoms dispose at tips of right pentagons and hexagons, covering surface of sphere or spheroid. These molecules can contain 28, 32, 50, 60, 70, 76, etc. atoms. First fullerene discovered by Kroto with co-workers [943], had molecule of 60 atoms (C_{60}). In 1991, Hebard with co-workers [657] discovered superconductivity in fullerene-based molecular solids, namely in potassium-doped fulleride salts at $T_c = 18$ K. The compounds A_3C_{60} (where A is the alkaline metal) are superconducting and their properties are given in Table A.2. The doping by electrons allowed one to reach the critical temperature of 33 K in the system $Rb_yCs_xC_{60}$ [1660].

The critical temperature in A_3C_{60} increases with growth of ion radius A^+ (that is with increasing distance between molecules of C_{60}) from 19 K at $A = $ K up to 33 K at $A = $ Rb. Due to maximal radius has ion of Cs^+, then maximal T_c would be attained in Cs_3C_{60}. This result has been reached in 2008 [513]: the synthesized bulk samples of Cs_3C_{60} demonstrated superconductivity under hydrostatic compression with non-monotonous dependence of T_c on pressure p (maximal $T_c = 38$ K at $p \approx 7$ kbar) that does not agree with BCS-model. In difference of K_3C_{60} and Rb_3C_{60}, demonstrating f. c. c. structure, in the Cs_3C_{60}, it has been observed a structure of A-15 type. In difference of HTSC in f. c. c. fullerides (fullerene-containing crystallite compounds), the dielectric state competing with superconductive one is absent. However this state is present in the fulleride Cs_3C_{60} with A-15 structure [1643]. Under atmospheric pressure, this fulleride is antiferromagnetic dielectric, by this antiferromagnetic and superconducting regions of phase diagram are not separated (as in HTSC), but overlapping, that is, superconducting state arises directly from antiferromagnetic one.

If there are defects in the lattice, ln T_c varies as the inverse square of the unit cell constant for C_{60}-fullerides (see Fig. A.1).

Table A.2 Critical temperature, T_c, and lattice parameter, a_0, for polycrystalline compounds C_{60} with atoms of alkali metals

Compound	T_c, K	a_0, Å
$RbCs_2C_{60}$	33	14.555 ± 0.007
Rb_2CsC_{60}	31.3	14.431 ± 0.006
Rb_3C_{60}	29.4	14.384 ± 0.010
KRb_2C_{60}	27	14.324 ± 0.010
K_2CsC_{60}	24	14.292 ± 0.010
K_2RbC_{60}	23	14.243 ± 0.010
K_3C_{60}	19.3	14.240 ± 0.006
Na_2CsC_{60}	12	14.134 ± 0.006
Li_2CsC_{60}	12	14.120 ± 0.021
Na_2RbC_{60}	2.5	14.028 ± 0.011
Na_2KC_{60}	2.5	14.025 ± 0.010

Fig. A.1 The dependence of ln T_c *vs* inverse square of the unit-cell constant [1569]

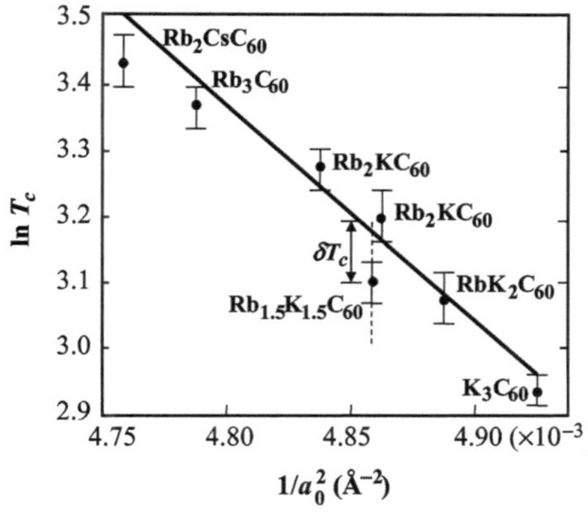

The research results have shown that superconductivity mechanism in fullerene-based materials is based on formation of Cooper couples as a result of electronic interaction with intramolecular oscillations in molecule of C_{60}. Nature of high T_c in fullerene-based materials is not understood in the whole. Herein, significant help could be rendered by experimental definition of the power, α, from the isotopic effect $T_c \sim M^{-\alpha}$, where M is the mass of atoms. In [500], ^{12}C has been substituted practically totally (on 99 %) by isotope ^{13}C in Rb_3C_{60} monocrystals with $T_c = 31$ K and has been found the value of power $\alpha = 0.21 \pm 0.012$. Based on these calculations, it is concluded that the phonon mechanism of electron pairing with "intermediate" force of electron–phonon interaction is realized in the doped fullerenes.

Then, an existence of Meissner's effect (repulsion of magnetic field from volume of superconductor) in the polycrystalline copper-containing fullerides at temperatures up to 110 K has been experimentally stated [1405]. The estimation of the experimental data showed that approximately *thousandth part* of the bulk sample has superconducting properties, causing the repulsion force.

In 2000, Prinzbach with co-workers [1406] discovered experimentally in gas phase C_{20} being smallest from possible fullerenes. At the surface of fullerene C_{20} chains C–C between adjacent atoms of carbon form only pentagons, but hexagons (in difference of fullerene C_{60}) are absent. After discovery of C_{20}, many attempts have been made to process on the base of this fullerene a solid, that is, fullerite (similar to fullerite from fullerene C_{60}). Interest to fullerite C_{20} is caused, in particular because of possible prospective to attain in it higher T_c compared with fullerites C_{60} (due to the predicted significant forcing electron–phonon interaction at decreasing fullerene sizes).

In 2000–2001, Schön et al. announced that, using an intercalation with $CHCl_3$ and $CHBr_3$ of C_{60} single crystals, directed to expansion of the lattice and formation of high densities of electrons and holes, they reached first the critical temperature, $T_c = 52$ K [1513], and then $T_c = 117$ K in hole-doped $C_{60}/CHBr_3$ [1514]. Figure A.2 presents the variation in critical temperature as a function of charge carrier

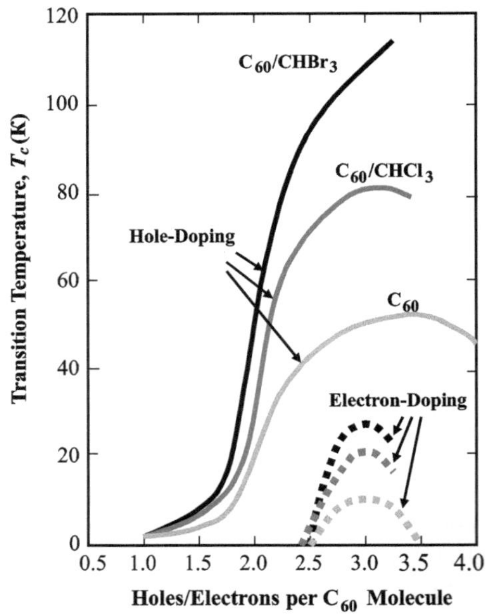

Fig. A.2 Variation in T_c as a function of charge carrier density for electron- and hole-doped C_{60} crystals in incorporation of $CHCl_3$ and $CHBr_3$ [1514]

density for electron- and hole-doped C_{60} crystals in different types of intercalation. Obviously, these results rendered doubtful.[1]

Superconductivity with $T_c = 7$ K has been found in C_{70} monocrystal with size of about 1 mm [546]. Obviously, carbon nanotubes (1D molecular conductors) are ideal candidates for research of 1D superconductivity. Today, maximum critical temperature, $T_c = 15$ K, for one-wall nanotubes [1658], and $T_{c,on} \sim 11.5$ K (beginning of resistive transition), T_c $(R = 0) = 7.8$ K, for multiwall nanotubes [1650] have been reached. Value of superconducting gap, $\Delta \sim 1.15$ meV, in multiwall nanotubes well agrees with expression of $\Delta = 1.76k_BT_c$ from BCS-model, if as T_c to select T_c $(R = 0)$; critical current behaves in accordance with Ginzburg–Landau theory, $I_c \sim [1 - (T/T_c)^2]^{3/2}$. Small value of H_c and comparatively great

[1] The critical papers [193, 1543, 1544] about Schön et al. results [1510–1512, 1515–1517] have been published in 2002. The papers [943–945, 1510–1512] have been devoted to the use of organic molecules, segregated on thin film, as "molecular switches", and also to the use of field transistors to change the charge carrier concentration in different organic matters, providing a regulation to their electric properties (from dielectric to semiconductor, from metal to superconductor and so on). Supposedly, this rendered to induce high-T_c superconductivity of the fullerenes C_{60} by electric field. However, first, total identity of the figures (right up to accidental noise inevitably accompanying any experiment) in [1517, 1516] has been found. The same figure has been also presented in [1510] devoted to another type of microelectronic devices. In this case, the coincidence has not been in the whole, but separate parts of the plots coincided totally with one another. Then, *eight* cases of the figure coincidence (total or fragmentary) have been found in six papers of Schön, devoted to different types of devices for various materials and temperatures. After that, 100 *scientific groups* in the world have attempted unsuccessfully to repeat the results of Schön et al.

penetration depth of magnetic flux (>10 nm) witness on that these carbon nanotubes relate to type-II superconductors. As one of possible ways to further growth of T_c, it is proposed to increase a concentration of free carriers by using doping of the multiwall nanotubes (for instance, by B or Ca). Note also an observation of Josephson effect in tunnel junction from graphen monolayer disposed between two superconducting electrodes Ti/Al, divided by distance of (100–500) nm one of other [660].

The attempts to switch graphite into superconducting state, using its doping by different chemical elements, led to superconductivity with $T_c < 1$ K for cases of K and Na [67, 633]. Introduction of ytterbium and calcium atoms between graphite layers led to the fabrication of superconducting compounds with $T_c = 6.5$ and 11.5 K, respectively [1788]. In this case, the role of Yb and Ca atoms is that they supply free charge carriers in the graphite layers.

Then, the oriented (111) boron-doped diamond thin films (with boron concentration 0.53 %) have been grown at (001) silicon substrates, using MPCVD technique (modification method of the chemical vapor deposition) [1648]. These films demonstrated $T_{c,on} = 7.4$ K and T_c $(R = 0) = 4.2$ K at $H = 0$. The linear extrapolation gives H_{c2} $(0) = 10.4$ T and H_{irr} $(0) = 5.12$ T, the critical current density $J_c = 200$ A/cm^2 at $H = 0$. First signs of superconductivity arose at boron concentration of 0.18 %. Finally, the materials, combining superconductivity, superhardness and high strength (these materials could be used for research of electric and superconducting properties under pressure) have been synthesized at high static pressures (up to 7.7 GPa) and temperatures (up to 2173 K) in the following systems [400]: (i) diamond/Nb, $T_c = 12.6$ K ($\Delta T = 1.5$ K), H_{c2} (4.2 K) $= 1.25$ T (diamond matrix and superconducting channels from niobium carbide); (ii) diamond/Mo, $T_c = 9.3$ K, $\Delta T = 5$ K (diamond matrix and superconducting channels from molybdenum carbide); (iii) composites with matrices from superhard materials (80 wt%) and MgB_2 channels, namely diamond/MgB_2 ($T_c = 37$ K) and cubic boron nitride/MgB_2 ($T_c = 36.1$ K), frame of these composites has microhardness in the range of 57–95 GPa.

A-*15 Compounds*

Chemical compounds of binary composition $A_{3-x}B_{1+x}$ crystallize into many different structures, depending on the value of x, temperature and pressure. One of the structures existing near $x = 0$, A_3B (where A = Nb, V, Ta, Zr and B = Sn, Ge, Al, Ga, Si) has the structure of beta-tungsten, designated in crystallography by the symbol A-15, and is superconducting. Hardy and Hulm first discovered A-15 superconductor (V_3Si) in 1954 [636]. Intermetallic compound Nb_3Ge has a critical temperature $T_c = 23.2$ K [526], while Nb_3Ga shows $T_c = 20.7$ K, Nb_3Al, $T_c = 19.1$ K and Nb_3Sn, $T_c = 18.3$ K. Figure A.3 shows the structure of the binary A_3B compound. The transition temperatures of several A-15 structures are given in Table A.3.

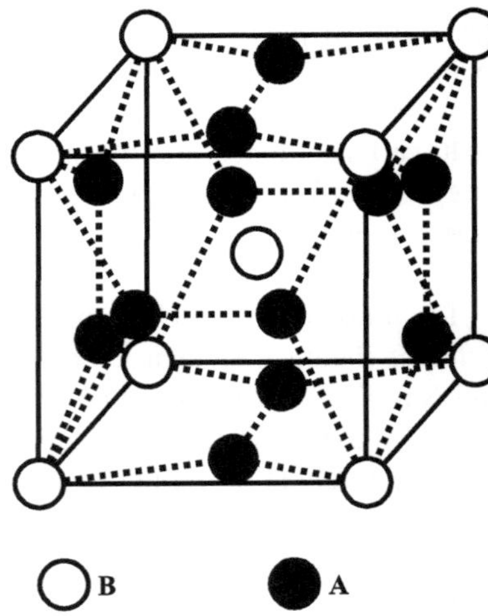

Fig. A.3 Crystalline structure of A_3B compound (A-15 superconductors). The atoms A form one-dimensional chains on each face of the cube. Chains on the opposite faces are parallel, while on the neighboring faces they are orthogonal to each other

The critical temperature and second critical field increase in Nb_3Al compound, adding Ge and Cu in Nb/Al at the initial stage of the *rapid heating, quenching and transforming* (RQHT) process [739], and they attain for Nb_3Al–Ge ($T_c = 19.4$ K, $H_{c2} = 39.5$ T) and for Nb_3Al–Cu ($T_c = 18.2$ K, $H_{c2} = 28.7$ T) [738]. Moreover, the superconductors Nb_3Al–Ge, Cu have highest critical current densities among all metallic multifilamentary superconductors at $H > 20$ T and $T = 4.2$ K. Only significantly more expensive HTSC, based on Bi-2223 and Bi-2212, have the near values of supercurrent at $T < 20$ K.

Magnetic Superconductors (Chevrel Phases)

In 1971, Chevrel [274] and co-workers discovered a new class of ternary molybdenum sulfides having the general chemical formula $M_xMo_6S_8$, where M stands for a large number of metals and rare earths. These superconductors have unusually high values of the upper critical field, B_{c2}, given in Table A.4 [1201]. The superconducting compounds REMo$_6X_8$ (where RE = Gd, Tb, Dy, Er and X = S, Se, Te), and RERh$_4$B$_4$ (where RE = Nd, Sm, Tm) are usually related to the Chevrel phases [1629]. Magnetic superconductors demonstrate some novel features not found in conventional type-I superconductors. Upon cooling from the superconducting phase ($T < T_c$), the material becomes normal again at a low temperature and the superconductivity is destroyed. Very often this normal phase is magnetically ordered. Upon cooling from the normal state the system becomes superconducting below T_c, and upon further cooling it becomes magnetically ordered below Neél temperature, T_N (where $T_N < T_c$). Thus, the superconducting phase occurs only in a limited range of temperatures, $T_N < T < T_c$.

Table A.3 The critical temperatures, T_c (K), of A-15 compounds

Ti_3Sb	6.5	Mo_3Ga	0.76
$Zr_{80}Sn^q_{20}$	0.92	$Mo_{77}Si_{23}$	1.7
Zr–Pb	0.76	$Mo_{77}Ge_{23}$	1.8
$Zr_{\sim 3}Bi^p$	3.4	Ti_3Ir	4.2
$V\text{–}Al^f$	11.8	Ti_3Pt	0.5
V_3Ga	15.9	$V_{29}Re_{71}$	8.4
V_3Si	17.0	$V_{50}Os_{50}$	5.7
$V_{\sim 3}Ge$	6.0–7.5	$V_{65}Rh_{35}$	≈ 1
$V_{\sim 3}Ge^f$	6.0–11.0	$V_{63}Ir_{37}$	1.7
$V_{\sim 79}Sn_{\sim 21}$	4.3	$V_{\sim 3}Pd$	0.08
$V\text{–}Sn^q$	7.0–17.0	V_3Pt	3.7
$V_{77}As_{23}$	0.2	$Nb_{75}Os_{25}$	1.0
$V_{76}Sb_{24}$	0.8	$Nb_{75}Rh_{25}$	2.6
Nb_3Al	19.1	$Nb_{72}Ir_{28}$	3.2
Nb_3Be	10.0	Nb_3Pt	11.0
Nb_3Ga	20.7	$Ta_{85}Pt_{15}$	0.4
Nb_3Pb	5.6	$Cr_{72}Ru_{28}$	3.4
$Nb_{\sim 3}In^p$	8.0–9.2	$Cr_{73}Os_{27}$	4.7
$Nb_{82}Si^q_{18}$	4.4	$Cr_{78}Rh_{22}$	0.07
$Nb\text{–}Si^f$	9.3	$Cr_{82}Ir_{18}$	0.75
$Nb\text{–}Si^f$	4.0–8.0	$Mo_{40}Tc_{60}$	13.4
$Nb\text{–}Si^f$	11.0–17.0	$Mo_{\sim 65}Re^f_{\sim 35}$	≈ 15 (A-15)
$Nb\text{–}Ge^q$	6.0–17.0	$Mo_{75}Os_{25}$	12.7
$Nb\text{–}Ge^f$	23.2	$Mo_{78}Ir_{22}$	8.5
Nb_3Sn	18.3	$Mo_{82}Pt_{18}$	4.6
Nb–Sb	2.0	$W_{\sim 60}Re^f_{\sim 40}$	11.0
$Nb_{\sim 3}Bi^p$	3.0	$Ta_{\sim 80}Au_{20}$	0.55
$Ta_{\sim 3}Ge^f$	8.0	Zr_3Au	0.9
$Ta_{\sim 3}Sn$	8.3	$V_{76}Au_{24}$	3.0
$Ta_{\sim 3}Sb$	0.7	$Nb_{\sim 3}Au$	11.5
Mo_3Al	0.58		

The left column has no transition metals; the right column has transition metals; three alloys of gold are also included. *Note* q = quenching p = pressure, f = film

Table A.4 The critical temperature and the upper critical magnetic field of Chevrel phases

No.	Compound	T_c (K)	B_{c2} (T)
1	$PbMo_6S_8$	15	60
2	$LaMo_6S_8$	7	44.5
3	$SnMo_6S_8$	12	63

The interaction of conduction electrons with magnetic atoms leads to the formation of a bound state of the electron with the magnetic atom lower a certain characteristic temperature called the *Kondo temperature*. It was found that the resistivity of a magnetic alloy, such as Fe impurities in Cu shows a minimum in the resistivity as a function of temperature due to the interaction of conduction electrons of the metal atom with the magnetic moment of the magnetic atom. Such a *Kondo effect* exists in superconductors containing magnetic impurity atoms at low temperatures, because the superconductivity is destroyed. In the case of magnetic superconductors, there is a sub-lattice of magnetic atoms in addition to the lattice of metallic atoms, so that magnetically ordered phase exists at low temperatures lower the superconducting phase, $T_N < T_c$.

While ferromagnetic superconductors (e.g., $ErRh_4B_4$ and $HoMo_6S_8$) demonstrate the above behavior, the antiferromagnetic superconductors (e.g., $\mathit{RE}Mo_6S_8$, where RE = Gd, Tb, Dy, Er and also $\mathit{RE}Rh_4B_4$, where RE = Nd, Sm, Tm) present examples of the co-existence of the two phases together with anomalous behavior of the second critical field. In $ErRh_4B_4$, possessing tetragonal structure with $a = 5.299$ Å and $c = 7.58$ Å, a plot of the *ac*-magnetic susceptibility (χ_{ac}) and electric resistance depicts a normal to superconducting transition at $T_{c1} = 8.7$ K, followed by a loss of susceptibility at $T_{c2} = 0.9$ K together with the appearance of the ferromagnetic long-range order. This result takes place at zero magnetic field. At non-zero finite field also, the resistance disappears at some temperature interval within the above domain. The compound $HoMo_6S_8$ becomes superconducting at $T_{c1} = 1.3$ K in zero field. On further cooling, it becomes normal at $T_{c2} = 0.6$ K together with the appearance of ferromagnetic long-range order.

The antiferromagnetic superconductors provide the most striking case of the co-existence of the two kinds of order. These systems (e.g., $\mathit{RE}Mo_6S_8$, where RE = Gd, Tb, Dy, Er and also $\mathit{RE}Rh_4B_4$, where RE = Nd, Sm, Tm) demonstrate near T_N the antiferromagnetic alignment of rare-earth magnetic moments in the superconducting state of the system. The most important result is the anomalous behavior of the upper critical field as a function of temperature near T_N. In particular, $\mathit{RE}Mo_6S_8$ (RE = Gd, Tb, Dy) demonstrate anomalous decreasing of H_{c2} near T_N. The crystal structure of $\mathit{RE}Mo_6S_8$ is presented in Fig. A.4.

An attempt of increasing T_c in $\mathit{RE}Mo_6S_8$ owing to change of *RE* ion radius leads to the structure transition, as a result of which the dielectric gap opens at Fermi level and the superconducting properties disappear together with metallic ones. Based on BCS model, this is explained that the growth of T_c is caused by the approaching Fermi level to the peak of the electron state density, $N(E)$. However, it is disadvantageous energetically for the structure therefore it suffers phase transition, which suppresses superconductivity.

The superconducting carbosulfide Nb_2SC_x [1465] with layered crystalline lattice has critical temperature $T_c = 5$ K at $x = 0.8$–1. The superconductivity demonstrates volume character, and Nb atom octahedron is the key structure element, in the center of which C atom locates.

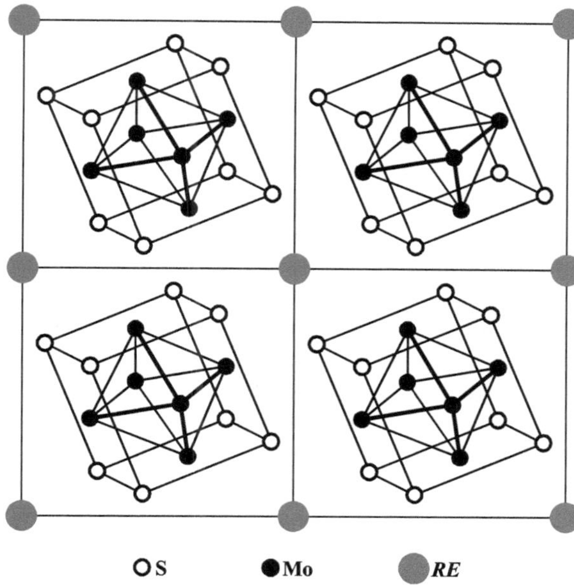

Fig. A.4 Structure of $REMo_6S_8$, which has Mo octahedron inside sulphur cube, which are inside a rare-earth cube [1569]

Heavy Fermion Superconductors

Heavy fermions are defined by state of electrons, in which they form Fermi-liquid of quasi-particles with anomalous great effective mass, m. The value of m in 10^2–10^3 times greater than effective mass of electrons in normal and transition metals, where as rule, $m \sim (1\text{–}10)m_0$ (where m_0 is the mass of free electron). Almost all heavy fermion compounds are to be antiferromagnetics with Neél temperature, $T_N \sim$ 5–20 K.

Superconductivity of heavy fermions has been discovered in 1979 by Steglich with co-workers in $CeCu_2Si_2$ [1613]. Co-existence of superconductivity and antiferromagnetic order is found in U-based heavy fermions (UPt_3, $U(Pt_{1-x}Pd_x)_3$, $U_{1-x}Th_xBe_{13}$, URu_2Si_2, UNi_2Al_3, UPd_2Al_3, U_6Co and U_6Fe) [729, 793, 840, 1487] and in the heavy fermions: $RERh_2Si_2$ (where $RE =$ La and Y), $Cr_{1-x}Re_x$, $CeRu_2$, $CeIn_3$, $CePd_2Si_2$, $CeNi_2Ge_2$, $CeCoIn_5$ and $Ce_nT_mIn_{3n+2m}$ (where $T =$ Rh or Ir, $n = 1$ or 2; $m = 1$) [1096, 1356, 1416, 1668].

In $CeCoIn_5$ critical temperature $T_c = 2.3$ K [1416], but there are evidences of existence in this compound of second superconductive phase (so-called, Q-phase) in strong (near to H_{c2}) magnetic field at the $T < 0.3$ K. In the work [846], it has been shown that compound $CeCoIn_5$ differs from other heavy fermion superconductors by that what in it superconductive and magnetic orders not only co-exist but superconductivity is necessary condition of magnetism in the Q-phase. At the increasing H up to H_{c2}, far magnetic order disappears together with superconductivity, demonstrating that the order parameter in the Q-phase presents itself a complex superposition of parameters of the superconductive and magnetic orders.

Co-existence of superconductivity and ferromagnetism is discovered in the compounds UGe_2 [1494], URhGe [53] and $ZrZn_2$ [1357]. Superconductor URhGe loses its resistance at the temperature $T < 9.5$ K. In $ZrZn_2$, the maximum critical temperature is slightly less than 3 K at ambient pressure and decreases with increasing pressure. The superconductor $YNi_{1.9}B_{1.2}$ demonstrates one of the maximum $T_c \approx 14$ K among known triple compounds with ferromagnetic component [1407]. Another superconducting phase $YNiB_3$ has $T_c \sim 4.5$ K (tetragonal lattice with parameters of $a = 0.3782$ nm, $c = 1.1347$ nm, spatial group of symmetry *P4/mmm*) [1408].

On superconductive properties the heavy fermion superconductors significantly differ from conventional superconductors. Power (but no exponential) dependence of $c_V(T)$, factors of ultra-sound absorption, thermal conductivity, relaxation time of NMR signal at the $T \sim T_c$ point on gapless character of heavy fermion superconductors. These data and also that what arising antiferromagnetic order in all known cases precedes to superconductive transition allow us to propose a possibility of superconductive state with no zero momentum of Cooper couple and no phonon pairing mechanism via antiferromagnetic spin fluctuations.

As it is known, the electronic heat capacity of a solid varies linearly with temperature at low temperatures, $c = \gamma T$, where the constant of proportionality, γ, determines the mass of the electron in the conduction band. It is found that in many compounds of Ce or U, this band mass is very large. Not only the effective mass but also Fermi rate differs by two orders of magnitude from the free electron values found in ordinary metals. For example in $CeCu_2Si_2$, the Fermi rate, $v_F = 8.7 \times 10^8$ m/s and the band mass, $m_{ef} = 220m_0$, where m_0 is the free electron mass. The critical temperature is $T_c \approx 0.5$ K and the London penetration depth approaches a high value of $\lambda_L = 0.2$ μm at zero temperature. The large specific heat is caused by the strong correlations between electrons. In UBe_{13}, a sharp superconducting transition occurs at 0.97 K. The variation in T_c from sample to sample is in the range of 0.88–0.97 K. When Np is added to UBe_{13}, the T_c reduces. For $Np_{0.68}U_{0.32}Be_{13}$, the transition temperature is about 0.8 K. The compound UPt_3 has a superconducting transition temperature of 0.54 K. The extreme low temperature indicates that there are strong magnetic correlations [1569].

The superconductor $PuCoGa_5$ with $T_c \sim 18$ K [1482] occupies intermediate place between heavy fermion superconductors with $T_c \sim 1$ K and HTSC with $T_c \sim 100$ K. Besides "exotic" superconducting state, the common property for all these materials is the very strong Coulomb correlations between electrons of atomic *f*-shells (in $CeCoIn_5$ and $PuCoGa_5$) or *d*-shells (in HTSC), which lead to sharp increase of the effective electron mass or to transition in the state of Mott's dielectric, respectively. Co-existence of the localized and delocalized states, which are close in energy, favors the magnetic mechanism of the electron pairing [321].

Iron is the ferromagnetic, which demonstrates superconductivity at very high pressure. The superconducting transition is registered at 15 GPa $< P <$ 30 GPa by both resistive and magnetic (Meissner's effect) methods [1565]. The dependence $T_c(P)$ is "bell-like" (similar to the dependence of T_c on hole concentration in HTSC) with maximum value of $T_c = 2$ K at $P = 21$ GPa.

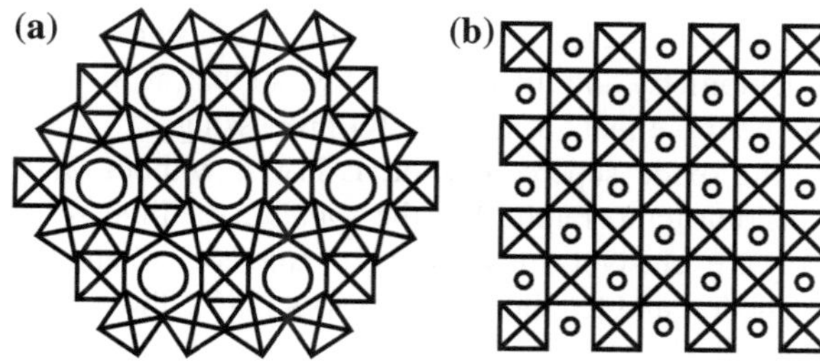

Fig. A.5 (**a**) Hexagonal structure of the tungsten bronze. One elementary layer presented, which is formed by the octahedrons MO_6 with large ions in cavities. (**b**) Perovskite structure [243]

Lithium is the metal at normal pressure. Under pressure of $P \approx 30$ GPa, it becomes a superconductor and T_c attains 20 K at $P = 48$ GPa. This is the highest T_c for "simple" one-element superconductors [1564].

Oxide Superconductors Without Copper

The superconductors of A_xWO_3 type have hexagonal structure of the *tungsten bronze*, where the large alkali ions of K, Rb and Cs are used as A more often (see Fig. A.5a). The compound $NaWO_3$ demonstrates *perovskite* structure with octahedral W^{5+} and cuboctahedral Na^+ coordination polyhedra. In perovskites, one cation shares the close-packed sites with the $3O^{2-}$ anions, and the other cation occupies exclusively octahedral sites coordinated by O [345]. Many superconducting materials from this family demonstrate $T_c \approx 2$–7 K. The monocrystals of perovskite dielectric WO_3, doped by Na, demonstrate in surface layer of $Na_{0.05}WO_3$ high-temperature superconductivity with $T_c = 91$ K [677]. In this case, Na segregates at the grain surface and takes the structure of the WO_3 surface layer.

The superconductor without copper, $Ba_{1-x}K_x$ (BKBO) demonstrates maximum critical temperature ($T_c = 31$ K) among "old" oxide superconductors [1201]. BKBO has been intensively studied owing to (i) it is non-Cu containing superconductor with high T_c, and (ii) it has isotropic three-dimensional structures without two-dimensional copper oxide planes of the layered HTSCs. Instead, it has the three-dimensional network of BiO_6 octahedra. Studies of symmetry role in superconductivity have caused the next research directions:

(i) the role of symmetry lowering tilting of the BiO_6 octahedra [688];
(ii) the low temperature BKBO-structure [957];
(iii) the rich structural phase diagram of the BKBO-system, including the sites of monoclinic, rhombohedral, orthorhombic and cubic structures [1346];
(iv) the role of isotropic superconductivity in intrinsic and extrinsic properties such as flux pinning [1115] and pair tunneling [108] through superconductor–insulator–superconductor (SNS) junctions.

Superconductivity in BKBO occurs at low temperatures over the range $0.37 \leq x \leq 0.50$. Into this range the lattice constant varies approximately linear from

0.4289 to 0.4270. The cubic lattice constant is quite sensitive to both the O stoichiometry and the amount of K, so that, it would be carefully to interpret these competing effects at the BKBO processing.

The oxide superconductor, $LiTi_2O_4$, without Bi and Cu, also possessing high transition temperature, has the *spinel* crystalline structure (AB_2O_4), in which f.c.c. array is exclusively made from O^{2-} anions and A and B cations occupy a fraction of the octahedral and tetrahedral interstices [345]. Two other oxide superconductors: NbO and TiO, showing $T_c \leq 1$ K, demonstrate clearly the expanded chains of the "metal–metal" type as the previous oxide. The oxide superconductor $BaPb_{1-x}Bi_xO_3$ (BPBO) has a single layer perovskite structure (ABO_3), depicted in Fig. A.5b. In BPBO, the single perovskite unit is tetragonal but no cubic and, unlike $BaTiO_3$, it has a large c/a ratio. This is differed from other tetragonal HTSCs, whose tetragonality occurs due to unidirectional stacking of cubic perovskite units. BPBO has the critical temperature T_c of about 13 K for $x = 0.3$. By this superconductivity is observed into range $0.05 \leq x \leq 0.30$. Combined P and K doping reduces T_c to about 12 K in BPBO. Substitution of alkali metals for Ba in $BaBiO_3$ also leads to superconductivity.

The layered superconductor $Na_xCoO_2 \bullet 1.3H_2O$ with $T_c = 4$ K has been fabricated using the chemical oxidation technique [1644]. The conductive layers (CoO_2) in this superconductor alternate with the dielectric buffer layers ($Na_x \bullet 1.3H_2O$), which carry out a function of reservoirs for electric charge. First and second critical fields reach values $H_{c1}(0) = 28.1$ Oe, $B_{c2}(0) = 61$ T. The corresponding values of superconductive coherence length, $\xi = 2.32$ nm, penetration depth, $\lambda = 568$ nm, and Ginzburg–Landau parameter, $\kappa = \lambda/\xi = 244$. Similar to HTSC, the value of T_c is highest at the optimum level of doping and decreases in overdoped and underdoped samples [1496].

Pyrochlore Oxides

The superconductors with the structure of pyrochlore oxide has general formula AOs_2O_6, where A = Cs ($T_c = 3.3$ K), Rb ($T_c = 6.3$ K) and K ($T_c = 9.6$ K). However, these compounds with the same chemical formula differ sharply in their superconducting properties. If $RbOs_2O_6$ is the BCS superconductor [192, 848], then KOs_2O_6 similar to HTSC relates to oxides of transitional metals, but does not crystallize in the perovskite structure. KOs_2O_6 crystallizes in the pyrochlore oxide structure, based on triangle lattice, which is a classic example of frustration effect in spin system, forming numerous spin structures [1841]. Generally pyrochlore oxides present a great group of titan-, tantalum- and niobium-containing minerals with cubic crystals and general formula $A_2B_2O_6O'$, where A is the large cation, B is the smaller cation (usually $5d$-transitional metal, i.e., Re, Os or Ir). First, superconductivity in this oxide class has been discovered in $Cd_2Re_2O_7$ compound ($T_c = 1$ K) [629]. The difference between $Cd_2Re_2O_7$ and KOs_2O_6 is in the number of d-electrons in B-cation. It is interesting that KOs_2O_6 compound with fractional degree of oxidation has the critical temperature 10 times higher than $Cd_2Re_2O_7$, possessing even number of 5*d*-electrons. In the compounds with the same structure like $RbOs_2O_6$ [1842] and $CsOs_2O_6$ [1840], Os ion has fractional degree of oxidation, +5.5, disposing in intermediate state between $5d^2$ and $5d^3$. In this case, the 5*d*-electrons of Os define transport and magnetic properties of these materials, simultaneously, a coupling of which in these oxide compounds is mostly interesting.

The effect of high pressure (up to 10 GPa) on superconductivity in the AOs_2O_6 compounds has been studied for A = Cs, Rb and K [1218]. The critical temperature for all three materials increased together with pressure up to a maximum value of $T_c = 7.6$ K (at 6 GPa), $T_c = 8.2$ K (at 2 GPa) and $T_c = 10$ K (at 0.6 GPa) for A = Cs, Rb and K, respectively, after that it diminishes down to total disappearance at 7 and 6 GPa for A = Rb and K, and above 10 GPa for A = Cs.

Rutheno-Cuprates

Sr_2RuO_4 compound ($T_c \sim 1$ K) presents itself layered perovskite without cooper, demonstrating superconductivity [1068]. This compound relates to the class of "self-doped" conductors due to small ratio, *U/W* (where *U* is the energy of Coulomb repulsion and *W* is the width of Brillouin's zone), that is, there the role of electron correlation is not important compared to cuprates [1068]. The *p*-type of pairing (spin-triplet) in Sr_2RuO_4 is realized. This system is also called the system with *ladder structure*. If to seek number of legs of the ladders per cell to infinity, then a transition to two-dimensional structure occurs. In cuprate HTSC, the ladder role could be played by stripes, then, a total analogy between cuprate and ruthenium systems is possible.

Triple perovskite 1212 phase ("hybrid" rutheno-cuprate superconductor) $RuSr_2GdCu_2O_8$ consists of both "superconducting" CuO_2 layers and "ferromagnetic" RuO_2-layers (see Fig. A.6) and to be isostructural with $YBa_2Cu_3O_7$ [121]. Herein, superconductivity co-exists with electronic ferromagnetism in microscopic scale (the temperature of ferromagnetic ordering 135 K and T_c = 50 K) [1110]. As it follows from Fig. A.6, Gd rare-earth ion occupies the Y-site (Gd, Sm, Eu, and Pr have been substituted on this site [69]) and Sr occupies the Ba-site of the YBCO superconductor. Re–O octahedra replace the Cu–O chains, with the higher coordination of Ru reflecting the Ru^{4+} charge state. Note, the investigations of the magnetization and magnetic resistance of the $RuSr_2GdCu_2O_8$ have demonstrated influence of the magnetic moments of Ru atoms on the electrons of conductivity.

The intragrain critical temperature, T_c, of HTCS cuprate $RuSr_2(Gd,Ce)_2Cu_2O_{10+d}$ has been studied depending on hole concentration (which is found by oxygen content) [1815]. In this case, T_c changed in very wide limits (17–40 K) with the change of *p* being only 0.03 holes/CuO_2. Into this range of *p*, the intragrain superfluid density (which is inversely proportional to square of the magnetic field penetration depth, $1/\lambda^2$) and the value of the diamagnetic jump

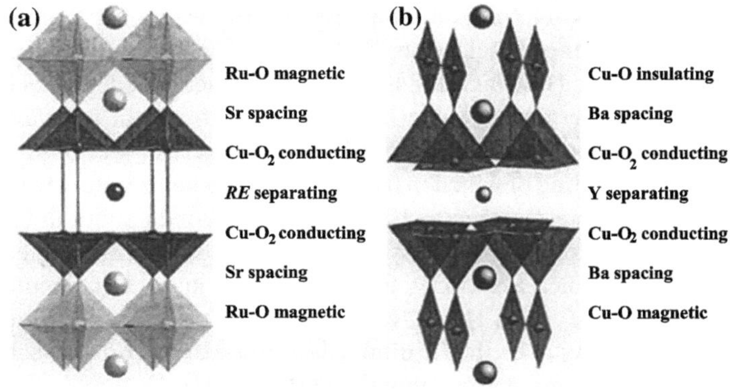

Fig. A.6 Comparison of structure of the (**a**) $RuSr_2GdCu_2O_8$ and (**b**) $YBa_2Cu_3O_7$ [345]

(found at the sample cooling in magnetic field) have increased more than 10 times. These results contradict to correlations between T_c, p and $1/\lambda^2$, observed in homogeneous HTSC. This is possible due to the phase de-lamination and granularity. Moreover, there is an effect of anomalous increasing of distance between CuO_2 planes during cooling of layered compound $RuSr_2Nd_{0.9}Y_{0.2}Ca_{0.9}Cu_2O_{10}$, doped up to level of the boundary "antiferromagnetism/superconductivity" at the phase diagram [1119]. This means negative value of the thermal expansion factor. Difference in volumes of the antiferromagnetic and superconducting phases may be caused by the phase segregation, observed often in weakly doped HTSC.

High-Temperature Superconductive Cuprates

Superconducting compound belongs to HTSC, if it has one or more CuO_2 planes. The cuprates possess a perovskite, strongly anisotropic crystalline structure, defining majority of their physical and mechanical properties. In conventional superconductors, there are no important structural effects, because the coherence length (ξ) is much longer than the penetration depth (λ). However, it is not the case for cuprates due to $\xi << \lambda$ for them. In general, the high-T_c materials are basically tetragonal (orthorhombic), and superconductivity in cuprates occurs in the copper-oxide planes. These layers are always separated by layers of other atoms such as Bi, O, Y, Ba, La, etc., which provide the charge carriers into CuO_2 planes. In the CuO_2 planes, each copper ion is strongly coupled to four ions, separated by a distance approximately 1.9 Å. At fixed doping level, by increasing the number of CuO_2 planes, T_c first increases, reaching the maximum at $n = 3$ (where n is the number of CuO_2 layers per unit cell), and then decreases. So, nuclear magnetic resonance (NMR) data show that the charge carriers are distributed heterogeneously between CuO_2 layers (into limits of one elementary cell) in

Table A.5 Transition temperature and number of CuO_2 planes for different cuprates

No.	Cuprate	CuO_2 planes	T_c (K)	Abbreviation
1	$La_{2-x}Sr_xCuO_4$	1	38	LSCO
2	$Nd_{2-x}Ce_xCuO_4$	1	24	NCCO
3	$YBa_2Cu_3O_{7-x}$	2	93	YBCO
4	$Bi_2Sr_2CuO_{6+x}$	1	~ 12	Bi-2201
5	$Bi_2Sr_2CaCu_2O_{8+x}$	2	95	Bi-2212
6	$(Bi, Pb)_2Sr_2Ca_2Cu_3O_{10+x}$	3	110	Bi-2223
7	$Tl_2Ba_2CuO_{6+x}$	1	95	Tl-2201
8	$Tl_2Ba_2CaCu_2O_{8+x}$	2	105	Tl-2212
9	$Tl_2Ba_2Ca_2Cu_3O_{10+x}$	3	125	Tl-2223
10	$TlBa_2Ca_2Cu_4O_{11}$	3	128	Tl-1224
11	$HgBa_2CuO_{4+x}$	1	98	Hg-1201
12	$HgBa_2CaCu_2O_{6+x}$	2	128	Hg-1212
13	$HgBa_2Ca_2Cu_3O_{8+x}$	3	135	Hg-1223
14	$HgBa_2Ca_3Cu_4O_{10+x}$	4	123	Hg-1234
15	$HgBa_2Ca_4Cu_5O_{12+x}$	5	110	Hg-1245
16	$(Ag, Cu)Ba_2Ca_3Cu_4O_{11-x}$	4	117	Ag-1234

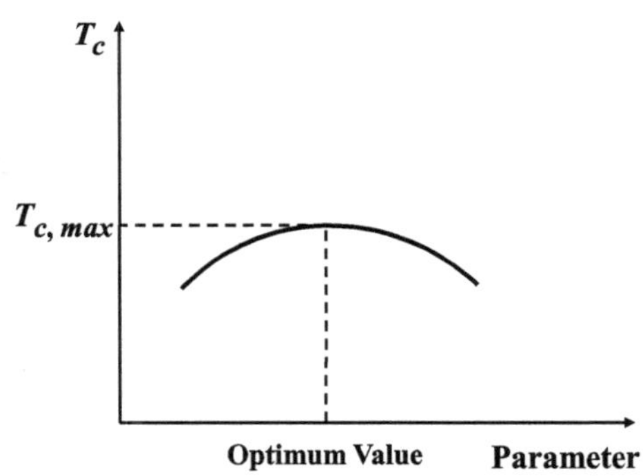

Fig. A.7 Critical temperature as a function of a parameter, which can be the doping level, the number of CuO_2 planes per unit cell, the unit-cell constants, the buckling angle of CuO_2 layers, etc.

mercurial HTSC at $n > 3$. Their concentration into "internal" layers is smaller than into "external" [925]. Based on the phenomenological approach, it has been shown [247] that due to lower concentration of carriers into "internal" layers of the multilayered elementary cell, a value of pseudogap in these layers is large and superconducting order is suppressed, therefore the following increasing of n does not lead to growth of T_c.

The critical temperature and the number of the CuO_2 planes are presented for different HTSC in Table A.5. In most superconducting cuprates, by changing the doping level at fixed n, the $T_c(p)$ dependence has the bell-like shape, where p is the hole concentration in CuO_2 planes (see Fig. A.7). This shape of $T_c(x)$ dependence is more or less universal in cuprates, where x can be p, n, the lattice constants a, b or c, the buckling angle of CuO_2 layers, etc. Thus, the maximum T_c value can be only attained when all the required parameters have their optimum values.

Table A.6 Composition and oxygen content dependence of critical temperature in the system $Bi_2Sr_2Ca_{1-x}Y_xCu_2O_{8+\delta}$ [1569]

No.	x	δ	Cu-valency	T_c (K)
1	0.0	0.23	2.23	70
2	0.1	0.23	2.18	90
3	0.3	0.30	2.15	86
4	0.35	0.30	2.12	74
5	0.4	0.29	2.09	64
6	0.5	0.33	2.08	40
7	0.75	0.41	2.03	<4
8	1.0	0.51	2.01	<4

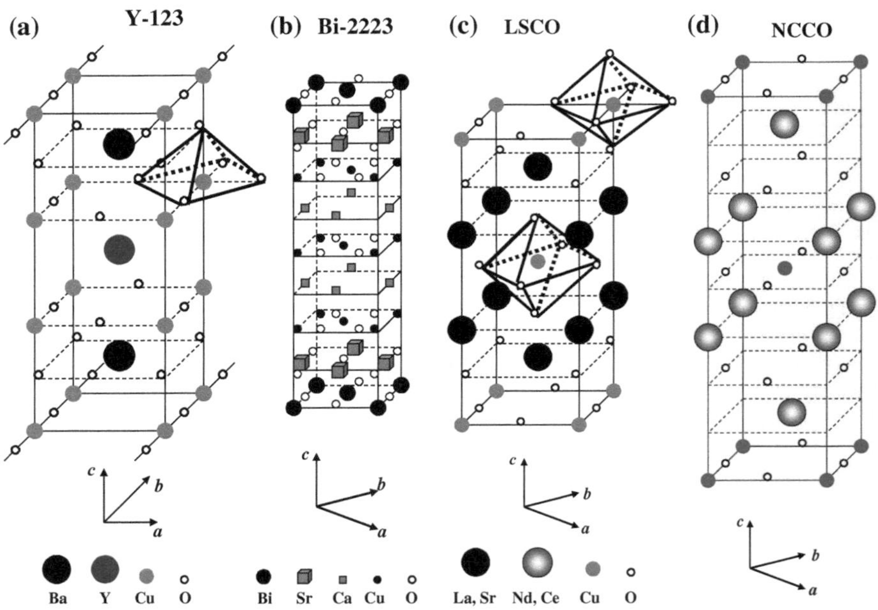

Fig. A.8 Crystal structures of HTSC: (**a**) YBCO, (**b**) Bi-2223, (**c**) LSCO, and (**d**) NCCO

In the HTSC also, as in conventional compounds, there is a change in valence due to the change in electron–phonon interaction. The oxygen content defines the transition temperature. As an example, the transition temperature, T_c, as a function of x and δ for the system $Bi_2Sr_2Ca_{1-x}Y_xCu_2O_{8+\delta}$ is presented in Table A.6. Crystal structures of high-temperature superconductors LSCO, YBCO, Bi-2212 and NCCO are shown in Fig. A.8. Search for superconductors with high values of T_c is continuing. Table A.7 shows a variety of superconducting cuprates, differing in their structural and chemical properties discovered beginning from 1986. These superconductors are grouped in compositional principle. The partial substitution of Y by Ca and Ba by La in $YBa_2Cu_4O_8$ does not have an influence in practice on T_c.

Table A.7 Superconducting cuprates discovered from 1986

$YBa_2Cu_3O_7$ + *RE* equivalents	$(X, Cu)(Eu, Ce)_2(Eu, Sr)_2Cu_2O_8$, where X = Pb, Ga
$(Y, Pr)Ba_2Cu_3O_7$	$Sr_{1-x}CuO_2$
$(Y, Ca)Ba_2(Cu, Zn)_3O_7$	$(Ca, Sr, La, Nd)CuO_2$
$(Sm, Nd)Ba_2[Cu_{1-y}(Ni, Zn)_y]_3O_7$	$(Cu, CO_2)(Ba, Sr)CuO_{3+\delta}$
$TbSr_2(Cu, Mo)_3O_7$	$(Cu, CO_2)(Ba, Sr)CaCu_2O_{5+\delta}$
$YBa_2Cu_4O_8$ + *RE* equivalents	$(Cu, CO_2)(Ba, Sr)Ca_2Cu_3O_{7+\delta}$
$Y_2Ba_4Cu_7O_{15}$	$(Ca, Na, Sr, K)_2CuO_2Cl_2$
$Bi_2(Sr, La)_2CuO_{6+x}$	$(Sr, K)_3Cu_2O_4Cl_2$
$Bi_{2+z}Sr_{2-x-z}La_xCuO_y$	$(Cu, X)Sr_2(Y, Ca)Cu_2O_7$,
$Bi_2Sr_2CaCu_2O_{8+x}$	where X = Pb, CO_2, SO_3, BO_3, Bi, Mo
$Bi_2Sr_2Ca_2Cu_3O_{10+x}$ + (Bi, Pb) equivalents	$GaSr_2(Y, Ca)Cu_2O_7$
$Bi_2Sr_2Ca_3Cu_4O_{12+x}$	$NbSr_2(Nd, Ce)_2Cu_2O_{10}$
$Bi_2Sr_2Ca_2Cu_3O_8F_4$	(Nd, Pr, Sm)CeCuO
$Tl_2(Ba, La)_2CuO_{6+x}$	$(La, Pr, Nd)_{2-x}Ce_xCuO_{4-y}$
$Tl_2Ba_2CaCu_2O_{8+x}$	$Sr_2CuO_3\ F_x$
$Tl_2Ba_2Ca_2Cu_3O_{10+x}$	$Nd_2CuO_{4-d}F_\delta$
$Tl_2Ba_2Ca_3Cu_4O_{12+x}$	$(Nd, Ce, Sr)_2CuO_{4-y}$
$Tl(Ba, La)_2CuO_5$	$(La, RE)_2CuO_4$, where *RE* = Y, Lu, Sm, Eu, Gd, Tb
$TlBa_2CaCu_2O_7$	$(La, M)_2CuO_4$, where M = Ba, Sr, Ca
$(Tl_{0.5}, Pb_{0.5})Sr_2(Ca, Y)Cu_2O_7$	$(La, Sr)_2CaCu_2O_6$
$(Tl_{0.5}Pb_{0.5})(CaSr_{2-x})(La_xCu_2)O_7$	$(Eu, Ce)_2(Eu, Sr)_2Cu_3O_9$
$(Tl_{0.5}Sn_{0.5})Ba_2(Ca_{0.5}Tm_{0.5})Cu_2O_x$	$ACuO_2$, where A is the alkaline-earth metal
$TlBa_2Ca_2Cu_3O_9$ + (Tl, Pb)Sr equivalents	$[CaCu_2O_3]_4$
$TlBa_2Ca_3Cu_4O_{11}$	$(Cu_{0.5}Cr_{0.5})Sr_2Ca_{n-1}Cu_nO_{2n+3+d}$
$(Cu, Tl)Ba_2Ca_2Cu_3O_y$	$CaBa_2Ca_2Cu_3O_{9-\delta}$
$TlO_x(CO_3)_y(Sr, Ca)_{n+1}Cu_nO_z$	$CuBa_2Ca_3Cu_4O_{12-d}$
$(Ag, Cu)Ba_2Ca_3Cu_4O_{11-x}$	$(Sr, Ca)_{14}Cu_{24}O_{41}$
$Hg(Ba, La)_2CuO_{4+\delta}$	$Pb_2(Sr, La)_2Cu_2O_6$
$HgBa_2CaCu_2O_{6+\delta}$	$PbSr_2(La, Ca)Cu_3O_8$
$HgBa_2Ca_2Cu_3O_{8+\delta}$	$Ba_2RE(Ru, Cu)O_6$, where *RE* = Y, Pr
$HgBa_2Ca_2Cu_3O_{8+\delta}\ F_x$	$Sr_2RE(Ru, Cu)O_6$, where *RE* = Y, Pr
$(Hg, Re)Ba_2Ca_2Cu_3O_{8+\delta}$	$RuSr_2RECu_2O_8$, where *RE* = Eu, Gd, Y
$HgBa_2Ca_3Cu_4O_{10+\delta}$	$RuSr_2(Nd, Y, Gd, Ca, Ce)_2Cu_2O_{10}$
$HgBa_2Ca_4Cu_5O_{12+\delta}$	
$Hg_2Ba_2CaCu_2O_{7+\delta}$	

At the same time, the substitution of Cu by Zn or Ni leads to rapid degradation of superconductivity [1800]. This result agrees totally with two-dimensional picture, in accordance of which high-temperature superconductivity is caused by pairing into CuO_2 layers. If intactness of these layers is violated (as in the case of substitution of the copper atoms) then T_c decreases. If atomic disorder arises outside these layers (as in the case of substitution of the yttrium and barium atoms), then T_c does not change.

It is interesting to note that T_c values exceeding 117 K have been reached in a homologous series of compounds $Ag_{1-x}Cu_xBa_2Ca_{n-1}Cu_nO_{2n+3-x}$. A sample with nominal composition $(Ag,Cu)Ba_2Ca_3Cu_4O_{11-x}$ (n = 4) has been processed by a high pressure solid state synthesis in 1994 [737]. This family does not contain the toxic components (Bi, Tl, Pb, Hg) determined in other HTSCs with $T_c > 100$ K. The structure of $AgBa_2Ca_3Cu_4O_{10}$ is identified as body-centered tetragonal one. Its primitive unit cell is similar to that of Tl-1234 but with Ag_2O layers. If Ag randomly occupies Cu sites then the crystal structure becomes primitive tetragonal (a = 0.38635 nm and c = 1.8111 nm). In the ordered structure, eight of unit cells are centered by an Ag ion in b. c. t. cell.

Search of scaling relationships, that is, the laws, stating systematic and universe dependencies between characteristics of complex system, has an important value for HTSCs. One of the first discovered scaling relationships between physical values characterizing normal and superconductive states of HTSC was Uemura correlation [1720]: a direct proportionality between density of superfluid component, ρ_s, and critical temperature, T_c. However, this correlation was no universe because comparatively well described only underdoped HTSC, but violated in optimally doped and overdoped samples. Attempts to connect ρ_s with static specific conductivity, σ_{dc}, at the $T \sim T_c$ also did not lead to success.

Universe scaling relationship, proposed by Homes with co-workers [701], has the form

$$\rho_s = A\sigma_{dc}T_c. \tag{A.1}$$

It carries out *for all* HTSCs independently on value of T_c, charge carrier type (holes or electrons), doping level, crystalline structure and current direction (parallel or perpendicular to CuO_2 planes). The proportionality factor, $A = 120 \pm 25$, if ρ_s is measured in s^{-2}, σ_{dc} defined in $(\Omega{\cdot}cm)^{-1}$, but T_c determined in K. To the direct dependence (A.1) even belong data for low-temperature superconductor Pb and Nb. This straight line covers a diapason above five orders of magnitude on each of axes. The discovered empiric dependence is very surprising, taking into account the principally different character of current transfer in various crystallographic directions (i.e., coherent in *ab*-plane and incoherent along *c*-axis). By this, the value of $T_c \sim \rho_s / \sigma_{dc}$ is higher with decreasing electric conductivity in normal state.

Methods of nanotechnology can provide significant increasing pinning of magnetic flow and corresponding growth critical current. The pinning features in nanocomposite films of $YBa_2Cu_3O_7$, processed by solution method and containing chaotically oriented non-superconducting nanodots of $BaZrO_3$, Y_2O_3, $BaCeO_3$, Ba_2YTaO_6 with mean radius (10–20) nm have been studied in [1035]. As a result, a record pinning force (above 20 GN/m^3) has been attained in films with content of $BaZrO_3$ about 10 mol.% (see Fig. A.9).

Based on the conducted investigations in [1035], it has been suggested a novel model of magnetic flux pinning, according to which a strain of HTSC-matrix in vicinity of nanodots leads to extension of Cu–O chains and, as a result, de-pairing of the dots occurs at the adjacent nodes of copper and normal regions being the pinning centers are formed. This model allows one to understand why, in spite of on itself heterogeneity and anisotropy, the strain in nanoscale leads to quasi-isotropic pinning.

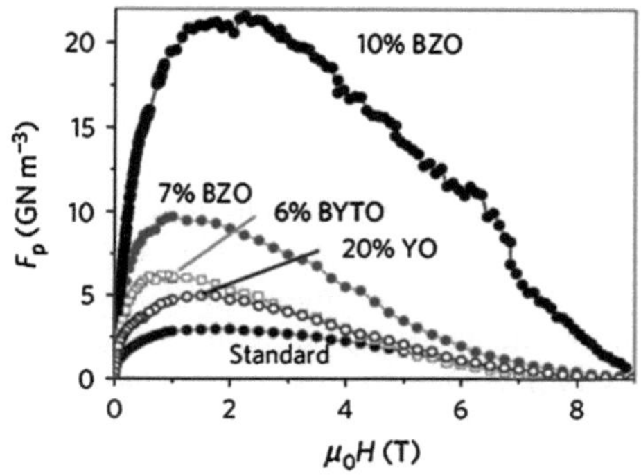

Fig. A.9 Pinning force at $T = 77$ K in different nanocomposites [1035]

High-temperature superconductivity in thin films [145] and in monocrystals [1888] of $PrBa_2Cu_3O_x$ has been discovered in 1996. The value of $T_c = 56.5$ K at pressure $P = 0$ in $PrBa_2Cu_3O_x$ monocrystal ($x = 6.6$) increases up to 105 K at $P = 9.3$ GPa [1834, 1889]. This result contrasts sharply with data for $YBa_2Cu_3O_x$, in which the value of $T_c \sim 90$ K at $x = (6.8–7)$ is almost independent of P up to $P = 10$ GPa. Obviously, the different responses of $PrBa_2Cu_3O_x$ and $YBa_2Cu_3O_x$ on high pressure is connected with various characters of distribution of the charge carriers between structure units of elementary cell, and with different re-distribution of the charge under pressure, respectively. The value of $T_c \approx 90$ K has been attained in the polycrystalline $PrBa_2Cu_3O_{7-d}$ samples [54]. Most probable cause of superconductivity in $PrBa_2Cu_3O_{7-d}$ is the partial substitution of Pr by Ba [1234].

HTSC $NdBa_2Cu_3O_x$ (Nd-123) holds set of records among *RE*-123 ($T_c \sim 95$ K, high irreversible line, etc.). The main advantages of Nd-123 consists in anomalous peak effect, leading to significant increasing of intragrain currents for account of formation of the effective pinning centers, which begin to work at liquid nitrogen temperature in magnetic field of some Tesla, that are most interesting for HTSC technical applications. Only in Nd-123, it is possible to attain the results, using *chemical methods*, which are comparable in effect to cumbersome, expensive and difficult-accessible methods of *physical* formation of the pinning centers (e.g., due to neutron irradiation or ion bombardment). Additional advantages of the Nd-123 phase are also connected with its higher chemical stability and higher rate of solidification. New pinning centers into Nd-123 are formed during de-lamination of re-saturated solid solution [691, 1215]. In this case, the sites of solid solutions form in the basic superconducting matrix. These sites are distributed homogeneously in the matrix and are coupled coherently with it, because they act as effective pinning centers in the external magnetic field. In the case of non-zero magnetic field, superconductivity in them is suppressed sharply, causing the peak effect. Due to these new pinning types, the irreversibility line in the Nd-123 samples displaces magnetic fields above 8 T at 77 K (the record value for *RE*-123 superconductors). Moreover, Nd and Ba can exchange places in the Nd-system due to favor combination of the ionic radii that forms the defects, serving strong pinning centers. Majority of HTSC have *p*-type of superconductivity ($YBa_2Cu_3O_7$,

$La_{2-x}Sr_xCuO_4$, $Bi_2Sr_2CaCu_2O_8$, etc.). $Nd_{2-x}Ce_xCuO_4$ is the most studied of no numerous HTSC with *n*-type conductivity. Note that antiferromagnetism is the main alternative to superconductivity as in electronic HTSC also as in hole-doped superconductors [809].

The structure of "infinite-layer" $ACuO_2$ (where *A* is the alkaline-earth metal) compounds presents a set of parallel CuO_2 layers, separated by the layers of the *A* atoms. They are simplest (in structure and chemical composition) copper-oxide superconductors and have no structure blocks, playing the role of the charge reservoirs. Due to this reason, they are ideal candidates for research of basic characteristics of CuO_2 plane, which is the main structure unit of copper-oxide HTSC. The critical temperature, $T_c \sim 60$ K, has been attained in $CaCuO_2$ monolayers, surrounded from both sides by $Ba_{0.9}Nd_{0.1}CuO_{2+x}$ layers, which carried out a function of the electric charge reservoirs. In this case, the measured "one-layer" critical current density, $J_c > 10^8$ A/cm^2 at $T = 4.2$ K [86].

The irreversibility line in the infinite-layer copper-oxide electronic HTSC $Sr_{0.9}La_{0.1}CuO_2$ samples without admixtures, synthesized under high pressure, disposed much higher than in $La_{1.85}Sr_{0.15}CuO_4$ and $Nd_{1.85}Ce_{0.15}CuO_4$ [800]. The critical current density was also much higher that proposed strong pinning of magnetic flux. Moreover, it has been stated that superconductivity of $Sr_{0.9}La_{0.1}CuO_2$ is three-dimensional in contrast to quasi-two-dimensional superconductivity of all HTSC cuprates [867].

Chalcogens

Superconductivity of sulfur with $T_c = 10$ K arises under $P = 93$ GPa, at $P = 200$ GPa, the value of T_c attains a maximum of 17.3 K, and at further increasing of the pressure again decreases down to 15 K. The similar dependences of T_c (P) are observed in Se and Te. For example, the superconducting transition takes place in Se at $T_c = (4–6)$ K in the range of $P = (15–25)$ GPa, while $T_c = 8$ K at $P > 150$ GPa [570].

The superconducting composites C/S, fabricated from press-powders of graphite and sulfur (23 wt%), which are synthesized in argon, have demonstrated $T_c = 35$ K (the superconducting transition has been found by Meissner's effect) [323]. In this case, the magnetization curves demonstrated hysteresis that is proper for type-II superconductors.

The controlled intercalation of copper into di-chalcogen $TiSe_2$ leads to initiation of the waves of charge density (WCD), which are periodical modulations of the density of conductive electrons, and also to arising of superconductivity at the copper content $x = 0.04$ in calculation per formal unit [1195]. The maximum $T_c = 4.15$ K has been observed at $x = 0.08$. Thus, a correlation of WCD with superconductivity could be studied in the example of Cu_xTiSe_2.

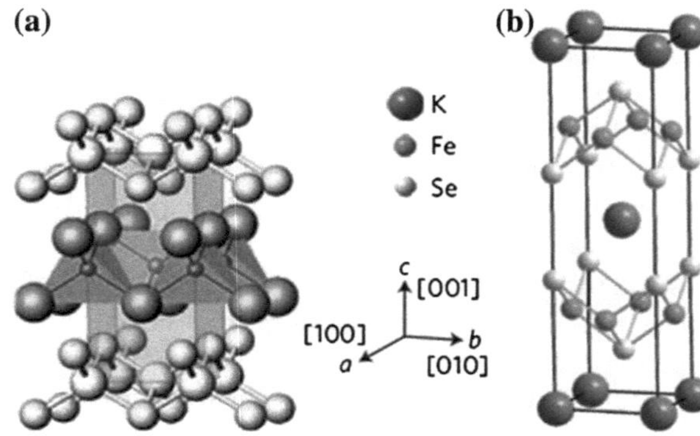

Fig. A.10 Crystalline structures of (**a**) $LaO_{1-x}F_xFeAs$; *red balls* present Fe atoms, *yellow* As, *white* O, and *blue* La, and (**b**) KFe_2Se_2

Superconductive Pnictides and Chalcogenides of Iron

In February 2008, the layered pnictides and chalcogenides of iron with high superconductive properties have been discovered [806] (structure of $LaO_{1-x}F_xFeAs$ is presented in Fig. A.10a). Today five families of the superconductors (FeSCs) have been determined and characterized [263, 591, 763, 767, 1238]:

(i) the 1111 family of *Ln*OFeAs (T_c = 56 K, where *Ln* = La, Ce, Pr, Sm, Nd);
(ii) the 122 family of $AEFe_2As_2$ (T_c = 38 K, where *AE* = Ca, Sr, Ba, Eu);
(iii) the 111 family of *A*FeAs (T_c = 21 K, where *A* = Li, Na);
(iv) the 011 family of FeSe (T_c = 14 K);
(v) the family of $A_{1-x}Fe_{2-y}Se_2$ (*A* = K, Cs, Rb, Rb/Tl) (T_c = 30–32 K).

Moreover, new materials are studying, for instance *M*FFeAs ($T_c \sim 56$ K, where *M* = Ca, Sr, Ba, Eu) with superconductivity arising due to modification with Ln^{3+} ions [269, 1809], $(Sr_4Sc_2O_6)(Fe_2P_2)$ with T_c = 17 K [1260], and $(Sr_4V_2O_6)(Fe_2As_2)$ with T_c = 37.2 K [1881].

The discovered in 2010 iron selenides with critical temperature up to 32 K [591], differ of other HTSCs without copper by presence of antiferromagnetic dielectric phases, very high Neél temperatures and ordered Fe-vacancies (structure of KFe_2Se_2 is presented in Fig. A.10b). In the chalcogenides $K_{0.8}Fe_{1.7}Se_2$, $K_{0.8}Fe_{1.78}Se_2$ and $Tl_{0.6}Rb_{0.4}Fe_{1.67}Se_2$ after total disappearance of superconductivity at pressure, $P \approx 9$ GPa, further increasing pressure leads not only to restoration of superconductivity, but growth of T_c up to 48 K, that is significantly higher than the first maximum 33 K at 1 GPa. The range of pressures in which is observed this "restored superconductivity" is comparatively short: from 11.5 up to 13.2 GPa. By this, superconductivity disappears also suddenly as it arises, that is, the usual "bell-like" dependence $T_c(P)$ is totally absent [1628].

While the chemical compositions of above materials are very different, all the superconductors based on pnictides and chalcogenides of iron have many similar properties from viewpoint of their crystalline structure due to the presence of layers of Fe atoms surrounded by ions of elements of V group of the Mendeleev Table, pnictogens (P, As) or elements of VI group, chalcogens (S, Se, Te) in tetrahedral configuration. These layers are divided by "blocking layers" from alkali, alkali-earth or rare-earth elements and oxygen/fluorine. The conducting FeAs or FeSe layers with an antifluorite structure are divided by positively charged layers, forming charge reservoirs, of various natures. Such layers into 1111 phases demonstrate the fluorite-like structure $(LnO)^+$, at the same time, in the 122 phases, these layers are consisted of alkali-earth metals AE^{2+}, and in the 111 phases from alkali metal ions (Li^+ or Na^+).

Inside the layers, the chains are predominantly of the covalent character, at the same time, the chains between the layers are ionic. In order to arising superconductivity, it is required to dope at least one type of the layers, which cause the hole or electron conductivity. This could be reached by partial substitution of fluorine for oxygen in the layers of *Ln*O superconductors of the 1111 family or owing to generation of a partial oxygen deficit at the electron conductivity. The same effect in the 122 structures is reached by partial substitution of alkali cations for the alkali-earth cations at the hole conductivity. Moreover, several modifications of the conducting layers are also possible.

So, the partial replacement of Fe atoms for Co or Ni leads to the superconductivity in the 122 phases. The pressure, as usually, also renders positive effect on the superconductivity of undoped 122 phases. The simplest chemical compositions containing the layers "transition metal–chalcogen" and demonstrating the superconducting properties are the $Fe_{1+x}Se$ compounds with the anti-PbO structure (T_c = 8 K) [286], with the critical temperature rising under the pressure to 37 K [690].

It is assumed that FeSCs superconductivity is caused by layers of Fe-pnictogen/chalcogen similarly to CuO_2 layers in cuprate HTSCs. Similarly to cuprates, a chemical substitution plays key role in superconductivity of FeSCs. However, there are significant differences, namely (i) atoms of pnictogens and chalcogens dispose over and under Fe-layers, at the same time, CuO_2 layers are practically plain, (ii) it is possible in FeSCs a chemical substitution of elements directly in "active layers" of Fe-pnictogen/chalcogen, but in cuprates—only in blocking buffer layers. Moreover, parent undoped state in FeSCs is metallic, but in cuprates—dielectric one.

Phase diagram of FeSCs is very similar to those in some other unconventional superconductors, including cuprates: during doping or under high pressure non-superconducting antiferromagnetic state is replaced by superconducting, non-magnetic one. In some FeSCs (for example, $LaFeAsO_{1-x}F_x$) the superconducting and antiferromagnetic phases are divided in phase diagram, but in other FeSCs they are overlapping (the last variant exists more often). Similarity of FeSCs phase diagrams at chemical substitution and compression (see Fig. A.11) together with the same character of changing lattice periods in both cases suggest that concentration of charge carriers is no unique factor (and maybe even no main one) defining disposition of interphase boundaries and in this case a greater role is played by features of crystalline structure which causes electronic one. Note also other feature of FeSCs, namely coincidence or proximity of temperatures of the antiferromagnetic and structural transitions (Fig. A.11).

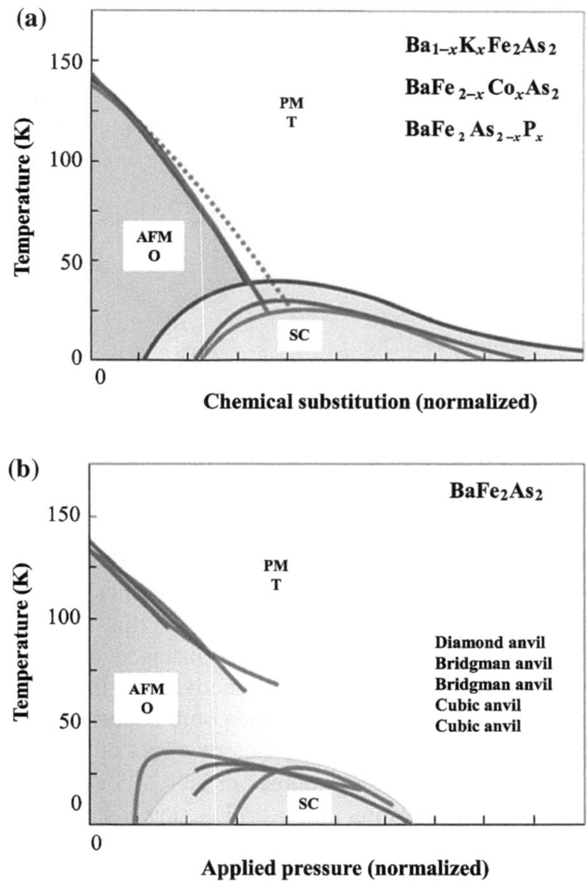

Fig. A.11 Phase diagrams of $BaFe_2As_2$ at chemical substitution (**a**), and hydrostatic compression (**b**). The next designations: *PM* paramagnetism, *AFM* antiferromagnetism, *SC* superconductivity, *T* tetragonal phase, *O* orthorhombic phase; the *dotted line* shows temperature of structural transition [1286]

The physical properties of the FeSCs superconductive phases are depended on distance between the conducting layers. For example, the maximal value of T_c increases together with basal spacing (the Fe–Fe distance) according to the superconductivity conception [720]. At increasing the depth of the charge reservoir, the distance between the conducting layers in iron-containing superconductors gradually increases from ~5.5 Å in the 011 phase up to 6.35–7.0 Å in the 111 and 122 phases and attains 8.4–8.7 Å in the 1111 family; in the superconducting $(Sr_4V_2O_6)(Fe_2As_2)$, this distance is equal to 15.7 Å [1881]. Other important structural parameter is the As–Fe–As angle which would be near to 109.5° in order to obtain highest T_c, which is possible for these families.

Fermi surface in FeSCs consists of hole pockets in the center of Brillouin zone and electronic ones in its corners (see Fig. A.12). The main contribution in the electronic state density is caused by *d*-orbitals of Fe atoms [1286]. In relation of superconductivity mechanism for FeSCs, majority researches assume non-phonon nature of pairing (for instance, for account of spin fluctuation exchange).

Fig. A.12 Fermi surface in $BaFe_{2-x}Co_xAs_2$. *Violet* and *blue colors* show hole and electronic pockets [1286]

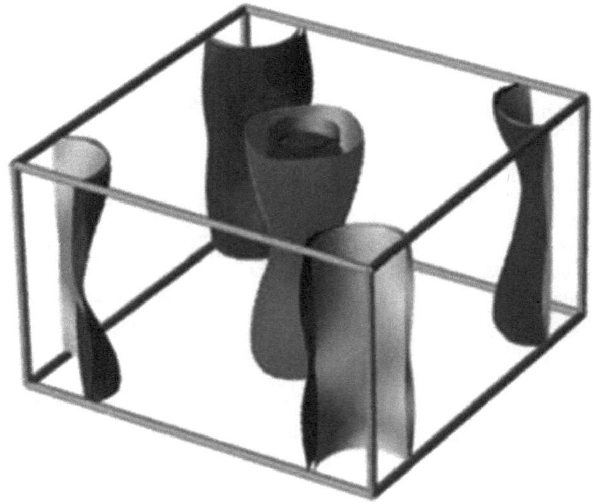

This could be argued by presence in all FeSCs (similar to cuprates) collective resonance magnetic mode at the $T < T_c$, whose energy increases proportionally to T_c with growth of T_c. However this mode maybe non-related with superconductivity, but arises, for instance due to decreasing rate of magnon relaxation on electrons of conductivity at the superconductive transition. It is possible in FeSCs, there is some combination of phonon and non-phonon pairing due to observation of strong isotope effect on iron.

Spin state of Cooper couples in FeSCs is usual singlet that follows from NMR data. At the same time, their orbital state rather no *s*-wave and no *d*-wave (as in cuprates), but $s_{\pm}$-wave, that is the superconductive order parameter has the same sign on each of pockets of Fermi surface, but different one on hole and electronic pockets. This allows one to explain majority experiments (including phase-sensitive ones). However, it is not excluded presence in FeSCs several states almost degenerated on energy of superconducting states with different symmetries of pairing. Then any smallest perturbation (for example, defects) could lead to "switching" from one symmetry type to other and to non-unique interpretation of experimental data.

Rare-Earth Borocarbides

The rare-earth borocarbide $LuNi_2B_2C$ has critical temperature, T_c, approximately 16 K, at the same time, LuNiBC is not superconducting. The superconductor is obtained, adding carbon in Lu plane. On the other hand, LuNiBC is obtained from $LuNi_2B_2C$, adding the other layer of Lu–C. Thus, the superconducting composition is fabricated from insulator by changing corresponding atom ratio. Similarly in $La_{2-x}Sr_xCuO_4$, superconducting phase is formed from insulating phase by changing

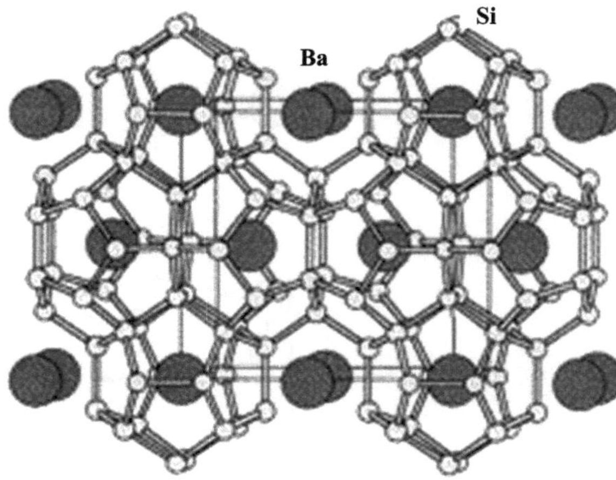

Fig. A.13 Crystal structure of silicon clathrate superconductor Ba_8Si_{46} [1830]

parameter x [1201]. The compounds $RENi_2B_2C$ with RE = Y, Lu, Tm, Er, Ho and Dy are superconducting with moderate high critical temperatures, $T_c \approx 16$ K. At low temperatures, there is antiferromagnetic phase, and the phase exists in which can co-exist superconductivity and magnetic long-range order. Based on the example of $Ho_{1-x}RE_xNi_2B_2C$ (RE = Y, Lu) system with different numbers of Ho(RE)C layers, a clear correlation of T_c with the density of states in Fermi level has been found. The various magnetic structures have been derived which co-exist with superconductivity. The highest $T_c = 23$ K is attained in YPd_2B_2C.

Silicon-Based Superconductors

Up to recent time, only one superconducting compound of silicon family, namely $ThSi_2$ with $T_c = 1.56$ K, has been known. However, recently, it has been stated that semimetal $CaSi_2$ becomes a superconductor with $T_c = 14$ K at pressure $P > 12$ GPa [1477].[2] In contrary to known crystalline silicon structures, the various cavities can form at slightly distorted angles between links. Clathrate crystals are the compounds consisting of molecules and atoms ("guests") into cavities of frame of the crystalline lattice ("host"). Clathrate structures form on the base of different matters with tetrahedral coordination of atoms. Two types of clathrates are known, namely the silicon compounds M_xSi_{46} and M_xSi_{136} (where M = Na, K, Rb, Cs). If part of the alkaline atoms in this "friable" (but right) clathrate structure of the silicon compound M_xSi_{46} (where M = Na, K) is substituted by Ba, then this structure becomes superconducting [834]. Clathrates $Na_2Ba_6Si_{46}$ and Ba_8Si_{46} have critical temperatures, $T_c = 4$ K [834] and 8 K [1825], respectively. Crystal structure of Ba_8Si_{46} is presented in Fig. A.13.

[2] Other silicon-based superconductors: CaAlSi ($T_c = 7.8$ K [1050]); SrAlSi ($T_c = 5.1$ K [1050].

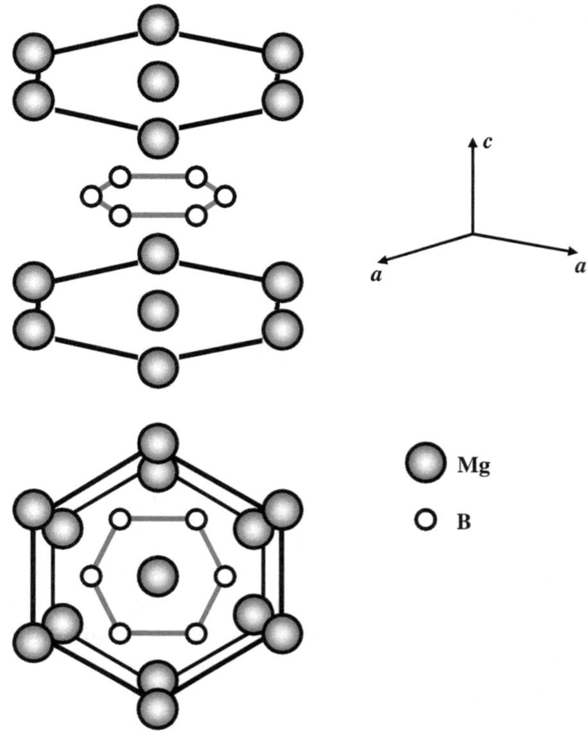

Fig. A.14 Crystal structure of MgB_2 [1224]

MgB_2 and Related Superconductors

The superconducting system MgB_2 discovered in 2001 demonstrates highest volume superconductivity (critical temperature, $T_c = 39$ K) among non-copper oxide conductors [1224], excluding superconductive oxipnictides discovered in 2008.[3] Crystal structure of MgB_2 (relates to Laves phases [991]), having spatial group of symmetry *P6/mmm*, is shown in Fig. A.14 (where $a = 3.086$ Å, and $c = 3.524$ Å are the parameters of hexagonal unit cell). It is most surprising that simplicity and accessibility of this material, known from the beginning of 1950 [792], combines with complexity of the superconductivity phenomenon.

Even in non-textured multiphase MgB_2 samples, superconducting currents flow in all volume of the sample and are not sensitive to weak magnetic field (which should suppress J_c at the existence of Josephson links between grains). The temperature scaling of the pinning strength practically in all superconducting H–T plane supports the concept that critical current density is found by the magnetic

[3] WO_3 monocrystals, doped by Na, demonstrate high-temperature superconductivity with $T_c = 91$ K in the *surface layer* with content of $Na_{0.05}WO_3$ [677]. The nature of the observed superconductivity shows that in this case, supercurrents flow on the crystal surface but not into its volume.

Table A.8 Characteristics of Nb_3Ge (*A*-15 compound), BKBO ($x \approx 0.4$) and MgB_2

No.	Compound	T_c (K)	v_F (10^7cm/s)	ξ(Å)	$2\Delta/k_BT_c$	B_{c2}(T)
1	Nb_3Ge	23	2.2	35–50	4.2	38
2	BKBO	31	3.0	35–50	4.5	32
3	MgB_2	39	4.8	35–50	4.5	39

T_c is the critical temperature; v_F is Fermi rate; ξ is the coherence length; Δ is the energy gap; k_B is Boltzmann constant; and B_{c2} is the upper critical magnetic field

flux pinning [982]. It has been shown in experiments that both the transport (i.e., intergranular) J_c and its dependence on H and T, totally coincide with those for inductive (i.e., intragranular) J_c. The test results prove that grain boundaries in MgB_2 are transparent for superconducting current [361]. It is observed small, but evident anisotropy of critical field, $H_{c2}^{ab}/H_{c2}^{c} = 1.1$ in the bulks [630]. At the same time, it is found that $H_{c2}^{ab}/H_{c2}^{c} = 1.8$–$2.0$ in MgB_2 films with *c*-axis, oriented perpendicular to the film surface [1343]. The high-qualitative-oriented MgB_2 films has critical current density, $J_c = 1.6 \times 10^7$ A/cm^2 at $T = 15$ K and $H = 0$ [864]. It is stated that T_c and residual electric resistivity, ρ_0, depend only on thickness, d, of the epitaxial MgB_2 films, but are independent of the film growth rate. An increase in d leads to increasing T_c and decreasing of ρ_0. Now, the values of $T_c = 41.8$ K and $\rho_0 = 0.28$ μΩ·cm at $d > 300$ nm are maximal and minimal magnitudes, respectively [1372]. The record value of $H_{c2} = 52$ T at $T = 4.2$ K has been reached in one sample of thin MgB_2 films in the field parallel to the film surface [469].

The irreversibility field, H_{irr}, in Nb_3Sn, Nb–Ti and MgB_2 is equal to 20, 10 and 14 T at $T = 4.2$ K, respectively [443]. Generally, the characteristics of the superconductors MgB_2, BKBO and *A*-15 (see Table A.8) [1201] show similar values. Therefore, this group of superconductors can be considered as intermediate between the conventional superconductors, subjected to BCS theory, and the superconductors with high critical temperature.

Today MgB_2 are usually processed by reaction of magnesium and boron precursor powders after their mixing and heating at a temperature above the melting point of pure magnesium ($T > 650$ °C). In this process MgB_2 powder are typically of 1–10 μm. The attained values of critical current density for MgB_2-coated conductors in Fe-sheath [784]: $J_c > 85$ kA/cm^2 ($T = 4.2$ K and $H = 0$) and $J_c = 23$ kA/cm^2 ($T = 20$ K and $H = 0$) are nearly sufficient for current-carrying cables.[4]

In order to improve properties long-length applications of MgB_2, it is necessary to use nanotechnology methods. With this aim low cost nanoboron powder could be used to process nano-size MgB_2 powder by application, for instance of techniques of the fluidized-bed for continuous processing and spark ablation for minimizing agglomeration. Due to nanoboron will be converted into MgB_2 at strictly controlled parameters (processing temperature, Mg pressure, heating rate,

[4] MgB_2-coated conductor (or tape), fabricated by laser evaporation on flexible metallic tape has $J_c = 1.1 \times 10^5$ A/cm^2 ($T = 4.2$ K and $H = 10$ T) [908].

etc.), avoiding unwanted nanoparticle size growth and reducing agglomeration. Moreover, the controlled surface passivation of the MgB_2 nanoparticle surface could be used to reduce the particle increasing at the final sintering stage of MgB_2 applied in processing superconductive wires. Additionally, at the processing MgB_2 nanoparticles, could be used the techniques of the nanoparticle size refinement, de-agglomeration, activation and high-energy planetary ball milling [612]. At the final stage, for MgB_2 long-length applications (wires, tapes, etc.) could be used the conventional Powder-In-Tube (PIT) method due to MgB_2 demonstrates well chemical and mechanical compatibility with majority metals which could be cold worked into wires, excepting copper.

Due to the overall critical current of a superconductor is caused in high degree by connectivity of cross-section in the sintered porous powder sample, then to attain high J_c performance, it is required to increase the packing density and attain higher connectivity between superconducting particles. In this sense, the beneficial effects of cold densification at reproducibly attaining a double increasing of the wire performance in short length wires have been demonstrated in [710]. Today the critical current density of MgB_2 does not exceed 10^6 A/cm^2 and is useful only at moderate magnetic fields (up to 2–4 T). Obviously, the increased critical current and field performance can be attained by using nano-dopants and inclusions during superconductor processing. The high compatibility of MgB_2 with many alloys also ensures a unique opportunity for processing low AC loss composite superconductor with fine filaments. In contrary to HTSCs which are only compatible with noble materials (silver, gold, platinum), the minimum filament size in MgB_2 is limited by the texture requirement and heterogeneity caused by residual weak links.

Several compounds, related to MgB_2, reach the following critical temperatures: BeB_2 (T_c = 0.79 K [1850]); Re_3B (T_c = 4.7 K [1619]); ZrB_2 (T_c = 5.5 K [524]); ReB_2 (T_c = 6.7 K [1619]); TaB_2 (T_c = 9.5 K [803]); $Nb_{0.76}B_2$ (T_c = 9.2 K at P = 5 GPa [1824]); $MgCNi_3$ (T_c = 8 K [656]); $MgAlB_4$ (T_c = 12 K [1012]). The calculations of electronic structure and constant of electron–phonon interaction predict that AgB_2 and AuB_2 should be high-temperature superconductors with T_c = 59 and 72 K, respectively [959].

Room-Temperature Superconductivity

There are a set of papers reporting superconductivity (or its possibility) at or above room temperature. In the first paper, superconductivity is observed in a thin surface layer of the complex compound $Ag_\beta Pb_6CO_9$ ($0.7 < \beta < 1$) at 240–340 K [380]. The second paper claims that superconductivity exists in carbon-based multiwall nanotubes at $T > 400$ K [1873]. Several papers and patents, published by Tripodi with co-workers, have been devoted a possible room-temperature superconductivity in palladium–hydrogen (PdH_x) samples for stoichiometric ratio $x = H/Pd \geq 1$ [1702–1705]. This superconductive phase, on opinion of the authors, has critical temperature into range 51 K $\leq T_c \leq$ 295 K, and measured critical current density,

$J_c \geq 6.1 \times 10^4$ A/cm^2 (at $T = 77$ K and $H_{DC} = 0$). The first results on that superconductivity in PdH_x have been published in 2003, however they have not been supported and repeated by scientific society up to date (in particular, a clear Meissner effect measurement for the PdH_x samples has not be stated). In a whole, today there is only one book devoted to room-temperature superconductivity [1202].

Appendix B
Finite Element Implementation of Carbon-Induced Embrittlement Model

Consider the finite element implementation of the governing equations (5.14) and (5.27) for carbon diffusion and non-mechanical energy flow, respectively [1309]. The next initial and boundary conditions supplement the governing equations:

$$C^{CT} = C_0^{CT}, \quad T = T_0, \quad \text{at } t = 0; \tag{B.1}$$

$$C^{CT} = C_b^{CT}, \ \text{ on } S_b; \quad J_k^C n_k = \varphi^C, \ \text{ on } S_\varphi; \tag{B.2}$$

$$T = T_s, \ \text{ on } S_T; \quad -k\frac{\partial T}{\partial x_i} n_i = \phi^E, \ \text{ on } S_F, \tag{B.3}$$

where C_0^{CT} and T_0 are the initial carbon concentration and temperature, which may vary within material volume *V*. If C_0^{CT} is larger than carbon terminal solid solubility, the initial carbon concentration in carbonate equals the terminal solubility of carbon in carbonate, and the initial carbon volume fraction is calculated according to (5.15). A similar comment is valid for C_b^{CT}, which is prescribed carbon concentration on S_b, a part of the bounding surface *S*. For the calculation of C^{TS}, stress and temperature distributions are taken into account. φ^C is the prescribed carbon flux on S_φ; φ^E is the prescribed heat flux on S_F, and T_s is the prescribed temperature on S_T. Note that $S_b \cup S_\varphi = S_T \cup S_F = S$. The quantities C_b^{CT}, φ^C, *T* and φ^E may vary with time.

The finite element equations are derived from variation descriptions of diffusion and energy flow. For this purpose variations of carbon concentration, δC^C, and temperature, δT, are considered, which satisfy the boundary conditions. Therefore,

$$\delta C^C = 0, \ \text{ on } S_b; \tag{B.4}$$

$$\delta T = 0, \ \text{ on } S_T. \tag{B.5}$$

I. A. Parinov, *Microstructure and Properties of High-Temperature Superconductors*, DOI: 10.1007/978-3-642-34441-1,

Relation (5.14) is multiplied by a carbon concentration variation satisfying (B.4). Subsequently it is integrated over the volume V. Then, taking into account (5.11), (5.12) and (5.31), the following expression is derived, which is valid at any time t:

$$\begin{aligned}\int_V \delta C^C \frac{dC^{CT}}{dt} dV = & -\int_{S_\varphi} \delta C^C \varphi^C dS - \int_V fD^C \frac{\partial C^C}{\partial x_k} \frac{\partial(\delta C^C)}{\partial x_k} dV \\ & - \int_V f\left(-\frac{D^C \overline{V}^C}{3RT} \frac{\partial \sigma_{mm}}{\partial x_k} + \frac{D^C Q^C}{RT^2} \frac{\partial T}{\partial x_k}\right) C^C \frac{\partial(\delta C^C)}{\partial x_k} dV.\end{aligned} \tag{B.6}$$

The above relation is simplified if, within a time increment, Δt, the change of carbonate volume fraction is included in C^C. Value of Δt is order of a characteristic time, introduced by diffusion and carbonate size, or smaller. Then

$$\frac{dC^{CT}}{dt} = f_t \frac{dC^C}{dt}. \tag{B.7}$$

Note that C^C is no longer the carbon concentration in carbonate, but the carbon concentration in a part of the material, which at time t is in the form of carbonate and has volume $f_t V$.

Spatial discretization is obtained by introducing the usual finite element interpolation for carbon concentration, carbonate volume fraction, temperature and stress trace. For example, carbon concentration is calculated from nodal values as

$$C^C = a_q C_q^C, \tag{B.8}$$

where a_q and C_q^C are the interpolation function and the nodal carbon concentration value for q-node, respectively. Substitution of (B.7) into (B.6) and use of spatial discretization lead to the finite element equations for carbon diffusion:

$$C_{pq} \frac{dC_q^C}{dt} + \left(D_{pq}^1 + D_{pq}^2\right) C_q^C = F_p, \tag{B.9}$$

where

$$C_{pq} = \int_V f_t a_p a_q dV, \quad D_{pq}^1 = \int_V f_t D^C \frac{\partial a_p}{\partial x_k} \frac{\partial a_q}{\partial x_k} dV; \tag{B.10}$$

$$D_{pq}^2 = \int_V f_t \left(-\frac{D^C \overline{V}^C}{3RT} \frac{\partial a_r}{\partial x_k} \sigma_{mm}^r + \frac{D^C Q^C}{RT^2} \frac{\partial a_s}{\partial x_k} T_s\right) a_q \frac{\partial a_p}{\partial x_k} dV; \quad F_p = -\int_{S_\varphi} a_p \varphi^C dS. \tag{B.11}$$

The time derivative of carbon concentration is approximated by

$$\frac{dC_q^{C,t+\Delta t}}{dt} = \frac{1}{\Delta t}\left(C_q^{C,t+\Delta t} - C_q^{C,t}\right). \tag{B.12}$$

Relation (B.9) is taken at time $t + \Delta t$. Equation (B.12) is substituted into (B.9) and the D^2_{pq}-term is transferred to the right hand, leading to the next relation (similar to the one developed in [1595] for hydrogen diffusion due to concentration and stress gradient):

$$\left(\frac{1}{\Delta t} C_{pq} + D^1_{pq}\right) C_q^{C,t+\Delta t} = F_p - D^2_{pq} C_q^{C,t} + \frac{1}{\Delta t} C_{pq} C_q^{C,t}. \tag{B.13}$$

The matrices C_{pq}, D^1_{pq} and D^2_{pq} are calculated by using the known nodal values of carbonate volume fraction, temperature and stress from the previous calculation step. The value of carbonate volume fraction corresponds to time t. However, the values of temperature and stress correspond to time $t + \Delta t$, according to the discussion at the end of Appendix *B*. Vector F_p is calculated from the boundary conditions at time $t + \Delta t$. Note that the solution of (B.13) provides a preliminary carbon concentration value, $C^{C,pr}_{t+\Delta t}$, which according to (B.7) may include carbon in carbonate. This preliminary value can be used for the determination of the total carbon concentration

$$C^{CT}_{t+\Delta t} = f_t C^{C,pr}_{t+\Delta t}, \tag{B.14}$$

as well as the new carbonate volume fraction, $f_{t+\Delta t}$:

$$f^{pr} = \frac{C^{CT}_{t+\Delta t} - C^{TS}_{t+\Delta t}}{C^C - C^{TS}_{t+\Delta t}}; \quad f_{t+\Delta t} = \begin{cases} 0, & f^{pr} < 0; \\ f^{pr}, & 0 \le f^{pr} \le 1, \end{cases} \tag{B.15}$$

where $C^{TS}_{t+\Delta t}$ is calculated based on the results for temperature and stress from the previous calculation step.

The new carbon concentration in the carbonate is derived, based on (B.15), as

$$C^C_{t+\Delta t} = \begin{cases} C^{C,pr}_{t+\Delta t}, & f^{pr} < 0, \\ C^{TS}_{t+\Delta t}, & 0 \le f^{pr} \le 1. \end{cases} \tag{B.16}$$

Then, the finite element implementation of the governing equations for non-mechanical energy flow will be obtained. Relation (5.27) is multiplied by a temperature variation, satisfying (B.5). Subsequently, it is integrated over volume V and the following expression is derived, which is valid at any time t:

$$\int_V \delta T \rho c_p \frac{dT}{dt} dV + \int_V \delta T \frac{\Delta \overline{H}^{car}}{\overline{V}^{car}} \frac{df}{dt} dV = -\int_{S_F} \delta T \varphi^E dS - \int_V k \frac{\partial T}{\partial x_i} \frac{\partial(\delta T)}{\partial x_i} dV - \int_V \delta T J_n^{CT} \frac{\partial \mu^C}{\partial x_n} dV. \tag{B.17}$$

As in the case of the carbon diffusion, spatial discretization is introduced and the following finite element equations are derived:

$$H_{qr}\frac{dT_r}{dt} + K_{qr}T_r = \Phi_q^1 + \Phi_q^2 - L_{qs}\frac{df_s}{dt}, \tag{B.18}$$

where

$$H_{qr} = \int_V \rho c_p a_q a_r dV; \quad K_{qr} = \int_V k\frac{\partial a_q}{\partial x_i}\frac{\partial a_r}{\partial x_i}dV; \tag{B.19}$$

$$\Phi_q^1 = -\int_{S_F} a_q \varphi^E dS; \quad \Phi_q^2 = -\int_V a_q J_n^C \frac{\partial \mu^C}{\partial x_n} dV; \quad L_{qs} = \int_V \frac{\Delta \overline{H}^{car}}{\overline{V}^{car}} a_q a_s dV. \tag{B.20}$$

Assuming that temperature time derivative is given by

$$\frac{dT_r^{t+\Delta t}}{dt} = \frac{1}{\Delta t}\left(T_r^{t+\Delta t} - T_r^t\right), \tag{B.21}$$

and following an approach similar to that for carbon diffusion, one may derive the next algebraic system from (B.18)

$$\left(\frac{H_{qr}}{\Delta t} + K_{qr}\right)T_r^{t+\Delta t} = \Phi_q^1 + \Phi_q^2 - L_{qs}\frac{df_s^t}{dt} + \frac{H_{qr}}{\Delta t}T_r^t, \tag{B.22}$$

where Φ_q^1 is calculated from the boundary conditions at time $t + \Delta t$. In order to estimate Φ_q^2 the nodal values of temperature, carbon concentration, carbonate volume fraction and stress from the previous calculation step are used.

A complete calculation cycle is as follows. At time t, all field quantities are known: $u_i^t, \varepsilon_{ij}^t, \sigma_{ij}^t, C_t^C, f_t, T_t, \frac{dC_t^C}{dt}, \frac{df_t}{dt}, \frac{dT_t}{dt}$, where u_i are the components of the displacement vector of a material particle. A time increment Δt is considered and the following calculation steps are performed:

(1) Material deformation problem is solved first. The boundary conditions of the applied traction and/or displacements are defined at time $t + \Delta t$. The isotropic expansion strain rate due to carbon dissolution, carbonate formation and thermal expansion is calculated by using the values and time rates of carbon concentration, carbonate volume fraction and temperature at time t. The parameters of the de-cohesion model are also derived from carbonate volume fraction and temperature distribution at time t. By performing calculation step (1), $u_i^{t+\Delta t}, \varepsilon_{ij}^{t+\Delta t}, \sigma_{ij}^{t+\Delta t}$ are calculated.

(2) The energy flow problem is solved next. The boundary conditions of the applied surface temperature and/or heat flux are defined at time $t + \Delta t$. Φ_q^2-term is calculated, based on the distributions of temperature, carbon concentration and carbonate volume fraction, at time t, as well as on the distribution of stress, $\sigma_{ij}^{t+\Delta t}$, calculated within step (1). The carbonate volume fraction rate, at time t, is used for the calculation of L_{qs}-term. By performing calculation step (2), $T_{t+\Delta t}$ is calculated.

(3) The carbon diffusion problem is solved at the end. The boundary conditions of the applied surface carbon concentration and/or carbon flux are defined at time $t + \Delta t$. In all terms, f_t is used. D^2_{pq}-term is calculated, based on the values $\sigma_{ij}^{t+\Delta t}$ and $T_{t+\Delta t}$, derived using steps (1) and (2), respectively. By performing calculation step (3), $C^C_{t+\Delta t}$ and $f_{t+\Delta t}$ are calculated.

The material deformation problem is solved, assuming a constant value of Young's modulus. The error in the calculations is further minimized, considering the value of E for the temperature in the crack-tip region, where embrittlement and fracture processes operate. When the variation of temperature, either in space or in time, is significant, its effect on elastic moduli should be taken into account (see, e.g., [1397]).

Adequate numerical results by using the finite element implementation can be obtained after carrying out preliminary tests, estimating properties of carbon, cuprate, carbonate and YBCO superconductor, which are necessary for calculation.

Appendix C
Macrostructure Modeling of Heat Conduction

Method of Summary Approximation for Quasi-Linear Equation of Heat Conduction

For macrostructure modeling of the heat front propagation during sintering and cooling of HTSC ceramic, the method of summary approximation (MSA) is used as a method for construction of economic schemes for quasi-linear non-stationary equations in the case of arbitrary region and any number of measurements p [1474].

The quasi-linear equation of heat conduction without heat sources (heat capacity, c_V, and material density, ρ, are suggested to be constant) has the form:

$$\begin{gathered}\frac{\partial u}{\partial t} = Lu; \quad x = \left(x_1, x_2, \ldots, x_p\right) \in G, \ t > 0; \\ Lu = \sum_{\alpha=1}^{p} L_\alpha u; \quad L_\alpha u = \frac{\partial}{\partial x_\alpha}\left[k_\alpha(u, x_\alpha)\frac{\partial u}{\partial x_\alpha}\right]; \quad k_\alpha \geq C_\alpha > 0\end{gathered} \tag{C.1}$$

with boundary and initial conditions:

$$u|_\Gamma = \mu(x,t); \quad u(x,0) = u_0(x); \quad x \in \overline{G}, \tag{C.2}$$

where u is the temperature; x_α are the spatial coordinates; t is the time and α is the fixed coordinate direction. A boundary Γ of a region G is sufficiently smooth, which is necessary for the existence of smooth solution, $u = u(x, t)$. It is assumed that smooth derivations are required.

As it is made in all economic schemes, the process of approximate solution for multidimensional problem is divided into several stages. Simple problem is solved at every stage. The operator L is presented by the operator sum of more simple structures:

I. A. Parinov, *Microstructure and Properties of High-Temperature Superconductors*, DOI: 10.1007/978-3-642-34441-1,

$$L = \sum_{\alpha=1}^{p} L_\alpha. \tag{C.3}$$

We consider multidimensional equation (C.1) and compare to problem (C.1), (C.2) a "chain" of equations:

$$\sum_{\alpha=1}^{p} P_\alpha u = 0; \quad P_\alpha u = \frac{1}{p}\frac{\partial u}{\partial t} - L_\alpha u. \tag{C.4}$$

At the interval, $0 \le t \le t_0$, an uniform lattice, $\varpi_\tau = \{t_j = j\tau; \quad j = 0, 1, \ldots, j_0\}$ with a step, $\tau = t_0/j_0$, is introduced. Every interval is divided into p parts, introducing points $t_{j+\alpha/p} = t_j + \alpha\tau/p$; $\alpha = 1, 2, \ldots, p - 1$. Successively (at $\alpha = 1, 2, \ldots, p$),

$$P_\alpha V_{(\alpha)} = 0; \quad x \in G; \quad t \in \left(t_{j+(\alpha-1)/p}; \, t_{j+\alpha/p}\right]; \quad \alpha = 1, 2, \ldots, p, \tag{C.5}$$

are solved suggesting

$$V_{(1)}(x, 0) = u_0(x); \; V_{(\alpha)}\left(x, t_{j+(\alpha-1)/p}\right) = V_{(\alpha-1)}\left(x, t_{j+(\alpha-1)/p}\right). \tag{C.6}$$

Also, it is assumed that the boundary condition of first type is given at Γ. The solution of this problem is named

$$V\left(x, t_j\right) = V_{(p)}\left(x, t_j\right); \quad j = 0, 1, \ldots, j_0. \tag{C.7}$$

Every of the equations, $P_\alpha V_{(\alpha)} = 0$, it is substituted by the difference scheme (approximating $\frac{\partial u}{\partial t}$ and L_α by corresponding difference relations at uniform lattice ω_h with steps $h_1, h_2, \ldots, h_p$)

$$\Pi_\alpha y_{(\alpha)} = 0; \quad \alpha = 1, 2, \ldots, p. \tag{C.8}$$

The scheme (C.8) approximates the equation $P_\alpha V_{(\alpha)} = 0$ in usual sense, that is

$$\Pi_\alpha u^{j+\alpha/p} - (P_\alpha u)^{j+\alpha/p} \to 0, \text{ at } \tau \to 0 \text{ and } h_\alpha \to 0. \tag{C.9}$$

The summary approximation of additive scheme (C.8) is attained due to the "chain" of differential equations (C.5), and (C.6) approximates corresponding equations (C.5) in usual sense.

Then, it is assumed that L_α consists of derivations only on variable x_α. Therefore, L_α is the one-dimensional operator, $P_\alpha V_{(\alpha)} = 0$ are the one-dimensional equations and additive scheme (C.8) is the local one-dimensional scheme (LOS). We write LOS, and with this aim the multidimensional equation is replaced by "chain" of one-dimensional equations:

$$\frac{1}{p}\frac{\partial V_{(\alpha)}}{\partial t} = L_\alpha V_{(\alpha)}, \tag{C.10}$$

at

$$t_{j+(\alpha-1)/p} < t \leq t_{j+\alpha/p};\ \alpha = 1, 2 \ldots p; \quad x \in G; \quad t_{j+\alpha/p} = (j + \alpha/p)\tau, \tag{C.11}$$

with conditions

$$V_{(1)}(x,0) = u_0(x);\ V_{(\alpha)}(x, t_{j+(\alpha-1)/p}) = V_{(\alpha-1)}(x, t_{j+(\alpha-1)/p});$$
$$V_{(\alpha)} = \mu(x,t), \quad \text{at}\, x \in \Gamma_\alpha. \tag{C.12}$$

For difference approximation of the operator L_α in node x_i, a three-point templet is used, consisting of the points: $x_i^{(-1\alpha)}, x_i, x_i^{(+1\alpha)}$, where $x_i^{(\pm 1\alpha)} = [x_1^{(i_1)}, \ldots, x_\alpha^{(i_\alpha)} \pm h_\alpha, \ldots, x_p^{(i_p)}]$; $x_\alpha^{(i_\alpha)} = h_\alpha\, i_\alpha; h_\alpha$ is the step of the lattice ω_h along α-direction. A number of internal nodes of the lattice ω_h consists of the points $(x_1, x_2, \ldots, x_p) \in G$ of crossing of the hyper-planes $x_\alpha = i_\alpha\, h_\alpha$; $i_\alpha = 0, \pm 1, \pm 2, \ldots$; $\alpha = 1, 2, \ldots, p$, but a number of boundary nodes γ_h consists of the points of crossing of the straight lines C_α, passing through all internal nodes $x \in \omega_h$, with boundary Γ. Also introduced are the next designations: $\gamma_{h,\alpha}$ is the number of boundary nodes along direction x_α; γ_h is the number of all boundary nodes $x \in \Gamma$.

Let $\overline{G} = \{0 \leq x_\alpha \leq l_\alpha\}$ is the parallelepiped, then Γ_α consists of the facets: $x_\alpha = 0$ and $x_\alpha = l_\alpha$. Approximating every heat conduction equation of number α at semi-interval $(t_{j+(\alpha-1)/p}; t_{j+\alpha/p}]$ by a two-layer scheme with weights, a "chain" of p one-dimensional schemes is stated that is called LOS:

$$\frac{y^{j+\alpha/p} - y^{j+(\alpha-1)/p}}{\tau} = \Lambda_\alpha[\sigma_\alpha y^{j+\alpha/p} + (1-\sigma_\alpha) y^{j+(\alpha-1)/p}], \tag{C.13}$$

where $\alpha = 1, 2, \ldots, p$; $x \in \omega_h$; $\sigma_\alpha \in [0, 1]$.

In regular nodes, Λ_α has second order of approximation, $\Lambda_\alpha u - L_\alpha u = O(h_\alpha^2)$, and in non-regular nodes, $\Lambda_\alpha u - L_\alpha u = O(1)$. Consider purely implicit LOS ($\sigma_\alpha \equiv 1$):

$$\frac{y^{j+\alpha/p} - y^{j+(\alpha-1)/p}}{\tau} = \Lambda_\alpha y^{j+\alpha/p}, \tag{C.14}$$

and join to this equation the boundary condition:

$$y^{j+\alpha/p} = \mu^{j+\alpha/p}, \text{ at } x \in \gamma_{h,\alpha};\ j = 0, 1, \ldots, j_0;\ \alpha = 1, 2, \ldots, p, \tag{C.15}$$

and initial condition:

$$y(x,0) = u_0(x). \tag{C.16}$$

Let us assume y^j is known. In order to define y^{j+1} at new layer from (C.14) and (C.15), the p equations (C.14) are required to solve together with the boundary condition (C.15), successively suggesting $\alpha = 1, 2, \ldots, p$. In order to define $y^{j+\alpha/p}$, we have the boundary-value problem:

$$A_{i_\alpha} y_{i_\alpha - 1}^{j+\alpha/p} - C_{i_\alpha} y^{j+\alpha/p} + A_{i_\alpha + 1} y_{i_\alpha + 1}^{j+\alpha/p} = 0, \quad \text{at } x \in \omega_h; \tag{C.17}$$

$$y^{j+\alpha/p} = \mu^{j+\alpha/p}, \text{ at } x \in \gamma_{h,\alpha};\ \alpha = 1, 2, \ldots, p. \tag{C.18}$$

Here, the lower indexes are pointed only, which change in calculations. The difference equation is written along the section of the straight line, the ends of which coincide with the nodes $\gamma_{h,\alpha}$. Eq. (C.17) is solved by the run method at fixed α-direction along corresponding sections. Successively suggesting α = 1, 2,..., p and changing the run directions, $y^{j+1/p}$, $y^{j+2/p}$, ..., y^{j+1} are calculated, expending $O(1)$ operations per lattice node. Thus, the LOS (C.14)–(C.16) is economic [1474]. Moreover, it can be shown that the LOS approximation error tends to zero at $\tau \to 0$ and $h_\alpha \to 0$, and from summary approximation, uniform convergence of the LOS with rate, $O\left(\tau + \max\limits_{1 \le \alpha \le p} h_\alpha^2\right)$ is followed [1474].

Heat Conduction of Heterogeneous Systems

A study of heat conduction of the heterogeneous systems is a sufficiently complex problem. In the case of HTSC ceramics, it is even more complicated due to significant porosity and different, compared to metal, mechanism of heat conduction. In metal, demonstrating a small porosity and high heat conduction of crystallites, the main mechanism of heat conduction is the convection. In oxide superconductors, possessing relatively greater porosity, together with heat transfer on solid component, the main sources of heat transfer are molecular and radiant components of heat conduction.

Effective Heat Conduction of Mixes and Composites

Existing mixes and composites can be presented by one of the models, depicted in Fig. C.1. A process of ceramic sintering can be described by using structure with mutual-penetrating components (see Fig. C.1b) and the loose granular material (see Fig. C.1e). All components of the structure with mutual-penetrating components are continuous in any direction and geometrically equivalent relative to effective heat conduction:

$$\lambda_{ef} = f_1(\lambda_1, \lambda_2) = f_2(\lambda_2, \lambda_1), \text{ at } m_1 = m_2, \tag{C.19}$$

where λ_i and m_i (i = 1, 2) are the heat conduction and concentration of the components.

Granular materials consist of monolithic particles (1) (see Fig. C.1e) and occupy an intermediate state between the structures with impregnation and the structures with mutual-penetrating components. The contacting particles and pores

(2), disposed between them, form continuous extent of solid components and cavities in any direction.

Following [408], define heat conduction of granular system. First, consider a structure with mutual-penetrating components. In order to take into account the distortion of the heat flux lines, we use the following formulae for effective heat conduction in the cases of adiabatic and isothermal division of elementary cell, respectively:

$$\lambda_{ad} = \lambda_1 \left[c^2 + \nu(1-c)^2 + \frac{2c(1-c)\nu}{1+c(\nu-1)} \right], \quad \nu = \lambda_2/\lambda_1; \tag{C.20}$$

$$\lambda_{is} = \lambda_1 \left[\frac{1-c}{c^2+\nu(1-c^2)} + \frac{c}{c(2-c)+\nu(1-c)^2} \right]^{-1}, \tag{C.21}$$

here c is the solution of the equation:

$$m_2 = 2c^3 - 3c^2 + 1; \quad m_2 = 1 - m_1. \tag{C.22}$$

Hence

$$c = 0.5 + A\cos(\varphi/3), \tag{C.23}$$

where

$$A = \begin{cases} -1; & \varphi = \arccos(1-2m_2), \ \text{at } 0 \le m_2 \le 0.5; \\ 1; & \varphi = \arccos(2m_2 - 1), \ \text{at } 0.5 \le m_2 \le 1, \end{cases} \tag{C.24}$$

and $3\pi/2 \le \phi \le 2\pi$. Then, effective heat conduction is defined as

$$\lambda_{ef} = (\lambda_{ad} + \lambda_{is})/2. \tag{C.25}$$

The approximate solution (C.25) gives values, which are near to numerical results and preserves property of component invariability.

Polystructural Model of Granular Material

A heat transfer into pores occurs due to molecular collisions and radiation. At the same time, convection is absent, as a rule. The molecular transfer of heat takes place due to interchange by kinetic energy at collisions of moving molecules with one another and with surface of solid or liquid component, limiting pores (grain and liquid surfaces). The heat transfer due to radiation occurs on account of the absorption, emission and dissipation of radiant energy. Both mechanisms can exist together and influence each other mutually.

The granular structure of ceramic powder consists of "frame" (1) (see Fig. C.1e), formed by the chaotic, but relatively dense package of continuously contacting grains (the structure of first order) and spatial lattice of more larger cavities (2), penetrating the powder, which together with the frame form the structure of second

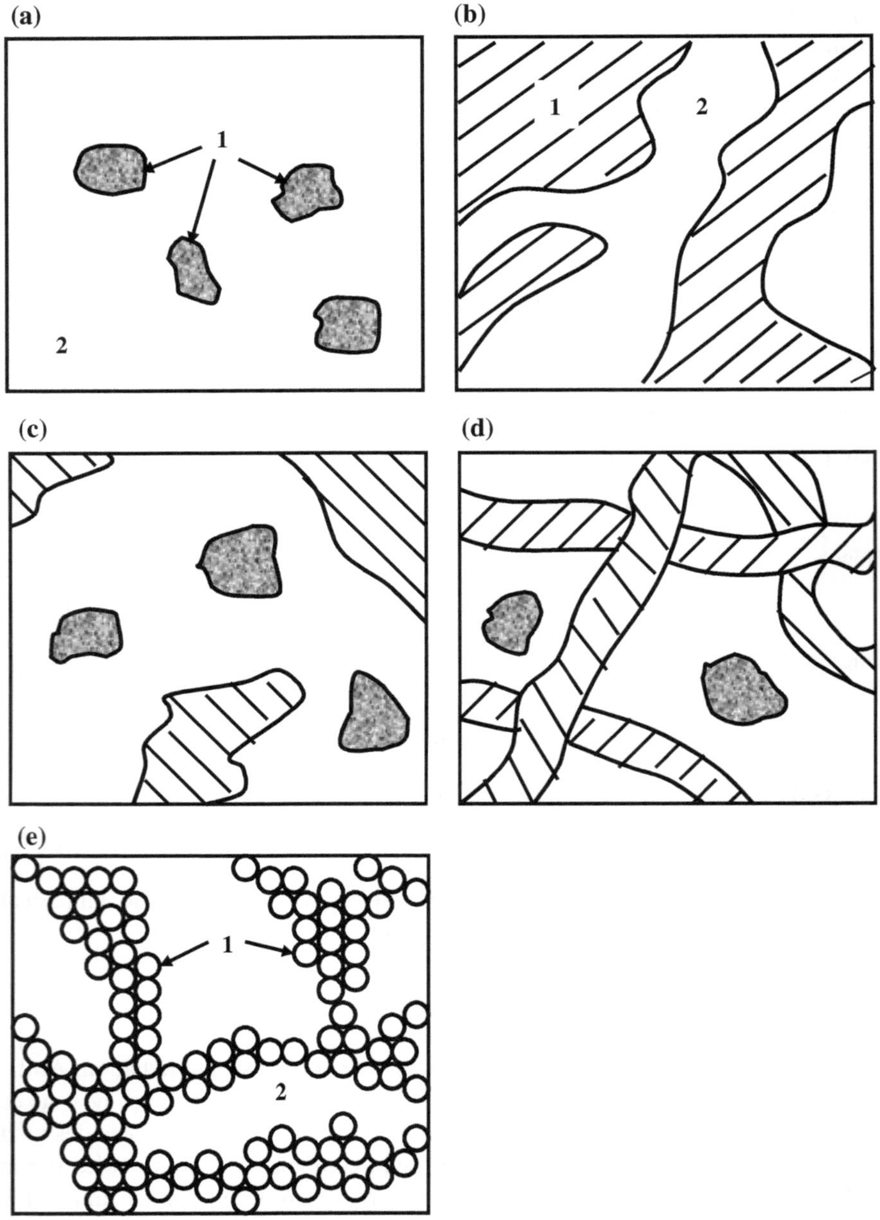

Fig. C.1 Heterogeneous systems with different structure: (**a**) impregnated structure; (**b**) structure with mutual-penetrating components; (**c**, **d**) combined structures with mutual-penetrating components and impregnation; (**e**) loose granular material

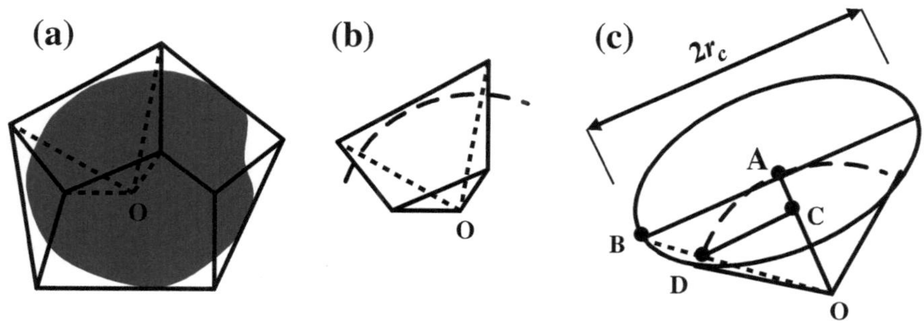

Fig. C.2 Substitution procedure used for definition of coordination number as function of porosity: (**a**) polyhedron formed due to crossing the planes tangential to particle in contact points; (**b**) pyramid is the element of polyhedron; (**c**) straight circular cone

order with mutual-penetrating continuous components. No destroying generality, it may be assumed that the structure of second order occurs at $m_2 \geq 0.4$ [409].

First, a dependence of coordination number, N_c (i.e., number of contacts per one particle) on the porosity m_2 is defined. For this, a granular system is considered, consisting of convex rounded particles. In all points of the particle contacts, we depict tangential planes. Then, volume of the system is divided into polyhedrons, circumscribed around any particle. In this case, the facet number of every polyhedron is equal to the number of contacts for given particle (Fig. C.2a).

The porosity of the system is presented through ratio of the difference between volumes of all polyhedrons and particles to volume of the polyhedrons. Determination of required dependence for the parameter N_c is carried out, using arbitrary polyhedron, circumscribed around particle with mean radius, r.

Then, N_c is the mean coordination number for all polyhedrons of the system. For chaotic actual structure, it is assumed that all contacts are uniformly distributed on particle surface. Then, it is suggested that the polyhedron consists of N_c identical pyramids with tops at the particle center (Fig. C.2b). In this case, mean porosity of pyramid is equal to mean porosity of the granular system. For simplicity, the pyramid is substituted by cone with the same solid angle and around-contact surfaces are replaced by spherical ones (Fig. C.2c).

Then, the porosity of this system is equal to

$$m_2 = 1 - V_{bs}/V_c; \quad V_c = 0.33\pi r_c^2 r, \tag{C.26}$$

where V_c is the cone volume, V_{bs} is the ball sector in the cone, r_c is the radius of the cone base, and r is the radius of the spherical surface.

In order to define r_c, we use equalities: $F_b/N_c = F_s = 2\pi r h_s$. Hence

$$h_s = 2r/N_{c,} \tag{C.27}$$

where F_b, F_s are the surface squares of the ball and ball segment and h_s is the segment height. Because ΔOAB is similar to ΔOCD, then from (C.27), we obtain

$$r_c = 2r(N_c - 1)^{1/2}/(N_c - 2). \tag{C.28}$$

The volume of the ball sector is smaller by N_c times than the volume of the ball

$$V_{bs} = 4\pi r^3/3N_c. \tag{C.29}$$

By using the formula for N_c and (C.26), (C.28) and (C.29), we obtain finally

$$N_c = \left[m_{2f} + 3 + \left(m_{2f}^2 - 10m_{2f} + 9\right)^{1/2}\right]/2m_{2f}. \tag{C.30}$$

Then, effective heat conduction of the structure of second order with mutual-penetrating components is calculated by using (C.20), reduced to the form

$$\lambda = \lambda_f\left[c_2^2 + \nu_f(1 - c_2)^2 + \frac{2\nu_f c_2(1 - c_2)}{\nu_f c_2 + 1 - c_2}\right]; \quad \nu_f = \lambda_{22}/\lambda_f. \tag{C.31}$$

Here, the geometrical parameter c_2 characterizes volume concentration of frame and is connected with the porosity, m_{22} (in the second order structure), through an equation of type (C.22):

$$m_{22} = 2c_2^3 - 3c_2^2 + 1. \tag{C.32}$$

The volume concentration of pores, m_2, in granular system with volume V and pore volume V_2 is equal to $m_2 = V_2/V$, but the porosities of the frame, m_{2f}, and the second order structure, m_{22}, are found as

$$m_{2f} = V_{2f}/\left(V_1 + V_{2f}\right); \quad m_{22} = V_{22}/V, \tag{C.33}$$

where V_1, V_{2f} and V_{22} are the volumes of particles, pores in the frame and pores in the second order structure. Moreover, there are next relations:

$$V = V_{22} + V_{2f} + V_1; \quad V_2 = V_{22} + V_{2f}. \tag{C.34}$$

Then, we obtain from (C.33) and (C.34)

$$m_{22} = \left(m_2 - m_{2f}\right)/\left(1 - m_{2f}\right). \tag{C.35}$$

In order to calculate heat conduction, using (C.31), it is necessary to know the heat conduction of component, filling pores of the second order structure, λ_{22}, and heat conduction of the frame, λ_f. The value of λ_{22} is the sum of molecular and radiant components. It depends both as on the physical properties of gas as on the geometrical and physical parameters of the pores. For structure of granular material, the molecular component is calculated as [408]

$$\lambda_{2m} = \frac{\lambda_g}{1 + B/(H\delta_{2c})}, \tag{C.36}$$

here $\delta_{2c} = 3d(1 - c_2)/c_2$ is the mean size of great cavities; $d = 2r$ is the grain diameter; $B = \frac{4\gamma(2-a)\Lambda_c H}{(\gamma+1)a\mathrm{Pr}}$; where λ_g is the heat conduction factor of gas in the infinite space at pressure H and temperature T; a is the accommodation factor of gas at uniform walls; Λ_c is the mean length of the gas molecule run; $\gamma = c_p/c_V$ is the adiabatic index, being a ratio of isobaric heat capacity to isochoric; $\Pr = \nu/k$ is Prandtl's criterion, being the ratio of kinematic viscosity, ν, to temperature conductivity of gas, k.

Relation for radiant component of the heat conduction factor in pores of the second order structure has the form [408]:

$$\lambda_{2r} \approx 0.23(T/100)^3 \frac{Yd}{c_2^2(1-c_2)(2-\varepsilon)}, \qquad \text{(C.37)}$$

where $Y = f(\tau, \varepsilon)$ is the function, taking into account influence of the optic thickness of sample, $\tau = \beta l_l$, and of the blackness degree, ε, limiting surfaces (walls); for "gray" approximation, $\beta = \alpha_\lambda + \gamma_\lambda$ is the spectrum factor of weakening; α_λ and γ_λ are the volume spectrum factors of absorption and scattering; l_l is the thickness of filling layer. The value of $Y \approx 1$ fulfills for granular systems with porosity $m_2 < 0.95$ [408]. We obtain from (C.36) and (C.37) relation for the heat conduction of gas component in pores of the second order structure:

$$\lambda_{22} = \lambda_g \left[1 + \frac{Bc_2}{3Hd(1-c_2)}\right]^{-1} + 0.23(T/100)^3 \frac{Yd}{c_2^2(1-c_2)(2-\varepsilon)}. \qquad \text{(C.38)}$$

Model of Granular System with Chaotic Structure

In order to model the granular system with chaotic structure, consider a system consisting of rounded absolute solid particles with heat conduction factors that are greater than corresponding parameter of component, occupying pores. Main fraction of heat flux passes through the regions surrounding point contacts of particles (the sizes of near-contact regions are much smaller than grain diameter). Then, we divide the heat flux into single flux tubes, so that the tube axis in every particle passes successively the near-contact regions at entry and exit of the flux (Fig. C.3a).

Assumption 1 Heat conduction of any tube is equal to effective heat conduction of all granular system.

It is assumed that the tube length is much greater than the cross-section length of particles with non-elongated shape, which fill the granular system volume chaotically. We divide the tube into elements: $i-1, i, i+1,\ldots$ Every element is limited by two planes perpendicular to the heat flow, namely the plane in contact point and the plane δ–δ, dividing particle in half (Fig. C.3). Lateral surface of the tube is formed by adiabatic surface.

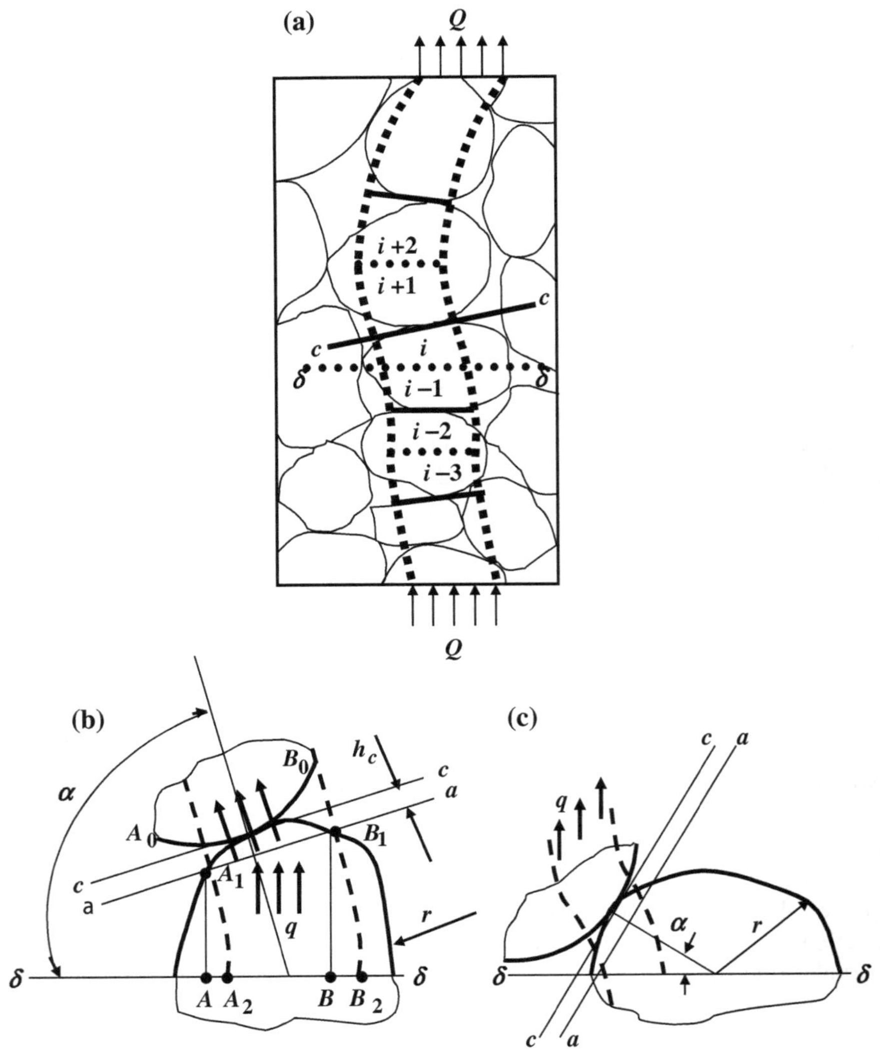

Fig. C.3 Account of chaotic character of particle package in frame: (**a**) flux tubes in the frame; (**b**) element of the flux tube of 1st type; (**c**) element of the flux tube of 2nd type

Thermal resistance of the tube is equal to the sum of thermal resistance of its elements, which are divided into two types. The elements without (first type) and with (second type) through pores are shown respectively in Fig. C.3b and *c*. The through pores are present in only those elements of the tube (second type) for which *a–a* plane (Fig. C.3c) contacts with δ–δ plane within tube.

The averaged element for case of the ordered cubic package of balls is shown in Fig. C.4a. Any ball contacts with six other balls in the points K, L, M, N, O, P.

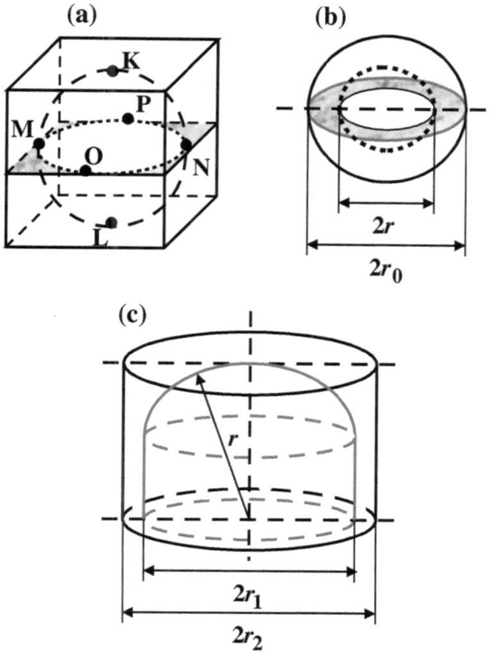

Fig. C.4 Definition of averaged geometrical parameters of elements of the flux tube in the frame: (**a**) disposition of contacts and through pores in cubic package of balls; (**b**) mean cross-section through pores; (**c**) element of chaotic structure with averaged parameters

Four contacts (points M, N, O, P) belong to through pores; the cross-section area of the through pores is shaded.

First, consider a heat transfer in the elements of first type (with distorted boundaries of elements). Let the thermal resistance of the distorted element ($A_2A_1A_0B_0B_1B_2$) (see Fig. C.3b) be the resistance of "straightened" element ($A_0A_1ABB_1B_0$) with the lateral adiabatic surfaces parallel to the direction of general heat flux, but cross-section square is the same as initial element square. Then, the distortion of flux lines in the "straightened" element occurs only in the plane of near-contact region.

Assumption 2 The near-contact regions of particle in any element of flux tube are formed by spherical surfaces with mean radius r.

The square of spherical surface per contact, S_c, is the equal to ratio of total square of the particle surface, $S = 4\pi r^2$, to the coordination number, N_c, that is, $S_c = 4\pi r^2/N_c$. A cross-section of grain part in the flux tube is presented in the form of circle with radius r_1 (Fig. C.4c). Dependence between parameters r_1 and N_c is stated, taking into account the following relations:

$$r_1^2 = h_{bs}(2r - h_{bs}); \quad S_c = 2\pi r h_{bs}, \tag{C.39}$$

where h_{bs} is the height of ball segment. Hence

$$y_1 = r_1/r = 2(N_c - 1)^{1/2}/N_c. \tag{C.40}$$

Using (C.30) and (C.40), the dependence $y_1 = y_1(m_2)$ may be stated. The heat resistance of the first type element is equal to the sum of straight ($A_1A_0B_0B_1$) and truncated (AA_1B_1B) cylinders (Fig. C.3b). Heat resistance of the truncated cylinder is equal to

$$R_{tc} = h_{tc}/(\lambda_1 S_{tc}), \tag{C.41}$$

where h_{tc} and S_{tc} is the mean height and base square of the truncated cylinder, respectively. Moreover

$$S_{tc} = S_{bs} \sin \alpha, \tag{C.42}$$

where S_{bs} is the base square of the straight cylinder;

$$h_{tc} = (r - h_{bs}) \sin \alpha, \tag{C.43}$$

where r is the ball radius and h_{bs} is the height of the straight cylinder. Then, we have from (C.41) to (C.43)

$$R_{tc} = (r - h_{bs})/(\lambda_1 S_{bs}). \tag{C.44}$$

Because geometrical parameters of the straight cylinder depend on coordination number only, then the heat resistance of the first type element does not depend on contact location. Therefore, during the study of heat transfer through elements of first type, it may be restricted by only considering the central element.

In the second type elements, contact points are disposed near interface plane and heat transfer depends significantly on fraction of through pores. Due to the existence of first, second and mixed type elements in chaotic structure, it is assumed that effective heat conduction of flux tube is equal to the heat conduction of combined element with averaged parameters, that is, the first type element with mean fraction of the through pores belonging to the second type element. In order to define mean fraction of the through pores, tangential planes, forming spatial polyhedron, are drawn about any particle in the points of contact with neighbors (Fig. C.2a). Average porosity of these polyhedrons and all systems are coincided.

Assumption 3 Irregular shape of particles (in the form of polyhedron) can be substituted by system of equal volume concentric balls with the same porosity (Fig. C.4b).

It is followed from Fig. C.4b that

$$r_0/r = (1 - m_2)^{-1/3} = r_2/r, \tag{C.45}$$

where $m_2 = (V_0 - V_1)/V_0$; V_0 and V_1 is the volume of external ball and particle, respectively and m_2 is the porosity. Quantitative estimation of the through pores is obtained by using the ratio of the through pore square per particle to the mean cross-section square of the particle.

Assumption 4 Through pores may be presented in the form of cylinders with annular base, enclosing central element of the tube (Fig. C.4c). In this case,

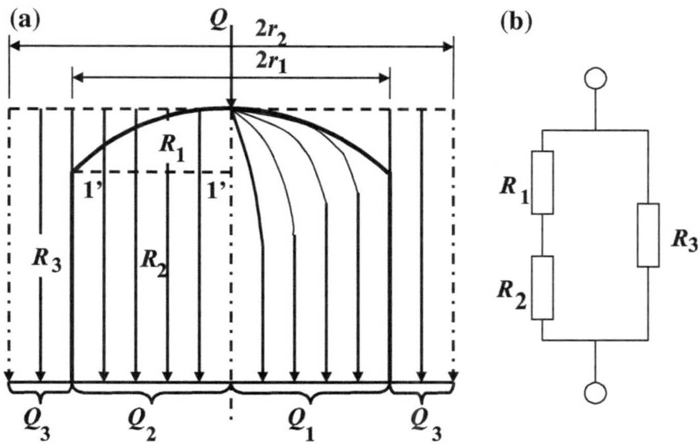

Fig. C.5 Definition of heat resistance of chaotic structure element of the granular system: (**a**) distribution of heat fluxes (actual flux spreading is shown at *right hand*, and idealized picture at *left hand*), (**b**) schematic junction of the heat resistances for single sections

relative square of the cylinder bases is the same as relative square of the through pores: $\pi(r_0^2 - r^2)/\pi r^2 = \pi(r_2^2 - r_1^2)/\pi r_1^2$.

We have from (C.40) and (C.45)

$$y_2 = r_2/r = y_1(1 - m_2)^{-1/3}. \tag{C.46}$$

Equations (C.40), (C.45) and (C.46) define finally the averaged geometrical parameters of the considered system (Fig. C.4). Presented model takes into account the existence of continuous contacts of particles in any direction (the stability condition) and isotropy of the granular system with chaotic structure. One is applied for porosity in the range of $0 \leq m_2 \leq 0.4$.

Heat Flux Through Averaged Element

After estimation of the averaged geometrical parameters of granular system, heat flux through an averaged element is investigated. As this element, we consider an element with bases presenting isothermal planes, and lateral surfaces are adiabatic (see Fig. C.5). A division by adiabatic surfaces parallel to the heat flux gives overestimated value for heat resistance, but a division by isothermal surfaces leads to underestimated value. A division of mixed type leads to more accurate results with error about 10 %.

General heat flux, Q, is presented by the sum: $Q = Q_1 + Q_2$, where fluxes Q_1 and Q_2 pass through particle and through pore in the averaged element, respectively. Then, heat resistances R_1, R_2 and R_3 are found. The resistance R_1 is related to spherical section of solid particle down to intermediate isotherm $1'$–$1'$ (Fig. C.5a). Then, the pointed region is divided into annular adiabatic planes.

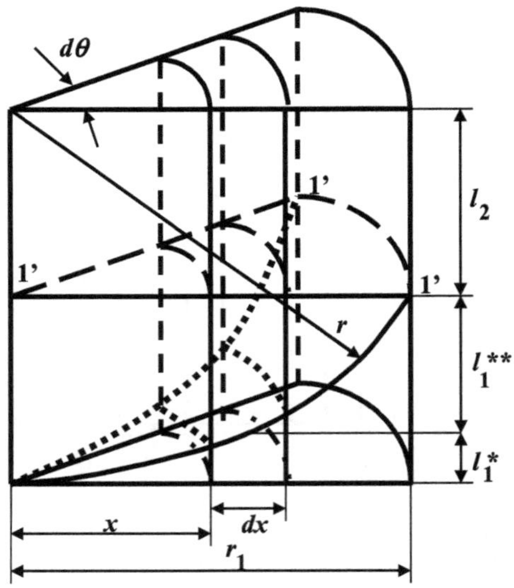

Fig. C.6 Definition of heat resistance of the element's sections with averaged parameters

Conductivity of single layer, $d\sigma_1$, with thickness, dx, is calculated as (see Fig. C.6) [408]

$$d\sigma_1 = \left(\frac{l_1^*}{\lambda_s dS_1} + \frac{l_1^{**}}{\lambda_1 dS_1}\right)^{-1} = dS_1\left(\frac{r - \sqrt{r^2 - x^2}}{\lambda_s} + \frac{\sqrt{r^2 - x^2} + \sqrt{r^2 - r_1^2}}{\lambda_1}\right)^{-1}, \tag{C.47}$$

where $dS_1 = xdxd\theta$.

Because the radiant component into gap between spheres is significantly smaller than the molecular one ($\lambda_{sr} << \lambda_{sm}$), then in integration of (C.47), we take into account only the molecular heat transfer. Hence, we have from (C.36)

$$\lambda_s \approx \lambda_{sm} = \lambda_g\left[1 + \frac{B}{2l_1^*(x)H}\right]^{-1}. \tag{C.48}$$

Total conductivity of the gap between spheres, taking into account only molecular transfer is

$$\sigma_{1m} = \int_0^{2\pi}\int_0^{r_1} d\sigma_{1m} = \frac{2\pi\lambda_g r}{1 - v_g}\left(D - 1 + \omega_m \ln\frac{\omega_m - D}{\omega_m - 1}\right), \tag{C.49}$$

where

$$D = \sqrt{1 - y_1^2};\ y_1 = r_1/r;\ \omega_m = [1 - v_g D + B/(Hd)]/(1 - v_g);\ v_g = \lambda_g/\lambda_1. \tag{C.50}$$

Now, we take into account a contribution of the radiant component of the heat conduction factor into gap between spheres, summarizing it with the molecular component. We obtain from (C.49) and (C.50)

$$\sigma_1 = \frac{2\pi\lambda_s r}{1 - \nu_s}\left(D - 1 + \omega_s \ln\frac{\omega_s - D}{\omega_s - 1}\right), \tag{C.51}$$

where

$$\omega_s = [1 - \nu_s D + B/(Hd)]/(1 - \nu_s); \quad \nu_s = \lambda_s/\lambda_1; \quad \lambda_s = \lambda_{sm} + \lambda_{sr}. \tag{C.52}$$

The value of λ_{sr} into gap between spheres is estimated approximately, replacing a complex shape of the gap through buffer layer with thickness δ_s, which is equal to the gap thickness between spheres, to being average-integral value on square, πr_1^2 (see Fig. C.6), that is,

$$\lambda_{sr} \approx 4\varepsilon_l \sigma_{S-B} T^3 \delta_s, \tag{C.53}$$

here

$$\delta_s = \frac{1}{\pi r_1^2}\int_0^{2\pi}\int_0^{r_1} dV_s, \tag{C.54}$$

where $dV_s = l_1^* dS_1 = (r - \sqrt{r^2 - x^2})x dx d\theta$; $\varepsilon_l = \varepsilon/(2 - \varepsilon)$; and $\sigma_{S\text{–}B}$ is Stefan–Boltzmann constant.

At integration of (C.54), it is taken into account that dependence $r_1 = f(N_c)$ is defined by relation (C.40). Then, we have from (C.54)

$$\delta_s = d/N_c. \tag{C.55}$$

(a) Finally, the heat resistances are found as

$$R_1 = 1/\sigma_1; \tag{C.56}$$

(b) the heat resistance, R_2, of the section between intermediate isotherm $1'-1'$ and upper base of the element (see Fig. C.5a) is

$$R_2 = \frac{l_2}{\lambda_1 \pi r^2} = \frac{\sqrt{1 - y_1^2}}{\lambda_1 \pi r y_1^2}; \tag{C.57}$$

(c) the heat resistance of the through pore, R_3, is

$$R_3 = \frac{r}{\lambda_{2p}\pi(r_2^2 - r_1^2)} = \frac{1}{\lambda_{2p}\pi r(y_2^2 - y_1^2)}; \quad y_2 = r_2/r, \tag{C.58}$$

where λ_{2p} is the heat conduction of gas into through pores with height equal to the particle radius; one is computed by using (C.38).

By using heat conduction, R_i, of single sections (C.51)–(C.58) and taking into account their schematic junction (see Fig C.5b), effective heat resistance, R, of element with averaged parameters is calculated as

$$R = f(R_i). \tag{C.59}$$

On other hand, we obtain by assuming that all volume of the element is filled by homogeneous matter with effective heat conduction, λ_f, which is equal to effective heat conduction of the granular system frame:

$$R = r/(\lambda_f \pi r_2^2). \tag{C.60}$$

Finally, we find effective heat conduction of the frame, λ_f, from (C.59) and (C.60):

$$\lambda_f = \frac{\lambda_1}{y_2^2}\left\{\left[\frac{D}{y_1^2} + \frac{(1-v)}{2v_s}\left(D - 1 + \omega_s \ln\frac{\omega_s - D}{\omega_s - 1}\right)^{-1}\right]^{-1} + \frac{\lambda_{2p}E}{\lambda_1}\right\}; \tag{C.61}$$

where

$$\begin{aligned} &D = (1-y_1^2)^{1/2}; E = y_2^2 - y_1^2; \omega_s = \frac{1 - v_s(1-y_1^2)^{1/2} + B/(Hd)}{1 - v_s};\\ &v_s = \frac{\lambda_s}{\lambda_1}; v_{2p} = \frac{\lambda_{2p}}{\lambda_1}; v_g = \frac{\lambda_g}{\lambda_1}; B = \frac{4\gamma(2-a)\Lambda_c H}{(\gamma+1)a\mathrm{Pr}}. \end{aligned} \tag{C.62}$$

This formula could be used to calculate effective heat conduction system at $m_2 < 0.5$.

Appendix D
Computational Algorithms for Definition of Structure-Sensitive Properties of Some HTSC Ceramics and Composites

Here we provide some computational algorithms for definition of microstructure, strength and conductive properties of some HTSC ceramics and composites of Y(*RE*)BCO and BSCCO compositions (see Figs. D.1, D.2, D.3, D.4, D.5, D.6, D.7, D.8) [1310, 1311].

I. A. Parinov, *Microstructure and Properties of High-Temperature Superconductors*,
DOI: 10.1007/978-3-642-34441-1,

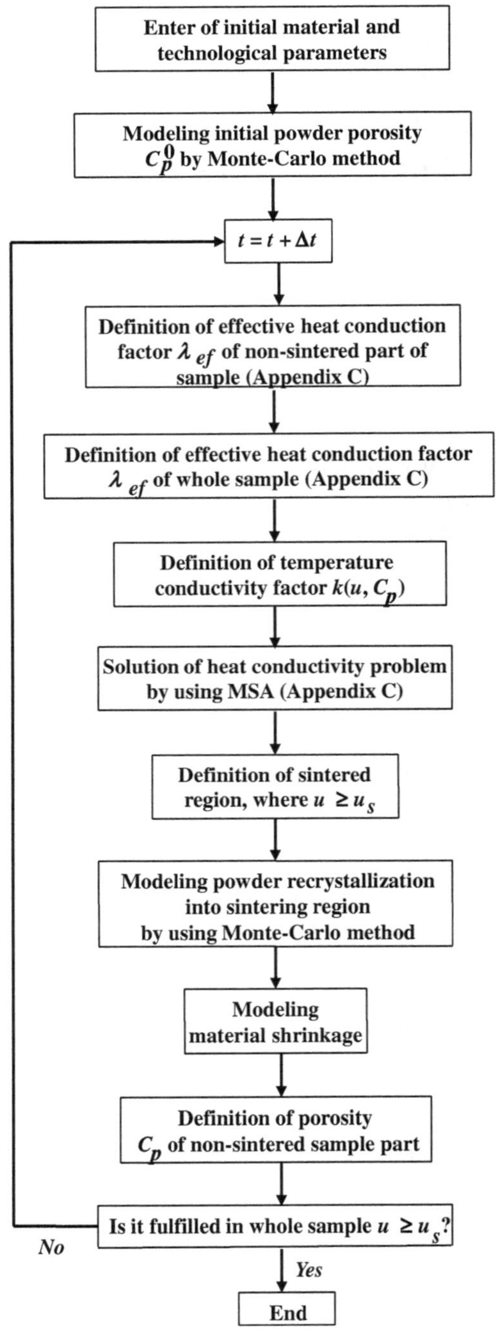

Fig. D.1 Block-scheme of algorithm describing microstructure formation during sintering

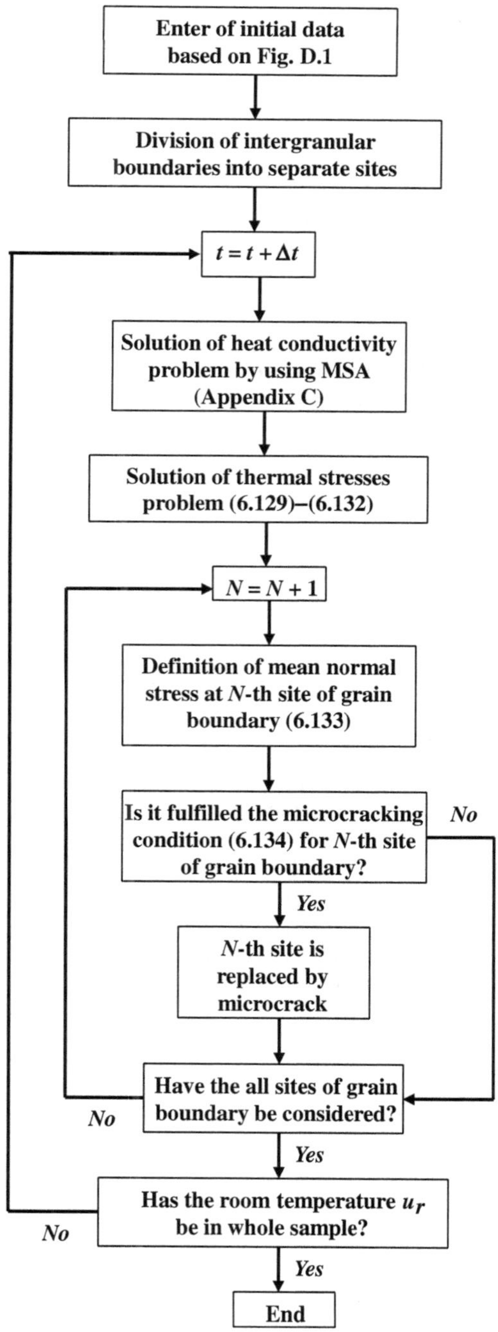

Fig. D.2 Block-scheme of algorithm describing micro-cracking during sample cooling

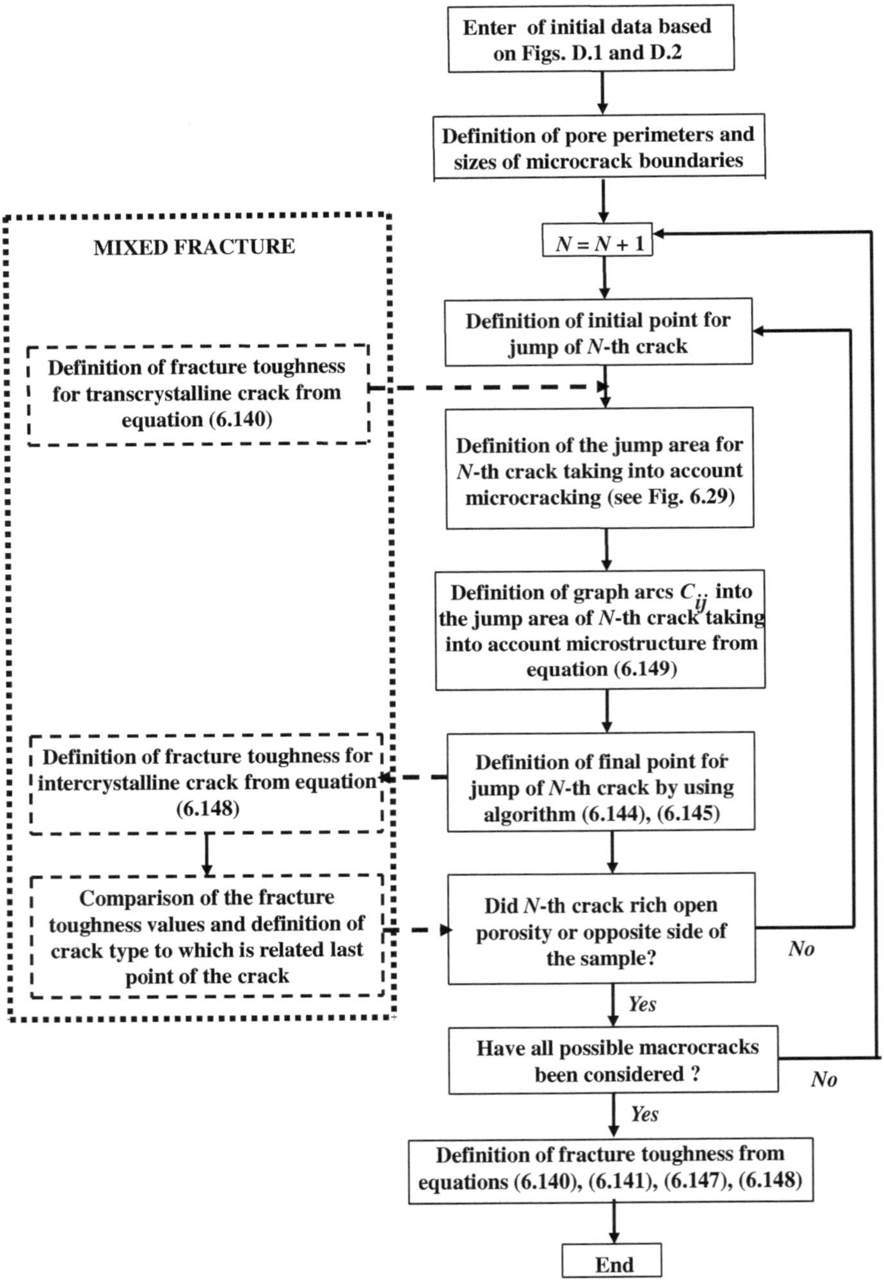

Fig. D.3 Block-scheme of algorithm describing propagation of macrocracks and estimating sample fracture toughness (cases of intercrystalline and mixed fracture)

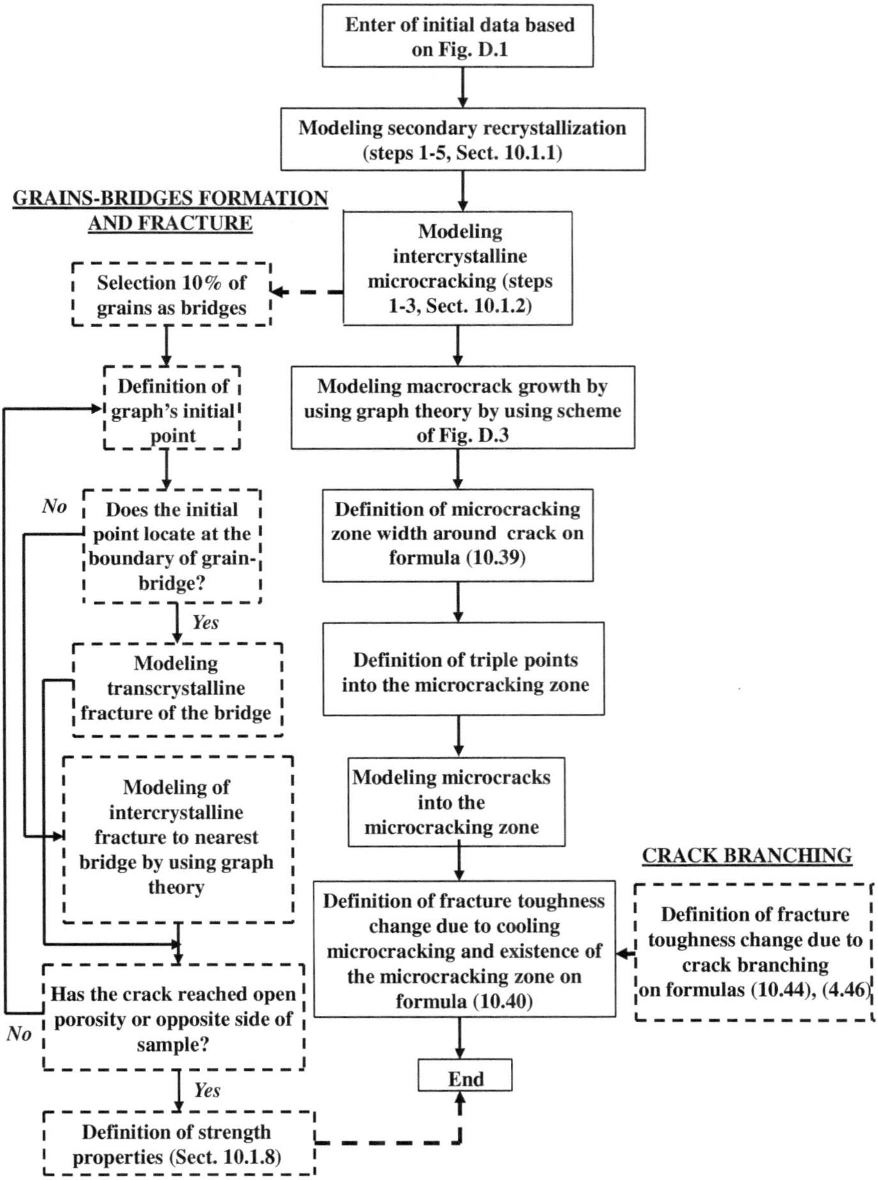

Fig. D.4 Block-scheme of algorithm describing grain growth, microcracking at cooling of YBCO ceramic and into process zone with corresponding estimation of fracture toughness; the algorithms of crack branching and also bridge formation and failure are presented, too

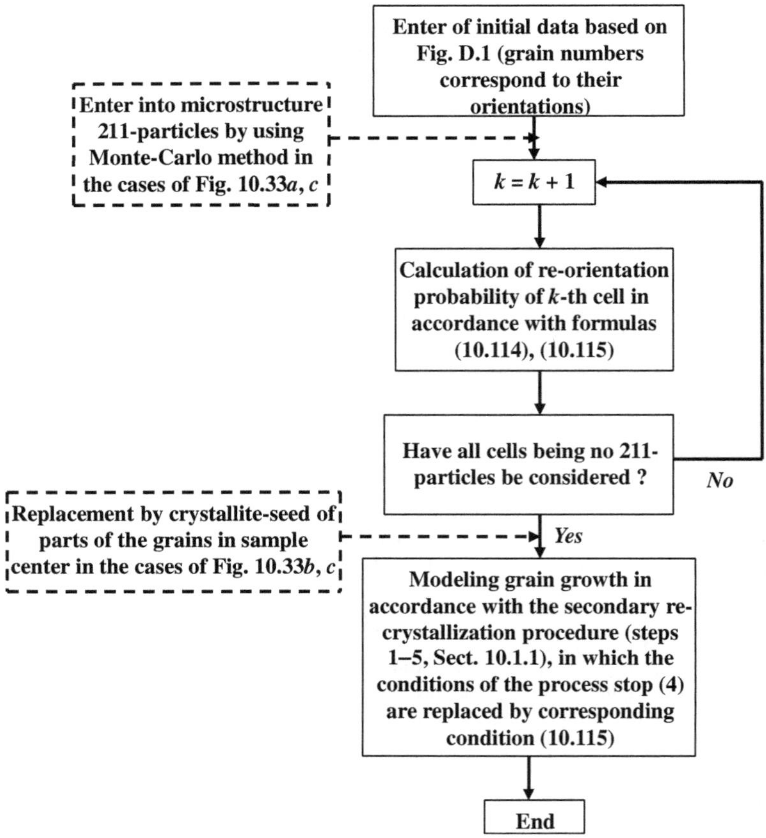

Fig. D.5 Block-scheme of algorithm describing three models of evolution of the YBCO microstructure presented in Fig. 10.33

Fig. D.6 Block-scheme of algorithm describing case of microstructure dissimilitude

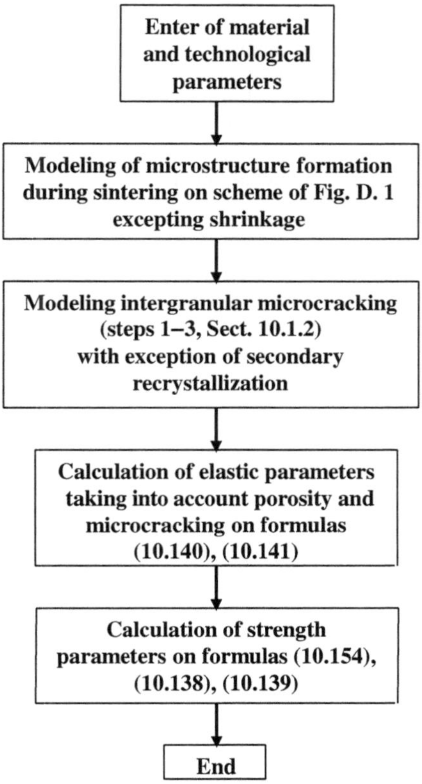

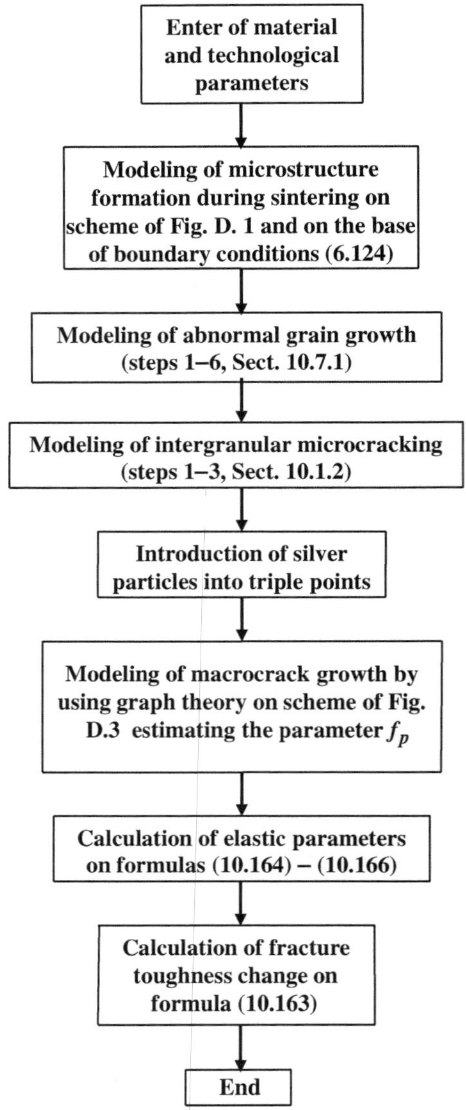

Fig. D.7 Block-scheme of algorithm describing microstructure formation, grain growth, micro- and macrofracture of Bi-2223/Ag ceramic with corresponding estimation of fracture toughness change

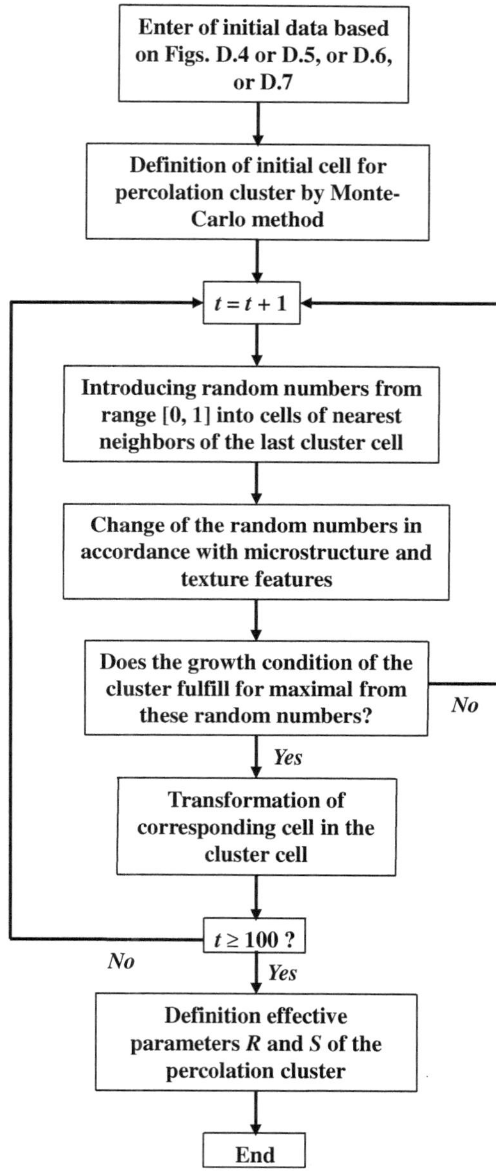

Fig. D.8 Block-scheme of algorithm describing growth of the percolation cluster with estimation of effective characteristics of electric conductivity of model structures presented in Figs. 12.29, 12.30, 12.31, 12.32

Appendix E
Eden Model

The Eden's model is a sequential model for the stochastic growth of compact clusters [420]. In this model, each new element of a cluster is added at growth site that is chosen with equal probability from a set of all possible growth sites. These sites are defined each time by a microscopic rule for cluster expansion, which, in general, identifies sites that are on the cluster edge. The set of possible growth sites reflects instantaneous (i.e., in the present time) shape of the cluster (see Fig. E.1).

Eden's clusters are compact in all space dimensions [1433], but their surfaces demonstrate self-affine fractal geometry [1079, 1475]. Computer simulation methods are used extensively to examine the interfacial properties of lattice-based Eden's clusters [798, 799, 1371]. In [799], three microscopically distinct versions of the Eden's growth process are considered and it is shown that they all have similar scaling behaviors with a finite size (or roughening) exponent, $\alpha = 0.50 \pm 0.02$, and a dynamic exponent, $\beta = 0.30 \pm 0.03$. The Eden's model describes a large class of irreversible, interfacial growth processes, which includes processes described by continuum equations.

The relaxation processes, which occur concurrently with the growth and disposition of one cell in a group of cells, are substantially unquantified. For example, in the extended Eden's model [93], relaxation is restricted to the synchronous motion of a train of neighboring cells. Computer simulation is performed on a square lattice in 2D strip with size L. Planes at $y = 0$ and $y = L$ are periodic boundaries, and the strip is infinite in the x-direction. Initially, all the lattice sites with $x \leq 0$ are occupied, and all those with $x > 0$ are empty. At each time step, an occupied lattice site $\boldsymbol{r}$ is chosen randomly. Then, if there are some unoccupied sites, $\boldsymbol{r}_0$, such that $\boldsymbol{r} - \boldsymbol{r}_0 = s\boldsymbol{k}$, where $\boldsymbol{k}$ is a lattice vector and $s \leq q$ (q is the maximum train length), the site $\boldsymbol{r}$ becomes, momentarily, doubly occupied. This double occupancy is relaxed by the occupation of a site chosen randomly from the set of unoccupied sites $\boldsymbol{r}_0$ for which $|\boldsymbol{r} - \boldsymbol{r}_0|$ is the minimum. In most cases, this set has only one member. The relaxation is constructed from the simultaneous motions of a train of neighboring cells in the direction of least resistance with the maximum train length, q. Note that for $q = 1$ this growth process has been modeled in [799].

I. A. Parinov, *Microstructure and Properties of High-Temperature Superconductors*,
DOI: 10.1007/978-3-642-34441-1,

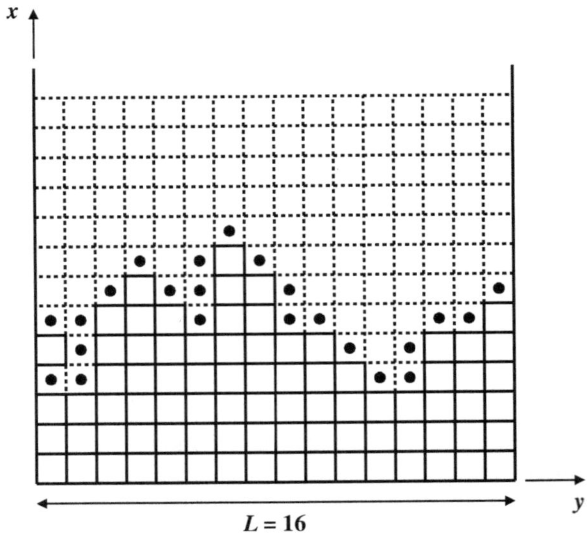

Fig. E.1 Typical cluster obtained, using Eden model. *Circles* denote sites located along the cluster perimeter

The growth process leads to a compact cluster with an irregular, rough surface. At any time, the surface of the cluster may be defined by a set of heights, $h(y_i)$, $1 \leq i \leq L$. The heights, $h(y_i)$, represent the extension of the cluster in the x-direction at a lateral point y_i. For a cluster with constant density, the mean height is found as

$$h_0 = L^{-1} \sum_{i=1}^{L} h(y_i). \tag{E.1}$$

For a particular value of h_0, the standard deviation of the heights

$$\sigma(L, h_0) = \sqrt{L^{-1} \sum_{i=1}^{L} [h(y_i) - h_0(L)]^2} \tag{E.2}$$

represents the surface roughness and has a scaling form [93]:

$$\begin{aligned} \sigma(L, h_0) &= L^{\alpha} f(h_0 L^{z}); \\ &\propto h_0^{\beta}, \quad \text{at } h_0 \ll L^{z}; \\ &\propto L^{\alpha}, \quad \text{at } h_0 \gg L^{z}, \end{aligned} \tag{E.3}$$

where $f(x)$ is a scaling function and $z = \alpha/\beta$. These equations describe a surface with a strip width that initially increases and then, after a time, which depends on the strip width L, reaches a saturated value.

The growth from beneath the surface of an expanding cluster leads to an interface that roughens slowly compared to one that is driven by a simple surface addition process. For the extended model $z \approx 2$ and the temporal spread of height fluctuations is diffusive [93]. This is in contrast to the super-diffusive behavior of pure Eden's model where $z < 2$. Moreover, in the steady state of the extended Eden's model, the fluctuations of the heights are less correlated along the surface than in the case of pure "deposition", that is, in the extended model the surface is less rough. The results are currently insufficient to establish the precise dependence of the exponents on the parameter q, but they establish a *qualitative* change, in an observable property of the cluster morphology, that arises as a result of the subsurface growth.

The averaged Green's function, $g(\boldsymbol{r} - \boldsymbol{r}_0)$, for the steady-state growth in the extended Eden's model is largely independent of the system size and geometry. It represents, quite generally, the probability that an unoccupied site at r_0 becomes occupied at the same time as a new particle is introduced at site r. The probability of growth increases with the distance from lower surface of the cluster because the number of equally weighted growth sites increases with depth through the surface profile. Simulations with $q > \sigma\ (L, \infty)$ confirm that $g(\boldsymbol{r}) \to$ const for $r > \sigma\ (L, \infty)$ [93]. Thus, the averaged Green's function may be used in a numerical scheme to construct, directly, clusters in different geometry.

References

1. G. Abbruzzese, Acta. Metall. **33**, 1329 (1985)
2. G. Abbruzzese, K. Lücke, Acta. Metall. **34**, 905 (1986)
3. A.A. Abrikosov, Doklady Akad. Nauk S.S.S.R **86**, 489 (1952)
4. A.A. Abrikosov, Zh. Eksp. Teor. Fiz. **32**, 1442 (1957) [Sov. Phys. JETP 5, 1174 (1957)]
5. P. Agarvala, M.P. Srivastava, P.N. Dheer et al., Phys. C **313**, 87 (1999)
6. T. Aichele, P. Görnert, R. Uecker, M. Mühlberg, IEEE Trans. Appl. Supercond. **9**(2), 1510 (1999)
7. I.I. Akimov, A.K. Shikov, T.P. Krinitsina et al., Phys. C **328**, 125 (1999)
8. V.A. Akopyan, A.N. Soloviev, I.A. Parinov, S.N. Shevtsov, *Definition of Constants for Piezoceramic Materials* (Nova Science Publishers, New York, 2010)
9. I. Aksay, C. Han, G.D. Maupin, et al., U.S. Patent No 5,061,682 (1991)
10. A.K.M. Alamgir, Y. Yamada, N. Harada et al., IEEE Trans. Appl. Supercond. **9**(2), 1864 (1999)
11. J.A. Alarco, E. Olsson, Z.G. Ivanov et al., Ultramicroscopy **51**, 239 (1993)
12. V.E. Alekhin, I.P. Miroshnichenko, A.G. Serkin, V.P. Sizov, *Optical Device for Measurement of Displacements of the Control Object Surfaces*. Russian Patent No. 2343402 (2009)
13. V.E Alekhin, I.P Miroshnichenko, A.G Serkin, V.P, Sizov *Registration Method of Displacements by Optical Transducers*. Russian Patent No. 2343403 (2009)
14. K.B. Alexander, A. Goyal, D.M. Kroeger et al., Phys. Rev. B **45**, 5622 (1992)
15. A.S. Alexandrov, cond-mat/9807185 (1998)
16. A.S. Alexandrov, in *Proceedings of International Conference on Basic Problem of HTSC*, Moscow, 18–22 Oct 2004
17. A.S. Alexandrov, N.F. Mott, Rep. Prog. Phys. **57**, 1197 (1994)
18. S. Alexandrov, F. Barlat, Acta. Mech. **133**, 57 (1999)
19. S. Alexandrov, G. Mishuris, W. Mishuris, Eur. J. Mech. A. Solids **19**, 811 (2000)
20. N. Alford, McN et al., J. Appl. Phys. **66**, 5930 (1989)
21. M.I. Ali, K. Ehata, S. Ohshima, IEEE Trans. Appl. Supercond. **9**(2), 3077 (1999)
22. C. Allgeier, J. Heise, W. Reith et al., Physica C **157**, 293 (1989)
23. C. Allgeier, J.S. Schilling, H.C. Ku et al., Solid State Commun. **64**, 227 (1987)
24. C.C. Almasan, S.H. Han, B.W. Lee et al., Phys. Rev. Lett. **69**, 680 (1992)

I. A. Parinov, *Microstructure and Properties of High-Temperature Superconductors*,
DOI: 10.1007/978-3-642-34441-1, © Springer-Verlag Berlin Heidelberg 2012

25. R.P. Aloysius, A. Sobha, P. Guruswamy, U. Syamaprasad, Supercond. Sci. Technol. **14**, 85 (2001)
26. G.A. Alvarez, J.G. Wen, F. Wang et al., IEEE Trans. Appl. Supercoud. **7**, 3017 (1997)
27. N. Amemiya, Z. Jiang, K. Yamagishi, T. Sasaoka, IEEE Trans. Appl. Supercond. **12**(1), 1599 (2002)
28. Amm K.M., Sastry P.V.P.S.S., Knoll D.C., Schwartz J., Adv. Cryog. Eng. (Mater.) 44b, 457 (1998)
29. T. Amrein, M. Seitz, D. Uhl et al., Appl. Phys. Lett. **63**, 1978 (1993)
30. D.M. Anderson, G.B. McFadden, A.A. Wheeler, Phys. Fluids **9**, 1870 (1997)
31. D.M. Anderson, G.B. McFadden, A.A. Wheeler, Ann. Rev. Fluid Mech. **30**, 139 (1998)
32. D.M. Anderson, G.B. McFadden, A.A. Wheeler, Physica D **135**, 175 (2000)
33. D.M. Anderson, G.B. McFadden, A.A. Wheeler, in *Interfaces for the Twenty-First Century*, ed. by M.R. Smith, M.J. Miksis, G.B. McFadden et al. (Imperial College Press, London, 2002)
34. G.W. Anderson, D.C. Webb, A.E. Spezio, J.N. Lee, Proc. IEEE **79**, 355 (1991)
35. J. Anderson, X.Y. Cai, M. Feldman et al., Supercond. Sci. Technol. **12**, 617 (1999)
36. M.P. Anderson, D.J. Srolovitz, G.S. Grest, P.S. Sahni, Acta. Metall. **32**, 783 (1984)
37. P.W. Anderson, in *Lectures on the Many-Body Problems*. ed. by E.R. Caianello (Academic Press, New York, 1964)
38. P.W. Anderson, Science **235**, 1196 (1987)
39. P.W. Anderson, *The Theory of Superconductivity in the High-T_c Cuprate Superconductors* (Princeton University Press, Princeton, 1997)
40. P.W. Anderson, Phys. Rev. Lett. **96**, 017001 (2005)
41. P.W. Anderson, Cond-mat/0603726 (2006)
42. P.W. Anderson, Nat. Phys. **3**, 160 (2007)
43. P.W. Anderson, Int. J. Mod. Phys. B **25**(1), 1 (2011)
44. P.W. Anderson, P.A. Lee, M. Randeria, et al., J. Phys. Cond. Mat. **16**, R754 (2004)
45. P.W. Anderson, N.P. Ong, J. Phys. Chem. Solids **67**, 1 (2006)
46. T.L. Anderson, *Fracture Mechanics. Fundamentals and Applications* (CRC Press, Texas, 1995)
47. I.V. Andrianov, J. Awrejcewicz, L.I. Manevitch, *Asymptotical Mechanics of Thin-Walled Structures: a Handbook* (Springer, Berlin, 2004)
48. A.O. Animalu, *Intermediate Quantum Theory of Crystalline Solids* (Prentice-Hall, New Jersey, 1977)
49. V.I. Anisimov, M.A. Korotin, I.A. Nekrasov et al., Phys. Rev. B. **66**, 100502 (2002)
50. J.F. Annett, M. Randeria, S.R. Renn, Phys. Rev. B **38**, 4660 (1988)
51. L. Antognazza, S.J. Berkowitz, T.H. Geballe, K. Char, Phys. Rev. B **51**, 8560 (1995)
52. L. Antognazza, B.H. Moeckly, T.H. Geballe, K. Char, Phys. Rev. B **52**, 4559 (1995)
53. D. Aoki, A. Huxley, E. Ressouche et al., Nature **413**, 613 (2001)
54. F.M. Araujo-Moreira, P.N.L. Filho, S.M. Zanetti, et al., cond-mat/9909086 (1999)
55. D. Ariando, H.-J.H. Darminto, V. Smilde, D.H.A. Leca, Phys. Rev. Lett. **94**, 167001 (2005)
56. G. Arlt, J. Mater. Sci. **25**, 2655 (1990)
57. N.P. Armitage, D.H. Lu, D.L. Feng et al., Phys. Rev. Lett. **86**, 1126 (2001)
58. R.A. Armstrong, *Basic Topology* (Springer, Berlin, 1983)
59. M.C. Aronson, S.B. Dierker, B.S. Dennis et al., Phys. Rev. B **44**, 4657 (1991)
60. T. Aselage, K. Keefer, J. Mater. Res. **3**, 1279 (1988)
61. M.F. Ashby, Acta. Metall. **37**, 1273 (1989)

62. M.F. Ashby, C. Gandhi, D.M.R. Taplin, Acta. Metall. **27**, 699 (1979)
63. N.W. Ashcroft, Phys. Rev. Lett. **21**, 1748 (1968)
64. S.P. Ashworth, B.A. Glowacki, M.P. James, IEEE Trans. Appl. Supercond. **5**(2), 1271 (1995)
65. H. Assmann, A. Guenther, K. Rigby, M. Wohlfart, Adv. Supercond. V,p. 685. Y. Bando, H. Yamauchi (eds.), (Springer, Tokyo, 1993)
66. M. Ausloos, N. Vandewalle, R. Cloots, Europhys. Lett. **24**, 629 (1993)
67. V. Avdeev et al., Pis'ma JETP **43**, 376 (1986)
68. B. Avitzar, *Metal Forming Processes and Analysis* (McGraw-Hill, New York, 1968)
69. V.P.S. Awama, J. Nakamura, M. Karppinen et al., Phys. C. **357–360**, 121 (2001)
70. J. Ayache, P. Odier, N. Pellerin, Supercond. Sci. Technol. **7**, 655 (1994)
71. T. Aytug, M. Paranthaman, A.A. Gapud et al., J. Appl. Phys. **98**, 114309 (2005)
72. S.E. Babcock, X.Y. Cai, D.L. Kaiser, D.C. Larbalestier, Nature **347**, 167 (1990)
73. S.E. Babcock, X.Y. Cai, D.C. Larbalestier et al., Phys. C **227**, 183 (1994)
74. S.E. Babcock, D.C. Larbalestier, Appl. Phys. Lett. **55**, 393 (1989)
75. S.E. Babcock, D.C. Larbalestier, J. Mater. Res. **5**, 919 (1990)
76. S.E. Babcock, D.C. Larbalestier, J. Phys. Chem. Solids **55**, 1125 (1994)
77. S.E. Babcock, J.L. Vargas, Annual Rev. Mater. Sci **25**, 193 (1995)
78. E. Backen, J. Hänisch, R. Hühne et al., IEEE Trans. Appl. Supercond. **17**, 3733 (2007)
79. A. Badia, C. López, Phys. Rev. Lett. **87**, 127004 (2001)
80. A. Badia, C. López, Phys. Rev. B. **65**, 104514 (2002)
81. A. Badia, C. López, J.Low Temp. Phys. **130**, 129 (2003)
82. R. Baetzold, Phys. Rev. B **42**, 56 (1990)
83. K.E. Bagnall, I.V. Grigorieva, J.W. Steeds et al., Supercond. Sci. Technol. **8**, 605 (1995)
84. G.A. Baker, P. Graves-Morris, *Pade Approximants Two Volume Set* (Addison-Wesley Publishing Company, New York, 1981)
85. U. Balachandran, M. Lelovic, B.C. Prorok et al., IEEE Trans. Appl. Supercond. **9**(2), 2474 (1999)
86. G. Balestrino, P.G. Medaglia, P. Orgiani et al., Phys. Rev. Lett. **89**, 156402 (2002)
87. V.P. Balkevich, *Technical Ceramics* (Stroyizdat, Moscow, 1984)
88. R.W. Balluffi, P.D. Bristowe, Surf. Sci. **144**, 28 (1984)
89. B.R. Balmer, C.R. Grovenor, R. Riddle, IEEE Trans. Appl. Supercond. **9**(2), 1888 (1999)
90. A.L. Barabasi, H.E. Stanley, *Fractal Concepts in Surface Growth* (Cambridge University Press, Cambridge, 1995)
91. J. Bardeen, L.N. Cooper, J.R. Schrieffer, Phys. Rev. **108**, 1175 (1957)
92. G.I. Barenblatt, Adv. Appl. Mech. **7**, 55 (1962)
93. G.C. Barker, M.J. Grimson, J. Phys. A: Math. Gen. **27**, 653 (1994)
94. P.N. Barnes, T.J. Haugan, C.V. Varanasi, T.A. Campbell, Appl. Phys. Lett. **85**, 4088 (2004)
95. E.S. Barr, Infrared Phys. **3**, 195 (1963)
96. J.W. Barrett, L. Prigozhin, Interface. Free Bound. **8**, 349 (2006)
97. S.E. Barret, D.J. Durand, C.H. Pennington et al., Phys. Rev. B **41**, 6283 (1990)
98. J. Bartholomew, Phys. Rev. B **28**, 5378 (1983)
99. G. Baskaran, Z. Zou, P.W. Anderson, Solid State Commun. **63**, 973 (1987)
100. D. Basov, T. Timusk, Rev. Mod. Phys. **77**, 721 (2005)
101. G.K.J. Batchelor, Fluid Mech. **52**, 245 (1972)

102. G.K. Batchelor, R.W. O'Brien, Proc Roy. Soc. Lond. A **355**, 313 (1977)
103. C.A. Bateman, L. Zhang, H. Chan et al., J. Am. Ceram. Soc. **75**, 1281 (1992)
104. A.S. Batrakov, M.M. Butusov, G.P. Grechka, *Laser Measuring Systems* (Radio i Svyaz, Moscow, 1981)
105. T. Bauch, T. Lindström, F. Tafuri et al., Science **311**, 57 (2006)
106. T. Bauch, F. Lombardi, F. Tafuri et al., Phys. Rev. Lett. **94**, 087003 (2005)
107. M. Bauer, R. Semerad, H. Kinder, IEEE Trans. Appl. Supercond. **9**(2), 1502 (1999)
108. B.A. Baumert, Structural, Chemical, and Electronic Properties of Superconducting Thin Films and Junctions. PhD thesis, Carnegie Mellon University, Pittsburg, USA, 1994
109. Z. Bazant, L. Cedolin, *Stability of Structures: Elastic, Inelastic, Fracture and Damage Theories* (Oxford University Press, Oxford, 2003)
110. C.P. Bean, Phys. Rev. Lett. **8**, 250 (1962)
111. C.P. Bean, Rev. Mod. Phys. **36**, 31 (1964)
112. F. Becca, A. Parola, S. Sorella, Phys. Rev. B. **61**, R16287 (2000)
113. J. Bech, M. Eriksen, F. Toussant et al., in *Proceeding Powder Metallurgy World Congress and Exhibition* (2000)
114. K. Bechgaard, K. Carneiro, M. Olsen et al., Phys. Rev. Lett. **46**, 852 (1981)
115. A. Beck, O.M. Froelich, D. Koelle et al., Appl. Phys. Lett. **68**, 3341 (1996)
116. C. Beckermann, H.-J. Diepers, I. Steinbach et al., Comp. Phys. **154**, 468 (1999)
117. J.G. Bednorz, K.A. Müller, Z. Phys. B **64**, 189 (1986)
118. V. Beilin, A. Goldgirsh, E. Yashchin et al., Phys. C **309**, 56 (1998)
119. J.F. Bell, *The Experimental Foundations of Solid Mechanics. Encyclopedia of Physics*, vol. 1 (Springer, Berlin, 1973)
120. A.V. Belyaev, V.N. Lebedev, E.G. Fisenko et al., *Problems of HTSC*, part 1, (Rostov Sate University, 1990), p. 158
121. C. Bernhard, J.L. Tailon, C.H. Niedermayer et al., Phys. Rev. B. **59**, 14099 (1999)
122. S.J. Bennison, B.R. Lawn, Acta Metall. **37**, 2659 (1989)
123. V.L. Berezinski, Zh. Eksp. Teor. Fiz. **59**, 907 (1970) [Sov. Phys. JETP, **32**, 493 (1971)]
124. D.J. Bergman, Phys. Rep. C. **43**, 377 (1978)
125. S.J. Berkowitz, C.F. Shih, W.H. Mallison et al., IEEE Trans. Appl. Supercond. **7**(2), 3056 (1997)
126. J.G. Berryman, J. Acoust. Soc. Am. **68**, 1809 (1980)
127. J.G. Berryman, P.A. Berge, Mech. Mater. **22**, 149 (1996)
128. J.G. Berryman, G.W. Milton, J. Phys. D: Appl. Phys. **21**, 87 (1988)
129. R. Besseling, R. Higgebrugge, P.H. Kes, Phys. Rev. Lett. **82**, 3144 (1998)
130. S. Beuven, M. Darula, O. Harnack et al., IEEE Trans. Appl. Supercond. **9**(2), 3769 (1999)
131. M.B. Bever, *Encyclopedia of Materials Science and Engineering* (Pergamon Press, Oxford, 1986), p. 3354
132. R.L. Bhattacharya, R.D. Blaugher, U.S. Patent No 5,413,987 (1995)
133. B. Bhushan, J.N. Israelachvili, U. Landman, Nature **374**, 6076 (1995)
134. B. Bhushan, V.N. Koinkar, J. Ruan, J. Eng. Tribol. **208**, 17 (1994)
135. B. Bhushan, O. Marti, in *Springer Handbook of Nanotechnology*, ed. by B. Bhushan, 2nd edn. (Springer, Berlin, 2007)
136. B. Bhushan, J. Ruan, ASME J. Tribol. **116**, 3896 (1994)
137. W. Bian, Y. Zhu, Y.L. Wang, M. Suenaga, Phys. C **248**, 119 (1995)
138. J. Biaxeras, G. Fcurnet, J. Phys. Chem. Solids **28**, 1541 (1967)

139. B.A. Bilby, R. Bullough, E. Smith, Proc. Roy. Soc. (London), A **231**, 263 (1955)
140. G. Binnig, A. Baratoff, H.E. Hoenig, J.G. Bednorz, Phys. Rev. Lett. **45**, 1352 (1980)
141. G. Binnig, C.F. Quate, Ch. Gerber, Phys. Rev. Lett. **56**, 930 (1986)
142. G. Binning, H. Rohrer, Rev. Mod. Phys. **59**, 615 (1987)
143. G. Binning, H. Rohrer, Ch. Gerber, E. Weibel, Phys. Rev. Lett. **49**, 57 (1982)
144. H.K. Birnbaum, M.L. Grosbeck, M. Amano, J. Less-Common Metals **49**, 357 (1976)
145. H.A. Blackstead, J.D. Dow, D.B. Chrisey et al., Phys. Rev. B **54**, 6122 (1996)
146. G. Blatter, M.V. Feigel'man, V.B. Geshkenbein et al., Rev. Mod. Phys. **66**, 1125 (1994)
147. R. Bleykhut, *Fast Algorithms of Digital Signal Processing* (Mir, Moscow, 1989)
148. W.J. Boettinger, J.A. Warren, Met. Trans. A **27**, 657 (1996)
149. W.J. Boettinger, J.A. Warren, J. Cryst. Growth. **200**, 583 (1999)
150. P.V. Bogdanov, A. Lanzara, X.J. Zhou et al., Phys. Rev. B **64**, 180505 (2001)
151. N.N. Bogolyubov, Pis'ma Zh. Eksp. Teor. Fiz. **34**, 58 (1958)
152. B.A. Boley, J.H. Wiener, *Theory of Thermal Stresses* (Wiley, New York, 1960)
153. V.V. Bolotin, *Stability Problems in Fracture Mechanics* (Wiley, New York, 1996)
154. V.I. Bol'shakov, I.V. Andrianov, V.V. Danishev'skyy, *Asymptotical Methods for Calculation of Composite Materials Taking into Account Internal Structure* (Porogi, Dneropetrovsk, 2008)
155. P.A.A. Booi, S.P. Benz, Appl. Phys. Lett. **64**, 2163 (1994)
156. N.E. Booth, Appl. Phys. Lett. **50**, 293 (1987)
157. S.V. Borisenko, A.A. Kordyuk, T.K. Kim et al., Phys. Rev. B **64**, 140509 (2002)
158. S.V. Borisenko, A.A. Kordyuk, A. Koitzsch et al., Phys. Rev. Lett. **92**, 207001 (2004)
159. A. Bossavit, IEEE Trans. Magn. **30**, 3363 (1994)
160. S. Boucher, J. Composite Mater. **8**, 82 (1974)
161. Ph. Boullay, B. Domenges, M. Hervieu, B. Raveau, Chem. Mater. **5**, 1683 (1995)
162. A.J. Bourdillon, N.X. Tan, C.L. Ong, J. Mater. Sci. Lett. **15**, 439 (1996)
163. S. Boutemy, J. Kessler, J. Schwartz, IEEE Trans. Appl. Supercond. **7**(2), 1552 (1997)
164. V.S. Boyko, V.M. Gorbatenko, L.F. Krivenko et al., Fiz. Nizk. Temper. **15**, 988 (1989)
165. V.S. Boyko, L.F. Krivenko, V.V. Demirsky, V.D. Natsik, Supercond. Phys. Chem. Tech. **4**, 1815 (1991)
166. I. Bozovich, IEEE Trans. Appl. Supercond. **11**(1), 2686 (2001)
167. I. Bozovic, G. Logvenov, M.A.J. Verhoeven et al., Nature **422**, 873 (2000)
168. I. Bozovic, G. Logvenov, M.A.J. Verhoeven et al., Phys. Rev. Lett. **93**, 157002 (2004)
169. M. Braden, O. Hoffels, W. Schnelle et al., Phys. Rev. B **47**, 12288 (1993)
170. E.H. Brandt, Int. J. Mod. Phys. B **5**, 751 (1991)
171. E.H. Brandt, Phys. B **169**, 91 (1991)
172. E.H. Brandt, Rep. Prog. Phys. **58**, 1465 (1995)
173. E.H. Brandt, Phys. Rev. B **54**, 3530 (1996)
174. E.H. Brandt, Phys. Rev. B **54**, 4246 (1996)
175. E.H. Brandt, Phys. Rev. B **55**, 14513 (1997)
176. E.H. Brandt, Phys. Rev. B **58**, 6506 (1998)
177. E.H. Brandt, Phys. Rev. B **58**, 6523 (1998)
178. E.H. Brandt, Phys. Rev. B **59**, 3369 (1999)
179. R.J. Braun, J.W. Cahn, G.B. McFadden et al., Acta Mater. **46**, 1 (1998)
180. R.J. Braun, B.T. Murray, J. Cryst. Growth **174**, 41 (1997)

181. K. Breuer, J. Levy, H. Paczkowski, Microwave J. **32**, 81 (1989)
182. H. Brezis, in *Contributions to Nonlinear Functional Analysis*, ed. by E. Zarantonello (Academic Press, Madison, 1971)
183. P.W. Bridgeman, *Studies in Large Plastic Flow and Fracture* (McGraw-Hill, New York, 1952)
184. C.J. Brinker, G.W. Sherer, *Sol Gel Science: The Physics and Chemistry of Sol-Gel Processing* (Academic Press, Boston, 1990)
185. L.S. Brizhik, A.S. Davydov, Fiz. Nizk. Temp. **10**, 358 (1984)
186. J. Brnic, M. Canadija, G. Turkalj, in *Proceedings 6th International Conference on Advanced Manufacturing Systems and Technology (AMST'02)*, ed. by E. Kuljanic (Springer, New York, 2002)
187. J.S. Brooks, in *Handbook of High-Temperature Superconductivity: Theory and Experiment*, eds. by J.R. Schrieffer, J.S. Brooks (Springer, New York, 2007)
188. R. Brout, Phys. Rev. **118**, 1009 (1960)
189. W.F. Brown, J. Chem. Phys. **23**, 1514 (1955)
190. N. Browning, M.F. Chisholm, S. Pennycook et al., J. Phys. C **212**, 185 (1993)
191. D. Bruggeman, Ann Physik (Liepzig) **24**, 636 (1935)
192. M. Brühwiler, S.M. Kazakov, N.D. Zhigadlo et al., Phys. Rev. B **70**, 020503 (2004)
193. G. Brumfiel, Nature **417**, 367 (2002)
194. E. Bruneel, F. Persyn, S. Hoste, Supercond. Sci. Technol. **11**, 88 (1998)
195. E. Bruneel, A.J. Ramirez-Cuesta, I. van Drische, S. Hoste, Int. Phys. Conf. Ser. **167**(1), 247 (1999)
196. O.P. Bruno, R. Proc, Soc. Lond. A. **433**, 353 (1991)
197. P. Brusov, *Mechanisms of High-Temperature Superconductivity* (Rostov State University Press, Rostov-on-Don, 1999)
198. W. Buckles, D. Driscoll, IEEE Power Eng. Rev. **20**(5), 16 (2000)
199. B. Budiansky, J. Mech. Phys. Solids. **13**, 223 (1965)
200. B. Budiansky, J.C. Amazigo, A.G. Evans, J. Mech. Phys. Solids. **36**, 167 (1988)
201. B. Budiansky, J.W. Hutchinson, J.C. Lambropoulos, Int. J. Solids Struct. **19**, 337 (1983)
202. L.N. Bulaevskii, Zh Eksp, Teor. Fiz. **64**, 2241 (1973)
203. L.N. Bulaevskii, J.R. Clem, L.I. Glazman, A.P. Malozemoff, Phys. Rev. B **45**, 2545 (1992)
204. A.A. Bul'bich, Pis'ma Zh. Tech. Fiz. **12**, 645 (1986)
205. N. Bulut, Adv. Phys. **51**, 1587 (2002)
206. B.C. Bunker, et al., in *High Temperature Superconducting Materials: Preparation, Property and Processing*, eds. by W.E. Hatfield, J.H. Miller, Jr (Marcel Dekker Inc., New York, 1988)
207. F.E. Buresch, in *Proceedings of International Conference on Residual Stresses Science Technology*. (Garmish-Partenkirkhen, 1986), Oberursel, (1987) **1**, p. 539
208. F.E. Buresch, E. Babilon, G. Kleist,*Proceedings of 2nd International Conferences on Residual Stress*. (ICRS2, Nancy, 1988), London, (1988), p. 1003
209. F.E. Buresch, O. Buresch, *Procedings of 14th International Conference on Science of Ceramics* - 14 (Canterbury, 1987), Stoke-on-Trent, (1988), p. 683
210. M.J. Burns, A.W. Kleinsasser, K.A. Delin et al., IEEE Trans. Appl. Supercond. **7**(2), 3564 (1997)
211. S. Burns, J. Phys. C **206**, 97 (1993)
212. S.J. Burns, Supercond. Sci. Technol. **7**, 337 (1994)
213. M. Cacciari, G. Mazzanti, G.C. Montanari, IEEE Trans. Dielectr. Electr. Instr. **3**(1), 18 (1996)

214. W.G. Cady, *Piezoelectricity* (Dover, New York, 1964)
215. C. Caginalp, J. Jones, Appl. Math. Lett. **4**, 79 (1991)
216. G. Caginalp, E. Socolovsky, SIAM J. Sci. Comput. **15**, 106 (1994)
217. C. Cai, J. Hänisch, R. Hühne et al., J. Appl. Phys. **98**, 123906 (2005)
218. C.B. Cai, J.C. Zhang, S.X. Cao et al., Physica C **460**, 1355 (2007)
219. X.Y. Cai, A. Gurevich, I.F. Tsu, et al., Phys. Rev. B **57**, 10951 (1998)
220. X.Y. Cai, A.A. Polyanskii, Q. Li et al., Nature **392**, 906 (1998)
221. X.Y. Cai, A. Polyanskii, Q. Li et al., Nature **393**, 909 (1998)
222. Y. Cai, J.S. Chung, M.F. Thorpe, S.D. Mahanti, Phys. Rev. B **42**, 8827 (1990)
223. Z.X. Cai, Y. Zhu, IEEE Trans. Appl. Supercond. **9**, 2714 (1999)
224. Z.X. Cai, Y. Zhu, D.O. Welch, Phys. Rev. B **52**, 13035 (1995)
225. R. Caillard, V. Garnier, G. Desgardin, Physica C **340**, 101 (2000)
226. C.G. Callan Jr, Phys. Rev. D **2**, 1541 (1970)
227. A.M. Campbell, J.E. Evetts, Adv. Phys. **21**, 1191 (1972)
228. H.C. Cao, M.D. Thouless, A.G. Evans, Acta Metall. **36**, 2037 (1988)
229. A.H. Cardona, H. Suzuki, T. Yamashita et al., Appl. Phys. Lett. **62**, 411 (1993)
230. D.A. Cardwell, M. Kambara, N. Hari Babu, et al., IEEE Trans. Appl. Supercond. **11**(1) (2001)
231. D.A. Cardwell, W. Lo, H.T. Leung, J.C.L. Chow, in *Proceedings of 9th International Symposium on Superconductivity* (*ISS*'96), (Sapporo, Japan, Oct 21–24, 1996), (Springer, Tokio, 1997), p. 725
232. A.H. Carim, T.E. Mitchell, Ultramicroscopy **51**, 228 (1993)
233. F.M.A. Carpay, in *Ceramic Microstructures* '76, eds. by R.M. Fulrath, J.A. Pask (Westview, Boulder, 1977) p. 261
234. F.M.A. Carpay, J. Am. Ceram. Soc. **60**, 82 (1977)
235. E. Cartan, Cont. Rend. Acad. Sci. **174**, 593 (1922)
236. E. Cartan, Ann. Ec. Norm. Sup. **40**, 325 (1922)
237. E. Cartan, Ann. Ec. Norm. Sup. **42**, 17 (1922)
238. E. Cartan, A. Einstein, *Letters on Absolute Parallelism* (Princeton University Press, Princeton, 1979)
239. C.B. Carter, Acta Metall. **36**, 2753 (1988)
240. S. Casalbuoni, L. von Sawilski, J. Kotzler, cond-mat/0310565 (2003)
241. P.A. Casey, PhD Thesis (Princeton, New Jersey, 2010)
242. P.A. Casey, J.D. Koralek, N.C. Plumb et al., Nat. Phys. **4**, 210 (2008)
243. R. Cava, J. Phys. C **282–287**, 27 (1997)
244. G. Celotti, A. Tampieri, D. Rinaldi, IEEE Trans. Appl. Supercond. **9**(2), 1779 (1999)
245. J. Chakhalian, J.W. Freeland, H.-U. Habermeier et al., Science **318**, 1114 (2007)
246. V. Chakrapani, D. Balkin, P. McGinn, Appl. Supercond. **1**, 71 (1993)
247. S. Chakravarty, H.Y. Kee, K. Völker, Nature **428**, 53 (2004)
248. B.K. Chakraverty, J. Ranninger, D. Feinberg, Phys. Rev. Lett. **81**, 433 (1998)
249. H. Chaloupka, N. Klein, M. Peiniger et al., IEEE Trans. Microwave Theory Tech. **39**, 1513 (1991)
250. K.S. Chan, J. Lankford, Acta Metall. **36**, 193 (1988)
251. H.W. Chandler, J. Mech. Phys. Solids **33**, 215 (1985)
252. H.W. Chandler, Solid. Mech. Appl. **39**, 235 (1995)
253. Y.M. Chang, D. Winkler, T. Claeson, Appl. Phys. Lett. **62**, 3195 (1993)

254. S.A. Chaplygin, *On Gas Jets* (Brown University, Providence, 1944)
255. S.J. Chapman, SIAM J. Appl. Math. **55**(5), 1233 (1995)
256. K. Char, L. Antognazza, T.H. Geballe, Appl. Phys. Lett. **63**, 2420 (1993)
257. K. Char, M.S. Colclough, S.M. Garrison et al., Appl. Phys. Lett. **59**, 733 (1991)
258. Ch. Charach, C.K. Chen, P.C. Fife, J. Stat. Phys. **95**, 1141 (1999)
259. Ch. Charach, P.C. Fife, SIAM J. Appl. Math. **58**, 1826 (1998)
260. Ch. Charach, P.C. Fife, Open. Syst. Inf. Dyn. **5**, 99 (1998)
261. P. Chaudhari, D. Dimos, J. Mannhart, in *Superconductivity*, eds. by J.G. Bednortz, K.A. Müller (Springer, Heidelberg, 1990)
262. F.K. Chen, Appl. Mech. Eng. **3**, 413 (1998)
263. G.F. Chen, W.Z. Hu, J.L. Luo, N.L. Wang, PRL. **102**, 227004 (2009)
264. L.Q. Chen, Scripta Metall. Mater. **32**, 115 (1995)
265. L.Q. Chen, A.G. Khachaturyan, Acta Metall. Mater. **39**, 2533 (1991)
266. N. Chen, D. Shi, K.C. Goretta, J. Appl. Phys. **66**, 2485 (1989)
267. T.Y. Chen, Z. Tesanovic, R.H. Liu et al., Nature **453**, 1224 (2008)
268. H.W. Cheng, L. Greengard, J.Comput. Phys. **136**, 629 (1997)
269. P. Cheng, B. Shen, G. Mu et al., EPL **85**, 67003 (2009)
270. G.P. Cherepanov, *Mechanics of Brittle Fracture*, eds. by R. de Wit, W.C. Cooley (McGraw Hill, New York, 1979)
271. K.S. Chernyavsky, *Stereology in Metallurgical Science* (Metallurgiaya, Moscow, 1977)
272. A. Cherradi, G. Desgardin, L. Mazo, B. Raveau, Supercond. Sci. Technol. **6**(799), 7 (1993)
273. I. Chevalier, J. Mat. Proc. Tech. **32**, 145 (1992)
274. R. Chevrel, M. Sergent, J. Prigent, J. Solid State Chem. **3**, 515 (1971)
275. M. Chiao, R. Hill, C. Lupien et al., Phys. Rev. B. **62**, 3554 (2000)
276. N.R. Chikara, A. Aleem, Int. J. Mech. Sci. **43**, 1685 (2001)
277. C.C. Chin, R.J. Lin, Y.C. Yu et al., IEEE Trans. Appl. Supercond. **7**(2), 1403 (1997)
278. I. Chiorescu, P. Bertet, K. Semba et al., Nature **431**, 159 (2004)
279. M.F. Chisholm, S.J. Pennycook, Nature **351**, 47 (1991)
280. J.H. Cho, M.P. Maley, J.O. Willis et al., Appl. Phys. Lett. **64**, 3030 (1994)
281. R. Choksi, S Conti, R. VKohn, F.Otto, Comm. Pure Appl. Math. LXI, 0595 (2008)
282. R. Choksi, R.V. Kohn, F. Otto, J. Nonlinear Sci. **14**, 119 (2004)
283. D.K. Christen, J.R. Thompson, H.R. Kerchner, et al., in *AIP Conference of Proceedings of Superconductivity and its Applications* **273**, 24 (1993)
284. R.M. Christensen, K.H. Lo, J. Mech. Phys. Solids. **27**, 315 (1979)
285. C.W. Chu, IEEE Trans. Appl. Supercond. **7**(2), 80 (1997)
286. C.W. Chu, F. Chen, M. Gooch et al., Physica C **469**, 326 (2009)
287. C.W. Chu, L. Gao, F. Chen et al., Nature **365**, 323 (1993)
288. C.Y. Chu, J.L. Routbort, N. Chen et al., Supercond. Sci. Technol. **5**, 306 (1992)
289. T.J. Chuang, J.R. Rice, Acta Metall. **21**, 1625 (1973)
290. Y.D. Chuang, A.D. Gromko, A. Fedorov et al., Phys. Rev. Lett. **87**, 117002 (2001)
291. A.V. Chubukov, D. Pines, J. Schmalian, in *The Physics of Conventional and Unconventional Superconductors*, eds. by K.H. Bennemann. J.B. Kettersen (Springer, New York, 2002)
292. M.J. Cima, M.C. Flemings, A.M. Figueredo et al., J. Appl. Phys. **72**, 179 (1992)
293. M.J. Cima, M.C. Flemings, A.M. Figueredo et al., J. Appl. Phys. **78**, 1868 (1992)
294. D.R. Clarke, K.T. Faber, J. Phys. Chem. Solids **48**, 1115 (1987)

295. J.R. Clem, Phys. Rev. B **43**, 7837 (1991)
296. J.R. Clem, M.W. Coffey, Phys. Rev. B **42**, 6209 (1990)
297. J.R. Clem, A. Perez-Gonzalez, Phys. Rev. B. **30**, 5041 (1984)
298. J.P. Clerk, G. Giraud, J.M. Laugier, Adv. Phys. **39**, 191 (1990)
299. J.-P. Cleuziou, W. Wernsdorfer, V. Bou et al., Nat. Nanotechnol. **1**(1), 53 (2006)
300. W.S. Coblenz, J.M. Dynys, R.M. Cannon, R.L. Coble, in *Sintering Processes*, ed. by G.C. Kuczynski (Plenum Press, New York, 1980), p. 141
301. L.B. Coleman, M.J. Cohen, D.J. Sandman et al., Sol. State Comm. **12**, 1125 (1973)
302. S. Coleman, E. Weinberg, Phys. Rev. D **7**, 1888 (1973)
303. *Conectus Global Market for Superconductivity*, December 2009. http://www.conectus.org/market.html
304. M. Conti, Phys. Rev. E **55**, 701 (1997)
305. M. Conti, Phys. Rev. E **55**, 765 (1997)
306. M. Conti, Phys. Rev. E **56**, 3197 (1997)
307. R.F. Cook, Acta Metall. Mater. **38**, 1083 (1990)
308. R.F. Cook, D.R. Clarke, Acta Metall. **36**, 555 (1988)
309. H.T. Cooper, W. Gao, S. Li et al., Supercond. Sci. Technol. **14**, 862 (2001)
310. L.N. Cooper, Phys. Rev. **104**, 1189 (1956)
311. E.F.C.R. Cosserat, Acad. Sci. Paris. **126**, 1089 (1898)
312. G.W. Crabtree, W.K. Kwok, U. Welp, et al., in *Proceedings NATO Advanced Study Institute on the Physics and Materials Science of Vortex States, Flux Pinning and Dynamics* eds. by S. Bose, R. Kossowski (Kluwer Academic Publishers, Dordrecht, 1999)
313. G.W. Crabtree, D.R. Nelson, Phys. Today **50**, 38 (1997)
314. A. Crisan, P. Badica, S. Fujiwara et al., Appl. Phys. Lett. **80**, 3566 (2002)
315. A. Crisan, P. Badica, S. Fujiwara et al., IEEE Trans. Appl. Supercond. **13**, 3726 (2003)
316. A. Crisan, V.S. Dang, P. Mikheenko et al., Physica C **470**, 55 (2010)
317. A. Crisan, S. Fujiwara, J.C. Nie et al., Appl. Phys. Lett. **79**, 4547 (2001)
318. A. Crisan, A. Sarkar, P. Mikheenko et al., J. Supercond. Nov. Magn. **22**, 631 (2009)
319. C. Cronstrom, M. Noga, cond-mat/9906360 (1999)
320. H.Y. Cui, cond-mat/0212059 (2002)
321. N.J. Curro, T. Caldwell, E.D. Bauer et al., Nature **434**, 622 (2005)
322. M.E. Curtin, D. Polignone, J. Vinals, Math. Models. Methods Apl. Sci. **6**, 815 (1996)
323. R.R. da Silva, J.H.S. Torres, Y. Kopelevich, et al., cond-mat/0105329 (2001)
324. M. Daemling, J.M. Seuntjens, D.C. Larbalestier, Nature **346**, 332 (1990)
325. E. Dagotto, Rev. Mod. Phys. **66**, 763 (1994)
326. E. Dagotto, T.M. Rice, Science **271**, 618 (1996)
327. C. Dahnken, M. Aichhorn, W. Hanke et al., Phys. Rev. B **70**, 245110 (2004)
328. A. Damascelli, Z. Hussain, Z.X. Shen, Rev. Mod. Phys. **75**, 474 (2004)
329. R.R. Daminov, M.F. Imayev, M. Reissner et al., Physica C **408–410**, 46 (2004)
330. Z. Damm, T.S. Orlova, B.I. Smirnov, V.V. Shpeizman, Fiz. Tverd. Tela. **36**(8), 2465 (1994)
331. G.A. Daniels, A. Gurevich, D.C. Larbalestier, Appl. Phys. Lett. **77**, 3251 (2000)
332. B.S. Danilin, *Reviews on High-Temperature Superconductivity*, vol. 4(8), (MCNTI, Moscow, 1992), p. 101
333. V.V. Danishevskyy, A. Gillet, P. Pareige, D. Weichert, Mech. Mater. **40**, 284 (2008)
334. C. Dasgupta, B.I. Halperin, Phys. Rev. Lett. **47**, 1556 (1981)

335. S. Daul, D.J. Scalapino, S.R. White, Phys. Rev. Lett. **84**, 4188 (2000)
336. R.W. Davidge, Acta. Metall. **29**, 1695 (1981)
337. R.W. Davidge, T.J. Green, J. Mater. Sci. **3**, 629 (1968)
338. A.S. Davydov, *Solitons in Molecular Systems* (Naukova Dumka, Kiev, 1988)
339. A.S. Davydov, Phys. Rep. **190**, 191 (1990)
340. A.S. Davydov, *Solitons in Molecular Systems* (Kluwer Academic, Dordrecht, 1991)
341. L. de Arcangelis, H.J. Herrmann, Phys. Rev. B **39**, 2678 (1989)
342. P.G. de Gennes, Solid State Comm. **3**, 127 (1965)
343. P.G. de Gennes, *Superconductivity of Metals and Alloys* (Benjamin, New York, 1966)
344. P.G. de Gennes, J. Phys. Lett. (Paris) **37L**, 1 (1976)
345. M. de Graaf, M.E. McHenry, *Structure of Materials. An Introduction to Crystallography, Diffraction, and Symmetry* (Cambridge University Press, Cambridge, 2007)
346. C. de la Cruz, Q. Huang, J.W. Lynn et al., Nature **453**, 899 (2008)
347. A.L. de Lozanne, Supercond. Sci. Technol. **12**, R43 (1999)
348. P. de Rango, M.P. Lees, P. Lejay et al., Nature **349**, 770 (1991)
349. R.S. Decca, H.D. Drew, E. Osquiguil et al., Phys. Rev. Lett. **85**, 3708 (2000)
350. M.P. Delamare, M. Hervien, J. Wang et al., Phys. C **262**, 220 (1996)
351. M.P. Delamare, I. Monot, J. Wang, G. Desgardin, J. Electron. Mater. **24**, 1739 (1995)
352. M.P. Delamare, I. Monot, J. Wang et al., Supercond. Sci. Technol. **9**, 534 (1996)
353. M.P. Delamare, H. Walter, S. Bringmann et al., Phys. C **323**, 107 (1999)
354. K.A. Delin, A.W. Kleinsasser, Supercond. Sci. Technol. **9**, 227 (1996)
355. K.G. Denbigh, *The Thermodynamics of the Steady State* (Methuen, London, 1951)
356. G. Desgardin, I. Monot, B. Raveau, Supercond. Sci. Technol. **12**, R115 (1999)
357. D. Deutsch, R. Jozsa, Proc. Roy. Soc. Lond. A **439**, 553 (1992)
358. G. Deutscher, O. Entin-Wohlmann, S. Fishman, Y. Shapira, Phys. Rev. B **21**, 5041 (1980)
359. H.E. Deve, A.G. Evans, Acta Metall. Mater. **39**, 1171 (1991)
360. D. Dew Hughes, Cryogenics **36**, 660 (1986)
361. M. Dhalle, P. Toulemonde, C. Beneduce, et al., cond-mat/0104395 (2001)
362. A. Díaz, L. Mechin, P. Berghuis et al., Phys. Rev. Lett. **80**, 3855 (1998)
363. L. DiCarlo, J.M. Chow, J.M. Gambetta et al., Nature **460**, 240 (2009)
364. P. Diko, Supercond. Sci. Technol. **11**, 68 (1998)
365. P. Diko, Supercond. Sci. Technol. **13**, 1202 (2000)
366. P. Diko, G. Fuchs, G. Krabbes, Phys. C **363**, 60 (2001)
367. P. Diko, W. Gawalek, T. Habisreuther et al., Phys. Rev. B **52**, 13658 (1995)
368. P. Diko, W. Gawalek, T. Habisreuther et al., J. Microscopy **184**, 46 (1996)
369. P. Diko, H. Kojo, M. Murakami, Phys. C **276**, 188 (1997)
370. P. Diko, N. Pellerin, P. Odier, Phys. C **247**, 169 (1995)
371. P. Diko, S. Takebayashi, M. Murakami, Phys. C **297**, 216 (1998)
372. P. Diko, V.R. Todt, D.J. Miller, K.C. Goretta, Phys. C **278**, 192 (1997)
373. S. Dimopoulos, L. Susskind, Nucl. Phys. B **155**, 237 (1979)
374. D. Dimos, P. Chaudhari, J. Mannhart, Phys. Rev. B **41**, 4038 (1990)
375. D. Dimos, P. Chaudhari, J. Mannhart, F.K. LeGoues, Phys. Rev. Lett. **61**, 219 (1988)
376. H. Ding, T. Yokoya, J.C. Campuzano et al., Nature **382**, 51 (1996)
377. R.J. Dinger, D.R. Bowling, A.M. Martin, IEEE Trans. Microwave Theory Tech. **39**, 1498 (1991)

378. P.A.M. Dirac, Proc. Roy. Soc. A **133**, 60 (1931)
379. U.S. Dixit, P.M. Dixit, J. Mat. Proc. Tech. **47**, 201 (1995)
380. D. Djurek, Z. Medunic, A. Tonejc, M. Paljevic, Phys C **351**, 78 (2001)
381. G.J. Dolan, G.V. Chandrashekhar, T.R. Dinger et al., Phys. Rev. Lett. **62**, 827 (1989)
382. L.A. Dorosinskii, M.I. Indenbom, V.I. Nikitenko et al., Phys. C **203**, 149 (1992)
383. L.A. Dorosinskii, M.V. Indenbom, V.I. Nikitenko et al., Phys. C **206**, 360 (1993)
384. L.A. Dorosinskii, V.I. Nikitenko, A.A. Polyanskii, V.K. Vlasko-Vlasov, Phys. C **219**, 81 (1994)
385. S.E. Dorris, J.T. Dusek, M.T. Lanagan et al., Ceram. Bull. **70**, 722 (1991)
386. S.E. Dorris, B.C. Prorok, M.T. Lanagan et al., Phys. C **212**, 66 (1993)
387. S.X. Dou, J. Horvat, X.L. Wang et al., IEEE Trans. Appl. Supercond. **7**(2), 2219 (1997)
388. S.X. Dou, H.K. Liu, S.J. Guo, Supercond. Sci. Technol. **2**, 274 (1989)
389. S.X. Dou, H.K. Liu, Y.C. Guo, D.L. Shi, IEEE Trans. Appl. Supercond. **3**(2), 1135 (1993)
390. K. Doverspike, C.D. Hubbard, R.K. Williams et al., Phys. C **172**, 486 (1991)
391. V.P. Dravid, H. Zhang, Y.Y. Wang, Phys. C **213**, 353 (1993)
392. D. Driscoll, V. Dombrovsky, B. Zhang, IEEE Power Eng. Rev. **20**(5), 39 (2000)
393. D. Driscoll, B. Zhang, IEEE Trans. Appl. Supercond. **11**(1), 615 (2001)
394. F. Drolet, K.R. Elder, M. Grant, J.M. Kosterlitz, Phys. Rev. E **61**, 6705 (2000)
395. M.D. Drory, M.D. Thouless, A.G. Evans, Acta Metall. **36**, 2019 (1988)
396. Y.N. Drozdov, S.V. Gaponov, S.A. Gusev et al., Supercond. Sci. Technol. **9**, 1 (1996)
397. Y.N. Drozdov, S.V. Gaponov, S.A. Gusev et al., IEEE Trans. Appl. Supercond. **7**(2), 1642 (1997)
398. D.C. Drucker, W. Prager, Quart. Appl. Math. **10**, 157 (1952)
399. W.J. Drugan, J.R. Rice, T.L. Sham, J. Mech. Phys. Solids **30**, 447 (1982)
400. G.A. Dubitzky, V.D. Blank, S.G. Buga et al., Pis'ma JETP **81**, 323 (2005)
401. I.N. Dubrovina, R.G. Zakharov, E.G. Kositsyn et al., Supercond. Phys. Chem. Tech. **3**, 3 (1990)
402. D.S. Dugdale, J. Mech. Phys. Solids **8**, 100 (1960)
403. E.A. Dul'kin, Supercond. Phys. Chem. Tech. **5**, 104 (1992)
404. E.A. Dul'kin, Supercond. Phys. Chem. Tech. **7**, 105 (1994)
405. E.A. Dul'kin, J. Supercond. **11**, 275 (1998)
406. E. Dul'kin, V. Beilin, E. Yashchin et al., Pis'ma Zh. Tech. Fiz. **27**(9), 79 (2001)
407. E. Dul'kin, V. Beilin, E. Yashchin, M. Roth, Supercond. Sci. Technol. **15**, 1081 (2002)
408. G.N. Dul'nev, Yu.P. Zarichnyak, *Thermal Conductivity of Mixtures and Composite Materials* (Energiya, Leningrad, 1974)
409. G.N. Dul'nev, Yu.P. Zarichnyak, B.L. Muratova, Inzh. Fiz. Zh. **16**, 1019 (1969)
410. J. Dunders, *Mathematical Theory of Dislocations* (The American Society of Mechanical Engineers, New York, 1969)
411. L. Duran, F. Kircher, P. Regnier et al., Supercond. Sci. Technol. **8**, 214 (1995)
412. D.J. Durand, J. Carpentter, E. Landizinsky et al., IEEE Trans. Appl. Supercond. **2**(1), 33 (1992)
413. E.N. Dvorkin, M.B. Goldschmit, M.A. Cavaliere et al., J. Mater. Process. Technol. **68**, 99 (1997)
414. A.I. D'yachenko, Fiz. Tech. Vys. Davl. **6**, 5 (1996)
415. A.M. Dykhne, Sov. Phys. JETP. **32**, 63 (1971)

416. A.V. Dyskin, L.M. Ring, R.L. Salganik, *Estimation of Damage of $YBa_2Cu_3O_{7-\delta}$ Superconducting Ceramic on Data about its Deformability, Preprint 417* (IPM AN USSR, Moscow, 1989)
417. E.A. Early, R.L. Steiner, A.F. Clark, K. Char, Phys. Rev. B **50**, 9409 (1994)
418. C.J. Eastell, B.M. Henry, C.G. Morgan et al., IEEE Trans. Appl. **7**(2), 2083 (1997)
419. H.S. Edelman, J.A. Parrell, D.C. Larbalestier, J. Appl. Phys. **81**, 2296 (1997)
420. M. Eden, in *Proceedings 4th Berkeley Symposium on Mathematics, Statistics and Probability* vol. **4**, 223 (1961)
421. A.L. Efros, *Physics and Geometry of Disorder: Percolation Theory* (Mir, Moscow, 1987)
422. K. Ehata, M.I. Ali, M. Kusunoki, S. Ohshima, IEEE Trans. Appl. Supercond. **9**(2), 3081 (1999)
423. H. Eichelkraut, G. Abbruzzese, K. Lücke, Acta Metall. **36**, 55 (1988)
424. J.W. Eischen, S. Torquato, J. Appl. Phys. **74**, 159 (1993)
425. J.W. Ekin, S.L. Bray, N. Cheggour et al., IEEE Trans. Appl. Supercond. **11**(1), 3389 (2001)
426. J.W. Ekin, D.K. Finnemore, Q. Li et al., Appl. Phys. Lett. **61**, 858 (1992)
427. J.W. Ekin, H.R. Hart Jr, A.R. Gaddipati, J. Appl. Phys. **68**, 2285 (1990)
428. K.R. Elder, F. Drolet, J.M. Kosterlitz, M. Grant, Phys. Rev. Lett. **72**, 677 (1994)
429. K.R. Elder, J.D. Gunton, M. Grant, Phys. Rev. E **54**, 6476 (1996)
430. C.M. Elliott, Y. Kashima, IMA J. Num. Anal. **27**, 293 (2007)
431. C.M. Elliott, D. Kay, V. Styles, J. Num. Anal. **42**, 1324 (2004)
432. C.M. Elliott, D. Kay, V. Styles, IMA J. Num. Anal. **25**, 182 (2005)
433. V. Emery, S. Kivelson, Nature **374**, 434 (1995)
434. V. Emery, S. Kivelson, O. Zachar, Phys. Rev. B **56**, 6120 (1997)
435. H. Emmerich, Phys. Rev. B **65**, 233406 (2002)
436. H. Emmerich, *The Diffuse Interface Approach in Materials Science Thermodynamic Concepts and Aplications of Phase-Field Models* (Springer, New York, 2003)
437. H. Emmerich, K. Kassner, T.J. Ihle, A. Wei*β*, J. Cryst. Growth **211**, 43 (2000)
438. H. Emmerich, K. Kassner, C. Misbah, T. Ihle, J. Phys. Condens. Mater. **11**, 9985 (1999)
439. A. Endo, S.H. Chauhan, T. Egi, Y. Shiohara, J. Mater. Res. **11**, 795 (1996)
440. A. Endo, S.H. Chauhan, Y. Nakamura, Y. Shiohara, J. Mater. Res. **11**, 1114 (1996)
441. K. Endo, H. Yamasaki, S. Milsawa et al., Nature **355**, 327 (1992)
442. F. Englert, R. Brout, Phys. Rev. Lett. **13**, 321 (1964)
443. C.B. Eom, M.K. Lee, J.H. Choi et al., Nature **411**, 558 (2001)
444. C.B. Eom, A.F. Marshall, Y. Suzuki et al., Nature **353**, 544 (1991)
445. M. Eriksen, J.I. Bech, N. Bay, et al., in *Proceedings 4th Europ Conference on Applied Superconductivity* (*EUCAS'*99), paper 12–75 (Barselona, 1999)
446. M. Eriksen, J.I. Bech, B. Seifi, N. Bay, IEEE Trans. Appl. Supercond. **11**(1), 3756 (2001)
447. R. Erlandsson, G.M. McClelland, C.M. Mate, S. Chiang, J.Vacuum Sci. Technol. A **6**, 266 (1988)
448. J.D. Eshelby, Proc. Roy. Soc. Lond. A. **241**, 376 (1957)
449. J.D. Eshelby, *Continuum Theory of Dislocation* (Inostrannaya Literatura, Moscow, 1963)
450. A.G. Evans, Acta Metall. **26**, 1845 (1978)
451. A.G. Evans, N. Burlingame, M. Drory, W.M. Kriven, Acta Metall. **29**, 447 (1981)
452. A.G. Evans, R.M. Cannon, Acta Metall. **34**, 761 (1986)
453. A.G. Evans, B.J. Dalgleish, M. He, J.W. Hutchinson, Acta Metall. **37**, 3249 (1989)
454. A.G. Evans, Y. Fu, Acta Metall. **33**, 1525 (1985)

455. A.G. Evans, A.H. Heuer, J. Am. Ceram. Soc. **63**, 241 (1980)
456. A.G. Evans, J.W. Hutchinson, Acta Metall. **37**, 909 (1989)
457. A.G. Evans, D.B. Marshall, Acta Metall. **37**, 2567 (1989)
458. K.T. Faber, A.G. Evans, Acta Metall. **31**, 565 (1983)
459. J. Fabrega, J. Fontcuberta, A. Serquis, A. Caneiro, Phys. C **356**, 254 (2001)
460. F. Falk, Acta Metall. **28**, 1773 (1980)
461. F.Z. Falk, Phys. B **51**, 177 (1983)
462. D. Favrot, M. Déchamps, J. Mater. Res. **6**, 2256 (1991)
463. D.M. Feldmann, J.L. Reeves, A.A. Polyanskii et al., Appl. Phys. Lett. **77**, 2906 (2000)
464. D.L. Feng, N.P. Armitage, D.H. Lu et al., Phys. Rev. Lett. **86**, 5550 (2001)
465. D.L. Feng, A. Damascelli, K.M. Shen et al., Phys. Rev. Lett. **88**, 107001 (2002)
466. S. Feng, B.I. Halperin, P.N. Sen, Phys. Rev. B. **35**, 197 (1987)
467. Y. Feng, K.E. Hautanen, Y.E. High, et al., Phys. C **192**. 293 (1992)
468. Y. Feng, D.C. Larbalestier, Interface Sci. **1**, 401 (1994)
469. C. Ferdeghini, V. Ferrando, C. Tarantini, et al., cond-mat/0411404 (2004)
470. P.J. Ferreira, H.B. Liu, J.B. Vander Sande, IEEE Trans. Appl. Supercond. **9**(2), 2231 (1999)
471. I.A. Parinov (ed.) ,*Ferroelectrics and Superconductors: Properties and Applications* (Nova Science Publishers, New York, 2011)
472. E.G. Fesenko, Dantsiger A. Ya., O.N. Razumovskaya, *New Piezoceramic Materials* (Rostov State University Press, Rostov-on-Don, 1983)
473. A.V. Fetisov, A.A. Fotiev, Supercond. Phys. Chem. Tech. **5**, 1071 (1992)
474. V.B. Fetisov, A.V. Fetisov, A.A. Fotiev, Supercond. Phys. Chem. Tech. **3**, 2627 (1990)
475. P.F. Fewster, *X-Ray Scattering from Semiconductors* (Imperial College Press, London, 2000)
476. M.B. Field, X.Y. Cai, S.E. Babcock, D.C. Larbalestier, IEEE Trans. Appl. Supercond. **3**(2), 1479 (1993)
477. M.B. Field, A. Pashitski, A. Polyanskii et al., IEEE Trans. Appl. Supercond. **5**(2), 1631 (1995)
478. J. Figueras, T. Puig, X. Obradors et al., Nature Phys. **2**, 402 (2006)
479. B. Fischer, S. Kautz, M. Leghissa et al., IEEE Trans. Appl. Supercond. **9**(2), 2480 (1999)
480. D.S. Fisher, M.A. Fisher, D.A. Huse, Phys. Rev. B **43**, 130 (1991)
481. K.H. Fischer, Phys. C **178**, 161 (1991)
482. K.H. Fischer, Supercond. Rev. **1**, 153 (1995)
483. R.A. Fisher, J.E. Gordon, N. Phillips, in *Handbook of High-Temperature Superconductivity: Theory and Experiment*, eds. by J.R. Schrieffer, J.S. Brooks (Springer, New York, 2007)
484. M.D. Fiske, Rev. Mod. Phys. **36**, 221 (1964)
485. H. Fjellvåg, P. Karen, A. Kjekshus et al., Acta Chem. Scand. A. **42**, 178 (1988)
486. M.C. Flemings, *Solidification Processing* (McGraw-Hill, New York, 1974)
487. S. Fleshler, M. Fee, S. Spreafico, A.P. Malozemoff, *Advanced Supercond. XII, 12th International Symposium on Superconductivity* (*ISS'1999*), (Oct 17–19, Morioka, Japan, 1999), p. 625
488. S. Fleshler, W.K. Kwok, U. Welp, et al., Phys. Rev. B **47**, 14448 (1993)
489. R. Flükiger, E. Giannini, M. Lomello-Tafin et al., IEEE Trans. Appl. Supercond. **11**(1), 3393 (2001)
490. R. Flükiger, G. Grasso, J.C. Grivel et al., Supercond. Sci. Technol. **10**, A68 (1997)
491. M. Fol'mer, *Initiation Kinetics of New Phase* (Nauka, Moscow, 1986)

492. F.P. Ford, Corrosion **52**, 375 (1996)
493. K. Fossheim, S.D. Tuset, T.W. Ebbesen et al., Phys. C **248**, 195 (1995)
494. G.A. Francfort, F. Murat, Arch. Rational Mech. Analysis. **94**, 307 (1986)
495. M. Friesen, A. Gurevich, in *Ext. Abstr. International Workshop on Critical Current (IWCC-99)*, Madison, Wisconsin, USA, July 7–10, 1999, (1999), p. 102
496. H. Fröhlich, Phys. Rev. **79**, 845 (1950)
497. H. Fröhlich, Proc. Roy Soc. (London) **A215**, 291 (1950)
498. Y. Fu, A.G. Evans, Acta Metall. **30**, 1619 (1982)
499. Y. Fu, A.G. Evans, Acta Metall. **33**, 1515 (1985)
500. M.S. Fuhrer, K. Cherrey, A. Zettl, M.L. Cohen, Phys. Rev. Lett. **83**, 404 (1999)
501. H. Fujimoto, H. Kamijo, T. Higuchi et al., IEEE Trans. Appl. Supercond. **9**(2), 301 (1999)
502. H. Fujimoto, M. Murakami, S. Gotoh et al., Adv. Supercond. **2**, 285 (1990)
503. H. Fujimoto, M. Murakami, N. Koshizuka, Phys. C **203**, 103 (1992)
504. S. Fujisawa, E. Kishi, Y. Sugawara, S. Morita, Jpn. J. Appl. Phys. **33**, 3752 (1994)
505. S. Fujisawa, M. Ohta, T. Konishi et al., Rev. Sci. Instrum. **65**, 644 (1994)
506. S. Fujita, S. Godoy, *Theory of High Temperature Superconductivity* (Kluwer Academic, Dordrecht, 2001)
507. S. Fujita, K. Ito, S. Godoy, *Quantum Theory of Conducting Matter, Superconductivity* (Springer, Berlin, 2009)
508. P. Fulde, R.A. Ferrell, Phys. Rev. **135**, A550 (1964)
509. A.M. Gabovich, A.I. Voytenko, Fiz. Nizk. Temp. **26**, 419 (2000)
510. P.L. Gammel, L.F. Schneemeyer, J.V. Waszczak et al., Phys. Rev. Lett. **61**, 1666 (1988)
511. C. Gandhi, M.F. Ashby, Acta Metall. **27**, 1565 (1979)
512. C.H.-A. Gandin, M. Rappaz, D. West, B.L. Adams, Metallurg. Mater. Trans. **26A**, 1543 (1995)
513. A.Y. Ganin, Y. Takabayashi, Y.Z. Khimyak et al., Nature Mater. **7**, 367 (2008)
514. F. Gao, T. Wang, J. Mater. Sci. Lett. **9**, 1409 (1990)
515. L. Gao, Y.Y. Xue, F. Chen et al., Phys. Rev. B **50**, 4260 (1994)
516. W. Gao, J. Chen, C.O. Yang et al., Phys. C **193**, 255 (1992)
517. Y. Gao, G. Bai, D.J. Lam, K.L. Merkle, Phys. C **173**, 487 (1991)
518. Y. Gao, K.L. Merkle, G. Bai et al., Phys. C **174**, 1 (1991)
519. Y. Gao, K.L. Merkle, C. Zhang et al., J. Mater. Res. **5**, 1363 (1990)
520. Z-S. Gao, L. Y-PWang Qi, et al., Supercond. Sci. Technol. **21**, 112001 (4 pp.) (2008)
521. J.L. Garcia-Munoz, M. Suaadi, J. Foncuberta et al., Phys. C **268**, 173 (1996)
522. J.C.M. Garnet, Phil. Trans. Roy. Soc. Lond. A **203**, 385 (1904)
523. V. Garnier, R. Caillard, A. Sotelo et al., Physica C **319**, 197 (1999)
524. V.A. Gasparov, N.S. Sidorov, I.I. Zver'kova, M.P. Kulakov, Pis'ma JETP **73**, 532 (2001)
525. P. Gautier-Picard, PhD Thesis (Grenoble, 1998)
526. J.R. Gavaler, Appl. Phys. Lett. **23**, 480 (1973)
527. B.M. Geerken, R. Griessen, L.M. Huisman, E. Walker, Phys. Rev. B **26**, 1637 (1982)
528. E.M. Gershenzon, G.N. Gol'tsman, I.G. Gogidze et al., Supercond. Phys. Chem. Tech. **3**, 1582 (1990)
529. L.V. Gibiansky, S. Torquato, Phys. Rev. Lett. **71**, 2927 (1993)
530. L.V. Gibiansky, S.J. Torquato, Mech. Phys. Solids. **43**, 1587 (1995)
531. L.V. Gibiansky, S. Torquato, Int. J. Eng. Sci. **34**, 739 (1996)
532. L.V. Gibiansky, S. Torquato, Proc. Roy. Soc. Lond. A **452**, 253 (1996)

533. D. Giller, A. Shaulov, R. Prozorov et al., Phys. Rev. Lett. **79**, 2542 (1997)
534. V.L. Ginzburg, Sov. Solid State **2**, 61 (1960)
535. V.L. Ginzburg, Sov. Phys. JETP **47**, 2318 (1964)
536. V.I. Ginzburg, L.D. Landau, Zh. Eksp. Teor. Fiz. **20**, 1064 (1950)
537. F. Gire, D. Robbes, C. Gonzales et al., IEEE Trans. Appl. Supercoud. **7**(2), 3200 (1997)
538. S.L. Glashow, Nucl. Phys. **22**, 519 (1961)
539. W. Goldacker, M. Quilitz, B. Obst, H. Eckelmann, Phys. C **310**, 182 (1998)
540. N. Goldenfeld, *Lectures on Phase Transitions and the Renormalization Group* (Addison-Wesley, Reading, 1992)
541. J. Goldstone, Nuovo Cim. **19**, 154 (1961)
542. J. Goldstone, A. Salam, S. Weinberg, Phys. Rev. **127**, 965 (1962)
543. R. Golopan, T. Roy, T. Rajasekharan et al., Phys. C **244**, 106 (1995)
544. E.A. Goodilin, D.B. Kvartalov, N.N. Oleynikov et al., Phys. C **235–240**, 449 (1994)
545. E.A. Goodilin, N.N. Oleynikov, Supercond. Res. Develop. **5/6**, 81 (1995)
546. S. Goodman, Nature **414**, 831 (2001)
547. K. Goretta, W. Delaney, J. Routbort et al., Supercond. Sci. Technol. **9**, 422 (1996)
548. K.C. Goretta, M. Jiang, D.S. Kupperman et al., IEEE Trans. Appl. Supercond. **7**(2), 1307 (1997)
549. K.C. Goretta, M.L. Kullberg, D. Bär et al., Supercond. Sci. Technol. **4**, 544 (1991)
550. K.C. Goretta, D.S. Kupperman, S. Majumdar et al., Supercond. Sci. Technol. **11**, 1409 (1998)
551. K.C. Goretta, M.E. Loomans, L.J. Martin et al., Supercond. Sci. Technol. **6**, 282 (1993)
552. C.J. Gorter, Physica **23**, 45 (1957)
553. C.J. Gorter, H.B.G. Casimir, Physica **1**, 306 (1934)
554. L.P. Gor'kov, Zh. Eksp. Teor. Fiz. **36**, 1918 (1959) [Sov. Phys. JETP **9**, 1364 (1959)]
555. L.P. Gor'kov, Zh Eksp, Teor. Fiz. **37**, 833 (1959)
556. L.P. Gor'kov, Zh Eksp, Teor. Fiz. **37**, 1407 (1959)
557. L.P. Gor'kov, A.V. Sokol, Pis'ma. Zh. Eksp. Teor. Fiz. **46**, 420 (1987)
558. T. Goto, T. Kuji, Y.S. Jiang, et al., IEEE Trans. Appl. Supercond. **9**(2), 1653 (1999)
559. H. Gould, J. Tobochnik, *An Introduction to Computer Simulation Methods. Applications to Physical Systems*, part 2 (Addison-Wesley, New York, 1988)
560. L.V. Govor, V.P. Dobrego, V.N. Golubev et al., Supercond. Phys. Chem. Tech. **5**, 2142 (1992)
561. A. Goyal, K.B. Alexander, D.M. Kroeger et al., Phys. C **210**, 197 (1993)
562. A. Goyal, P.D. Funkenbusch, D.M. Kroeger, S. Burns, Phys. C **182**, 203 (1991)
563. A. Goyal, D.P. Norton, J.D. Budai et al., Appl. Phys. Lett. **69**, 1975 (1996)
564. A. Goyal, D.P. Norton, D.M. Kroeger et al., J. Mater. Res. **12**, 2924 (1997)
565. A. Goyal, W.C. Oliver, P.D. Funkenbusch et al., Phys. C **183**, 221 (1991)
566. A. Goyal, E.D. Specht, D.M. Kroeger, T.A. Manson, Appl. Phys. Lett. **68**, 711 (1996)
567. D.Z. Grabko, Yu.S. Boyarskaya, R.P. Zhitaru et al., Supercond. Phys. Chem. Tech. **2**, 67 (1989)
568. G. Grasso, A. Jeremie, R. Flükiger, Supercond. Sci. Technol. **8**, 827 (1995)
569. M. Grayson, L.B. Rigal, D.C. Schmadel et al., Phys. Rev. Lett. **89**, 037003 (2002)
570. E. Gregoryanz, V.V. Struzhkin, R.J. Hemley, et al., cond-mat/0108267 (2001)

571. A.A. Grekov, N. Egorov, Ya, D.N. Karpinsky, in *Proceedings International Sympos. on Strength of Materials and Construction Elements at Sound and Ultrasound Frequencies of Loading*, (Naukova Dumka, Kiev, 1987), p. 187
572. S.A. Gridnev, O.N. Ivanov, Supercond. Phys. Chem. Tech. **5**, 1143 (1992)
573. S.A. Gridnev, O.N. Ivanov, O.V. Dybova, Supercond. Phys. Chem. Tech. **3**, 1449 (1990)
574. S.A. Gridnev, O.N. Ivanov, O.V. Dybova, in *Collection of Abstracts International Conference on Transparent Ferroelectric Ceram*ics (TFC'91), (Riga, Oct 2–6, 1991), p. 77
575. R. Griessen, Phys. Rev. Lett. **60**, 1674 (1990)
576. A.A. Griffith, Phil. Trans., R. Soc. Lond. A. **221**, 163 (1920)
577. M.L. Griffith, R.T. Huffman, J.W. Halloran, J. Mater. Res. **9**, 1633 (1994)
578. L.D. Grineva, V.P. Zatsarinny, V.A. Aleshin et al., Strength Mater. **4**, 34 (1993)
579. J.C. Grivel, A. Jeremie, R. Flükiger, Supercond. Sci. Technol. **8**, 41 (1995)
580. J.C. Grivel, A. Jeremie, B. Hansel, R. Flükiger, Supercond. Sci. Tech. **6**, 730 (1993)
581. R. Gross, in *High Temperature Superconducting Systems*, eds. by S.L. Shindé, D. Rudman (Springer, New York, 1992), p. 176
582. R. Gross, in *Interfaces in High-T_c Superconducting Systems*, eds. by S.L. Shinde, D.A. Rudman (Springer, New York, 1994). p. 174
583. R. Gross, L. Alff, A. Beck et al., IEEE Trans. Appl. Supercond. **7**(2), 2929 (1997)
584. R. Gross, P. Chaudhari, M. Kawasaki, A. Gupta, Phys. Rev. B **42**, 10735 (1992)
585. K.R. Grossmann, K.R. Elder, M. Grant, J.M. Kosterlitz, Phys. Rev. Lett. **71**, 3323 (1993)
586. L.K. Grover, in *Proceedings of 28th Annual ACM Symposium on the Theory of Computing*, Philadelphia, USA, May 22–24, 1996
587. J.E. Gubernatis, J.A. Krumhansl, J. Appl. Phys. **46**, 1875 (1975)
588. D.U. Gubser, Phys. C **341–348**, 2525 (2000)
589. F. Gugenberger, C. Meingast, G. Roth et al., Phys. Rev. B **49**, 13137 (1994)
590. S. Guillaume, A.E. Siam, J.Qualitative Theory Diff. Equ. **11**, 1 (2005)
591. J. Guo, S. Jin, G. Wang et al., Phys. Rev. B. **82**, 180520 (2010)
592. Y.C. Guo, H.K. Liu, S.X. Dou, Phys. C **235–240**, 1231 (1994)
593. Y.C. Guo, H.K. Liu, S.X. Dou et al., Supercond. Sci. Technol. **11**, 1053 (1998)
594. G.S. Guralnik, C.R. Hagen, T.W.B. Kibble, Phys. Rev. Lett. **13**, 585 (1964)
595. A. Gurevich, Fiz. Tverd. Tela **30**, 1384 (1988)
596. A. Gurevich, Phys. Rev. B **42**, R4857 (1990)
597. A. Gurevich, Phys. Rev. B **46**, 3187 (1992)
598. A. Gurevich, Phys. Rev. B **46**, 3638 (1992)
599. A. Gurevich, *Ext. Abstr. Int. Workshop on Critical Current (IWCC-99)*, Madison, Wisconsin, USA, July 7–10, 1999), p. 2
600. A. Gurevich, M. Benkraouda, J. R. Clem, Phys. Rev. B **54**, 13196 (1996)
601. A. Gurevich, L.D. Cooley, Phys. Rev. B **50**, 13563 (1994)
602. A. Gurevich, M. Friesen, Phys. Rev. B **62**, 4004 (2000)
603. A. Gurevich, M. Friesen, V. Vinokur, Phys. C **341–348**, 1249 (2000)
604. A. Gurevich, H. Küpfer, C. Keller, Europhys. Lett. **15**, 789 (1991)
605. A. Gurevich, J. McDonald, Phys. Rev. Lett. **81**, 2546 (1998)
606. A. Gurevich, R.G. Mints, A.L. Rakhmanov, *The Physics of Composite Superconductors* (Begell House, New York, 1997)
607. A. Gurevich, E.A. Pashitskii, Phys. Rev. B **56**, 6213 (1997)
608. A. Gurevich, E.A. Pashitskii, Phys. Rev. B **57**, 13878 (1998)

609. A. Gurevich, V.M. Vinokur, Phys. Rev. Lett. **83**, 3037 (1999)
610. A.L. Gurson, D.W. Yuan, *Net Shape Processing of Powder Materials*, vol. 216 (ASME-AMD, Metals Park, 1995)
611. M. Gurvitch, J.M. Valles Jr, A.M. Cucolo et al., Phys. Rev. Lett. **63**, 1008 (1989)
612. A.A. Gusev, E.G. Avvakumov, V.P. Isupov, et al., in *Piezoelectric Materials and Devices*, ed. by Ivan A. Parinov (Nova Science Publishers, New York, 2012)
613. J. Gutierrez, A. Llordes, J. Gazquez et al., Nat. Mater. **6**(5), 367 (2007)
614. A.N. Guz', YuN Nemish, *Perturbation Method of Boundary Shapes in Mechanics of Solid Media* (Vishcha Shkola, Kiev, 1989)
615. E.M. Gyorgy, R.B. Van Dover, K.A. Jackson et al., Appl. Phys. Lett. **55**, 283 (1989)
616. R. Hackl, M. Lambacher, A. Erb, et al., in *Advances in Solid State Physics*, ed. by B. Kramer **45** (Springer, New York, 2005)
617. G. Hager, G. Wellein, E. Jeckelmann, H. Fehske, Phys. Rev. B. **71**, 75108 (2005)
618. B.I. Halperin, T.C. Lubensky, S.K. Ma, Phys. Rev. Lett. **32**, 292 (1974)
619. M. Hamasumi, *Nonferrous Metals and Alloys*, 3rd edn. (Uchidarouhokaku-shinsya, Tokyo. 1984), p. 210
620. N.M. Hamdan, P.V.P.S.S. Sastry, J. Schwartz, Phys. C **341–348**, 513 (2000)
621. N.M. Hamdan, P.V.P.S.S. Sastry, J. Schwartz, IEEE Trans. Appl. Supercond. **12**(1), 1132 (2002)
622. N.M. Hamdan, A. Ziq Kh, A.S. Al-Harthi, Phys. C **314**, 125 (1999)
623. N.M. Hamdan, A. Ziq Kh, J. Shirokoff, Supercond. Sci. Technol. **7**, 118 (1994)
624. G. Hammeri, H. Bielefeldt, B. Goetz et al., IEEE Trans. Appl. Supercond. **11**(1), 2830 (2001)
625. G. Hammeri, A. Schmehl, R.R. Schulz et al., Nature **407**, 162 (2000)
626. B. Han, G.-H. Chen, L.-H. Zhang, Chin. Phys. Lett. **17**(11), 847 (2000)
627. Z. Han, T. Freltoft, Appl. Supercond. **2**, 201 (1994)
628. Z. Han, P. Skov-Hansen, T. Freltoft, Supercond. Sci. Technol. **10**, 371 (1997)
629. M. Hanawa, Y. Muraoka, T. Tayama, Phys. Rev. Lett. **87**, 187001 (2001)
630. A. Handstein, D. Hinz, G. Fuchs, et al., cond-mat/0103408 (2001)
631. M. Hangyo, S. Shikii, M. Yamashita et al., IEEE Trans. Appl. Supercond. **9**(2), 3038 (1999)
632. J. Hänisch, C. Cai, V. Stehr et al., Supercond. Sci. Technol. **19**, 534 (2006)
633. N.B. Hannay, T.H. Geballe, B.T. Matthias et al., Phys. Rev. Lett. **14**, 225 (1965)
634. K. Harada, T. Matsuda, J. Bonevich et al., Nature **360**, 51 (1992)
635. K. Harada, T. Matsuda, H. Kasai et al., Phys. Rev. Lett. **71**, 3371 (1993)
636. G.F. Hardy, J.K. Hulm, Phys. Rev. **93**, 1004 (1954)
637. N. Hari Babu, M. Kambara, J. McCrone, et al., IEEE Trans. Appl. Supercond. **11**(1), 2838 (2001)
638. J.M. Harris, P.J. White, Z.-X. Shen et al., Phys. Rev. Lett. **79**, 143 (1997)
639. H.R. Hart Jr, F.E. Luborsky, R.H. Arendt et al., IEEE Trans. Magn. **27**, 1375 (1991)
640. R. Hartley, *Linear and Nonlinear Programming* (Wiley, New York, 1985)
641. D.P. Harvey, M.I. Jolles, Metall. Trans. **21A**, 1719 (1990)
642. K. Hasegawa, N. Yoshica, K. Fujinao, et al., in *Proceedings 16th ICEC/ICMC* (Elsevier Science, Amsterdam, 1997), p. 1413
643. T. Hasegawa, Y. Hikichi, T. Koizumi et al., IEEE Trans. Appl. Supercond. **7**(2), 1703 (1997)
644. T. Hasegawa, T. Kitamura, H. Kobayashi et al., Phys. C **190**, 81 (1991)

645. Z. Hashin, J. Appl. Mech. **29**, 143 (1962)
646. Z. Hashin, J. Mech. Phys. Solids. **13**, 119 (1965)
647. Z. Hashin, J. Appl. Phys. **89**, 2261 (2001)
648. Z. Hashin, B.W. Rosen, J. Appl. Mech. **31**, 223 (1964)
649. Z. Hashin, S. Shtrikman, J. Appl. Phys. **33**, 1514 (1962)
650. Z. Hashin, S. Shtrikman, J. Mech. Phys. Solids. **10**, 335 (1962)
651. Z. Hashin, S. Shtrikman, J. Mech. Phys. Solids. **11**, 127 (1963)
652. R. Haslinger, R. Joynt, Phys. Rev. B **61**, 4206 (2000)
653. T. Haugan, P.N. Barnes, R. Wheeler et al., Nature **430**, 867 (2004)
654. B. Hawsey, D. Peterson, in *Proceedings of HTS Wire Development Workshop* (St. Petersburg, Florida USA, 20 February 2005)
655. R.M. Hazen, L.W. Finger, R.J. Angel et al., Phys. Rev. Lett. **60**, 1657 (1988)
656. T. He, Q. Huang, A.P. Ramirez, et al., cond-mat/0103296 (2001)
657. A.F. Hebard, M.J. Rosseinsky, R.C. Huddon et al., Nature **350**, 600 (1991)
658. R.W. Heckel, Trans. Metal. Soc. AIME **221**, 671 (1961)
659. R. Hedderich, T. Schuster, H. Kuhn et al., Appl. Phys. Lett. **66**, 3215 (1995)
660. H.B. Heersche, P. Jarillo-Herrero, J.B. Oostinga et al., Nature **446**, 56 (2007)
661. R.H. Heffner, D.E. Mac Laughlin, J.E. Sonier, et al., cond-mat/0102137 (2001)
662. K. Heine, J. Tenbrink, M. Thöner, Appl. Phys. Lett. **55**, 2441 (1989)
663. N.F. Heinig, R.D. Redwing, I.F. Tsu, et al., Appl. Phys. Lett. **69**, 577 (1996)
664. E.E. Hellstrom, J. Miner, J. Met. Mater. Soc. **44**, 44 (1992)
665. E.E. Hellstrom, MRS Bull. **17**, 45 (1992)
666. E.E. Hellstrom, in *High-Temperature Superconducting Materials Science and Engineering. New Concepts and Technology*, ed. by D. Shi (Elsevier Science, London, 1995), p. 383
667. E.E. Hellstrom, R.D. Ray, *II. HTS Materials, Bulk Processing and Bulk Applications* (World Scientific, Singapore, 1992), p. 354
668. J. Helsing, J. Stat. Phys. **90**, 1461 (1998)
669. B. Hensel, J.C. Grivel, A. Jeremie, et al., Phys. C **205**, 329 (1993)
670. H.F. Hess, R.B. Robinson, R.C. Dynes et al., Phys. Rev. Lett. **62**, 214 (1989)
671. H.F. Hess, R.B. Robinson, J.V. Waszczak, Phys. Rev. Lett. **64**, 2711 (1990)
672. A.H. Heuer, J. Am. Ceram. Soc. **62**, 317 (1979)
673. A. Hibino, R. Wantabe, J. Mat. Sci.: Mater. Electronics **1**, 13 (1990)
674. P. Higgs, Phys. Lett. **12**, 132 (1964)
675. P. Higgs, Phys. Rev. Lett. **13**, 508 (1964)
676. *High Temperature Superconductivity*, eds. by V.L. Ginzburg, D.A. Kirzhnitz (Consultants Bureau, New York, 1982)
677. High-T_c Update. **13**(9), May 1 (1999)
678. H. Hilgenkamp, J. Mannhart, Supercond. Sci. Technol. **10**, 880 (1997)
679. H. Hilgenkamp, J. Mannhart, Appl. Phys. Lett. **73**, 265 (1998)
680. H. Hilgenkamp, J. Mannhart, Rev. Mod. Phys. **74**, 485 (2002)
681. H. Hilgenkamp, J. Mannhart, B. Mayer, Phys. Rev. B **53**, 14586 (1996)
682. R. Hill, J. Mech. Phys. Solids. **5**, 229 (1957)
683. R. Hill, J. Mech. Phys. Solids. **11**, 357 (1963)
684. R. Hill, J. Mech. Phys. Solids **13**, 213 (1965)
685. M. Hillert, Acta Metall. **13**, 227 (1965)

686. M. Hillert, *Solidification and Casting of Metals* (The Metals Society, New York, 1977), p. 81
687. D.A. Hills, P.A. Kelly, D.N. Dai, A.M. Korsunsky, *Solution of Crack Problems. The Distributed Dislocation Technique* (Kluwer Academic Publish., Dordrecht, 1996)
688. D.G. Hinks, B. Dabrowski, J.D. Jorgenson et al., Nature **333**, 836 (1988)
689. E. Hinrichsen, S. Roux, A. Hansen, Phys. C **167**, 433 (1990)
690. H. Hiramatsu, T. Katase, T. Kamiya, et al., Cond. Mat. arXiv:0903.3710 (2009)
691. T. Hirayama, Y. Ikuhara, M. Nakamura et al., J. Mater. Res. **12**, 293 (1997)
692. J.P. Hirth, B. Carnahan, Acta Metall. **26**, 1795 (1978)
693. J.P. Hirth, J. Lothe, *Theory of Dislocations* (Wiley, New York, 1982)
694. H. Banno, Am. Ceram. Soc. Bull. **66**, 1332 (1987)
695. N. Hjashi, P. Diko, K. Nagashima, M. Murakami, J. Mater. Sci. Eng. B. **53**, 104 (1998)
696. O. Hoffman, G. Sachs, *Introduction to the Theory of Plasticity for Engineers* (McGraw-Hill, NY, 1953)
697. M.J. Hogg, F. Kahlmann, E.J. Tarte et al., Appl. Phys. Lett. **78**, 1433 (2000)
698. W.L. Holstein, C. Wilker, D.B. Laubacher et al., J. Appl. Phys. **74**, 1426 (1993)
699. R.L. Holtz, IEEE Trans. Appl. Supercond. **11**(1), 3238 (2001)
700. R.L. Holtz, S. Fleshler, D.U. Gubser, Adv. Eng. Mater. **3**(3), 131 (2001)
701. C.C. Homes, S.V. Dordevic, M. Strongin et al., Nature **430**, 539 (2004)
702. C. Honerkamp, M. Salmhofer, N. Furukawa, T.M. Rice, Phys. Rev. B **63**, 35109 (2001)
703. G-W. Hong, K-B. Kim, I-H. Kuk, et al., IEEE Trans. Appl. Supercond. **7**(2), 1945 (1997)
704. T. Honma, P.H. Hor, cond-mat/0602452 (2006)
705. B. Hopzapfel et al., IEEE Trans. Appl. Supercond. **11**(1), 3872 (2001)
706. P.H. Hor, R.L. Meng, Y.Q. Wang et al., Phys. Rev. Lett. **58**, 1891 (1987)
707. P.M. Horn, D.T. Keane, G.A. Held et al., Phys. Rev. Lett. **59**, 2772 (1987)
708. J. Horvat, R. Bhasale, Y.C. Guo et al., Supercond. Sci. Technol. **10**, 409 (1997)
709. W.F. Hosford, R.M. Caddel, *Metal Forming* (PTR Prentice Hall, Englewood Cliffs, 1993)
710. M.S.A. Hossain, C. Senatore, R. Flükiger et al., Supercond. Sci. Technol. **22**(8), 7 (2009)
711. A. Hosseini, R. Harris, S. Kamal et al., Phys. Rev. B **60**, 1349 (1999)
712. F.C. Hsu, J.Y. Luo, K.W. Yeh et al., Proc. Nat. Acad. Sci. USA **105**, 14262 (2008)
713. C.H. Hsueh, A.G. Evans, Acta Metall. **29**, 1907 (1981)
714. C.H. Hsueh, A.G. Evans, Acta Metall. **31**, 189 (1983)
715. C.H. Hsueh, A.G. Evans, J. Am. Ceram. Soc. **68**, 120 (1985)
716. C.H. Hsueh, A.G. Evans, R.L. Coble, Acta Metall. **30**, 1269 (1982)
717. M.S. Hu, A.G. Evans, Acta Metall. **37**, 917 (1989)
718. M.S. Hu, M.D. Thouless, A.G. Evans, Acta Metall. **36**, 1301 (1988)
719. Q.Y. Hu, H.K. Liu, S.X. Dou, M. Apperley, IEEE Trans Appl. Supercond. **7**(2), 1849 (1997)
720. X. Huang arXiv: 0808.4093v3 (27 November, 2008)
721. Y.B. Huang, G. Grasso, F. Marti et al., Inst. Phys. Conf. Ser. **158**, 1385 (1997)
722. Y.B. Huang, F. Marti, G. Witz et al., IEEE Trans. Appl. Supercond. **9**(2), 2722 (1999)
723. Y.K. Huang, B. ten Haken, H.H.J. ten Kate, IEEE Trans. Appl. Supercond. **9**(2), 2702 (1999)
724. J. Hubbard, Phys. Rev. Lett. **3**(2), 77 (1959)
725. R.P. Huebener, *Magnetic Flux Structures in Superconductors* (Springer, Berlin, 2001)
726. S.F. Hulbert, J. Br. Ceram. Soc. **6**, 11 (1969)

727. B.D. Hunt, M.C. Foote, W.T. Pike et al., Phys. C **230**, 141 (1994)
728. B.D. Hunt, M.G. Forrester, J. Tolvacchio et al., IEEE Trans. Appl. Supercond. **7**(2), 2936 (1997)
729. M. Hunt, M. Jordan, in *Advances in Solid State Physics*, ed. by B. Kramer **39** (Vieweg, Braunschweig/Wiesbaden, 1999), p. 351
730. F. Hunte, J. Jaroszynski, A. Gurevich et al., Nature **453**, 903 (2008)
731. I. Hušek, P. Kováč, L. Kopera, Supercond. Sci. Technol. **9**, 1066 (1996)
732. I. Hušek, P. Kováč, W. Pachla, Supercond. Sci. Technol. **8**, 617 (1995)
733. N.E. Hussey, in *Handbook of High-Temperature Superconductivity: Theory and Experiment*, eds. J.R. Schrieffer, J.S. Brooks (Springer, New York, 2007)
734. N.E. Hussey, M. Abdel-Jawad, A. Carrington et al., Nature **425**, 814 (2003)
735. J.W. Hutchinson, J. Mech. Phys. Solids **16**, 337 (1968)
736. B.B. Hwang, S. Kobayashi, Int. J. Mach. Tools Manufact. **30**, 309 (1990)
737. H. Ihara, K. Tokiwa, H. Ozawa et al., Jpn. J. Appl. Phys. **33**, L300 (1994)
738. Y. Iijima, A. Kikuchi, K. Inoue, Cryogenics **40**, 345 (2000)
739. Y. Iijima, M. Kosuge, T. Takeuchi, K. Inoue, Adv. Cryog. Eng. Mater. **40**, 899 (1994)
740. Y. Iijima, K. Onabe, N. Futaki et al., J. Appl. Phys. **74**, 1905 (1993)
741. H. Ikuta, A. Mase, T. Hosokawa et al., Adv. Supercond. **11**, 657 (1999)
742. A.Y. Ilyushechkin, T. Yamashita, B. Williams, I.D.R. Mackinnon, Supercond. Sci. Technol. **12**, 142 (1999)
743. T. Imai, K. Ahilan, F.L. Ning et al., Phys. Rev. Lett. **102**, 177005 (2009)
744. M.F. Imayev, R.R. Daminov, D.B. Kabirova, et al., in *Ferroelectrics and Superconductors: Properties and Applications*, ed. by I.A. Parinov (Nova Science Publishers, New York, 2011)
745. M.F. Imayev, R.R. Daminov, V.A. Popov et al., Physica C **422**, 27 (2005)
746. M.F. Imayev, V.A. Daminov, V.A. Popov, O.A. Kaibyshev, Inorg. Mater. **41**(5), 1 (2005)
747. M.F. Imayev, R.R. Daminov, M. Reissner et al., Physica C **467**, 14 (2007)
748. M.F. Imayev, D.B. Kabirova, A.N. Korshunova, et al., in *Proceedings of the 4th International Conference on Recrystallization and Related Phenomena* (The Japan Institute of Metals, Tokio, 1999)
749. M.F. Imayev, R.O. Kaibyshev, F.F. Musin et al., Mater. Sci. Forum **170–172**, 445 (1994)
750. M.I. Indenbom, N.N. Kolesnikov, M.P. Kulakov et al., Phys. C **166**, 486 (1990)
751. M.V. Indenbom, V.I. Nikitenko, A.A. Polyanskii, V.K. Vlasko-Vlasov, Cryogenics **30**, 747 (1990)
752. S.E. Inderhees, M.B. Salamon, N. Goldenfeld et al., Phys. Rev. Lett. **60**, 1178 (1988)
753. S.E. Inderhees, M.B. Salamon, J.P. Rice, D.M. Ginsberg, Phys. Rev. Lett. **66**, 232 (1991)
754. T. Ingrafea, et al. *FRANC3D, 3D FRacture ANalysis Code.* (The Cornell University Fracture Group, Cornell University, Ithaca, 1996)
755. A. Ino, C. Kim, T. Mizokawa et al., J. Phys. Soc. Jpn. **68**, 1496 (1999)
756. K. Inomata, S. Sato, K. Nakajima et al., Phys. Rev. Lett. **95**, 107005 (2005)
757. L.B. Ioffe, V.B. Geshkenbein, M.V. Feigel'man et al., Nature **398**, 679 (1999)
758. M. Ionescu, S.X. Dou, E. Babic, et al., in *Proceedings of International Workshop on Superconductivity*, Maui, Hawaii, USA, June 18–21 (1995), p. 44
759. M. Ionescu, A.H. Li, Y. Zhao et al., J. Phys. D Appl. Phys. **37**, 1824 (2004)
760. G.R. Irwin, J. Appl. Mech. **24**, 361 (1957)
761. K. Itoh, O. Ishii, Y. Nagai et al., J. Supercond.**5**, 485 (1991)
762. Z.G. Ivanov, P.Å. Nilsson, S. Winkler et al., Appl. Phys. Lett. **59**, 3030 (1991)

763. A.L. Ivanovskii, Physics Uspekhi. **51**(12), 1229 (2008)
764. T. Izumi, Y. Nakamura, Y. Shiohara, J. Mater. Res. **7**, 1621 (1992)
765. T. Izumi, Y. Nakamura, Y. Shiohara, J. Cryst. Growth **128**, 757 (1993)
766. T. Izumi, Y. Nakamura, T.H. Sung, Y. Shiohara, *Proceedings 4th International Symp on Superconductivity*, ed. by T. Ishiguro (Tokyo, Oct 14–17, 1991)
767. Y.A. Izumov, E.Z. Kurmaev, Physics Uspekhi. **51**(12), 1261 (2008)
768. J.D. Jackson, *Classical Electrodynamics* (Wiley, New York, 1990)
769. K.A. Jackson, *Solidification* (American Society for Metals, New York, 1971), p. 121
770. J. Jaklic, P. Prelovsek, Adv. Phys. **49**, 1 (1999)
771. Y.A. Jee, G-W. Hong, C-J. Kim, T.H. Sung, IEEE Trans. Appl. Supercond. **9**(2) 2093 (1999)
772. Y.A. Jee, G-W. Hong, C-J. Kim, T.H. Sung, IEEE Trans. Appl. Supercond. **9**(2), 2097 (1999)
773. Y.A. Jee, S-J.L. Kang, H-S. Jung, J. Mater. Res. **13**, 583 (1998)
774. D.J. Jeffrey, Proc. Roy. Soc. Lond. A **335**, 355 (1973)
775. D.J. Jeffrey, Proc. Roy. Soc. Lond. A **338**, 503 (1974)
776. H.M. Jensen, Acta Metall. **38**, 2637 (1990)
777. P.A. Jensen, J.W. Barnes, *Network Flow Programming* (Wiley, New York, 1980)
778. D. Jérome, A. Mazaud, M. Ribault, K. Bechgaard, J. Phys. Lett. (Paris). **41**, L195 (1980)
779. J. Jiang, T.C. Shields, J.S. Abell et al., IEEE Trans. Appl. Supercond. **9**(2), 1812 (1999)
780. Y-S. Jiang, T. Kobayashi, T. Goto, IEEE Trans. Appl. Supercond. **9**(2), 1657 (1999)
781. M. Jimenez-Melendo, A. Dominguez-Rodriguez, J.L. Routbort, Scr. Met. Mater. **32**, 621 (1995)
782. S. Jin, G.W. Kammlott, T.H. Tiefel et al., Phys. C **181**, 57 (1991)
783. S. Jin, G.W. Kammlott, T.H. Tiefel et al., Phys. C **198**, 333 (1992)
784. S. Jin, H. Mavori, C. Bower, R.B. van Dover, Nature **411**, 563 (2001)
785. S. Jin, T.H. Tiefel, R.C. Sherwood et al., Phys. Rev. B **37**, 7850 (1996)
786. X.Y. Jin, J. Lisenfeld, Y. Koval et al., Phys. Rev. Lett. **96**, 177003 (2006)
787. T.H. Johansen, Phys. Rev. B **60**, 9690 (1999)
788. D.H. John, St., Acta Metall. Mater. **38**, 631 (1990)
789. D.H. John, St., Hogan L. M., Acta Metall. **25**, 77 (1977)
790. F. John, Commun. Pue. Appl. Math. **13**, 239 (1960)
791. S.M. Johnson, M.I. Gusman, D.J. Rowcliffe, Adv. Ceram. Mater. **2**, 337 (1987)
792. M. Jones, R. Marsh, J. Am. Chem. Soc. **76**, 1434 (1954)
793. M. Jordan, M. Hunt, H. Adrian, Nature **398**, 47 (1999)
794. J.D. Jorgensen, M.A. Beno, D.G. Hinks et al., Phys. Rev. B **36**, 3608 (1987)
795. J.D. Jorgensen, B.W. Veal, W.K. Kwok et al., Phys. Rev. B **36**, 5731 (1987)
796. B.D. Josephson, Phys. Lett. **1**, 251 (1962)
797. D.T. Jover, R.J. Wijngaarden, R. Griessen et al., Phys. Rev. B **54**, 10175 (1996)
798. R. Jullien, R. Botet, J. Phys. A: Math. Gen. **18**, 2279 (1985)
799. R. Jullien, R. Botet, Phys. Rev. Lett. **54**, 2055 (1985)
800. C.U. Jung, J.Y. Kim, M-S. Kim, et al., Phys. C **366**, 299 (2002)
801. A. Junod, A. Erb, C. Renner, Physica C **317–318**, 333 (1999)
802. B. Kabius, J.W. Seo, T. Amrein et al., Phys. C **231**, 123 (1994)
803. D. Kaczorowski, A.J. Zaleski, O.J. Zogał, J. Klamut, cond-mat/0103571 (2001)

804. J. Kameda, Acta Metall. **34**, 867 (1986)
805. J. Kameda, Acta Metall. **34**, 883 (1986)
806. Y. Kamihara, T. Watanabe, M. Hirano, H. Hosono, J. Am. Chem. Soc. **130**(11), 3296 (2008)
807. A. Kaminski, S. Rosenkranz, H.M. Fretwell et al., Nature **416**, 610 (2002)
808. G.W. Kammlott, T.H. Tiefel, S. Jin, Appl. Phys. Lett. **56**, 2459 (1990)
809. H.J. Kang, P. Dai, J.W. Lynn et al., Nature **423**, 522 (2003)
810. K.-Y. Kang, I. Song, Y.S. Ha et al., IEEE Trans. Appl. Supercond. **9**(2), 3074 (1999)
811. A. Kanigel, U. Chatterjee, M. Randeria et al., Phys. Rev. Lett. **99**, 157001 (2007)
812. Y. Kantor, D.J. Bergman, J. Mech. Phys. Solids **32**, 41 (1984)
813. A. Kapitulnik, M.R. Beasley, C. Castellani, C. DiCastro, Phys. Rev. B **37**, 537 (1988)
814. M.A. Kaplitsky, *Energetic Method in Dynamical Theory of Brittle Fracture, Preprint 511–B96* (VINITI, Moscow, 1996)
815. B.S. Karasik, M.C. Gaids, W.R. McGrath et al., IEEE Trans. Appl. Supercond. **7**(2), 3580 (1997)
816. A. Karma, Phys. Rev. E **49**, 2245 (1994)
817. A. Karma, Phys. Rev. Lett. **87**, 115701 (2001)
818. A. Karma, W.-J. Pappel, Phys. Rev. E **53**, R3017 (1996)
819. A. Karma, W.-J. Pappel, Phys. Rev. E **57**, 4323 (1998)
820. A. Karma, M. Plapp, Phys. Rev. Lett. **81**, 4444 (1998)
821. D.N. Karpinsky, C.O. Kramarov, A.N. Orlov, Strength Mater. **1**, 97 (1981)
822. D.N. Karpinsky, I.A. Parinov, Glass Ceram. **3**, 27 (1991)
823. D.N. Karpinsky, I.A. Parinov, in *Proceedings of Jt. FEFG/ICF Int. Conf. Fract. of Eng. Mater. and Struct.* (Singapore, 1991), London (1991), p. 327
824. D.N. Karpinsky, I.A. Parinov, Strength Mater. **7**, 34 (1991)
825. D.N. Karpinsky, I.A. Parinov, J. Appl. Mech. Techn. Phys. **1**, 150 (1992)
826. D.N. Karpinsky, I.A. Parinov, Ferroelectric Lett. **19**, 151 (1995)
827. D.N. Karpinsky, I.A. Parinov, L.V. Parinova, Ferroelectrics **133**, 265 (1992)
828. M. Karuna, J.A. Parrell, D.C. Larbalestier, IEEE Trans. Appl. Supercond. **5**(2). 1279 (1995)
829. J. Kase, T. Morimoto, K. Togano et al., IEEE Trans. Mag. **27**, 1254 (1991)
830. J.A. Kash, J. C.-H. Tsang. *Noninvasive optical method for measuring internal switching and other dynamic parameters of CMOS circuits.* US Patent No 5,940,545 (1999)
831. Y. Kashima, ESAIM: Mathematical modelling numerical. Analysis **42**, 333 (2008)
832. A.S. Katz, A.G. Sun, R.C. Dynes, K. Char, Appl. Phys. Lett. **66**, 105 (1995)
833. W. Kautek, B. Roas, L.H. Schultz, Thin Solid Films **191**, 317 (1990)
834. H. Kawaji, H. Horie, S. Yamanaka, M. Ishikawa, Phys. Rev. Lett. **74**, 1427 (1995)
835. M. Kawasaki, E. Sarnelli, P. Chaudhari et al., Appl. Phys. Lett. **62**, 417 (1993)
836. J. Kawashima, Y. Yamada, I. Hirabayashi, Phys. C **306**, 114 (1998)
837. P.E. Kazin, V.V. Poltavets, Y.D. Tretyakov et al., Physica C **280**, 253 (1997)
838. P.E. Kazin, Y.D. Tretyakov, V.V. Lennikov et al., J. Mater. Chem. **11**, 168 (2001)
839. S. Keil, R. Straub, R. Gerber et al., IEEE Trans. Appl. Supercond. **9**(2), 2961 (1999)
840. R.J. Keizer, A. de Visser, A.A. Menovsky et al., cond-mat/9903328 (1999)
841. J.B. Keller, J. Appl. Phys. **34**, 991 (1963)
842. J.B. Keller, J. Math. Phys. **5**, 548 (1964)
843. J.B. Keller, L. Rubenfeld, J. Molyneux, J.Fluid Mech. **30**, 97 (1967)

844. M.G. Kendall, A. Stuart, *The Advanced Theory of Statistics, 1–3* (Griffin Publ, London, 1983)
845. N. Kenmochi, *Solvability of Nonlinear Evolution Equations with Time-Dependent Constraints and Applications, The Bulletin of the Faculty of Education*, vol. 30. (Chiba University, Chiba, Japan, 1981)
846. M. Kenzelmann, Th Strässle, C. Niedermayer et al., Science **321**, 1652 (2008)
847. P.H. Kes, Phys. C **185–189**, 288 (1991)
848. R. Khasanov, D.G. Eshchenko, J. Karpinski et al., Phys. Rev. Lett. **93**, 157004 (2004)
849. D.I. Khomskii, *Basic Aspects of the Quantum Theory of Solids: Order and Elementary Excitations* (Cambridge University Press, Cambridge, 2010)
850. T.W.B. Kibble, J. Math. Phys. **2**, 212 (1961)
851. C-J. Kim, G-W. Hong, Supercond. Sci. Technol. **12**, R27 (1999)
852. C-J. Kim, H-J. Kim, J-H. Joo, G-W. Hong, Phys. C **354**, 384 (2001)
853. C-J. Kim, K-B. Kim, G-W. Hong, H-Y. Lee, J. Mater. Res. **10**, 1605 (1995)
854. C-J. Kim, K-B. Kim, I-H. Kuk, G-W. Hong, Phys. C **281**, 244 (1997)
855. C-J. Kim, K-B. Kim, I-H. Kuk, G-W. Hong, J. Mater. Res. **13**, 269 (1998)
856. C-J. Kim, K-B. Kim, H-W.Park, et al., Supercond. Sci. Technol. **9**, 76 (1996)
857. C-J. Kim, K-B. Kim, D-Y. Won, J. Mater. Res. **8**, 699 (1993)
858. C-J.Kim, K-B. Kim, D-Y. Won, et al., J. Mater. Res. **9**, 1952 (1994)
859. C-J.Kim, K-B. Kim, D-Y. Won, G-W. Hong, Phys. C **228**, 351 (1994)
860. C-J. Kim, I-H. Kuk, G-W. Hong, et al., Mater. Res. Lett. **34**, 392 (1998)
861. C.-J. Kim, S.H. Lai, P.J. McGinn, J. Mater. Lett. **19**, 185 (1994)
862. C-J. Kim, Y-S. Lee, H-W. Park, et al., Phys. C **276**, 101 (1997)
863. C.-J. Kim, H.-W. Park, K.-B. Kim et al., Mater. Lett. **29**, 7 (1996)
864. H-J. Kim, W.N. Kang, E–M. Choi, et al., cond-mat/0105363 (2001)
865. I.C. Kim, S. Torquato, J. Appl. Phys. **68**, 3892 (1990)
866. I.C. Kim, S. Torquato, J. Appl. Phys. **69**, 2280 (1991)
867. M-S. Kim, C.U. Jung, J.Y. Kim, et al., cond-mat/0102420 (2001)
868. Y.B. Kim, C.F. Hempstead, A.R. Strnad, Phys. Rev. Lett. **9**, 306 (1962)
869. Y.B. Kim, C.F. Hempstead, A.R. Strnad, Phys. Rev. **129**, 528 (1963)
870. Y.T. Kim, N. Provatas, N. Goldenfeld, J. Dantzig, Phys. Rev. E **59**, R2546 (1999)
871. S.A.J. Kimber, A. Kreyssig, Y.Z. Zhang et al., Nat Mater. **8**, 471 (2009)
872. A.H. King, A. Singh, J.Y. Wang, Interface Sci. **4**, 347 (1993)
873. A.M. Kini, U. Geiser, H.H. Wang et al., Inorg. Chem. **29**, 2555 (1990)
874. J.R. Kirtley, M.B. Ketchen, K.G. Stawiasz et al., Appl. Phys. Lett. **66**, 1138 (1995)
875. J.R. Kirtley, F. Tafuri, in *Handbook of High-Temperature Superconductivity: Theory and Experiment*, eds. by J.R. Schrieffer, J.S. Brooks (Springer, New York, 2007)
876. J.R. Kirtley, C.C. Tsuei et al., Ariando, Verwijs C. J. M., et al. Nat. Phys. **2**, 190 (2006)
877. T. Kiss, K. Oda, S. Nishimura et al., Phys. C **357–360**, 1123 (2001)
878. T. Kiss, H. van Eck, B. ten Haken, H.H.J. ten Kate, IEEE Trans. Appl. Supercond. **11**(1), 3888 (2001)
879. T.J. Kistenmacher, J. Appl. Phys. **64**, 5067 (1988)
880. H. Kitaguchi, K. Itoh, H. Kumakura et al., Phys. C. **357–360**, 1193 (2001)
881. S. Kivelson, cond-mat/0109151 (2001)
882. S.A. Kivelson, E. Fradkin, cond-mat/0507459 (2005)
883. S. Kiryu, S. Kohjiro, A. Shoji, et al. *Ext. Abstr. ISEC'95* (Nagoya, September, 1995)

884. S. Kiryu, W. Zhang, S. Han et al., IEEE Trans. Appl. Supercond. **7**(2), 3107 (1997)
885. A.-K. Klehe, A.K. Gangopadhyay, J. Diederichs et al., Physica C **223**, 313 (1994)
886. R. Kleiner, P. Müller, Phys. Rev. B **49**, 1327 (1994)
887. R. Kleiner, F. Steinmeyer, G. Kunkel, P. Müller, Phys. Rev. Lett. **68**, 2394 (1992)
888. H. Kleinert, Phys. Lett. **89A**, 246 (1982)
889. H. Kleinert, Phys. Lett. **91A**, 295 (1982)
890. H. Kleinert, *Gauge Fields in Condensed Matter, Electromagnetism*, vol. II, Stresses and Defects (World Scientific, Singapore, 1989)
891. H. Kleinert, Phys. Lett. B **246**, 127 (1990)
892. H. Kleinert, Int. J. Mod. Phys. A **7**, 4693 (1992)
893. H. Kleinert, Phys. Lett. B **293**, 168 (1992)
894. H. Kleinert, *Multivalued Fields in Condensed Matter, Electromagnetism, and Gravitaion* (World Scientific, Singapore, 2008)
895. A.W. Kleinsasser, K.A. Delin, IEEE Trans. Appl. Supercoud. **7**(2), 2964 (1997)
896. R.F. Klie, J.P. Buban, M. Varela et al., Nature **435**, 475 (2005)
897. J.K. Knowles, E.J. Sternberg, Elasticity. **8**, 329 (1978)
898. H.A. Knudsen, A. Hansen, Phys. Rev. B **61**, 11336 (2000)
899. R. Kobayashi, J.A. Warren, W.C. Carter, Physica D **140**, 141 (2000)
900. M.R. Koblischka, Supercond. Sci. Technol. **9**, 271 (1996)
901. M.R. Koblischka, T.H. Johansen, H. Bratsberg, Supercond. Sci. Technol. **10**, 693 (1997)
902. M.R. Koblischka, R.J. Wijngaarden, Supercond. Sci. Technol. **8**, 199 (1995)
903. M.R. Koblischka, R.J. Wijngaarden, D.G. de Groot et al., Phys. C **249**, 339 (1995)
904. R.H. Koch, V. Foglietti, W.J. Gallagher et al., Phys. Rev. Lett. **63**, 1511 (1989)
905. A. Kofman, *Introduction in Applied Combinatoric* (Nauka, Moscow, 1975)
906. K. Kojima, S. Suda, X. Kong, H. Itozaki, Physica C **445–448**, 979 (2006)
907. A. Kolleck, G.A. Schneider, F. Meschke, Acta Mater. **48**, 4099 (2000)
908. K. Komori, K. Kawagishi, Y. Takano et al., Appl. Phys. Lett. **81**, 1047 (2002)
909. V.G. Kompantseva, K.V. Rusanov, Supercond. Res. Develop. **3**(4), 41 (1994)
910. K. Kondo, in *Proceedings The II Japan National Congress on Applied Mechanics* (Tokyo, 1952)
911. N. Kondo, E. Sato, F.J. Wakai, Am. Ceram. Soc. **81**, 3221 (1998)
912. L. Kopera, P. Kováč, I. Hušek, Supercond. Sci. Technol. **11**, 433 (1998)
913. A.A. Kordyuk, S.V. Borisenko, A.N. Yaresko et al., Phys. Rev. B **70**, 214525 (2004)
914. T. Koretsune, M. Ogata cond-mat/0505618 (2005)
915. G.A. Korn, T.M. *Korn, Mathematical Handbook for Scientists and Engineers. Definitions, Theorems and Formulas for Reference and Review* (McGraw Hill, New York, 1961)
916. Y. Korneeva, I. Florya, A. Semenov, Appl. Phys. Lett. **79**(3), 705 (2001)
917. A.N. Korshunov, A.A. Shevtshionok, V.I. Ovtshinnikov, et al., *Collect. Abstr. Int. Conf. Transparent Ferroelectric*. Ceram. (TFC'91) (Riga, October 2–6, 1991), p. 79
918. D.A. Korzekwa, J.F. Bingert, E.J. Rodtburg, P. Miles, Appl. Supercond. **2**, 261 (1994)
919. V.P. Koshelets, S.V. Shitov, A.M. Baryshev et al., Inst. Phys. Conf. Ser. **148**, 731 (1995)
920. V.P. Koshelets, S.V. Shitov, A.V. Shchukun et al., Appl. Phys. Lett. **69**, 699 (1996)
921. V.P. Koshelets, S.V. Shitov, A.V. Shchukin et al., IEEE Trans. Appl. Supercond. **9**(2), 4133 (1999)
922. J.M. Kosterlitz, D.J. Thouless, J. Phys. C. **5**, L124 (1972)
923. J.M. Kosterlitz, D.J. Thouless, J. Phys. C. **6**, 1181 (1973)

924. J.M. Kosterlitz, D.J. Thouless, J. Phys. C. **7**, 1046 (1974)
925. H. Kotegawa, Y. Tokunaga, K. Ishida et al., Phys. Rev. B **64**, 064515 (2001)
926. J. Kotzler, L. von Sawilski, S. Casalbuoni, Phys. Rev. Lett. **92**, 067005 (2004)
927. K. Kourtakis, M. Robbins, P.K. Gallagher, J. Sol. State Chem. **84**, 88 (1990)
928. S.P. Kovalev, V.A. Kuz'menko, G.G. Pisarenko, *Computer Simulation of Microstructure Processes in Ceramic Materials* (Preprint AN UkrSSR, IPS, Kiev, 1983)
929. P. Kováč, I. Hušek, L. Cesnak, Supercond. Sci. Technol. **7**, 583 (1994)
930. P. Kováč, I. Hušek, F. Gömöry et al., Supercond. Sci. Technol. **13**, 378 (2000)
931. P. Kováč, P.E. Richens, P. Bukva et al., Supercond. Sci. Technol. **12**, 168 (1999)
932. T.K. Koyama, N. Takezawa, M. Tachiki, Phys. C **172**, 501 (1992)
933. Y.A. Kozinkina, I.A. Parinov, Adv. Cryog. Eng. (Mater.) **44**b, 449 (1998)
934. S.M. Kozlov, Uspekhi Math. Nauk **44**(2), 79 (1989)
935. S.O. Kramarov, Y.V. Dashko, Strength Mater. **10**, 52 (1987)
936. W.S. Kreher, J. Mech. Phys. Solids. **50**, 1029 (2002)
937. W. Kreher, W. Pompe, J. Mat. Sci. **16**, 694 (1981)
938. V. Kresin, Solid State Comm. **63**, 725 (1987)
939. V. Kresin, S.A. Wolf, Solid State Comm. **63**, 1141 (1987)
940. V.Z. Kresin, S.A. Wolf, *Fundamentals of Superconductivity* (Plenum Press, New York, 1990)
941. D.M. Kroeger, A. Choudhury, J. Brynestad et al., J. Appl. Phys. **64**, 331 (1988)
942. E. Kröner, in *The Physics of Defects*, ed. by R. Balian, et al. (North Holland, Amsterdam, 1981)
943. H.W. Kroto, J.R. Heath, S.C. O'Bbrien et al., Nature **318**, 1622 (1985)
944. V.D. Krstic, J. Mater. Sci. **23**, 259 (1988)
945. L. Krusin-Elbaum, J.R. Thompson, R. Wheeler et al., Appl. Phys. Lett. **64**, 3331 (1994)
946. T. Kumagai, H. Yokota, K. Kawaguchi et al., Chem. Lett. **16**, 1645 (1987)
947. H. Kumakura, H. Kitaguchi, K. Togano et al., IEEE Trans. Appl. Supercond. **9**(2), 1804 (1999)
948. H. Kumakura, K. Togano, H. Maeda et al., Appl. Phys. Lett. **58**, 2830 (1991)
949. P. Kumar, V. Pillai, D.O. Shah, Appl. Phys. Lett. **62**, 765 (1993)
950. M.F. Kupriyanov, G.M. Konstantinov, in *Proceedings International Conf. Electr. Ceram. Product. Propert.*, part I (Riga, April 30–May 2, 1990), p. 36
951. V.V. Kurin, A.V. Yulin, Phys. Rev. B **55**, 11659 (1997)
952. V. Kurin, A. Yulin, E. Goldobin et al., IEEE Trans. Appl. Supercond. **9**(2), 3733 (1999)
953. K. Kuroda, T. Takami, K. Nishi et al., Jpn. J. Appl. Phys. **40**, L609 (2000)
954. L.S. Kuzmin, in *Proceedings Sympos. Nano and Giga Challenges in Microelectronics Research and Opportunities in Russia* (Moscow, Sept 10–13, 2002)
955. L. Kuzmin, D. Chouvaev, M. Tarasov et al., IEEE Trans. Appl. Supercond. **9**(2), 3186 (1999)
956. L. Kuzmin, I. Devyatov, D. Golubev, in *Proceeedigs The 4th International Conference on mm and submm Waves* (San Diego, CA, 1998)
957. G.H. Kwei, J.A. Goldstone, A.C. Lawson et al., Phys. Rev. B. **39**, 7378 (1989)
958. W.K. Kwok, G.W. Crabtree, A. Umezawa et al., Phys. Rev. B. **37**, 106 (1988)
959. S.K. Kwon, S.J. Youn, K.S. Kim, B.I. Min, cond-mat/0106483 (2001)
960. P.V. Lade, Int. J. Numer. Anal. Methods Geomech. **12**, 351 (1988)
961. M. Lahtinen, J. Paasi, J. Sarkaniemi et al., Phys. C **244**, 115 (1995)

962. B.M. Lairson, S.K. Streiffer, J.C. Bravman, Phys. Rev. B **42**, 10067 (1990)
963. R.S. Lakes, Science **235**, 1038 (1987)
964. R.S. Lakes, J. Mater. Sci. **26**, 2287 (1991)
965. R.S. Lakes, Adv. Mater. **5**, 293 (1993)
966. R.S. Lakes, Phys. Rev. Lett. **86**, 2897 (2001)
967. R.S. Lakes, W.J. Drugan, J. Mech. Phys. Solids. **50**, 979 (2002)
968. R.S. Lakes, T. Lee, A. Bersie, Y.C. Wang, Nature **410**, 565 (2001)
969. L.D. Landau, Sov. Phys. JETP **7**, 371 (1937)
970. L.D. Landau, E.M. Lifshitz, *Electrodynamics of Continuous Media*, 2nd edn. (Nauka, Moscow, 1972) [Pergamon Press, Oxford, 1968]
971. L.D. Landau, E.M. Lifshitz, *Statistical Physics*, 3rd edn., part 1 (Nauka, Moscow, 1976) [Pergamon Press, Oxford, 1980]
972. L.D. Landau, E.M. Lifshits, *Fluid Mechanics* (Pergamon, Oxford, 1987)
973. R.J. Landauer, Appl. Phys. **23**, 779 (1952)
974. R. Landauer, in *Electrical, Transport and Optical Properties of Inhomogeneous Media*, eds. by J.C. Garland, D.B. Tanner (AIP, New York, 1978)
975. K.M. Lang, V. Madhavan, J.E. Hoffman et al., Nature **415**, 412 (2002)
976. A. Lanzara, P.V. Bogdanov, X.J. Zhou et al., Nature **412**, 510 (2001)
977. D.C. Larbalestier, Science **274**, 736 (1996)
978. D.C. Larbalestier, IEEE Trans. Appl. Supercond. **7**(2), 90 (1997)
979. D.C. Larbalestier, S.E. Babcock, X.Y. Cai et al., Phys. C **185–189**, 315 (1991)
980. D.C. Larbalestier, X.Y. Cai, Y. Feng et al., Phys. C **221**, 299 (1994)
981. D.C. Larbalestier, M.P. Maley, MRS Bull. **17**, 50 (1993)
982. D.C. Larbalestier, M.O. Rikel, L.D. Cooley, et al., cond-mat/0102216 (2001)
983. A.I. Larkin, Y.N. Ovchinnikov, Sov. Phys. JETP **20**, 762 (1965)
984. A.I. Larkin, Y.N. Ovchinnikov, J. Low Temp. Phys. **34**, 409 (1979)
985. U.D. Larsen, O. Sigmund, S.J. Bouwstra, Vibroelectromech. Syst. **6**, 99 (1997)
986. A. Lauder, D.W. Face, W.L. Holstein et al., Adv. Supercond. **5**, 925 (1993)
987. R.B, Laughlin, cond-mat/0209269 (2002)
988. J.Y. Laval, T.S. Orlova, Supercond. Sci. Technol. **15**, 1244 (2002)
989. J.Y. Laval, T.S. Orlova, Supercond. Sci. Technol. **16**, 1139 (2003)
990. J.Y. Laval, W. Swiatnicki, Phys. C **221**, 11 (1994)
991. F. Laves, K. Löhberg, P. Rahlfs, Nachr. Ges. Wiss. Göttingen Fachgr. IV **1**, 67 (1934)
992. W.E. Lawrence, S. Doniach, in *Proceedings* 12*th International Conf on Low Temperature Physics*, ed. by E. Kanda (Academic Press, Kyoto, 1971), p. 361
993. I.D. Lawrie, Phys. Rev. B **50**, 9456 (1994)
994. N. Laws, J.R. Brockenbrough, Trans. ASME. J. Eng. Mater. Technol. **110**, 101 (1988)
995. N. Laws, J.C. Lee, J. Mech. Phys. Solids **37**, 603 (1989)
996. P.L. Leath, W. Tang, Phys. Rev. B **39**, 6485 (1989)
997. P.L. Leath, W. Xia, Phys. Rev. B **44**, 9619 (1991)
998. H. Ledbetter, M. Lei, A. Hermann, Z. Sheng, Phys. C **255**, 397 (1994)
999. D.F. Lee, V. Selvamanickam, K. Salama, Physica C **202**, 83 (1992)
1000. H. Lee, W. Siu, K.W. Yu, Phys. Rev. B **52**, 4217 (1995)
1001. P. Lee, Phys. Rev. Lett. **71**, 1887 (1993)
1002. A. Leenders, M. Ullrich, H.C. Freyhardt, IEEE Trans. Appl. Supercond. **9**(2), 2074 (1999)

1003. A.J. Leggett, Phys. Rev. Lett. **83**, 392 (1999)
1004. R.G. Leisure, in *Encyclopedia of Materials: Science and Technology* (Elsevier, London, 2004), p. 1
1005. P.W. Leung, Phys. Rev. B. **62**, R6112 (2000)
1006. L.-P. Lèvy, *Magnetism and Superconductivity* (Springer, Berlin, 2000)
1007. O. Lèvy, D. Bergman, J. Phys.: Condens. Mater. **5**, 7095 (1993)
1008. D.J. Lew, Y. Suzuki, A.F. Marshall et al., Appl. Phys. Lett. **65**, 1584 (1994)
1009. J.C. Li, R.A. Oriani, L.S. Darken, Zeitschrift für Physikalische Chemie Neue Folge **49**, 271 (1966)
1010. J.C. Li, S.C. Sanday, Acta Metall. **34**, 537 (1986)
1011. J.N. Li, A.A. Menovsky, J.J.M. Franse, Phys. Rev. B **48**, 6612 (1993)
1012. Q. Li J, Li L., Liu F. M., et al., cond-mat/0104320 (2001)
1013. L. Li, J.G. Checkelsky, S. Komiya et al., Nat. Phys. **3**, 311 (2007)
1014. Li L., Y. Wang, M.J. Naughton, et al., Europhys. Lett. **72**, 451 (2005) (cond-mat/0507617)
1015. S. Li, M. Bredehöft, W. Gao et al., Supercond. Sci. Technol. **11**, 1011 (1998)
1016. S. Li, W. Gao, H. Cooper et al., Phys. C **356**, 197 (2001)
1017. Y. Li, J.A. Kilner, M. Dhallé et al., Supercond. Sci. Technol. **8**, 764 (1995)
1018. Y. Li, G.K. Perkins, A.D. Caplin et al., Phys. C **341–348**, 2037 (2000)
1019. Li Y., Z-X. Zhao ., Phys. C **351,** 1 (2001)
1020. I.M. Lifshitz, V.V. Slyozov, J. Phys. Chem. Solids **19**, 35 (1961)
1021. K.K. Likharev, Rev. Mod. Phys. **51**, 101 (1979)
1022. K.K. Likharev, B.T. Ulrich, *Systems with Josephson Junctions. The Basics of the Theory* (Moscow State University Press, Moscow, 1978)
1023. N.C.H. Lin, S.P. Yukon, IEEE Trans. Appl. Supercond. **9**(2), 4320 (1999)
1024. R-J. Lin, L-J. Chen, L-J. Lin, et al., Jpn. J. Appl. Phys. **35**, 5805 (1996)
1025. L.E. Lindgren, J. Edberg, J. Mater. Process. Technol. **24**, 85 (1990)
1026. J.A. Lipa, D.R. Swanson, J.A. Nissen et al., Phys. Rev. Lett. **76**, 944 (1996)
1027. P. Lipavský, J. Kolaček, K. Morawetz, E.H. Brandt, T.-J. Yang, *Bernoulli Potential in Superconductors: How the Electrostatic Field Helps to Understand Superconductivity* (Springer, Berlin, 2007)
1028. D.A. Lisachenko, Supercond. Phys. Chem. Tech. **6**, 1757 (1993)
1029. W.A. Little, Phys. Rev. A **134**, 1416 (1964)
1030. H.K. Liu, S.X. Dou, Phys. C **250**, 7 (1995)
1031. H.K. Liu, J. Horvat, R. Bhasale et al., IEEE Trans. Appl. Supercond. **7**(2), 1841 (1997)
1032. H.K. Liu, A. Polyanskii, W.M. Chen et al., IEEE Trans. Appl. Supercond. **11**(1), 3764 (2001)
1033. H.K. Liu, R.K. Wang, S.X. Dou, Phys. C **229**, 39 (1994)
1034. R.H. Liu, T. Wu, G. Wu et al., Nature **459**, 64 (2009)
1035. A. Llordés, T. Puig, X. Obradors et al., Nat. Mater. **11**, 329 (2012)
1036. S.-W. Lo, Y.-H. Lu, J. Mat. Proc. Tech. **123**, 212 (2002)
1037. W. Lo, D.A. Cardwell, J.C.L. Chow, J. Mater. Res. **13**, 1141 (1998)
1038. W. Lo, D.A. Cardwell, C.D. Dewhurst, S-L. Dung, J. Mater. Res. **11**, 786 (1996)
1039. W. Lo, D.A. Cardwell, S-L. Dung, R.G. Barter, IEEE Trans. Appl. Supercond. **5**(2), 1619 (1995)
1040. W. Lo, D.A. Cardwell, S-L. Dung, R.G. Barter, J. Mater. Sci. **30**, 3995 (1995)
1041. W. Lo, D.A. Cardwell, S-L. Dung, R.G. Barter, J. Mater. Res. **11**, 39 (1996)

1042. J.-P. Locquet, J. Perret, J. Fompeyrine et al., Nature **394**, 453 (1998)

1043. F. Lombardi, F. Tafuri, F. Ricci et al., Phys. Rev. Lett. **89**, 207001 (2002)

1044. F. London, *Superfluids* (Wiley, New York, 1950)

1045. F. London, H. London, Physica **2**, 341 (1935)

1046. F. London, H. London, Proc. Roy. Soc. **A149**, 71 (1935)

1047. J.W. Loram, J.L. Tallon, Y. Liang, Phys. Rev. B **69**, 060502R (2004)

1048. L. Kelvin (Thomson, W.) Philos. Mag. **26**, 414 (1888)

1049. L. Rayleigh, Phil. Mag. **34**, 481 (1892)

1050. B. Lorenz, J. Lenzi, J. Cmaidalka, et al., cond-mat/0208341 (2002)

1051. M. Lorenz, H. Hochmuth, D. Natusch et al., Appl. Phys. Lett. **68**, 3332 (1996)

1052. F.K. Lotgering, J.Inorg. Nucl. Chem. **9**, 113 (1959)

1053. I.M. Low, R.D. Skala, H.G. Mohazzab, J. Mater. Sci. Lett. **13**, 1340 (1994)

1054. D.H. Lu, D.L. Feng, N.P. Armitage et al., Phys. Rev. Lett. **86**, 4370 (2001)

1055. S.V. Lubenets, V.D. Natsik, L.S. Fomenko, Fiz. Nizk. Temp. **21**, 475 (1995)

1056. G.B. Lubkin, Phys. Today **49**, 48 (1996)

1057. P. Lukas, M. Klesnil, J. Krejci, Phys. Stat. Sol. **27**, 545 (1968)

1058. J.S. Luo, N. Merchant, V.A. Maroni et al., Appl. Supercond. **1**, 101 (1993)

1059. K.A. Lurie, A.V. Cherkaev, Proc. R. Soc. Edinburgh. **99A**, 71 (1984)

1060. K.A. Lurie, A.V. Cherkaev, Proc. R. Soc. Edinburgh. **104A**, 21 (1986)

1061. J.T. Lynch, A.C. Anderson, R.S. Withers, et al., in *Proceedings IEEE MTTS International Microwave Symposium Digest*, 524 (1982)

1062. W.G. Lyons, R.S. Withers, J.M. Hamm et al., IEEE Trans. Magn. **27**, 2932 (1991)

1063. Y.-W. Ma, Z.-S. Gao, Y.-P. Qi et al., Physica C **469**, 651 (2009)

1064. M. Maciejewski, A. Baiker, K. Conder et al., Phys. C **227**, 343 (1994)

1065. J.L. MacManus-Driscoll, S.R. Foltyn, Q.X. Jia et al., Nat. Mater. **3**(7), 439 (2004)

1066. J. Maeda, Y. Nakamura, T. Izumi, Y. Shiohara, Supercond. Sci. Technol. **12**, 563 (1999)

1067. H. Maeda, Y. Tanaka, M. Fukutomi, T. Asano, Jpn. J. Appl. Phys. **27**, L209 (1988)

1068. Y. Maeno, H. Hashimoto, K. Yoshida et al., Nature **372**, 532 (1994)

1069. A. Magerl, B. Berre, G. Alefeld, Phys. Stat. Sol. A **36**, 161 (1976)

1070. I. Maggio-Aprile, Ch. Renner, A. Erb et al., Phys. Rev. Lett. **75**, 2754 (1995)

1071. M.A. Magil, F.J. Zwerneman, Engng. Fract. Mech. **56**, 9 (1997)

1072. Y.-W. Mai, B.R. Lawn, Annu Rev. Mater. Sci. **16**, 415 (1986)

1073. T.A. Maier, M. Jarrell, T.C. Schulthess et al., Phys. Rev. Lett. **95**, 237001 (2005)

1074. P. Majewski, H. Bestengen, S. Eschner, F. Aldinger, Z. Metall. **86**, 563 (1995)

1075. M.V. Makarova, P.E. Kazin, YuD Tretyakov et al., Physica C **419**, 61 (2005)

1076. M.T. Malachevsky, D.A. Esparza, Phys. C **324**, 153 (1999)

1077. M. Malberg, J. Bech, N. Bay et al., IEEE Trans. Appl. Supercond. **9**(2), 2577 (1999)

1078. L.E. Malvern, *Introduction to the Mechanics of a Continuous Medium* (Prentice-Hall, Englewood Cliffs, 1969)

1079. B.B. Mandelbrot, *The Fractal Geometry of Nature* (Freeman, San Francisco, 1982)

1080. J. Mannhart, P. Chaudhari, D. Dimos et al., Phys. Rev. Lett. **61**, 2476 (1988)

1081. S.S. Manson, *Thermal Stress and Low-Cycle Fatigue* (McGrow-Hill, New York, 1973)

1082. S. Marinel, I. Monot, J. Provost, G. Desgardin, Supercond. Sci. Technol. **11**, 563 (1998)

1083. K.R. Marken, W. Dei, L. Cowey et al., IEEE Trans. Appl. Supercond. **7**(2), 2211 (1997)

1084. R.S. Markiewicz, J. Phys. Chem. Solids **58**, 1179 (1997)

1085. L.K. Markov, T.S. Orlova, N.N. Peschanskaya et al., Fiz. Tverd. Tela. **45**(9), 1551 (2003)
1086. D.B. Marshall, B.N. Cox, A.G. Evans, Acta Metall. **33**, 2013 (1985)
1087. D.B. Marshall, M. Drory, A.G. Evans, Fract. Mech. Ceram. **6**, 289 (1983)
1088. D.S. Marshall, D.S. Dessau, A.G. Loeser et al., Phys. Rev. Lett. **76**, 4841 (1996)
1089. J.S. Martens, A. Pance, K. Char et al., Appl. Phys. Lett. **63**, 1681 (1993)
1090. F. Marti, M. Daumpling, R. Flukiger, IEEE Trans. Appl. Supercond. **5**, 1884 (1995)
1091. F. Marti, M. Dhallé, R. Flükiger et al., IEEE Trans. Appl. Supercond. **9**(2), 2766 (1999)
1092. F. Marti, G. Grasso, J.-C. Grivel, R. Flükiger, Supercond. Sci. Technol. **11**, 485 (1998)
1093. F. Marti, Y.B. Huang, G. Witz et al., IEEE Trans. Appl. Supercond. **9**(2), 2521 (1999)
1094. J.W. Martin, *Micromechanisms in Particle-Hardened Alloys* (Cambridge University Press, New York, 1980)
1095. L. Martini, Supercond. Sci. Technol. **11**, 231 (1998)
1096. N.D. Martur, F.M. Grosche, S.R. Julian et al., Nature **394**, 39 (1998)
1097. Y. Masuda, R. Ogawa, Y. Kawate et al., J. Mater. Res. **8**, 693 (1993)
1098. L.J. Masur, J. Kellers, F. Li et al., IEEE Trans. Appl. Supercond. **12**(1), 1145 (2002)
1099. C.M. Mate, G.M. McClelland, R. Erlandsson, S. Chiang, Phys. Rev. Lett. **59**, 1942 (1987)
1100. Y. Matsui, H. Maeda, Y. Tanaka, S. Horiuchi, Jpn. J. Appl. Phys. **27**, L361 (1988)
1101. K. Matsumoto, T. Horide, K. Osamura et al., Physica C **412–414**, 1267 (2004)
1102. K. Matsumoto, S.B. Kim, J.G. Wen et al., Adv. Supercond. **11**, 773 (1999)
1103. K. Matsunaga, A. Nishimura, S. Satoh, O. Motojima, Adv. Cryog. Eng. (Mater.) **46**, 691 (2000)
1104. M. Maul, B. Schulte, J. Appl. Phys. **74**, 2942 (1993)
1105. E. Maxwell, Phys. Rev. **78**, 477 (1950)
1106. J.C. Maxwell, *Treatise on Electricity and Magnetism* (Clarendon Press, Oxford, 1873)
1107. B. Mayer, L. Alff, T. Träuble et al., Appl. Phys. Lett. **63**, 996 (1993)
1108. I.I. Mazin, M.D. Johannes, Nat Phys. **5**, 141 (2009)
1109. B.W. McConnel, IEEE Power Eng. Rev. **20**(6) (2000)
1110. J.E. McCrone, J.R. Cooper, J.L. Tallon, cond-mat/9909263 (1999)
1111. K. McElroy, R.W. Simmonds, J.E. Hoffman et al., Nature **422**, 592 (2003)
1112. P. McGinn, M.A. Black, A. Valenzuela, Phys. C **156**, 57 (1988)
1113. P. McGinn, W. Chen, N. Zhu et al., Appl. Phys. Lett. **59**, 120 (1991)
1114. J. McHale, G.H. Myer, R.E. Salomon, J. Mater. Res. **10**, 1 (1995)
1115. M.E. McHenry, M.P. Maley, G.H. Kwei, J.D. Thompson, Phys. Rev. B. **39**, 7339 (1989)
1116. P. McIntyre, M.J. Cima, M.F. Ng, J. Appl. Phys. **68**, 4183 (1990)
1117. D.R. McKenzie, R.C. McPhedran, G.H. Derrick, Proc. Roy. Soc. Lond. A **362**, 211 (1978)
1118. R. McLaughlin, Int. J. Eng. Sci. **15**, 237 (1977)
1119. A.C. McLaughlin, F. Sher, J.P. Attfield, Nature **436**, 829 (2005)
1120. R.C. McPhedran, D.R. McKenzie, Proc. Roy. Soc. Lond. A. **359**, 45 (1978)
1121. R.C. McPhedran, G.W. Milton, Appl. Phys. A **26**, 207 (1981)
1122. P. Meakin, Phys. Rep. **235**, 189 (1993)
1123. S. Medvedev, T.M. McQueen, I.A. Troyan et al., Nat. Mater. **8**, 630 (2009)
1124. P. Mele, K. Matsumoto, T. Horide et al., Supercond. Sci. Technol. **19**, 44 (2006)
1125. E.Z. Meilikhov, Phys. C **271**, 277 (1996)
1126. C. Meingast, A. Junod, E. Walker, Phys. C **272**, 106 (1996)
1127. C. Meingast, O. Kraut, T. Wolf et al., Phys. Rev. Lett. **67**, 1634 (1991)

1128. W. Meissner, R. Ochsenfeld, Naturwissenschaften **21**, 787 (1933)
1129. K.S. Mendelson, J. Appl. Phys. **46**, 917 (1975)
1130. E. Mendoza, T. Puig, E. Varesi et al., Phys. C. **334**, 7 (2000)
1131. R.L. Meng et al., J. Supercond. **11**, 181 (1998)
1132. N. Merchant, J.S. Luo, V.A. Maroni et al., Appl. Phys. Lett. **65**, 1039 (1994)
1133. K.L. Merkle, Ultramicroscopy **37**, 130 (1991)
1134. F. Meschke, O. Raddatz, A. Kolleck, G.A. Schneider, J. Am. Ceram. Soc. **83**, 353 (2000)
1135. J. Maruzen (ed.),*Metal Data Book*, 3rd edn. (Institute of Metals, Tokyo, 1993)
1136. V.V. Metlushko, G. Guntherodt, P. Wagner et al., Appl. Phys. Lett. **63**, 2821 (1993)
1137. G. Meyer, N.M. Amer, Appl. Phys. Lett. **57**, 2089 (1990)
1138. H. Miao, H. Kitaguchi, H. Kumakura, K. Togano, Cryogenics **38**, 257 (1998)
1139. H. Miao, H. Kitaguchi, H. Kumakura et al., Physica C **303**, 81 (1998)
1140. H. Miao, F. Lera, A. Larrea, G.F. de la Fuente, R. Navarro, IEEE Trans. Appl. Supercond. **7**(2), 1833 (1997)
1141. O. Michikami, A. Yokosawa, H. Wakana, Y. Kashiwaba, Jpn. J. Appl. Phys. **36**, 2646 (1997)
1142. B.P. Mikhailov, G.S. Burkhanov, G.M. Leytus et al., Neorg. Mater. **32**, 1225 (1996)
1143. P. Mikheenko, V.S. Dang, M.M. Awang Kechik et al., IEEE Trans. Appl. Supercond. **21**(3), 3184 (2011)
1144. P. Mikheenko, A. Sarkar, V.S. Dang et al., IEEE Trans. Appl. Supercond. **18**, 3491 (2009)
1145. P. Mikheenko, A. Sarkar, V.S. Dang et al., Physica C **469**, 798 (2009)
1146. C.A. Miller, S. Torquato, J. Appl. Phys. **68**, 5486 (1990)
1147. D.J. Miller, S. Sengupta, J.D. Hettinger et al., Appl. Phys. Lett. **61**, 2823 (1992)
1148. K. Miller, J. Mat. Sci. Technology. **9**, 453 (1993)
1149. L.M. Milne-Thompson, *Theoretical Aerodynamics* (Dover, New York, 1973)
1150. G.W. Milton, Phys. Rev. Lett. **46**, 542 (1981)
1151. G.W. Milton, J. Mech. Phys. Solids. **30**, 177 (1982)
1152. G.W. Milton, in *Physics and Chemistry of Porous Media*, eds. by D.L. Johnson, P.N. Sen (AIP, New York, 1984)
1153. G.W. Milton, J. Mech. Phys. Solids. **40**, 1105 (1992)
1154. G.W. Milton, *The Theory of Composites* (Cambridge University Press, Cambridge, 2002)
1155. G.W. Milton, Commun. Pure Appl. Math. **43**, 63 (1990)
1156. G.W. Milton, N. Phan-Thien, Proc. Roy. Soc. Lond. A. **380**, 305 (1982)
1157. N. Minami et al., Jpn. J. Appl. Phys. **31**, 784 (1992)
1158. M. Mironova, D.F. Lee, V. Selvamanickam et al., Philos. Mag. A **71**, 855 (1995)
1159. I.P. Miroshnichenko, I.A. Parinov, E.V. Rozhkov, A.G. Serkin, Metallurgist **50**(7–8), 408 (2006)
1160. I.P. Miroshnichenko, I.A. Parinov, E.V. Rozhkov, A.G. Serkin, *Optical Device for Measurement of Displacements*. Russian Patent No. 2373492 (2009)
1161. I.P. Miroshnichenko, I.A. Parinov, E.V. Rozhkov, A.G. Serkin, *Device for Testing Thin Specimens on Bending*. Russian Patent No. 2376567 (2009)
1162. I.P. Miroshnichenko, I.A. Parinov, E.V. Rozhkov, A.G. Serkin, *Optical Device for Measurement of Displacements*. Russian Patent No. 2407988 (2010)
1163. I.P. Miroshnichenko, I.A. Parinov, E.V. Rozhkov, et al. in *Piezoelectrics and Related Materials: Investigations and Applications*, ed. by I. A. Parinov (Nova Science Publishers, New York, 2012)
1164. I.P. Miroshnichenko, A.G. Serkin, Russ. J. Nondestr. Test. **43**(4), 234 (2007)

1165. D.R. Mishra, H.S. Zadeh, D. Bhattacharya, R.G. Sharma, Phys. C **341–342**, 1931 (2000)
1166. T. Mito, H. Chikaraishi, S. Hamaguchi, IEEE Trans. Appl. Supercond. **12**(1), 606 (2002)
1167. R. Mitsuhashi, Y. Suzuki, Y. Yamanari et al., Nature **464**, 76 (2010)
1168. K. Miyake, S. Schmitt-Rink, C.M. Varma, Phys. Rev. B **34**, 6554 (1986)
1169. T. Miyamoto, J. Katagari, K. Nagashima, M. Murakami, IEEE Trans. Appl. Supercond. **9**(2), 2066 (1999)
1170. T. Miyamoto, K. Nagashima, N. Sakai, M. Murakami, Supercond. Sci. Technol. **13**, 816 (2000)
1171. T. Miyamoto, K. Nagashima, N. Sakai, M. Murakami, Phys. C **349**, 69 (2001)
1172. A. Miyase, Y.S. Yuan, M.S. Wong et al., Supercond. Sci. Technol. **8**, 626 (1995)
1173. G.S. Mkrtychan, V.V. Schmidt, Zh. Eksp. Teor. Fiz. **61**, 367 (1971)
1174. S. Mo, J. Hove, A. Sudbø, Phys. Rev. B **65**, 104501 (2002)
1175. B.H. Moeckly, K. Char, Appl. Phys. Lett. **71**, 2526 (1997)
1176. B.H. Moeckly, K. Char, Phys. Rev. B **65**, 174504 (2002)
1177. B.H. Moeckly, D.K. Lathrop, R.A. Buhrman, Phys. Rev. B **47**, 400 (1993)
1178. O. Mohr, Zeit. Ver. Deut. Ing. **44**, 1524 (1900)
1179. W. Mönch, *Semiconductor Surfaces and Interfaces* (Springer, Berlin, 2001)
1180. I. Monot, T. Hihuchi, N. Sakai, M. Murakami, Supercond. Sci. Technol. **7**, 783 (1994)
1181. I. Monot, J. Wang, M.P. Delamare et al., Phys. C **267**, 173 (1996)
1182. G.C. Montanari, I. Ghinello, L. Gherardi, P. Caracino, Supercond. Sci. Technol. **9**, 385 (1996)
1183. G.C. Montanari, I. Ghinello, L. Gherardi, R. Mele, IEEE Trans. Appl. Supercond. **7**(2), 1303 (1997)
1184. M. Monteverde, M. Núñez-Regueiro, C. Acha, et al, Phys. C **408–410,** 23 (2004)
1185. G. Moore, S. Kramer, G. Kordas, Mater. Lett. **7**, 415 (1989)
1186. J.C. Moore, M.I. Bisset, D.C. Knoll et al., IEEE Trans. Appl. Supercond. **9**(2), 1692 (1999)
1187. J.C. Moore, M.J. Naylor, C.R.M. Grovenor, IEEE Trans. Appl. Supercond. **9**(2), 1787 (1999)
1188. J.C. Moore, P. Shiles, C. Eastell et al., J.Inst. Phys. Conf. Ser. **158**, 901 (1997)
1189. P.E.D. Morgan, R.M. Hously, J.R. Porter, J. Ratto, J. Phys. C **176**, 279 (1991)
1190. T. Mori, K. Tanaka, Acta Metall. **21**, 571 (1973)
1191. M. Morita, K. Nagashima, S Takebayashi, et al., *Proceedings Internaional Workshop on Superconductivity* (Okinawa, Japan, 1998), p. 115
1192. M. Morita, M. Sawamura, S. Takebayashi et al., Phys. C **235–240**, 209 (1994)
1193. M. Morita, S. Takebayashi, M. Tanaka et al., Adv. Supercond. **3**, 733 (1991)
1194. M. Morita, M. Tanaka, S. Takebayashi, Jpn. J. Appl. Phys. **30**, L813 (1991)
1195. E. Morosan, H.W. Zandbergen, B.S. Dennis et al., Nat Phys. **2**, 544 (2006)
1196. A. Mourachkine, J. Low Temp. Phys. **117**, 401 (1999)
1197. A. Mourachkine, Europhys. Lett. **55**, 86 (2001)
1198. A. Mourachkine, Europhys. Lett. **55**, 559 (2001)
1199. A. Mourachkine, J. Supercond. **14**, 375 (2001)
1200. A. Mourachkine, Supercond. Sci. Technol. **14**, 329 (2001)
1201. A. Mourachkine, *High-Temperature Superconductivity in Cuprates. The Nonlinear Mechanism and Tunneling Measurements* (Kluwer Academic, Dordrecht, 2002)
1202. A. Mourachkine, *Room-Temperature Superconductivity* (Cambridge International Science Publishing, Cambridge, 2004)

1203. G. Mozurkewich, M.B. Salamon, S.E. Inderhees, Phys. Rev. B **46**, 11914 (1992)
1204. R. Mulet, O. Diaz, E. Altshuler, Supercond. Sci. Technol. **10**, 758 (1997)
1205. W.W. Mulins, R.F. Sekerka, J. Appl. Phys. **34**, 323 (1963)
1206. K.A. Müller, M. Takashige, J.G. Bednorz, Phys. Rev. Lett. **58**, 1143 (1987)
1207. R. Müller, C. Janowitz, M. Schneider et al., J. Supercond. **15**, 147 (2002)
1208. H. Müller-Krumbhaar, H. Emmerich, E. Brener, M. Hartmant, Phys. Rev. E **63**, 026304 (2001)
1209. M. Murakami, *Melt Processed High-Temperature Superconductors* (World Scientific Publ, Singapore, 1992)
1210. M. Murakami, Supercond. Sci. Technol. **5**, 185 (1992)
1211. M. Murakami, H. Fujimoto, S. Gotoh et al., Phys. C **185–189**, 321 (1991)
1212. M. Murakami, S. Gotoh, N. Koshizuka et al., Cryogenics **30**, 390 (1990)
1213. M. Murakami, M. Morita, K. Doi, K. Miyamoto, Jpn. J. Appl. Phys. **28**, 1189 (1989)
1214. M. Murakami, N. Sakai, T. Higuchi, S.I. Yoo, Supercond. Sci. Technol. **9**, 694 (1996)
1215. M. Murakami, N. Sakai, T. Higuchi, S.I. Yoo, Supercond. Sci. Technol. **9**, 1015 (1996)
1216. M. Murakami, K. Yamaguchi, H. Fugimoto et al., Cryogenics **32**, 930 (1992)
1217. M. Murakami, S.I. Yoo, T. Higuchi et al., Jpn. J. Appl. Phys. **33**, L715 (1994)
1218. T. Muramatsu, N. Takeshita, C. Terakura, et al., cond-mat/0502490 (2005)
1219. J.M. Murduck, J. Burch, R. Hu et al., IEEE Trans. Appl. Supercond. **7**(2), 2940 (1997)
1220. K.G. Murty, *Linear Programming* (Wiley, New York, 1983)
1221. K.E. Myers, D.W. Face, D.J. Kountz et al., IEEE Trans. Appl. Supercond. **5**(2), 1684 (1995)
1222. K.E. Myers, Lijie Bao. J. Supercond. **11**, 129 (1998)
1223. T. Nabatame, S. Koike, O.B. Hyun et al., Appl. Phys. Lett. **65**, 776 (1994)
1224. J. Nagamatsu, N. Nakagawa, T. Muranaka et al., Nature **410**, 63 (2001)
1225. M. Nahum, J.M. Martinis, Appl. Phys. Lett. **63**, 3075 (1993)
1226. M. Nahum, P.L. Richards, C.A. Mears, IEEE Trans. Appl. Supercond. **3**, 2124 (1993)
1227. S. Nakahara, G.J. Fisanick, M.F. Yan et al., J. Crystal Growth **85**, 639 (1987)
1228. N. Nakamura, G.D. Gu, K. Takamuku et al., Appl. Phys. Lett. **61**, 3044 (1992)
1229. N. Nakamura, Y. Nakai, Y. Kanai et al., Rev. Sci. Instrum. **75**, 3034 (2004)
1230. Y. Nakamura, Y. Shiohara, J. Mater. Res. **11**, 2450 (1996)
1231. Y. Nambu, Phys. Rev. **117**, 648 (1960)
1232. Y. Nambu, Phys. Rev. D **10**, 4262 (1974)
1233. Y. Nambu, G. Jona-Lasinio, Phys. Rev. **124**, 246 (1961)
1234. V.N. Narozhnyi, D. Eckert, K.A. Nenkov, et al., cond-mat/9909107 (1999)
1235. Nature 414, 6861 (2001)
1236. M.J. Naylor, C.R.M. Grovenor, IEEE Trans. Appl. Supercond. **9**(2), 1860 (1999)
1237. A.A. Needleman, J. Appl. Mech. **54**, 525 (1987)
1238. I.A. Nekrasov, M.V. Sadovskii, JETP Lett. **93**(3), 166 (2011)
1239. J.J. Neumeier, Physica C **233**(3&4), 354 (1994)
1240. K.C. Numan, J.B. Keller, J. Mech. Phys. Solids. **32**, 259 (1984)
1241. J.J. Neumeier, H.A. Zimmermann, Phys. Rev. B **47**, 8385 (1993)
1242. M.S. Newson, D.T. Ryan, M.N. Wilson, H. Jones, IEEE Trans. Appl. Supercond. **12**(1), 725 (2002)
1243. S. Nhien, P.L. Langlois, G. Desgardin, in *Proceedings 4th Journées d'Études SEE*, ed. by G. Desgardin, AP18 (Caen, March, 1997)

1244. M. Nideröst, A. Suter, P. Visani et al., Phys. Rev. B **53**, 9286 (1996)
1245. J.C. Nie, H. Yamasaki, H. Yamada et al., Supercond. Sci. Technol. **17**, 845 (2004)
1246. F.S. Nogueira, H. Kleinert, Phys. Rev. B **77**, 045107 (2008)
1247. S. Nomura, C. Suzuki, N. Watanabe et al., IEEE Trans. Appl. Supercond. **12**(1), 788 (2002)
1248. A.N. Norris, Mech. Mat. **4**, 1 (1985)
1249. D.P. Norton, A. Goyal, J.D. Budai et al., Science **274**, 755 (1996)
1250. J.G. Noudem, J. Beille, D. Bourgault et al., Phys. C **230**, 42 (1994)
1251. J.G. Noudem, J. Beille, A. Draperi et al., Supercond. Sci. Technol. **6**, 795 (1993)
1252. J.G. Noudem, E.S. Reddy, M. Tarka et al., Phys. C **366**, 93 (2002)
1253. M. Nunez-Regueiro, J-L. Tholence, E.V. Antipov, et al., Science **262**, 97 (1993)
1254. H.M. O'Bryan, P.K. Gallagher, Adv. Ceram. Mater. **2**, 640 (1987)
1255. O.O. Oduleye, S.J. Penn, N.McN. Alford, Supercond. Sci. Technol. **11**, 858 (1998)
1256. O.O. Oduleye, S.J. Penn, N.McN. Alford, et al., IEEE Trans. Appl. Supercond. **9**(2), 2621 (1999)
1257. N. Ogawa, I. Hirabayashi, S. Tanaka, Phys. C **177**, 101 (1991)
1258. R.W. Ogden, *Non-Linear Elastic Deformation* (Dover, New York, 1997)
1259. R.A. Ogg Jr, Phys. Rev. **69**, 243 (1946)
1260. H. Ogino, Y. Matsumura, Y. Katsura, et al., arXiv: 0903.3314 (2009)
1261. T.S. Oh, J. Rodel, R.M. Cannon, R.O. Ritchie, Acta Metall. **36**, 2083 (1988)
1262. S. Ohshima, T. Ogasawara, K. Ehata, IEEE Trans. Appl. Supercond. **7**(2), 3060 (1997)
1263. T. Oka, Y. Itoh, Y. Yanagi et al., Phys. C **200**, 55 (1992)
1264. M. Okada, K. Tanaka, T. Wakuda et al., IEEE Trans. Appl. Supercond. **9**(2), 1904 (1999)
1265. M. Okaji, K. Nara, H. Kato et al., Cryogenics **34**, 163 (1994)
1266. N. Okazaki, T. Hasegawa, K. Kishio et al., Phys. Rev. B **41**, 4296 (1990)
1267. C.G. Olson, R. Liu, A.-B. Yang et al., Science **245**, 731 (1989)
1268. H.K. Onnes, Proc. Sect. Sci. **XII**, 1107 (1911)
1269. J. Orenstein, in *Handbook of High-Temperature Superconductivity: Theory and Experiment*, eds. by J.R. Schrieffer, J.S. Brooks (Springer, New York, 2007)
1270. T.S. Orlova, J.Y. Laval, Fiz. Tverd. Tela. **49**(11), 1964 (2007)
1271. T.S. Orlova, J.Y. Laval, A. Dubon et al., Supercond. Sci. Technol. **11**, 467 (1998)
1272. T.S. Orlova, J.Y. Laval, C. Nguyen-van-Huong, A. Dubon, Supercond. Sci. Technol. **12**, 1156 (1999)
1273. T.S. Orlova, J.Y. Laval, C. Nguyen-van-Huong, A. Dubon, Supercond. Sci. Technol. **14**, 59 (2001)
1274. T.S. Orlova, L.K. Markov, B.I. Smirnov et al., Fiz. Tverd. Tela. **33**(12), 3595 (1991)
1275. T.S. Orlova, B.I. Smirnov, Fiz. Tverd. Tela. **43**(6), 974 (2001)
1276. T.S. Orlova, B.I. Smirnov, V.V. Shpeizman et al., Fiz. Tverd. Tela. **32**(4), 1031 (1990)
1277. E.Z. Orowan, Phys. **89**(605), 634 (1934)
1278. K. Osamura, S. Nanaka, M. Matsui, Phys. C **257**, 79 (1996)
1279. K. Osamura, K. Ogawa, T. Horita, et al., in*Ext. Abstr. Int. Workshop on Critical Current* (*IWCC*-99) , Madison, Wisconsin, USA, July 7–10, 1999, p. 206
1280. K. Osamura, K. Ogawa, T. Thamizavel, A. Sakai, Phys. C **335**, 65 (2000)
1281. K. Osamura, M. Sugano, Phys. C **357–360**, 1128 (2001)
1282. K. Osamura, M. Sugano, T. Wada, S. Ochiai, Adv. Cryog. Eng. (Mater.) **46**, 639 (2000)
1283. I.V. Ostrovsky, I.N. Selivonov, Fiz. Nizk. Temper. **24**, 67 (1998)
1284. W. Pachla, P. Kováč, H. Marciniak et al., Phys. C **248**, 29 (1995)

1285. W. Pachla, H. Marciniak, A. Szulc et al., IEEE Trans. Appl. Supercond. **7**(2), 2090 (1997)
1286. J. Paglione, R.L. Greene, Nat Phys. **6**, 645 (2010)
1287. S.V. Pak, A.N. Slipenyuk, V.N. Varyukhin, A.V. Samelyuk, Metal. Phys. Surf. Technol. **19**(3), 46 (1997)
1288. T.T.M. Palstra, B. Batlogg, L.F. Schneemeyer et al., Phys. Rev. Lett. **61**, 1662 (1988)
1289. T.T.M. Palstra, O. Zhou, Y. Iwasa et al., Solid State Commun. **93**, 327 (1995)
1290. S.H. Pan, E.W. Hudson, K.M. Lang et al., Nature **403**, 746 (2000)
1291. S.H. Pan, J.P. O'Neal, R.L. Badzey et al., Nature **413**, 282 (2001)
1292. V.M. Pan, A.L. Kasatkin, V.L. Svetchnikov et al., Cryogenics **33**, 21 (1993)
1293. V.M. Pan, *Study of High Temperature Superconductors*, ed. by A. Narlikar (Nova, New York, 1990), p. 319
1294. V.G. Panteleev, K.S. Ramm, L.S. Agroskin, L.V. Solovieva, Glass Ceram. **2**, 24 (1989)
1295. A.S. Parikh, B. Meyer, K. Salama, Supercond. Sci. Technol. **7**, 455 (1994)
1296. I.A. Parinov, Ferroelectrics **131**, 131 (1992)
1297. I.A. Parinov, Ferroelectrics **172**, 253 (1995)
1298. I.A. Parinov, Ferroelectric Lett. **19**, 157 (1995)
1299. I.A. Parinov, J. Surface Invest. X-Ray, Synch. Neutron Techn. **11**, 39 (1996)
1300. I.A. Parinov, Supercond. Res. Develop. **9**(10), 16 (1998)
1301. I.A. Parinov, in *Ext. Abstr. Int. Workshop on Critical Current* (*IWCC*-99) , Madison, Wisconsin, USA, July 7–10, 1999, p. 166
1302. I.A. Parinov, IEEE Trans Appl. Supercond. **9**(2), 4304 (1999)
1303. I.A. Parinov, Strength Mater. **4**, 92 (1999)
1304. I.A. Parinov, Compos. Mech. Des. **6**, 445 (2000)
1305. I.A. Parinov, in *Proceedings of 10th International Congress of Fracture*, ICF100470PR (CD-ROM), Honolulu, Hawaii, USA, Dec 2–6, 2001
1306. I.A. Parinov, Compos. Mech. Des. **8**, 172 (2002)
1307. I.A. Parinov, *Microstructure and Properties of High-Temperature Superconductors* (Rostov State University Press, Rostov-on-Don, 2004)
1308. I.A. Parinov, Compos. Mech. Des. **11**, 242 (2005)
1309. I.A. Parinov, Compos. Mech. Des. **11**, 357 (2005)
1310. I.A. Parinov, *DrSc thesis* (South-Russian State Technical University, Novocherkassk, 2007)
1311. I.A. Parinov, Int. J. Pure Appl. Math. **49**(3), 421 (2008)
1312. I.A. Parinov, *Superconductors and Superconductivity, Handbook, Three-Volume Set: Vol. 2 Theory and Properties* (Southern Federal University Press, Rostov-on-Don, 2008)
1313. I.A. Parinov, *Superconductors and Superconductivity. Handbook, Three-Volume Set: Vol. 3, Applications and Prospective* (Southern Federal University Press, Rostov-on-Don, 2010)
1314. I.A. Parinov, in *Piezoceramic Materials and Devices*, ed. by Ivan A. Parinov (Nova Science Publishers, New York, 2010)
1315. I.A. Parinov, S.H. Chang, in *Ferroelectrics and Superconductors: Properties and Applications*, ed. by Ivan A. Parinov (Nova Science Publishers, New York, 2011)
1316. I.A. Parinov, L.I. Parinova, Ferroelectrics **211**, 41 (1998)
1317. I.A. Parinov, L.I. Parinova, E.V. Rozhkov, Phys. C **377**, 114 (2002)
1318. I.A. Parinov, L.V. Parinova, in *Proceedings International Conference on Struct. and Proper. Brittle and Quasiplast. Mater.* (SPM'94) (Latv. Acad. Sci., Riga, 1994), p. 98
1319. I.A. Parinov, L.V. Parinova, Supercond. Phys. Chem. Tech. **7**, 79 (1994)
1320. I.A. Parinov, L.V. Parinova, Supercond. Phys. Chem. Tech. **7**, 1382 (1994)

1321. I.A. Parinov, L.V. Parinova, Strength Mater. **1**, 113 (1997)

1322. I.A. Parinov, A.V. Popov, E.V. Rozhkov, A.G. Prygunov, Russ. J. Non-Destruct. Control **1**, 66 (2000)

1323. I.A. Parinov, A.G. Prygunov, E.V. Rozhkov, et al., *Russian Patent* No 2,169,348 (2001)

1324. I.A. Parinov, E.V. Rozhkov, IEEE Trans. Appl. Supercond. **9**(2), 2058 (1999)

1325. I.A. Parinov, E.V. Rozhkov, Compos. Mech. Des. **10**, 355 (2004)

1326. I.A. Parinov, E.V. Rozhkov, C.E. Vassil'chenko, IEEE Trans. Appl. Supercond. **7**(2), 1941 (1997)

1327. I.A. Parinov, E.V. Rozhkov, C.E. Vassil'chenko, Adv. Cryog. Eng. (Mater.) **44b,** 639 (1998)

1328. I.A. Parinov, Yu.S. Vasil'eva, Strength Mater. **8**, 77 (1994)

1329. H-W. Park, K-B. Kim, K-W. Lee, Supercond. Sci. Technol. **9**, 694 (1996)

1330. F. Parmigiani, G. Chiarello, N. Ripamonti, Phys. Rev. B **36**, 7148 (1987)

1331. J.A. Parrell, S.E. Dorris, D.C. Larbalestier, Adv. Cryogenic Eng. **40**, 193 (1994)

1332. J.A. Parrell, S.E. Dorris, D.C. Larbalestier, Phys. C **231**, 137 (1994)

1333. J.A. Parrell, Y. Feng, S.E. Dorris, D.C. Larbalestier, J. Mater. Res. **11**, 555 (1996)

1334. J.A. Parrell, D.C. Larbalestier, S.E. Dorris, IEEE Trans. Appl. Supercond. **5**(2), 1275 (1995)

1335. J.A. Parrell, D.C. Larbalestier, G.N. Riley Jr et al., J. Mater. Res. **12**, 2997 (1997)

1336. J.A. Parrell, A.A. Polyanskii, A.E. Pashitski, D.C. Larbalestier, Supercond. Sci. Technol. **9**, 393 (1996)

1337. E.A. Pashitskii, A. Gurevich, A.A. Polyanskii et al., Science **275**, 367 (1997)

1338. E.A. Pashitskii, A.A. Polyanskii, A. Gurevich et al., Phys. C **246**, 133 (1995)

1339. Y.A. Pashkin, T. Yamamoto, O. Astafiev et al., Nature **421**, 823 (2003)

1340. K.V. Pasmussen, O.B. Pederson, Acta Metall. **14**, 1467 (1980)

1341. R. Passerini, M. Dhallé, G. Witz et al., IEEE Trans. Appl. Supercond. **11**(1), 3018 (2001)

1342. S. Patel, S. Chen, T. Haugan et al., Cryogenics **35**, 257 (1996)

1343. S. Patnaik, L.D. Cooley, A. Gurevich, et al., cond-mat/0104562 (2001)

1344. S. Pavard, D. Bourgault, C. Villard et al., Physica C **316**, 198 (1999)

1345. M. Pechini, U.S. Patent No 3,330,697 (1967)

1346. S. Pei, J.D. Jorgenson, B. Dabrowski et al., Phys. Rev. B. **41**, 4126 (1990)

1347. R. Peierls, Proc. Roy. Soc. (London) **A155**, 613 (1936)

1348. H. Peisl, in *Hydrogen in Metals* , eds. by G. Alefeld, J. Volkl, Basic Properties, vol. 1 (Springer, New York, 1978), p. 53

1349. A. Perez-Gonzalez, J.R. Clem, Phys. Rev. B. **31**, 7048 (1985)

1350. A. Perez-Gonzalez, J.R. Clem, J. Appl. Phys. **58**, 4326 (1985)

1351. A. Perez-Gonzalez, J.R. Clem, Phys. Rev. B. **32**, 2909 (1985)

1352. W.T. Perrins, D.R. McKenzie, R.C. McPhedran, Proc. Roy. Soc. Lond. A **369**, 207 (1979)

1353. N.A. Pertsev, G. Arlt, Fiz. Tverdogo Tela **33**, 3077 (1991)

1354. J.M. Peterson, J.J. Hermans, J. Composite Mater. **3**, 338 (1969)

1355. R.L. Peterson, J.W. Ekin, Phys. Rev. B **37**, 9848 (1988)

1356. C. Petrovic, R. Movshovich, M. Jaime, et al., cond-mat/0012261 (2000)

1357. C. Pflelderer, M. Uhlarz, S.M. Hyden et al., Nature **412**, 58 (2001)

1358. D.T. Philips, A. Garcia-Diaz, *Fundamentals of Network Analysis* (Prentice-Hall, Englewood Cliffs, 1981)

1359. I.S. Grigoriev, E.Z. Meilikhov (eds.), *Physical Values. Handbook*, vol. 223 (Energoatomizdat, Moscow, 1991), p. 47

1360. Phys. Today (July, 1994), p. 11

1361. S.W. Pierson, O.T. Valls, Z. Tesanovic, M.A. Lindemann, Phys. Rev. B **57**, 622 (1998)

1362. I.A. Parinov (ed.), *Piezoceramic Materials and Devices* (Nova Science Publishers, New York, 2010)

1363. I.A. Parinov (ed.), *Piezoelectric Materials and Devices* (Nova Science Publishers, New York, 2012)

1364. I.A. Parinov (ed.), *Piezoelectrics and Related Materials: Investigations and Applications* (Nova Science Publishers, New York, 2012)

1365. A.B. Pippard, Proc. Roy. Soc. **A216**, 547 (1953)

1366. A.B. Pippard, Phil. Trans. Roy. Soc. A: Math. Phys. Eng. Sci. **248**(941), 97 (1955)

1367. G.G. Pisarenko, *Fracture Resistance of Piezoelectric Ceramics* (Preprint AN UkrSSR, Kiev, 1984)

1368. G.G. Pisarenko, *Strength of Piezoceramics* (Naukova Dumka, Kiev, 1987)

1369. J. Pitel, P. Kováč, K. Hense, H. Kirchmayr, IEEE Trans. Appl. Supercond. **12**(1), 1475 (2002)

1370. V. Plecháček, IEEE Trans. Appl. Supercond. **9**(2), 2078 (1999)

1371. M. Plischke, Z. Racz, Phys. Rev. Lett. **53**, 415 (1984)

1372. A.V. Pogrebnyakov, J.M. Redwing, J.E. Jones, et al., cond-mat/0304164 (2003)

1373. D. Poilblanc, Phys. Rev. B. **48**, 3368 (1993)

1374. D. Poilblanc, Phys. Rev. B. **49**, 1477 (1994)

1375. J.P. Poirier, *Plasticité á Haute Température des Solides Cristallins* (Eyrolles, Paris, 1976)

1376. J. Pokluda, F. Kroupa, L. Obdražálek, *Mechanical Properties and Structure of Solids* (PCdir, Brno, 1994)

1377. J. Pokluda, P. Šandera, *Random Micromechanisms of Fracture and Fatigue In a Multiscale Context* (Springer, London, 2010)

1378. J. Pokluda, P. Šandera, J. Horníková, Fat. Fract. Engng. Mater. Struct. **27**, 141 (2004)

1379. J. Pokluda, P. Šandera, J.J. Horníková, ASTM Int. **2**, 1 (2005)

1380. J. Polák, *Plasticity and Low Cycle Fatigue Life of Metals* (Academia, Prague, 1991)

1381. J. Polák, T. Lepistö, P. Keitunnen, Mater. Sci. Eng., A **74**, 85 (1985)

1382. M. Polak, J.A. Parrell, A.A. Polyanskii et al., Appl. Phys. Lett. **70**, 1034 (1997)

1383. M. Polak, A.A. Polyanskii, W. Zhang et al., Adv. Cryog. Eng. (Mater.) **46**, 793 (2000)

1384. M.Z. Polany, Physics **89**, 660 (1934)

1385. J. Polchinski, Nucl. Phys. B **231**, 269 (1984)

1386. A.M. Polyakov, JEPT Lett. **20**, 894 (1974)

1387. A.A. Polyanskii, Private communication (2004)

1388. A. Polyanskii, V. Beilin, E. Yashchin et al., IEEE Trans. Appl. Supercond. **11**(1), 3736 (2001)

1389. A.A. Polyanskii, X.Y. Cai, D.M. Feldman, Larbalestier. in *NATO Science Series*, vol. 3. eds. by I. Nedkov, M. Ausloos, High Technology (Kluwer Academic, Netherlands 1999), pp. 72, 353

1390. A.A. Polyanskii, D.M. Feldmann, D.C. Larbalestier, in *Handbook of Superconducting Materials*, eds. by D. Cardwell, D Ginley (Institute of Physics Publishing, Bristol, 1999), p. 236

1391. A.A. Polyanskii, A. Gurevich, E.A. Pashitskii et al., Phys. Rev. B **53**, 8687 (1996)

1392. A.A. Polyanskii, A.E. Pashitski, A. Gurevich et al., Adv. Supercond. **9**, 469 (1997)

1393. C.P. Poole Jr, T. Datta, H.A. Farach et al., *Copper Oxide Superconductors* (Wiley, New York, 1988)
1394. D.L. Porter, A.G. Evans, A.H. Heuer, Acta Metall. **27**, 1649 (1979)
1395. J.R. Porter, W. Blumenthal, A.G. Evans, Acta Metall. **29**, 1899 (1981)
1396. M. Potthoff, Eur. Phys. J. **B32**, 429436 (2003)
1397. G.L. Povirk, A. Needleman, S.R. Nutt, Mater. Sci. Eng. **A125**, 129 (1990)
1398. D. Prall, R.S. Lakes, Int. J. Mech. Sci. **39**, 305 (1997)
1399. M.R. Presland et al., Phys. C **176**, 95 (1991)
1400. M. Prester, Phys. Rev. B **54**, 606 (1996)
1401. R. Preuss, W. Hanke, C. Grober, H.G. Evertz, Phys. Rev. Lett. **79**, 1122 (1997)
1402. L. Prigozhin, Eur. J. Appl. Math. **7**, 237 (1996)
1403. L. Prigozhin, J. Comput. Phys. **129**, 190 (1996)
1404. T.A. Prikhna, W. Gawalek, V. Moshchil et al., Phys. C **354**, 415 (2001)
1405. A.V. Prikhod'ko, O.I. Kon'kov, Semiconductors **35**, 687 (2001)
1406. H. Prinzbach, A. Weiler, P. Landenberger et al., Nature **407**, 60 (2000)
1407. A.M. Prokhorov, N.P. Lyakishev, G.S. Burkhanov et al., Inorg. Mater. **32**, 854 (1996)
1408. A.M. Prokhorov, N.P. Lyakishev, G.S. Burkhanov et al., Dokl. Phys. **363**, 771 (1998)
1409. A.V. Pronin, M. Dressel, A. Pimenov et al., Phys. Rev. B **57**, 14416 (1998)
1410. N. Provatas, N. Goldenfeld, J. Dantzig, Phys. Rev. Lett. **80**, 3308 (1998)
1411. N. Provatas, N. Goldenfeld, J. Dantzig, Comp. Phys. **148**, 265 (1999)
1412. L.P. Pryadko, S.A. Kivelson, O. Zachar, Phys. Rev. Lett. **92**, 67002 (2004)
1413. M.P. Puls, Acta Metall. **29**, 1961 (1981)
1414. M. Putti, I. Pallecchi, E. Bellingeri, et al. Sci. Technol. **23**, 134003 (10 pp.) (2010)
1415. M. Rábara, N. Sekimura, H. Kitagushi et al., Supercond. Sci. Technol. **12**, 1129 (1999)
1416. H.A. Radovan, N.A. Fortune, T.P. Murphy et al., Nature **425**, 51 (2003)
1417. H.D. Ramsbottom, H. Ito, T. Horita, K. Osamura, Appl. Supercond. **5**, 157 (1997)
1418. C. Ratsch, P. Ruggerone, M. Scheffler, in *Surface Diffusion: Atomistic and Collective Processes*, ed. by M.C. Tringides (Plenum, New York, 1997)
1419. R. Ravinder Reddy, M. Muralidhar, V. Hari Babu, P. Venugopal Reddy, Supercond. Sci. Technol. **8**, 101 (1995)
1420. R.T. Read, *Dislocations in Crystals* (McGraw-Hill, New York, 1953)
1421. E.S. Reddy, G.J. Schmitz, Supercond. Sci. Technol. **15**, L21 (2002)
1422. Y. Ren, R. Weinstein, J. Liu et al., Phys. C **251**, 15 (1995)
1423. Y. Ren, R. Weinstein, R. Sawh, J. Liu, Phys. C **282–287**, 2301 (1997)
1424. Z.F. Ren, J.Y. Lao, L.P. Gao et al., J. Supercond. **11**, 159 (1998)
1425. Ch. Renner, B. Revaz, K. Kadowaki et al., Phys. Rev. Lett. **80**, 3606 (1998)
1426. C.A. Reynolds, B. Serin, W.H. Wright, L.B. Nesbitt, Phys. Rev. **78**, 487 (1950)
1427. N. Reyren, S. Thiel, A.D. Caviglia et al., Science **317**, 1196 (2007)
1428. J. Rhyner, Phys. C. **212**, 292 (1993)
1429. J. Rhyner, G. Blatter, Phys. Rev. B **40**, 829 (1989)
1430. J. Rice, in *Fracture. An Advanced Treatise Mathematical Fundamentals*, vol. 204, ed. by H. Liebowitz (Academic Press, New York, 1968), p. 2
1431. J.R. Rice, M.A. Johnson, in *Inelastic Behavior of Solids*, eds. by M.F. Kanninen, W.F. Adler, A.R. Rosenfield, R.I. Jaffee (McGraw-Hill, New York, 1970), p. 641
1432. P.L. Richards, J. Clarke, R. Leoni et al., Appl. Phys. Lett. **54**, 283 (1989)
1433. D. Richardson, Proc. Cambr. Phil. Soc. **74**, 515 (1973)

1434. O. Richmond, S. Alexandrov, J. Mech. Phys. Solids. **48**, 1735 (2000)
1435. S.C. Riggs, O. Vafek, J.B. Kemper, et al., cond-mat/1008.1568 (2010)
1436. M.O. Rikel, J.L. Reeves, N.A. Scarbrough, E.E. Hellstrom, Phys. C **341–348**, 2573 (2000)
1437. M.O. Rikel, R.K. Williams, X.Y. Cai et al., IEEE Trans. Appl. Supercond. **11**(1), 3026 (2001)
1438. T. Ring, *Fundamentals of Ceramic Powder Processing and Synthesis* (Academic Press, New York, 1996)
1439. R.O. Ritchie, Mat. Sci. Eng. A **103**, 15 (1988)
1440. D. Rodgers, K. White, V. Selvamanickam et al., Supercond. Sci. Technol. **5**, 640 (1992)
1441. M.A. Rodriquez, B.J. Chen, R.L. Snyder, Phys. C **195**, 185 (1992)
1442. M.A. Rodriquez, R.L. Snyder, B.J. Chen et al., Phys. C **206**, 43 (1993)
1443. N.B. Romalis, V.P. Tamuzh, *Fracture of Structure-Heterogeneous Solids* (Zinatne, Riga, 1989)
1444. F. Ronnung, M. Danerud, M. Lindgren, D. Winkler, IEEE Trans. Appl. Supercond. **7**, 2599 (1997)
1445. P. Rosakis, A. Ruina, R.S. Lakes, J. Mater. Sci. **28**, 4667 (1993)
1446. K.H. Roscoe, Geotechnique **20**, 129 (1970)
1447. L.R.F. Rose, Int. J. Fract. **31**, 233 (1986)
1448. L.R.F. Rose, J. Amer. Ceram. Soc. **69**, 212 (1986)
1449. L.R.F. Rose, J. Mech. Phys. Solids **35**, 383 (1987)
1450. C.H. Rosner, IEEE Trans. Appl. Supercond. **11**(1), 39 (2001)
1451. J. Rossat-Mignod, L.P. Regnault, C. Vettier et al., Phys. C **185–189**, 86 (1991)
1452. V. Rouessac, G. Poullain, G. Desgardin, B. Raveau, Supercond. Sci. Technol. **11**, 1160 (1998)
1453. V. Rouessac, J. Wang, J. Provost, G. Desgardin, Phys. C **268**, 225 (1996)
1454. J.L. Routbort, K.C. Goretta, D.J. Miller et al., J. Mater. Res. **7**, 2360 (1992)
1455. L.A. Rubenfeld, J.B. Keller, SIAM J. Appl. Math. **17**, 495 (1969)
1456. N.A. Rutter, B.A. Glowacki, IEEE Trans. Appl. Supercond. **11**(1), 2730 (2001)
1457. N.A. Rutter, B.A. Glowacki, J.E. Evetts, Supercond. Sci. Technol. **13**, L25 (2000)
1458. V.V. Sagaradze, V.I. Zeldovich, V.G. Pushin et al., Supercond. Phys. Chem. Tech. **3**, 1309 (1990)
1459. M. Sahimi, *Applications of Percolation Theory* (Taylor & Frances, London, 1994)
1460. D. Saint James, de P.G. Gennes, Phys. Rev. Lett. **7**, 306 (1963)
1461. N. Sakai, A. Mase, H. Ikuta et al., Supercond. Sci. Technol. **13**, 770 (2000)
1462. N. Sakai, K. Ogasawara, K. Inoue et al., IEEE Trans. Appl. Supercond. **11**(1), 3509 (2001)
1463. N. Sakai, S-J. Seo, K. Inoue, et al., Adv. Supercond. **11**, 685 (1999)
1464. N. Sakai, S-J. Seo, K. Inoue, et al., Phys. C **335**, 107 (2000)
1465. K. Sakamaki, H. Wada, H. Nozaki et al., Solid State Commun. **112**, 323 (1999)
1466. A. Salam, in *Elementary Particle Theory: Relativistic Groups and Analyticity*, ed. by N. Svarholm, Nobel Symp. No. 8 (Almquist & Wiksell, Stockholm, 1968)
1467. A. Salam, J.C. Ward, Phys. Lett. **13**, 168 (1964)
1468. K. Salama, V. Selvamanickam, L. Gao, K. Sun, Appl. Phys. Lett. **54**, 2352 (1989)
1469. H. Salamati, A.A. Babaei-Brojeny, M. Safa, Supercond. Sci. Technol. **14**, 816 (2001)
1470. M.B. Salamon, J. Shi, N. Overend, M.A. Howsend, Phys. Rev. B. **47**, 5520 (1993)
1471. S. Salib, M. Mironova, C. Vipulanandan, K. Salama, Supercond. Sci. Technol. **9**, 1071 (1996)

1472. E. Salje, *Phase Transitions in Ferroelastic and Co-elastic Crystals* (Cambridge University Press, Cambridge, 1990)
1473. M. Salmhofer, Commun. Math. Phys. **194**, 249 (1998)
1474. A.A. Samarsky, *Theory of Difference Schemes* (Nauka, Moscow, 1983)
1475. L.M. Sander, *Solids Far from Equilibrium*, ed. by C. Godreche (Cambridge University Press, Cambridge, 1992)
1476. F. Sandiumenge, N. Vilalta, S. Pinol et al., Phys. Rev. B. **51**, 6645 (1995)
1477. S. Sanfilippo, H. Elsinger, M. Núñez-Regueiro et al., Phys. Rev. B **61**, 3800 (2000)
1478. A. Sarkar, P. Mikheenko, V.S. Dang et al., Physica C **469**, 1550 (2009)
1479. A. Sarkar, P. Mikheenko, V.S. Dang et al., J. Supercond. Nov. Magn. **24**, 173 (2011)
1480. E. Sarnelli, P. Chaudhari, J. Lacy, Appl. Phys. Lett. **62**, 777 (1993)
1481. E. Sarnelli, P. Chaudhari, W.Y. Lee, E. Esposito, Appl. Phys. Lett. **65**, 362 (1994)
1482. J.L. Sarrao, L.A. Morales, J.D. Thompson et al., Nature **420**, 297 (2002)
1483. P.V.P.S.S. Sastry, J. Schwartz, IEEE Trans. Appl. Supercond. **9**(2), 1684 (1999)
1484. T. Sata, K. Sakai, S. Tashiro, J. Am. Ceram. Soc. **75**, 805 (1992)
1485. K. Sato, T. Hikata, H. Mukai et al., IEEE Trans. Magn. **27**(2), 1231 (1991)
1486. N. Sato, M. Kawachi, K. Noto et al., Phys. C **357–360**, 1019 (2001)
1487. S. Sato, N. Aso, K. Miyake et al., Nature **410**, 340 (2001)
1488. T. Sato, T. Kamiyama, Y. Naitoh et al., Phys. Rev. B **63**, 132502 (2001)
1489. T. Sato, T. Kamiyama, T. Takahashi et al., Science **291**, 1517 (2001)
1490. T. Sato, H. Matsui, S. Nishina et al., Phys. Rev. Lett. **89**, 67005 (2002)
1491. M. Satoh, S. Haseyama, M. Kojima, et al. Adv. Cryog. Eng. (Mater.) **44b**, 405 (1998)
1492. T. Satoh, M. Hidaka, S. Tahara, IEEE Trans. Appl. Supercoud. **7**(2), 3001 (1997)
1493. M. Sauzay, P. Gilormini, Theor. Appl. Fract. Mech. **38**, 53 (2002)
1494. S.S. Saxena, P. Agarwal, K. Ahilan et al., Nature **406**, 587 (2000)
1495. D.J. Scalapino, in *Handbook of High-Temperature Superconductivity: Theory and Experiment*, eds. by J.R. Schrieffer, J.S. Brooks (Springer, New York, 2007)
1496. R.E. Schaak, T. Klimczuk, M.L. Foo, R.J. Cava, cond-mat/0305450 (2003)
1497. P. Schätzle, G. Krabbes, G. Stöver et al., Supercond. Sci. Technol. **12**, 69 (1999)
1498. J.S.Schilling, in *Handbook of High-Temperature Superconductivity: Theory and Experiment*, eds. by J.R. Schrieffer, J.S. Brooks (Springer, New York, 2007)
1499. J.S. Schilling, S. Klots, in *Physical Properties of High Temperature Superconductors*,ed. D.M. Ginsberg (World Scientific, Singapore, 1992), pp. 3, 59
1500. Z. Schlesinger et al., Phys. C **235–240**, 49 (1994)
1501. A. Schmehl, B. Goetz, R.R. Schulz et al., Europhys. Lett. **47**, 110 (1999)
1502. V.V. Schmidt, *The Physics of Superconductors*, eds. by P. Müller, A.V. Ustinov (Springer, Berlin, 1997)
1503. G.J. Schmitz, J. Laakmann, CH. Wolters, et al., J. Mater. Res. **8**, 2774 (1993)
1504. G.J. Schmitz, B. Nestler, M. Seeβellberg, J. Low Temp. Phys. **105,** 1451 (1996)
1505. G.J. Schmitz, M. Seeβelberg, B. Nestler, R. Terborg, Phys. C **282–287**, 519 (1997)
1506. T. Schneider, H. Keller, Physica C **207**, 306 (1993)
1507. W. Schnelle, N. Knauf, J. Bock et al., Physica C **209**, 456 (1993)
1508. R.J. Schoelkopf, S.H. Moseley, C.M. Stahle et al., IEEE Trans. Appl. Supercond. **9**(2), 2935 (1999)
1509. S.E. Schoenfeld, R.J. Asaro, S. Ahzi, Phil. Mag. A **73**, 1591 (1996)
1510. J.H. Schön, S. Berg, Ch. Kloc, B. Batlogg, Science **287**, 1022 (2000)

1511. J.H. Schön, A. Dodabalapur, C.H. Kloc, B. Batlogg, Science **290**, 963 (2000)
1512. J.H. Schön, C.H. Kloc, B. Batlogg, Appl. Phys. Lett. **77**, 3776 (2000)
1513. J.H. Schön, C.H. Kloc, B. Batlogg, Nature **408**, 549 (2000)
1514. J.H. Schön, C.H. Kloc, B. Batlogg, Science **293**, 2432 (2001)
1515. J.H. Schön, C.H. Kloc, R.C. Haddon, B. Batlogg, Science **288**, 656 (2000)
1516. J.H. Schön, H. Meng, Z. Bao, Nature **413**, 713 (2001)
1517. J.H. Schön, H. Meng, Z. Bao, Science **294**, 2138 (2001)
1518. E. Schrödinger, Proc. Roy. Ir. Acad. A **49**, 135 (1943)
1519. E. Schrödinger, Proc. Roy. Ir. Acad. A **52**, 1 (1948)
1520. E. Schrödinger, Proc. Roy. Ir. Acad. A **54**, 79 (1951)
1521. L.W. Schubnikow, W.I. Chotkewitsch, J.D. Schepelew, J.N. Rjabinin, Phys. Z. Sowiet., **10**(2), 165 (1936) [Sov. Phys. JETP, **7**, 221 (1937)]
1522. L.W. Schubnikow, I.E. Nakhutin, Nature **139**, 589 (1937)
1523. T.H. Schuster, M.V. Indenbom, M.R. Koblischka et al., Phys. Rev. B **49**, 3443 (1994)
1524. T.H. Schuster, T. Kuhn, E.H. Brandt et al., Phys. Rev. B **52**, 10375 (1995)
1525. T.H. Schuster, T. Kuhn, E.H. Brandt et al., Phys. Rev. B **54**, 3514 (1996)
1526. T.H. Schuster, T. Kuhn, E.H. Brandt et al., Phys. Rev. B **56**, 3413 (1997)
1527. T.H. Schuster, H. Kuhn, M.V. Indenbom, Phys. Rev. B **52**, 15621 (1995)
1528. J. Schwartz, J.K. Heuer, K.C. Goretta et al., Appl. Supercond. **2**, 271 (1994)
1529. J. Schwinger, Phys. Rev. **144**, 1087 (1966)
1530. D.W. Sciama, Rev. Mod. Phys. **36**, 463 (1964)
1531. *Second-Generation HTS Conductors*. ed. by A. Goyal (Springer, London, 2005)
1532. L.I. Sedov, *Two-Dimensional Problems in Hydrodynamics and Aerodynamics* (Interscience, New York, 1965)
1533. M. Seeβellberg, G.J. Schmitz, M. Nestler, I. Steinbach, IEEE Trans. Appl. Supercond. **7**(2), 1739 (1997)
1534. R.F. Sekerka, Z. Bi, in *Interfaces for the Twenty-First Century*, eds. by M.R. Smith, M.J. Miksis, G.B. McFadden, et al. (Imperial College Press, London, 2002)
1535. J. Selsing, Am Ceram. Soc. **44**, 419 (1961)
1536. G. Selvaduray, C. Zhang, U. Balachandran et al., J. Mater. Res. **7**, 283 (1992)
1537. P. Send, S.K. Bandyopadhyay, P. Barat et al., Phys. C **255**, 306 (1995)
1538. S. Sengupta, JOM, 19 (October, 1998)
1539. S. Sengupta, D. Shi, Z. Wang et al., Phys. C **199**, 43 (1992)
1540. S. Sengupta, V.R. Todt, K.C. Goretta et al., IEEE Trans. Appl. Supercond. **7**(2), 1727 (1997)
1541. OYu. Serdobol'skaya, G.P. Morozova, Fiz. Tverd. Tela **31**(8), 280 (1989)
1542. A.G. Sergeev, M.V. Latyshev, V.V. Teregerya, *Metrology, Standardization, Sertification* (Logos, Moscow, 2003)
1543. R.F. Service, Science **296**, 1376 (2002)
1544. R.F. Service, Science **296**, 1584 (2002)
1545. M.W. Shafer, T. Penny, B. Olsen, Phys. Rev. B **36**, 4047 (1987)
1546. R. Shah, S. Tangrila, S. Rachakonda, M. Thirukkonda, J. Electron. Mater. **241**, 1781 (1995)
1547. S. Shapiro, Phys. Rev. Lett. **11**, 80 (1963)
1548. T.M. Shaw, D. Dimos, P.E. Batson et al., J. Mater. Res. **5**, 1176 (1990)
1549. S. Sheka, P. Venugopal Reddy, Mod. Phys. Lett. B **7**, 935 (1993)

1550. Z.-X. Shen, D.S. Dessau, Phys. Rep. **253**, 1 (1995)

1551. Z.-X. Shen, D.S. Dessau, B.O. Wells et al., Phys. Rev. Lett. **70**, 1553 (1993)

1552. P. Sheng, R.V. Kohn, Phys. Rev. B **26**, 1331 (1982)

1553. Z.Z. Sheng, A.M. Hermann, Nature **332**, 55 (1988)

1554. Z.Z. Sheng, A.M. Hermann, A. El Ali et al., Phys. Rev. Lett. **60**, 937 (1988)

1555. P.G. Shewmon, Trans. Am. Inst. Min. Engrs. **230**, 1134 (1964)

1556. P.G. Shewmon, *Diffusion in Solids* (The Minerals, Metals & Materials Society, Pennsylvania, 1989)

1557. D. Shi, Appl. Supercond. **1**, 61 (1993)

1558. D. Shi, S. Sengupta, J.S. Lou et al., Phys. C **213**, 179 (1993)

1559. K. Shibutani, T. Egi, S. Hayashi et al., IEEE Trans. Appl. Supercond. **3**, 935 (1993)

1560. T.W. Shield, K.S. Kim, T.T. Shield, Trans. ASME. J. Appl. Mech. **61**, 231 (1994)

1561. T.C. Shields, J.B. Langhorn, S.C. Watcham et al., IEEE Trans. Appl. Supercond. **7**(2), 1478 (1997)

1562. A. Shilling, M. Cautoni, J.D. Gao, H.R. Ott, Nature **363**, 56 (1993)

1563. H. Shimakage, A. Kawakami, Z. Wang, IEEE Trans. Appl. Supercond. **9**(2), 1645 (1999)

1564. K. Shimizu, H. Ishikawa, D. Takao, Nature **419**, 597 (2002)

1565. K. Shimizu, T. Kimura, S. Furomoto et al., Nature **412**, 316 (2001)

1566. J. Shimoyama, J. Kase, T. Morimoto et al., Jpn. J. Appl. Phys. **31**, L1167 (1992)

1567. Y. Shiohara, N. Hobara, Phys. C **341–348**, 2521 (2000)

1568. B.I. Shklovskii, A.L. Efros, S. Luryi, *Electronic Properties of Doped Semiconductors* (Nauka, Moscow, 1979)

1569. K.N. Shrivastava, *Superconductivity. Elementary Topics* (World Scientific Publ. Corp., Singapore, 2000)

1570. L.V. Shubnikov, W.J. de Haas, Commun. Phys. Lab. Univ. Leiden. **207**, 210 (1930)

1571. L.S. Sigl, P.A. Mataga, B.J. Dalglish et al., Acta Metall. **36**, 945 (1988)

1572. F.B. Silsbee, J. Wash. Acad. Sci. **6**, 597 (1916)

1573. D.J. Singh, W.E. Pickett, Physica C **233**, 237 (1994)

1574. H.K. Singh, A.K. Saxena, O.N. Srivastava, Phys. C **273**, 181 (1997)

1575. J.P. Singh, J. Joo, N. Vasanthamohan, R.B. Poeppel, J. Mater. Res. **8**, 2458 (1993)

1576. J.P. Singh, H.J. Leu, R.B. Poeppel et al., J. Appl. Phys. **66**, 3154 (1989)

1577. J. Singleton, Rep. Prog. Phys. **63**, 1111 (2000)

1578. J. Singleton, J. Solid, State Chem. **168**, 675 (2002)

1579. G.J. Sizoo, H.K. Onnes, Commun. Phys. Lab. Univ. Leiden, No. **180**b (1925)

1580. P. Skov-Hansen, Z. Han, R. Flükiger et al., Inst. Phys. Conf. Ser. **167**, 623 (1999)

1581. P. Skov-Hansen, M.R. Koblischka, P. Vase, IEEE Trans. Appl. Supercond. **11**(1), 3740 (2001)

1582. V.P. Skripov, V.P. Koverda, *Spontaneous Solidification of Re-Cooled Liquids* (Nauka, Moscow, 1984)

1583. L.I. Slepyan, Y.S. Yakovlev, *Integral Transformations in Non-stationary Problems of Mechanics* (Sudostroenie, Leningrad, 1980)

1584. C.P. Slichter, in *Handbook of High-Temperature Superconductivity: Theory and Experiment*, eds. by J.R. Schrieffer, S. Brooks (Springer, New York, 2007)

1585. H.-J.H. Smilde, D.H.A. Blank Ariando, G.J. Gerritsma et al., Phys. Rev. Lett. **88**, 057004 (2002)

1586. H.J.H. Smilde, A.A. Golubov, H. Ariando, G. Rijnders et al., Phys. Rev. Lett. **95**, 257001 (2005)

1587. H.-J.H. Smilde, H. Hilgenkamp, G. Rijnders et al., Appl. Phys. Lett. **80**, 4579 (2000)
1588. B.I. Smirnov, YuM Baikov, A.N. Kudymov et al., Fiz. Tverd. Tela. **37**(6), 1794 (1995)
1589. B.I. Smirnov, Yu.M. Baikov, L.K. Markov, T.S. Orlova, Pisma ZhTF. **21**(12), 64 (1995)
1590. P.J. Smith, D.A. Cardwell, N. Hari Babu, Y. Shi, *Applied Superconductivity*, vol. 2. eds. by X. Obrados, F. Sandiumenge, J. Fontcuberta (2000) p. 75
1591. A. Sobha, R.P. Aloysius, P. Guruswamy et al., Phys. C **307**, 277 (1998)
1592. A. Sobha, R.P. Aloysius, P. Guruswamy et al., Phys. C **316**, 63 (1999)
1593. A. Sobha, R.P. Aloysius, P. Guruswamy, U. Syamaprasad, Supercond. Sci. Technol. **13**, 1487 (2000)
1594. I.M. Sobol, *Numerical Monte-Carlo Methods* (Nauka, Moscow, 1973)
1595. P. Sofronis, R.M. McMeeking, J. Mech. Phys. Solids **37**, 317 (1989)
1596. V.V. Sokolovsky, *Theory of Plasticity* (Gostekhizdat, Moscow, 1950)
1597. L. Solymar, *Superconductive Tunneling and Applications* (Chapman and Hall, London, 1972)
1598. C. Song, F. Liu, H. Gu et al., J. Mater. Sci. **26**, 11 (1991)
1599. S. Sorella, G. Martins, R. Becca et al., Phys. Rev. Lett. **88**, 117002 (2002)
1600. S. Souma, Y. Machida, T. Sato et al., Nature **423**, 65 (2003)
1601. M.A. Spears, A.G. Evans, Acta Metall. **30**, 1281 (1982)
1602. E.D. Specht, A. Goyal, D.M. Kroeger, Phys. Rev. B **53**, 3585 (1996)
1603. E.D. Specht, A. Goyal, D.M. Kroeger, Supercond. Sci. Technol. **13**, 592 (2000)
1604. A.J.M. Spencer, in *Mechanics of Solids. The Rodney Hill 60th Anniversary Volume*, eds. by H.G. Hopkins, M.J. Sewell (Pergamon Press, Oxford, 1982), p. 607
1605. B.J. Spencer, S.H. Davis, P.W. Voorhees, Phys. Rev. B. **47**, 9760 (1993)
1606. D.J. Srolovitz, M.P. Anderson, G.S. Grest, P.S. Sahni, Acta Metall. **32**, 1429 (1984)
1607. D.J. Srolovitz, G.S. Grest, M.P. Anderson, Acta Metall. **33**, 2233 (1985)
1608. M.P. Staines, N.E. Flower, Supercond. Sci. Technol. **4**, 232 (1991)
1609. D.M. Stamp, B. Budiansky, Int. J. Solids Struct. **25**, 635 (1989)
1610. D. Stauffer, *Introduction to the Percolation Theory* (Taylor and Frances, London, 1985)
1611. N. Stefanakis, N. Flytzanis, Supercond. Sci. Technol. **14**, 16 (2001)
1612. D.M. Stefanescu, B.K. Dhindaw, S.A. Kacar, A. Moitra, Metall. Trans. A **19**, 2847 (1988)
1613. F. Steglich, J. Aarts, C.D. Bredl et al., Phys. Rev. Lett. **43**, 1892 (1979)
1614. I. Steinbach, F. Pezzolla, M. Nestler et al., Phys. D **94**, 135 (1996)
1615. P.P. Stoev, I.I. Papirov, V.A. Finkel, Fiz. Nizk. Temper. **23**, 1019 (1997)
1616. J. Straley, S. Kenkel, Phys. Rev. B **29**, 6299 (1984)
1617. R.L. Stratonovich, Sov. Phy. Doklady **2**, 416 (1958)
1618. P. Strobel, J.C. Toledano, D. Morin et al., Phys. C **201**, 27 (1992)
1619. G.K. Strukova, V.F. Degtyareva, D.V. Shovkun, et al., cond-mat/0105293 (2001)
1620. A. Subedi, L. Zhang, D. Singh, M. Du, Phys. Rev. B. **78**, 134514 (2008)
1621. M.A. Subramanian et al., Science **239**, 1015 (1988)
1622. E. Sudhakar Reddy, T. Rajasekharm, Phys. Rev. B **55**, 14160 (1997)
1623. M. Sugano, K. Osamura, S. Ochiai, IEEE Trans. Appl. Supercond. **11**(1), 3022 (2001)
1624. T. Sugiura, H. Hashizume, K. Miya, Int. J. Appl. Electromagn. Mater. **2**, 183 (1991)
1625. N. Sugiyama et al., J. Jpn. Inst. Metals. **61**, 985 (1997)
1626. M. Sumida, Y. Nakamura, Y. Shiohara, T. Umeda, J. Mater. Res. **12**, 1979 (1997)
1627. F. Sumiyoshi, R. Kinoshita, Y. Miyazono et al., IEEE Trans. Appl. Supercond. **9**(2), 2549 (1999)

1628. L. Sun, X.J. Chen, J. Guo et al., Nature **483**, 67 (2012)

1629. O. Fisher, M.B. Maple (eds.) *Superconductivity in Ternary Compounds II* (Springer, Berlin, 1982)

1630. N. Suzuki, K. Itoh, O. Michikami, IEEE Trans. Appl. Supercond. **6**, 13 (1996)

1631. N. Suzuki, Y. Nagai, K. Itoh, O. Michikami, *Proceedings 6th Internatinal Symposium. on Superconductivity, ISS'93* (1993)

1632. S. Svechnikov, O.V. Okunev, P.A. Yagoubov et al., IEEE Trans. Appl. Supercond. **7**(2), 3548 (1997)

1633. M.V. Swain, J. Mater. Sci. Lett. **5**, 1313 (1986)

1634. A.C. Switendick, in *Hydrogen in Metals I. Topics Applied Physics*, vol. 28. eds. by G. Alefeld, J. Volkl (Springer, Berlin, 1978), p. 101

1635. K. Symanzik, Commun. Math. Phys. **18**, 227 (1970)

1636. K. Symanzik, Commun. Math. Phys. **23**, 49 (1971)

1637. Y. Syono, M. Kikuchi, K. Ohishi et al., Jpn. J. Appl. Phys. **26**, L498 (1987)

1638. K. Tachikawa, T. Watanabe, T. Inoue, K. Shirasu, Jpn. J. Appl. Phys. **30**, 639 (1991)

1639. H. Tada, P. Paris, G. Irwin, *The Stress Analysis of Cracks. Handbook* (Del Research Corp., Hellertown, 1973)

1640. F. Tafuri, J.R. Kirtley, F. Lombardi et al., Sov. J. Low Temp. Phys. **30**, 785 (2004)

1641. F. Tafuri, F. Miletto Granozio, F. Carillo, et al. Phys. Rev. B **59**, 11523 (1999)

1642. Yu.M. Tairov, V.F. Tzvetkov, *Technology pf Semiconducting and Dielectric Materials* (Vysshaya Shkola, Moscow, 1990)

1643. Y. Takabayashi, A.Y. Ganin, P. Jeglič et al., Science **323**, 1585 (2009)

1644. K. Takada, H. Sakurai, E. Takayama-Muromachi et al., Nature **422**, 53 (2003)

1645. H. Takahashi, N. Mori, in *Studies of High Temperature Superconductors*, vols. 16, 1, ed. by A.V. Narlikar (Nova Science, New York, 1995)

1646. N. Takahishi, N. Takeda, A. Koukitu et al., Jpn. J. Appl. Phys. **37**, L379 (1998)

1647. M. Takano, J. Takada, K. Oda et al., Jpn. J. Appl. Phys. **27**, L1041 (1988)

1648. Y. Takano, M. Nagao, I. Sakaguchi et al., Appl. Phys. Lett. **85**, 2851 (2004)

1649. H. Takashima, M. Tsuchimoto, T. Onishi, Jpn. J. Appl. Phys. **40**, 3171 (2001)

1650. I. Takesue, J. Haruyama, N. Kobayashi et al., Phys. Rev. Lett. **96**, 057001 (2006)

1651. I. Takeuchi, P.A. Warburton, Z. Trajanovic et al., Appl. Phys. Lett. **69**, 112 (1996)

1652. J.L. Tallon, J.R. Cooper, Adv. Supercond. **5**, 339 (1993)

1653. J. Talvacchio, M.G. Forrester, J.R. Gavaler, T.T. Braggins, IEEE Trans. Mag. **27**(2), 978 (1991)

1654. A. Tampieri, G. Celotti, G. Calestani, IEEE Trans. Appl. Supercond. **9**(2), 2010 (1999)

1655. H. Tamura, T. Mito, A. Iwamoto et al., IEEE Trans. Appl. Supercond. **12**(1), 1319 (2002)

1656. S. Tanaka, *Superconductivity Research Laboratory Publication* (Tokyo, Japan, 1998)

1657. W.H. Tang, Supercond. Sci. Technol. **13**, 580 (2000)

1658. Z.K. Tang, L. Zhang, N. Wang et al., Science **292**, 2462 (2001)

1659. S. Tangrila, R. Shah, S. Rachakonda, in*Net Shape Processing of Powder Materials, 216* (ASME-AMD, New York, 1995)

1660. K. Tanigaki, T.W. Ebbesen, S. Saito et al., Nature **352**, 222 (1991)

1661. H. Taniguchi, M. Miyashita, K. Uchiyama et al., Phys. Soc. Jpn. **72**, 468 (2003)

1662. G.I. Taylor, Proc. Roy. Soc. A **145**, 362 (1934)

1663. B. ten Haken, A. Beuink, H.H.J. ten Kate, IEEE Trans. Appl. Supercond. **7**(2), 2034 (1997)

1664. B. ten Haken, H.H.J. ten Kate, J. Tenbrink, IEEE Trans. Appl. Supercond. **5**(2), 1298 (1995)

1665. K. Terashima, H. Matsui, D. Hashimoto et al., Nature Phys. **2**, 27 (2006)
1666. J. Thomas, P. Verges, P. Schatzle et al., Phys. C **251**, 315 (1995)
1667. J. Thomasson, C. Ayache, I.L. Spain, M. Villedieu, J. Appl. Phys. **68**, 5933 (1990)
1668. J.D. Thompson, R. Movshovich, Z. Fisk, et al., cond-mat/0012260 (2000)
1669. J.R. Thompson, Y.R. Sun, H.R. Kerchner et al., Appl. Phys. Lett. **60**, 2306 (1990)
1670. E.G. Thomsen, C.T. Yang, S. Kobayashi, *Mechanics of Plastic Deformation in Metal Processing* (McMillan, New York, 1965)
1671. M.F. Thorpe, W. Jin, S.D. Mahanti, Phys. Rev. B **40**, 10294 (1989)
1672. G. 't Hooft, Nucl. Phys. B, **35**, 167 (1971)
1673. G. 't Hooft, Nucl. Phys. B, **79,** 276 (1974)
1674. M.D. Thouless, J. Am. Ceram. Soc. **73**, 2144 (1990)
1675. M.D. Thouless, A.G. Evans, Acta Metall. **36**, 517 (1988)
1676. M.D. Thouless, A.G. Evans, M.F. Ashby, J.W. Hutchinson, Acta Metall. **35**, 1333 (1987)
1677. M.D. Thouless, E. Olsson, A. Gupta, Acta. Metal. Mater. **40**, 1287 (1992)
1678. T.R. Thurston, U. Wildgruber, N. Jisrawi et al., J. Appl. Phys. **79**, 3122 (1996)
1679. S.P. Timoshenko, J.N. Goodier, *Theory of Elasticity, 3-d edn* (McGraw Hill, New York, 1970)
1680. S.M. Ting, K.R. Marken, L. Cowey et al., IEEE Trans. Appl. Supercond. **7**(2), 700 (1997)
1681. M. Tinkham, *Introduction to Superconductivity* (McGraw-Hill, New York, 1996)
1682. J.E. Tkaczyk, J.A. DeLuca, P.L. Karas et al., Appl. Phys. Lett. **62**, 3031 (1993)
1683. V.R. Todt, S. Sengupta, D. Shi et al., J. Electron. Mater. **23**, 1127 (1994)
1684. K. Togano, H. Kumakura, K. Kadowaki et al., Adv. Cryog. Eng. (Mater.) **38**, 1081 (1992)
1685. M. Tomita, M. Murakami, Phys. C **341–348**, 2443 (2000)
1686. M. Tomita, M. Murakami, Supercond. Sci. Technol. **13**, 722 (2000)
1687. M. Tomita, M. Murakami, Phys. C **354**, 358 (2001)
1688. M. Tomita, M. Murakami, Nature **421**, 517 (2003)
1689. M. Tomita, M. Murakami, K. Sawa, Y. Tachi, Phys. C **357–360**, 690 (2001)
1690. M. Tomita, K. Nagashima, M. Murakami, T. Herai, Phys. C **357–360**, 832 (2001)
1691. A. Tonomura, H. Kasai, O. Kamimura et al., Nature **397**, 308 (1999)
1692. R. Törnhardt, G. Amberg, J.Cryst. Growth **194**, 406 (1998)
1693. S. Torquato, J. Appl. Phys. **58**, 3790 (1985)
1694. S. Torquato, J. Chem. Phys. **83**, 4776 (1985)
1695. S. Torquato, in *Macroscopic Behavior of Heterogeneous Materials from the Microstructure*, vol. 147. eds. by S. Torquato, D. Krajcinovic (Am. Soc. Mech. Eng.–AMD, NY, 1992)
1696. S. Torquato, J. Mech. Phys. Solids. **45**, 1421 (1997)
1697. S. Torquato, *Radom Heterogeneous Materials Microstructure and Macroscopic Properties* (Springer, New York, 2002)
1698. S. Torquato, J.D. Beasley, Int. J. Eng. Sci. **24**, 415 (1986)
1699. S. Torquato, J. Rubinstein, J. Appl. Phys. **69**, 7118 (1991)
1700. J.M. Tranquada, B.J. Sternlieb, J.D. Axe et al., Nature **375**, 561 (1995)
1701. C. Træholt, J.G. Wen, H.W. Zandbergen et al., Phys. C **230**, 425 (1994)
1702. P. Tripodi, *High T_c Palladium Hydride Superconductor*. Patents WO0167525, EP1269550, AU7436301, US2003144151
1703. P. Tripodi, D. Di Gioacchino, R. Borelli, J.D. Vinko, Physica C **388–389**, 571 (2003)
1704. P. Tripodi, D. Di Gioacchino, J.D. Vinko, Physica C **408–410**, 350 (2004)

1705. P. Tripodi, D. Di Gioacchino, J.D. Vinko, J. Brazilian, Physics, **34**(3B), 1177 (2004)
1706. Y. Tsabba, S. Reich, Phys. C **254**, 21 (1995)
1707. I-F. Tsu, S.E. Babcock, D.L. Kaiser, J. Mater. Res. **11**, 1383 (1996)
1708. I-F. Tsu, D.L. Kaiser, S.E. Babcock, in *Proceedings MSA Annual Meeting* (1996), p. 338
1709. I-F. Tsu, J-L. Wang, S.E. Babcock, et al., Phys. C **349**, 8 (2001)
1710. I-F. Tsu, J-L. Wang, D.L. Kaiser, S.E. Babcock, Phys. C **306**, 163 (1998)
1711. M. Tsuchimoto, H. Takashima, Jpn. J. Appl. Phys. **39**, 5816 (2000)
1712. M. Tsuchimoto, H. Takashima, Phys. C **341–348**, 2467 (2000)
1713. M. Tsuchimoto, H. Takashima, T. Onishi, Phys. C **357–360**, 759 (2001)
1714. C.C. Tsuei, J.R. Kirtley, Rev. Mod. Phys. **72**, 969 (2000)
1715. H. Tsunakawa, R. Aoki, Powder Tech. **33**, 249 (1982)
1716. M. Turchinskaya, D.L. Kaiser, F.W. Gayle et al., Phys. C **216**, 205 (1993)
1717. M. Turchinskaya, D.L. Kaiser, F.W. Gayle et al., Phys. C **221**, 62 (1994)
1718. V. Tvergaard, J.W. Hutchinson, J. Mech. Phys. Solids **40**, 1377 (1992)
1719. A. Ubbelode, *Melting and Crystalline Structure* (Mir, Moscow, 1969)
1720. Y.J. Uemura, V.J. Emery, A.R. Moodenbaugh et al., Phys. Rev. B. **38**, 909 (1988)
1721. M. Ueyama, T. Hikata, T. Kato, K. Sato, Jpn. J. Appl. Phys. **30**, L1384 (1991)
1722. D.R. Uhlmann, B. Chalmers, K.A. Jackson, J. Appl. Phys. **35**, 2986 (1964)
1723. S. Uji, H. Shinagawa, T. Terashima et al., Nature **410**, 908 (2001)
1724. M. Ullrich, A. Leenders, H.C. Freyhardt, IEEE Trans. Appl. Supercond. **7**(2), 1813 (1997)
1725. M. Ullrich, A. Leenders, J. Krelaus et al., Mater. Sci. Eng., B **53**, 143 (1998)
1726. T. Umeda, H. Kozuka, S. Sakka, Adv. Ceram. Mater. **3**, 520 (1988)
1727. A. Umezawa, Y. Feng, H.S. Edelman et al., Phys. C **198**, 261 (1992)
1728. A. Umezawa, Y. Feng, H.S. Edelman et al., Phys. C **219**, 378 (1993)
1729. N. Uno, N. Enomoto, Y. Tanaka, H. Takami, Jpn. J. Appl. Phys. **27**, L1003 (1988)
1730. R. Utiyama, Phys. Rev. **101**, 1597 (1956)
1731. D. Vaknin, S.K. Sinha, C. Stassis et al., Phys. Rev. B **41**, 1926 (1990)
1732. T. Valla, P.D. Johnson, Z. Yusof et al., Nature **417**, 627 (2002)
1733. M. Vallet-Regi, J. Ramirez, C.V. Ragel, J.M. Gonzales-Calbet, Phys. C **230**, 407 (1994)
1734. C.J. van der Beek, P.H. Kes, Phys. Rev. B **43**, 13032 (1991)
1735. C.J. van der Beek, P.H. Kes, M.R. Maley et al., Phys. C **195**, 307 (1992)
1736. N. Vandewalle, R. Cloots, M. Ausloos, J. Mater. Res. **10**, 268 (1995)
1737. C. Varanasi, M.A. Black, P.J. McGinn, J. Mater. Res. **11**, 565 (1996)
1738. C. Varanasi, P. McGinn, J. Phys. C **207**, 79 (1993)
1739. J.L. Vargas, N. Zhang, D.L. Kaiser, S.E. Babcock, Phys. C **292**, 1 (1997)
1740. A.G. Varias, Comput. Mech. **21**, 316 (1998)
1741. A.G. Varias, A.R. Massih, Eng. Fract. Mech. **65**, 29 (2000)
1742. A.G. Varias, A.R. Massih, J. Nucl. Mater. **279**, 273 (2000)
1743. A.G. Varias, A.R. Massih, J. Mech. Phys. Solids **50**, 1469 (2002)
1744. A.G. Varias, N.P. O'Dowd, R.J. Asaro, C.F. Shih, Mater. Sci. Eng. **A126**, 65 (1990)
1745. C.M. Varma, Phys. Rev. B. **55**, 14554 (1997)
1746. C.M. Varma, Phys. Rev. Lett. **83**, 3538 (1999)
1747. G. Vekinis, M.F. Ashby, P.W. Beaumont, Acta Metall. Mater. **38**, 1151 (1990)
1748. P.A. Vermeer, in *Physics of Dry Granular Media* eds. by H.J. Herrmann, J-P. Hovi, S. Luding (Kluwer Academic Publishers, Dordrecht, 1998), p. 163

1749. N. Vilatla, F. Sandiumenge, S. Pinol, X. Obrados, J. Mater. Res. **12**, 38 (1997)
1750. N. Vilalta, F. Sandiumenge, J. Rabier et al., Philos. Mag. A **76**, 837 (1997)
1751. N. Vilalta, F. Sandiumenge, E. Rodriguez et al., Philos. Mag. B **75**, 431 (1997)
1752. L.Y.A. Vinnikov, L.A. Gurevich, G.A. Emel'chenko, YuA Osip'yan, Pisma Zh. Eksp. Teor. Fiz. **47**, 109 (1988)
1753. Y. Viouchkov, J. Schwartz, IEEE Trans. Appl. Supercond. **11**(1) (2001)
1754. Y. Viouchkov, H.W. Weijers, Q.Y. Hu et al., Adv. Cryog. Eng. (Mater.) **46**, 647 (2000)
1755. G.F. Virshup, M.E. Klusmeier-Brown, I. Bozovic, J.N. Eckstein, Appl. Phys. Lett. **60**, 2288 (1992)
1756. V.K. Vlasko-Vlasov, G.W. Crabtree, U. Welp, V.I. Nikitenko, in *Physycs and Materials Science of Vortex States, Flux Pinning and Dynamics*, eds. by R. Kossowsky et al., NATO ASI, Ser. E 356, 205 (Kluwer, Dordrecht, 1999)
1757. V.K. Vlasko-Vlasov, V.N. Goncharov, V.I. Nikitenko et al., Phys. C **222**, 367 (1994)
1758. A.F. Volkov, Phys. C **183**, 177 (1991)
1759. B.V. Vuchic, K.L. Merkle, K.A. Dean et al., Appl. Phys. Lett. **67**, 1013 (1995)
1760. D.A. Vul, E.K.H. Salje, Phys. C **253**, 231 (1995)
1761. A. Wallraff, D.I. Schuster, A. Blais et al., Nature **431**, 162 (2004)
1762. A. Wallraff, D.I. Schuster, A. Blais et al., Phys. Rev. Lett. **95**, 060501 (2005)
1763. Wang Chongmin, Zhe Xiaoli, Zhang Hongtu, J. Mater. Sci. Lett. **7**, 621 (1988)
1764. J. Wang, I. Monot, G. Desgardin, J. Mater. Res. **11**, 2703 (1996)
1765. J. Wang, I. Monot, M. Hervieu et al., Supercond. Sci. Technol. **9**, 69 (1996)
1766. J.L. Wang, X.Y. Cai, R. Kelley et al., J. Phys. C **230**, 189 (1994)
1767. J.L. Wang, I-F. Tsu, X.Y. Cai, et al., J. Mater. Res. **11,** 868 (1996)
1768. R.K. Wang, J. Horvat, H.K. Liu, S.X. Dou, Supercond. Sci. Technol. **9**, 875 (1996)
1769. R.K. Wang, H.K. Liu, S.X. Dou, Supercond. Sci. Technol. **8**, 168 (1995)
1770. R.K. Wang, H.T. Ren, L. Xiao et al., Supercond. Sci. Technol. **3**, 344 (1990)
1771. S.-L. Wang, R.F. Sekerka, Phys. Rev. E **53**, 3760 (1996)
1772. W.G. Wang, Z. Han, P. Skov-Hansen et al., IEEE Trans. Appl. Supercond. **9**(2), 2613 (1999)
1773. W.G. Wang, H.K. Liu, Y.C. Guo et al., Appl. Supercond. **3**, 599 (1995)
1774. Y. Wang, L. Li, N.P. Ong, Phys. Rev. B. **73**, 024510 (2006)
1775. Z.L. Wang, A. Goyal, D.M. Kroeger, Phys. Rev. B **47**, 5373 (1993)
1776. W.H. Warnes, D.C. Larbalestier, Cryogenics **26**, 643 (1986)
1777. J.A. Warren, W.J. Boettinger, Acta Metall. Mater. A **43**, 689 (1995)
1778. J.A. Warren, B.T. Murray, Mod. Simul. Mater. Sci. Eng. **4**, 215 (1996)
1779. G. Wassermann, J. Grewen, *Texturen Metallischer Werkstoffe* (Springer, Berlin, 1962)
1780. N. Watanabe, K. Fukamachi, Y. Ueda et al., Phys. C **235–240**, 1309 (1994)
1781. J. Weertman, Acta Metall. **26**, 1731 (1978)
1782. J.J. Weertman, J.R. Weertman, *Elementary Dislocation Theory* (Oxford University Press, New York, 1992)
1783. S. Weinberg, Phys. Rev. Lett. **18**, 188 (1967)
1784. S. Weinberg, Phys. Rev. Lett. **19**, 1264 (1967)
1785. S. Weinberg, Phys. Rev. **5**, 1412 (1972)
1786. S. Weinberg, Phys. Rev. D **19**, 1277 (1979)
1787. S. Weinberg, Int. J. Mod. Phys. A **23**(11), 1627 (2008)
1788. T.E. Weller, M. Ellerby, S.S. Saxena et al., Nat. Phys. **1**, 39 (2005)

1789. U. Welp, M. Grimsditch, S. Fleshler et al., J. Supercond. **7**, 159 (1994)

1790. U. Welp, D.O. Gunter, G.W. Crabtree et al., Nature **376**, 44 (1995)

1791. G.J. Weng, Int. J. Eng. Sci. **22**, 845 (1984)

1792. A.A. Wheeler, W.J. Boettinger, G.B. Mc Fadden, Phys. Rev. A **45**, 7424 (1992)

1793. A.A. Wheeler, W.J. Boettinger, G.B. McFadden, Phys. Rev. E **47**, 1893 (1993)

1794. A.A. Wheeler, G.B. McFadden, W.J. Boettinger, Proc. Roy. Soc. Lond. A **452**, 495 (1996)

1795. A.A. Wheeler, B.T. Murray, R.J. Schaefer, Physica D **66**, 243 (1993)

1796. R. White, D.J. Scalapino, Phys. Rev. Lett. **81**, 3227 (1998)

1797. R.J. Wijngaarden, R. Griessen, in *Studies of High Temperature Superconductors*, vols. 2, 29. ed. by A.V. Narlikar (Nova Science, New York, 1989)

1798. R.J. Wijngaarden, D.T. Jover, R. Griessen, Phys. B **265**, 128 (1999)

1799. J.P. Wikswo, Jr. in *SQUID Sensors: Fundamentals, Fabrication, and Applications*, ed. by H. Weinstock (Kluwer Academic Publishers, Dordrecht, 1996), p. 629

1800. G.V.M. Williams, J.L. Tallon, Phys. Rev. B **57**, 10984 (1998)

1801. R.K. Williams, K.B. Alexander, J. Brynestad et al., J. Appl. Phys. **70**, 906 (1991)

1802. J. RWillis, in *Mechanics of Solids: The R. Hill Anniversary Volume*, eds. by H.G. Hopkins, M.J. Sewell (Pergamon Press, Oxford, 1982)

1803. K.G. Wilson, Phys. Rev. D **10**, 2445 (1974)

1804. T.A. Witten, L.M. Sander, Phys. Rev. B **27**, 5686 (1983)

1805. D.A. Wollman, D.J. Van Harlingen, W.C. Lee et al., Phys. Rev. Lett. **71**, 2134 (1993)

1806. M.S. Wong, A. Miyase, Y.S. Yuan, S.S. Wang, J. Am. Ceram. Soc. **77**, 2833 (1994)

1807. R. Wördenweber, Rep. Prog. Phys. **62**, 187 (1999)

1808. C. Cm. Wu, S.M. Freiman, R.W. Rice, J.J. Mecholsky, J. Mater. Sci. **13**, 2659 (1978)

1809. J.G. Wu, Y.L. Xie, H. Chen, et al. J. Phys.: Condens. Mater. **21**, 142203 (2009)

1810. M.K. Wu, J.R. Ashburgn, C.J. Torng et al., Phys. Rev. Lett. **58**, 908 (1987)

1811. X.D. Wu, S.R. Foltyn, P. Arendt et al., Appl. Phys. Lett. **65**, 1961 (1994)

1812. B. Xu, F. Arias, S.T. Brittain et al., Adv. Mater. **11**, 1186 (1999)

1813. J-A. Xu, Supercond. Sci. Technol. **7**, 1 (1994)

1814. Y. Xu, A. Matsuda, M.R. Beasley, Phys. Rev. B **42**, 1492 (1990)

1815. Y.Y. Xue, B. Lorenz, A. Baikalov, et al., cond-mat/0205487 (2002)

1816. Y. Yamada, Y. Hishinuma, F. Yamashita et al., IEEE Trans. Appl. Supercond. **7**(2), 1715 (1997)

1817. Y. Yamada, K. Itoh, K. Wada, K. Tachikawa, IEEE Trans. Appl. Supercond. **9**(2), 1868 (1999)

1818. K. Yamada, C.H. Lee, K. Kurahashi et al., Phys. Rev. B **57**, 6165 (1998)

1819. Y. Yamada, M. Satou, S. Murase, et al., in *Proceedings 5th International Symposium on Superconductivity* (*ISS'92*) (Kobe, Japan, 1992), p. 717

1820. Y. Yamada, Y. Shiohara, Phys. C **217**, 182 (1993)

1821. Y. Yamada, F. Yamashita, K. Wada, K. Tachikawa, Adv. Cryog. Eng. (Mater.) **44b**, 547 (1998)

1822. K. Yamafuji, T. Kiss, Phys. C **258**, 197 (1996)

1823. K. Yamaguchi, M. Murakami, H. Fujimoto et al., Jpn. J. Appl. Phys. **29**, L1428 (1990)

1824. A. Yamamoto, C. Takao, T. Masui, et al., cond-mat/0208331 (2002)

1825. S. Yamanaka, E. Enishi, H. Fukuoka, M. Yasukawa, Inorg. Chem. **39**, 56 (2000)

1826. T. Yamashita, J.A. Alarco, A.J. Ilyushechkin et al., IEEE Trans. Appl. Supercond. **9**(2), 2581 (1999)

1827. M.F. Yan, R.M. Cannon, H.K. Bowen, in *Ceramic Microstructures* '76, eds. by R. M. Fulrath, J.A. Pask (Westview, Boulder, 1977), p. 276

1828. C.N. Yang, R. Mills, Phys. Rev. **96**, 191 (1954)

1829. C.S. Yang, P.M. Hui, Phys. Rev. B **44**, 12559 (1991)

1830. K.-Y. Yang, T.M. Rice, F.-C. Zhang, Phys. Rev. B. **73**, 174501 (2006)

1831. I.K. Yanson, V.M. Svistunov, I.M. Dmitrenko, Zh. Eksp. Teor. Fiz. **48**, 976 (1965) [Sov. Phys. JETP 21, 650 (1965)]

1832. X. Yao, K. Furuya, Y. Nakamura et al., J. Mater. Res. **10**, 3003 (1995)

1833. X. Yao, Y. Shiohara, Supercond. Sci. Technol. **10**, 249 (1997)

1834. J. Ye, Z. Zou, A. Matsushita et al., Phys. Rev. B **58**, 619 (1998)

1835. F. Yeh, K.W. White, J. Appl. Phys. **70**, 4989 (1991)

1836. Y. Yeshurin, A.P. Malozemoff, Phys. Rev. Lett. **60**, 2202 (1988)

1837. Y. Yeshurin, A.P. Malozemoff, A. Shaulov, Rev. Mod. Phys. **68**, 911 (1996)

1838. T. Yokoya, T. Kiss, A. Chainani et al., Science **294**, 2518 (2001)

1839. F. Yonezawa, M.H. Cohen, J. Appl. Phys. **54**, 2895 (1983)

1840. S. Yonezawa, Y. Muraoka, Z. Hiroi, J. Phys. Soc. Jpn. **73**, 1655 (2004)

1841. S. Yonezawa, Y. Muraoka, Y. Matsushita, Z. Hiroi, J. Phys.: Condens. Matter **16**, L9 (2004)

1842. S. Yonezawa, Y. Muraoka, Y. Matsushita, Z. Hiroi, J. Phys. Soc. Jpn. **73**, 819 (2004)

1843. J. Yoo, H. Chung, J. Ko, H. Kim, IEEE Trans. Appl. Supercond. **7**(2), 1837 (1997)

1844. S.H. Yoo, V.W. Wong, Y. Xin, Phys. C **273**, 189 (1997)

1845. S.I. Yoo, N. Sakai, K. Segawa et al., Adv. Supercond. **7**, 705 (1995)

1846. J. Yoshida, H. Katsuno, K. Nakayama, T. Nagano, Phys. Rev. B **70**, 54511 (2004)

1847. M. Yoshida, A. Tsuzuk, I. Hirabayashi, *Proceedings International Workshop on Superconductivity* (Honolulu, HI, 1992), p. 270

1848. H. Yoshino, M. Yamazaki, H. Fuke et al., Adv. Supercond. **6**, 759 (1994)

1849. S.J. Yotsutani, Math. Soc. Japan. **31**, 623 (1978)

1850. D.P. Young, P.W. Adams, J.Y. Chan, F.R. Fronczek, cond-mat/0104063 (2001)

1851. C.W. Yuan, Z. Zheng, A.L. de Lozanne et al., J. Vac. Sci. Technol., B **14**, 1210 (1996)

1852. H.Q. Yuan, J. Singleton, F.F. Balakirev et al., Nature **457**, 565 (2009)

1853. Y.S. Yuan, M.S. Wong, S.S. Wang, Phys. C **250**, 247 (1995)

1854. Y.S. Yuan, M.S. Wong, S.S. Wang, J. Mater. Res. **11**, 8 (1996)

1855. Y.S. Yuan, M.S. Wong, S.S. Wang, J. Mater. Res. **11**, 18 (1996)

1856. Y.S. Yuan, M.S. Wong, S.S. Wang, J. Mater. Res. **11**, 1645 (1996)

1857. K.S. Yugvesson, T.L. Korzenovski, Y.S. Kim et al., IEEE Trans. Microwave Theory Tech. **37**, 365 (1989)

1858. S.P. Yukon, N.C. Lin, in *Macroscopic Quantum Phenomena and Coherence in Superconductivity Networks* (World Scientific, Singapore, 1995)

1859. A. Yurgens, D. Winkler, T. Claeson, N.V. Savaritsky, Appl. Phys. Lett. **70**, 1769 (1997)

1860. V.B. Zabolotnyy, D.S. Inosov, D.V. Evtushinsky et al., Nature **457**, 569 (2009)

1861. N.V. Zavaritski, S.S.S.R. Doklady Akad. Nauk, **86**, 501 (1952)

1862. B. Zeimetz, N.A. Rutter, B.A. Glowacki, J.E. Evetts, Supercond. Sci. Technol. **14**, 672 (2001)

1863. E. Zeldov, N.M. Amer, G. Koren, A. Gupta, Appl. Phys. Lett. **56**, 1700 (1990)

1864. E. Zeldov, A.I. Larkin, V.B. Geshkenbein et al., Phys. Rev. Lett. **73**, 1428 (1994)

1865. E. Zeldov, D. Majer, M. Konczykowsky et al., Nature **375**, 373 (1995)

1866. J. Zhai, Z. Xing, S. Dong et al., Appl. Phys. Lett. **88**, 062510 (2006)

1867. P.X. Zhang, H. Maeda, L. Zhou et al., IEEE Trans. Appl. Supercond. **9**(2), 2770 (1999)

1868. P.X. Zhang, L. Zhou, P. Ji et al., Supercond. Sci. Technol. **8**, 15 (1995)

1869. W. Zhang, E.E. Hellstrom, Phys. C **234**, 137 (1994)

1870. W. Zhang, M. Polak, A. Polyanskii, et al., Adv. Cryog. Eng. (Mater.) **44b**, 509 (1998)

1871. Y. Zhang, M. Mironova, D.F. Lee et al., Jpn. J. Appl. Phys. **34**, 3077 (1995)

1872. Y. Zhang, V. Selvamanickam, D.F. Lee et al., Jpn. J. Appl. Phys. **33**, 3419 (1994)

1873. G.-M. Zhao, Y.S. Wang, cond-mat/0111268 (2001)

1874. J. Zhao, D.T. Adroja, D.-X. Yao et al., Nat. Phys. **5**, 555 (2009)

1875. J. Zhao, Q. Huang, C. de la Cruz et al., Nat. Mater. **7**, 953 (2008)

1876. Z. Xi, Z. Lian, Supercond. Sci. Technol. **7**, 908 (1994)

1877. L. Zhou, P. Zhang, P. Ji et al., Supercond. Sci. Technol. **3**, 490 (1990)

1878. X.J. Zhou, T. Cuk, T. Devereaux, et al. in *Handbook of High-Temperature Superconductivity: Theory and Experiment*, eds. by J.R. Schrieffer, J.S. Brooks (Springer, New York, 2007)

1879. X.J. Zhou, T. Yoshida, D.-H. Lee et al., Phys. Rev. Lett. **92**, 187001 (2004)

1880. W. Zhu, P.S. Nicholson, J. Mater. Res. **7**, 38 (1992)

1881. X. Zhu, F. Han, G. Mu, et al, arXiv: 0904.1732v4 (2009)

1882. Y. Zhu, M. Corcoran, M. Suenaga, Interface Sci. **1**, 361 (1993)

1883. Y. Zhu, Z.L. Wang, M. Suenaga, Philos. Mag. A **67**, 11 (1993)

1884. Y. Zhu, J.M. Zuo, A.R. Moodenbaugh, M. Suenaga, Philos. Mag. A **70**, 969 (1994)

1885. A.A. Zhukov, H. Küpfer, V.A. Ryabchuk et al., Phys. C **219**, 99 (1994)

1886. M. Ziese, Phys. Rev. B **55**, 8106 (1996)

1887. J. Ziman, *Models of Disorder* (Cambridge University Press, Cambridge, 1979)

1888. Z. Zou, K. Oka, T. Ito, Y. Nishihara, Jpn. J. Appl. Phys. **36**, L18 (1997)

1889. Z. Zou, J. Ye, K. Oka, Y. Nishihara, Phys. Rev. Lett. **80**, 1074 (1998)

Index

I. A. Parinov, *Microstructure and Properties of High-Temperature Superconductors*,
DOI: 10.1007/978-3-642-34441-1, © Springer-Verlag Berlin Heidelberg 2012

Z

MIX
Papier aus verantwortungsvollen Quellen
Paper from responsible sources
FSC® C105338

If you have any concerns about our products,
you can contact us on
ProductSafety@springernature.com

In case Publisher is established outside the EU,
the EU authorized representative is:
Springer Nature Customer Service Center GmbH
Europaplatz 3, 69115 Heidelberg, Germany

Printed by Libri Plureos GmbH
in Hamburg, Germany